Acoustical Imaging

Volume 25

Acoustical Imaging

Recent Volumes in This Series:

Volume 10	Proceedings of the Tenth International Symposium October 12–16, 1980, edited by Pierre Alais and Alexander F. Metherell
Volume 11	Proceedings of the Eleventh International Symposium May 4–7, 1981, edited by John P. Powers
Volume 12	Proceedings of the Twelfth International Symposium July 19–22, 1982, edited by Eric A. Ash and C. R. Hill
Volume 13	Proceedings of the Thirteenth International Symposium October 26–28, 1983, edited by M. Kavch, R. K. Mueller, and J. F. Greenleaf
Volume 14	Proceedings of the Fourteenth International Symposium April 22–25, 1985, edited by A. J. Berkhout, J. Ridder, and L. F. van der Wal
Volume 15	Proceedings of the Fifteenth International Symposium July 14–16, 1986, edited by Hugh W. Jones
Volume 16	Proceedings of the Sixteenth International Symposium June 10–12, 1987, edited by Lawrence W. Kessler
Volume 17	Proceedings of the Seventeenth International Symposium May 31–June 2, 1988, edited by Hiroshi Shimizu, Noriyoshi Chubachi, and Jun-ichi Kusibiki
Volume 18	Proceedings of the Eighteenth International Symposium September 18–20, 1989, edited by Hua Lee and Glen Wade
Volume 19	Proceedings of the Nineteenth International Symposium April 3–5, 1991, edited by Helmut Ermert and Hans-Peter Harjes
Volume 20	Proceedings of the Twentieth International Symposium September 12–14, 1992, edited by Yu Wei and Benli Gu
Volume 21	Proceedings of the Twenty-First International Symposium March 28–30, 1994, edited by Joie Pierce Jones
Volume 22	Proceedings of the Twenty-Second International Symposium September 3–7, 1995, edited by Piero Tortoli and Leonardo Masotti
Volume 23	Proceedings of the Twenty-Third International Symposium April 13–16, 1997, edited by Sidney Lees and Leonard A. Ferrari
Volume 24	Proceedings of the Twenty-Fourth International Symposium September 23–25, 1998, edited by Hua Lee
Volume 25	Proceedings of the Twenty-Fifth International Symposium March 19–22, 2000, edited by Michael Halliwell and Peter N. T. Wells

Acoustical Imaging
Volume 25

Edited by

Michael Halliwell

and

Peter N. T. Wells

Centre for Physics and Engineering Research in Medicine
and
Department of Medical Physics and Bioengineering
Bristol General Hospital
Bristol, United Kingdom

Springer Science+Business Media, LLC

The Library of Congress cataloged the first volume of this series as follows:

International Symposium on Acoustical Holograpy
Acoustical holography; proceedings; v. 1–
Springer Science+Business Media, LLC.

v. illus. (part col.), ports. 24 cm.
Editors, 1967– , A. F. Metherell and L. Larmore (1967 with H. M. A. el-Sum)
Symposium for 1967 held at the Douglas Advanced Research Laboratories, Huntington Beach, Calif.

1. Acoustic holography—Congresses—Collected works. I. Metherell, Alexander A., ed. II. Larmore, Lewis, ed. III. el-Sum, Hussein Mohammed Amin, ed. IV. Douglas Advanced Research Laboratories. V. Title.

QC244.5.1.5 69-12533

Proceedings of 25th International Symposium on Acoustical Imaging, held March 19–22, 2000, in Bristol, United Kingdom.

ISSN 0270-5117

ISBN 978-1-4757-8794-8 ISBN 978-0-306-47107-0 (eBook)
DOI 10.1007/978-0-306-47107-0

Originally published by Kluwer Academic/Plenum Publishers, New York in 2000
Softcover reprint of the hardcover 1st edition 2000

http://www.wkap.nl/

10 9 8 7 6 5 4 3 2 1

A C.I.P. record for this book is available from the Library of Congress

25th International Acoustical Imaging Symposium

Chairmen: Peter N T Wells
Michael Halliwell

International Advisory Board
(invited members)

PREFACE

The origin of the International Acoustical Imaging Symposium series can be traced to 1967, when a meeting on acoustical holography was held in California. In those days, acoustical holography was at the leading edge of research but, as the importance of this subject waned, so the title of the series was changed from Acoustical Holography to Acoustical Imaging in 1978. The early Symposia were held at various venues in the United States. In 1980, the series became international, with the Symposium that year taking place in Cannes in France. The pattern now is to try to met alternately in the USA and in another part of the world so that active researchers everywhere can conveniently attend at a reasonably high frequency.

It was a great privilege for us in Bristol in the United Kingdom to be chosen to host the 25th Symposium, which convened on 19 March 2000 and spread over four days. We were blessed not only by good weather, but also by the attendance of nearly 100 participants who came from 17 countries. A large number of papers were accepted for presentation, either orally or as posters. Whether an oral presentation or a poster, all were considered to have equal merit, and no distinction is made between them in the published proceedings. There were no parallel sessions, so every participant could attend every presentation. The resultant disciplinary cross-fertilisation maintained the tradition of past Symposia. It stimulated discussion and seemed very much to be appreciated.

We thank the members of the International Advisory Board for their contributions to the success of the Symposium. Many of them helped to publicise the event; they advised on the contents of the programme; and most of them attended and participated in the Symposium. We acknowledge the excellence of the three keynote speakers, Mr. Gil Devey, Dr. Tom Szabo, and Professor Han Thijssen. We are also grateful to the Department of Health, Imasonic and the Institution of Electrical Engineers, who generously sponsored the Symposium.

The overall success of any Symposium is due in no small part to the meticulous planning and careful attention to detail. It is usually invidious to single out individuals in relation to these duties but several members of the Department of Medical Physics and Bioengineering in Bristol are worthy of our thanks. Mrs. Jan Carne, however, does deserve special mention, for it was she who held the whole event together, and who coordinated the production of these Proceedings.

There is no formal structure to sustain the International Acoustical Imaging Symposium series. Everyone who attended the Symposium in Bristol was invited to a meeting to discuss the future of the event. The acceptance of the invitation from Dr. Roman Maev for the 26th

Symposium to convene in Windsor, Ontario, Canada, in the fall of 2001 was enthusiastically confirmed. Following that event, the 27th Symposium will be held in Saarbrücken, Germany, at the invitation of Professor Walter Arnold.

Michael Halliwell
Peter Wells

CONTENTS

Keynote Papers

Mathematics and Physics of Acoustical Imaging

Transducers and Arrays

Nondestructive Evaluation

Geophysical and Underwater Ultrasonics

Microscopy and Microscanning

Scattering by Blood and Tissue

Medical and Biological Image Formation

Tissue Characterisation

Tissue Motion and Blood Flow

Elasticity Imaging

Hard Tissues

Novel and Emerging Methods

TURNING POINTS IN DIAGNOSTIC ULTRASOUND

Thomas L. Szabo

Agilent Technologies
Imaging Systems Division
3000 Minuteman Road
Andover, MA 01180 USA

BEGINNINGS

Critical changes in the evolution of diagnostic ultrasound imaging were made possible by the parallel growth of key enabling technologies. Turning points in imaging can be understood in the context of other scientific developments available at the time. In some cases imagination preceded the means needed for its realization. In other cases, allied technologies greatly accelerated the growth of diagnostic imaging.

Diagnostic imaging using ultrasound has been in existence for about fifty to sixty years, but its origins may have begun in ancient times. During a search for shipwrecks on the sea floor near Greece with SONAR, Harold Edgerton (1986) and his crew found a "sounder," the remains of a cord with knots tied at regular intervals with a lead weight on its end. This ancient Greek method of measuring water depth by throwing the weight overboard and by counting knots was used well into the nineteenth century. Henry Thoreau employed a sounder to measure the depth of Walden Pond.

The discovery of piezoelectricity by the Curie brothers in 1880 and the invention of the triode amplifier by Lee De Forest set the stage for further advances in depth-measurement (O'Brien, 1998). The sinking of the Titanic in 1912 (Wells,1979; Woo, 1998) motivated L. F. Richardson (1912) to invent underwater echo-ranging with sound waves to detect icebergs within a month of the disaster. At the time, however, there was no easy means to implement the method. By 1918, C. Chilowsky and P. Langevin (Biquard, 1972) took advantage of the further development of the enabling technologies of piezoelectric quartz manufacture and vacuum tube amplifiers to make echo-ranging in water practical and widespread.

F. Firestone's invention of the supersonic reflectoscope in 1941 (Woo, 1998) used the pulse echo-ranging principle for the location of defects in metals in the form of a reasonably compact instrument. When devices like this were applied to the human body in Japan, the United States and Sweden in the late forties and early fifties (Goldberg and

Kimmelman, 1988), a new world of possibility for medical diagnosis was born. Rokoru Uchida in Japan was one of the first to use flaw detectors for medical "A-line" pulse-echo ranging. Dr. J. J. Wild used A-Mode pulse echoes for medical diagnosis in 1949. In Sweden in 1953, Dr. I. Edler (1991) and Professor C. H. Hertz detected heart motion with a flaw detector and began what later was called "echocardiography."

A basic echo-ranging system is shown in Figure 1. In 1949, Dr. D. Howry in the United States found that the soft tissues and organs of the body, because of their high water content, have similar acoustic velocities and impedances which allowed the penetration of elastic waves through multiple tissue interfaces (Erickson et al, 1974). During 1947-49, Dr .G. Ludwig, characterized the properties of tissues and found the average speed of sound in tissues to be 1.54mm/us, the standard value still used today on imaging systems (Goldberg and Kimmelman, 1988).

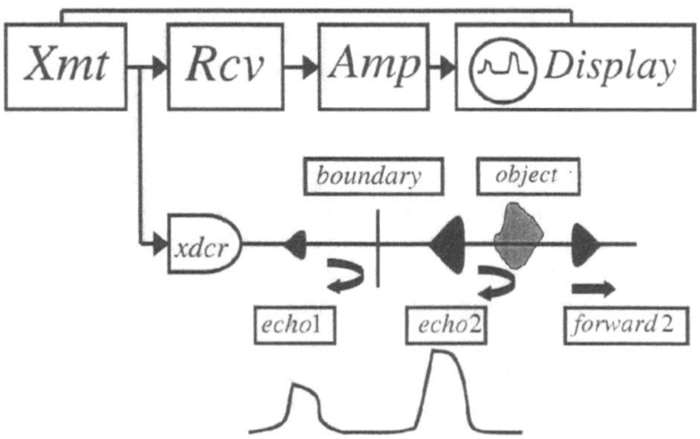

Figure 1. Basic echo-ranging system showing multiple reflections.

PORTRAIT PHOTOGRAPHERS

The imaginative leap from echo-ranging to imaging came in 1942 in Austria when Dr. Karl Dussik and his brother (Goldberg and Kimmelman, 1988) produced a through transmission ultrasound attenuation image of the brain which they called a "hyperphonogram." This transcranial method was not adopted widely because of difficult refraction artifacts in the skull but it inspired many others to work on imaging with ultrasound. Their work is even more remarkable because it preceded the widespread development of RADAR imaging. RADAR, which also used the pulse–echo principle but with electromagnetic waves, produced images of remote targets in the form of a plan-position indicator , PPI , a radial range display image. Echo-ranging with sound, evolved based on new RADAR principles, into SONAR with its own PPI images.

The work of the Dussiks and war developments in pulse-echo imaging motivated others to make acoustic images of the body. Dr. Howry was able to show that highly detailed pulse echo tomographic images of cross-sections of the body correlated well with anatomical features (Goldberg and Kimmelman, 1988; Holmes, 1980).

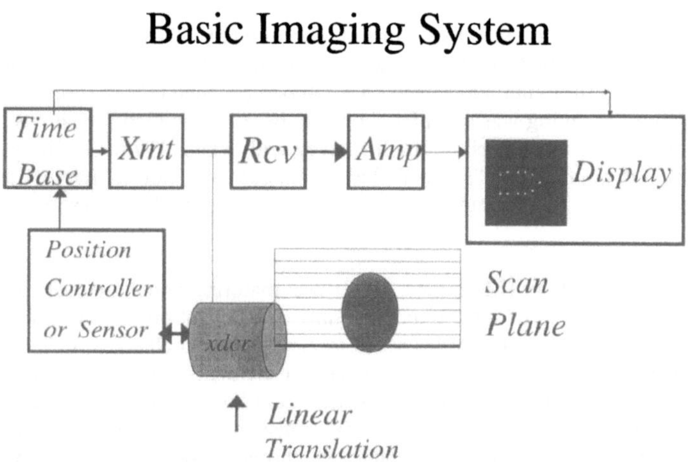

Figure 2. Basic imaging system block diagram showing transducer position control or sensing.

The nineteen-fifties were a period of active experimentation with imaging methods and ways of making contact with the body. The key element that makes an imaging system (Figure 2.) different from an echo-ranging system is a method of either scanning the transducer either in a freehand form with the detection of transducer position in space or of controlling the motion of the transducer. Scanning methods are shown in Figure 3. The most popular method was freehand compound scanning which involved both translation and rocking; this became known as static B-scanning. In the early fifties, Dr. J.J. Wild and Dr. J. Reid were working on an alternative method, a real-time compound scanner, in which they used mechanically scanned (controlled position) transducers (Goldberg and Kimmelman, 1988). Their vision of true real-time scanning did not become a reality for another twenty years because of the unavailability of enabling technologies.

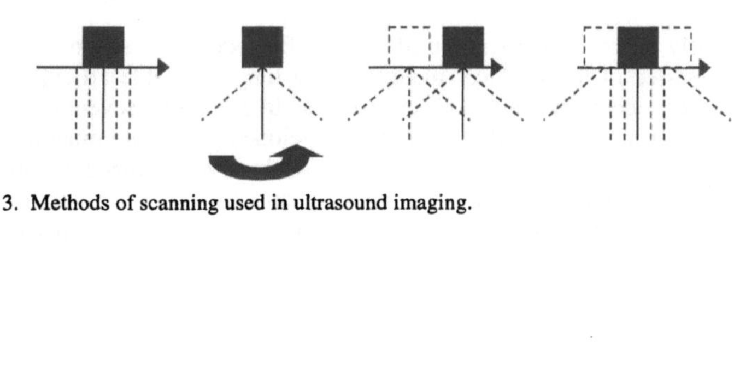

Figure 3. Methods of scanning used in ultrasound imaging.

Sonography in this time period was like portrait photography. Different patterns of freehand scanning were developed to achieve the "best picture." For each position of the transducer a corresponding time-line was traced on a cathode ray tube (CRT) screen so the image was not seen until scanning was completed or later because the picture was usually in the form of a long time exposure photograph (Devey and Wells, 1978; Goldberg and Kimmelman, 1988). Of course, the "subject" being imaged was not to move during scanning. In 1959, the situation was improved by the introduction of the Polaroid camera scope camera.

A seemingly unrelated technology (EE Times, 1997) was maturing simultaneously which would revolutionize the practice of ultrasound imaging. The inventions of the transistor and of the digital computer in the late forties set in motion profound changes. In 1958 the integrated circuit including several transistors and circuit elements accelerated the pace. Moore in 1964 predicted that microprocessors growth would be exponential as illustrated in Fig. 4 (Santo and Wollard, 1978). By 1971, 2300 transistors on a single chip had as much computational power as the ENIAC computer which twenty-five years before was a big as a boxcar and weighed 60,000 pounds. Hand calculators such as the hp-35 sped up chip miniaturization. Digital memories and programmable chips were produced.

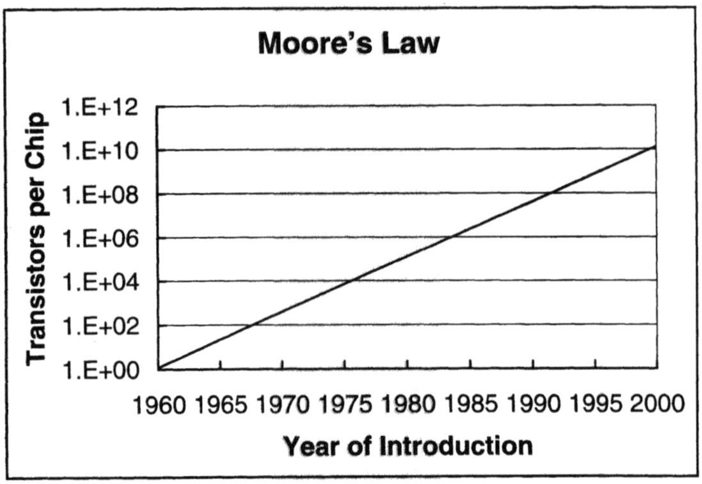

Figure 4. Moore's law predicting the exponential increase in transistor density over time (adapted from Santo and Wollard, 1978).

Eventually by the early sixties, commercialized contact B-mode static scanners became available. These consisted of a transducer mounted on a long moveable articulated arm with spatial position encoders, display and electronics. In the early seventies, Dr. G. Kossoff (1974) and his group in Australia and others developed grey scale imaging. Until then, images were in biphase: black and white. Grey scale revolutionized imaging by making possible a greater dynamic range for viewing and by revealing subtle tissue features and textures not seen before. Grey scale became widespread because of programmable digital memories and microprocessors which allowed the ultrasound image to be stored and scan-converted to the rectangular format of cathode ray tubes at video rates.

Ultrasound was shown to be a safe noninvasive methodology for the diagnosis of diseased tissue, the location of cysts, fetal abnormalities and heart irregularities. By the late seventies, millions of clinical exams had been performed with these instruments (Devey and Wells, 1978).

CINEMATOGRAPHERS

Despite these successes, researchers were attempting to make real-time images. RADAR and SONAR images benefited from the maturing of electronically scanned and focused phased array technology in the late fifties and sixties. 1965 marked the appearance of Vidoson, the first real-time mechanical scanner. J. C. Somer (1968) designed a phased array for medical ultrasound imaging. In 1971 Professor N. Bom's group in Rotterdam was building linear arrays for real-time imaging. These efforts resulted in the Minivisor (Ligtvoet et al, 1978), the first portable ultrasound system including a built in linear array, display and a two-hour battery. By the early seventies, real-time contact mechanical scanners were beginning to replace the static B-scanners. By 1974 Professor Thurstone and Dr. von Ramm obtained live images of the heart with their sixteen channel phased array imaging system, the Thaumascan (von Ramm and Thurstone, 1975).

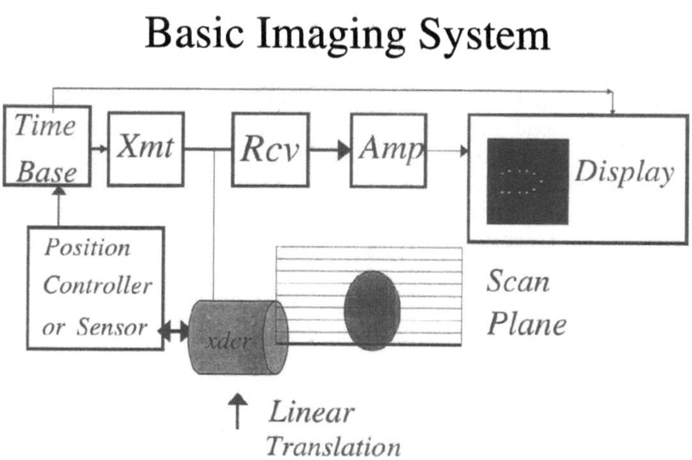

Figure 5. Basic imaging system with linear array.

By 1980, commercial real-time phased array imaging systems were made possible by recent developments in video, microprocessors, digital memory, and the miniaturization offered by programmable integrated circuits. The Hewlett Packard 70020A phase array system was a forerunner of future phased array systems which had modular architecture and programmable capabilities, the means for their future growth (HP Journal, 1983). The appearance of real-time systems with good image quality marked the beginning of the end of the static B-scanners (Klein, 1981).

During the decade of the eighties, array systems became the dominant imaging modality. Several electronic advancements (EE Times, 1997) made possible the rapid improvements in imaging during this decade: application specific integrated circuits (ASIC's),digital signal processing chips (DSP's), and the computer-aided design (CAD) of very large scale integration (VLSI) circuits. Added to imaging were new signal processing modalities which included M-mode, developed in 1954; and Doppler ultrasound, invented in 1955, but through the use of arrays. In 1985, color flow imaging made its debut on these systems, providing real-time visualization of blood flow. Concurrently, SONAR systems evolved to the point where Captain Ballard discovered the Titanic at the bottom of the sea with SONAR and video equipment in 1986 (Murphy, 1986).

Also during the eighties, transducer technology underwent tremendous growth. Based on the Mason equivalent circuit model and the waveguide and matching layer design technology and PZT piezoelectric materials developed during and after World War 2, ultrasonic phased array design evolved rapidly. Specialized phased and linear arrays were developed for specific cardiogy, radiology, endoscopic, transesophageal, and intraoperative applications. With improved materials and piezoelectric composites, arrays with several hundred elements and higher frequencies became available. Wider transducer bandwidths allowed the imaging and the operation of other modes within the same transducer at several frequencies selectable by the user.

By the nineties, developments in more powerful microprocessors, high density gate arrays, surface mount technology, and the availability of low cost analog to digital(A/D) chips made available higher computation and faster processing in smaller volumes and at lower costs. Imaging systems incorporating these advances evolved into digital architectures and beamformers. Broadband communication enabled the live transfer of images for telemedicine. Transducers appeared with even wider bandwidths and in 1.5D and matrix array configurations. New second harmonic imaging modes exploited the nonlinear characteristics of contrast agents and tissues. Special effect image processing techniques developed for movies such as John Cameron's Titanic, enabled nearly real time three dimensional imaging including surface-rendered images of fetuses. The continued miniaturization of electronics, especially ASIC's, made possible truly portable imaging systems for arrays with full imaging capabilities.

CONCLUSION

From the tables chronicling ultrasound imaging developments and enabling technologies, it is evident that there is often a time lag between the appearance of a technology and its effect. The most dramatic changes have been through the continual miniaturization of electronics in accordance with Moore's law. Smaller size components led to the first commercially available phased array imaging systems as well as new portable imaging systems which weigh only a few pounds. Moore's law is apparently approaching physical limits, and a second Moore's law predicts rapidly increasing production costs with reduced chip size. Because of the time lag of technology implementation, the full effect of the latest developments have not had their full effect on ultrasound imaging.

Ultrasound imaging is inherently linked to the context of current allied technologies. Change is in the direction of higher complexity at reduced costs.

Table 1. Chronology of ultrasound imaging developments and enabling technologies.

Time	Ultrasound	Enablers
Pre-WWII	Echo-Ranging	Piezoelectricity Vacuum Tube Amplifiers
1940's	Dussik Image of Brain PPI Images	RADAR, SONAR Supersonic Reflectoscope ENIAC computer Transistor
1950's	A-Line Compound Scanning Doppler Ultrasound M-Mode	Integrated Circuits Phased Array Antennas
1960's	Contact Static B-Scanner Real-time mechanical scanner	Moore's Law Microprocessors VLSI Hand-Held Calculators
1970's	Real-time imaging Scan-Conversion Grey Scale Linear & Phased Arrays	RAM EPROM ASIC HP-35 Scientific Calculator Altair PC
1980's	Commercial Array Systems Pulsed Wave Doppler Color Flow Imaging Wideband & Specialized Transducers	Gate Arrays Digital Signal Processing Chips Surface Mount Components Computer Aided Design of VLSI Circuits
1990's	Digital Systems 1.5D & Matrix Arrays Harmonic Imaging Commercialized 3D Imaging	Low Cost A/D Converters Powerful PC's 3D Image Processing 0.1um Fabrication Linewidths for Electronics

REFERENCES

Biquard, P.,Paul Langevin, Ultrasonics 10:213-214 (1972).

Birnbaum, J. and Williams, R. S., Physics and the information revolution, Physics Today 53:38-42 (2000).

Bom, N.,Lancee, C. T., van Zwieten, G.,Kloster, F. E.,and Roelandt, J., Multiscan echocardiography, CirculationXLVIII:1066-1074 (1973).

Devey, G. B., and Wells, P. N.T., Ultrasound in medical diagnosis, Scientific American 238:98-106+ (1978).

Edler, I.,Early echocardiography, Ultrasound in Med.& Biol. 17:425-431 (1991).

Edgerton, H. G., 1986,*Sonar Images*, Prentice hall, Englewood Cliffs.

Electronic Engineering Times, (October 30,1997).

Erickson, K. R., Fry, F. J., and Jones, J.P., Ultrasound in medicine-a review, IEEE Trans. Sonics Ultrasonics SU-21:144-170 (1974).

Goldberg, B. B. and Kimmelman, B. A., *Medical Diagnostic Ultrasound: A Retrospective on Its 40th Anniversary*, Eastman Kodak Company, New York, (1988).

Hewlett Packard Journal 34;3-40, 1983

Holmes, J. H. ,Diagnostic ultrasound during the early years of the A. I. U. M., J. Clin. Ultrasound 8:299-308 (1980).

Klein, H. G., Are B-scanners' days numbered in abdominal diagnosis? Diagnostic Imaging 3:10-11 (1981).

Kossoff, G.,Display techniques in ultrasound pulse echo investigations: a review,Journal of Clinical Ultrasound 2:61-72 (1974).
Ligtvoet C., Rijsterborgh, H., Kappen L., and Bom N., Real time ultrasonic imaging with a hand-held scanner, Ultrasound in Med. & Biol. 4:91-92 (1978).
Murphy, J. , Down into the deep, Time 128:48-54 (1986).
O'Brien, Assessing the risks for modern diagnostic ultrasound imaging, Japanese J. of Applied Physics 37:2781-2788 (1998).
Richardson, L.F., British Patent No. 11,125 (filed May 10, 1912; issued March 27,1913.
Santo, B. and Wollard, K., The world of silicon: it's dog eat dog, IEEE Spectrum 25:30-39 (1978).
Somer,J. C., Electronic sector scanning for ultrasonic diagnosis, Ultrasonics 153-159 (1968).
Von Ramm, O. T. and Thurstone, F. L., Thaumascan: Design considerations and performance characteristics, Ultrasound in Medicine1:373-378 (1975).
Wells, P. N. T., 1979,Historical review of echo-instrumentation, *Echocardiology*, C. T. Lancee, ed.,Martinus Nijhoff, The Hague,
Woo, J. ,/www/ob-ulrasound.net.

ULTRASONIC TISSUE CHARACTERIZATION

Johan M. Thijssen
Clinical Physics Laboratory-435
University Children's Center
6500 HB Nijmegen, the Netherlands
E-mail: J.Thijssen@ckskg.azn.nl

INTRODUCTION

Ultrasonic Tissue Characterization (UTC) is a well-established field of scientific research since the first publication on computer analysis of radiofrequency signals by Mountford and Wells (1972). After another milestone, i.e. the first symposium at the National Bureau of Standards, Gaithersburg, MD, in 1975 (Linzer 1976), followed by a similar European meeting in 1979 (Thijssen, 1980), an ever lasting stream of scientific papers has been devoted to this topic. The industrial interest has been relatively low, partly because of a few disappointments, partly because the technological demands of developed UTC methods were too high. Presently, modern digital scanners may be considered to be "ready" to accept these methods!

A definition of UTC could be" The assessment by ultrasound of quantitative information about the histological characteristics, and pathological changes thereof, of biological tissues". This quantitative information is extracted form the echographic data at two different levels.

First, the raw data, i.e. the radiofrequency signals corresponding to single transmit-receive echo lines in an image. Second, the 2-D echographic images, which are constructed by the scanner from the rf-signals generally, have been not only demodulated but also, in addition, compressed and filtered. A basic problem inherent in both ways of data analysis is the so-called beam diffraction: due to the limited transmit- and receive apertures of the US transducer and additional transmit/ receive focussing regimes, the spatial and spectral beam characteristics are depth dependent. This means that even in an isotropic and homogeneous medium the backscattered signals are depth dependent. The position and width of the corresponding amplitude spectrum, but also the "speckle" pattern in images, changes from the transducer face to greater depth. Furthermore, amplitude decay due to attenuation in the examined tissues is compensated by time gain compensation (TGC) setting of the receiving amplifier. These problems can be accounted for by adequate calibration measurements when rf-signals are processed (e.g. Huisman and Thijssen, 1996). In case of image analysis, the derived UTC parameters have to be corrected, or a posteriori calibrated (c.f. Valckx et al., 1996).

Acoustical Imaging, Volume 25, Edited by Halliwell and Wells.
Kluwer Academic/Plenum Publishers, 2000.

The information that can be extracted by analysis of the rf-data concerns primarily the acoustic characteristics of tissues: the attenuation coefficient and backscattering coefficient, both as function of frequency. Because of the relatively limited frequency range comprised by the transmitted US pulse, generally a linear with frequency relation is assumed for soft tissues. The speed of sound cannot be estimated from single and coincident transmit/ receive scan lines. For this reason, techniques were developed using linear array transducers, where separate transmit and receive angles were employed, directed to the same region-of-interest (ROI). Alternatively, through-transmission techniques are used for bone characterization, where the travel time between two transducers is measured through the heel bone or a finger phalange and compared to a reference measurement in water. For this clinical application, also the attenuation coefficient is measured by US transmission (cf. Langton, 1990).

Analysis of rf-data is also performed in "elastography" (c.f. Ophir and Cespedes 1991). Here the backscattered rf-signals are acquired prior to and after a slight compression of the tissues. The rf-signals are correlated and the relative compression, i.e. the tissue strain, is estimated vs. depth. This strain is related to the elasticity of the tissue (i.e. Young's modulus), hence the terms elastography, elasticity imaging, or sonoelasticity. The latter technique is using a low frequency vibrator (<1000Hz) and imaging is performed with conventional color Doppler techniques (Levinson, 1988).

Finally, it may be remarked that the rf-based techniques are used to estimate the discussed acoustic tissue parameters locally ("sliding window") and to produce so-called parametric images. This might be identified as the ultimate goal of innovation in medical ultrasound!

A second part of UTC is based on the analysis of the texture in 2-D echographic images. It was shown by realistic simulations, that the statistical and speckle characteristics of echographic texture are dependent on the "number density" of the scattering sites within a medium (Oosterveld et al., 1985). In other words, the gray level statistics and the speckle size changed continuously when the number of scatterers, on average, within the resolution cell corresponding to the transmitted beam width and pulse length increased from one to approximately ten. In the latter case "fully developed" speckle occurs. This result shows that image analysis could in principle reveal the effective number density of tissues, as well as pathological changes of this parameter. The number density may be considered as a histological characteristic of tissues.

Another method was devised to reveal eventual spatial regularity in the histology, e.g. the lobular structure of the liver parenchyma (Wagner et al. 1986). Based on second order statistical texture analysis some evidence for its usefulness could be shown.

Further texture analysis methods are not making use of any a priori assumptions about tissue histology but merely use methods developed in other, generally technical, fields of image analysis.

Although, success has been claimed in several clinical studies, the patient numbers are generally too low and the calibrations, or corrections of transducer and equipment performance too limited, to be able to generalize and find optimal strategies from existing literature. Adequate preconditioning and calibration should preferably be incorporated within commercial equipment thus enabling exchange of data and of results between institutions.

PHYSICAL BACKGROUNDS

Acoustic Tissue model

A general model of biological tissues is shown in Figure 1. The model, suitable for most parenchymal tissues (i.e., constituting organs like liver, spleen etc.), is assuming homogeneity and isotropy. Acoustically it is characterized by a fixed speed of sound (on the order of 1540 m/s in soft tissues) and by absorption and scattering. Absorption is due to relaxation phenomena of translational and rotational vibration modes of biological macromolecules. Scattering is considered to obey the born approximation (i.e., single scattering, undisturbed wave propagation) and is related to small inhomogeneities in acoustic impedance, which are randomly

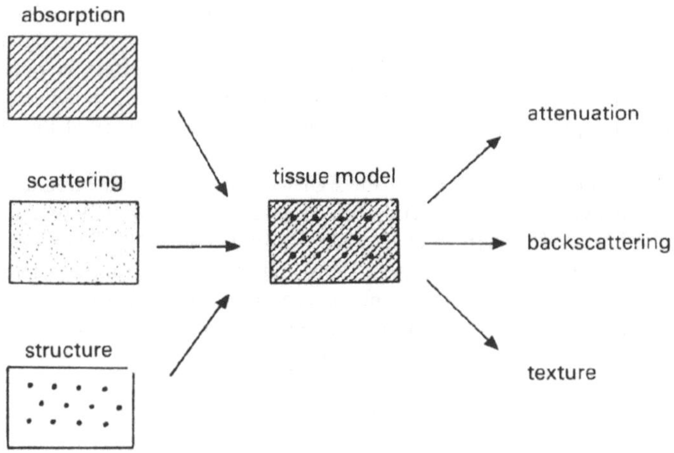

Figure 1. Block diagram of "acoustic tissue model" and kind of information that can be derived with ultrasonic tissue characterization.

distributed in 3D space. For instance, the microvasculature (arterioles), and the collagen meshwork of parenchymal tissues are producing relatively strong scattering contributions as compared to individual cells (e.g., blood cells). The overall attenuation due to absorption and scattering is accessible for estimation as is shown in Figure 1.In terms of backscattered echo amplitude e (t, z).

$$e(t,z) = e(t,o)\exp\{-\mu(f)2z\} \tag{1}$$

Where: f= frequency

z= depth (i.e., 2z= two-way travel path of scattered wave)

μ= attenuation coëfficient [nepers/cm].

In addition to the random (= diffuse) scattering, in some tissues small range order is present which is related to structural regularity on a millimeter scale, e.g., the triads of Kiernan constituting the portal system of the liver. So, in addition to diffuse scattering, a structural scattering component is present which is revealed in the texture of 2D echographic images (Jacobs and Thijssen, 1991).

Preprocessing

The characteristics of the ultrasound beam are depth dependent, both in continuous wave- and in pulsed mode. The limited dimensions of the transducer cause this. The resulting intensity distribution is called the "beam diffraction pattern", or "directivity function", which applies both to the transmission- and to the reception mode of the transducer employment (reciprocity principle). Generally, the directivity effects and the additional depth effects of transmit- and receive focusing are summarized in the same term: beam diffraction. Furthermore, it was shown experimentally (Thijssen et al., 1981; Cloostermans and Thijssen, 1983) that the received echographic spectra, calculated from the radiofrequency data, are highly depth dependent and, therefore, influence the estimates of the frequency dependence of the attenuation- and backscattering coefficients (see also Fink et al., 1984; Robinson et al., 1984; Verhoef et al., 1985). Finally, it was shown by experiments and simulations (Oosterveld et al., 1985) that also the texture features estimated from echographic images display a significant depth dependence (see also section 2.3 Processing: Texture analysis).

The tissue attenuation yields both an amplitude decay of the echographic signals, as well as an effective downshift of the spectral information. The latter effect is not shown in echographic images, due to the amplitude demodulation. It is, however, used in acoustospectrographic approaches to estimate the frequency dependence of the attenuation coefficient.

It can be concluded that prior to UTC the signals have to be corrected, in the frequency domain, for the influence of beam diffraction (cf. Thijssen et al., 1981; Cloostermans et al., 1983; Fink and Cardoso, 1984; Verhoef et al., 1985; Romijn et al., 1989; Romijn et al., 1991; Oosterveld et al., 1991; Huisman and Thijssen, 1996) and, in addition, for the employed TGC.

Equation (1) is obviously too simple to describe the real world conditions in echography. After transforming Eq (1) to the frequency domain and insertion of transducer performance, the backscattered spectrum E (f, z) becomes:

$$E(f,z) = P^2(f)D^2(f,z)T^2(f,z)S(f) \tag{2}$$

Where: P(f) = electro-acoustic transfer function of transducer
D(f,z) = beam diffraction transfer function (spectrogram)
T(f,z)= tissue transfer function (attenuation)
S(f)= backscatter transfer function.

In Eq. (2) it is assumed that the time gain compensation amplification has been corrected for. The diffraction correction consists of the estimation of P(f) from a reflection in focus in a medium with known acoustic characteristics and of $D^2(f,z)$ from the average backscatter vs. depth of a medium with known backscatter characteristics (e.g., Romijn et al., 1989; Huisman and Thijssen, 1996).

The next step is to analyse the corrected spectra vs. depth:

$$E_c(f,z) = T^2(f,z)S(f) \tag{3}$$

By taking the logarithmic spectra (in dB) and differentiating with respect to depth z:

$$\frac{d}{dz}\{\log_{10}[E_c(f,z)]\} = 2\frac{d}{dz}\{log_{10}[T(f,z)]\} \tag{4}$$

In this way, the tissue transfer function is obtained, which is then used to compensate the spectrogram for the effects of attenuation and resulting is:

$$E_c^*(f,z) = S(f) \tag{5}$$

So, the homogeneous backscattering vs. frequency is obtained over the whole depth range that is being analysed.

Finally, the attenuation-compensated data can be software demodulated and a corrected echographic image can be constructed ready for texture analysis (cf. Romijn et al., 1991; Oosterveld et al., 1991).

When radiofrequency signals are not available, a pragmatic approach for pre-processing the video images prior to analysis is the use of a reference measurement on a tissue-mimicking phantom with known acoustic characteristics. This method allows for correction of the diffraction and attenuation effects in the axial direction, i.e. along the scan lines and for the equipment settings of overall gain and TGC (cf. Valckx et al., 1996; 2000). In the past, many authors neglected position dependencies of the areas in the image selected for analyse, or they restricted the analysis to a single Region of Interest (ROI) in the focal transmit zone of the transducer.

ACOUSTIC SPECTROSCOPY

Acoustic spectroscopy is concerned with the analysis of the frequency dependence of the attenuation- and backscattering coefficient.

Attenuation coefficient

The attenuation as given by Eq. (1) can be written in the frequency domain, while replacing the echo by the tissue transfer function:

$$T(f,z) = \exp\{-2\mu(f)z\} \tag{6}$$

Transferring to decibells:

$$20\log_{10}[T(f,z)] = -2\alpha(f)z \tag{7}$$

Where: $\alpha(f) = [20\log_{10}(e)]\mu(f) \simeq 8.68\mu(f)$.

Insertion of Eq. (7) into Eq. (4) yields the attenuation coefficient from the measured logarithmic echo spectrogram.

It has been shown in many studies that in the low megahertz (1-10MHz) frequency range the attenuation coefficient α (f) is linearly dependent on frequency (c.f., Figure 2):

$$\alpha(f) = \alpha_1(f - f_c) + \alpha_0 \tag{8}$$

Where: α_0 = midband value (i.e., at $f=f_c$)

α_1 = slope.

The attenuation coefficient can be estimated in various ways, generally by using a "sliding" window technique. The rf-lines of an echographic image are segmented by suitable window functions (e.g., Hamming windows) of length D samples and which are overlapping each other by 50 percent. The amplitude spectra of the windows are calculated (yielding the Spectrogram) at discrete frequencies f_i. Eq. (7) is calculated at each frequency, yielding a set of values $\alpha(f_i)$ by linear regression analysis. Finally, the linear regression according to Eq. (8) is calculated of α (f_i) vs. f_i yielding α_0 and α_1. This method is called the "multi-narrow band" technique (Cloostermans and Thijssen, 1983).

An alternative method is making use of the property given by Eq. (8) that the higher frequencies of the received spectrum of the echo signals are attenuated more than the low frequencies. This results in an effective "down ward" shift of the spectrum with increasing depth. The so-called "centroid shift" method (Kak and Dines, 1978; Fink et al., 1983) can applied in case the spectrum corresponding to the transmitted ultrasound pulse is Gaussian shaped and α_o in Eq. (8) is zero. Then:

$$\alpha_1 = \frac{f_c' - f_c}{2z\sigma_f^2} \tag{9}$$

Where: f_c'= centroid frequency at depth z.
σ_f = "standard deviation" of (Gaussian) spectrum.

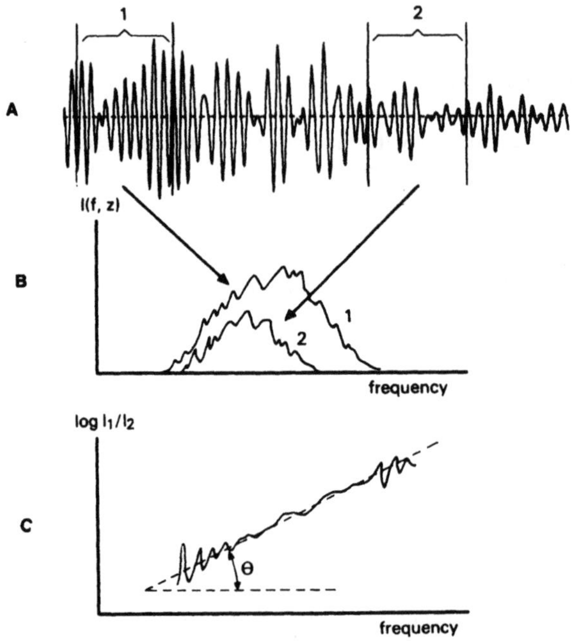

Figure 2. Plot of a radiofrequency signal with attenuation (A). Spectra are calculated of time windows 1 and 2 (B). Plot of difference of log-spectra, corresponding to 1 and 2, vs. frequency (C). Linear fit yields two parameters: attenuation coefficient intercept at central frequency (α_0) and slope constant (α_1), cf. Eq. (8).

Backscatter coefficient

The spectrographic data are corrected for the attenuation, by using the estimated α(f, z) and subsequently averaged over all windows in the ROI, yielding one single backscatter (amplitude-) spectrum: S (f). Again, this can be approximated by a straight line when plotting the log-spectral amplitude vs. frequency (see also Fig. 3):

$$S(f) = b_1(f - f_c) + b_0 \tag{10}$$

The estimated slope of the backscatter coefficient may be used to assess the (effective) size of the scatterers in the tissue. As shown in Fig. 3, the backsattering coefficient vs. frequency curve changes both in position and shape with the size of the scatterer when calculated for a particular model. The model used in Fig. 3 was the spherical scatterer (cf. Ueda and Ozawa, 1985; Romijn et al., 1989).

The slope of the linear approximation decreases in a monotonous fashion when changing the size from 20 to 500 μm. So the slope estimate yields a unique size estimate. This idea was successfully applied in the assessment of an animal model of melanomas (Romijn et al., 1989), in clinical studies of eye tumours (Feleppa et al., 1986; Romijn et al., 1991; Thijssen et al., 1991) and for estimating the size of renal glomeruli (Hall et al., 1996).

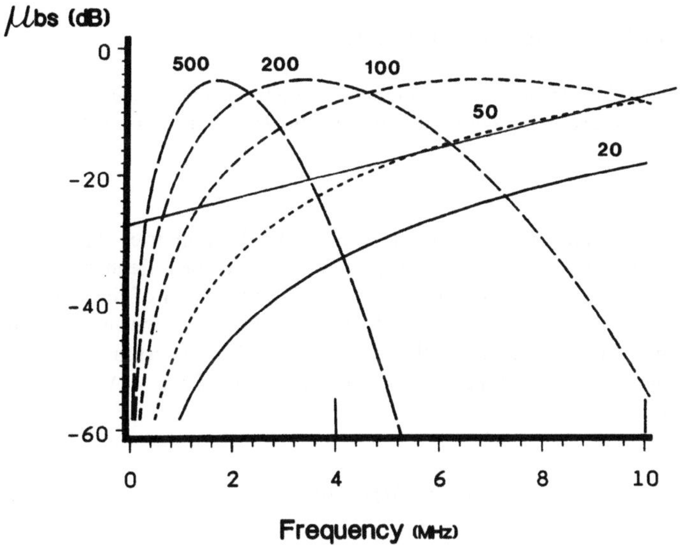

Figure 3. Backscattering coefficient vs. frequency for scatterers of indicated size (micrometers). Linear fit yields two parameters: backscattering at central frequency (b_0) and slope constant (b_1), cf. Eq. (10).

Accuracy and precision

The accuracy and precision of the estimated parameters have recently been reconsidered (Huisman and Thijssen, 1996, 1998[a]; Lizzi et al., 1997). Apart from the inherent uncertainty of echographic signals (i.e. "speckle noise"), the number of scan lines N, the window size D, number of windows, L along scan lines, frequency bandwidth W, of the employed transducer, play an important role.

It could be shown (Huisman and Thijssen, 1996) that mean of the logarithmic spectrum (i.e. in dB) according to Eq. (7) and the standard deviation at each discrete (sampled) frequency equal:

$$\mu_T = 2z\alpha(f_i)$$
$$\sigma_T = \frac{4.54}{\sqrt{N}} \tag{11}$$

The midband value and slope of the attenuation coefficient can be estimated with a precision (c.f., Huisman and Thijssen, 1996):

$$\sigma_{\alpha_0} = \frac{7.86}{D\sqrt{KLN}} = \frac{5.56\sqrt{c}}{\sqrt{D^3 NW}} \tag{12a}$$

$$\sigma_{\alpha_1} = \frac{33.24}{DW\sqrt{KLN}} = \frac{19.26\sqrt{c}}{\sqrt{D^3 NW^3}} \tag{12b}$$

Where: c = speed of sound

K= number of discrete frequencies within band W.

Similar expressions could be derived for the midband value b_o, and the slope b_1 of the linear approximation of the backscatter coefficient (Eq.9) (c.f., Huisman and Thijssen, 1996; Lizzi et al., 1997).

Elastography

Elastography, or Elasticity Imaging is a rather new field of UTC (Ophir et al., 1991, 1996; Svoroda et al., 1994; Gao et al. 1996). The basic principle is, that a radiofrequency echogram is acquired two times: once before and once after a small indentation of the tissue by the transducer (e.g., 0.2 mm). By comparing the signals at two depths within the analysis window, the relative change in mutual distance due to the external force yields an estimate of the strain. When denoting the distance prior to force: l, and after applying the force: $l - \Delta l$, the strain s is given by: (cf. Fig. 4)

$$s = \frac{\{l-(l-\Delta l)\}}{l} = \frac{\Delta l}{l} \qquad (13)$$

Precompression A-line

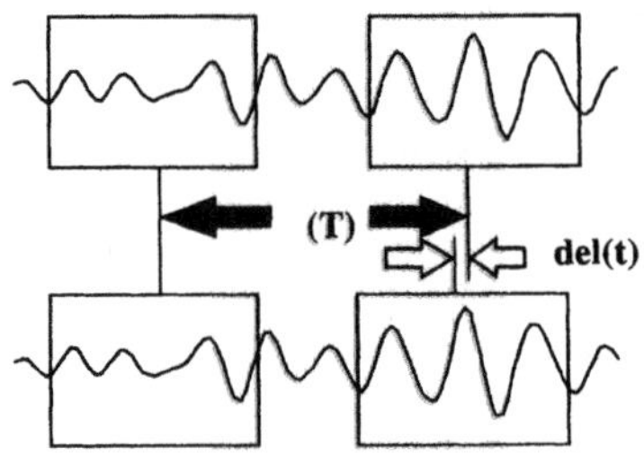

Postcompression A-line

Figure 4. Radiofrequency signals before (top) and after (bottom) mechanical compression of medium. Time windows are crosscorrelated to estimate the "time shifts" due to deformation of both windows separately. Change in time distance, del(t), multiplied by the speed of sound yields the compression (Δl in Eq. 13). (Courtesy: J.Ophir, Univ.Texas, Houston).

A problem with the use of rf-data for the estimation is the deformation, even within an analysis window, of the signal at larger compressions (>>1%). The correlation algorithm is not finding the correct time shift any more. This so-called signal decorrelation imposes a theoretical limit on the accuracy of strain estimation as well. Recently, Ophir and colleagues investigated two new methods for strain estimation. The use of the signal envelope (i.e., demodulated rf-signals) in the correlation resulted in improved accuracy for larger compressions as compared to rf-based methods, and the reverse for small compressions (Varghese and Ophir, 1998).

The second method employs the power spectrum estimate vs. depth, where it could be shown that the compression results in a small upward shift of the mean frequency (centroid) due to compression. This method is far more robust than time domain correlation methods (Konofagu et al., 1999). The authors anticipate to enhance clinical applicability of elastography by these techniques, such as enabling “free hand” elastography.

The promising aspect of elastography is, that it yields 2D images with relatively large contrasts and high signal-to-noise ratio (Belaid et al., 1994; Varghese and Ophir, 1998). Moreover, the range of tissue stiffness' is relatively large) e.g., Krouskop et al., 1998). So, it seems evident that elastography has a potential of becoming a quantitative method of "palpation", where the grey level in the images is indicative of the hardness of a lesion.

A rather innovative technique was introduced recently by Heimdal et al., (1998): strain rate imaging (SRI). Basically, it is derived from a correlation technique of ultrasound velocity estimation, which is called Tissue Doppler Imaging. The low velocity, high amplitude Doppler signals obtained from the heart walls, or from arterial walls, are used to calculate the local velocity gradient along scan lines in 2D velocity images. This local velocity gradient can be rewritten:

$$SRI = \frac{\partial v_{z_i}}{\partial z} = \frac{\partial(\frac{dz}{dt})_{z=z_i}}{\partial z} = \frac{ds_{z_i}}{dt} \tag{14}$$

Where: v_{zi} = velocity of (heart wall) movement at depth z
s_{zi}= strain at z_i

The technique is being evaluated for the assessment of regional myocardial disturbances.

TEXTURE

Speckle formation

The scattering due to small (subwavelength) inhomogeneities yields an image of "soft" tissues. Since, the echoes from nearby scatterers arrive at the transducer face with small time differences interference occurs. The radiofrequency A-lines constituting a 2-D image are displaying random amplitude vs. time, as well as small phase jumps. When considering amplitude demodulation as estimation of the envelope of the analytical rf signals, the interference may be described as a random walk in 2-D space. If the number of scatterers within the resolution cell of the transducer is high (>10) then it follows from the central limit theorem, that the analytical signal has a 2-D Gaussian probability distribution function (p.d.f.). It can be shown that in this case the magnitude (envelope) has a Rayleigh p.d.f. and the phase a uniform p.d.f. (c.f. Papoulis, 1965).

In 2-D imaging, the sound beam is moved, which means a gradual decorrelation of the axial speckle pattern. This gradual change results in a pattern of "banana" shaped speckles. Each speckle has some similarity with the 2-D image of a single point reflector (i.e. the point-spread-function, PSF). It should be realised, however, that the lateral speckle size is highly dependent on the distance to the transducer and not a direct image of any structure!

Texture statistics

Texture can be defined as the arrangement of grey levels in an image. The most simple quantification of this arrangement is the grey level histogram, which is a first order statistic. This means that the texture is measured pixel wise, without considering the relations among pixels (first order statistics of image texture).

For large scatterer number density the histogram of echographic images, when the amplitude of the echostrength is processed linearly, obeys the Rayleigh probability distribution function (pdf; e.g. Burckhardt, 1978; Wagner et al., 1983):

$$p(\hat{e}) = \frac{\hat{e}}{\sigma^2}\exp\{-\frac{\hat{e}^2}{2\sigma^2}\} \tag{15}$$

Where: $\hat{e}$ = demodulated signal ("envelope")
σ^2 = signal power.
It can be shown (c.f., Papoulis, 1965) that:

$$\mu_{\hat{e}}^2 = \frac{\pi\sigma^2}{2}$$
$$\sigma_{\hat{e}}^2 = \sigma^2(2-\frac{\pi}{2}) \tag{16}$$

Hence, the so-called "signal-to-noise" ratio (SNR) becomes:

$$SNR = \frac{\mu}{\sigma} = (\frac{\pi}{4-\pi})^{1/2} \tag{17}$$

The general expression (irrespective of scatterer density) for the pdf of the signal intensity I is given by (Jakeman, 1984):

$$p(I) = ½\int_0^\infty udu J_0(u\sqrt{I}) < J_0^N(u\beta) >_{N,\beta} \tag{18}$$

Where: J_0 = Bessel function of zero order, first kind
β = scattering amplitude of individual scatterers
N = number of scatterers within resolution cell.
This equation can not be analytically solved, but expressions were obtained for the moments of the pdf. For large N (i.e., N>10) this pdf turns into the exponential pdf.
It was shown (Jakeman, 1984) that the "signal-to-noise ratio"(SNR) equals:

$$SNR_I = \frac{\mu_I}{\sigma_I} = (1+\frac{<\beta^4>}{N<\beta^2>^2})^{-½} \tag{19}$$

Taking the scatterer number density n and the volume corresponding to the transmitted ultrasound pulse (in focus) equal to V, then N=nV. The number density can then be estimated from Eq. (18) if β is constant. In soft biological tissues $\beta \approx 3$ (cf. Sleefe and Lele, 1988).

More recently, it was shown that the pdf of the envelope amplitude for low (effective) scatterer density is given by the K-distribution (Jakeman and Tough, 1987):

$$p(\hat{e}) = \frac{2b}{\Gamma(\alpha)}(\frac{b\hat{e}}{2})^\alpha K_{\alpha-1}(b\hat{e}) \tag{20}$$

Where: $b = 2\sqrt{\frac{\alpha}{E(\hat{e}^2)}}$
Γ = Gamma function
$\alpha = (\mu+1)$ N= effective number of scatterers within resolution cell.
μ = constant, depending on scatterer – and transducer characteristics.
K_x = modified Bessel function of second kind, and order x.
The moments of the K-distribution have been analytically derived and may be used to calculate

the signal-to-noise ratio of the envelope (Dutt and Greenleaf, 1995):

$$SNR_{\hat{e}} = \frac{\sqrt{\pi}\Gamma(\alpha + \frac{1}{2})}{\sqrt{4\Gamma(\alpha+1) - \pi\Gamma^2(\alpha + \frac{1}{2})}} \tag{21}$$

It can be shown that in the limit case, where α goes to infinity (in practice $\alpha \geq 10$) the K-distribution approaches the Rayleigh pdf ("fully developed speckle").

It should be realised, that in general terminology of echographic scanners a 'linear' processing means that a linear look-up table is used to encode the image gray levels. The systems use nevertheless a logarithmic compression amplifier prior to digitisation, which yields an overall histogram that is considerably different from the Rayleigh pdf (Kaplan and Ma, 1994; Dutt and Greenleaf, 1996). This pdf of the logarithmic echosignals yields a fixed standard deviation (i.e., additive, rather than multiplicative, speckle noise).

Recently, it has been argued that when a structural component is present in the location of backscattering sites in tissue (i.e. a more or less regular pattern, like in the distribution of liver triads), the Rayleigh pdf should be replaced by the Rician pdf (e.g., Goodman, 1975; Insana et al., 1986; Thijssen, 1992).

A method has been developed to unravel the contributions of the diffuse and structural scattering components (Wagner et al., 1986; Insana et al., 1986). The characteristic distance of the structural scatterers is revealed in the power spectrum by spectral peaks superimposed on the spectrum corresponding to the diffuse scattering.

Taking the characteristic structural distance Δz_s, i.e. a time difference

$$2\Delta\tau = \frac{2\Delta z_s}{c} \tag{22}$$

which equals the two-way travel time in pulse-echo mode. Then it can be shown, that F(f), the backscattered power spectrum (in the temporal frequency domain, 1D case), is given by (Mesdag, 1985; Jacobs and Thijssen, 1991):

$$|F(f)| = \left|\frac{\sin N2\pi f\tau}{\sin 2\pi f\tau}\right| |S(f)| \tag{23}$$

Where: N= number of scatterers within duration of transmitted ultrasound pulse, τ.

S= power spectrum, corresponding to the envelope of diffusely backscattered echo signal.

The spectral peaks can be analysed by estimating the first peak at $f > 0$, or more sophisticated methods like for instance cepstral analysis (Varghese and Donehue, 1995) could be applied. Till now the analysis, initiated by Fellingham and Sommer (1984), has not been highly successful in clinical applications. The simulation study by Jacobs and Thijssen (1991) indicated that if the structural distance variations throughout the tissue volume are larger that 10 percent of the mean, the distance estimate becomes inaccurate.

Texture Analysis

The first order statistics are quantified by calculating the moments and derived parameters. A useful derived histogram parameter is the signal-to-noise ratio (SNR) (Eq. 17).

This parameter was analysed as an UTC parameter (Oosterveld et al., 1985; Dutt en Greenleaf 1995) and depends on the density of scatterers (Eqs. 19,21). The SNR was also used for parametric imaging (Verhoeven et al., 1991).

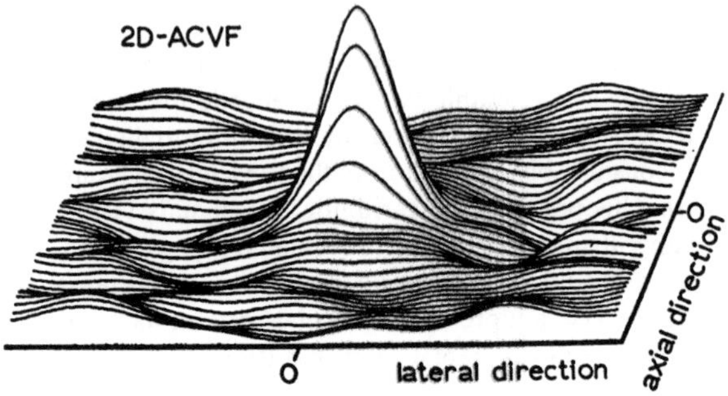

Figure 5. Two-dimensional graph of the autocorrelation function of an echographic image of a homogenous, isotropic scattering medium (Oosterveld et al., 1985).

The second order statistics are used to describe and quantify the relations between neighbouring pixels. At first, the autocovariance function (ACVF) was introduced (Wagner et al., 1983; Oosterveld et al., 1985; cf. Fig. 5) which can be considered as a measure of the average size of the speckles in echographic images:

$$ACVF(x,z) = \iint_{ROI} d\xi d\zeta \{\hat{e}(\xi + x, \zeta + z) - <\hat{e}>\}\{\hat{e}(\xi,\zeta) - <\hat{e}>\} \tag{24}$$

Where: ROI = "region of interest" area

$<\hat{e}>$ = mean of signal envelope in ROI

x,z = lateral, axial coordinates, respectively.

The average size of the speckles in the focal zone of the transducer (i.e., plane wave conditions) can be quantified by the full-width-at-half-maximum (FWHM) of the ACVF (Wagner et al., 1983; Oosterveld et al., 1985):

$$FWHM_x = 0.80\frac{\lambda f}{D} \qquad FWHM_z = \frac{0.61}{\Delta f} \tag{25}$$

Where: $\lambda = c/f_c$ = wavelength of transmitted ultrasound at center frequency.

F = focal distance

D' = effective diameter of transducer (=D/1.08)

Δf= halfwidth (-6dB width) of spectrum corresponding to transmitted vs. pulse.

Eq. (21) applies in case of fully developed speckle (i.e. high scatterer number density) in the focal zone. It may be remarked, that the single scatterer case, corresponding to the point-spread-function (PSF) of the used transducer can be written in formulas similar to Eq. (18), and yield 10 to 20 percent larger values (Thijssen, 1992).

The texture analysis based on a tissue model containing both diffuse and structural scattering (Fig. 1) was undertaken by Wagner et al. (1986) and Insana et al. (1986). These authors, among other parameters, estimated the "Rician" variance ($\sigma_r^2(I)$) of the backscattered intensity from the Fourier transform of the ACVF. This is equivalent to the area under the curve fitted through the minima of the estimated power spectrum. This curve is modelled (or measured) as the power spectrum derived form a plane reflector in focus. Subtraction of the Rician power from the total area under the spectrum yields the variance of the structurally backscattered power, $\sigma_s^2(I)$.

These variances can be used as tissue parameters, or to calculate the backscattered diffuse and structural intensities. The latter procedure yielded algebraic instabilities in a study of diffuse liver diseases (Oosterveld et al., 1991).

The speckle size is not only dependent on the transducer characteristics, but varies with the scatterer number density as well (Oosterveld et al., 1985). A serious problem with the ACVF as a texture parameter is that in particular the lateral size is strongly dependent on the

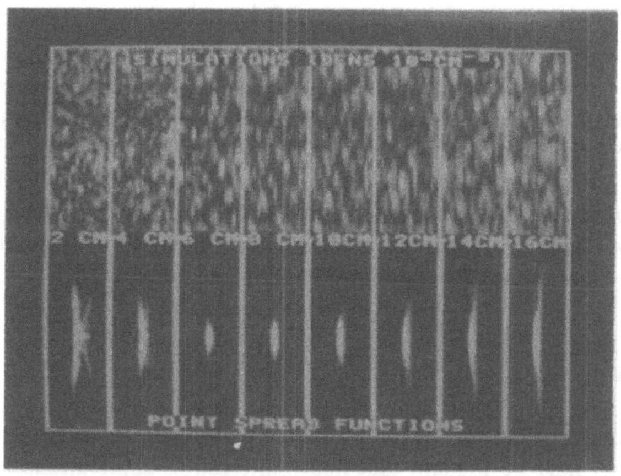

Figure 6. Top: Simulated echograms with increasing depth from left to right; note fine texture near to the transducer (left) and increasing speckle size at larger depth (right). Bottom: Echograms of point-like reflector at same depths; note narrowing of this so-called "point-spread-function" in the focal zone.

depth and also on the attenuation within the tissue (Oosterveld et al., 1985, Huisman and Thijssen, 1998[a]). These dependencies have often been overlooked, or neglected, in literature. The only way out seems to be to analyse the second order statistics in the axial direction exclusively. The problem remains that even after the diffraction, attenuation corrections (see Section 2.1 Pre-processing) and software demodulation, the lateral speckle size is still strongly depth dependent. An example of the pure speckle pattern vs. depth obtained by simulations is shown in Fig. 6, together with the images of a point-like reflector at the same depths as the "tissue" images.

Further methods of second order statistics analysis have been taken from the general image processing literature. An often used statistic is the co-occurrence matrix (Haralick et al., 1973). This is actually a second order (i.e. 2-dimensional-) histogram approximating the probability of occurrence of all combinations of two grey levels for a set of two pixels at a certain spatial distance (or "pixel displacement"). Many parameters can be derived from this matrix, some of which are often used in UTC: contrast, correlation, entropy and angular second moment (cf. Thijssen et al., 1993; Valckx and Thijssen, 1997).

Several other second order statistics applied to UTC can be found in literature of the last ten years. It is questionable whether these contribute in essentially different ways as compared to the ACF and the co-occurrence matrix. In any case, at the scale of the size of the speckles (i.e. small displacements), all methods will be equivalent. At a larger scale (supra-speckle scale) macroscopic changes in tissue structure should be present which can produce a kind of modulation of (local-mean-) grey levels, e.g. as is found in a fibrotic liver. In that case, the displacement used in the estimation of the second order statistic should be much larger than most authors have considered sofar. An alternative to these methods has recently been presented by Huisman (Huisman and Thijssen, 1998[b]) in the way of an adaptive feature extraction algorithm based on higher order neural networks (HONN). This approach does not make assumptions about the scale and it can be based on first and/or second order statistics without further restrictions. Their results, as worked out for the first order case, are rather promising.

Classification

This is the final goal of UTC: labelling an analysed window, or a ROI, as normal or, alternatively, as diseased. The number of estimated parameters used for the classification can be relatively high, but the number of independently proven cases of a disease is however very often relatively low. This means, that learning the "classifier", i.e. the algorithm to decide, is based on a limited set of clinical cases. The classifier can be e.g. a discriminate function (Bayesian classifier; Oosterveld et al., 1993; Wagner et al., 1995), a neural network (Klein Gebbink et al., 1993; Huisman and Thijssen, 1998[b]; Schmitz and Ermert, 1999), or a decision tree (Giessen et al., 1996). In any case, it has been shown by Foley (1972) that the number of parameters that can be taken into account in a linear classification algorithm should not exceed the smallest number of cases, in any of the classes, divided by three (or more, in case of non-linear classifiers). This rule is often not considered in literature!

CONCLUSIONS

Although not all clinical fields of UTC are covered by this review (e.g. skin, skeletal muscle, eye), it nevertheless may be evident that many interesting applications are emerging. The tendency of Industry to incorporate high performance, very powerful computing hardware into the echographic equipment might lead very soon to feasibility studies in which UTC is not only *on-line*, but even *in real time* available to the diagnostician. The author believes these will be the conditions that may induce clinical interest and thereby will bring about more and more clinical evidence and, in the end, broad acceptance. One of the major challenges still is to find the optimal set of UTC parameters to be used in "(multi)parametric imaging" (e.g. Schmitz and Ermert 1999; Valckx et al., 1996;2000). Indeed this set has to be established by analysing a high number of clinical cases and by comparison with some kind of gold standard. Further progress is to be expected from new techniques, which are being, introduced such as: elastography, contrast echography and second harmonic imaging. The introduction of these techniques has proceeded to a more or less clinical phase of explorations. Applications are proceeding towards the fields of tissue differentiation and tumour diagnosis. The elastography by assessing the hardness of tissues, the contrast echography and second harmonic imaging by improved contrast of (neo)vascularization of tissues and by dynamic flow and perfusion studies of tissues.

The future of ultrasound is still bright, but as before, we can state that future has begun!

REFERENCES

Belaid, N., Cespedes, I., Thijssen, J.M., and Ophir, J., 1994, Lesion detection in elastography versus echography: a quantitative analysis, *Ultrasound Med. Biol.*, 20: 877-91.

Burckhardt, C.B.,1978, Speckle in ultrasound B-mode scans., *IEEE Trans. Son. Ultrason.*, SU-25: 1-6.

Cespedes, I., Huang,Y., Ophir, J., and Spratt, S., 1995, Methods for the estimation of sub sample time delays of digitized echo signals, *Ultrasonic Imag.* 17: 142-71.

Cloostermans, M.J.T.M., and Thijssen, J.M., 1983, A beam corrected estimation of the frequency dependent attenuation of biological tissues from backscattered ultrasound, *Ultrasonic Imag.* 5: 136-47.

Dutt,V., and Greenleaf, J.F., 1995, Speckle analysis using signal to noise ratios based on fractional order moments, *Ultrasonic Imag.* 17: 251-68.

Dutt,V., and Greenleaf, J.F., 1996, Statistics of the log-compressed echo envelope, *J. Acoust. Soc. Am.* 99: 3817-25.

Feleppa, E.J., Lizzi, F.L.,Coleman, D.J., and Yaremko, M.M., 1986, Diagnostic spectrum analysis in ophthalmology, *Ultrasound Med. Biol.* 12: 623-31.

Foley, D., 1972, Considerations of sample and feature size. *IEEE Trans. Inform. Theory* IT-18: 618-26.

Fellingham, L.L., and Sommer, F.G., 1984, Ultrasonic characterization of tissue structure in the in vivo human liver and spleen, *IEEE Trans. Son. Ultrason.* SU-31: 418-28.

Fink, M., Hottier, F., and Cardoso, J.F., 1983, Ultrasonic signalprocessing for invitro attenuation measurement: short time Fourcer analysis, *Ultrasonic Imag.* 5: 117-35.

Fink, M.A., and Cardoso, J.F., 1984, Diffraction effects in pulse- echo measurements, *IEEE Trans. Son. Ultrason.* SU-31: 313-29.

Gao, L., Parker, K.J., Lerner, R.M., and Levinson, S.F., 1996, Imaging of elastic properties of tissue, a review, *Ultrasound Med. Biol.* 72: 959-77.

Giessen, R.J.B., Huynen, A.L., Aarnink, R.G., de la Rosette, J.J.M.C.H., Debruyne, F.M.J., and Wijkstra, H., 1996, Construction and application of hierarchical decision tree for classification of ultrasonic prostate images, *Med. Biol. Eng. Comp.* 15: 105-9.

Goodman, J.W., 1975, Statistical properties of laser speckle, in: *Laser Speckle and Related Phenomena,* J.C. Dainty, ed., Springer Verlag, Berlin. P. 9-75.

Hall, T.J., Insana, M.F., Harrison, L.A., and Cox, G.G., 1996, Ultrasonic measurement of glomerular diameters in normal adult humans, *Ultrasound. Med. Biol.* 22: 987-97.

Haralick, R.M., Shanmugam, K., and Dinastein, I., 1973, Textural features foe image classification. *IEEE Trans. Syst. Man. Cyber.* TSMC-3: 614-24.

Heimdal, A., Stoylen, A., Torp, H., and Skaerpe, T., 1998, Real time strain rate imaging of the left ventricle by ultrasound, *J. Am. Soc. Echocardiogr.* 11: 1013-9.

Huisman, H.J., and Thijssen, J.M., 1996, Precision and accuracy of acoustospectrographic parameters, *Ultrasound Med. Biol.* 22: 855-71.

Huisman, H.J., and Thijssen, J.M., 1998[a], An in vivo ultrasonic model of liver parenchyma, *IEEE Trans. Ultrason. Ferroel. Freq. Contr.* UFFC-45: 739-50.

Huisman, H.J., and Thijssen, J.M., 1998[b], Adaptive feature extraction with application to ultrasonic image analysis, *Ultrasonic Imag.* 20: 132-48.

Insana, M.F., Wagner, R.F., Garra, B.S., Brown, D.G., and Shawker, T.H., 1986, Analysis of ultrasound image texture via generalized Rician statistics, *Opt Eng* . 25: 743-8.

Jacobs, E.M.G.P., and Thijssen, J.M., 1991, A simulation study of echographic imaging of structurally scattering media, *Ultrasonic Imag.* 13: 316-33.

Jakeman, E.,1984, Speckle statistics with a small number of scatterers. *Opt. Eng.* 23: 453-61.

Jakeman, E., and Tough, R.J.A., 1987, Generalized k-distribution; a statistical model for weak scattering, *J. Opt. Soc. Am.* 4: 1764-72.

Kak, A., and Dines, K.A., 1978, Signal processing of broad band pulsed ultrasound measurement of attenuation in soft tissues, *IEEE Trans. Biomed. Eng.* BME-25: 321-44.

Kaplan, D., and Ma, Q., 1994, On the statistical characterization of log-compressed Rayleigh signals, *Proceedings IEEE Ultrasonics Symposium.* P. 961-74.

Klein Gebbink, M.S., Verhoeven, J.T.M., Thijssen, J.M., and Schouten, T.E., 1993, Application of neural networks for the classification of diffuse liver disease by quantitative echography, *Ultrasonic Imag.* 15: 205-17.

Konofagou, E.E., Varghese, T., Ophir, J., and Alem, S.K., 1999, Power spectral strain estimators in elastography, *Ultrasound Med. Biol.* 25: 1115-29.

Krouskopf, T.A., Wheeler, T.M., Kallel, F., Garra, B.S., and Hall, T., 1998, Elastic moduli of breast and prostate tissues under compression. *Ultrasonic Imag.* 20: 260-74.

Langton, C.M., Evans, G.P., and Riggs, C.M., 1990,A contact method for the assessment of ultrasonic velocity and broadband attenuation in cortical and cancellous bone, *Clin. Phys. Physiol. Meas.* 11: 177-80

Levinson, S.F., 1988, The ultrasonic investigation of dynamic muscle elasticity in vivo, *J. Ultrasound Med.* 7: 127 (abstract).

Linzer, M., 1976, ed., *Ultrasonic Tissue Characterization I.*, NBS Spec Publ, Gaithersburg 453.

Lizzi, F.L., Feleppa, E.J., Astor, M., and Kalisz, A., 1997, Statistics of ultrasonic spectral parameters for prostate and liver examinations, *IEEE Trans. Ultrason. Ferroel. Freq. Contr.* UFFC-44: 935-42.

Mesdag, P.R., 1985, *Estimation of Medium Parameters by Acoustic Echo Measurements,* PhD Thesis, Technological University, Delft, The Netherlands.

Mountford, R.A., and Wells, P.N.T., 1972, Ultrasonic liver scanning: the A-scan in the normal and cirrhosis, *Phys. Med. Biol.* 17: 261-9.

Oosterveld, B.J., Thijssen, J.M., and Verhoef ,W.A, 1985, Texture in B-mode echograms: 3-D simulations and experiments of the effects of diffraction and scatterer density, *Ultrasonic Imag* 7: 142-60.

Oosterveld, B.J., Thijssen, J.M., Hartman, P.C., Romijn, R.L., and Rosenbusch, G.J.E., 1991, Ultrasound attenuation and B-mode texture analysis of diffuse liver disease: methods and preliminary results, *Phys. Med. Biol.* 36: 1031-64.

Oosterveld, B.J., Thijssen, J.M., Hartman, P.C., and Rosenbusch, G.J.E., 1993, Detection of diffuse liver disease by quantitative echography; dependence on a priori choice of parameters, *Ultrasound Med Biol.* 19: 21-5.

Ophir, J., Cespecles, I., Ponnekant, H., Yazdi, Y., and Li, X., 1991, Elastography: a method for imaging the elasticity in biological tissues, *Ultrasonic Imag.* 13: 111-34.

Ophir, J., Cespedes, I., Garra, B., Ponnekanti, H., Huang, Y., and Maklad, N., 1996, Elastography: ultrasound imaging of tissue strain and elastic modulus in vivo, *Eur. J. Ultrasound* 3: 49-70.

Papoulis, A., 1965, *Probability, Random Variables and Stochastic Processes*, McGrawHill New York .

Robinson, D.A., Wilson, L.S., and Bianchi, T., 1984, Beam pattern (diffraction) correction for ultrasonic attenuation measurement, *Ultrasonic Imag.*6: 293-303.

Romijn, R.L., Thijssen, J.M., van Delft, J.L., de Wolff-Rouendaal, D., van Best, J., and Oosterhuis, J.A., 1989 In nivo ultrasound backscattering estimation for tumor diagnosis: an animal study, *Ultrasound Med. Biol.* 15: 471-9.

Romijn, R.L., Thijssen, J.M., Oosterveld, B.J., and Verbeek, A.M., 1991, Ultrasonic differentiation of intracular melanoma: parameters and estimation methods, *Ultrasonic Imag.* 13: 27-55.

Schmitz, G., and Ermert, H., 1999, Tissue characterization of the prostate using radiofreuency ultrasonic signals, *IEEE Trans. Ultrasonics Feroel. Freq. Control* 46: 126-38.

Sleefe, G.E., and Lele, P.P., 1988, Tissue characterization based on scatterer number density estimation, *IEEE Trans. Ultrasonics Ferroel. Freq. Control* UFFC-35: 749-57.

Svoroda, A.R., Emilianow, S.Y., Lubinski, M.A., Sarvazyan, A.P., and O'Donnell, M., 1994, Theoretical analysis and verification of ultrasound displacement and strain imaging, *IEEE Trans. Ultrason. Ferroel. Freq. Contr.* UFFC-41: 302-13.

Thijssen, J.M., ed., 1980, *Ultrasonic Tissue Characterization: Clinical Achievements and Technological Potentials,* Stafleu, Alphen aan de Rijn.

Thijssen, J.M., Cloostermans, M.J.T.M., and Verhoef ,W.A., 1981, Measurement of ultrasonic attenuation in tissues from scattering reflections: in vitro assessment of applicability, in: *Ultrasonography in Ophthalmology 8,* Thijssen, J.M., Verbeek, A.M., ed., Dr. W. Junk Publishers, The Hague. 431-9.

Thijssen, J.M., Oosterveld, B.J., Hartman, P.C., and Rosenbusch, G.J.E., 1993, Correlations between acoustic and texture parameters from rf- and B-mode liver echograms, *Ultrasound Med. Biol.* 19: 13-23.

Thijssen, J.M., 1992, Echographic image processing, In: *Advances in Electronics and Electron Physics Vol. 84,* Hawkins, P.W.,ed., Academic Press, Boston, pp. 317-49.

Thijssen, J.M., Verbeek, A.M., Romijn R.L., De Wolff-Rouzendaal, D., and Oosterhuis, J.A., 1991, Echographic differentiation of histological types of intraocular melanoma., *Ultrasound Med. Biol.* 17: 127-38.

Ueda, M., and Ozawa, Y., 1985, Spectral analysis of echoes for backscattering measurement, *J. Acoust. Soc. Am.* 77: 38-47.

Valckx, F.M.J., van Geemen, A.J., and Thijssen, J.M., 1996, Online parametric ultrasound imaging, *Ultrasonic Imag.* 18: 51.

Valckx, F.M.J., and Thijssen, J.M., 1997, Characterization of echographic image texture by cooccurrence matrix parameters, *Ultrasound Med Biol.* 23: 559-71.

Valckx, F.M.J., Thijssen, J.M., van Geemen, A.J., and Rotteveel, J.J., 2000, Calibrated parametric ultrasound imaging, *Ultrasonic Imag* : in press.

Varghese, T., and Donehue, K.D., 1995, Estimating mean scatterer spacing with the frequency-smoothed spectral autocorrelation method, *IEEE Trans. Ferroel. Freq. Contr.* UFFC-42: 451-63.

Varghese, T., and Ophir, J., 1998, Characterization of elastographic noise using the envelope of echo signals, *Ultrasound Med. Biol.* 24: 543-55.

Verhoef, W.A., Cloostermans, M.J.T.M., and Thijssen, J.M., 1985, Diffraction and dispersion effects on the estimation of ultrasound attentuation and velocity in biological tissues, *IEEE Trans. Biomed. Eng.* BME-32: 521-9.

Verhoeven, J.T.M., Thijssen, J.M., and Teeuwes, A.G.M., 1991, Improvement of lesion detection by echographic image processing; signal-to-noise ratio imaging, *Ultrasonic Imag.* 13: 238-51.

Wagner, R.F., Smith, S.W., Sandrik, J.M., and Lopez, H., 1983, Statistics of speckle in ultrasound B-scans, *IEEE Trans. Son. Ultrason.* SU-30: 156-63.

Wagner, R.F., Insana, M.F., and Brown, D.G., 1986, Unified approach to the detection and classification of speckle texture in diagnostic ultrasound, *Opt. Eng.* 25: 738-42.

Wagner, R.F., Wear, K.A., Perez, J.E., McGill, J.B., Schechtman, K.B., and Miller, J.G., 1995, Quantitative assessment of myocardial ultrasound tissue characterization through receiver operating characteristic analysis of Baysian classifiers, *J. Am. Coll. Cardiol.* 25: 1706-11.

DESIGN-DIRECTED RESEARCH: *A NEW FUNDING PARADIGM*

Gilbert B. Devey

Program Director
Bioengineering and Environmental Sciences Division
National Science Foundation
Arlington, VA 22230

INTRODUCTION

The 20th century delivered enormous advances in medical science and, especially during the period 1950-1999, produced a myriad of revolutionary medical technologies; advances in medical imaging are arguably the most remarkable. Early in the last century the art and science of medicine was largely *experience-based* but by mid-century it had become a *science-based* activity. During the latter half of the century the extensive R&D programs conducted by the aerospace, defense, telecommunications, and computer industries produced important new technologies, a number of which were adapted for medical purposes. The new technologies resulted from *design-directed* (problem-solving) investigations, which contrasts with the biological/biomedical sciences where research is *hypothesis-driven*. The classical research structure that served so well in yesteryear is now being challenged: emerging medical technologies require systemic and cultural changes in the research management paradigm if they are to flourish. Research in medical ultrasound would benefit.

EXPERIENCE-BASE TO SCIENCE-BASE MEDICINE

The schema for medical education and medical research established by the founders of the Johns Hopkins University (1876) had been widely adopted in the United States by the first quarter of the 20th century. Modeled on the German system's emphasis on the scientific basis of medicine, "the essential element...was the idea that the hospital should form a part of the medical school...a single individual headed both the hospital department

Acoustical Imaging, Volume 25, Edited by Halliwell and Wells.
Kluwer Academic/Plenum Publishers, 2000.

and the corresponding department in the medical school...the hospital contained laboratories and classrooms in close proximity to the wards."[1] The Hopkins model worked well until the mid-1950s when government support for training and research substantially altered its features.

There were few regulatory or economic barriers to the introduction of new medical technologies early in the 20th century but the paucity of scientific knowledge available to the basic and the clinical sciences was a major barrier that impeded clinically useful medical technology developments. The art and science of medicine by and large was *experience-based,* one result of which was the misuse of otherwise bona fide devices and drugs. When *x-ray imaging* began in clinical settings the lack of knowledge of the effects of ionizing radiation led to unnecessary death and suffering, and *thalidomide* caused birth defects due to incomplete information gleaned during the testing process. By the mid 20th century the *experience-base* nature of biomedical research was rapidly being transformed into a *science-base* activity. But the health care enterprise lacked the infrastructure essential for technological developments to reach their full potential. A number of social programs (e.g., health insurance), public policy concepts (e.g., government support of medical research and training; drug/device regulation; provider accreditation; technology assessment), and business concepts (e.g., academic researchers collaborating with industry; large-scale industrial research programs) remained to be established during the latter half of the century.

Medical technologies (devices) typically emerge as derivatives of technologies developed for other purposes; the massive R&D programs conducted by the aerospace, defense, telecommunications, and computer industries during the latter half of the 20th century produced important new technologies, a number of which were adapted for medical purposes[2]. The new technologies derived from investigations that are *problem-solving* in nature, in contrast with the biological/biomedical sciences where research is *hypothesis-driven*. This was pointed out in 1965 by NIH (National Institutes of Health) Director James A. Shannon who reported "to develop a practical device...requires the know-how of the engineering industry...*These* (sic) *are not the common characteristics of the scientists we support through our grant procedures*" [emphasis added][3]. But, even today, the NIH grant application Form 398 calls for the applicant to state the hypotheses to be tested. The dominance of the *hypothesis-driven* concept is believed by many to have impeded the development of medical technologies[4].

HARBINGERS OF CHANGE

U.S. National Institutes of Health (NIH)

Bioengineering Consortium (BECON). In 1997 NIH established BECON to address current needs and prepare for the future of this emerging area. BECON has initiated three tangible and important funding initiatives to support *design-directed* (problem-solving) research.

1. Bioengineering Research Grants (PAR-00-009) to support basic bioengineering research. An application for support should propose basic bioengineering *design-directed* or *hypothesis-driven* research.

2. Bioengineering Research Partnerships (PAS-00-006) support a multidisciplinary research team applying an integrative, systems approach to develop knowledge and/or methods to prevent, detect, diagnose, and treat disease and understand health and behavior. An application may propose *design-directed* or *hypothesis-driven* research.
3. Bioengineering Nanotechnology Initiative (PA-00-018) is directed to small businesses under the Small Business Innovation Research (SBIR) program. A major objective is to encourage team approaches to research.

Center for Scientific Review (CSR). The CSR is in the second and final phase of a comprehensive review of the grant application review process. The review is to "foster the expanded research opportunities created by the stunning success of the biomedical research enterprise as well as to permit the review system to keep pace with the accelerating rate of change in the way biomedical research is now performed"[4].

The DRAFT Phase 1 Report notes that:

"Although "hypothesis-driven" research could be interpreted broadly to include...generating the knowledge needed to solve important problems and developing novel techniques...the practice of NIH study sections has been to interpret it narrowly ...Under these conditions, exploratory research using and developing new technology suffers....Concerning the role of preliminary data...For new ideas, little or no preliminary data may be required"[4].

Office of Bioimaging/Bioengineering. The FY2000 Appropriations Bill for the NIH states that research in the fields of biomedical imaging and engineering does not fit the organizational structure of the NIH. "For these reasons, NIH is urged to establish the Office of Bioimaging/Bioengineering and to review the feasibility of establishing an *Institute of Biomedical Imaging and Engineering*"[5]. Currently the NIH is in the process of establishing the Office.

National Cancer Institute (NCI). Three recent actions taken by NCI contribute to a shift in the research funding paradigm, a shift of considerable importance to the ultrasound research community.

1) The first NCI-Industry Forum and Workshop on Biomedical Imaging in Oncology was convened in September 1999 in response to the rapid pace of innovation, discovery, and new developments in cancer research, basic science, and imaging technology.

The meeting had four goals:

1. Bring together the people who fund and conduct the research, regulate products of the research and reimburse the use of the products.
2. Expand the role of anatomic, functional, and molecular imaging.
3. Develop strategies to apply advances in imaging to unmet clinical challenges in cancer.
4. Understand better how organizations involved or interested in imaging make decisions. how NCI identifies scientific opportunities; how industry develops and commercializes technology; how the FDA conducts its regulatory process; and how HCFA (Medicare) conducts and decides on reimbursement.

The Meeting Summary is at http://dino.nci.nih.gov/dctd/forum/summary.htm.

2) In June 2000 the NCI Biomedical Imaging Program (BIP) [URL: http://www.nci.nih.gov/dip/] expects to publish a Request for Proposals (RFP) for the development of an Ultrasound Interface to develop powerful ultrasound research interfaces that allow extensive control over a wide range of system parameters. These interfaces will be software add-ons to scanner control software that is continuously evolving. The

interface software could run either on the current ultrasound computer systems, or on stand-alone computer systems (PC's) interfaced to the ultrasound scanner control. Although the creation of these interfaces necessarily relies on access to proprietary information, such access can be partly restricted by protective software shells, while still accessing the control parameters. The intent of this concept is to allow research interfaces to be developed by several leading ultrasound manufacturers to maximize the research impact for the ultrasound imaging community.

3) The BIP is preparing a Program Announcement *Development of Novel Imaging Technologies*. Projects to be supported would be directed at the development of imaging methods and image enhancing materials, and limited evaluation or feasibility studies using either pre-clinical or clinical models. The intent is to stimulate: (a) the development of highly innovative imaging methods and enhancement methods, including high risk/high gain technologies that exploit our expanding knowledge of the molecular basis of cancer, and (b) the integration of these emerging and more traditional technologies for more effective solutions for cancer.

National Science Foundation (NSF)

In 1985 NSF began an Engineering Research Centers (ERC) Program to provide an integrated (multi-disciplinary; multi-institutional; multi-sector) environment for academe and industry to focus on next-generation advances in complex engineered systems. Activity within ERCs lies at the interface between the *discovery-driven* culture of science and the *innovation-driven* culture of engineering, creating a synergy between science, engineering, and industrial practice. ERCs provide the intellectual foundation for industry to collaborate with faculty and students on resolving generic, long-range challenges to produce the knowledge base for steady advances in technology and a speedy transition to the marketplace. Eight bioengineering-related and biotechnology-related ERCs have been funded* and an ERC *Best Practices Manual* is available[6].

CONCLUSION

The medical research enterprise has migrated from an experience-base to a science-base largely dominated by an hypothesis-driven culture. The stunning successes of the biomedical research endeavor, coupled with the enormous progress made by design-directed (problem-solving) investigations in the physical sciences and engineering have produced knowledge that requires a new paradigm of investigation. The new knowledge has produced the emerging technologies of tissue engineering, artificial organs, and image-guided procedures, plus nascent technologies such as MEMS (micro-electromechanical systems), WIMS (wireless-integrated micro-systems), bio-nanotechnology, smart drug

*
1. Neuromorphic Systems Engineering.
2. Center for Engineered Biomaterials
3. Biotechnology Process Engineering Center
4. Center for Biofilm Engineering
5. Center for Emerging Cardiovascular Technologies
6. Center for Engineering of Living Tissues
7. Center for Computer Assisted Surgical Systems and Technology
8. Center for Bioengineering Educational Technology

delivery systems, and genome mining. These new technology developments require systemic and cultural changes in the research management paradigm if they are to flourish.

Three of the developments mentioned herein are critical to progress in medical technology in the 21st century:

1. *Design-directed* research should be on the same level of importance as *hypothesis-driven* research.
2. A research/development management scheme based on the NSF ERC model should be adopted.
3. The incipient NCI vision for a discovery-to-application systemic model including the researcher, the developer (industry), the regulator, and the payer should supplant the existing disconnected process.

The NCI vision to integrate all elements of the discovery-to-application continuum is an innovative and important concept. The NCI approach would account for the major scientific, technical, social, and political factors necessary for medical technology development to flourish. But, installing such a model is likely to be a bureaucratic nightmare.

REFERENCES

1. Harvey, AM, GH Brieger, SL Abrams, VA McKusic. *A Model of its Kind.* Vol. 1. *a centennial history of medicine at Johns Hopkins.* The Johns Hopkins University Press, Baltimore and London (1989).
2. Wells, PNT. Healthcare and the medical engineering Phoenix. *Gresham College Lecture*, February 15, 1999.
3. U.S. Congress. House Subcommittee of the Committee on Appropriations, *Hearings, Department of Health, Education, and Welfare*, Part 3, NIH, 69th Congress, 1st Session, 1965; p. 180.
4. NIH. Recommendations for change at the NIH's Center for Scientific Review. *Cultural Changes*, p. 5. URL http://www.csr.nih.gov/bioopp/intro.htm
5. Congress directs NIH to create Office for Bioengineering. *The AIMBE News*, Vol. 7, No.:4, 1.
6. URL: http://www.erc-assoc.org/.

COMPUTATION OF LINEAR ULTRASOUND FIELDS USING 2D FOURIER-BESSEL SERIES

Paul D Fox

Department of Informatics
University of Oslo
P.O. Box 1080, Blindern
N-0316 Oslo, Norway

INTRODUCTION AND 1D FOURIER-BESSEL SERIES

A method for modelling 2D arrays and computing their propagated fields in linear media will be discussed. The field computation technique represents an extension from recent 1D Fourier-Bessel modelling of annular array surface pressure profiles to the 2D Fourier-Bessel modelling of 2D arrays. For ease of discussion, the main principles of the technique will be discussed initially in terms of 1D Fourier-Bessel series and then extended to 2D series in the latter sections. In[1,2] an N-ring annular array spatial pressure profile $q(r)$ corresponding to a time-varying transducer surface pressure $q(r)e^{-j\omega t}$ was modelled using the 1D Fourier-Bessel series

$$q(r) = \sum_{i=1}^{\infty} A_i J_0(\alpha_i r) \quad where \quad A_i = q_1 C_{i,1} + q_2 C_{i,2} + \dots + q_N C_{i,N} = \sum_{p=1}^{N} q_p C_{i,p} \tag{1}$$

in which r is the radial distance from the cylindrical centerline of the transducer and

$$\alpha_i = x_i/a \ , \quad J_0(x_i) = 0 \quad with \quad C_{i,p} = \frac{2\left[r_p^+ J_1(\alpha_i r_p^+) - r_p^- J_1(\alpha_i r_p^-) \right]}{\alpha_i a^2 J_1^2(\alpha_i a)} \tag{2}$$

The index p represents the pth annulus in the transducer, q_p the corresponding quantisation amplitude, r_p^-, r_p^+ the inner and outer radii of the pth annulus respectively. The distance $a \geq R$, where R is the radius of the transducer, is the spatial range over which we choose to apply the series. The values x_i are the monotonically increasing set of known roots to $J_0(x_i) = 0$. Time delays may be incorporated into the system by writing the quantisation levels q_p as

$$q_p = \gamma_p + j\delta_p = |q_p| e^{j\theta_p} \quad where \quad |q_p| = \sqrt{\gamma_p^2 + \delta_p^2} \ , \quad \theta_p = \tan^{-1}[\delta_p/\gamma_p] \tag{3}$$

in which γ_p and δ_p, $p = 1 \dots N$ are real and imaginary parts of q_p respectively; the surface

Acoustical Imaging, Volume 25, Edited by Halliwell and Wells.
Kluwer Academic/Plenum Publishers, 2000.

pressure $q(r)e^{-j\omega t}$ then becomes

$$q(r)e^{-j\omega t} = |q_p|e^{-j\omega(t-\tau_p)} \qquad where \qquad \tau_p = \theta_p/\omega \tag{4}$$

which corresponds to implementation of quantisation amplitudes $|q_p| = (\gamma_p^2 + \delta_p^2)^{1/2}$ and time delays $\tau_p = (\tan^{-1}[\delta_p/\gamma_p])/\omega$ on each annulus p respectively. Note also that the A_i coefficients in (1) become complex in this case. With or without time delays, the transducer surface pressure $q(r)e^{-j\omega t}$ then propagates a field which in a linear medium is a sum of weighted limited diffraction zero-order Bessel beams

$$q(r,z,t) = \sum_{i=1}^{\infty} A_i J_0(\alpha_i r) e^{j\beta_i z} e^{-j\omega t} \quad : \quad \beta_i = \sqrt{k^2 - \alpha_i^2} \quad , \quad k = \omega/c \tag{5}$$

where z is the propagation distance perpendicular to the transducer surface. The β_i term becomes imaginary for all values $\alpha_i > k$ $(x_i > ka)$, which always occurs[2] for a certain value $i = l(a) \approx ka/\pi + 1/4$ since the roots $x_i \approx (4i-1)\pi/4$ are monotonically increasing with respect to i. The magnitude of β_i therefore also increases with i and the corresponding terms $e^{j\beta_i z}$ rapidly become evanescent in the propagation direction z. The infinite sum in (5) is therefore effectively truncated to the finite sum

$$q(r,z,t) = \sum_{i=1}^{l(a)} A_i J_0(\alpha_i r) e^{j\beta_i z} e^{-j\omega t} \qquad where \qquad l(a) \approx ka/\pi + 1/4 \tag{6}$$

In addition, each component basis function in the sum has the form of a zero order Bessel beam[3-6]

$$p(r,z,t) = J_0(\alpha r) e^{j\beta z} e^{-j\omega t} \qquad where \qquad \beta = \sqrt{k^2 - \alpha^2} \quad , \quad k = \omega/c \tag{7}$$

which is a nondiffracting solution to the wave equation only when realised over an infinite aperture at $z = 0$ since r must tend to infinity before the amplitude of $J_0(\alpha r) \approx \sqrt{2/\pi\alpha r}.\cos(\alpha r - \pi/4)$ drops to zero. When implemented on a finite aperture, this condition is not satisfied and the beam becomes a limited diffraction beam which is no longer an exact solution to the wave equation. The difference between these two conditions is key in computing the propagated field $f(r,z,t)$ from the limited diffraction field of equation (6).

FIELD COMPUTATION

We now allow the basis distance a in (6) to tend to infinity such that each basis function $A_i J_0(\alpha_i r) e^{j\beta_i z} e^{-j\omega t} \equiv A_i J_0(x_i r/a) e^{j\beta_i z} e^{-j\omega t}$ becomes a valid Bessel beam solution to the wave equation, this is so since its realisation at $z = 0$ takes place over an infinite aperture $a \to \infty$ as required. This means that each component in (6) now propagates in a nondiffracting manner in the form of (7). Combining $a \to \infty$ with the notation for A_i given in (1) then gives the field $f(r,z,t)$ as

$$f(r,z,t) = \lim_{a\to\infty} \sum_{i=1}^{l(a)} \sum_{p=1}^{N} q_p C_{i,p} J_0(x_i r/a) e^{j\beta_i z} e^{-j\omega t} \qquad where \qquad \beta_i = \sqrt{k^2 - (x_i/a)^2} \tag{8}$$

In practice we are unable to compute this infinite limit since we note from (6) that the number of nonevanescent terms $l(a)$ included in the sum also tends to infinity as $a \to \infty$. However, we are able to a to increase gradually in (6) from relatively low to relatively high values and at each iteration compute the sum in (6) until it converges to a steady state value representing (8). Various iterative schemes for the increase in a may be employed, although the simplest is merely to initialise it with $a = R$ and increase it by a constant factor at each iteration. As an example, take the 15mm diameter 4-ring ($N = 4$) equal-area transducer discussed in[7] and calculate the field at 3.5MHz assuming a water propagation speed of $c = 1540$ m/s, for which the wavenumber is $k = 14280 m^{-1}$. The quantisation levels are taken as $q_1 = +0.61$, $q_2 = -0.21$, $q_3 = -0.39$, $q_4 = -0.32$, or equivalently $|q_1| = 0.61$, $|q_2| = 0.21$, $|q_3| = 0.39$, $|q_4| = 0.32$ with half-cycle relative delays $\tau_1 = 0$, $\tau_2 = 143$, $\tau_3 = 143$, $\tau_4 = 143$ nanoseconds. Inner and outer radii for the transducer are $r_1^- = 0$, $r_1^+ = 3.75$, $r_2^- = 3.75$, $r_2^+ = 5.3 = r_3^-$, $r_3^+ = 6.5 = r_4^-$ and $r_4^+ = R = 7.5$ (mm). We then apply (6) along with (2) for the iterative aperture ratios $a = R, 3R, 9R, 27R, 81R, 243R$ and plot the resulting fields $q(r,z,t)$ in Figures 1-6 (normalised[7] relative to 0dB along the transducer centerline $r = 0$ for all values of z). Relative contours of -6dB (inner), -12dB (middle), and -18dB (outer) are indicated in each case. The generated field patterns $q(r,z,t)$ are poor in the farfield for relatively low values of aperture ratio (Figures 1-3), but converge towards the true field $f(r,z,t)$ for higher aperture ratios (Figures 4-6). Independent field calculation software has also been used to verify that the field $q(r,z,t) \approx f(r,z,t)$ shown in Figure 6 does indeed give a correct representation of the propagated field for the given problem. Hence we now have a Fourier-Bessel based method for field computation. However, the 1D series discussed above is restricted in its application to circular symmetric (annular) arrays since it does not provide any allowance for the polar field variations which occur in 2D arrays. These arrays have spatial variations of the form $q(r,\varphi)$, in which φ is the polar angle perpendicular to the z axis around the centerline of the transducer. To analyse these fields it is necessary to extend the analysis from 1D to 2D Fourier-Bessel series.

2D FOURIER-BESSEL SERIES

A 2D function $q(r,\varphi)$ may be modelled[8] by the infinite 2D Fourier-Bessel series

$$q(r,\varphi) = \sum_{n=0}^{\infty}\sum_{i=1}^{\infty} A_{n,i} J_n(\alpha_{n,i} r)\cos(n\varphi) + \sum_{n=0}^{\infty}\sum_{i=1}^{\infty} B_{n,i} J_n(\alpha_{n,i} r)\sin(n\varphi) \tag{9}$$

where

$$\alpha_{n,i} = x_{n,i}/a \quad , \quad J_n(x_{n,i}) = 0 \quad , \quad q(a,\varphi) = 0 \tag{10}$$

and the orthogonal coefficients $A_{n,i}$ and $B_{n,i}$ are given by

$$A_{0,i} = \frac{1}{\pi a^2 J_1^2(\alpha_{0,i} a)} \int_{r=0}^{r=a}\int_{\varphi=0}^{\varphi=2\pi} q(r,\varphi) J_0(\alpha_{0,i} r)\, r\, dr\, d\varphi \quad , \quad B_{0,i} = 0 \quad : \quad n = 0 \tag{11}$$

$$\begin{bmatrix} A_{n,i} \\ B_{n,i} \end{bmatrix} = \frac{2}{\pi a^2 J_{n+1}^2(\alpha_{n,i} a)} \int_{r=0}^{r=a}\int_{\varphi=0}^{\varphi=2\pi} q(r,\varphi) J_n(\alpha_{n,i} r) \begin{bmatrix} \cos(n\varphi) \\ \sin(n\varphi) \end{bmatrix} r\, dr\, d\varphi \quad : \quad n > 0 \tag{12}$$

This series is also valid over an interval $0 < r \leq a$, where a is any value subject to $q(a,\varphi) = 0$. Hence we may choose a to be any distance greater than or equal to that from the centre of the transducer to its furthest edge, since then $q(a,\varphi) = 0$ for any shape transducer by definition. By analogy to $q(r)$ in (1) for the 1D case, $q(r,\varphi)$ in (9) for the 2D case then corresponds to a source $q(r,\varphi)e^{-j\omega t}$ generating a set of limited diffraction Bessel beams

$$q(r,\varphi,z,t) = \sum_{n=0}^{\infty}\sum_{i=1}^{\infty} A_{n,i} J_n(\alpha_{n,i} r)\cos(n\varphi)\, e^{j\beta_{n,i} z} e^{-j\omega t} \quad , \quad \beta_{n,i} = \sqrt{k^2 - \alpha_{n,i}^2}$$

$$+ \sum_{n=0}^{\infty}\sum_{i=1}^{\infty} B_{n,i} J_n(\alpha_{n,i} r)\sin(n\varphi)\, e^{j\beta_{n,i} z} e^{-j\omega t} \tag{13}$$

in which each of the two sums in (13) is seen to comprise a weighted sum of nth order nonrotating limited diffraction Bessel beams[5]

$$p(r,\varphi,z,t) = J_n(\alpha r)\cos(n\varphi - \varphi_0)\,e^{j\beta z}e^{-j\omega t} \quad : \quad \beta = \sqrt{k^2 - \alpha^2} \quad , \quad k = \omega/c \qquad (14)$$

Here n is the order of the beam, $J_n(\alpha r)$ is the Bessel function of the first kind of order n, and φ_0 is an arbitrary offset polar angle. The first sum in (13) is thus a set of beams (14) with coefficients $A_{n,i}$, basis values $\alpha_{n,i}$, and offset angles $\varphi_0 = 0$, whilst the second has coefficients $B_{n,i}$, basis values $\alpha_{n,i}$ and offset angles $\varphi_0 = \pi/2$ (due to $\sin(n\varphi) \equiv \cos(n\varphi - \pi/2)$ for all $n\varphi$). Note also that the terms corresponding to $n = 0$ in the 2D expressions (9,10,12) collapse to the 1D expressions (1,2) already encountered for annular arrays. Hence the 2D series is a consistent expansion of the 1D series and similar analysis techniques apply for both. This includes incorporation of time delays on different elements by writing $q(r,\varphi)$ in complex form

$$q(r,\varphi) = \gamma(r,\varphi) + j\delta(r,\varphi) = |q(r,\varphi)|\,e^{j\theta(r,\varphi)} \qquad (15)$$

since then a similar analysis to that of (3) and (4) may be applied. In this case the $A_{n,i}$ and $B_{n,i}$ coefficients in (11,12) also become complex and this is reflected as time delays in the limited diffraction field (13) by writing $A_{n,i} = |A_{n,i}|e^{j\theta(A_{n,i})}$, $B_{n,i} = |B_{n,i}|e^{j\theta(B_{n,i})}$ and extracting the relevant delays accordingly.

EVANESCENCE IN 2D SERIES

Similarly to the 1D case, the majority of the terms in (13) may also be neglected due to $\beta_{n,i}$ being imaginary for all values $\alpha_{n,i} > k$ ($x_{n,i} > ka$). In addition the Bessel function $J_n(x)$ may be approximated (with increasing accuracy as $x \to \infty$) by

$$J_n(x) \simeq \sqrt{2/\pi x}\,.\cos(\,x - (2n+1)\pi/4\,) \qquad (16)$$

and since the ith positive root of a cosine function is $(2i-1)\pi/2$, the ith positive root $x_{n,i}$ of $J_n(x_{n,i}) = 0$ becomes approximately

$$x_{n,i} \simeq (\,4i + 2n - 1)\pi/4 \qquad (17)$$

which increases monotonically with both i and n. Hence for all orders n there exists a limiting value $i = l(n,a)$ after which evanescence occurs due to $x_{n,i} > ka$. Setting $i = l(n,a)$ and $x_{n,i} = ka$ in (17) then gives $l(n,a) \simeq ka/\pi + 1/4 - n/2$, which decreases with increasing Bessel order n. It then drops to a limiting value of zero when $n = m(a) \simeq 2(ka/\pi + 1/4)$. In fact these expressions are only approximate due to the approximate nature of (16), and an empirical adjustment in the term from $n/2$ to $n/3$ may also be made to obtain revised estimates

$$l(n,a) \simeq ka/\pi + 1/4 - n/3 \quad , \quad m(a) \simeq 3(ka/\pi + 1/4) \qquad (18)$$

which can be more accurate for higher frequencies than the original expressions. In fact the exact limits $l(n,a)$ and $m(a)$ usually lie somewhere between the two extremes, and may always be found numerically by testing either for when $x_{n,i} > ka$ or for the sudden convergence of (13) with respect to i for a given a. Thus precise expressions are not required for implementation of the scheme; their purpose is merely to give an insight into the 2D evanescence mechanism. The result is that both the infinite series in (13) may be truncated to the finite sums

$$\begin{aligned} q(r,\varphi,z,t) = &\sum_{n=0}^{m(a)}\sum_{i=1}^{l(n,a)} A_{n,i}\,J_n(x_{n,i}r/a)\cos(n\varphi)\,e^{j\beta_{n,i}z}e^{-j\omega t} \\ &+ \sum_{n=0}^{m(a)}\sum_{i=1}^{l(n,a)} B_{n,i}\,J_n(x_{n,i}r/a)\sin(n\varphi)\,e^{j\beta_{n,i}z}e^{-j\omega t} \end{aligned} \qquad \begin{aligned} l(n,a) &\simeq ka/\pi + 1/4 - n/3 \\ m(a) &\simeq 3(ka/\pi + 1/4) \\ \beta_{n,i} &= \sqrt{k^2 - (x_{n,i}/a)^2} \\ J_n(x_{n,i}) &= 0 \end{aligned} \qquad (19)$$

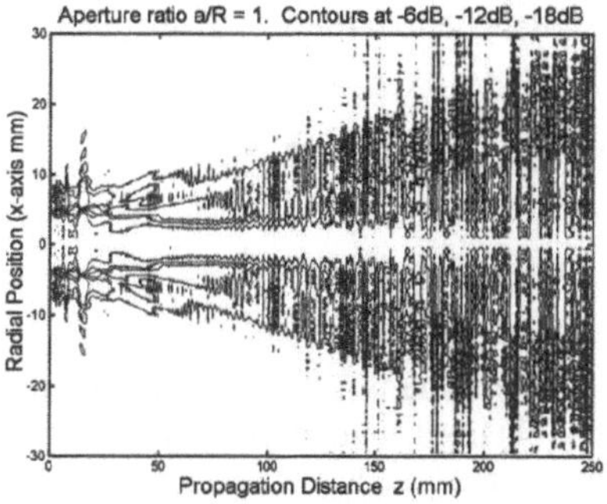

Figure 1 Aperture ratio : $a = R$

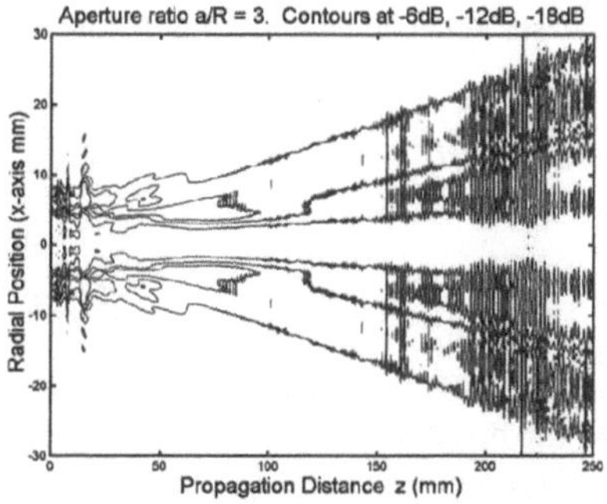

Figure 2 Aperture ratio : $a = 3R$

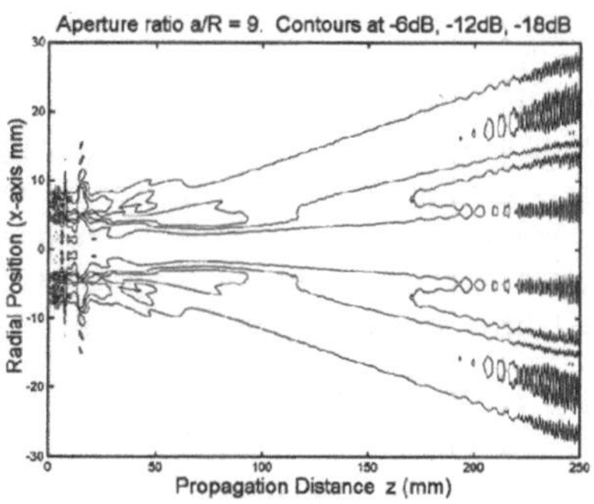

Figure 3 Aperture ratio : $a = 9R$

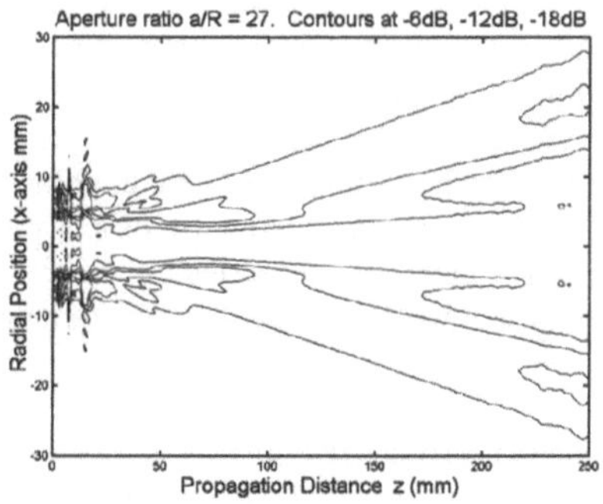

Figure 4 Aperture ratio : $a = 27R$

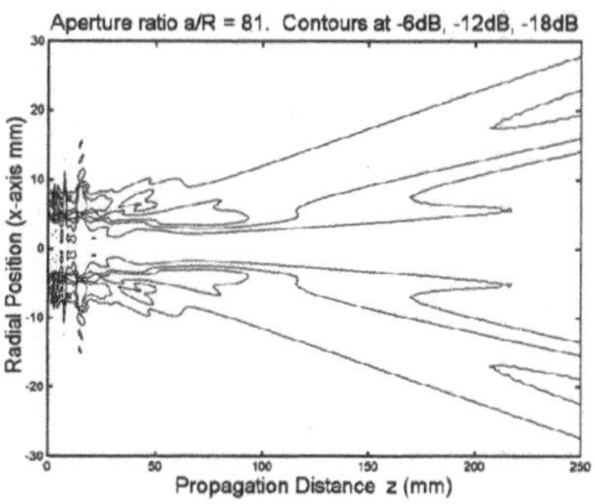

Figure 5 Aperture ratio : $a = 81R$

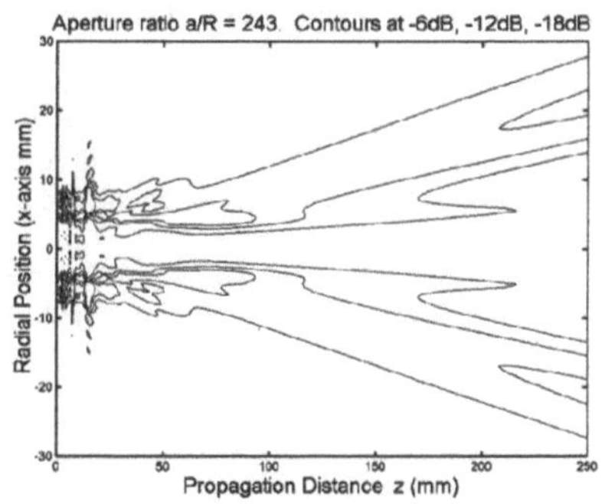

Figure 6 Aperture ratio : $a = 243R$

in which $\alpha_{n,i} = x_{n,i}/a$ has been substituted from (10). This means that we are able to characterise the nonevanescent portion of any 2D transducer surface pressure $q(r,\varphi)$ in terms of a finite basis of known limited diffraction J_n beams similarly to the finite set of J_0 beams for the 1D case. The corresponding fields are then also calculated in the same manner as per the 1D case, which is by allowing the basis distance a to tend to infinity to give $f(r,\varphi,z,t)$ as

$$\begin{aligned} f(r,\varphi,z,t) = \lim_{a\to\infty} q(r,\varphi,z,t) &= \lim_{a\to\infty} \sum_{n=0}^{m(a)} \sum_{i=1}^{l(n,a)} A_{n,i} J_n(x_{n,i} r/a) \cos(n\varphi)\, e^{j\beta_{n,i} z} e^{-j\omega t} \\ &+ \lim_{a\to\infty} \sum_{n=0}^{m(a)} \sum_{i=1}^{l(n,a)} B_{n,i} J_n(x_{n,i} r/a) \sin(n\varphi)\, e^{j\beta_{n,i} z} e^{-j\omega t} \end{aligned} \tag{20}$$

For numerical purposes as for the 1D case previously, the value a is then increased steadily from low to high values until convergence of the field is obtained, with similar field plots being obtained for a given choice of viewing plane. (Plots omitted here due to space restrictions).

CONCLUSIONS AND FURTHER WORK

A method for analysing the field distribution from 2D transducers has been discussed. 2D Fourier-Bessel series have been used to characterise the quantised field in terms of a known set of limited diffraction basis functions, from which the propagated field is evaluated by performing an iterative evaluation until steady state field convergence is achieved. Although the method has been developed here in polar coordinates, the equations may also be converted into Cartesian coordinates for ease of analysis of rectangular 2D arrays. Similar to previous work on 1D Fourier-Bessel series for annular arrays[7], it is also hoped to extend the present 2D Fourier-Bessel work to least-squares tuning of limited diffraction beams on 2D arrays.

ACKNOWLEDGEMENT
This work was sponsored by the European Union under TMR contract ERBFMBICT 982917.

REFERENCES

1. S. Holm and P.D. Fox, "Analysis of Bessel beam quantisation in annular arrays," *Proc. 22nd Scandinavian Symposium on Physical Acoustics*, Ustaoset, Norway, pp. 43-44. (Jan. 31 - Feb. 3, 1999). ISSN 1501-6773.
2. P.D. Fox and S. Holm, "Decomposition of acoustic fields in quantised Bessel beams," Poster at joint event Ultrasonics International 99 / 1999 World Congress on Ultrasonics, Lyngby (Copenhagen), Denmark. (Jun. 29 - Jul. 1, 1999). Full paper to appear in "Ultrasonics".
3. J. Durnin, "Exact solutions for nondiffracting beams. I. The scalar theory," *J. Opt. Soc. Amer.*, vol. 4, pp. 651-654. (Apr. 1987).
4. J.-Y. Lu and J. F. Greenleaf, "Ultrasonic nondiffracting transducer for medical imaging," *IEEE Trans. Ultrason., Ferroelect., Freq. Cont.*, vol. 37, no. 5, pp. 438-447. (Sept. 1990).
5. J-Y. Lu and J.F. Greenleaf, "Sidelobe reduction for limited diffraction pulse-echo systems," *IEEE Trans. Ultrason., Ferroelect., Freq. Cont.*, vol. 40, no. 6, pp. 735-746. (Nov. 1993).
6. S. Holm, "Bessel and conical beams and approximation with annular arrays," *IEEE Trans.*

Ultrason., Ferroelect., Freq. Cont., vol. 45, no. 3, pp. 712-718. (Sept. 1998).

7. P.D. Fox and S. Holm, "Tuning of Bessel beams in annular arrays," *Proc. 25th International Acoustical Imaging Symposium*, Bristol, United Kingdom. (Mar. 19-22, 2000).
8. F. Bowman, *Introduction to Bessel Functions*, Dover Publications Incorporated., New York. (1958). ISBN 0-486-60462-4.

FIELD ANALYSIS WITH A NEW FIELD SIMULATION AND PROPAGATION TECHNIQUE

Sidney Leeman[1] and Andrew J. Healey[2]

[1]Department of Medical Engineering and Physics
King's College School of Medicine and Dentistry
London SE22 8PT, U.K.
[2]Nycomed Imaging AS
N0401 Oslo, Norway

INTRODUCTION

The development of an exact and practicable 3-D simulation technique that allows the full time-space complexity of a wideband ultrasound field to be displayed as an image is presented, and the theory explained. The input to the method is the excitation of the transducer face, and the field may be calculated at any distance from the transducer. Arbitrary excitation waveforms and apodisation may be fully taken into account. All common transducer geometries may be incorporated, and both planar or focused transducers may be handled with ease. All diffraction effects are completely taken into account. The method is directly applicable to transient fields and allows for efficient computation of such fields. Visualisation of the field, presented here, is as a spatial pressure distribution at any temporal instant, but other options are possible. The phase of the field is shown. This new method for ultrasound field simulation and visualisation is shown to be eminently practicable and represents a substantial improvement over conventional methods. The technique is computationally rapid, and a less than state-of-the-art desktop personal computer is adequate for rapid results.

The method is a generalisation of the angular spectrum[1] technique, since it handles transient fields with ease, and provides a novel approach towards eliminating evanescent waves from the theory. In contrast to the Tupholme-Stepanishen method[2,3] (which turns out to be cumbersome in practice), the technique presented here provides an image of the field that does not have to be laboriously built up on a point-by-point basis. Moreover, the method is applicable to a strongly curved bowl transducer, without the approximations inherent in the impulse response method.

The simulation technique is capable of calculating pulsed fields from a number of different (ideal) transducer geometries, but aspects of real transducer behaviour, such as he lateral modes that give rise to head waves, may also be incorporated, as well as apodisation and arbitrary excitation voltages. The extension to multielement arrays, with or without phasing, is straightforward.

In this communication, the development of the technique is restricted to fields from single element transducers, propagating in a lossless environment, but it is briefly indicated how the theory may be extended to more complicated media.

THE DIRECTIVITY SPECTRUM

Any spatially varying, time-dependent ultrasound (pressure) field may be expressed as a Fourier transform:

$$p(\boldsymbol{r},t)=\frac{1}{(2\pi)^4}\iiiint d^3\boldsymbol{k}\,d\omega\, F(\boldsymbol{k},\omega)\exp(i[\boldsymbol{k}\bullet\boldsymbol{r}-\omega t]) \tag{1}$$

where $\boldsymbol{r}=(x,y,z)$ denotes the spatial location vector, and t the time. F is the 4-dimensional Fourier transform of the 4-dimensional field. Note that, at this stage, the values of $\boldsymbol{k}$ and ω are totally unrelated. A constraint that may be placed on p is that it obey the canonical wave equation,

$$\nabla^2 p-(1/C^2)\ddot{p}=0 \tag{2}$$

This additional constraint may be incorporated into Equation (1) by demanding that the k-integration be carried out on the 'energy shell', viz such that

$$k_x^2+k_y^2+k_z^2=\boldsymbol{k}\cdot\boldsymbol{k}=\omega^2/C^2 \tag{3}$$

Formally, this may be ensured by introducing a delta function, $\delta(\boldsymbol{k}^2-\omega^2/C^2)$ into the Fourier integral. On effecting the ω-integration, the following expression for a forward travelling wave results:

$$p(\boldsymbol{r},t)=\iiint d^3\boldsymbol{k}D(\boldsymbol{k})e^{-i\omega t}\exp(i\boldsymbol{k}\cdot\boldsymbol{r}) \tag{4}$$

This involves only a *three* dimensional integration (on the energy shell), and it is easily checked, by direct substitution, that p is, indeed, a solution of the canonical wave equation. For reasons that are made clear below, we have called $D(\boldsymbol{k})$ the directivity spectrum: it is, in fact, a generalisation of the angular spectrum, but is immediately applicable to transient fields.

The directivity spectrum may be obtained from the initial value (t=0) of the field:

$$D(\boldsymbol{k})=\int d^3\boldsymbol{r}p(\boldsymbol{r},0)\exp(-i\boldsymbol{k}\cdot\boldsymbol{r}) \tag{5}$$

Thus, once the time-zero field is known, the forward travelling field at any other time may be established, in principle, via the directivity spectrum and Equation (4). Since the field is constrained to be a solution of the canonical wave equation, it will be shifted to a different location at later times. The full 4-D space-time complexity of the field is thus encoded in the 3-D directivity spectrum. Moreover, this description is valid for any field, transient or continuous, focused or not: we prefer, here, to consider only transient fields, displayed at discrete time instants in a spatial plane containing the geometric beam axis.

In practice, predicting an arbitary field at discrete times by the above scheme would be computationally intensive, but assuming that the field exhibits certain symmetries, as is almost certainly the case for 'real' fields, simplifies matters. A major simplification, however, is gained by considering the behaviour of the projections of the field, as outlined below.

FIELD PROJECTIONS

Consider a coordinate system where the beam travels in the z-direction, and is axially symmetric. A projection of interest, is that of the spatial extent of the time-zero pulse onto the (x,z) plane,

$$P(x,z) \equiv \int_{\infty}^{\infty} dy\, p(x,y,z,0)$$

$$= \frac{1}{(2\pi)^3} \int dy \iiint d^3\boldsymbol{k}\, D(k_x,k_y,k_z) \exp(i\boldsymbol{k}\cdot\boldsymbol{r})$$

$$= \frac{1}{(2\pi)^2} \int dk_x dk_z\, D(k_x,0,k_z) \exp(i[k_x x + k_z z]) \equiv FT_2\{D(k_x,0,k_z)\} \qquad (6)$$

with FT_2 denoting the two-dimensional Fourier transform.

Clearly, the *projection* of the field, $P(x,z)$ is related to the value of the directivity spectrum in the (k_x,k_z) plane by Fourier transformation. At some later time, when the pulse will have propagated some distance, L, into the ideal medium, the associated projection is given by

$$\int_{\infty}^{\infty} dy\, p(x,y,z,L/C) = \frac{1}{(2\pi)^3} \int dy \iiint d^3\boldsymbol{k}\, D(k_x,k_y,k_z) \exp\left(i\boldsymbol{k}\cdot\boldsymbol{r} - i\sqrt{\boldsymbol{k}\cdot\boldsymbol{k}}\,L\right)$$

$$= FT_2\{D(k_x,0,k_z)\exp(-i\sqrt{k_x^2+k_z^2}\,L\} \equiv FT_2\{D(k_x,0,k_z)\times\text{"}shift\ factor\text{"}\} \qquad (7)$$

Thus, provided the appropriate values of the directivity spectrum (known via Equation (6) from the time-zero pulse) are modified by a shift factor, as indicated in Equation (7), then the (two-dimensional) projection of the pulse may be recovered at some later time, when it it is located at distance L into the medium. This is a relatively rapid calculation, because it involves only two two-dimensional Fourier transformations and the evaluation of the shift factor.

The technique described above affords an opportunity to simulate methods such as Schlieren photography. More informative, however, is the display of the transient field in the (x,z) plane. The pulse visualisation desired is

$$p(x,0,z,L/C) = \frac{1}{(2\pi)^3} \iiint d^3\boldsymbol{k}\, D(k_x,k_y,k_z) \exp(i[k_x x + k_z z] - i\sqrt{\boldsymbol{k}\cdot\boldsymbol{k}}\,L)$$

$$= FT_2\{\mathrm{Proj}[D(k_x,k_y,k_z)\exp\,(-i\sqrt{\boldsymbol{k}\cdot\boldsymbol{k}}\,L)]\} \qquad (8)$$

where Proj[....] denotes projection onto the (k_x,k_z) plane. Since axial symmetry is assumed for the fields considered here, the values of the projection required in Equation (8) may be recovered from the values of $D(k_x,0,k_z)$ (and an appropriate shift factor) as obtained from the time-zero pulse.

FIELD PROPAGATION AND VISUALISATION

The results derived are sufficient to compute a cross-sectional view (in the (x,y) plane) of a transient field from the time-zero pulse, which is the input to the simulation, and is

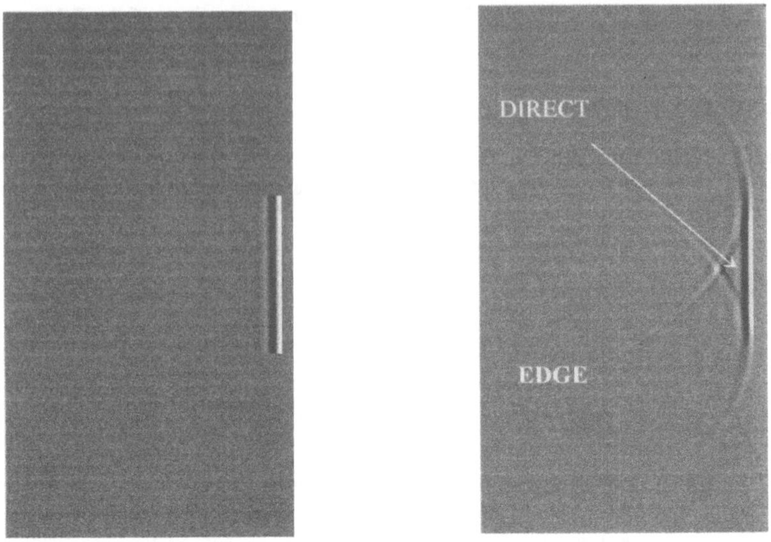

Figure 1. Cross sectional views of pulse from planar circular transducer, with no apodisation: (left) at transducer face, and (right) after propagation into a lossless medium. The direct and edge waves are clearly seen. Inthese, and in other images, a linear gray scale is employed, and is normalised for each image for better visualisation. The horizontal dimension is stretched 4× with respect to the vertical.

Figure 2. Propagation of unipolar pulse from transducer as in Figure (1): at transducer face (left) and some distance into ideal medium (right). The edge wave develops opposite polarity in the transducer 'shadow' zone. R.Dantas is acknowledged for modifying existing programmes to provide these images.

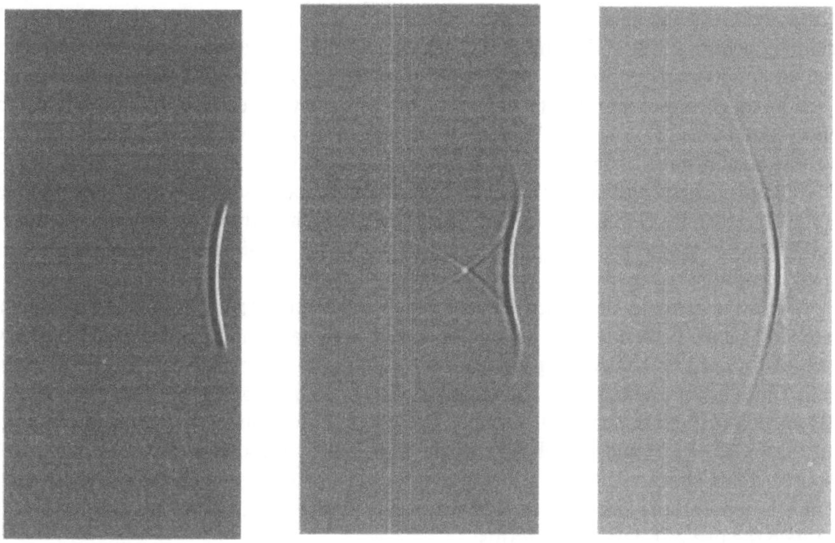

Figure 3. Cross sectional view of pulse from a focused bowl transducer: (left) at the transducer face, (middle) pre-focus, (right) post focus. The wave fronts change curvature as they progress through the focus.

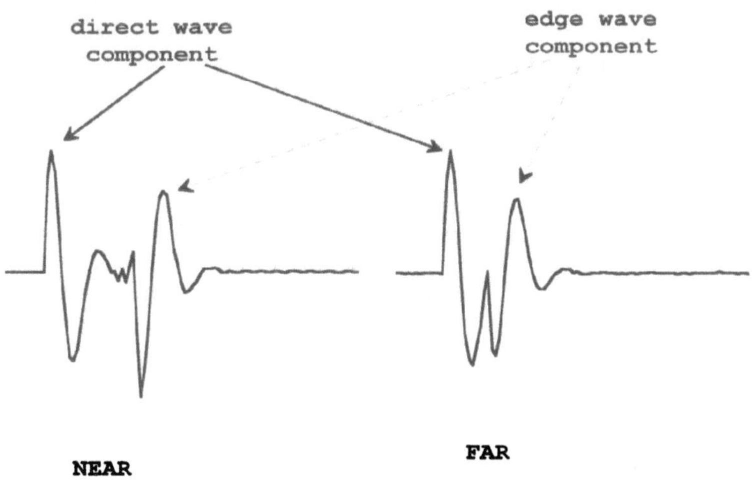

Figure 4. Point hydrophone measurements showing direct and edge wave components of the axial pulse, at two different distances from the transducer face. The edge wave component appers to travel faster than the direct wave, thus exhibiting apparently super-luminal behaviour.

taken to represent the excitation of the 'transducer', or, more correctly, the aperture. The technique consists of the following steps: (i) assume the form of the time-zero pulse (this is the initial condition), (ii) form its projection onto the (x,z) plane, (iii) Fourier transform the projection to give the appropriate values of the directivity spectrum, (iv) propagate these Fourier components to a new z- location by means of an appropriate shift factor (v) reconstruct the spatial form of the pulse, in the (x,z) plane, at the new location (Equation (8)). The whole propagation calculation can be performed in under a minute on a somewhat aged Pentium 90 personal computer, running Matlab 4.2c, with without any attempt to optimise the efficiency of the code. Computation times would be significantly reduced with a more modern computer.

It is of interest to note that the technique advocated here obviates the problematic evanescent waves present in the angular spectrum approach, and can be used equally well for either forward or backward propagation. Although only cross-sectional field views at specific times are presented here, the technique has been used by us for other field visualisations, including three dimensional views. The computations involve nothing more complicated than a 2D FFT, and the physics of the problem is encoded in the shift matrix. The computed fields show well-known diffractive features of pulses, even though these are not specifically fed into the calculations, and this suggests that such effects are enforced by little more then the use of Fourier transforms!

Figure (1) shows cross-sectional views of the initial and propagated pulses from a planar, uniformly excited, disc transducer. The phase of the field is clearly visible, both direct- and edge- waves are shown, and the phase changes of the latter, in the 'shadow' zone of the transducer are seen. Apodisation is easily handled (not shown), and shows the expected result that the edge wave component is considerably reduced. Of somewhat more interest, is the case of a positive unipolar excitation (Figure (2)). Perhaps unexpectedly, the edge wave component nonetheless develops a negative-going portion in the shadow zone. In fact, it can be seen that this result is in accord with a theoretical prediction, by Stokes[4], that a spherical wave (but *not* a progressive plane wave) with condensation of only a single sign cannot exist. Transducer geometries that are not easily computed by other methods, such as a strongly focused bowl transducer, are equally easily treated by the directivity spectrum technique. Figure (3) shows the exact field in such a case, the usual approximation of an almost planar radiating surface not having to be invoked.

Other transducer geometries are also possible. In particular, pulses from square transducers have been calculated (not shown). In practice, transducers always exhibit some geometrical symmetries, which can be exploited to simplify the computations. An arbitrary geometry, with no symmetries, can also be treated, but at the cost of an increase in computational complexity.

SUPERLUMINAL FIELD COMPONENTS

The on-axis pulse as measured by an ideal point hydrophone may also be computed. However, it is sufficient, and quicker, for our purposes to obtain a very good approximation to such measurements directly from representations as shown in Figure (2), say, with results as in Figure (4). This shows the on-axis pulse at from a planar, circular transducer at two distances from the transducer face. The direct- and edge- wave components are unambiguously shown, and it is clear that the gap between the two narrows with distance from the transducer face. The edge-wave component thus appears to travel faster than the direct-wave component! There can be no doubt as to the identification of these two components, since that is made on the basis of a two dimensional image. Moreover, since the field is calculated, with no approximations, from the canonical wave equation, there can also be no doubt that the wave speed has a single value, C. The cross sectional images of the field make it clear that both the direct and edge waves travel at the same speed, although their geometries are rather different. Hence, the apparent superluminal behaviour of the edge wave component, as revealed by a point hydrophone measurement, is an artefact caused by the mesurement technique and the geometry of that wave component. It is not an unexpected feature of the physics of wave propagation. The existence of superluminal components in more complicated pulsed fields may also be ascribed to geometrical factors, as above.

THE DIRECTIVITY FUNCTION

The directivity function is an important feature when characterising transducer fields. It is essentially a far field concept, valid for mono-frequency excitation only, but is encoded directly by the directivity spectrum, which is determined by the time-zero pulse. A cross section through the directivity spectrum, $D(k_x, 0, k_z)$, expressed in polar coordinates, contains the whole spectrum of directivity functions, for all the frequencies present in the pulse. This is shown in Figure (5). The directivity spectrum thus provides an appropriate vehicle for the description and understanding of the directivity function for wideband pulses. Moreover, the directivity function is seen to be an intrinsic feature of the pulsed field, not necessarily determined by its asymptotic behaviour only.

For a planar, circular transducer, excited at frequency f, the directivity function has the form $d(\theta, f) = \dfrac{|J_1(ka \sin\theta)|}{|ka \sin\theta|}$, where θ is the angle to the acoustic axis, a is the transducer radius, and $k = 2\pi/\lambda = 2\pi f / C$. Zeros of the monofrequency directivity function occur at $ka \sin\theta = z_N$, where z_N denotes the zeros of the first order Bessel function, J_1. This implies that the polar representation of the directivity spectrum, which is relevant for a whole range of frequencies, will exhibit parallel lines of zeros, spaced at virtually identical intervals determined by the transducer radius. This feature is indeed discernible in the directivity spectra shown in Figure (6).

To the best of our knowledge, the directivity function for a focused bowl transducer has not been calculated analytically, without invoking simplifying approximations. The considerations outlined above suggest that it could be obtained directly from the directivity spectrum, without the necessity to resort to approximate methods. It is to be hoped that the values thus obtained could provide a good fit to relatively uncomplicated standard functions.

FREQUENCY NULLS

Inspection of the directivity spectra in Figure (6) reveals a totally unexpected feature. The directivity spectrum for a focused bowl shows, remarkably, discrete zeros, or 'nulls', at certain locations. Frequency nulls are not seen for planar transducers, and their location, in the focused case, depends only on the curvature of the bowl. Their presence is thus not related to the physics of the problem, but is dictated by geometrical considerations.

Frequency nulls will influence the directivity function at certain frequencies only. Under such circumstances, the directivity function will not be peaked along the beam axis, but will be double humped - a feature which has been ascribed to faulty transducer manufacture. When activated at other frequencies, such 'faulty' tranducers will regain the expected peak output along the beam axis!

Another consequence of the geometric origin of the frequency nulls is that their

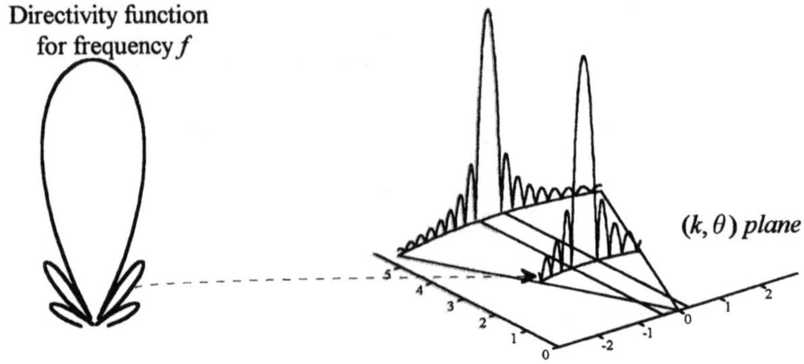

Figure 5. The directivity function, at frequency f, is given by the values of $D(k,\theta)$ along an appropriate arc. The zeros of the former, for all frequencies present in the pulse, lie along straight, parallel lines in the (k,θ) plane.

presence should not affect the development of nonlinearities in the field. Thus, it is to be expected that an apparent deficit in a particular frequency (because of a frequency null) will not be expected to reduce the rate at which higher harmonics develop in a nonlinear field, unless, quite fortuitously and unexpectedly in general, other frequency nulls should exist also at the harmonic frequencies. On the basis of this expectation, it is to be cautioned that experimental verification of nonlinear theories should always be carried out with planar transducers.

HEAD WAVES

When a real transducer is activated in thickness mode, the resulting dimension change causes an accompanying lateral change as well -- so that a radial wave is also generated. As the latter sweeps across the transducer, pressure waves are launched into the ambient medium. These 'head waves', as they have been called, can interfere with the edge wave component of the field, and may confuse the interpretation of point hydrophone measurements of a transient field[5]. One advantage of the simulation technique advocated here, is that the head waves may be modelled independently, so that their behaviour may be more closely studied. One such investigation is shown in Figure (7). It is clear that the head wave component in the field may mimic the edge wave behaviour.

The radial wave travels across the transducer face at a speed characteristic of the transducer material. Hence, incorporation of the head wave into the simulation requires a knowledge of this feature. The simulation shown adopts a typical radial wave speed for a transducer constructed from PZT.

PULSE PROPAGATION IN MORE COMPLICATED MEDIA

The absorptive and dispersive properties of the medium in which the (simulated) pulse propagates are encoded in the shift matrix, and, in principle, do not present any new problems for the simulation algorithm. Such simulations have already been carried out (not shown). Even propagation in uniform solids may be studied, although some modification is necessary because the shear wave propagation cannot be modelled as a pressure wave.

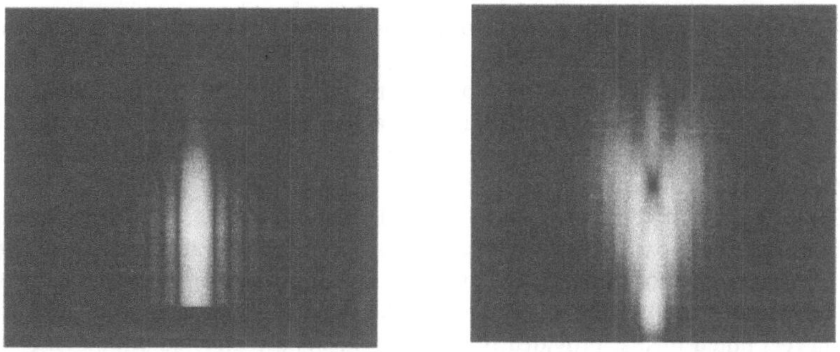

Figure 6. Magnitude of the directivity spectrum, in a cross sectional plane in $k-\theta$ space, for (left) a planar and (right) a focused transducer. The focused transducer characteristically exhibits the existence of frequency nulls. Both directivity spectra show striations, with lateral spacing related to transducer size. For the focused case, the degree of curvature of the striations relates to the degree of curvature of the focused bowl.

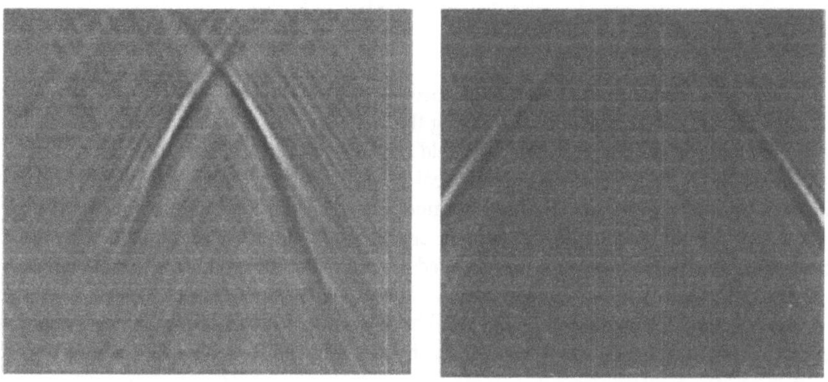

Figure 7. Simulated head wave, travelling upwards. A square transducer was assumed, and the view is of a projection of the field onto the (x,z) plane. (Left) The two visible components cross on axis at roughly the same point as the edge wave components, and interference between these two wave types is possible, leading to difficulty in interpreting axial point hydrophone measurements. (Right) At further distances from the transducer face, the two head wave components no longer cross, and confusion with edge waves is avoided.

Results comparable with finite element techniques may be obtained, but at a considerable saving in computation time. Nonlinear propagation offers more of a challenge, because of harmonic pumping, but recourse may be had to other methods of describing nonlinear effects. Of course, in all such simulations, diffraction effects are fully taken into account, but all the nuances of real transducer behaviour have to be tagged onto the ideal sources incorporated into the simulation.

The extension of the technique to multi-element transducers is straightforward, and simulations have been carried out for linear arrays, albeit it with a relatively small number of square elements only (not shown). The incorporation of phasing offers no difficulty.

The methods used in the simulation may be duplicated for laboratory field measurements, in real liquid media, provided that measurements are effected with a large aperture hydrophone, which measures field projections directly. Preliminary measurements, with typical circular (plane and focused) and square transducers, have already been carried out, and have verified that simulations of the above type indeed provide a very good approximation to transient fields in water and in the (few) lossy oils that have been investigated. The projections of the field may be measured at any convenient distance from the transducer face, and the reconstructed field may be propagated backwards, to the transducer face, without the complications of evanescent waves, which are absent from the theory. This provides the opportunity to make a number of statements about the source transducer, from the field measurements. Such data include: effective radiating area, apodisation function, excitation waveform, and even type of transducer material (from the measured head wave component).

CONCLUSIONS

A new field simulation technique has been proposed, and has been found to be computationally efficient, and capable of displaying the full complexity of diffraction effects. Idealised transducer behaviour, as well as field symmetries (circularly symmetric or square transducers) have been assumed. Non-ideal transducer behaviour, such as lateral modes giving rise to head waves, may be incorporated. The assumption of field symmetries may be dropped, but would lead to an increase in computation time. A significant feature of the technique is that pulsed fields from focused bowl transducers may be modelled, without approximation, even for strong curvatures. The extension to lossy, dispersive, and even solid, media may be made relatively straightforwardly. Propagation in nonlinear media offers more of a challenge, and the method is being extended to deal with that case.

REFERENCES

1. C.J. Vecchio, M.E. Schafer, and P.A. Lewin, Prediction of ultrasonic field propagation through layered media using the extended angular spectrum method, *Ultrasound Med. Biol.* 20:611 (1994)

2. G.E. Tupholme, Generation of acoustic pulses by baffled plane pistons, *Mathematika* 16:209 (1969)

3. P.R. Stepanishen, Transient radiation from pistons in an infinite planar baffle, *J. Acoust. Soc. Am.* 49:1629 (1971)

4. J.W.S. Rayleigh, *The Theory of Sound* II, 2nd ed., Dover,New York (1945)

5. A.J. Hayman and J.P. Weight, Transmission and reception of short ultrasonic pulses by circular and square transducers, *J. Acoust. Soc. Am.* 66/4:945 (1979)

TUNING OF BESSEL BEAMS ON ANNULAR ARRAYS

Paul D Fox and Sverre Holm

Department of Informatics
University of Oslo
P.O. Box 1080, Blindern
N-0316 Oslo, Norway

INTRODUCTION

Bessel beams[1-4] are a family of nondiffracting solutions to the linear wave equation. One is the circular-symmetric J_0 Bessel beam

$$p(r,z,t) = J_0(\alpha r)e^{j\beta z}e^{-j\omega t} \qquad where \qquad \beta = \sqrt{k^2 - \alpha^2} \quad , \quad k = \omega/c \tag{1}$$

where $p(r,z,t)$ is the field pressure relative to steady state. $J_0(\alpha r)$ is the Bessel function of the first kind of order zero, r is the radial distance from the cylindrical centerline, z is the propagation distance perpendicular to the transducer surface, and $k = \omega/c$ is the wavenumber (ω is frequency in rads/sec and c is speed of sound). This is a nondiffracting beam with infinite depth of field since its radial profile $J_0(\alpha r)$ is independent of the distance z from the transducer surface. If the transducer is placed at $z = 0$ and centred around $r = 0$, the required surface pressure to generate (1) is $p(r,z{=}0,t) = J_0(\alpha r)e^{-j\omega t}$. However, this requires a transducer of infinite radius and infinite radial resolution since $J_0(\alpha r) \propto \sqrt{2/\pi\alpha r}.\cos(\alpha r - \pi/4)$ extends to infinity before the amplitude $\sqrt{2/\pi\alpha r}$ decays to zero. In practice the profile must be truncated to zero at the transducer radius R, and the beam then assumes a limited depth of field given by[1] approximately $R\sqrt{(k/\alpha)^2 - 1}$, $\alpha \le k$. In addition, the idealised Bessel beam pressure distribution also needs to be quantised over the transducer surface due to the limited set of annular rings available. In[2,3] the annuli quantisation levels were generated as the mean and maximum values of $J_0(\alpha r)$ over each annulus respectively. In this article we use Fourier-Bessel theory to develop a least-squares design for selecting the quantisation levels to obtain an improved beam profile with lower sidelobes.

FOURIER-BESSEL THEORY FOR ANNULAR ARRAYS

As discussed in[5,6], an N-ring annular quantisation profile $q(r)$ may be modelled by the Fourier-Bessel series[7] :

Acoustical Imaging, Volume 25, Edited by Halliwell and Wells.
Kluwer Academic/Plenum Publishers, 2000.

$$q(r) = \sum_{i=1}^{\infty} A_i J_0(\alpha_i r) \quad where \quad A_i = q_1 C_{i,1} + q_2 C_{i,2} + \dots + q_N C_{i,N} = \sum_{p=1}^{N} q_p C_{i,p} \tag{2}$$

with

$$\alpha_i = x_i/a \quad , \quad J_0(x_i) = 0 \quad , \quad C_{i,p} = \frac{2\left[r_p^+ J_1(\alpha_i r_p^+) - r_p^- J_1(\alpha_i r_p^-)\right]}{\alpha_i a^2 J_1^2(\alpha_i a)} \tag{3}$$

Here the index p represents the pth annulus in the transducer, q_p the corresponding quantisation amplitude, and r_p^-, r_p^+ the inner and outer radii of the pth annulus respectively. This expansion is valid over an interval $0 < r \leq a$, where a is any value subject to $q(a) = 0$, we shall therefore select $a \geq R$ such that the pressure distribution over entire surface of the transducer may be modelled. The term $J_1(\alpha_i a)$ is a first-order Bessel function and the monotonically increasing basis values α_i are $\alpha_i = x_i/a$, where $x_i \approx (4i-1)\pi/4$ are the infinite set of known positive solutions to $J_0(x_i) = 0$. The quantised sum $q(r)$ in (2) then corresponds to a transducer surface pressure $q(r)e^{-j\omega t}$ generating a set of nondiffracting beam components

$$q(r,z,t) = \sum_{i=1}^{\infty} p_i(r,z,t) = \sum_{i=1}^{\infty} A_i J_0(\alpha_i r) e^{j\beta_i z} e^{-j\omega t} \quad : \quad \alpha_i = x_i/a \quad , \quad \beta_i = \sqrt{k^2 - \alpha_i^2} \tag{4}$$

which in practice are limited diffraction beams due to their truncation to zero at $r = a$ in the transducer plane at $z = 0$. Figure 1 (upper plot) shows the form of the Bessel function $J_0(x)$, where the zero crossings indicate the first few solutions x_i to $J_0(x_i) = 0$. The lower plot then shows the first five basis functions $J_0(\alpha_i r)$; note that the definition $\alpha_i = x_i/a$ scales the Bessel function such that the ith root of $J_0(\alpha_i r)$ is placed at $r = a$ for all i and all a.

FIELD COMPUTATION

However, if we allow the basis distance a to tend to infinity then each basis function $A_i J_0(\alpha_i r) e^{j\beta_i z} e^{-j\omega t} \equiv A_i J_0(x_i r/a) e^{j\beta_i z} e^{-j\omega t}$ in (4) becomes a fully valid Bessel solution (1) to the wave equation since its realisation in the transducer plane at $z = 0$ is taking place over an infinite realisation aperture $a \to \infty$. (Note that the case $a > R$ corresponds to an "extra" annulus on the transducer, whose inner radius is R, outer radius is a, and quantisation amplitude is $q_{N+1} = 0$; this means that although the sum for A_i in (2) should theoretically be extended from $p = 1 \dots N$ to $p = 1 \dots N{+}1$, it remains truncated in practise to $p = 1 \dots N$ due to $q_{N+1} = 0$). Combining this with the notation given in (2) then generates the field $f(r,z,t)$ for the limited diffraction sum ($a \neq \infty$) in (4) as

$$f(r,z,t) = \lim_{a\to\infty} q(r,z,t) = \lim_{a\to\infty} \sum_{i=1}^{\infty}\sum_{p=1}^{N} q_p C_{i,p} J_0(x_i r/a) e^{j\beta_i z} e^{-j\omega t} \quad , \quad \beta_i = \sqrt{k^2 - (x_i/a)^2} \tag{5}$$

In addition β_i becomes imaginary for all root values $x_i > ka$, and $e^{j\beta_i z}$ generally becomes evanescent in the propagation direction z due to the magnitudes of β_i involved in ultrasound. The number $l(a)$ of remaining non-evanescent basis functions[6] α_i is of order $l(a) \approx ka/\pi + 1/4$ and hence the nonevanescent field is given by the truncated sum

$$f(r,z,t) = \lim_{a\to\infty} \sum_{i=1}^{l(a)}\sum_{p=1}^{N} q_p C_{i,p} J_0(x_i r/a) e^{j\beta_i z} e^{-j\omega t} \qquad where \qquad l(a) \approx ka/\pi + 1/4 \tag{6}$$

Furthermore, if we also allow the quantisation levels q_p to become complex in the forms

$$q_p = \gamma_p + j\delta_p = |q_p| e^{j\theta_p} \qquad where \qquad |q_p| = \sqrt{\gamma_p^2 + \delta_p^2} \quad , \quad \theta_p = \tan^{-1}[\delta_p/\gamma_p] \tag{7}$$

where γ_p and δ_p, $p = 1 \dots N$ are real and imaginary parts respectively, then (6) becomes

$$f(r,z,t) = \lim_{a\to\infty} \sum_{i=1}^{l(a)} \sum_{p=1}^{N} |q_p| C_{i,p} J_0(x_i r/a) e^{j\beta_i z} e^{-j\omega(t-\tau_p)} \qquad where \qquad \tau_p = \theta_p/\omega \tag{8}$$

which corresponds to implementation of quantisation amplitudes $|q_p| = (\gamma_p^2 + \delta_p^2)^{½}$ and time delays $\tau_p = (\tan^{-1}[\delta_p/\gamma_p])/\omega$ on each annulus p respectively.

LEAST-SQUARES BEAM DESIGN

Consider the alternative form $q_p = \gamma_p + j\delta_p$ substituted from (7) into (6), and rearrange terms to obtain the field $f(r,z,t)$ in the form $f(r,z,t) = e^{-j\omega t} f(r,z)$ as

$$f(r,z,t) = e^{-j\omega t} \lim_{a\to\infty} \sum_{i=1}^{l(a)} \sum_{p=1}^{N} J_0(x_i r/a) \,.\, C_{i,p} e^{j\beta_i z} \,.\, (\gamma_p + j\delta_p) = e^{-j\omega t} f(r,z) \tag{9}$$

where $f(r,z)$ is the spatial field component postmultiplying $e^{-j\omega t}$. The real and imaginary components $\Re_f(r,z)$, $\Im_f(r,z)$ of $f(r,z)$ may then be written as

$$\begin{bmatrix} \Re_f(r,z) \\ \Im_f(r,z) \end{bmatrix} = \left[\lim_{a\to\infty} \sum_{i=1}^{l(a)} J_0(x_i r/a) \begin{pmatrix} +M_{i,1}^{\Re}, -M_{i,1}^{\Im}, \dots, +M_{i,N}^{\Re}, -M_{i,N}^{\Im} \\ +M_{i,1}^{\Im}, +M_{i,1}^{\Re}, \dots, +M_{i,N}^{\Im}, +M_{i,N}^{\Re} \end{pmatrix} \right] . \begin{bmatrix} \gamma_1 \\ \delta_1 \\ \vdots \\ \gamma_N \\ \delta_N \end{bmatrix} \tag{10}$$

where the matrix entries $M_{i,p}^{\Re}$ and $M_{i,p}^{\Im}$ are the real and imaginary parts of $M_{i,p} \quad C_{i,p} e^{j\beta_i z}$ respectively. This means that we now have all components of the radiated field $f(r,z)$ for all combinations $r = r_u$ and $z = z_v$ of interest, with $u = 1 \dots n_u$ and $v = 1 \dots n_v$ where and n_u, n_v are the respective numbers of points, as a linear function of the real and imaginary annular quantisation components γ_1, δ_1, ... γ_N, δ_N :

$$\begin{bmatrix} \Re_f(r_1,z_1) \\ \Im_f(r_1,z_1) \\ \vdots \\ \Re_f(r_{n_u},z_{n_v}) \\ \Im_f(r_{n_u},z_{n_v}) \end{bmatrix} = \lim_{a\to\infty} \begin{bmatrix} \sum_{i=1}^{l(a)} J_0(x_i r_1/a) \begin{pmatrix} +M_{i,1,1}^{\Re}, -M_{i,1,1}^{\Im}, \dots, +M_{i,N,1}^{\Re}, -M_{i,N,1}^{\Im} \\ +M_{i,1,1}^{\Im}, +M_{i,1,1}^{\Re}, \dots, +M_{i,N,1}^{\Im}, +M_{i,N,1}^{\Re} \end{pmatrix} \\ \vdots \quad \vdots \quad \vdots \quad \vdots \quad \vdots \\ \sum_{i=1}^{l(a)} J_0(x_i r_{n_u}/a) \begin{pmatrix} +M_{i,1,n_v}^{\Re}, -M_{i,1,n_v}^{\Im}, \dots, +M_{i,N,n_v}^{\Re}, -M_{i,N,n_v}^{\Im} \\ +M_{i,1,n_v}^{\Im}, +M_{i,1,n_v}^{\Re}, \dots, +M_{i,N,n_v}^{\Im}, +M_{i,N,n_v}^{\Re} \end{pmatrix} \end{bmatrix} . \begin{bmatrix} \gamma_1 \\ \delta_1 \\ \vdots \\ \gamma_N \\ \delta_N \end{bmatrix} \tag{11}$$

in which $M_{i,p,v}^{\Re}$ and $M_{i,p,v}^{\Im}$ represent the real and imaginary parts of $M_{i,p,v} = C_{i,p} e^{j\beta_i z_v}$ respectively, and where the quantisation components γ_1, δ_1, ... γ_N, δ_N define the quantisation amplitudes $|q_p|$ and delays τ_p of each annulus according to (7) and (8). The least squares design objective is then to select these components so as to obtain a high quality beam with low sidelobes and adequate depth of field. Defining a desired field $d(r,z,t)$ in the form

$$d(r,z,t) = e^{-j\omega t} d(r,z) \qquad where \qquad d(r,z) = \Re_d(r,z) + j\Im_d(r,z) \tag{12}$$

where $d(r,z)$ is the desired spatial distribution, then gives an error $e(r,z,t) = f(r,z,t) - d(r,z,t)$ between actual and desired fields as

$$e(r,z,t) = e^{-j\omega t}([\Re_f(r,z) - \Re_d(r,z)] + j[\Im_f(r,z) - \Im_d(r,z)]) \tag{13}$$

Hence, taking its magnitude squared $|e(r,z,t)|^2$ and summing over a set of sampled spatial locations $r = r_u$ and $z = z_v$ of interest generates a total squared error sum S as :

$$S = \sum_{u=1}^{n_u} \sum_{v=1}^{n_v} |e(r_u,z_v,t)|^2 = \sum_{u=1}^{n_u} \sum_{v=1}^{n_v} ([\Re_f(r_u,z_v) - \Re_d(r_u,z_v)]^2 + [\Im_f(r_u,z_v) - \Im_d(r_u,z_v)]^2) \quad (14)$$

which is independent of time t. This expression may also be written in block form as

$$S = [F-D]'.[F-D] = [MX-D]'.[MX-D] \quad (15)$$

in which the operator $'$ represents the transpose, M is the large matrix containing the terms $M^{\Re}_{i,p,v}$ and $M^{\Im}_{i,p,v}$ on the right hand side of (11), and F, D and X are the vectors

$$F = \begin{bmatrix} \Re_f(r_1,z_1) \\ \Im_f(r_1,z_1) \\ \vdots \\ \Re_f(r_{n_u},z_{n_v}) \\ \Im_f(r_{n_u},z_{n_v}) \end{bmatrix} \quad , \quad D = \begin{bmatrix} \Re_d(r_1,z_1) \\ \Im_d(r_1,z_1) \\ \vdots \\ \Re_d(r_{n_u},z_{n_v}) \\ \Im_d(r_{n_u},z_{n_v}) \end{bmatrix} \quad , \quad X = \begin{bmatrix} \gamma_1 \\ \delta_1 \\ \vdots \\ \gamma_N \\ \delta_N \end{bmatrix} \quad (16)$$

The vector X contains all the quantisation components γ_p, δ_p, $p = 1 \ldots N$ and the objective is then to minimise S in (15) with respect to all γ_p, δ_p. This is a standard linear least squares problem format, with solution $X = X_{ls}$ and corresponding field $F = F_{ls}$ as

$$X_{ls} = [M'M]^{-1}M'D \quad , \quad F_{ls} = MX_{ls} = M[M'M]^{-1}M'D \quad (17)$$

(subject to $n_u n_v \geq N$ and M being full rank for the inverse $[M'M]^{-1}$ to exist). In principle it then remains only to define the desired field vector D to obtain X_{ls}. In practice we also need to cater for the limit $a \to \infty$ in the matrix M in (11). We achieve this by allowing the basis distance a to increase from small to increasingly large values until all entries of X_{ls} converges to steady state values. Typically one may begin with setting $a = R$ and then either double or treble the value of a at each successive iteration, although more complex adaptive schemes may also be employed if desired. The final effect is to obtain steady state least squares values of X_{ls}, which then in turn generate corresponding quantisation amplitudes $|q_p| = (\gamma_p^2 + \delta_p^2)^{1/2}$ and time delays $\tau_p = (\tan^{-1}[\delta_p/\gamma_p])/\omega$ according to (7) and (8) for implementation on each annulus p.

NUMERICAL EXAMPLE

Consider a 15mm diameter, 4-ring (N =4) equal-area transducer discussed previously in[2]. This has inner and outer radii $r_1^- = 0$mm, $r_1^+ = 3.75$mm, $r_2^- = 3.75$mm, $r_2^+ = 5.30$mm $= r_3^-$, $r_3^+ = 6.50$mm $= r_4^-$ and $r_4^+ = R = 7.50$mm (Figure 2). At 3.5MHz with a (water) propagation speed c = 1540 m/s, the wavenumber is $k = 14279.97\ m^{-1}$. In[2], the aim was to approximate a Bessel beam whose first zero coincided with the edge $r_1^+ = 3.75$mm of the first annulus. This corresponds to $\alpha = 641.2869\ m^{-1}$ since first zero x_1 of $J_0(.) = 0$ is 2.404826 and α must satisfy $\alpha r_1^+ = x_1$. The quantisation levels were then taken as the mean values of $J_0(\alpha r)$ over each annulus (Figure 3) yielding $q_1 = +0.61$, $q_2 = -0.21$, $q_3 = -0.39$, $q_4 = -0.32$, (or equivalently $|q_1| = 0.61$, $|q_2| = 0.21$, $|q_3| = 0.39$, $|q_4| = 0.32$ with half-cycle relative delays $\tau_1 = 0$, $\tau_2 = 143$, $\tau_3 = 143$, $\tau_4 = 143$ nanoseconds). The resulting field is given in Figure 4, normalised at all values of z relative to 0dB along the centerline $r = 0$. Relative contours of -6dB (inner), -12dB (middle), and -18dB (outer) are indicated. A completely nondiffracting

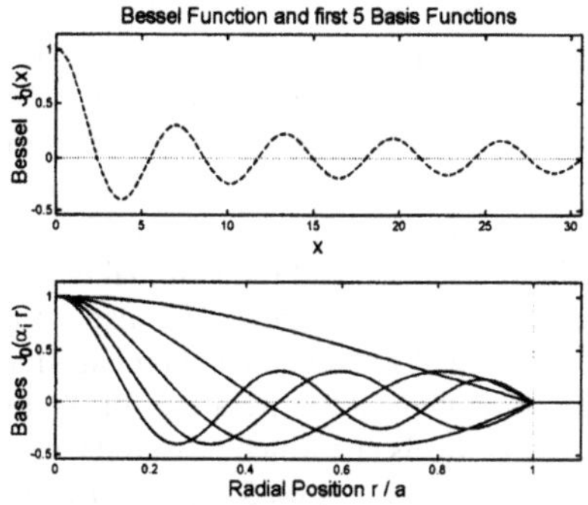

Figure 1 Bessel function and basis functions

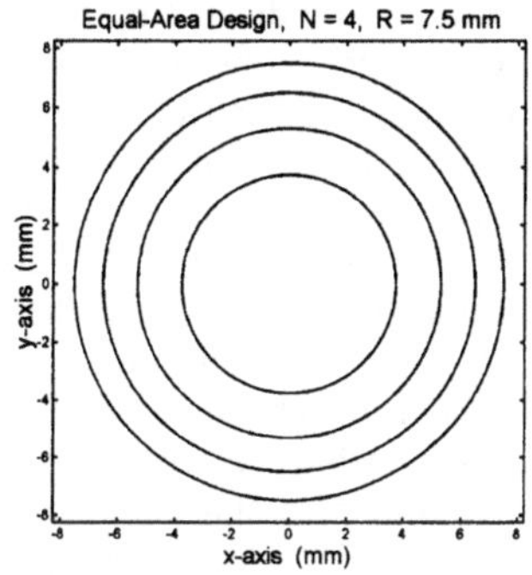

Figure 2 Topview of transducer

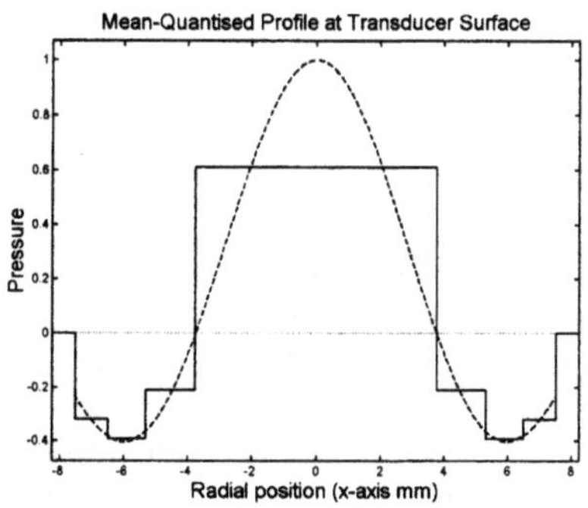

Figure 3 Mean-level quantisation

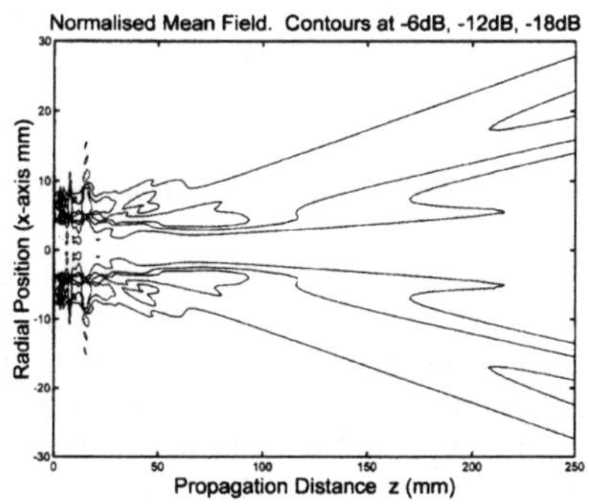

Figure 4 Field for mean levels

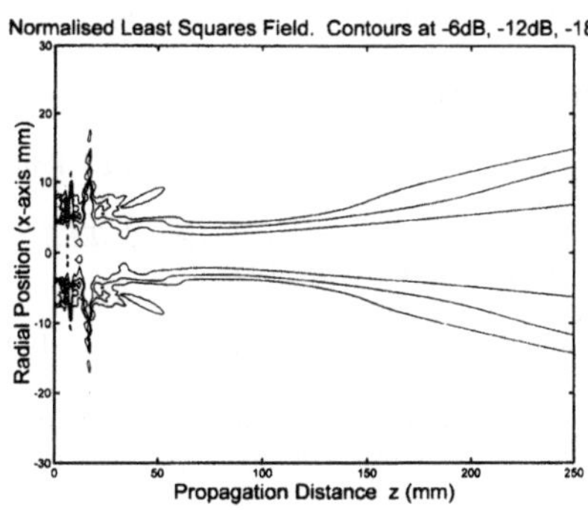

Figure 5 Field for least-squares levels 0-167mm

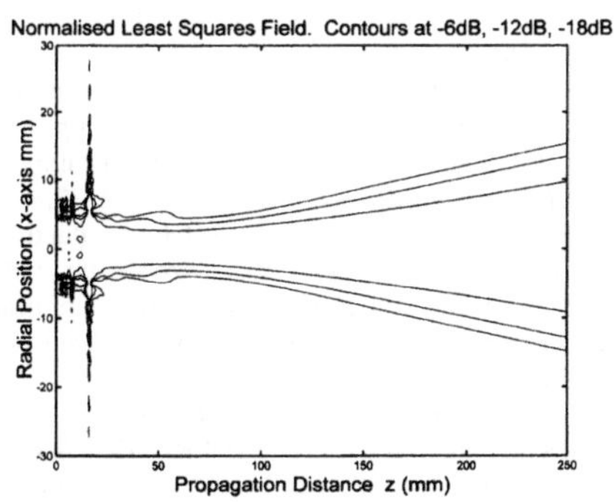

Figure 6 Field for least-squares levels 25-75mm

beam would thus be indicated by a set of parallel contours running in the z direction. In practice the -6dB mainlobe is almost nondiffracting between approx 50-170mm, although sidelobes are present and considerable diffraction is evident beyond 170mm. We then perform a least-squares design, in which we choose the design objective as a beam composed of only the mainlobe of $J_0(\alpha r)$, with zero sidelobes elsewhere. This gives $d(r,z) = J_0(\alpha r)e^{j\beta z}$ for $0 \le r \le r_1^+$ and $d(r,z) = 0$ for $r_1^+ < r$. We also define a region over which the error sum is $J_0(\alpha r)$ to be evaluated, and take 1.5 times the radius r in the r direction and the nominal depth of field $R\sqrt{(k/\alpha)^2-1} \approx$ 167mm in the z direction as summation boundaries. We also sample spatially at a high enough rate to avoid aliasing in both the radial r direction and propagation z direction. This enables all terms in D and M to be computed. After iterating for X_{ls} over several cycles with increasing values of a, the quantisation levels constructed from the steady state components of X_{ls} emerge as $|q_1| = 0.51$, $|q_2| = 0.15$, $|q_3| = 0.25$, $|q_4| = 0.23$ with relative delays $\tau_1 = 0$, $\tau_2 = 71$, $\tau_3 = 103$, $\tau_4 = 112$ nanoseconds. The corresponding field is given in Figure 5, exhibiting improved sidelobe characteristics and an improvement in diffraction qualities for the -6dB mainlobe in the farfield. The mainlobe is still however somewhat irregular below approx $z = 75$mm, and a second least-squares design computation considering only the region $z = 25$mm to $z = 75$mm is then performed in an attempt to improve this. Figure 6 gives the result, showing an improvement in the region of interest but at the cost of a wider mainlobe in the farfield. However, both least squares designs show improvements over the mean design.

CONCLUSIONS AND FURTHER WORK

A least-squares method for tuning flat annular arrays using Fourier-Bessel series has been discussed. The resultant design shows an improvement in sidelobe levels over a previous quantisation design when applied to the generation of Bessel beams. The present results are limited to continuous wave annular array studies in lossless media; the next phase is to extend the work to tuning to lossy media and then to consider the pulsed case. In addition work has also recently begun on modelling 2D arrays with 2D Fourier-Bessel series[8], and it is hoped that similar tuning techniques may be developed for these arrays too.

ACKNOWLEDGEMENT

This work was sponsored by the European Union under TMR contract ERBFMBICT 982917.

REFERENCES

1. J. Durnin, "Exact solutions for nondiffracting beams. I. The scalar theory," *J. Opt. Soc. Amer.*, vol. 4, pp. 651-654. (Apr. 1987).
2. S. Holm, "Bessel and conical beams and approximation with annular arrays," *IEEE Trans. Ultrason., Ferroelect., Freq. Cont.*, vol. 45, no. 3, pp. 712-718. (Sept. 1998).
3. J.-Y. Lu and J. F. Greenleaf, "Ultrasonic nondiffracting transducer for medical imaging," *IEEE Trans. Ultrason., Ferroelect., Freq. Cont.*, vol. 37, no. 5, pp. 438-447. (Sept. 1990).
4. J-Y. Lu and J.F. Greenleaf, "Sidelobe reduction for limited diffraction pulse-echo systems," *IEEE Trans. Ultrason., Ferroelect., Freq. Cont.*, vol. 40, no. 6, pp. 735-746. (Nov. 1993).
5. S. Holm and P.D. Fox, "Analysis of Bessel beam quantisation in annular arrays," *Proc. 22nd Scandinavian Symposium on Physical Acoustics*, Ustaoset, Norway, pp. 43-44. (Jan. 31 - Feb. 3, 1999). ISSN 1501-6773.
6. P.D. Fox and S. Holm, "Decomposition of acoustic fields in quantised Bessel beams," Poster at joint event Ultrasonics International 99 / 1999 World Congress on Ultrasonics, Lyngby (Copenhagen), Denmark. (Jun. 29 - Jul. 1, 1999). Full paper to appear in "Ultrasonics".

7. F. Bowman, *Introduction to Bessel Functions,* Dover Publications Incorporated., New York. (1958). ISBN 0-486-60462-4.
8. P.D. Fox, "Computation of linear ultrasound fields using 2D Fourier-Bessel series," *Proc. 25th International Acoustical Imaging Symposium,* Bristol, United Kingdom. (Mar. 19-22, 2000).

FINITE ELEMENT MODELLING OF PULSED BESSEL BEAMS AND X-WAVES USING DIFFPACK

Åsmund Ødegård, Paul D Fox, Sverre Holm, Aslak Tveito

Department of Informatics
University of Oslo
P.O.Box 1080, Blindern
N-0316 Oslo, Norway

INTRODUCTION

This article reports some preliminary results in the development of a finite element simulator for ultrasonic fields using the numerical programming library Diffpack. In our present work, we are combining the study of linearly–based limited diffraction beams (Bessel beams and X–Waves) with the development of a finite element package for analysis of linear and nonlinear media in order to compare how such waves behave in the two cases. The medium may be either homogeneous or inhomogeneous. The main content of this paper is on the linear model, but we also include some recent developments on propagation in nonlinear medium.

PROPAGATION MODELS, BESSEL BEAMS AND X–WAVES

Bessel beams (Durnin, 1987) and X-Waves (Lu and Greenleaf, 1992) are limited diffraction solutions to the lossless linear wave equation

$$\nabla^2 p - \frac{1}{c^2}\frac{\partial^2 p}{\partial t^2} = 0, \quad (x,t) \in \Omega \times \mathbb{R}^+ \tag{1}$$

in which c is the speed of sound. One variant of Bessel beams and X-Waves are those of order zero

$$p(r,z,t) = J_0(\alpha r)e^{j(\beta z - \omega t)} \qquad \textit{Bessel Beam}, \tag{2}$$

$$p(r,z,t) = \frac{a_0}{\sqrt{r^2 \sin^2\zeta + [a_0 - j(z\cos\zeta - ct)]^2}} \qquad X - Wave. \tag{3}$$

These exhibit circular symmetry and are therefore well suited to implementation on annular arrays. Here a_0 is a constant, ω is the frequency, r is the distance from the center line of the transducer, z is the propagation distance from the transducer surface, and ζ, α are design

Acoustical Imaging, Volume 25, Edited by Halliwell and Wells.
Kluwer Academic/Plenum Publishers, 2000.

parameters with $\beta = \sqrt{(\omega/c)^2 - \alpha^2}$. See (Lu, 1997; Holm, 1998; Fox and Holm, 1999) for details of design and implementation. For the purposes of simulation, we use the real parts of (2), (3) to generate the transducer surface pressure excitation at $z = 0$. The domain of computation is taken to be a box $\Omega \in \mathbb{R}^3$ over positive time and with initial condition

$$p(x, t_0) = p_0, \quad x \in \Omega, \tag{4}$$

$$\frac{\partial p}{\partial t}(x, t_0) = 0, \quad x \in \Omega, \tag{5}$$

where p_0 is the steady state (atmospheric) pressure. The boundary $\partial\Omega$ of the domain Ω is then divided into two parts, $\partial\Omega_T$ and $\partial\Omega_{T^c}$ such that $\partial\Omega_T$ is the part of the boundary on which the ultrasound transducer is located, and $\partial\Omega_{T^c}$ is the remainder. The waves in (2,3) are generated on the transducer Ω_T, and in the latter region we adopt a first order non-reflecting boundary (Engquist and Majda, 1977)

$$\frac{\partial p}{\partial n} = -\frac{1}{c}\frac{\partial p}{\partial t}, \quad x \in \partial\Omega_{T^c} \tag{6}$$

The longterm objective of our work is then to compare how the Bessel beams and X–Waves behave under the idealised lossless linear conditions of (1) and the more complex lossy nonlinear propagation conditions defined in (Makarov and Ochmann, 1996; Makarov and Ochmann, 1997) by the system

$$\nabla^2\varphi - \frac{1}{c^2}\frac{\partial^2\varphi}{\partial t^2} + \frac{1}{c^2}\frac{\partial}{\partial t}\left[(\nabla\varphi)^2 + \frac{B/A}{2c^2}\left(\frac{\partial\varphi}{\partial t}\right)^2 + b\nabla^2\varphi\right] = 0, \tag{7}$$

$$p - p_0 = \rho_0\frac{\partial\varphi}{\partial t} + V_a - E_k + \rho_0 b\nabla^2\varphi, \tag{8}$$

$$V_a = \frac{\rho_0}{2c^2}\left(\frac{\partial\varphi}{\partial t}\right)^2, \tag{9}$$

$$E_k = \frac{1}{2}\rho_0(\nabla\varphi\cdot\nabla\varphi) \tag{10}$$

Here, φ is the velocity potential, ρ_0 is the density, b is the absorption parameter and B/A is the nonlinearity parameter. The problem is solved on some domain $\Omega \in \mathbb{R}^3$ for positive time.

By ignoring the nonlinear terms and the lossy term in the relation (8), we get the simpler relation

$$p - p_0 = \rho_0\frac{\partial\varphi}{\partial t}, \tag{11}$$

which may be used as a replacement for the pressure — velocity potential relation (8). We assume that the system is at rest initially. Further, we may combine the initial condition for the pressure (4) and the pressure — velocity potential relation (11) to obtain a second initial condition for the velocity potential,

$$\varphi(x, t_0) = \varphi_0, \quad x \in \Omega, \tag{12}$$

$$\frac{\partial\varphi}{\partial t}(x, t_0) = 0, \quad x \in \Omega, \tag{13}$$

On the transducer boundary $\partial\Omega_T$, we combine the relation (11) and the wave–expressions (2,3) to set the boundary condition in terms of the velocity potential φ. Since the nonlinear model is an extension of the linear model, we also adopt the absorbing boundary condition (6) on the rest of the boundary, $\partial\Omega_{T^c}$, for the nonlinear model, i.e

$$\frac{\partial\varphi}{\partial n} = -\frac{1}{c}\frac{\partial\varphi}{\partial t}, \quad x \in \partial\Omega_{T^c} \tag{14}$$

NUMERICAL METHOD

Our approach is to use a Galerkin finite element method in the space domain, combined with a finite difference approximation of the time derivatives. Consider the linear model (1). Assume that $p^n = p^n(x)$ is the pressure at time $t_n = n \cdot \Delta t$. Then a semi-discrete scheme for the problem is given by

$$\nabla^2 p^n - \frac{1}{c^2 \Delta t^2}(p^{n+1} - 2p^n + p^{n-1}) = 0, \quad x \in \Omega. \tag{15}$$

Assume now that $\{N_i\}_{i=1}^m$ is a finite element basis for the solution space such that $p^n(x) \approx \sum_i p_i N_i$. Multiply (15) with N_j and apply integration by parts to the second order derivative in space, using the non–reflecting boundary condition (6). This yields a set of m equations for the unknowns:

$$\begin{aligned} &\sum_{i=1}^{m} \left[\frac{1}{c\Delta t} \int_{\partial\Omega_{Tc}} N_i N_j dn + \frac{1}{c^2 \Delta t^2}(N_i, N_j) \right] p_i = \\ &\quad \frac{1}{c\Delta t} \int_{\partial\Omega_{Tc}} p^n N_j dn - (\nabla p^n, \nabla N_j) + \frac{1}{c^2 \Delta t^2}(2p^n - p^{n-1}, N_j), \quad j = 1, \ldots, m, \end{aligned} \tag{16}$$

Here the inner product (u, v) is defined as the integral of the functions u and v over the solution domain

$$(u, v) = \int_\Omega uv dx$$

For the linear case, the problem then formulates into a linear problem of the type $Ax = b$ at each time step, which is solved with e.g. a Krylov subspace method such as Conjugate Gradients. For the nonlinear case, we obtain a nonlinear discrete problem which is solved with an iterative method, e.g. Newton–Raphson iterations, at each time step.

Consider now the nonlinear model (7). The same approach as for the linear case are used. For the time–derivatives, we use centred differences. Then a semi–discrete scheme for the problem is given by

$$\begin{aligned} &\nabla^2 \varphi^n - \frac{1}{c^2 \Delta t}(\varphi^{n+1} - 2\varphi^n + \varphi^{n-1}) \\ &\quad + \frac{1}{c^2 \Delta t} \left\{ \left[(\nabla\varphi)^2 + \frac{B/A}{2c^2} \left(\frac{\partial \varphi}{\partial t} \right)^2 + b\nabla^2 \varphi \right]^{n+1/2} \right. \\ &\quad \left. - \left[(\nabla\varphi)^2 + \frac{B/A}{2c^2} \left(\frac{\partial \varphi}{\partial t} \right)^2 + b\nabla^2 \varphi \right]^{n-1/2} \right\} = 0. \end{aligned} \tag{17}$$

The $n \pm 1/2$–power on the brackets, means that all expression inside the bracket should be evaluated at $t = t_{n \pm 1/2}$. This is performed by taking the mean value of the space–derivatives and using centred differences for the derivatives of time.

Assume now that $\{N_i\}_{i=1}^m$ is a finite element basis for the solution space such that $\varphi^n(x) \approx \sum_i \varphi_i^n N_i$,. Expand the brackets in (17), multiply with N_j and apply integration by parts to the second order derivatives in space, using the non–reflecting boundary condition (14). If we assume that φ^n and φ^{n-1} is computed, this yields a nonlinear problem for φ^{n+1}

$$F(\varphi^{n+1}; \varphi^n, \varphi^{n-1}) = 0,$$

where F is the function with components

$$\begin{aligned}
F_j = & -\frac{1}{2c}\int_{\partial\Omega}(\varphi^{n+1}-\varphi^{n-1})N_j dn - \Delta t(\nabla\varphi^n, \nabla N_j) \\
& -\frac{1}{c^2\Delta t}\Big([\varphi^{n-1}-2\varphi^n+\varphi^{n+1}], N_j\Big) \\
& +\frac{1}{c^2}\Bigg\{\frac{1}{4}\Big([(\nabla\varphi^{n+1})^2 + 2\nabla\varphi^n(\nabla\varphi^{n+1}-\nabla\varphi^{n-1}) - (\nabla\varphi^{n-1})^2], N_j\Big) \\
& +\frac{B/A}{2c^2\Delta t^2}\Big([(\varphi^{n+1})^2 - 2\varphi^n(\varphi^{n+1}-\varphi^{n-1}) - (\varphi^{n+1})^2], N_j\Big) \\
& +\frac{b}{\Delta tc}\int_{\partial\Omega}[\varphi^{n+1}+2\varphi^n-\varphi^{n-1}]N_j dn \\
& -\frac{b}{2}\Big([\nabla\varphi^{n+1}-\nabla\varphi^{n-1}], \nabla N_j\Big)\Bigg\} = 0, \quad j=0,\ldots,m.
\end{aligned} \tag{18}$$

Using the initial conditions (12, 13), we can initialise φ^{-1} and φ^0. Hence, assume that φ^{n-1} and φ^n are computed. If we set $\varphi^{n+1,0} = \varphi^n$, a hopefully better approximation to φ^{n+1} can be computed by Newton–Raphson iterations

$$J^k r^k = -F^k, \quad \text{where } r^k = \varphi^{n+1,k+1} - \varphi^{n+1,k}, \quad k = 0,1,2,\ldots. \tag{19}$$

Here J is the Jacobi matrix with components $J_{i,j} = \partial F_j/\partial\varphi_i^{n+1}$. Further, J^k and F^k are the respective expressions evaluated at $(\varphi^{n+1,k};\varphi^n,\varphi^{n-1})$. When some convergence criterion on the residual r^k is fulfilled, we set $\varphi^{n+1} = \varphi^{n+1,k+1}$.

DIFFPACK SIMULATOR

The simulator is implemented using the numerical Finite Element computational library Diffpack (Langtangen, 1999). This package is well suited to the simulation of ultrasound fields due to its versatility in structure and adaptability to simulation of different propagation conditions. The simulator is implemented as a class in C++, derived from the Finite Element solver class in Diffpack, and handles both pulsed and continuous wave 3D simulations. Inhomogeneous media are handled by introducing the speed of sound c as a field over the finite element grid.

The Diffpack library also includes support for parallel computations, through the mpi–system. We have used this system to implement a parallel version of the linear simulator.

RESULTS, CONCLUSION AND FURTHER WORK

As examples of the linear work we illustrate small scale nearfield simulations for pulsed Bessel beams (7-lobes) and X-Waves from a 6mm diameter transducer. The fields are solved in an $\mathbb{R}^3$ box measuring 10mm wide x 10mm high x 7mm deep. For the X–Wave, we use $a_0 = 0.05mm$ (Lu and Greenleaf, 1992) and $\zeta = 15°$. A 2.5MHz transducer in water (speed of sound $c = 1540m/s$) produces a wavelength $\lambda = 0.61mm$. The finite element grid was defined to provide 10 times oversampling of this wavelength, needing approximately $2.25 \cdot 10^6$ nodes. The solver then requires 1140MB of memory to solve this problem. The time step is set such that the wave is sampled at 20 times the frequency, which gives $\Delta t = 2.0 \cdot 10^{-8}$ for the given nominal frequency of 2.5MHz.

In Figures 1a and 1b we show snapshots of the pulsed Bessel beams and X-Waves travelling through a homogeneous medium. The figures depict x–y plane cuts through the center of

(a) Bessel beam

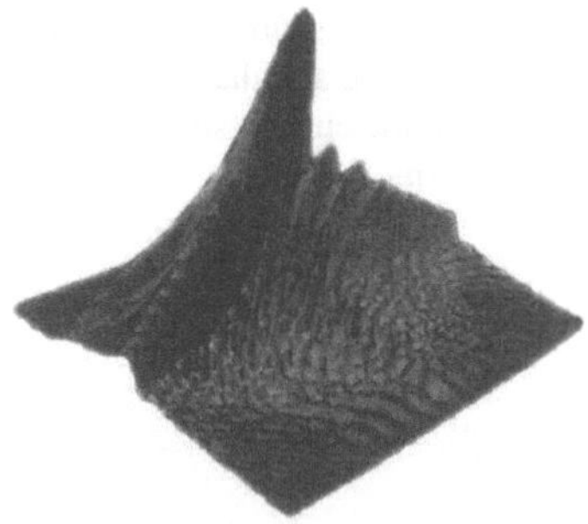

(b) X–Wave

Figure 1: Snapshots of pulsed beams. Direction of travel : bottom right to top left

the transducer for each case. The CPU time used to solve the problem for the pulsed Bessel beam travelling through the domain, was 1.66 hours and for the X–Wave was 4.35 hours. These computations were done on an SGI/Cray Origin 2000 computer with the scalar linear solver. The same computations were also carried out on a Beowulf cluster[1], using 18 processors. The CPU time used to solve the problem for the pulsed Bessel beam was then 3.04 minutes and for the X–Wave it was 3.33 minutes. This leads us to conclude that the CPU time used to solve the model for linear acoustic waves, using finite element methods, is not a problem anymore, because the method is suitable for implementation on parallel computers. The problem with the huge memory demands is still not solved.

In Figures 2a and 2b, the maximum field intensity at each point in space for the same pulses having passed through a medium is given. Here, a change in medium from human fat ($c = 1427m/s$) to muscle ($c = 1571m/s$) is included in the form of an abrupt boundary change angled at 45 degrees from bottom right to top left. In both cases the diffraction effects at the boundary are visible. Notably we observe the nondiffracting property of the Bessel beam close to the transducer surface degrading considerably upon impact with the boundary.

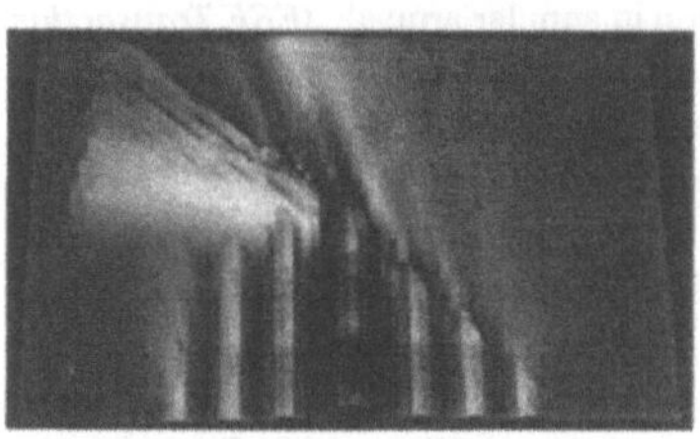

(a) Bessel beam

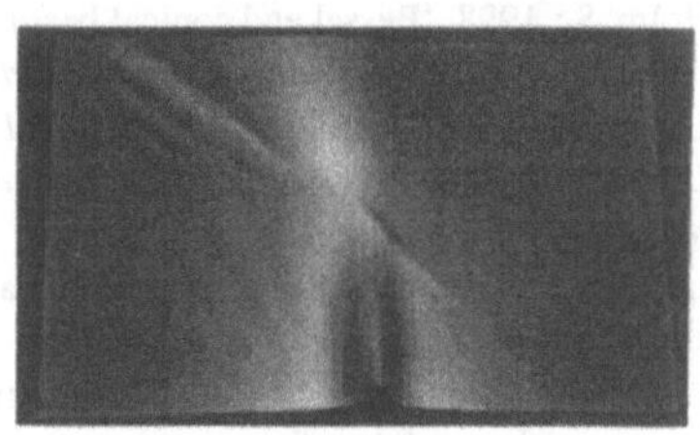

(b) X–Wave

Figure 2: Maximum intensities in inhomogeneous media. Direction of travel : bottom to top

Figure (3)a shows a snapshot of the pulsed Bessel beam in nonlinear medium, while

[1]A cluster of 24 Dual 500MHz Pentium-III computers with 512MB memory each and running the operating system Linux

Figure (3) shows the maximum field intensity. This is a 2D simulation in a $10mm \times 7mm$ domain, with a 6mm wide transducer. For the parameters in the model we have used $B/A = 5.0$, $b = 4.3 \cdot 10^{-6} m^2/s$ and $\rho = 1000 kg/m^3$.

Our initial results therefore suggest that finite elements exhibit strong potential for flexible and realistic simulation of ultrasound pulses, although techniques to reduce the memory demands of the current solver also need to be investigated in order to enable solution of large scale systems.

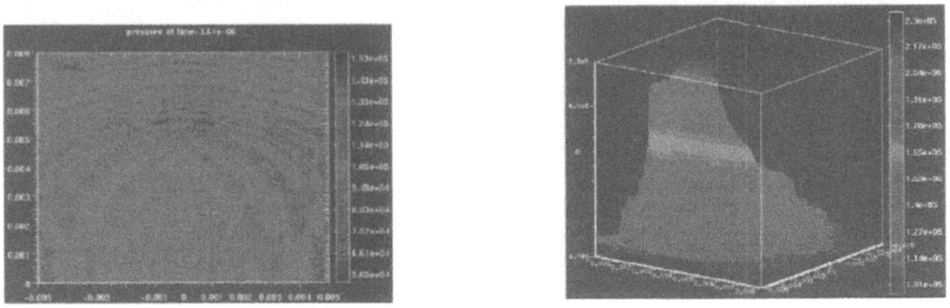

(a) Snapshot

(b) Maximum intensities

Figure 3: 2D simulation in a nonlinear medium.

ACKNOWLEDGEMENT

This work was partially sponsored by the European Union under TMR contract ERBFMBICT 982917.

REFERENCES

Durnin, J.: 1987, 'Exact solutions for nondiffracting beams'. *J. Opt. Soc. Am* **4**(4).

Engquist, B. and A. Majda: 1977, 'Absorbing boundary conditions for the numerical simulation of waves'. *Mathematics of computation* **31**(139).

Fox, P. and S. Holm: 1999, 'Effects of parameter mismatch in ultrasonic Bessel transducers'. *Proc. IEEE Norwegian Signal Processing Symposium NORSIG'99.*

Holm, S.: 1998, 'Bessel and conical beams and approximation in annular arrays'. *IEEE Transaction on Ultrasonics, Ferroelectrics and Frequency control* **45**(3).

Langtangen, H. P.: 1999, *Computational Partial Differential Equations, Numerical methods and Diffpack programming*, Vol. 2 of *Lecture notes in computational science and engineering.* Springer–Verlag.

Lu, J.-Y.: 1997, 'Designing limited diffraction beams'. *IEEE Transaction on Ultrasonics, Ferroelectrics and Frequency control* **44**(1).

Lu, J.-Y. and J. Greenleaf: 1992, 'Nondiffracting X–waves — exact solutions to free–space scalar wave equation and their finite aperture realizations'. *IEEE Transaction on Ultrasonics, Ferroelectrics and Frequency control* **39**(1).

Makarov, S. and M. Ochmann: 1996, 'Nonlinear and thermoviscous phenomena in Acoustics, Part I'. *ACUSTICA - acta acustica* **82**.

Makarov, S. and M. Ochmann: 1997, 'Nonlinear and thermoviscous phenomena in Acoustics, Part II'. *ACUSTICA - acta acustica* **83**.

NUMERICAL REALIZATION OF ALGORITHM FOR EXACT SOLUTION OF TWO-DIMENSIONAL MONOCHROMATIC INVERSE PROBLEM OF ACOUSTICAL SCATTERING

Alexei V. Bogatyrev [1], Valentin A. Burov [1], Sergei A. Morozov [1], Olga D. Rumyantseva [1], and Evgenii G. Sukhov [2]

[1] Moscow State University, Faculty of Physics, Dept. of Acoustics
Moscow, 119899, Russia
[2] Institute of Control Sciences of Russian Academy of Sciences

The present report is devoted to a detailed investigation of practical possibilities provided by a numerical realization of the Grinevich-Novikov algorithm[1,2]. The algorithm allows to reconstruct a distribution of two-dimensional scatterer of an arbitrary form

$$v(\vec{r}) \equiv \omega^2\left(c_o^{-2} - c^{-2}(\vec{r})\right) - 2i\omega\alpha(\vec{r},\omega)/c(\vec{r}) \tag{1}$$

by means of the scattering amplitude $f(\vec{k},\vec{l}) \equiv f(\varphi,\varphi')$ of plane waves: wave vectors $\vec{k} = \{k_o,\varphi\}$ and $\vec{l} = \{k_o,\varphi'\}$ characterize the falling direction and the scattering one. Here k_o and c_o are wave number and phase velocity, respectively, for the background medium without absorption; $c(\vec{r})$ and $\alpha(\vec{r},\omega)$ are velocity and amplitude absorption coefficient inside the scattering domain. The scatterer itself may have an absorption and be sufficiently strong because the rescatterings effects are rigorously mathematically taken into consideration. A description of the algorithm, advantages and convenience of its practical realization and also the first results of numerical modeling are presented in the previous report[3]. It should be remembered that the main moment of the algorithm is an extension of the values $f(\vec{k},\vec{l})$ to "non physical" domain of complex wave vectors with an infinitesimal imaginary part being orthogonal to a real part. The extension is realized by the generalized Green function introduced by L.D.Faddeev.

An area of a guarantee validity of the algorithm is defined by two conditions. The first condition is a localization of space spectrum $T(\vec{\xi},\vec{k})$ of the secondary sources (induced into the scatterer) inside $2k_o$-radius circle with the center at point $\vec{k}$ for each fixed $\vec{k}$:

$$T(\vec{\xi},\vec{k}) \approx 0, \text{ if } \left|\vec{\xi} - \vec{k}\right| \geq 2k_o. \tag{2}$$

The secondary sources are $T(\vec{r},\vec{k}) \equiv v(\vec{r}) \cdot U(\vec{r},\vec{k})$, where $U(\vec{r},\vec{k})$ is classical full field inside the scattering domain. The condition (2) provides rigour of the algorithm. The second condition concerns a norm of the scattering amplitude:

$$\|f(\varphi,\varphi')\| \equiv \sqrt{\int_0^{2\pi} d\varphi \int_0^{2\pi} d\varphi' \, |f(\varphi,\varphi')|^2} < \frac{1}{3\pi}. \tag{3}$$

It is a sufficient condition to provide solution uniqueness. The reconstruction of the scatterers with different power and different complication of their space structure has given results discussed below. At the stage of the direct problem solution, the scattering data $f(\varphi,\varphi')$ for central-symmetrical scatterers were simulated by an iterative solution of the Lippmann-Schwinger equation. However, it should be emphasized that at the stage of the scatterer reconstruction by the Grinevich-Novikov algorithm, the fact of the scatterer's central symmetry was not used at all.

Acoustical Imaging, Volume 25, Edited by Halliwell and Wells.
Kluwer Academic/Plenum Publishers, 2000.

Relationship between the scatterer power and a reconstruction quality is analyzed for the scatterers, for which the both conditions (2,3) are met. The result of the scatterer reconstruction appears to practically agree with the true scatterer, if a volume of the scattering data samples is larger than a minimal necessary volume V_{min}. The latter volume is evaluated as

$$V_{min} \cong \frac{S_T}{(2\pi/L)^2}. \tag{4}$$

Here S_T is an area in $\vec{\xi}$-space, inside of which the space spectrum of the secondary sources for the given scatterer is localized for all $\vec{k}$; L is a linear size of the scattering domain in $\vec{r}$-space. The value V_{min} is defined by the power of the given scatterer (the power depends on a relative contrast of the velocity inhomogeneity and on the absorption intensity) and by a width of the scatterer space spectrum[3]. It is important to point out the following fact. For any fixed scatterer having a finite size L and meeting the requirement (2) of the algorithm rigour (with an accuracy of negligible components of the secondary sources space spectrum), the value S_T and, consequently, the value V_{min} are finite. Then the data volume V_{min} can always be provided. However, as V_{min} increases, a conditionality of the inverse problem solution becomes worse, that will be demonstrated below by some scatterers.

Relationship between the scatterer power, on the one hand, *and a uniqueness and a stability of the inverse problem solution*, on the other hand, is followed. For this purpose a character and degree of a solution distortion are researched when the condition (3) for $\|f\|$ is no longer met and, consequently, a passage out the area of the guarantee validity takes place. The algorithm itself remains rigorous, because the condition (2) is met.

For example, *absorbing scatterer* is reconstructed. The function $\nu(\vec{r}) \equiv \mathrm{Re}\,\nu(\vec{r})$ (see (1): $c(\vec{r}) \equiv c_o$) has the Gauss shape with the half-width $d \approx 0.82\lambda_o$ on e^{-1} level (fig.1*a*). Its amplitude corresponds to the maximal absorption coefficient $\alpha \approx 0.3$ Nep/a length unit. (A length unit is a spatial scale conventionally defined. Now it is equal to $(\lambda_o/8) = 1$.) Summarized amplitude absorption along the central section of the scatterer is ≈ 3.6 Nep, that is very significant. The norm $\|f\| \approx 10/(3\pi)$ is noticeable more than the threshold value (3). Central sections of the true scatterer ν and its estimation $\hat{\nu}$ reconstructed by the Grinevich-Novikov algorithm when noisy interference is absent are shown by fig.1*b*. At the same time, the reconstruction in the Born (a single scattering) approximation testifies the significant power of the scatterer: the Born estimation $\hat{\nu}_{born}$ and the true value ν are even not of the same order (fig.1*c*). Thus, when the multiple scattering effects are taken into consideration, the estimation $\mathrm{Im}\,\hat{\nu}$ is close to the truth ν (fig.1*b*). However, a small false value $\mathrm{Re}\,\hat{\nu}$ of the real part appears because of bad conditionality of the solution for the given scatterer. So, the minimal inverse conditionality of the algorithm's matrices is $\approx 10^{-2}$. As a result, the reconstructed scatterer estimation appears to be very sensitive to numerical errors caused by sampling processes.

Another example is the reconstruction of *refractive-absorbing scatterer*. Its function $\nu(\vec{r})$ is pictured in fig.2*a,b*. Function $\mathrm{Re}\,\nu(\vec{r})$ (fig.2*a*) is caused by inhomogeneities of the phase velocity. Its central "peak" has the Gauss shape with the half-width $d \approx 0.58\lambda_o$ on e^{-1} level and with the maximal value corresponding to the velocity contrast $\Delta c/c_o = 0.15$. Any radial section of surrounding "wall" has the Gauss shape with the same d. The scatterer is not weak: summarized additional phase shift along the central section of the scatterer is $\Delta\psi \approx 0.22\pi$. Function $\mathrm{Im}\,\nu(\vec{r})$ (fig.2*b*) is caused by inhomogeneities of the amplitude absorption coefficient with the maximal value ≈ 0.32 Nep/a length unit (now the length unit is equal to $(\lambda_o/8) = 1$). Summarized absorption along the central section of the scatterer is ≈ 1.15 Nep. Though the norm $\|f\| \approx 12/(3\pi)$ is essentially more than the threshold value (3), the scatterer reconstruction is practically ideal (fig.2*c*). The minimal inverse conditionality of the algorithm's matrices is ≈ 0.2. At the same time, the reconstruction in a single scattering approximation gives inadequate results (fig.2*d*).

Thus, the scatterers, for which the Grinevich-Novikov algorithm remains rigorous, are reconstructed with the high accuracy (when the noisy interference and the errors of the numerical calculation are absent), even thought the norm $\|f\|$ is some ten times higher than the threshold value (3). The gradual increasing of the scatterer power (under the provision of the minimal volume of the scattering data samples) results only in monotonous deterioration of the problem conditionality, i.e. the solution becomes more sensitive to interference. In future, a regularization of the solution might be included in the algorithm.

Finally, the *reconstruction* quality *of the scatterers, for which the Grinevich-Novikov algorithm is not rigorous because of a presence of high-frequency components* in the space spectra of the scatterer and of its secondary sources is investigated. In other words, the first condition (2) of the guarantee validity of the algorithm is not met. A contribution to the scattering data by such the high-frequency components is, with respect to the algorithm, a certain interference. As an example, *refractive-absorbing scatterer* is considered. Its external view (fig.3*a,b*) is similar to that of the preceding scatterer (fig.2*a,b*), but their wave space scales are different. Now $d \approx 0.22\lambda_o$; maximal value of central Gauss "peak" for $\mathrm{Re}\, v(\vec{r})$ corresponds to $\Delta c/c_o = 0.3$; maximal amplitude absorbing coefficient is ≈ 0.027 Nep/a length unit (now the length unit is equal to $(\lambda_o/16) = 1$). The additional phase shift is $\Delta\psi \approx 0.27\pi$; the summarized absorption is ≈ 1.3 Nep.

Thus, wave sizes d/λ_o of characteristic space details of the given scatterer are small. Because of this, the space scatterer spectrum $\tilde{v}(\vec{\xi})$ has the high-frequency components $|\vec{\xi}| \geq 2k_o$, which can not be negligible. As a result, the secondary source space spectrum $T(\vec{\xi},\vec{k})$ has the high-frequency components $|\vec{\xi}-\vec{k}| \geq 2k_o$. At the same time, such the components for the preceding scatterer were negligibly small. Magnitude $|T(\vec{\xi},\varphi = 0)|$ for the considered scatterer is shown in fig.3*c*. Sections $|T(\xi_x,\xi_y = 0;\varphi = 0)|$ (fig.3*d*) and $|T(\xi_x = k_o,\xi_y;\varphi = 0)|$ (fig.3*e*) essentially differ from the corresponding sections T_{born} in the Born approximation. Significance of the multiple scattering effects is also confirmed by the Born scatterer estimation $\hat{v}_{born}$, which is far from the true value v (fig.3*f*).

The data norm is $\|f\| \approx 9/(3\pi)$; the minimal inverse conditionality of the matrices is ≈ 0.2. The scatterer estimation $\hat{v}$ is obtained without the noisy interference by the Grinevich-Novikov algorithm (fig.3*g,h*). A cause of reconstruction errors is presence of the high-frequency components in $\tilde{v}(\vec{\xi})$ and $T(\vec{\xi},\vec{k})$. These components appear as an interference for the algorithm. Really, as mentioned, the components $|\vec{\xi}| \geq 2k_o$ play a part in $\tilde{v}(\vec{\xi})$. It is confirmed by a comparison of the true scatterer $v(\vec{r})$ with the scatterer $v^{cut}(\vec{r})$ having the space spectrum restricted by the circle $|\vec{\xi}| < 2k_o$ (fig.3*g,h*). However, the Grinevich-Novikov algorithm is not able to reconstruct the components of the scatterer space spectrum with $|\vec{\xi}| \geq 2k_o$. Moreover, the components with $|\vec{\xi}| \leq 2k_o$ are reconstructed with a distortion because of the presence of the high-frequency components in $T(\vec{\xi},\vec{k})$ and accounting the multiple scattering effects. The more $|\vec{\xi}|$, the more distortion. This fact explicates a difference between $\hat{v}$ and v^{cut}. Space spectra of the central sections $x = 0$ for the true scatterer — $\tilde{v}(x = 0,\xi_y)$ — and for the reconstructed one — $\hat{\tilde{v}}(x = 0,\xi_y)$ — were calculated. A comparison of them (fig.3*i,j*) shows that relative mistake for the reconstruction $\hat{\tilde{v}}(x = 0,\xi_y)$ is $\approx 10\%$ at $|\vec{\xi}| \approx 0$ and it increases with increasing $|\vec{\xi}|$.

By such a manner, an influence of the high-frequency components of the space spectra on the reconstruction quality is comparable to that of the accidental noisy interference. An intensity of the latter interference is equal to intensity of the scattering data's part which is produced by such the spectral components.

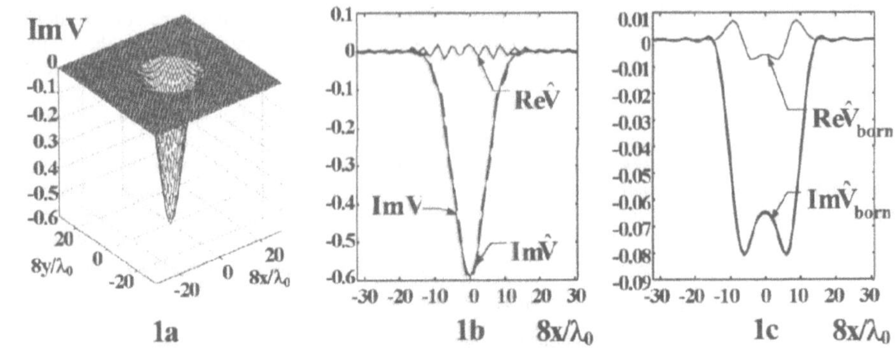

Figure 1. Absorbing scatterer (absorption inside the scatterer is 3.6 Nep):
- general view of the true scatterer (*a*);
- central sections of the true scatterer v (*b* – thin line) and of the scatterer $\hat{v}$ reconstructed when multiple scattering effects are taken into account (*b* – thick dashed line) and in Born approximation (*c*).

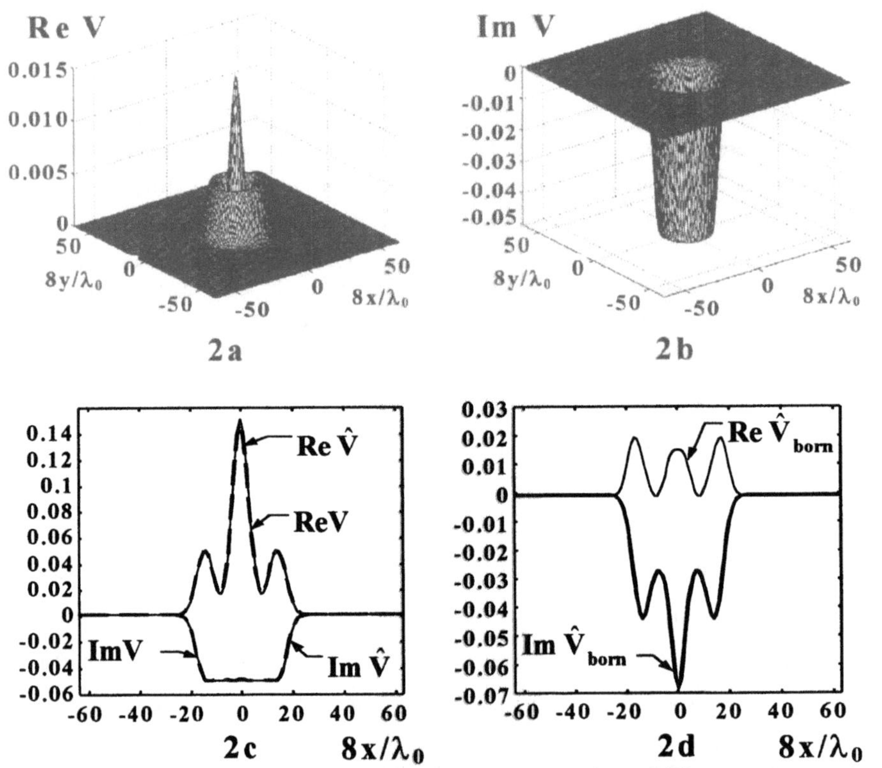

Figure 2. Refractive-absorbing scatterer (additional phase shift is $\Delta\psi \approx 0.22\pi$; absorption inside the scatterer is 1.15 Nep):
- general view of real (*a*) and imaginary (*b*) parts of the true scatterer;
- central sections of the true scatterer v (*c* – thin line) and the scatterer $\hat{v}$ reconstructed when multiple scattering effects are taken into account (*c* – thick dashed line) and in Born approximation (*d*).

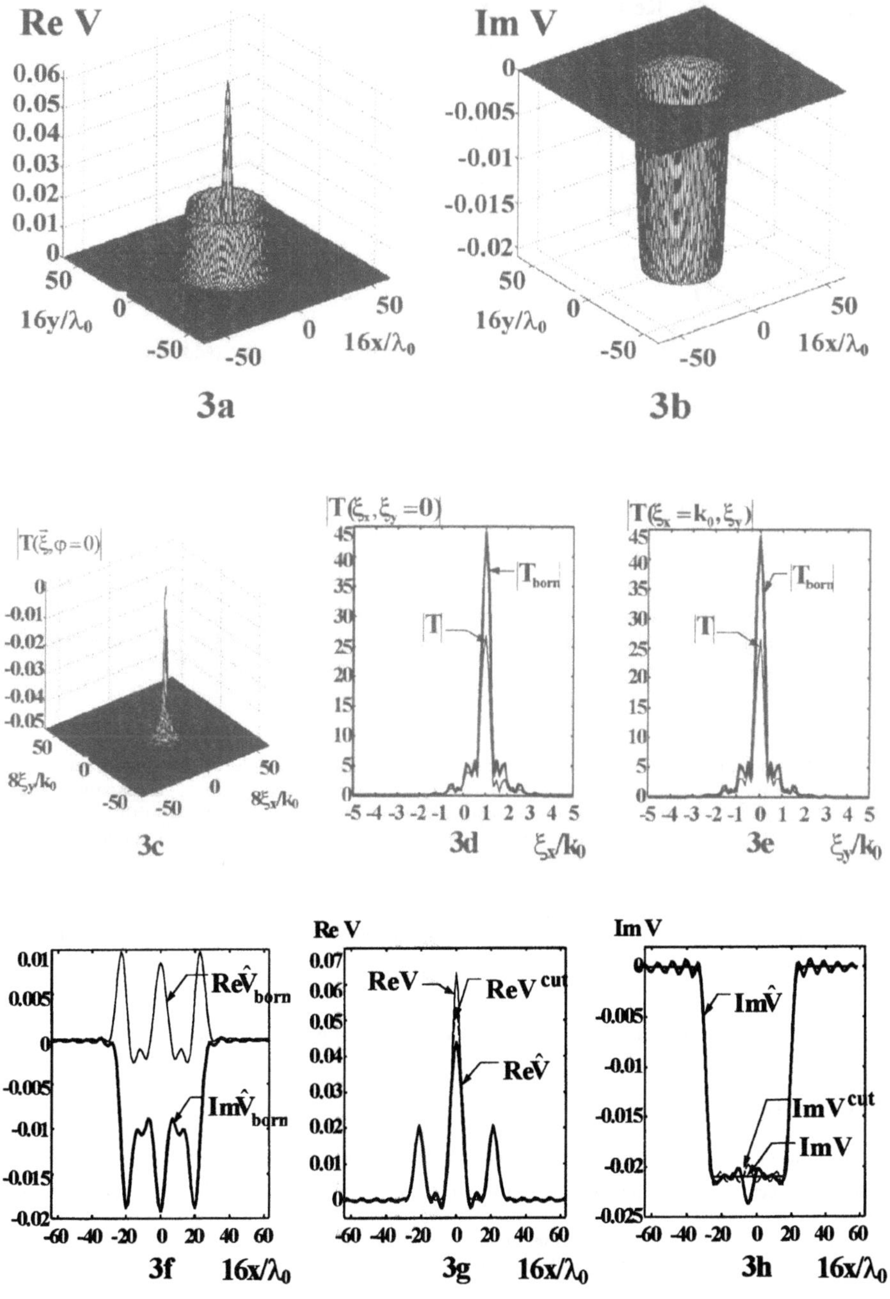

Figure 3. (See next page for details.)

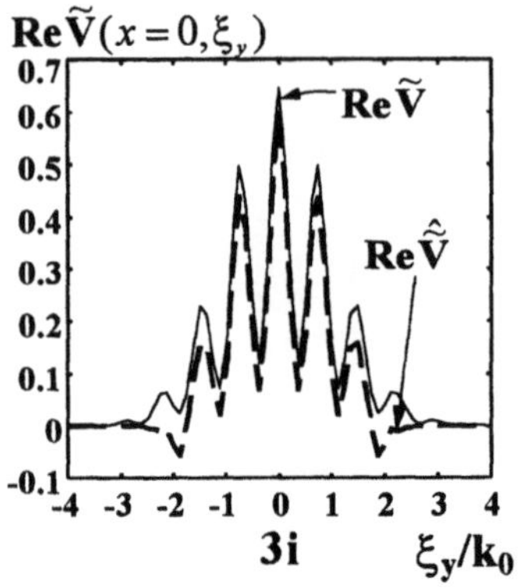

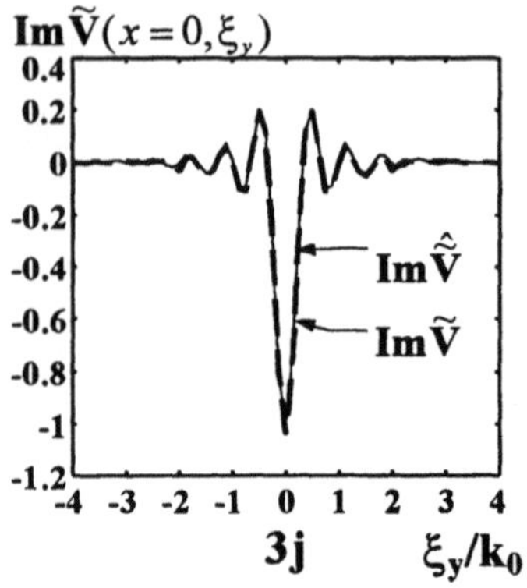

Figure 3. Refractive-absorbing scatterer creating the back scattering (additional phase shift is $\Delta\psi \approx 0.27\pi$; absorption inside the scatterer is 1.3 Nep):
- general view of real (*a*) and imaginary (*b*) parts of the true scatterer;
- magnitude of space spectrum of the secondary sources for fixed falling direction $\varphi = 0$: general view (*c*); section $\xi_y = 0$ (*d*) and section $\xi_x = k_o$ (*e*) when multiple scattering effects are taken into account (thin line) and in Born approximation (thick line);
- central section of the scatterer reconstructed in Born approximation (*f*);
- central sections of the real (*g*) and imaginary (*h*) parts of the scatterer: true scatterer (thin line) and its form after restriction of scatterer space spectrum by $2k_o$-radius circle (dashed line); scatterer reconstructed when multiple scattering effects are taken into account (thick line);
- real (*i*) and imaginary (*j*) parts of linear space spectrum of central section $x = 0$ for true (solid line) and for reconstructed (dashed line) scatterers.

The summarized results lead to the main conclusion: the Grinevich-Novikov algorithm is very perspective for practical problem. It has high computing efficiency conditioned by a comparatively small volume of computing operations. The data (the scattering amplitude) required for the algorithm and a circle scheme for obtaining the tomography data with different directions in real experiment go well together. The scatterer function can be found at any fixed point independently of its values at other ones. Such a locality of the problem is convenient for practical applications, in particular, for medical problems of the acoustical tomography. However, the algorithm imposes the restrictions (connected with the solution stability to the noisy interferences) on the limiting scatterer power and the width of the secondary sources spectrum, in particular, on a respective part of the back scattering.

Acknowledgments. This work was supported by the Russian Foundation for Fundamental Research.

REFERENCES

1. P. G. Grinevich, S. V. Manakov, The inverse problem of the scattering theory for two-dimensional Schrödinger operator, $\bar{\partial}$-method and nonlinear equations, *Func. Analysis and Its Applic.* **20** (2), 14-24 (1986).
2. R. G. Novikov, The inverse scattering problem on a fixed energy level for the two-dimensional Schrödinger operator, *J.Funkt. Anal.* **103**, 409-463 (1992).
3. V. A. Burov, S .A. Morozov, O. D. Rumiantseva, E. G. Sukhov, S. N. Vecherin, and A. Yu. Zhukovets, Exact solution of two-dimensional monochromatic inverse scattering problem and secondary sources space spectrum, in: *Acoustical Imaging-24*, edited by H. Lee (Plenum Press, New York , in press).

MODELING INVERSE PROBLEMS FOR STATISTICAL ESTIMATION OF SCATTERER CHARACTERISTICS

Valentin A.Burov[1], Elena E.Kasatkina[1], Evgenii G.Sukhov[2], and Olga D.Rumyantseva[1]

[1] Moscow State University, Faculty of Physics, Dept. of Acoustics
Moscow, 119899, Russia
[2] Institute of Control Sciences of Russian Academy of Sciences

1. ESTIMATION OF MAXIMUM A POSTERIORI PROBABILITY OF SCATTERER CHARACTERISTICS BY NOISED DATA

This report is devoted to a statistical approach to reconstruction problem of acoustical parameters of scattering medium inhomogeneities (for example, phase sound velocity and absorption coefficient) on the base of noised scattering data measured experimentally. The considered inverse scattering problem is *ill-posed.* Thus, regularizing methods should be applied to get a stable solution. However, up to this time, these are methods for a solution of linear ill-posed problems that have been developed in detail. In the statistical linear problems the optimal regularization based on use of a priori statistical information [1] is reduced to some kind of a filtration. On the other hand, regularizing methods for a solution of nonlinear ill-posed problems have not been sufficiently investigated and their investigation is being actively developed [2,3]. A nonlinearity of the solution with respect to input scattering data and, probably, a simultaneous nonlinearity with respect to unknown functions (these functions describe the scatterer and its secondary sources or inner fields) make the problem complicated by many aspects.

The general statistical approach for setting up a solution of the inverse problem, as the problem of an optimal estimation (in terms of maximum a posteriori probability method) of a scatterer function $\varepsilon(\mathbf{r})$, is used. The function $\varepsilon(\mathbf{r})$ is connected with the scatterer's characteristics as follows:

$$\varepsilon(\mathbf{r}) = \omega^2(c_0^{-2} - c^{-2}(\mathbf{r})) - 2i\omega\alpha(\mathbf{r},\omega)/c(\mathbf{r}) . \tag{1}$$

Here c_0, $c(\mathbf{r})$ are a sound velocity in a background medium without an absorption and inside a scattering domain $\mathfrak{R}$, correspondingly; $\alpha(\mathbf{r},\omega)$ is an amplitude absorption coefficient inside $\mathfrak{R}$. An algorithm developed for the problem solution is based on a minimization of a certain functional. This functional includes the following: a discrepancy between the experimental scattering data and estimated values, limiting equations and, finally, a priori information about correlation properties both of a noisy interference and the scatterer characteristics looked for. This algorithm gives regularized estimations of the velocity and the absorption coefficient inside the space-localized scatterer. *Processes of rescatterings are taken into account* in *maximally rigorous* mathematical formulation. A payment for the optimality of the approach is that a resulting system of nonlinear equations is very cumbersome (however, as numerical modeling tests have shown, a volume of computing operations is acceptable for modern computers).

In the previous paper [4] the functional has been constructed in the coordinate representation, wherein $\varepsilon(\mathbf{r})$ and full inner fields $u(\mathbf{r},\mathbf{k}_0)$ are estimated ($\mathbf{k}_0$ is a wave vector of an incident plane monochromatic wave). In the present paper a space frequency representation is considered, wherein a space spectrum of the scatterer $\tilde{\varepsilon}(\mathbf{k}')$ and a space spectrum $T(\mathbf{k}',\mathbf{k}_0)$ of the scatterer's secondary sources $\varepsilon(\mathbf{r})u(\mathbf{r},\mathbf{k}_0)$ (induced inside $\mathfrak{R}$) are estimated. Wave vectors $\mathbf{k}' \in K$, $\mathbf{k}_0 \in K_0$, where K and K_0 are spaces of wave vectors of any directions, but a wave vector length is arbitrary for K and is fixed for K_0 ($|\mathbf{k}_0| \equiv k_0$ is wave number in the background medium). The experimental scattering data are values of a complex scattering amplitude $T_{exp.}(\mathbf{k},\mathbf{k}_0)$; $\mathbf{k},\mathbf{k}_0 \in K_0$, $\mathbf{k}$ is wave vector of plane scattered wave in far field. The values $T_{exp.}$ and $\tilde{\varepsilon}$ are connected by the equation:

$$T_{exp.}(\mathbf{k},\mathbf{k}_0) = \tilde{\varepsilon}(\mathbf{k}-\mathbf{k}_0) + \frac{1}{(2\pi)^D}\int_K \tilde{\varepsilon}(\mathbf{k}-\mathbf{k}')\tilde{G}(|\mathbf{k}'|,k_0)T(\mathbf{k}',\mathbf{k}_0)d\mathbf{k}' + n(\mathbf{k},\mathbf{k}_0). \quad (2)$$

Here $\tilde{G}$ is space spectrum of the Green function for background medium; term $n(\mathbf{k},\mathbf{k}_0)$ describes the interference; D is the problem dimension. It is suggested that n and $\tilde{\varepsilon}$ are realizations of the Gaussian accidental processes with correlation matrices N and $\mathcal{E}$, respectively. Then the optimal estimation is reduced to minimizing the functional F:

$$F(\tilde{\varepsilon},\tilde{\varepsilon}^+,T,T^+) = F_0(N^{-1};\tilde{\varepsilon},\tilde{\varepsilon}^+,T,T^+) + F_a(\mathcal{E}^{-1};\tilde{\varepsilon},\tilde{\varepsilon}^+), \quad \text{where:} \quad (3)$$

$$F_0 = \int\limits_{K_0}\cdots\int [T^+_{exp.}(\mathbf{k}_{01},\mathbf{k}_1) - \tilde{\varepsilon}^+(\mathbf{k}_1-\mathbf{k}_{01}) - \frac{1}{(2\pi)^D}\int_K T^+(\mathbf{k}_{01},\mathbf{k}_1')\,\tilde{G}^+(|\mathbf{k}_1'|,k_0)\times$$

$$\times\tilde{\varepsilon}^+(\mathbf{k}_1-\mathbf{k}_1')\,d\mathbf{k}_1']\; N^{-1}(\mathbf{k}_1,\mathbf{k}_{01};\mathbf{k}_2,\mathbf{k}_{02})\;[T_{exp.}(\mathbf{k}_2,\mathbf{k}_{02}) - \tilde{\varepsilon}(\mathbf{k}_2-\mathbf{k}_{02}) - \quad (3a)$$

$$-\frac{1}{(2\pi)^D}\int_K \tilde{\varepsilon}(\mathbf{k}_2-\mathbf{k}_2')\tilde{G}(|\mathbf{k}_2'|,k_0)T(\mathbf{k}_2',\mathbf{k}_{02})\,d\mathbf{k}_2']\,d\mathbf{k}_1 d\mathbf{k}_{01} d\mathbf{k}_2 d\mathbf{k}_{02};$$

$$F_a = \iint_K \tilde{\varepsilon}^+(\mathbf{k}_1')\beta(\mathbf{k}_1',\mathbf{k}_2')\mathcal{E}^{-1}(\mathbf{k}_1',\mathbf{k}_2')\tilde{\varepsilon}(\mathbf{k}_2')d\mathbf{k}_1'd\mathbf{k}_2';\; \mathbf{k}_1,\mathbf{k}_{01},\mathbf{k}_2,\mathbf{k}_{02} \in K_0;\; \mathbf{k}_1',\mathbf{k}_2' \in K. \quad (3b)$$

Here, "+" stands for the Hermitian conjugation. In its physical sense, the term F_0 is a quadratic form of a discrepancy between the scattering amplitudes which are experimentally measured $T_{exp.}$, on the one hand, and are estimated by (2) under n≡0, on the other hand. The term F_a contains a priori statistical information about the scatterer characteristics; $\beta(\mathbf{k}_1',\mathbf{k}_2')$ is a penalty functional multiplier; average (over scatterers ensemble) value $\langle\tilde{\varepsilon}(\mathbf{k}')\rangle = 0$.

The functional (3) is complemented by limiting connection equations of the Lippmann-Schwinger-type for unknowns $T(\mathbf{k}',\mathbf{k}_0)$:

$$T(\mathbf{k}',\mathbf{k}_0) = \tilde{\varepsilon}(\mathbf{k}'-\mathbf{k}_0) + \frac{1}{(2\pi)^D}\int_K \tilde{\varepsilon}(\mathbf{k}'-\mathbf{k}'')\tilde{G}(|\mathbf{k}''|,k_0)T(\mathbf{k}'',\mathbf{k}_0)d\mathbf{k}'', \quad \mathbf{k}' \in K. \quad (4)$$

because only the noised values $T(\mathbf{k}',\mathbf{k}_0)$ at $|\mathbf{k}'| = k_0$ are experimentally known. The equations (4) could be included into F with the help of the functional Lagrange multipliers [4].

The functional F is minimized by an unknown function $\tilde{\varepsilon}^+(\vec{\chi})dV_{\bar{\chi}}$, where $dV_{\bar{\chi}}$ is an elementary volume in Cartesian coordinate system; a center of this volume is in an arbitrary fixed point $\vec{\chi} \in K$. Obtained equations system:

$$\begin{cases} \delta F(\tilde{\varepsilon},\tilde{\varepsilon}^+,T,T^+)/\,\delta\left[\tilde{\varepsilon}^+(\vec{\chi})dV_{\bar{\chi}}\right] = 0; \\ \text{Equations (4) for all } \mathbf{k}_0 \end{cases} \quad (5)$$

serves to estimating the unknowns $\tilde{\varepsilon}$ and T. Further, a final expression will be written for *two-dimensional problem* and the noisy interference which is δ-correlated by directions:

$$N^{-1}(\mathbf{k}_1,\mathbf{k}_{01};\mathbf{k}_2,\mathbf{k}_{02}) \approx (1/n_0)\cdot\delta(\mathbf{k}_2-\mathbf{k}_1)\cdot\delta(\mathbf{k}_{02}-\mathbf{k}_{01}), \quad \mathbf{k}_1,\mathbf{k}_{01},\mathbf{k}_2,\mathbf{k}_{02} \in K_0. \quad (6)$$

A sampling analog of (6) for M directions of the incident field and M directions of the scattered one is:

$$N^{-1}(\mathbf{k}_1,\mathbf{k}_{01};\mathbf{k}_2,\mathbf{k}_{02}) \approx (1/W_0)\cdot\delta_{\mathbf{k}_1,\mathbf{k}_2}\cdot\delta_{\mathbf{k}_{01},\mathbf{k}_{02}}, \quad \Delta\varphi = 2\pi/M,\; \Delta k = k_0\cdot\Delta\varphi,$$

$$W_0 \equiv n_0\cdot(\Delta k)^2\;;\quad \delta_{\mathbf{k}_1,\mathbf{k}_2} = \{1, \text{ if } \mathbf{k}_1 = \mathbf{k}_2;\; 0, \text{ if } \mathbf{k}_1 \neq \mathbf{k}_2\}.$$

Then the eq.(5) may be rearranged to the view:

$$\sum_{\vec{\eta}} A(\vec{\chi},\vec{\eta})\tilde{\varepsilon}(\vec{\eta}) = T'_{exp}(\vec{\chi}) + \Im_1(\vec{\chi}) + \Im_2(\vec{\chi}) \; . \tag{7}$$

The system of these equations (7) and the equations (4) allows to reconstruct $\tilde{\varepsilon}(\vec{\eta})$ for $|\vec{\eta}| \le 2k_0$, where independent samples are given in irregular coordinate net:

$$\vec{\eta} \in \left\{0 \cup (\mathbf{k}^n - \mathbf{k}_0^m\right\}, \text{ where} \tag{7a}$$

$$\mathbf{k}_0^m = \left\{k_0, \varphi_0^m\right\}, \quad \varphi_0^m = \Delta\varphi\cdot(m-1); \quad \Delta\varphi = 2\pi/M, \quad m = \overline{1, M};$$

$$\mathbf{k}^n = \left\{k_0, \varphi^n\right\}, \quad \varphi^n = \Delta\varphi\cdot(n-1); \quad n = \overline{(m+1),\ (m+M/2)}\,.$$

Similarly, samples $\vec{\chi}$ which are parameters of the eqs. (7), is chosen: $\vec{\chi} \in \left\{0 \cup (\mathbf{k}^{n_\alpha} - \mathbf{k}_0^{m_\beta}\right\}$, where $\mathbf{k}_0^{m_\beta} = \left\{k_0, \varphi_\beta = \varphi_0^{m_\beta}\right\}$, $\mathbf{k}^{n_\alpha} = \left\{k_0, \varphi_\alpha = \varphi^{n_\alpha}\right\}$. The matrix acting on the unknowns $\tilde{\varepsilon}(\vec{\eta})$ is:

$$A(\vec{\chi},\vec{\eta}) = E(\vec{\chi},\vec{\eta}) + \beta(\vec{\chi},\vec{\eta})K(\vec{\chi},\vec{\eta})W_0\mathcal{E}^{-1}(\vec{\chi},\vec{\eta}) \; ; \tag{7b}$$

$E(\vec{\chi},\vec{\eta})$ is the unit matrix, $K(\vec{\chi},\vec{\eta}) = S(\vec{\chi})\cdot S(\vec{\eta})C(\vec{\eta})/[2\cdot C(\vec{\chi})]$.

Coefficients S are: $S(\vec{\chi}) = \left\{|\sin(\varphi_\alpha - \varphi_\beta)|, \text{ at } |\vec{\chi}| \ne 0,\ 2k_0;\ \ \Delta\varphi/3, \text{ at } \vec{\chi} = 0 \text{ or } |\vec{\chi}| = 2k_0;\right\}$; for $S(\vec{\eta})$ the vector $\vec{\chi}$ is substituted by $\vec{\eta}$ and the argument $(\varphi_\alpha - \varphi_\beta)$ is substituted by $(\varphi^n - \varphi_0^m)$. These coefficients appear because of the different mechanism of forming the correlation functions $\mathcal{E}$ and N. Namely, for $\mathcal{E}(\mathbf{k}'_2,\mathbf{k}'_1) \equiv \left\langle \tilde{\varepsilon}^+(\mathbf{k}'_2)\cdot\tilde{\varepsilon}(\mathbf{k}'_1)\right\rangle$ the vectors $\mathbf{k}'_1, \mathbf{k}'_2 \in K$, whereas for $N(\mathbf{k}_2,\mathbf{k}_{02};\mathbf{k}_1,\mathbf{k}_{01}) \equiv \left\langle n^+(\mathbf{k}_2,\mathbf{k}_{02})\cdot n(\mathbf{k}_1,\mathbf{k}_{01})\right\rangle$ the vectors $\mathbf{k}_1,\mathbf{k}_2,\mathbf{k}_{01},\mathbf{k}_{02} \in K_0$. Coefficient $C(\vec{\eta})$ is the number of possible variants to form the fixed vector $\vec{\eta} = \mathbf{k} - \mathbf{k}_0$ by a pair of sampling vectors $\mathbf{k}, \mathbf{k}_{01} \in K_0$: $C(\vec{\eta}) = \left\{M, \text{ at } \vec{\eta} = 0;\ \ 2, \text{ at } |\vec{\eta}| \ne 0,\ 2k_0;\ \ 1, \text{ at } |\vec{\eta}| = 2k_0\right\}$; similarly for $C(\vec{\chi})$.

Values $T'_{exp}(\vec{\chi})$ are the scattering data averaged over these variants:

$$T'_{exp}(\vec{\chi}) = \left\{\frac{1}{M}\sum_{m_2=1}^{M} T_{exp}(\mathbf{k}_0^{m_2}, \mathbf{k}_0^{m_2}), \text{ at } \vec{\chi} = 0; \quad [T_{exp}(\mathbf{k}^{n_\alpha}, \mathbf{k}_0^{m_\beta}) + T_{exp}(-\mathbf{k}_0^{m_\beta}, -\mathbf{k}^{n_\alpha})]/2, \right.$$

$$\left. \text{at } \vec{\chi} = \mathbf{k}^{n_\alpha} - \mathbf{k}_0^{m_\beta},\ |\vec{\chi}| \ne 0,\ 2k_0; \quad T_{exp}(-\mathbf{k}_0^{m_\beta}, \mathbf{k}_0^{m_\beta}), \text{ at } \vec{\chi} = -2\cdot\mathbf{k}_0^{m_\beta}\ \ (\mathbf{k}^{n_\alpha} = -\mathbf{k}_0^{m_\beta})\right\}.$$

Term $\Im_1(\vec{\chi})$ has the similar structure, wherein T_{exp} is substituted by $(\tilde{\varepsilon}(\vec{\chi}) - T)$. Term $\Im_2(\vec{\chi})$ is:

$$\Im_2(\vec{\chi}) = \frac{S(\vec{\chi})}{C(\vec{\chi})}\cdot\frac{(\Delta k)^2}{(2\pi)^2}\sum_{m_1=1}^{M}\ \sum_{n_1=m_1+1}^{m_1+M} T^+(\mathbf{k}_0^{m_1}, \mathbf{k}^{n_1} - \vec{\chi})\cdot\tilde{G}^+(|\mathbf{k}^{n_1} - \vec{\chi}|, k_0)[T_{exp}(\mathbf{k}^{n_1}, \mathbf{k}_0^{m_1}) - T(\mathbf{k}^{n_1}, \mathbf{k}_0^{m_1})];$$

$$\Delta k = 2\pi k_0/M; \quad |\mathbf{k}_0^{m_1}| = |\mathbf{k}^{n_1}| = k_0\,.$$

The scatterer (1) may consist of velocity c-component and absorbing α-component, simultaneously: $\varepsilon(\mathbf{r}) = \varepsilon_c(\mathbf{r}) + i\cdot\varepsilon_\alpha(\mathbf{r})$ and, consequently, $\tilde{\varepsilon}(\mathbf{k}') = \tilde{\varepsilon}_c(\mathbf{k}') + i\cdot\tilde{\varepsilon}_\alpha(\mathbf{k}')$. In this case a separate estimation of the components is expedient. Then the term (3b) should be substituted by a sum of similar terms concerning each of the components. The equation (5) should be substituted by equations $\delta F/\delta(\tilde{\varepsilon}_c^+(\vec{\chi})dV_{\bar{\chi}}) = 0$, $\delta F/\delta(\tilde{\varepsilon}_\alpha^+(\vec{\chi})dV_{\bar{\chi}}) = 0$.

2. NUMERICAL RECONSTRUCTION OF SCATTERERS

The experimental scattering data $T^0_{exp}(\mathbf{k}, \mathbf{k}_0)$ in the absence of the noisy interference are simulated by interpolation of values $T(\mathbf{k}', \mathbf{k}_0)$ (founded for $\mathbf{k}' \in K$ by an iterative solution of the eq.(4)) to the circle $|\mathbf{k}'| = k_0$ at each fixed $\mathbf{k}_0$. Then the values T^0_{exp} are noised. Root-mean-square amplitude deviation σ_n of the noisy interference for both $\operatorname{Re} n(\mathbf{k}, \mathbf{k}_0)$ and $\operatorname{Im} n(\mathbf{k}, \mathbf{k}_0)$ is determined as $\sigma_n = q \cdot T_{max}$ $(q \geq 0)$, where $T_{max} \equiv \max\limits_{\mathbf{k},\, \mathbf{k}_0 \in K_0} |T^0_{exp}(\mathbf{k}, \mathbf{k}_0)|$. An amplitude ratio "interference/signal" is $I_n = \sqrt{\sum_{\mathbf{k},\mathbf{k}_0} |n(\mathbf{k}, \mathbf{k}_0)|^2} \,/\, \sqrt{\sum_{\mathbf{k},\mathbf{k}_0} |T^0_{exp}(\mathbf{k}, \mathbf{k}_0)|^2}$. In numerical modeling *two independent (by noisy interference) sets* of the noised data T_{exp} with the same σ_n are used. Then the effective ratio "interference/signal" is $I'_n = I_n / \sqrt{2}$.

During the solution of the inverse problem the values $\tilde{\varepsilon}(\vec{\eta})$ are estimated by the system (7), (4). The solution is interpolated from the irregular net (7a) to the regular Cartesian net. Then the final estimation $\hat{\varepsilon}(\mathbf{r})$ in $\mathbf{r}$-space is formed by the Fourier-transformation. A relative root-mean-square error in this estimation is a ratio of a discrepancy between $\hat{\varepsilon}(\mathbf{r})$ and the true function $\varepsilon(\mathbf{r})$ to a norm of $\varepsilon(\mathbf{r})$ over samples $\mathbf{r}$: $v_\varepsilon = \sqrt{\sum_{\mathbf{r}} \left|\hat{\varepsilon}(\mathbf{r}) - \varepsilon(\mathbf{r})\right|^2} \,/\, \sqrt{\sum_{\mathbf{r}} |\varepsilon(\mathbf{r})|^2}$. Efficiency of the algorithm (7), (4) has been confirmed by reconstruction of a number of modeling scatterers. The scattering data for them correspond to an active tomography for different interference levels.

2.1 Reconstruction of Weak Scatterers

The case of the Born approximation (of *a single scattering*) is considered. The terms $\Im_1(\vec{\chi})$ and $\Im_2(\vec{\chi})$ at the right part (7) are negligible, and the connecting eqs. (4) are not required. Then the Born estimation $\tilde{\varepsilon}_{born}(\vec{\eta})$ is founded for samples $\vec{\eta}$ (see (7a)) by the system of equations:

$$\sum_{\vec{\eta}} A(\vec{\chi}, \vec{\eta}) \cdot \tilde{\varepsilon}_{born}(\vec{\eta}) = T'_{exp}(\vec{\chi}) \,. \tag{8}$$

This estimation agrees with a priori information about the unknown scatterer and the interference. If the interference is small ($W_0 \to 0$, $A(\vec{\chi}, \vec{\eta}) \to E(\vec{\chi}, \vec{\eta})$), then $\tilde{\varepsilon}_{born}(\vec{\eta}) = T'_{exp}(\vec{\chi})$.

A regularization in the considered problem has two aspects. First, $\tilde{\varepsilon}(\vec{\eta})$ are founded only for $|\vec{\eta}| \leq 2k_0$. Secondly, it is a filtration of $\tilde{\varepsilon}(\vec{\eta})$ that reduces the interference influence. In the Born approximation the regularizing filtration is similar to the classical Wiener-type filtration. Really, the matrix $\left[A(\vec{\chi}, \vec{\eta})\right]^{-1}$ (see (7b)) plays a role of a filter, the value $W_0 \mathcal{E}^{-1}(\vec{\chi}, \vec{\eta})$ corresponding to the ratio "interference/signal". Then a choice of the penalty multiplier , as

$$\beta(\vec{\chi}, \vec{\eta}) = 1 / \kappa(\vec{\chi}, \vec{\eta}) \,, \tag{9}$$

leads to the regularizing filtration in (8) becoming the Wiener-type filtration.

An example of a weak scatterer is a refractive scatterer of the Gauss shape (fig.1*a*). Characteristic parameters are maximal velocity contrast $\Delta c / c_0 = 0.01$ and the space half-width (on e^{-1} level) $a / \lambda_0 \approx 0.82$. Then maximal additional phase shift within the scatterer is $\Delta\psi \approx 0.03\pi$.

The interference level is high: $\sigma_n = 0.15 \cdot T_{max}$ and $I_n \approx 75\%$ for each of the two noised sets $T_{exp}(\mathbf{k}, \mathbf{k}_0)$. The correlation function $\mathcal{E}(\vec{\eta}, \vec{\chi})$ is evaluated as $\mathcal{E}(\vec{\eta}, \vec{\chi}) \equiv \left\langle \tilde{\varepsilon}^+(\vec{\eta}) \cdot \tilde{\varepsilon}(\vec{\chi}) \right\rangle \approx \mathcal{E}_0((\vec{\eta} + \vec{\chi})/2) \cdot \Gamma(\vec{\eta} - \vec{\chi})$. Function $\mathcal{E}_0$ is a root-mean-square spectral density of scatterer energy. Function Γ gives a correlation area of the spectral components $\tilde{\varepsilon}^+(\vec{\eta})$ and $\tilde{\varepsilon}(\vec{\chi})$: $\Gamma(\vec{\eta} - \vec{\chi}) \approx \exp[-(\vec{\eta} - \vec{\chi})^2 (L/\pi)^2]$, where L is an effective linear size of the scattering domain.

Reconstruction results, in comparison with the true function $\varepsilon(\mathbf{r})$, are illustrated by fig.1*b*. During reconstruction *without the regularizing filtration* the correlation noisy matrix is formed in a proposition of zero noisy level ($W_0 = 0$, $N \equiv 0$) instead of the true level. Consequently, a priori information is not taken into account (see (7b)). The reconstruction error $\nu_\varepsilon \approx 50\%$ is approximately equal to the effective ratio "interference/signal" $I'_n \approx 53\%$. During reconstruction *with the help of the regularizing filtration* the matrix N is formed for the true noisy level ($W_0 = (\sigma_n)^2$), and the multiplier $\beta(\vec{\chi}, \vec{\eta})$ is given by (9). The reconstruction quality in the presence of two previous noisy sets is significantly better: $\nu_\varepsilon \approx 9\%$.

Two moments should be noted. First, the error ν_ε is proportional to $\cdot I'_n$ at the high noisy level, whereas ν_ε decreases slower than I'_n decreases at the low level because of sampling errors. Secondly, the regularized scatterer estimation $\hat{\varepsilon}(\mathbf{r})$ appears to be weakly dependent on β. For example, the choice $\beta(\vec{\chi}, \vec{\eta}) \equiv 1$ is possible.

2.2 Scatterer Reconstruction in Presence of Multiple Scattering Processes

The estimation of scatterers of a mean or strong power requires a solution of the combined system (7), (4). This system is nonlinear with respect to the unknowns ($\tilde{\varepsilon}, T$) and can be iteratively solved. During the numerical modeling an iterative scheme of an consistent estimation [5,6] of $\tilde{\varepsilon}$ and T has been realized. At each iterative step with the number l an estimation $\tilde{\varepsilon}_l(\vec{\eta})$ is the solution of the subsystem (7) in which the terms $\mathfrak{I}_1(\vec{\chi})$ and $\mathfrak{I}_2(\vec{\chi})$ are formed by means of $\tilde{\varepsilon}_{(l-1)}$ and $T_{(l-1)}$. An initial estimation is the Born one $\tilde{\varepsilon}_{born}(\vec{\eta})$ (see (8)). The use of the two noised data sets T_{exp} makes the subsystem (7) redundant. Its least mean square solution additionally stabilizes a convergence of the iterative procedure. The next is a solution of the subsystem (4) for estimation of $T_l(\mathbf{k}', \mathbf{k}_0)$ by $\tilde{\varepsilon}_l$, and the iterative cycle is repeated. The convergence of the procedure is controlled by a discrepancy $\mu_T(l)$ between $T_{exp}(\mathbf{k}, \mathbf{k}_0)$ and $T_l(\mathbf{k}, \mathbf{k}_0)$ at $|\mathbf{k}| = k_0$: $\mu_T(l) = \sqrt{\sum_{\mathbf{k},\mathbf{k}_0} |T_l(\mathbf{k}, \mathbf{k}_0) - T_{exp}(\mathbf{k}, \mathbf{k}_0)|^2}$. The iterative process is finished, when $\mu_T(l)$ appears about the noisy norm $\sqrt{\sum_{\mathbf{k},\mathbf{k}_0} |n(\mathbf{k}, \mathbf{k}_0)|^2}$.

The numerical modeling has been carried out for scatterers of two types. The first scatterer $\varepsilon_I(\mathbf{r})$ is refractive. It differs from the weak scatterer (fig.1*a*) only by a larger contrast: $\Delta c / c_0 = 0.2$. The maximal additional phase shift is $\Delta\psi \approx 0.55\pi$. The second scatterer is refractive-absorbing $\varepsilon_{II}(\mathbf{r}) = \varepsilon_I(\mathbf{r}) - i\varepsilon_I(\mathbf{r})$: its components $\varepsilon_c(\mathbf{r}) \equiv \mathrm{Re}\,\varepsilon_{II}(\mathbf{r})$ and $-\varepsilon_\alpha(\mathbf{r}) \equiv -\mathrm{Im}\,\varepsilon_{II}(\mathbf{r})$ are identical and coincide with $\varepsilon_I(\mathbf{r})$. Then a summarized absorption along the central section is ≈ 1.44 Nep. Reconstruction results illustrate both a necessity of taking into account the multiple scattering processes and the fundamental role of a nonlinear filtration for obtaining the regularized solution. So, *the Born approximation is not sufficient.* Even thought the noisy interference is absent, the estimation $\hat{\varepsilon}_{born}(\mathbf{r})$ essentially differs from the true scatterer function: the error is $\nu_\varepsilon \approx 70\%$ for the refractive scatterer (fig.2*c*) and $\nu_\varepsilon \approx 64\%$ for the refractive-absorbing one (fig.3*e*).

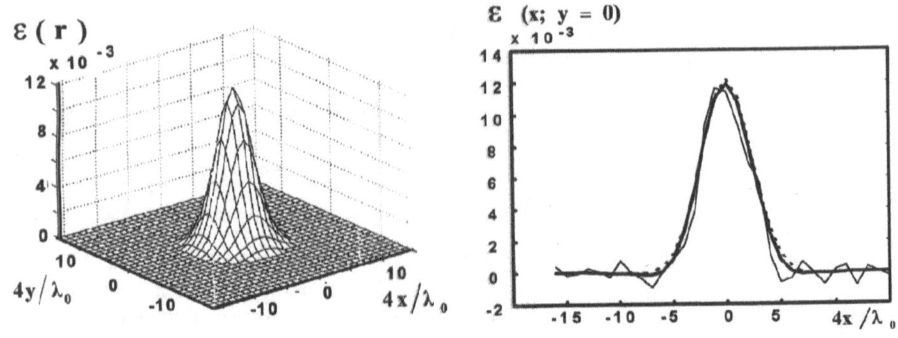

1*a* 1*b*

Figure 1. Weak refractive scatterer (velocity contrast $\Delta c/c_0 = 0.01$, maximal phase shift $\Delta\Psi = 0.033\pi$): (*a*) – general view; (*b*) – central sections: true scatterer (dashed line); scatterer reconstructed by the noised scattering data having root-mean-square noisy deviation $\sigma_n = 0.15 \cdot T_{max}$, with the help of regularizing filtration (thick line) and without it (thin line).

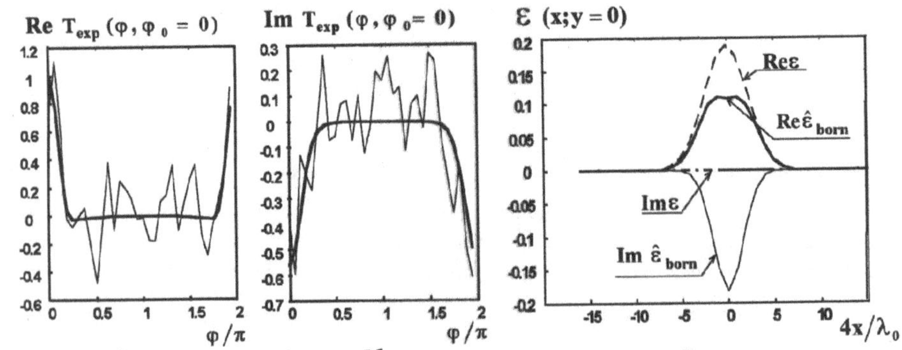

2*a* 2*b* 2*c*

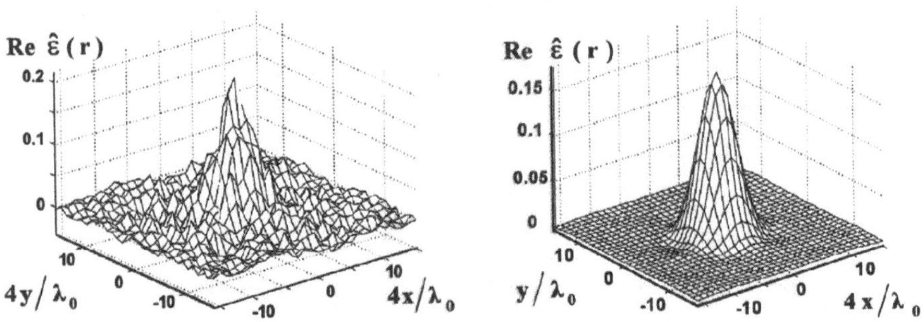

2*d* 2*e*

Figure 2. Refractive scatterer of a mean power ($\Delta c/c_0 = 0.2, \Delta\Psi = 0.55\pi$):

- Real (*a*) and imaginary (*b*) parts of the scattering data normalized by the maximal value of real part without noise: the data without noise (thick line) and in presence of noise with root-mean-square deviation $\sigma_n = 0.15 \cdot T_{max}$ (thin line).
- Central sections (*c*): true scatterer (dashed line); scatterer reconstructed in Born approximation without noise (solid line).
- Scatterer reconstructed by the noised data with $\sigma_n = 0.15 \cdot T_{max}$: (*d*) – case of «weak» regularizing filtration (proposing $\sigma_n = 0.001 \cdot T_{max}$), (*e*) – case of «full» filtration.

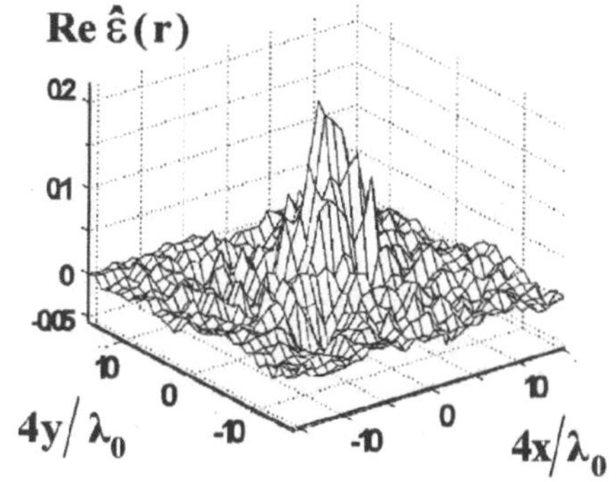

3a

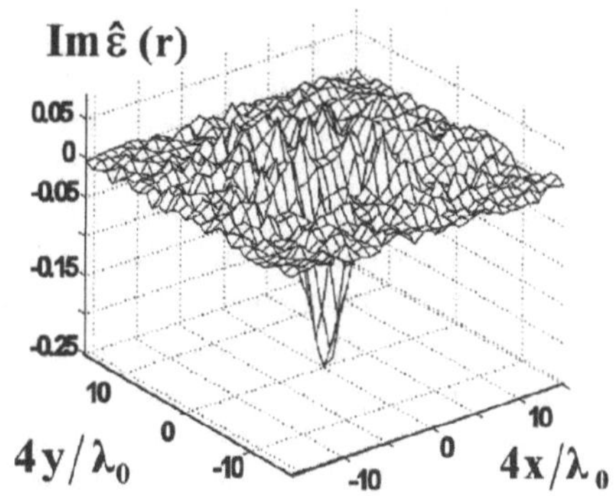

3b

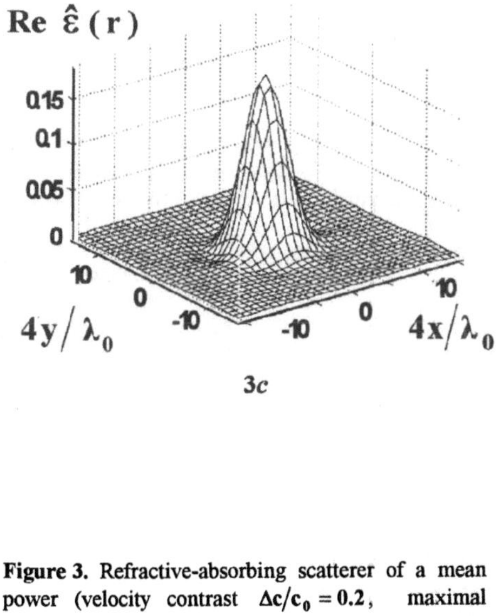

3c

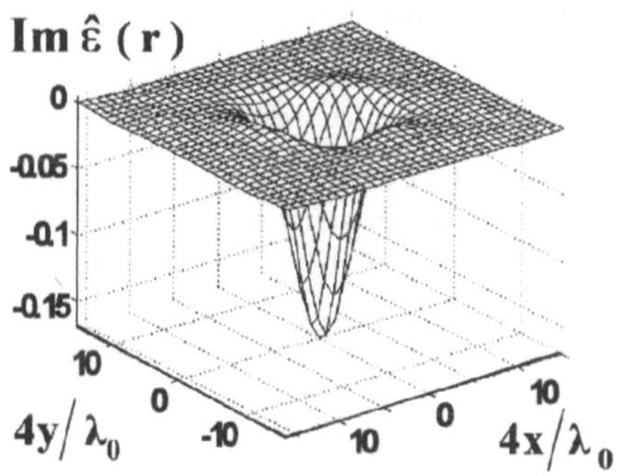

3d

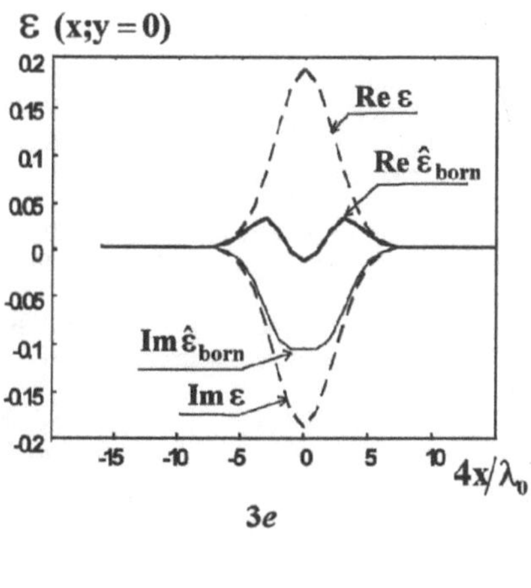

3e

Figure 3. Refractive-absorbing scatterer of a mean power (velocity contrast $\Delta c/c_0 = 0.2$, maximal phase shift $\Delta\Psi = 0.55\pi$; absorption inside the scatterer 1,44 Nep):

- Scatterer reconstructed by the noised data with root-mean-square noisy deviation $\sigma_n = 0.15 \cdot T_{max}$:

(*a,b*) – case of «weak» regularizing filtration (proposing $\sigma_n = 0.001 \cdot T_{max}$),

(*c,d*) – case of «full» filtration.

- Central sections (*e*): true scatterer (dashed line); scatterer reconstructed in Born approximation by the data without noise (solid line).

An influence of the regularizing filtration on a quality of the scatterer reconstruction has been researched under strong noisy interference: each of two the interference sets has $\sigma_n = 0.15 \cdot T_{max}$. The strong interference in the scattering data is shown in fig.2*a,b* for the refractive scatterer. Now the reconstruction by (7), (4) takes into account the multiple scattering effects. However, the used noisy matrix N is built under a proposition of two different noisy levels. Namely, the true level ($W_0 = (0.15 \cdot T_{max})^2$) and a level ($W_0 = (10^{-3} \cdot T_{max})^2$) being 150^2 times decreased by energy are proposed in cases of "full" regularizing filtration and of "weak" filtration, respectively.

The reconstruction with the "weak" filtration leads to an unacceptable result: $\nu_\varepsilon \approx 67\%$ at $I'_n \approx 52\%$ for the refractive scatterer (fig.2*d*); $\nu_\varepsilon \approx 66\%$ under $I'_n \approx 56\%$ for the refractive-absorbing scatterer (fig.3*a,b*). In distinction to the case of the weak scatterer without the filtration, the error ν_ε becomes more than I'_n because the inverse problem solution gets more ill-posed by means of the multiple scattering effects. The solution ill-position is also a reason why the "weak" filtration is necessary to provide the convergence of the iterative procedure.

The use of the "full" regularizing filtration leads to a sufficiently high quality of the reconstruction: $\nu_\varepsilon \approx 12\%$ for the refractive scatterer (fig.2*e*); $\nu_\varepsilon \approx 15\%$ for the refractive-absorbing scatterer (fig.3*c,d*). The reconstruction errors ν_ε for the considered nonweak scatterers are greater than for the weak scatterers with the same I'_n.

It is important that the regularizing filtration is always of the same type, independently of the iterative step number *l*. It is due to the fact that the matrix $A(\vec{\chi}, \vec{\eta})$ in (7), (7b) remains invariable for all the steps. Therefore, at each the step the regularization is reduced to the Wiener-type filtration being analogous to the case of the weak scatterers. The solution $\tilde{\varepsilon}_l(\vec{\eta})$ is the Born-type estimation, but with the correction of the right part of the eq. (8). In such a manner, the nonlinear generalization of the Wiener-type filtration has been obtained for sufficiently strong scatterers to provide the solution being close to the optimal one.

Advantages of the considered algorithm have been discussed in [4]. To extend a region of the iteration procedure convergence, it is expedient to use the simultaneous [7] evaluation of the unknown values with applying the filtration to each of them. However, volumes of computer memory and of computing operations needed for this purpose require involving the most modern computers. Moreover, in case of the scatterers of a strong power, a data redundancy is necessary to provide the solution uniqueness. Otherwise, the iterative solution might converge to a false one [8].

Acknowledgments. This work was supported by the Russian Foundation for Fundamental Research.

REFERENCES

1. A. N. Tikhonov and V. Y. Arsenin, *Solutions of Ill-Posed Problems* (Winston, Washington, 1977).
2. A. N. Tikhonov, A. V. Goncharskii, V. V. Stepanov, and A. G. Yagola, *Numerical Methods for the Solution of Ill-Posed Problems* (Kluwer, Dordrecht, 1995).
3. A. N. Tikhonov, A. S. Leonov, and A. G. Yagola, *Nonlinear Ill-Posed Problems* (Chapman & Hall, London, 1998).
4. V. A. Burov, E. E. Kasatkina, and O. D. Rumiantseva, Statistical estimations in inverse scattering problems, in: *Acoustical Imaging-22*, edited by P. Tortoli and L. Masotti (Plenum Press, New York, 1996), pp.113-118.
5. S. A. Johnson, Y. Zhou, M. L. Tracy, M. J. Berggren, and F. Stanger, Inverse scattering solutions by a sinc basis, multiple source, moment method. Pt.III: Fast algorithms, *Ultrason. Imaging*, **6** (4), 103-106 (1984).
6. V. A. Burov, M. N. Rychagov, and A. V. Saskovets, Account of multiple scattering in acoustic inverse problems of tomographic type, in: *Acoustical Imaging-19*, edited by H.Ermert and H.-P.Harjes (Plenum Press, New York, 1992), pp.35-39.
7. V. A. Burov, A. A. Goryunov, A. V. Saskovets, and T. A. Tikhonova, Inverse scattering problems in acoustics (review), *Sov. Phys. Acoust.*, **32**(4), 273-282 (1986).
8. V. A. Burov and O. D. Rumiantseva, Influence of the scattering data redundancy on uniqueness and stability in reconstruction of strong and complicated scatterers, in: *Acoustical Imaging-22*, edited by P. Tortoli and L. Masotti (Plenum Press, New York, 1996), pp.107-112.

BORN ULTRASONIC TOMOGRAPHY: SOME LIMITS AND IMPROVEMENTS

J.P. Lefebvre, P. Lasaygues, S. Mensah, S. Delamare, A. Wirgin

CNRS Laboratoire de Mécanique et d'Acoustique
31 Chemin Joseph Aiguier
13402 Marseille cedex 20
France
email : lefebvre@lma.cnrs-mrs.fr

SUMMARY

The goal of Ultrasonic Tomography (UT) is to reconstruct the spatial distribution of some physical parameter of an object from scattered ultrasonic measures. The measurements are made for more or less dense sets of emitter and receiver positions and of frequencies of the interrogating wave.

We solved this inverse scattering problem by using a Born approximation which leads to a particularly simple and attractive linear relation between the object function (OF) and the scattered field, particularly in the far-field (2D or 3D Fourier transform), making it possible, in principle, to reconstruct the OF in near real time for a sufficiently large set of scattering data.

We investigated wide-band Born UT both numerically and experimentally. Numerical simulations, using ideal measures with ideal objects, allow to examine in detail the influence of various parameters as object's dimension and contrast, transducers bandwidth, etc. It allows to analyse what happens when the Born approximation is no more valid (high frequencies, high contrasts), to find limits of quantitative and qualitative imagery, to imagine various improvement procedures (artefacts elimination, superresolution procedures leading to high resolution with low frequencies). Experiments (with a mechanical and an antenna-based system) show the applicability of the basic method and of its various improvements for medical and materials applications.

INTRODUCTION

We deal with ultrasonic imagery of objects, i.e. with reconstruction of objects from scattering measures, typically along a Σ sphere around the object. For simplicity, we consider 2D objects, the surface of measures Σ being a circle. For simplicity also we consider that the object is in the far field of the emitter, so that the impinging field can be approximated by a plane wave. Let assume that the object lies in the D space region, and is

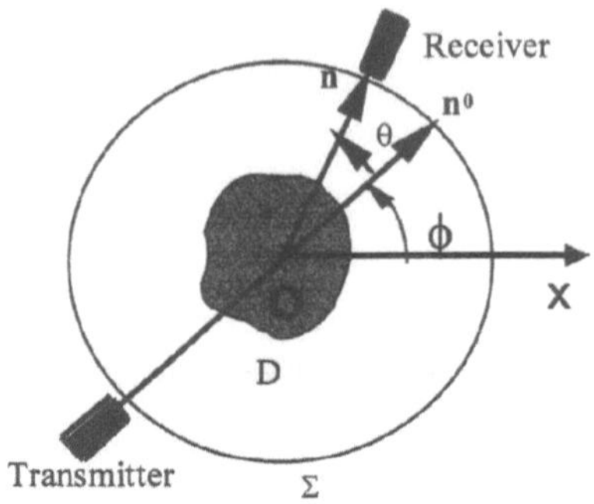

Figure 1: Acquisition geometry

characterised by its compressibility χ, space-variable parameter (for simplicity, the density is assumed constant and equal to the density of the surrounding medium). Let us note χ_0 the compressibility of the homogeneous host medium (coupling gel in *in vivo* imaging or water tank in *in vitro* imaging), one can show that the acoustic pressure is driven by the wave equation (1): $-\frac{1}{c_0^2}\frac{\partial^2 p}{\partial t^2}+\nabla^2 p=\frac{\mu_\chi}{c_0^2}\frac{\partial^2 p}{\partial t^2}$

where $\mu_\chi=\frac{\chi-\chi_0}{\chi_0}$ is the relative fluctuation of compressibility of the medium ($\mu_\chi=0$ outside D). Let us consider a plane wave excitation of pulsation ω, amplitude $P_i^0(\omega)$, direction of propagation $\mathbf{n_0}$ (unit vector) and of wave number $k_0=\frac{\omega}{c_0}$ in the homogeneous surrounding medium: $p_i=P_i^0(\omega)e^{-i(\omega t-k_0\mathbf{n_0}.\mathbf{x})}=P_i e^{-i\omega t}$ with $P_i=P_i^0(\omega)e^{ik_0\mathbf{n_0}.\mathbf{x}}$.

We seek for a solution of the form $p=Pe^{-i\omega t}$. Then P is solution of the Lippman-Schwinger integral equation:

$\mathbf{x}\in R^2$, $P(\mathbf{x},\omega)=P_i(\mathbf{x},\omega)+\int_D G_0(\mathbf{x},\mathbf{x}',\omega)\left[-k_0{}^2\mu_\chi(\mathbf{x}')P(\mathbf{x}',\omega)\right]d^2x'$ (state equation)

where G_0 is the 2D free space Green function: $G_0(\mathbf{x},\mathbf{x}',\omega)=-\frac{i}{4}H_0^{(1)}(k_0|\mathbf{x}-\mathbf{x}'|)$.

Peculiarly, the measures on Σ satisfy:

$\mathbf{y}\in\Sigma$, $P(\mathbf{y},\omega)=P_i(\mathbf{y},\omega)+\int_D G_0(\mathbf{y},\mathbf{x}',\omega)\left[-k_0{}^2\mu_\chi(\mathbf{x}')P(\mathbf{x}',\omega)\right]d^2x'$ (observation equation).

Born Approximation

In the weak scattering approximation (low contrasts) $P_d<<P_i$, one can use the Born Approximation: $\mathbf{y}\in\Sigma$ $P_d(\mathbf{y},\omega)\approx\int_D G_0(\mathbf{y},\mathbf{x}',\omega)\left[-k_0{}^2\mu(\mathbf{x}')_\chi P_i(\mathbf{x}',\omega)\right]d^2x'$

For far-field measures (great distance with respect to the wavelength and the size of the object: $\forall x'\in D$, $k_0|\mathbf{y}-\mathbf{x}'|>>1$, and $\forall x'\in D$, $|\mathbf{y}|>>|\mathbf{x}'|$ so that $|\mathbf{y}-\mathbf{x}'|\approx|\mathbf{y}|-\mathbf{n}.\mathbf{x}'$ where $\mathbf{n}\equiv\frac{\mathbf{y}}{|\mathbf{y}|}$ is the observation direction), one find the asymptotic scattered field:

$$P_d^\infty(\mathbf{y},\omega)\approx P_i^0(\omega)\left(\frac{i}{4}\sqrt{\frac{2}{\pi|\mathbf{y}|}}e^{i\left(k_0|\mathbf{y}|-\frac{\pi}{4}\right)}\right)h(\mathbf{n_0},\mathbf{n},\omega),\quad \text{with}\quad h(\mathbf{n_0},\mathbf{n},\omega)=|k_0|^{3/2}\hat{\mu}_\chi(\mathbf{K})\Big|_{\mathbf{K}=k_0(\mathbf{n}-\mathbf{n_0})},$$

where $\hat{\mu}_\chi(\mathbf{K})\equiv TF(\mu_\chi(\mathbf{x}))\equiv\int_{R^2}\mu_\chi(\mathbf{x})e^{-i\mathbf{K}.\mathbf{x}}d^2x$, space (2D) Fourier Transform of $\mu_\chi(\mathbf{x})$.

$h(\mathbf{n_0},\mathbf{n},\omega)$ is the transfer function in diffraction of the medium (Fourier transform of the impulse response) for directions of incidence $\mathbf{n_0}$ and observation $\mathbf{n}$. Strictly speaking, with the chosen convention for progressive and exploding waves, the corresponding impulse response is: $H(\mathbf{n_0},\mathbf{n},t)=TF^{-1}[h(\mathbf{n_0},\mathbf{n},\omega)]_{(-t)}$; i.e. $h(\mathbf{n_0},\mathbf{n},\omega)=\hat{H}(\mathbf{n_0},\mathbf{n},-\omega)$.

So: $\hat{H}(\mathbf{n}_0,\mathbf{n},-\omega)=h(\mathbf{n}_0,\mathbf{n},\omega)=|k_0|^{3/2}\hat{\mu}_\chi(\mathbf{K})\big|_{\mathbf{K}=k_0(\mathbf{n}-\mathbf{n}_0)}$. One can also write:

$\hat{H}(\mathbf{n}_0,\mathbf{n},-\omega)=h(\mathbf{n}_0,\mathbf{n},\omega)=[2(1-\mathbf{n}_0.\mathbf{n})]^{-3/4}|\mathbf{K}|^{3/2}\hat{\mu}_\chi(\mathbf{K})$, with $\mathbf{K}=k_0(\mathbf{n}-\mathbf{n}_0)$

Or, if one chooses as object function $O(\mathbf{x})$ such that $\hat{O}(\mathbf{K})\equiv|\mathbf{K}|^{3/2}\hat{\mu}_\chi(\mathbf{K})$:

$\hat{H}(\mathbf{n}_0,\mathbf{n},-\omega)=h(\mathbf{n}_0,\mathbf{n},\omega)=[2(1-\mathbf{n}_0.\mathbf{n})]^{-3/4}\hat{O}(\mathbf{K})$, with : $\mathbf{K}=\dfrac{\omega}{c_0}(\mathbf{n}-\mathbf{n}_0)$.

With an angular parametrization (ϕ, θ) with respect to the Cartesian frame $(\mathbf{i},\mathbf{j})$, where $\phi\equiv(\mathbf{i},\mathbf{n}_0)$ is the incidence angle and $\theta\equiv(\mathbf{n}_0,\mathbf{n})$ the diffraction angle, and setting $h(\mathbf{n}_0,\mathbf{n},\omega)\equiv h(\phi,\theta,\omega)$:

$\hat{H}(\phi,\theta,-\omega)=h(\phi,\theta,\omega)=|2\sin(\theta/2)|^{-3/2}|\mathbf{K}|^{3/2}\hat{\mu}_\chi(\mathbf{K})$, with $\mathbf{K}=K\boldsymbol{\xi}$, $K=2\dfrac{\omega}{c_0}\sin(\theta/2)$ and $\boldsymbol{\xi}$ unit vector such that $(\mathbf{i},\boldsymbol{\xi})\equiv\psi=\left(\phi+\dfrac{\theta}{2}+\dfrac{\pi}{2}\right)$; or, with respect to the chosen object function:

$\hat{H}(\phi,\theta,-\omega)=h(\phi,\theta,\omega)=|2\sin(\theta/2)|^{-3/2}\hat{O}(\mathbf{K})$, with $\mathbf{K}=K\boldsymbol{\xi}$, $K=2\dfrac{\omega}{c_0}\sin(\theta/2)$ and $\boldsymbol{\xi}$ unit vector such that $(\mathbf{i},\boldsymbol{\xi})\equiv\psi=\left(\phi+\dfrac{\theta}{2}+\dfrac{\pi}{2}\right)$.

Use of broadband pulses

One sees that a broad-band pulse of frequency-band $\omega\in[\omega_{\min},\omega_{\max}]$(ideally $\omega\in[0,\infty[$), emitted within direction $\mathbf{n}_0$ and received within direction $\mathbf{n}$ gives a *slice of the object in the spatial frequency plane* along the bisecting line of the two directions; i.e. *a projection of the object in the real space* along this line (slice-projection theorem /2/).

BORN TOMOGRAPHY WITH A FIXED DIFFRACTION ANGLE

Let us fill the Fourier space of the object $\hat{O}(\mathbf{K})$ with broadband ($\omega\in[\omega_{\min},\omega_{\max}]$, ideally $\omega\in[0,\infty[$) diffraction measures $h(\phi,\theta,\omega)$ at fixed diffraction angle θ, by rotating the object or the emitter-receiver system around the object over $\phi\in[-\pi,+\pi]$.

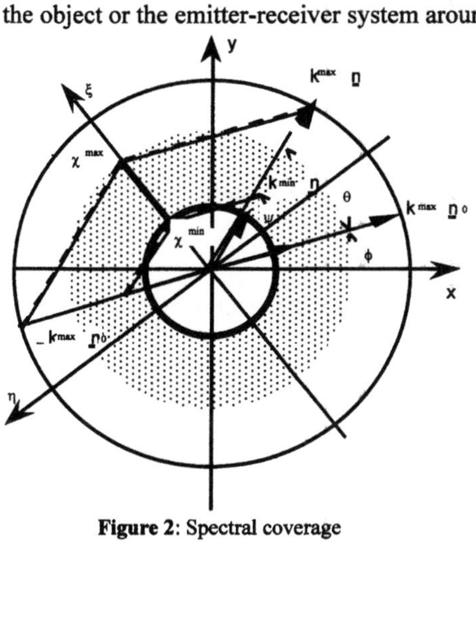

Figure 2: Spectral coverage

Then, in order to reconstruct the object $O(\mathbf{x})$, one has to compute the inverse Fourier Transform:

$$O(\mathbf{x})=\frac{1}{(2\pi)^2}\int_{\mathbb{R}^2}\hat{O}(\mathbf{K})e^{i\mathbf{K}.\mathbf{x}}d^2K,$$

or, in the polar frame (K,ψ) associated with the acquisition geometry:

$$O(\mathbf{x})=\frac{1}{(2\pi)^2}\int_0^{2\pi}d\psi\int_0^{+\infty}K\,\hat{O}(K,\psi)e^{i\mathbf{K}.\mathbf{x}}dK,$$

where $\hat{O}(\mathbf{K})\equiv\hat{O}(K,\psi)$; or, after some calculus:

$O(\mathbf{x}) \equiv O(x,y) = \frac{1}{2}\frac{1}{(2\pi)^2}\int_{-\pi}^{\pi} d\psi \int_{-\infty}^{+\infty} |K| \hat{O}(K,\psi) e^{iK(x\cos\psi + y\sin\psi)} dK$. It is the classical tomographic reconstruction algorithm named "summation of back-projections of filtered projections" /2/:

$$O(\mathbf{x}) \equiv O(x,y) = \frac{1}{2}\frac{1}{2\pi}\int_{-\pi}^{\pi} d\psi\, P_\psi^F(x\cos\psi + y\sin\psi),$$

with $\hat{P}_\psi^F(K) \equiv |K| \hat{P}_\psi(K)$ and $\hat{P}_\psi(K) = \hat{O}(K,\psi)$. $P_\psi(\xi) \equiv TF^{-1}(\hat{P}_\psi(K)) \equiv \frac{1}{2\pi}\int_{-\infty}^{+\infty} P_\psi(\xi) e^{iK\xi} d\xi$, whose Fourier Transform is a slice of the 2-D Fourier Transform of the object function O is a *projection* (slice-projection theorem /2/) of this object function.

With respect to acquisitions, i.e. Impulse Responses of the medium $H(\mathbf{n_0},\mathbf{n},t) \equiv H(\phi,\theta,t)$):

$$\hat{P}_\psi(K) = \hat{O}(K,\psi) = |2\sin(\theta/2)|^{3/2} \hat{H}\left(\psi - \frac{\theta}{2} - \frac{\pi}{2}, \theta, -\frac{c_0 K}{2\sin(\theta/2)}\right), \text{ or:}$$

$$\hat{P}_\psi(K) = \hat{O}(K,\psi) = |2\sin(\theta/2)|^{3/2} \hat{H}\left(\psi - \frac{\theta}{2} + \frac{\pi}{2}, \theta, \frac{c_0 K}{2\sin(\theta/2)}\right), \text{ and:}$$

$$O(\mathbf{x}) \equiv O(x,y) = \frac{1}{2}\frac{1}{2\pi}\int_0^{2\pi} d\phi\, P^F_{\phi+\frac{\theta}{2}-\frac{\pi}{2}}\left(x\cos\left(\phi + \frac{\theta}{2} - \frac{\pi}{2}\right) + y\sin\left(\phi + \frac{\theta}{2} - \frac{\pi}{2}\right)\right), \text{ with}$$

$$P^F_{\phi+\frac{\theta}{2}-\frac{\pi}{2}}(\xi) = |2\sin(\theta/2)|^{7/2} c_0^{-2} H^F\left(\phi, \theta, \frac{2}{c_0}\sin(\theta/2)\xi\right); \text{ and } \hat{H}^F(\phi,\theta,\omega) = |\omega| \hat{H}(\phi,\theta,\omega)$$

Numerical simulations with canonical objects

Numerical simulations with canonical objects as circular cylinders, allow to generate ideal exact data, and so to examine in detail and independently the influence of various parameters as object contrast, frequency-band of the transducers, and so on.

Simulations with circular cylinders

For a plane wave insonification $P_i(\mathbf{y},\omega) = P_i^0(\omega) e^{ik_0 \mathbf{n_0}\cdot\mathbf{y}}$, the field scattered by a homogeneous fluid circular cylinder (indexed by 1) can be written:

$$P(\mathbf{y},\omega) = P_i(\mathbf{y},\omega) + \sum_{n=-\infty}^{\infty} a_n H_n^{(1)}(k_0 r) e^{in\varphi} \text{ with}$$

$$a_n = \frac{(k_1/\rho_1) J_n(k_0 a) \dot{J}_n(k_1 a) - (k_0/\rho_0) \dot{J}_n(k_0 a) J_n(k_1 a)}{-(k_1/\rho_1) \dot{J}_n(k_1 a) H_n^{(1)}(k_0 a) + (k_0/\rho_0) \dot{H}_n^{(1)}(k_0 a) J_n(k_1 a)} P_i^0 e^{-i(\phi+\frac{\pi}{2})}$$

where $\phi = (\mathbf{i},\mathbf{n_0})$, $\varphi = (\mathbf{i},\mathbf{y})$, $r = |\mathbf{y}|$, $\dot{Z}_n(z) \equiv dZ_n(z)/dz$, $\xi_j = k_j a$ and $\delta_j = \frac{k_j}{\rho_j}$

This gives the far-field approximation:

$$P_d^\infty(\mathbf{y},\omega) = P^\infty(\mathbf{y},\omega) - P_i(\mathbf{y},\omega) \approx P_i^0(\omega)\left(\frac{i}{4}\sqrt{\frac{2}{\pi|\mathbf{y}|}} e^{i\left(k_0|\mathbf{y}| - \frac{\pi}{4}\right)}\right) h(\mathbf{n_0},\mathbf{n},\omega), \text{ with } \mathbf{n} \equiv \frac{\mathbf{y}}{|\mathbf{y}|}, \ \varphi = (\mathbf{i},\mathbf{n})$$

where $h(\mathbf{n_0},\mathbf{n},\omega) = -\frac{4i}{k_0}\sum_{n=-N=-\infty}^{N=\infty} a_n e^{in(\varphi-\pi/2)}$ can be calculated with much accuracy:

$$h(\mathbf{n_0},\mathbf{n},\omega) \approx -4i\sum_{n=-N}^{N} a_n e^{in(\varphi-\pi/2)} \text{ with } N = O(ka), \ k \equiv Max(k_0,k_1).$$

Image filtering induced by the transducers

Since transducers behave a limited frequency band $[\omega_{min}, \omega_{max}]$, they induce, within the inverse Born approximation, a spatial filtering of band: $[K_{min}, K_{max}] = \left[2\frac{\omega_{min}}{c_0}\sin\frac{\theta}{2}, 2\frac{\omega_{max}}{c_0}\sin\frac{\theta}{2}\right]$. So highest spatial frequencies are induced by highest frequencies of the signal and are highest for $\theta = \pi$, i.e. in back scattering. So back-scattering gives the best resolution. It is Reflection Tomography.

Image artefacts induced by high contrast

With increasing the contrast of the object, one finds more and more artefacts of reconstruction: numerous rings appear around and/or inside the object.

This is because the Born Approximation more and more fails and multiple reflection inside the object occurs.

We found an accurate method to attenuate these artefacts when imaging simple objects, based on the elimination of late signals (mainly multiples) /3/.

Figure 3 shows results of numerical simulations of back-scattering imaging (i.e. reflection tomography) experiments with a medium contrast canonical object (simulation of a Plexiglas cylinder in water). One sees clearly how inside ring artefacts are eliminated by our procedure.

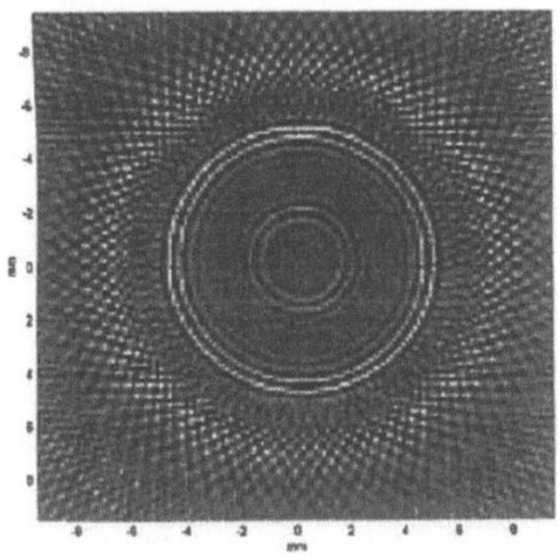

Figure 3-a: Image of a synthetic cylinder

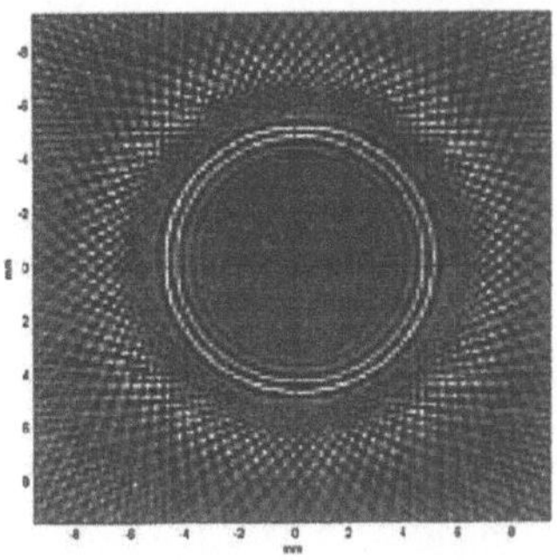

Figure 3-b: Elimination of ring artefacts

Experiments with a mechanical set-up

Experiments were conducted with using a mechanical system, which allows spatial sweeping of 2 transducers independently around an object immersed in a water tank (in-vitro imaging for biomedical applications). Main experiments were conducted in back scattering (reflection tomography), since that configuration gives the best image resolution.

Figure 4 shows a result got in back-scattering configuration with an alginate biomedical ghost (c=1529 ms^{-1}, $\chi = 4.437 \times 10^{-10} m^2N^{-1}$) riddled with hollows filled with water (c_0=1489 ms^{-1}, $\chi_0 = 4.533 \times 10^{-10} m^2N^{-1}$), immersed in water ($c_0$=1489 ms^{-1}, $\chi_0 = 4.533 \times 10^{-10} m^2N^{-1}$), with using a wide-band ultrasonic 2.25 MHz transducer (band [1.25 MHz – 3.75 MHz] at -15 dB) and an angular sampling rate of 2° (180 acquisitions over 360°). The contrast is very low ($\mu_\chi = 2.11 \times 10^{-2}$, representative of biomedical tissues) so that the Born approximation is efficient.

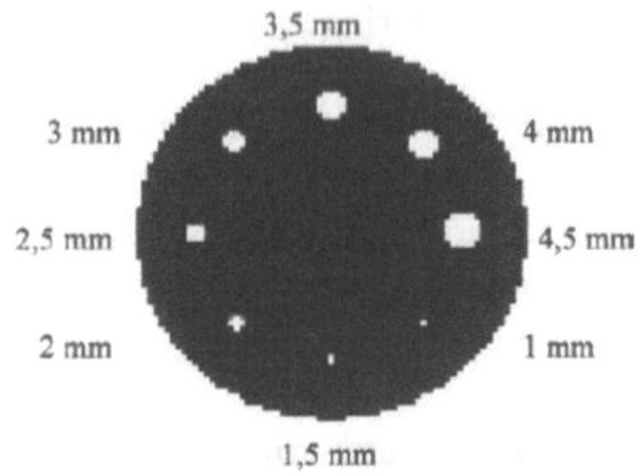

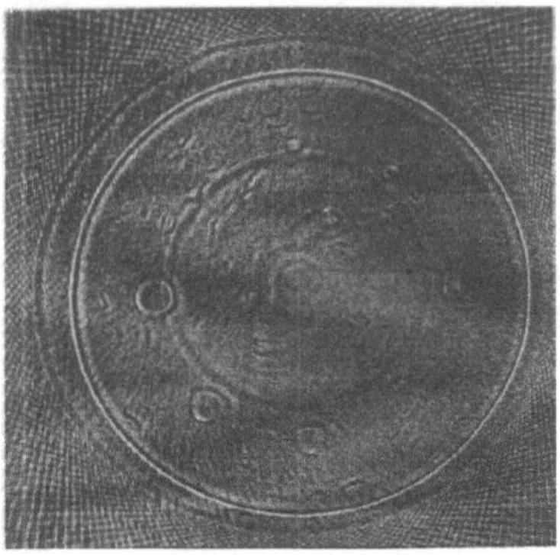

Figure 4-a: geometry of the alginate ghost

Figure 4-b: reflection imaging

Figure 5 shows artefact elimination with a real experiment on a Plexiglas cylinder (7 mm diameter) in water with using a wide-band ultrasonic 2.25 MHz transducer (band [1.25 MHz – 3.75 MHz] at -15 dB) and an angular sampling rate of 4° (90 acquisitions over 360°).

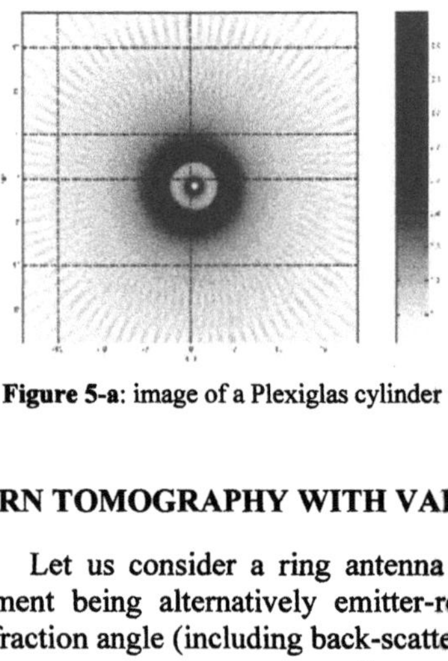

Figure 5-a: image of a Plexiglas cylinder

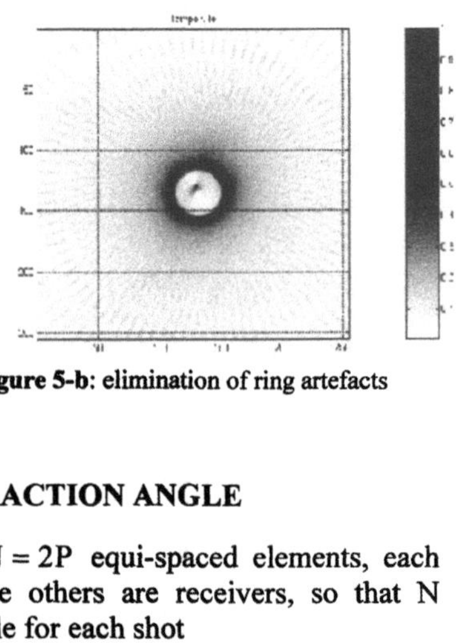

Figure 5-b: elimination of ring artefacts

BORN TOMOGRAPHY WITH VARIABLE DIFFRACTION ANGLE

Let us consider a ring antenna composed of N = 2P equi-spaced elements, each element being alternatively emitter-receiver while the others are receivers, so that N diffraction angle (including back-scattering) are available for each shot

So one get NxN wide-band diffraction measures.

three strategies can be used to solve the wide-band Born tomography problem:

- Consider each frequency, build an image for each frequency and take a mean image as the final one.

At each frequency one has a classical "diffraction tomography" problem, the simplest solution of which being the summation of filtered back-propagation algorithm /4/.

- Consider each diffraction angle, build an image for each angle and take a mean image as the final one.

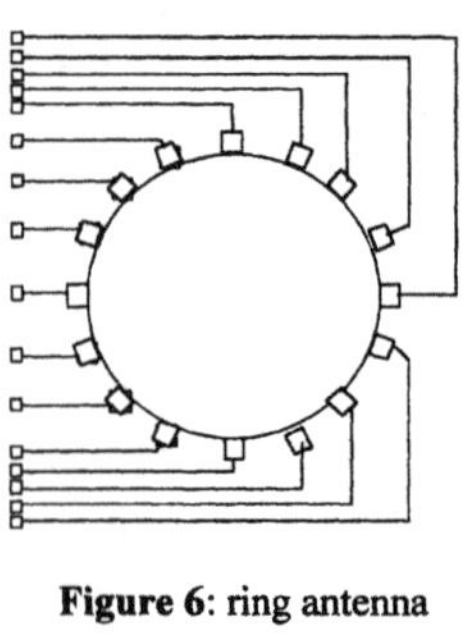

Figure 6: ring antenna

- Consider all the NxN acquired projections, remark that these projections fall only on 2N directions since they are along the bisecting lines of the emitter-receiver angles, take the mean of projections behaving same directions, and reconstruct an image with those 2N mean projections. We adopted that strategy, for its simplicity.

Simulation of 180 elements rings antenna by a mechanical system

It is easy (but time consuming) to simulate a 180 elements antenna with a mechanical system: proceed to a 180 steps angular sweep with the transmitter and for each position of the transmitter proceed to a 180 steps angular sweep of the receiver.

Figure 7 shows the result that would give a 180 elements antenna on the same biomedical ghost that the one of figure 4.
One sees the enhancement of image quality with regard to simple back-scattering measures (Figure 7-b = Figure 4-b).

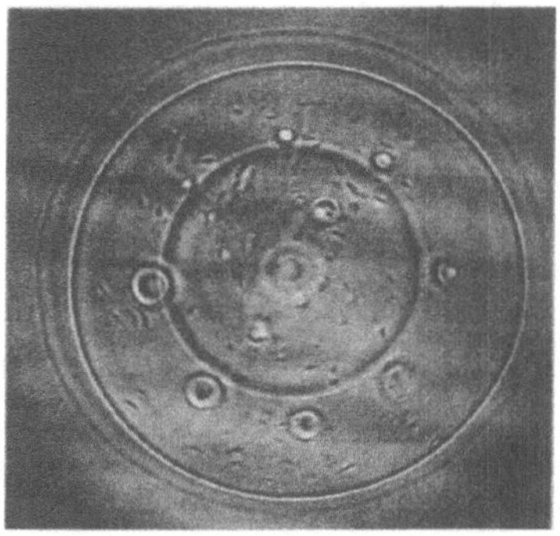

Figure 7-a: 180 x 180 diffraction measures

Figure 7-b: 180 reflection measures

Results with an 8 elements ring antenna

We built an 8 elements 2.25 MHz ultrasonic ring antenna tomograph, working in sequential mode (next step will be a 128 elements antenna working in parallel mode). A complete 8 x 8 = 64 acquisitions sequence (giving only 2 x 8 = 16 independent projections) takes about 500 ms.

Figure 8 shows results with using such a poor antenna.

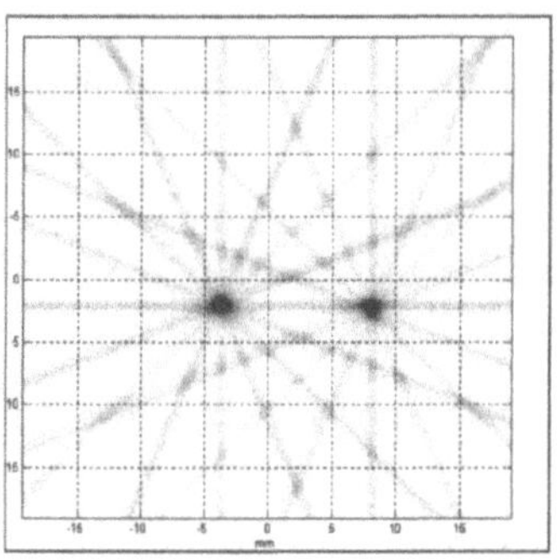

Figure 8-a: image of 2 0.1 mm nylon rods 10 mm spaced

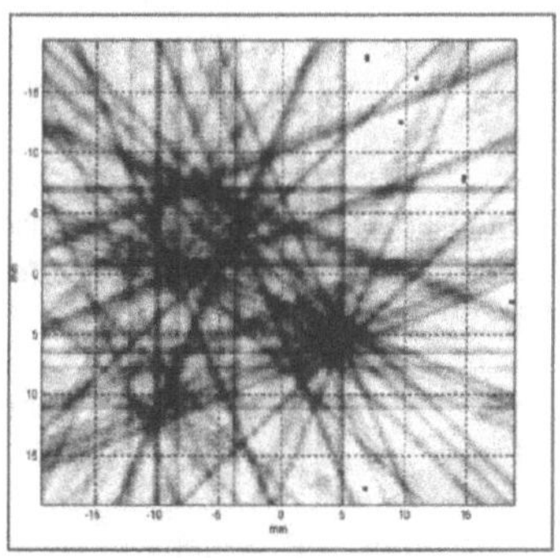

Figure 8-b: image of 3 copper rods (3 , 6, 9 mm)

CONCLUSION

Within the Born approximation, i.e. for small contrast objects like biomedical tissues, we have shown how slices of those objects in the space frequency plane, i.e. projections of those objects in the real space, can be obtained through wide-band ultrasonic diffraction measures. We also shown how classical tomographic algorithms can then be used in order to reconstruct the object from those projections.

Numerical experiments show limits of the technique with using band-limited transducers and significant contrast objects, but also how to improve the technique in that case.

Real experiments with using mechanical and electronic steering systems show the applicability of the method. For electronic steering with using a poor ring antenna tomograph, images clearly show limits of the classical summation of filtered back-projections algorithm for small number of projections.

Various ways of improvement will be explored before increasing the number of elements of the antenna:

- angular smoothing in order to increase the number of projections
- irregular repartition of transducers in order to increase the number of independent projections
- proceed to a few number of rotations of the system in order to increase the number of independent projections
- other algorithms, like algebraic algorithms, known to be more suitable for low number of projections

Those last algorithms are also known to be suitable for non-linear inversion, necessary for main quantitative imaging problems, when contrast is over few percents.

REFERENCES

1. P.M. Morse, K.U. Ingard, Theoretical Acoustics, 409-413, Mac Graw-Hill, NY, (1968).

2. S. R. Deans, The Radon transform and some of its Applications, Wiley-Interscience Publication, John Wiley & sons, New-York, (1983).

3. S. Delamare, Une amélioration de la tomographie ultrasonore en réflexion, *C. R. Acad. Sci. Paris*, t. 327, Série Iib, pp. 475-480,(1999)

4. A. J. Devaney, A filtered backpropagation algorithm for diffraction tomography. *Ultrasonic Imaging*, vol. 4, 336-350, (1982).

5. S. Mensah, J.P. Lefebvre, Ultrasonic reflection tomography, Specific problems and adapted solutions, *Mathematical Methods in Medical Imaging* III, SPIE Vol. 2299, pp.264-269, (1995).

6. S. Mensah, J.P. Lefebvre, Enhanced compressibility tomography, *IEEE Trans. UFFC*, vol. 44, n°6, pp. 1245-1252, (1997)

7. P. Lasaygues, J.P. Lefebvre, S. Mensah, High Resolution Low Frequency Ultrasonic Tomography, *Ultrasonic Imaging*, vol. 19, p 278, (1997).

A NEW CIRCULAR-SCANNING ULTRASONIC DIFFRACTION TOMOGRAPHY

Toyokatsu Miyashita* and Akio Honda

Dept. Electronics and Informatics
Ryukoku University
Seta Ohe-cho, Ohtsu, 520–2194 Japan

INTRODUCTION

Diffraction tomography images small structures of an object with respect to the wavelength of the probing wave in the presence of diffraction, reflection, and refraction phenomena. Theoretical method of ultrasonic diffraction tomography was first discussed by Mueller et al. [1] Although the wave equation is solved with an object as an inhomogeneous medium by a perturbation technique, the reconstruction scheme of the diffraction tomography is fundamentally similar to the X ray CT. That is, the measured data as a function of wide-range spatial coordinates is transformed into a two-dimensional Fourier-transform space of the object, and then the object is reconstructed by inverse Fourier transform. Several methods have been invented to get a wide-range Fourier transform of the cross sectional structure of the object, scanning the angle of the incident plane wave and also the position of the receiver over a wide range, combining a linear scanning and a rotation of the measurement system. The wide angular scanning of the incident plane wave is replaced by a multi-frequency measurement of the scattered wave using a short acoustic pulse, and a transmitter and receiver is simply moved along a line [2] as shown in Fig. 1(a). Experimental results of good quantitative reconstruction were presented by Yamada and Kurahashi[3] and Kuroiwa and Yamada[4]. In practice, the length of the linear scanning is limited and the direction from the transmitter and receiver to the object has also a limited angular range.

An alternative and natural method is to receive the scattered wave on a closed curve around the object as shown in Fig. 1(b). "Wave CT" has been investigated by Miyashita[5] using this type of alignment of VHF receiver antennas for the imaging of the complex electric susceptibility, i.e., refractive index and absorption coefficient of the object, directly solving the wave equation. This method include no weak scattering approximation and can reconstruct a cross-section of general objects which have a large contrast of the physical quantities.

*E-mail address: miya@rins.ryukoku.ac.jp

Acoustical Imaging, Volume 25, Edited by Halliwell and Wells.
Kluwer Academic/Plenum Publishers, 2000.

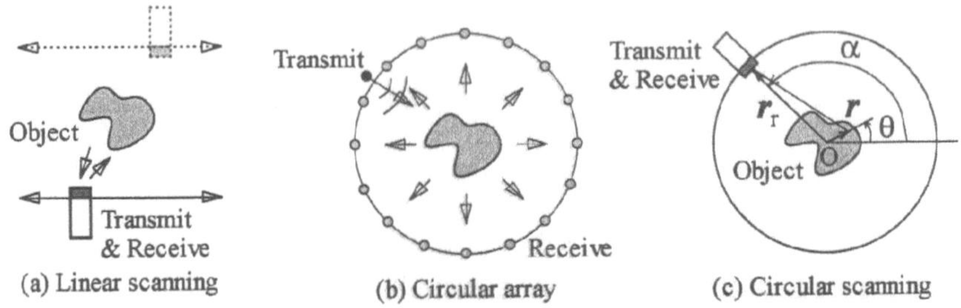

Fig. 1: Scanning types of diffraction CT.

A great demerit of "Wave CT" is to require a long computation time. An experimental study was reported by Joachimowicz et al.[6]

Circular scanning of the transmitter and/or receiver has been adopted indeed in several investigations about tomographic and holographic imagings, for example, by Higgins[7] and by Lewis and Lee[8]. They include no diffraction phenomena.

In this paper, first we present a theory of a circular-scanning ultrasonic diffraction tomography. Secondly we investigated simple experiments together with practical problems. Finally we show a theoretical investigation about key parameters introducing new numerical phantoms for evaluation of the circular scanning diffraction tomography.

MERITS AND DEMERITS OF THE CIRCULAR SCANNING

When the transmitter and receiver can be placed all around the object with the aid of water bag if necessary, a circular scanning type of acoustic diffraction tomography as shown in Fig. 1(c) is very attractive, even for a microwave tomography. The merits are as follows.

1. Circular scanning with a short pulse is simple, uniform, and fast compared with the linear scanning. A fixed alignment of the sensors is also possible using a water bag.
2. The most sensitive and uniform directivity of the sensors is used always facing toward an object during the circular scanning. With a narrow slit in front of the acoustic transducer, a high-quality cylindrical wave is transmitted and received.
3. Backscattered wave is detected completely separated from the incident wave and directly passed to the reconstruction processing.

Most biggest demerit is that any reconstruction formula of a closed form or of an orthogonal transform like Fourier transform for the linear scanning diffraction tomographies [2] was not invented for the circular scanning so far. To break through the difficulty, we have invented a new reconstruction method which has a closed-form formula composed of Fourier series and Hankel transforms, restricting the object in the far field of the cylindrical wave.

WEAK SCATTERING APPROXIMATION

Sound propagation within a lossless medium is described by the following equations.

$$\rho\frac{\partial u}{\partial t} = -\nabla p, \quad \frac{\partial p}{\partial t} = -K\nabla\cdot u. \tag{1}$$

Here, u, p, ρ and K are, respectively, particle velocity, sound pressure, density and bulk modulus of the medium. We introduce normalized contrasts $\chi_\rho = (\rho-\rho_0)/\rho$ and $\chi_K = (K-K_0)/K$ of the inhomogeneous object, where ρ_0 and K_0 denote, respectively, density and bulk modulus of the surrounding homogeneous medium. In the following consideration, wave propagation of the Fourier component $p(k)$ of the sound pressure is considered, where the wave number k is given by $k = \omega\sqrt{\rho_0/K_0}$ for the homogeneous medium. Following the well-known method of Born's approximation [1], The following formula of the scattered wave $p_s(k, r_r)$ for the ultrasonic diffraction tomography is derived [2].

$$p_s(k, r_r) = \iint_S \{-\chi_K(r)k^2 g(k, r_r|r)p_i(r) + \chi_\rho(r)\nabla_r g(k, r_r|r)\cdot\nabla_r p_i(r)\}\, ds. \tag{2}$$

Here, vector r_r denotes the position of the receiver, $p_i(r)$ the incident wave, and $g(k, r_r|r)$ Green's function for a two-dimensional homogeneous medium.

THEORY OF A CIRCULAR-SCANNING DIFFRACTION TOMOGRAPHY

We place the object at the center of the circular scanning for the diffraction tomography as shown in Fig. 1(c). The transmitter radiates a uniform cylindrical wave toward the object and the backscattered wave is detected by the same sensor which acts now as a receiver. Namely the transmitter's position r_t is always equal to r_r. An alternative circular scanning will have a constant nonzero angle between the transmitter and receiver. However, it is clear that the zero-angle scanning, i.e., pure backscattering detection, makes the longest path difference between adjacent points in the object.

The scanning angle is denoted by α, the scattered wave by $p_s(k, \alpha)$ hereafter, the object coordinate by $r = (r, \theta)$. The incident wave $p_i(r)$ and Green's function $g(k, r_r|r)$ in the two-dimensional homogeneous medium which propagate outwards with the temporal factor $e^{+i\omega t}$ are given by the second kind of 0th order Hankel function $H_0^{(2)}(z)$ as follows:

$$p_i(r) = g(k, r_r|r) = -\frac{i}{4}H_0^{(2)}(k|r_r - r|). \tag{3}$$

The object region is assumed to be so far away from the transmitter and receiver that the asymptotic formula of $H_0^{(2)}(z)$ is valid. This region is called "far field of the cylindrical wave" hereafter. Furthermore, the scope of the object region is assumed to be small compared with the scanning radius. Namely

$$|r_r - r| > 100\lambda, \quad |r| \ll |r_r| \tag{4}$$

Under these conditions, the above Hankel function is approximated as follows.

$$H_0^{(2)}(k|r_r - r|) \simeq H_0^{(2)}(kr_r)e^{-ik(\sqrt{r_r^2+r^2-2r_r r\cos(\theta-\alpha)}-r_r)} \simeq H_0^{(2)}(kr_r)e^{ikr\cos(\theta-\alpha)}. \tag{5}$$

The gradient of the Hankel function is also approximated as follows.

$$\nabla_r H_0^{(2)}(k|r_r - r|)\cdot\nabla_r H_0^{(2)}(k|r_t - r|) \simeq -k^2\{H_0^{(2)}(kr_r)\}^2 e^{2ikr\cos(\theta-\alpha)}. \tag{6}$$

Substituting these approximated expressions into eq. (2), we obtain the following formula for the diffracted and backscattered wave to be detected by the receiver.

$$p_s(k, \alpha) = \frac{k^2}{16}\left\{H_0^{(2)}(kr_r)\right\}^2 \int\int_S f(r)e^{2ikr\cos(\theta-\alpha)}\, r\,dr\,d\theta. \tag{7}$$

Hereafter, $f(r)$ defined by $f(r) = \chi_K(r) + \chi_\rho(r)$ represents the object.

First we calculate the Fourier coefficients $F_n(k)$ of the scattered wave $p_s(k, \alpha)$ with respect to the scanning coordinate α. This corresponds to the Fourier transform of the projection along the linear scanning coordinate in the first step of the image reconstruction of X ray CT.

$$F_n(k) = \frac{1}{2\pi}\int_0^{2\pi} p_s(k, \alpha)e^{-in\alpha}d\alpha = K_n(k)\int_0^R \left[\frac{1}{2\pi}\int_0^{2\pi} f(r, \theta)e^{-in\theta}d\theta\right] J_n(2kr)\, r\,dr, \tag{8}$$

where,

$$K_n(k) = i^n \frac{\pi k^2}{8} \left\{ H_0^{(2)}(kr_{\mathrm{r}}) \right\}^2. \tag{9}$$

Here, R denotes the maximum radius of the object region. An integral representation of the Bessel functions is used in the above derivation.

Next, for the inverse scattering, we make use of the following Hankel transform pair.

$$\phi(k) = \int_0^\infty \psi(r) J_n(kr) r dr, \qquad \psi(r) = \int_0^\infty \phi(k) J_n(kr) k dk. \tag{10}$$

From the above equations, it is found that the Fourier coefficients of the object function $f(r,\theta)$ with respect to the θ axis is equal to the Hankel transform of $F_n(k)/K_n(k)$ which is calculated from the measured data of the scattered wave. Therefore, it is derived that the object function is reconstructed by the following transform (11). In the practical applications, the range of n will be limited to the number of probing angles N, and the wave number k is also bounded within the passband of the sensor between k_1 and k_2 or between f_1 and f_2 in frequency.

$$\hat{f}(r,\theta) = 4 \sum_{n=-\infty}^{\infty} e^{in\theta} \int_0^\infty \frac{F_n(k)}{K_n(k)} J_n(2kr) k dk \cong 4 \sum_{n=-N}^{N} e^{in\theta} \int_{k_1}^{k_2} \frac{F_n(k)}{K_n(k)} J_n(2kr) k dk. \tag{11}$$

RECONSTRUCTION

A Simple Experimental Study

A simple experimental investigation was conducted to obtain cross-sectional images of an acrylic pipe, whose outer radius is 9.05 mm and internal radius is 8.10 mm. The radius of the circular scanning was 14.565 cm. A short ultrasonic pulse with a duration of about 1.5 cycles at 2.25 MHz was transmitted towards the acrylic pipe with a transducer (Panametrics, M306-SU) and a pulser/receiver (Panametrics, Model 5072). With a vertical slit of 3 mm width in front of the transducer whose effective diameter is 12.7 mm, cylindrical-wave radiation and receiving were achieved. The detected signal which was quantized in an 8 bit accuracy was Fourier-transformed and the passband components from $f_1 = 0.95$ MHz to $f_2 = 4.05$ MHz were extracted. The number of probing directions in the circular scanning was one for the symmetrical alignment of the object, and 32 for the asymmetrical one. The cross-sectional profile of the reconstructed image shows clearly and exactly the shape and size of the acrylic pipe as shown in Fig. 2. The internal hollow space of the pipe has a very small value except an artifact at the center of the object region. However the reconstructed cross section of the acrylic pipe has an undesirable two-peak structure. It was found that these phenomena were not due to inaccurate detection of the ultrasonic scattered wave but due to a lack of low frequency component in the short ultrasonic pulse, as shown in the next chapter.

Theoretical investigation of the experimental results using numerical phantoms

The scattered and diffracted wave by any circular cylinder of a uniform material parameter $f(r,\theta)$ or by a superposition of uniform cylinders can be described in an analytical form under the far field assumption. The waves which should be detected in the experiments are calculated numerically and the object is reconstructed as shown in Fig. 3. The first case has a wide frequency band from 24 kHz to 4 MHz, and the second case has an experimentally used frequency band from 1 MHz to 4 MHz. These results explain that the experimental results are insufficient mainly due to a lack of low frequency component, and also shows

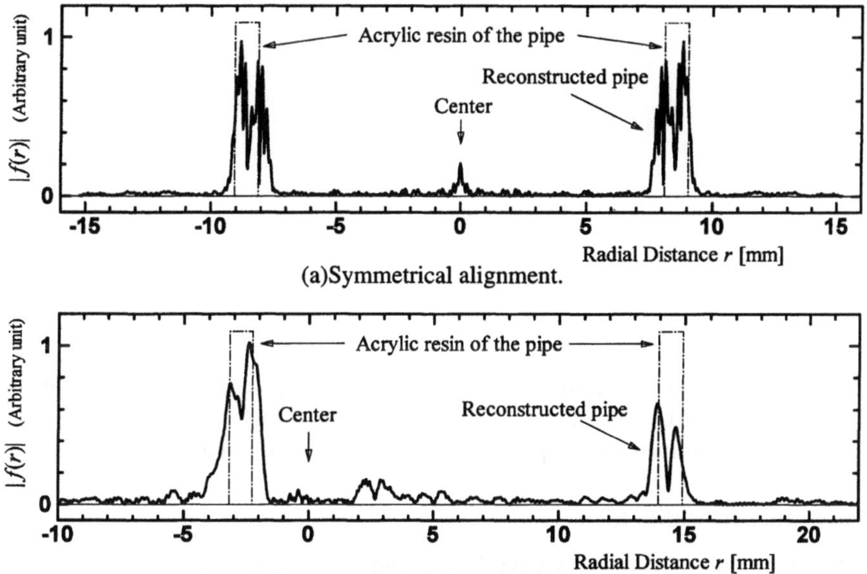

(a)Symmetrical alignment.

(b)Asymmetrical displaced alignment.

Fig. 2. Reconstructions of a pipe from the experimentally measured scattering.

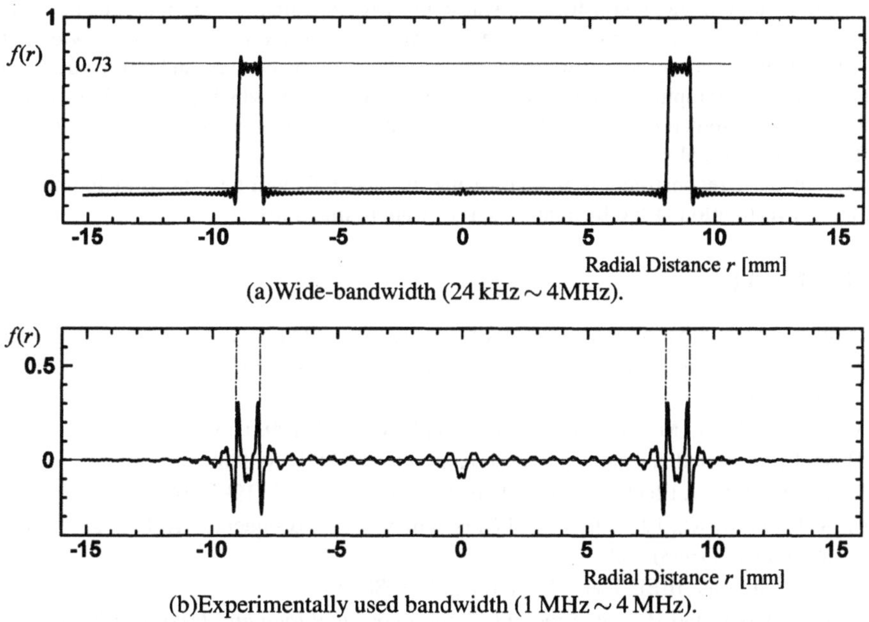

(a)Wide-bandwidth (24 kHz ~ 4MHz).

(b)Experimentally used bandwidth (1 MHz ~ 4 MHz).

Fig. 3. Reconstructions of a pipe from the numerically simulated scattering.

that quantitatively exact tomographic image should be given from a wide-band scattering measurement.

The phantom can be composed of eccentric cylinders for an object of relatively general structure. Reconstructions of such a phantom are shown in Fig. 4 with two different bandwidth of the probing wave. The diameter of the reconstruction area is 23.8 mm.

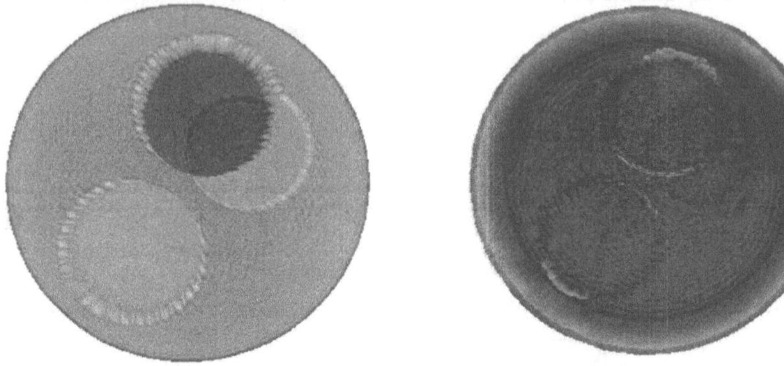

(a)Wide-bandwidth (24 kHz ~ 4MHz). (b)Experimentally used bandwidth (1 MHz ~ 4 MHz).

Fig. 4. Reconstructions of a phantom from the numerically simulated scatterings.

CONCLUDING REMARKS

Reconstruction formula for a circular scanning diffraction tomography was given in a compact closed form† with Fourier series and Hankel transform, making clear the required assumption that the object is placed in the far field of the illuminating cylindrical wave.

Tomographic image reconstruction was confirmed by a simple experiment using a short ultrasonic pulse in water. From a numerical investigation it was shown that a wide-band measurement including very low frequency component is necessary and sufficient for the quantitative tomographic reconstruction. Numerical phantoms which can simulate complex objects were presented for further detailed quantitative investigaions of the circular-scanning diffraction tomography.

Quantitative reconstruction of weak-scattering materials will be studied including comparative investigation of the Born and the Rytov approximations.

REFERENCES

[1] R. K. Mueller, M. Kaveh and G. Wade: Proc. IEEE **67** (1979) 567.

[2] K. Nagai: *Ultrasonic Holography* (Nikkan Kogyo Shinbun, Tokyo, 1986) Chap. 3 [in Japanese].

[3] A. Yamada and M. Kurahashi: Jpn. J. Appl. Phys. **32** (1993) 2507.

[4] M. Kuroiwa and A. Yamada: Jpn. J. Appl. Phys. **37** (1998) 3088.

[5] T. Miyashita: J. Electromagnetic Waves & Applications **7** (1993) 1421.

[6] N. Joahimowicz, J.J. Mallorqui, J.C. Bolomey and A. Broquetas: IEEE Trans. Medical Imaging **17** (1998) 562.

[7] W. E. Higgins: IEEE Trans. Med. Imag. **7** (1988) 59.

[8] W. R. Lewis and H. Lee, *Proc. Int. Symp. Acoustical Imaging, Firenze*, 1995 (Plenum Press, New York, 1996) Acoustical Imaging, Vol. 22, p. 37.

[9] K. Nagai: in Proc. Spring Meet. Acoust. Soc. Jpn. (1991) 779 [in Japanese].

†The formula was found recently to be very similar to a presentation by Nagai[9]. Our coefficient $K_n(k)$ given by eq. (9) is different from that of Nagai's.

DETERMINATION OF SURFACE WAVE VELOCITES IN INITIALLY DEFORMED, ANISOTROPIC LAYERED SOLIDS

Zuliang Yu, Siegfried Boseck

Department of Physics, University of Bremen
D-28359 Bremen, Germany

INTRODUCTION

This paper provides a theory to determine the velocities of surface waves in initially deformed, anisotropic layered solids. This theory is of great significance for stress measurement in thin films with acoustic microscopy. The stress cannot be measured directly when using acoustic microscopy but must be inferred from the measured surface wave velocity with inversion approach. Therefore, it is necessary to establish a basic theory to describe how the change of the surface wave velocities in deformed layered solids depends on the predeformation states due to stress and strain. We have solved this problem in the framework of acoustoelasticity for surface waves [1]. That is, the constitutive equations and the equations of motion have been derived by introducing the nonlinearity of elasticity and reexamining the theory of small (infinitesimal) motions superimposed on a finite elastic deformation. Furthermore, in order to solve the forward calculation problem, plane stress system was introduced in this work to describe the initial stress-strain field distributions in thin layered solids. Finally, this paper also presents the representative calculated examples from which some of the important conclusions will be extracted.

CONSTITUTIVE EQUATIONS AND EQUATIONS OF MOTION

Consider the configuration of a thin film/substrate combination and the coordinate system shown in Figure 1. Both of the film and the substrate are anisotropic whose crystalline axes do not coincide with in the general case. No generality is lost and much simplicity is gained,

Acoustical Imaging, Volume 25, Edited by Halliwell and Wells.
Kluwer Academic/Plenum Publishers, 2000.

we assume the propagation direction of surface wave to be the X_1 axis direction, but the X_1 axis may be inclined at an angle to the crystalline axis.

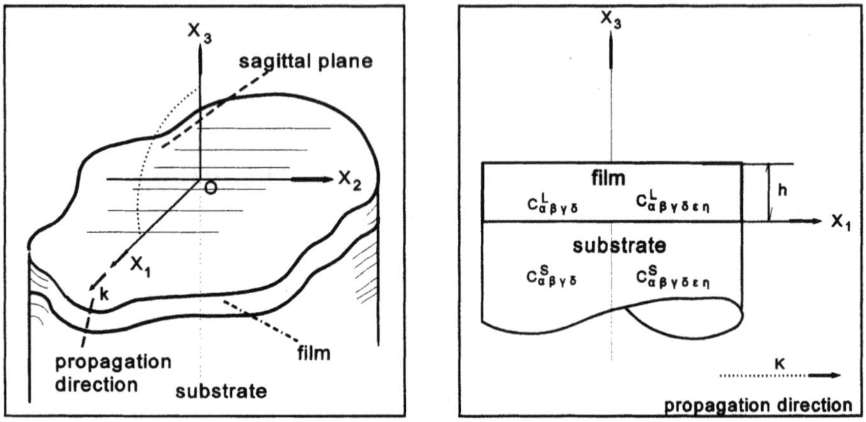

(a) (b)

Figure 1. Configuration of a layered solid and coordinate system for wave propagation.
(a) Wave propagation direction and sagittal plane.
(b) Geometry of the layered solid in the sagittal plane and material parameters.

We start by studying the surface wave propagation in anisotropic layered solids without deformation. The acoustic field equations and the elastic constitutive equations in each elastic medium in the absence of body forces can be written as [2]

$$
\begin{aligned}
\eta_{kl} &= \frac{1}{2}\left(\frac{\partial u_k}{\partial X_l} + \frac{\partial u_l}{\partial X_k}\right) && (a) \\
c_{ijkl}\frac{\partial^2 u_k}{\partial X_i \partial X_l} &= \rho \frac{\partial^2 u_j}{\partial t^2} && (b) \qquad (1) \\
\sigma_{ij} &= c_{ijkl}\eta_{kl} && (c)
\end{aligned}
$$

Equation 1(a) is an expression of the linearized strain-displacement relation where η_{kl} is the infinitesimal Langrangian strain tensor and $\partial u_k/\partial X_l$ are the displacement gradients, Equation 1(b) is the equation of motion for small-amplitude waves in which c_{ijkl} are the stiffness constants and ρ is the mass density, and Equation 1(c) is the elastic constitutive equation, i.e., Hooke's law, where σ_{ij} is the Cauchy stress tensor. It was known that surface waves may propagate in layered solids in many different independent modes [3]. The lowest mode is the Rayleigh wave mode. Higher modes are called Sezawa wave modes (generalized Lamb waves). All such surface waves become dispersive in the sense that (1) the phase velocities are a function of the normalized layer thickness, and (2) the phase velocities are of directional dependence due to anisotropy of materials. When taking account to stress or strain which occurs in the film and the substrate, the propagation of surface waves exhibits not only dispersive in the same sense as that described above, but their velocities are weakly affected by the deformation due to stress and strain. This effect is called acoustoelasticity. For examining the surface wave propagations in initially deformed, anisotropic layered solids the crucial aspect is how to able to characterize the deformed body due to stress or strain in terms of new exact physical quantities that can be substituted directly into the constitutive equations and the equations of motion to extract the velocities of surface waves travelling through it. This problem was solved through three development steps: (1) to determine the relation between stress and deformation by considering the nonlinearity of elasticity, (2) to derive the linearized governing equations relating the stress increments to the displacement gradients by defining the effective elastic constants, and (3) to establish the linearized equations of motion.

First, we consider that the deformation of a body is finite or large. In this case Hooke's law is no longer appropriate, and a nonlinear theory of elasticity is required in order to establish the theoretical basis in the change of elastic wave velocity. The nonlinear connection between the second Piola-Kirchhoff stress tensors, $S^{i}_{\alpha\beta}$ and $S^{f}_{\alpha\beta}$, and the Lagrangian finite strain tensors, $E^{i}_{\alpha\beta}$ and $E^{f}_{\alpha\beta}$, in the initial and final states can be given, respectively, as

$$\begin{aligned} S^{i}_{\alpha\beta} &= c_{\alpha\beta\gamma\delta} E^{i}_{\gamma\delta} + \frac{1}{2} c_{\alpha\beta\gamma\delta\epsilon\eta} E^{i}_{\gamma\delta} E^{i}_{\epsilon\eta} \qquad (a) \\ S^{f}_{\alpha\beta} &= c_{\alpha\beta\gamma\delta} E^{f}_{\gamma\delta} + \frac{1}{2} c_{\alpha\beta\gamma\delta\epsilon\eta} E^{f}_{\gamma\delta} E^{f}_{\epsilon\eta} \qquad (b) \end{aligned} \tag{2}$$

where $c_{\alpha\beta\gamma\delta}$ and $c_{\alpha\beta\gamma\delta\epsilon\eta}$ are the second-order and third-order elastic constants of a material in the natural state. Equation 2 characterizes the nonlinear stress-strain relations for a finitely deformed body in the first-order approximation.

Next, on the assumption that surface waves have small amplitude and, hence, the dynamic displacements due to wave motion are infinitesimal, the governing equations relating the increments of stresses, S_{IJ}, to the dynamic displacement gradients can be linearized as

$$S_{IJ} = \hat{C}_{IJKL} \frac{\partial u_K}{\partial X_L} \tag{3}$$

where $\partial u_K/\partial X_L$ are the displacement gradients due to wave motion. The coefficients $\hat{C}_{IJKL}$ are called the effective elastic constants in this paper and given as follows

$$\begin{aligned} \hat{C}_{IJKL} = {} & c_{IJKL}(1 - e^{i}_{NN}) + c_{IJKLMN}\eta^{i}_{MN} \\ & + c_{MJKL}\frac{\partial u^{i}_{I}}{\partial X_M} + c_{IMKL}\frac{\partial u^{i}_{J}}{\partial X_M} + c_{IJML}\frac{\partial u^{i}_{K}}{\partial X_M} + c_{IJKM}\frac{\partial u^{i}_{L}}{\partial X_M} \end{aligned} \tag{4}$$

where c_{IJKL} and c_{IJKLMN} are the second- and third-order elastic constants of a material in the initial state, η^{i}_{MN} is the Lagrangian infinitesimal strain tensor in the initial state and e^{i}_{NN} is the cubic dilatation, i.e., the ratio of volume change due to deformation to initial volume.

Finally, since the dynamic displacements of wave motion are small (infinitesimal), then the equations of motion relating to the dynamic displacements of surface waves can also be linearized as

$$A_{IJKL}\frac{\partial^2 u_K}{\partial X_I \partial X_L} = \rho^i \frac{\partial^2 u_J}{\partial t^2} \tag{5}$$

with

$$A_{IJKL} = \hat{C}_{IJKL} + \delta_{JK}\sigma^i_{JL} \tag{6}$$

where δ_{IJ} is the Kronecker delta, ρ^i and σ^i_{IJ} are the mass density and the Cauchy stress in the initial state, respectively. A_{IJKL} are called the modified effective elastic constants.

DETERMINATION OF SURFACE WAVE VELOCITIES

Having established the equations of motion and defined the new physical quantities, the determination of surface wave velocities can directly follow the Farnell's analytical formalism in undeformed layered solids [4]. The particle displacements due to surface waves in each medium can be assumed to be linear combinations of terms of the form

$$u_J = \alpha_J \exp(iKL_3X_3)\exp[iK(X_1 - Vt)] \tag{7}$$

where α_J are the relative amplitudes of the different components of each partial wave, K is the magnitude of the propagation vector, V is the phase velocity measured along the propagation vector, and L_3 is the decay constant in the surface wave case. The decay constant L_3 and the amplitude factors α_J are solved from the following system of the secular equations

$$[K^2\Delta_{IK} - \rho^i\omega^2\delta_{IK}]\,\alpha_K = 0 \tag{8}$$

with

$$\Delta_{IK} = A_{IJKL}L_J L_L \tag{9}$$

where $\omega = KV$ is the angular frequency, and Δ_{IK} are the acoustic tensor. In general, there are three pairs of complex-conjugate roots of L_3 to Equation 8 for each given velocity value, V. All six values of L_3 for the film region are maintained. However, only three of the six solutions of L_3 are maintained for the substrate to ensure that the solutions decay to zero with depth below the interface. For each combination of root L_3 and α_J, the corresponding term, Equation 7, satisfies the equations of motion Equation 5. These waves should be combined into linear combinations with weighting factors chosen in an attempt to satisfy the boundary conditions. Thus, in the film and the substrate regions we have

$$\begin{aligned} u_J^L &= \left\{\sum_{n=1}^{6} B_n \alpha_J^{(n)} \exp(iKL_3^{(n)}X_3)\right\}\exp[iK(X_1 - Vt)] \quad &(a)\\ u_J^S &= \left\{\sum_{m=1,3,5} D_m \alpha_J^{(m)} \exp(iKL_3^{(m)}X_3)\right\}\exp[iK(X_1 - Vt)] \quad &(b)\end{aligned} \tag{10}$$

Here and hereafter the superscripts L and S denote the quantities in the layer and the substrate, respectively. The superscripts (n) and (m) denote the nth and mth solutions of the eigenvalue problems derived from the film and substrate secular equations, respectively. B_n and D_m are unknown weighting constants which are determined from the boundary condition equations. Finally, the surface wave velocities can be determined with iterative approach by vanishing the following 9×9 "boundary condition determinant"

$$|ma_{bp}| = 0 \tag{11}$$

The elements of the boundary condition determinant consist of two parts: the "film" columns with the subscript $p = 1,2,3\ldots6$ and the "substrate" columns with the subscript $p = 7,8,9$. For the "film" columns, the elements ma_{bp} are

$$\begin{aligned}
ma_{1p} &= -\alpha_1^{(n)} \qquad\qquad p = n = 1,2,\ldots,6\\
ma_{2p} &= -\alpha_2^{(n)}\\
ma_{3p} &= -\alpha_3^{(n)}\\
ma_{4p} &= -(A^L_{33KI} + A^L_{33K3}L_3^{(n)})\alpha_K^{(n)}\\
ma_{5p} &= -(A^L_{31KI} + A^L_{31K3}L_3^{(n)})\alpha_K^{(n)}\\
ma_{6p} &= -(A^L_{32KI} + A^L_{32K3}L_3^{(n)})\alpha_K^{(n)}\\
ma_{7p} &= [(A^L_{33KI} + A^L_{33K3}L_3^{(n)})\alpha_K^{(n)}]\exp(iKL_3^{(n)}h)\\
ma_{8p} &= [(A^L_{31KI} + A^L_{31K3}L_3^{(n)})\alpha_K^{(n)}]\exp(iKL_3^{(n)}h)\\
ma_{9p} &= [(A^L_{32KI} + A^L_{32K3}L_3^{(n)})\alpha_K^{(n)}]\exp(iKL_3^{(n)}h)
\end{aligned} \tag{12}$$

where h is the thickness of the film. For the "substrate" columns, the elements ma_{bp} are

$$\begin{aligned}
ma_{1p} &= \alpha_1^{(m)} \qquad\qquad p = 7,8,9 \dashrightarrow m = 1,3,5\\
ma_{2p} &= \alpha_2^{(m)}\\
ma_{3p} &= \alpha_3^{(m)}\\
ma_{4p} &= (A^s_{33RI} + A^s_{33R3}L_3^{(m)})\alpha_R^{(m)}\\
ma_{5p} &= (A^s_{31RI} + A^s_{31R3}L_3^{(m)})\alpha_R^{(m)}\\
ma_{6p} &= (A^s_{32RI} + A^s_{32R3}L_3^{(m)})\alpha_R^{(m)}\\
ma_{7p} &= 0\ ,\ ma_{8p} = 0\ ,\ ma_{9p} = 0
\end{aligned} \tag{13}$$

PLANAR STRESS STATE IN THIN FILMS

For the forward calculation of surface wave velocities, it is necessary to have the knowledge of the initial deformations due to stress and strain in samples. In the most general case, stress and strain tensors have six independent components, where three are normal components (σ_{11}, σ_{22}, σ_{33}; η_{11}, η_{22}, η_{33}) and others are shear components (σ_{12}, σ_{13}, σ_{23}; η_{12}, η_{13}, η_{23}). This number can become smaller, if one considers a thin film deposited on a thick substrate. In this case the stress state in the thin film can be regarded as a planar stress system, and the substrate really is free of stress and strain. A planar stress system is where one of the three mutually perpendicular components of stress is taken to be constant or zero.

Furthermore, from the consideration of the boundary conditions in thin films we have

$$\sigma_{33} = \sigma_{23} = \sigma_{13} = 0 \tag{14}$$

Thus the maximum number of the independent stress components becomes three: σ_{11}, σ_{22}, and σ_{12}. The analysis of the equations of differential equilibrium in thin films shows that these components have the same value at all depths in the film. In other words, the planar stress system essentially presents a two-dimensinal problem. Figure 2 shows the geometry for a planar stress state, in which two normal stress components, σ_{11} and σ_{22}, are parallel to the (X_1, X_2) coordinate axes. Thus, σ_{11} and σ_{22} are also called axial stresses. The two principal stresses, designated by σ_1 and σ_2, are deviated from the (X_1, X_2) coordinates by an angle Ω. By "principal stress" we mean that the shear stress components are all zero.

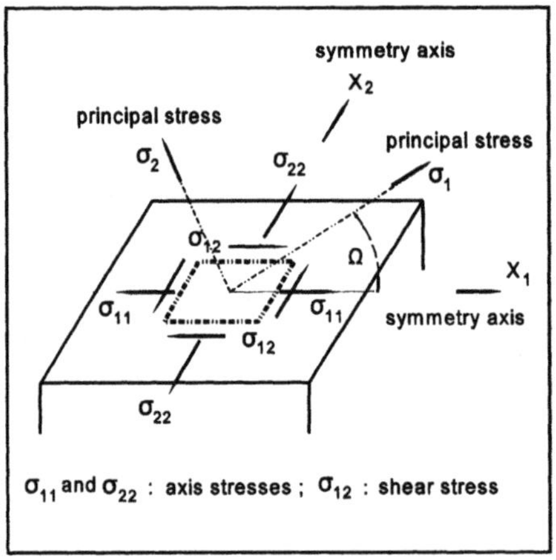

Figure 2. Geometry for a planar stress state.

The number of the nonzero independent components of stress and strain tensors may be further reduced. For this purpose, we have proposed some models of the stress-strain field distributions in thin films, namely, uniaxial-strain and uniaxial-stress models and biaxial-strain and biaxial-stress models. Biaxial models are often used in the evaluation of residual stresses in thin films, and uniaxial models in the evaluation of applied stresses. By "uniaxial model" we mean that only one principal stress or strain is nonzero. On the other hand, biaxial model means that two nonzero principal stresses or strains occur in the system.

NUMERICAL EXAMPLES

A series of the numerical experiments has been carried out in this work. Here the representative calculated examples are shown for a thin anisotropic layered Cu/Si(001) sample with the normalized layer thickness of $f \cdot h$ = 0.25 GHz$\times\mu$m. The single-crystal film and substrate materials, Cu(001) and Si(001), exhibit cubic symmetry. Figures 3(a) and 3(b) respectively show the velocity dispersions of the Rayleigh wave and the pseudo-Sezawa wave due to an uniaxial stress, σ_{11}, applied in the [100] direction. The dispersion curves of both the surface waves do not exhibit the symmetry about any angle. From Figure 3(a) it can be seen that the tension (with positive sigh) increases the velocities of the Rayleigh wave near the [100] direction and decreases those near the [010] axis. On the contrary, the compression (with negative sigh) decreases the Rayleigh wave velocities near the [100] direction and increases those near the [010] direction. Figure 3(b) shows that the velocities of the pseudo-Sezawa wave near the [110] direction are increased monotonically as the compression is applied, while the response to the tension is opposite to the compression with slightly complicated characteristic. The tension first monotonically decreases the velocities of the pseudo-Sezawa wave but, then, the dispersion curves may be overlapped as the tension exceeds +2 GPa.

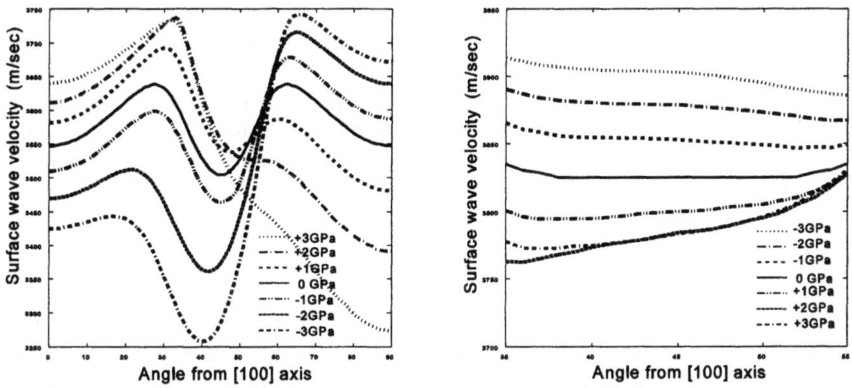

(a) for the Rayleigh wave (b) for the pseudo-Sezawa wave.

Figure 3. Directional dispersions of the surface waves for Cu/Si(001) under uniaxial stress, σ_{11}, applied in the [100] direction. The tested magnitudes of the stress, σ_{11}, are illustrated as parameter. The full lines are the dispersions of the surface waves with no stress.

Figures 4(a) and 4(b) respectively show the velocity responses for the Rayleigh wave and the pseudo-Sezawa wave, caused by the biaxial stresses, σ_{11} and σ_{22}, in the [100] and [010] directions with the equal value of σ_0. Each of the dispersion curves exhibits the mirror symmetry with respect to the [110] direction. In Figure 4(a) the tension monotonically decreases the Rayleigh wave velocities near the [100] and [010] axes. Above $\sigma_0 = +2$ GPa, the bell-shaped (unimodal-distribution) dispersion curves of the Rayleigh wave appear in the figure. While the trend of the velocity change due to the compression is opposite to that in the tension, i.e., the compression monotonically increases the Rayleigh wave velocities near the [100] and [010] axes. Figure 4(b) shows that the velocities of the pseudo-Sezawa wave are increased monotonically as the compression is applied. While the dependence on the tension is complicated. The velocities of the pseudo-Sezawa wave are decreased first monotonically with the increment of the tension. Then, if the tension exceeds +2 GPa, the velocity curves come back toward to that of the unstressed state and may be overlapped in the data space.

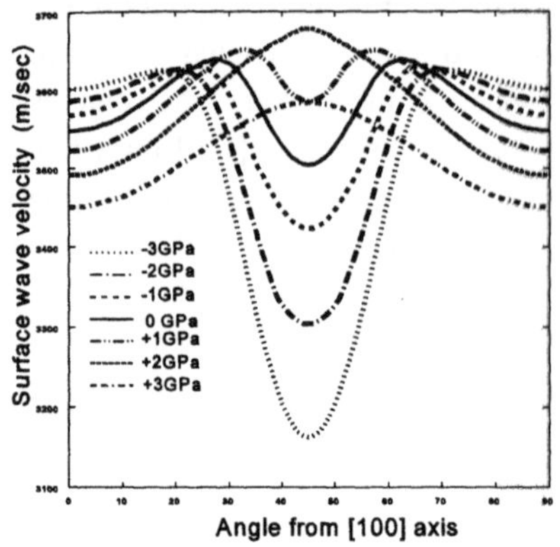

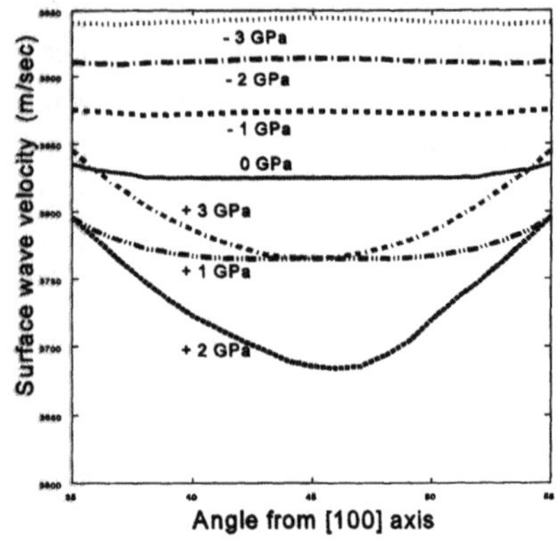

(a) for the Rayleigh wave **(b)** for the pseudo-Sezawa wave.

Figure 4. Directional dispersions of the surface waves for Cu/Si(001) under biaxial stress, $\sigma_{11} = \sigma_{22} = \sigma_0$, and $\sigma_{12} = 0$. The tested magnitudes of the stress, σ_0 , are illustrated as parameter. The full lines are the dispersions of the surface waves with no stress.

CONCLUDING REMARKS

In general, the acoustoelastic effect of surface waves is very small for the most crystalline materials. From Figure 3(a) it can be seen that the relative change of the Rayleigh wave velocity (change of the Rayleigh wave velocity per wave velocity in an undeformed state) in the Cu/Si(001) sample is approximately equal to 1.5×10^{-2}/GPa. Therefore, a precision of 1 part in 100 or higher is required for the measurement of wave velocities. A most important result obtained from the numerical studies is that the stress- and strain-induced velocity changes of surface waves exhibit the monotonic feature at the specified propagation directions. That is, the velocities of the Rayleigh wave near the [100] and [010] directions and the pseudo-Sezawa wave near the [110] direction vary always monotonically, if the deformations are not very large. This result is of great significance in stress measurement with acoustic microscopy. Owing to the monotonic characteristic in the velocity change the approach to examine the unknown stress or strain by inferring from the measured surface wave velocities will be simplified because the inversion solutions may be obtained uniquely and certainly.

REFERENCES

1. Y.H.Pao, Acoustoelasticity and ultrasonic measurement of stress, in: W.P.Mason, R.N.Thuston (Eds.), *Physical acoustics*. **17**:61, Academic Press, New York (1984).
2. B.A.Auld, *Acoustic field and waves in solids*. John Wiley & Sons Inc., New York (1973).
3. Z.Yu and S.Boseck, Scanning acoustic microscopy and its applications to material characterization, *Review of Modern Physics*. **67**:863 (1995).
4. G.W.Farnell and E.L.Adler, Elastic wave propagation in thin layers, in: W.P.Mason and R.N.Thuston (Eds.), *Physical acoustics*. **9**:35, Academic Press, New York (1972).

COMPARISON BETWEEN TIME REVERSAL AND INVERSE FILTER FOCUSING

Jean-François Aubry, Mickaël Tanter, Jean-Louis Thomas and Mathias Fink

Laboratoire Ondes et Acoustique, ESPCI, Université Paris VII,
U.R.A. C.N.R.S. 1503, 10 rue Vauquelin, 75005 Paris, France

INTRODUCTION

In most applications of ultrasound, medical therapy or diagnostic, non destructive testing, the first step consists in focusing an ultrasound beam deeply inside the medium to be investigated. The variations of speed of sound are usually unknown and the method has to be adaptive. Thus, to focus ultrasonic waves in an heterogeneous and absorbing medium using a phased array, one has to calculate the optimal set of signals (for monochromatic waves, amplitude and phase) to be applied on the transducers of the array. The solution obtained by Time Reversal Processing (TRP) corresponds to the spatial and temporal filter matched to an initial source available in the medium. However, this process doesn't ensure that the field amplitude at other locations is as small as possible. We have illustrated this problem previously using the example of focusing through the human skull, which is an absorbing medium. We found that by combining the TRP with an amplitude compensation, the focusing quality could be significantly improved [1]. This technique was aimed to recover an inverse spatial filtering from the matched filtering conducted by the TRP. Another approach consists in using an entire array of transducers on both sides of the skull. In that case, the whole set of Green's functions characterizing the propagation between the different source locations and the transducers array is acquired, giving now the whole information needed to characterize the medium. A real spatial inverse filtering technique can then be achieved.

In the first part of this paper, a matrix formalism of the propagation operator relating the array transducers to a set of control points located in the medium is introduced. Each Fourier component of this spatio-temporal filter is a complex matrix **H**. For a given Fourier component, the inverse filter process is then easily described by the matrix inverse diffraction operator $\mathbf{H}^{-1}$. The physical meaning of such an inverse diffraction operator is studied by using the singular value decomposition (S.V.D) of **H**. In the same way, time reversal processing is then described by introducing a time reversal operator $\mathbf{H}\tilde{\mathbf{H}}$, where $\tilde{\mathbf{H}}$ is the transpose conjugate of **H**. This time-reversal operator is also described by using the singular value decomposition of **H** and finally compared to $\mathbf{H}^{-1}$. A first experiment is conducted in homogeneous medium and put in evidence the close similarity of inverse filter processing and time reversal focusing in that particular case. In the second part of this paper, the matrix operator characterizing the propagation through the skull is experimentally acquired. The behaviour of the two operators $\mathbf{H}\mathbf{H}^{-1}$ and $\mathbf{H}\tilde{\mathbf{H}}$, respectively characterizing time reversal and inverse filter focusing processes is described and compared. In that experimental case, losses of information due to the bone absorption occur in the medium and involve a degradation of time reversal focusing. We measure and compare the focal quality obtained in the different cases: time reversal, inverse filter and time reversal combined with amplitude compensation.

I. THEORITICAL APPROACH : A MATRICIAL FORMALISM

I. 1. The Propagation Operator H

To compare inverse filter focusing and time reversal processing, we need to define the linear operator relating the elements of the transducer array to a set of control points. The signals applied to the transducer elements

Acoustical Imaging, Volume 25, Edited by Halliwell and Wells.
Kluwer Academic/Plenum Publishers, 2000.

are the input of the filter and the output consists of the signals measured on the control points. For sake of simplicity, we restrict ourselves to a one dimensional array of transducers and a spatial sampling of the field, the set of control points, along a line parallel to the array aperture, Fig. 1. We define an impulse response, $h_{mj}(t)$, for each couple (m,j) comprising a control point and a transducer element. This impulse response $h_{mj}(t)$ is the signal received on the mth control point after a temporal Dirac's function is applied on the jth transducer of the array. As the transformations are supposed to be linear and invariant under a time shift, the MxJ temporal functions $h_{mj}(t)$ for $1 \le m \le M$ and $1 \le j \le J$ can describe any transmit-receive operation for the same arrangement. In the Fourier domain, it gives the transfer matrix representing the propagation of each monochromatic component between the array and the set of control points.

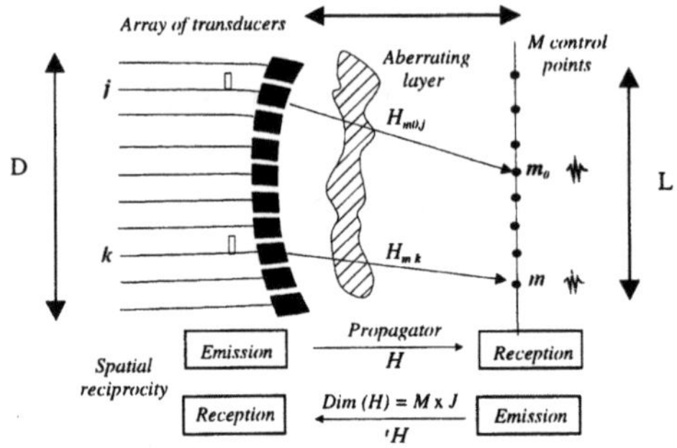

Figure 1. Source-transducer impulse response.

Indeed, let $e_j(t)$, $1 \le j \le J$, be the J input signals on the array. The output signals $f_m(t)$, $1 \le m \le M$, received in the control domain are given by :

$$f_m(t) = \sum_{j=1}^{J} h_{mj}(t) \underset{t}{*} e_j(t), \quad 1 \le m \le M \qquad (1)$$

where * is a temporal convolution operator. A temporal Fourier transform leads to the following relation :

$$F_m(\omega) = \sum_{j=1}^{J} \mathbf{H}_{mj}(\omega) E_j(\omega), \quad 1 \le m \le M \qquad (2)$$

Eq. 2 leads to the simplified matrix form $F(\omega) = \mathbf{H}(\omega)E(\omega)$ where $E(\omega) = (E_j(\omega))_{1 \le j \le J}$ is the column vector of the Fourier transform of the transmitted signals and $F(\omega) = (F_m(\omega))_{1 \le m \le M}$ is the column vector of the Fourier transform of the received signals. The transfer matrix $\mathbf{H}(\omega) = (H_{mj}(\omega))_{1 \le m \le M, 1 \le j \le J}$ describes the propagation in the medium from the array to the set of control points and thus, is called the ***propagation operator***.

I. 2. Inverse filter focusing: the inverse diffraction operator $\mathbf{H}^{-1}$

In this case, we try to deal with an inverse diffraction experiment : the aim is to determine the desired field distribution E on a boundary surface (the transducers array) that would give rise after propagation to the field distribution F on another surface (the image plane or control plane). As we want to focus in the image plane, F is chosen to correspond to the optimal focusing obtained in homogeneous medium. To solve this problem, we have to acquire the direct wavefield propagator $\mathbf{H}$ between the time-reversal mirror and the image plane. Thus, E and F are related by :

$$F = \mathbf{H} E \quad \Leftrightarrow \quad E = \mathbf{H}^{-1} F \qquad (3)$$

In order to regularize the problem, a Singular Value Decomposition is applied to $\mathbf{H}$ before inversion :

$$\mathbf{H} = \mathbf{U}\mathbf{D}\tilde{\mathbf{V}} \qquad (4)$$

where dim(**H**)= M x J, **U** is an M x J column-orthogonal matrix, **V** is an J x J orthogonal matrix, **D** is an J x J diagonal matrix ($\mathbf{D}=[\lambda_i\delta_{i,j}]_{i,j}$) and $\delta_{i,j}$ is the Krönecker symbol. The inverse of **H** is directly related to the inverse of **D** with :

$$\mathbf{H}^{-1} = \mathbf{V}\,\mathbf{D}^{-1}\,\tilde{\mathbf{U}} = \mathbf{V}\left[\frac{1}{\lambda_i}\delta_{i,j}\right]_{i,j}\tilde{\mathbf{U}} \qquad (5)$$

As **H** is close to singular, some eigenvalues λ_i become very small and so, sensitive to noise. During inversion, $1/\lambda_i$ tends to infinity and produces very large errors. Indeed, only the main eigenvalues of the decomposition have to be used during inversion. In order to regularize the problem, the matrix inversion is only applied to the main eigenvectors of the singular value decomposition of **H**, giving rise to a noise-filtered approximation $\hat{\mathbf{H}}^{-1}$ of the inverse matrix $\mathbf{H}^{-1}$:

$$\hat{\mathbf{H}}^{-1} = \mathbf{V}\,\hat{\mathbf{D}}^{-1}\,\tilde{\mathbf{U}} = \mathbf{V}\begin{bmatrix} \frac{1}{\lambda_1} & 0 & \cdots & & 0 \\ 0 & \ddots & & & \\ & & \frac{1}{\lambda_N} & \ddots & \vdots \\ \vdots & & \ddots & 0 & \\ & & & \ddots & 0 \\ 0 & & \cdots & 0 & 0 \end{bmatrix}\tilde{\mathbf{U}} \qquad (6)$$

where N is the number of eigenvalues used in the inversion. In a previous paper [2], we explained the number N of eigenvalues corresponding to the rank of **H**, but also the physical meaning of the associated eigenvectors [3]. In fact, the number N of spatial degrees of freedom necessary to define the field received in the control plane is given by :

$$N = \frac{L}{\delta_x} = 2\frac{L}{\lambda}\,\mathbf{sin}(\,\mathbf{tan}^{-1}\frac{D}{2F}\,) \qquad (7)$$

where D is the emitting array aperture, F is the focal distance and L is the lateral exploration width in the focal plane, see Fig. 1. This definition of N (valid when L<D) remains valid in the near field of the array. In the Fraunhofer assumption, when D/2F >>1, it gives the simple relation :

$$N \approx \frac{DL}{\lambda F} \qquad (8)$$

In that final case, *N* can be seen as the number of independant focal spots (lateral resolution $\lambda F/D$) that the array is able to generate in the aperture *L* of the image plane. In order to illustrate these results, the propagation matrix **H** has been acquired experimentally. Each term of the matrix is acquired by moving a single transducer on each location of the control points. This single transducer works as a receiver and records successively the signal emitted by each transducer of the array. We only keep the amplitude and phase of the received signal at the central frequency of the transducers. The *N* main eigenvectors of the singular value decomposition of **H** define the whole information that can be conveyed by the field to the control points and so, are kept for the inversion of the matrix. The others are only sensitive to noise and so, will not be taken into account. The inverse propagation matrix $\hat{\mathbf{H}}^{-1}$ is then deduced from Eq (6). Therefore, according to Eq (3), it allows to deduce the field *E* to be emitted on the array from the field *F* arising on the control points after propagation in the medium. As we try to deal with a focusing problem, we choose as an objective the ideal focusing pattern F_{m0} on the control point m_0. This ideal focusing pattern is defined by :

$$F = F_{m0} = {}^{t}\{\mathbf{0} \;\; \ldots \;\; \mathbf{0} \;\; \mathbf{1} \;\; \mathbf{0} \;\; \ldots \;\; \mathbf{0}\} \qquad (9)$$

According to Eq (3) and (6), the emission vector E_{inv} is directly deduced from F_{m0} :

$$E_{inv} = \hat{\mathbf{H}}^{-1} F_{m0} \qquad (10)$$

This vector calculated by the inverse filter technique can now be reemitted by the array . It will give rise to the focusing pattern F_{inv} in the control plane defined by :

$$F_{inv} = \mathbf{H}E_{inv} = \mathbf{H}\hat{\mathbf{H}}^{-1} F_{m0} \qquad (11)$$

This operator $\mathbf{H}\hat{\mathbf{H}}^{-1}$ will be referred to as ***the inverse focusing operator*** in the next part of this paper. Due to the limited array aperture, $\mathbf{H}\hat{\mathbf{H}}^{-1}$ is not equal to the Identity matrix $\mathbf{I}_{\mathbf{MxM}}$, Fig. 3a. Thus, F_{inv} is not exactly equal to F_{m0} :

$$F_{inv} = \mathbf{H}\hat{\mathbf{H}}^{-1} F_{m0} = \mathbf{UD}\tilde{\mathbf{V}}\mathbf{V}\hat{\mathbf{D}}^{-1}\tilde{\mathbf{U}} F_{m0} = \mathbf{UD}\hat{\mathbf{D}}^{-1}\tilde{\mathbf{U}} F_{m0} = \mathbf{U}\begin{bmatrix} \mathbf{I}_{N\times N} & 0 & \cdots & 0 \\ 0 & 0 & & \vdots \\ \vdots & & \ddots & \\ 0 & \cdots & & 0 \end{bmatrix}\tilde{\mathbf{U}} F_{m0} \qquad (12)$$

Eq (12) clearly shows that F_{inv} only corresponds to the projection of the objective vector F_{m0} on the N main eigenvectors of the singular decomposition of **H.** Figure 2 presents the initial objective F_{m0} of this inverse filter process, its projection F_{inv} on the main eigenvectors of the singular value decomposition of the experimentally acquired matrix **H**, and finally the experimental focusing obtained by this inverse filter process. Note that the inverse filter is not able to reconstruct F_{m0} because of the limited rank of **H** and that we obtain the classical point spread sinc function of the linear array. Moreover, one can notice on figure 2 that the experimental focusing obtained by emitting E_{inv}, solution given by the inverse filter process, is almost equal to the projection of F_{m0} on the N main eigenvectors of the singular value decomposition of **H**.

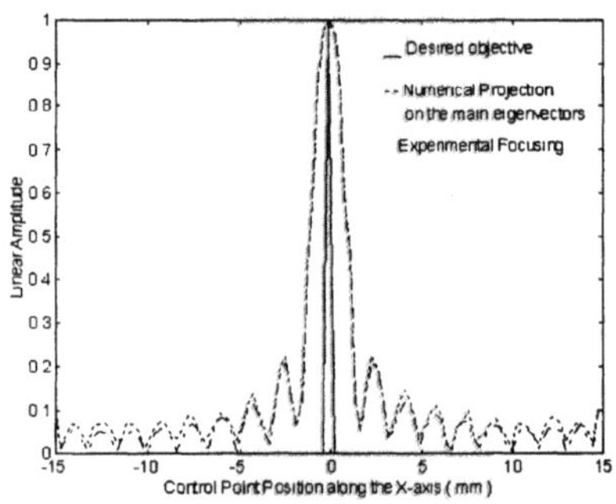

Figure 2. Spatial distribution in the control plane, on the one hand, of the objective F_{m0}, and on the other hand, of its numerical projection F_{inv} on the 18 main eigenvectors of the SVD of **H**. Finally, these vectors are compared to the experimental focusing achieved by inverse filter processing.

I. 3. Time reversal focusing: the time reversal operator $\mathbf{H}.{}^t\mathbf{H}^*$

Contrary to the inverse filter technique, a time reversal processing does not require knowledge of the whole propagation operator **H**. In the case of time-reversal focusing, one uses as inputs the time-reversed version of the set of signals measured by the array elements when one of the control points behaves as an acoustic source. According to the spatial reciprocity relation, $h_{mj}(t)$ is also the signal measured by the jth transducer element (working as a receiver) when a temporal delta function is applied on the mth control point. Thus spatial reciprocity implies that the transpose of **H**, ${}^t\mathbf{H}$ corresponds to the propagation matrix between the control points and the array elements, Fig. 1. In the first step of a time reversal experiment, one of the control points behaves like a source. Let be this source located at point m_0 in the image plane, defined by the column vector $F_{m0} = {}^t\{0, ...,0,1,0,...,0\}$. The wavefield E_0 received by the array when F_{m0} is transmitted writes :

$$E_0 = {}^t\mathbf{H}F_{m0} \qquad (13)$$

where ${}^t\mathbf{H}(\omega) = (h_{jm}(\omega))_{1\le j\le J, 1\le m\le M}$ is the transfer matrix relating the control points to the transducer elements. Thus, E_0 can be seen as $E_0 = \{\mathbf{H}\}_{m0} = {}^t\mathbf{H}F_{m0}$, where $\{\mathbf{H}\}_{m0}$ is the m_0 th row of **H**. A time reversal operation corresponds in the Fourier domain to a phase conjugaison. We apply the time reversal operation to the vector E_0 received on the array and we deduce the new vector to be emitted by the array :

$$E_0^* = \left({}^t\mathbf{H}F_{m0}\right)^* = \{\mathbf{H}\}^*_{m0} \quad (14)$$

where $\tilde{\mathbf{H}}$ be the adjoint operator of $\mathbf{H}$. By definition, $\tilde{\mathbf{H}} = {}^t\mathbf{H}^*$ and $*$ is the conjugation operation. The focusing obtained in the image plane is then given by :

$$F = \mathbf{H}\left({}^t\mathbf{H}F_{m0}\right)^* = \mathbf{H\tilde{H}}F_{m0} \quad (15)$$

where $\mathbf{H\tilde{H}}$ is called the time-reversal operator [4] and can be compared to the inverse focusing operator defined in Eq (11) for the inverse filter technique. In spite of time reversal invariance of the wave equation in lossless media, the time reversal operator is not the Identity matrix $\mathbf{I}_{\mathbf{M\times M}}$. In fact, each column m_0 of this matrix corresponds to the directivity pattern obtained in the image plane, by applying the time reversal process to an initial source located at m_0 in this image plane. As the TRM has a finite aperture, this directivity pattern is a point spread function and its beamwidth is linked to the ratio aperture/focal length of the mirror. The high spatial angular frequencies of the objective are lost. Thus, $\mathbf{H\tilde{H}}$ in homogeneous lossless medium, as well as the inverse filter operator $\mathbf{H\hat{H}}^{-1}$ in the first subpart, only tends towards the identity matrix, see Fig. 3b : The diagonal of this matrix remains spread because of the limited angular aperture of the mirror. Moreover, according to Eq (15) and (4), the time reversal focusing F_{TR} can be rewritten :

$$F_{\mathbf{TR}} = \mathbf{H\tilde{H}}\,F_{m0} = \mathbf{UD\tilde{V}V\tilde{D}\tilde{U}}\,F_{m0} = \mathbf{UD\tilde{D}\tilde{U}}F_{m0} = \mathbf{U}\left\lfloor \lambda_i^{\,2}\delta_{i\,j} \right\rfloor \tilde{\mathbf{U}}F_{m0} \quad (16)$$

As well as for the inverse filter operator, the focusing F_{TR} obtained by the TRP can be seen as the projection of the desired objective F_{m0} , but now on the whole set of eigenvectors of the SVD of $\mathbf{H}$ weighted by the root of their corresponding eigenvalue. Note that according to their predominant weight, only the N main eigenvectors have a non negligible contribution to the construction of F_{TR}.

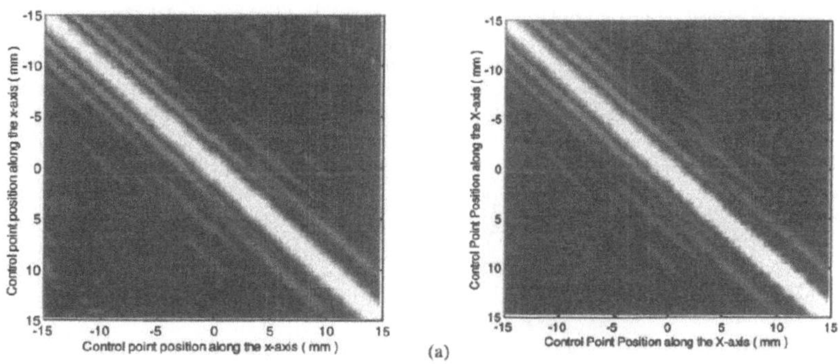

(b)

Figure 3. Amplitude representation of the matrix (a) $\mathbf{HH}^{-1}$ and (b) $\mathbf{H\tilde{H}}$ deduced from $\mathbf{H}$ experimentally acquired in water.

Thus, F_{TR} is only nearly equal to F_{inv} when the N main eigenvalues of the SVD of $\mathbf{H}$ are equal. A comparison between these two focusing patterns can be made in the case of the propagation through an homogeneous medium. In that case, as seen in **I.b**, the transducers directivity introduces a quite slow decreasing in the main eigenvalue distribution of $\mathbf{H}$ and F_{TR} is no more equivalent to F_{inv}. Physically, during a time reversal experiment, this transducers directivity introduces on the array aperture an apodization of the wavefield coming from the initial source. The time reversed wavefield reemitted by the time reversal mirror suffers this amplitude modulation a second time and the time reversal operation no longer achieves an inverse filter.

II. EXPERIMENTAL APPROACH : FOCUSING THROUGH THE SKULL

In the first theoretical part, we have introduced two different operators $\mathbf{H\hat{H}}^{-1}$ and $\mathbf{H\tilde{H}}$ allowing to quantify the optimal focusing quality that can be reached with two different techniques, respectively the inverse filter technique and the time reversal processing. We present now an experiment in order to compare these two processes. Note however that the inverse filter technique requires the acquisition of the whole matrix **H**, as the TRP only requires the knowledge of one single row of this matrix.

II-1 Experimental acquisition of the propagation operator through the skull

In that experiment, a skull bone is located between the image plane and the TRM. Sound absorption in bone (varying approximately between 0.3 and 2 $dB.mm^{-1}.Mhz^{-1}$) breaks the time-reversal invariance of the wave equation and so degrades the time-reversal focusing. Figure 4 presents the set-up of this first experiment. The 1D array is made of 128 transducers located on a cylindrical surface and working at a central frequency of 1.5 MHz. The radius of curvature of the cylindrical surface is equal to 100 mm. The transducer width and height are respectively equal to 1 mm and 30 mm. The array pitch is equal to 1mm. The skull is located closed to the array. The control points are disseminated in the medium as stated on page 4. In our case, we have chosen to place them in a plane called control plane or image plane. Indeed, we wanted to plot focusing patterns in this plane. The control plane is made of 100 controls points located along a line orthogonal to the axial direction and containing the array geometrical focus. The control points pitch is equal to 0.3 mm and thus, the control points are located between x = -15 mm and x = +15 mm in the control plane.

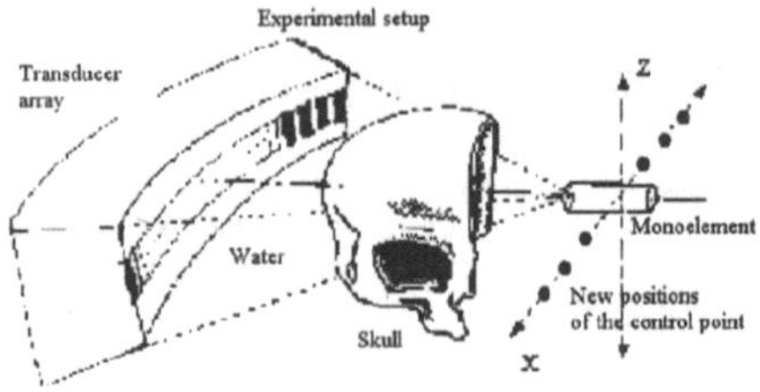

Figure 4. First experimental set-up : A skull is located between a 1D cylindrical array and the image plane.

The matrix propagator is experimentally acquired both in water and through the skull following the procedure explained in the first part. Note that R.Seip *et al* proposed to achieve dynamic focusing in ultrasound hyperthermia treatments by acquiring this propagation operator **H** with implantable hydrophone arrays [5].

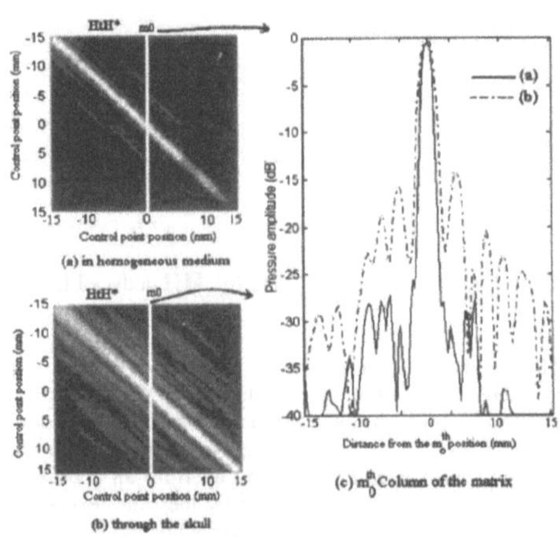

Figure 5. Matrix $\mathbf{H}^t\mathbf{H}^*$ (a) in homogeneous medium and (b) through the skull, and the corresponding directivity pattern obtained by time-reversal processing at a position m_0 in the image plane in both cases.

The amplitude of the whole time-reversal operator $\mathbf{H}\tilde{\mathbf{H}}$ is presented in figure 5 when (a) the medium is homogeneous and (b) when the skull is located between the array and the control points. As explained previously, the m_0^{th} column of the time-reversal operator corresponds to the focusing obtained by time reversal processing when a single initial acoustic source is available at the position m_0 in the image plane. Thus, one can notice on figure 5 (c) that the directivity pattern obtained by time reversal processing through the skull is degraded in comparison with the same experiment in water.

As the TRP achieves a spatial matched filter in spite of the absorption, there is no spatial shift of the focus. The amplitude of the field is always maximized on the initial source location. However, the TRP imposes no constraint on the other points in the image plane : Especially, the focal beam is spread in comparison with the one obtained in homogeneous medium and sidelobes appear. An improvement of the time-reversal process has been proposed in a previous paper to recover an optimal focusing through the skull. It consists in a numerical amplitude compensation of the wavefront received on the mirror [1]. This amplitude compensation allows to take into account the effects of skull absorption : The amplitude modulation of the wavefront is inverted before applying the time reversal operation. Thus, this new synthesized field corresponds to the field that would have propagated through a medium with gain. This virtual medium is exactly the same as the real one, except for the acoustic losses that are replaced by gains. Hence, it tends to re-create the inverse of the real medium. In the next subpart, we compare this solution to the optimal one obtained by the inverse filter process when the whole propagation operator **H** has been acquired.

II-2. Inverse filter process applied to focusing through the skull

We try to compare the solution vector E to be emitted on the array in order to focus at m_0 with the two different methods : Time reversal processing combined with amplitude compensation and the inverse filter solution $E = \mathbf{H}^{-1}F$. The numerical inversion is realized on a subspace defined by the eigenvectors associated to the most significant eigenvalues. The eigenvalues distribution of the experimental matrix H acquired through the skull is presented in figure 6.

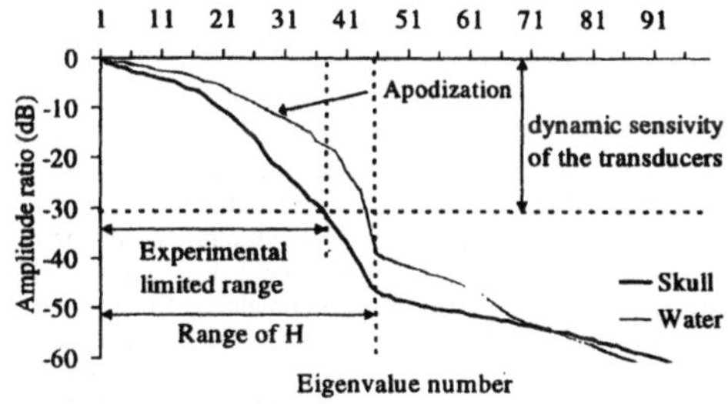

Figure 6. Eigenvalue ω_i distribution of **H** in homogeneous medium (gray line) and through the skull (black line).

The inverse filtering solution $E = \mathbf{H}^{-1}F$ obtained with a pertinent number of 38 eigenvalues is compared with the phase and amplitude (Fig. 7) of the solution obtained by time reversal and amplitude compensation. The inverse solution converges towards the time-reversal solution improved by amplitude compensation. The standard deviation between the phase variations of the two solutions is very small ($\approx \pi/13$). Moreover, figure 7b can be seen as a proof that the algorithm of amplitude compensation applied to the TRP gives good results, as there is a very good agreement with the amplitude modulation of the inverse filtering solution.

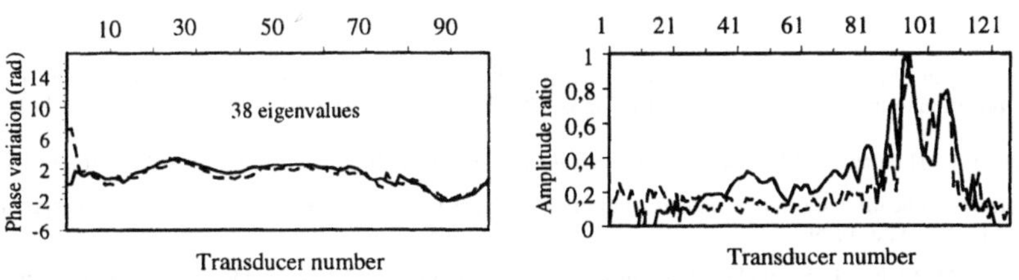

Figure 7. (a) Phase and (b) amplitude variation applied on the TRM by time reversal and amplitude compensation (black line) compared with the inverse filtering solution obtained by SVD of the matrix propagator **H**, (gray line).

These different solution vectors are then experimentally reemitted by the TRM, and the resulting experimental directivity patterns are compared with the optimal focusing obtained in homogeneous medium, Fig. 8. The focusing obtained by improved time-reversal processing and inverse filtering are optimal up to -20 dB. Under this limit, sidelobes are more important for the improved time-reversal focusing than for the focusing obtained by inverse filtering. This difference can be explained by the restricting assumption of very thin absorbing layer made in the case of time-reversal and amplitude compensation. In the case of the inverse filtering solution, no particular assumption has been made as the whole matrix characterizing the medium is known.

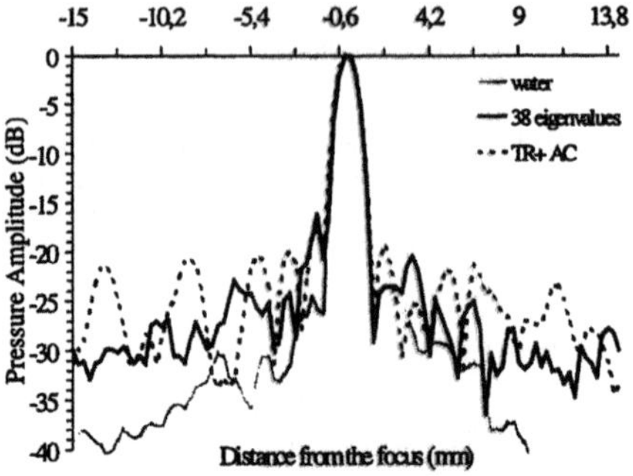

Figure 8. Experimental directivity patterns obtained in the focal plane. Optimal focusing in water (gray solid line), time-reversal focusing through the skull improved by amplitude compensation (dotted line), and focusing obtained by the inverse filtering process (black solid line).

CONCLUSION

The TRP is based on the time-reversal invariance of the wave equation in a lossless medium. In theory, it achieves an inverse wavefield propagator limited only by the evanescent waves. But in many practical situations, the wavefield radiated by the source can not be completely recorded and this loss of information generates an irreversibility. In particular, this loss in the information content conveyed by the field arises in an absorbing medium. But another fundamental symmetry of physics, the spatial reciprocity relation, is usually valid. Combined with the linearity and invariance under a time shift of the wavefield propagator, it affords to the TRP its strong robustness. Thus, in these conditions, the TRP is a spatio-temporal matched filter.

However a matched filter is not always enough. Such a filter maximizes the amplitude of the field at a given time and location but does not constrain the side lobe level. In particular to focus through the skull, it has been shown that better results could be obtained if a correction of the amplitude is added to the TRP. This amplitude correction tends to recover a spatial inverse filtering from the time-reversal solution by compensating the spatial frequencies of the field that suffered the most important absorption. This amplitude compensation may be improved since an inversion of the transfer matrix gives even better results. Each configuration (aberrations of the propagation medium, location, size and number of the tranducers and control points) can be described for each monochromatic component by a propagation matrix. The optimal focusing ability in that configuration is then defined as the projection of the desired objective on the main eigenvectors of the singular value decomposition of this matrix.

REFERENCES

1. J.-L. Thomas and M. Fink.« Ultrasonic Beam Focusing Through Tissue Inhomogeneities with a time reversal mirror : Application to transskull therapy », IEEE Trans. Ultr. Fer. Freq. Ctrl, vol 43, Nov. 1996, pp 1122- 1129.

2. M. Tanter, J.-L. Thomas and M. Fink. « Time reversal and the inverse filter », Accepted for publication in JASA.

3. M. Tanter "Application du retournement temporel à l'hyperthermie ultrasonore du cerveau". thesis of the university Paris VII, May 1999.

4. C Prada and M. Fink, « Eigenmodes of the time reversal operator : a solution to selective focusing in multiple target media », Wave Motion, Vol 20, p 151-163, 1994.

5. R. Seip, P. VanBaren and E. Ebbini, "Dynamic focusing in ultrasound hyperthermia using implantable hydrophone arrays", IEEE Trans. Ultrason. Ferro-elec. Freq. Control **41**, 5, 706-713 (1994).

INCREASING RESOLUTION CAPABILITY OF TWO-DIMENSIONAL TOMOGRAPH OVER THIRD COORDINATE: SEPARATING RECONSTRUCTION OF $c(\mathbf{r})-$, $\rho(\mathbf{r})-$, $\alpha(\mathbf{r},\omega)-$ SCATTERER CHARACTERISTICS

V.A.Burov, A.L.Konjushkin, and O.D.Rumyantseva

Moscow State University, Faculty of Physics, Dept. of Acoustics
Moscow, 119899, Russia

1. QUASI-THREE DIMENSIONAL TOMOGRAPHIC SCHEMES WITH INCLINED TRANSDUCERS

Systems of two-dimensional acoustical tomography in frequency range from about 1 MHz to some MHz presume that probing waves have a plane front over the third coordinate (i.e. in a direction being perpendicular to a tomography plane) along a desired length of a parallel beam zone. The usual object length of 10÷20 cm, wave length in soft biological tissues of 0.5÷1 mm and the mentioned frequency range define a thickness of the parallel beam zone of about 1÷2 cm. Then a resolution capability over the third coordinate is more than on one order inferior to a resolution capability in the tomography plane [1,2]. A scheme of complete three-dimensional tomography [3] provides the high resolution capability of an reconstructed imaging over all three dimensions. However, its realization demands an essential complication both of the tomography system and of a reconstruction algorithm, in comparison with the two-dimensional tomography. Because of these difficulties a variant by compromise to improve the two-dimensional tomograph is proposed below. It does not demand the essential complication of transmitter-receiver antenna array and the reconstruction algorithm, but it allows increasing the resolution capability over the third coordinate.

Theoretical researches have been carried out on base of an array scheme for acoustical ultrasonic tomograph being now developed in Acoustic Department of Moscow State University [4]. Transducers of the circular antenna array have sensitive surfaces of a convex cylindrical shape. Each of the transducers can work both as a transmitter and as a receiver. Arranging the transducers on the antenna circle differs essentially from arranging proposed in [1,2]. In our case the antenna array is sparsed in a special manner, and the transducers are arranged irregularly (non equidistantly). The number of the transducers is only 26. However, the antenna array can be rotated by a sampled angle $\Delta\varphi = 2\pi/256$. Thanks to the both facts - the sparsed antenna and its rotation, an obtained volume of the experimental scattering data is equivalent to a volume which could be measured in case of immovable antenna array having 256 transmitter-receiver transducers. Moreover, pulse-mode is used. The sparsed array permits to have transducers with large radiating and sensitive curved surface and, consequently, low impedance.

For the reconstruction of a scatterer (an investigated organ) a *two-step algorithm* is used. The two steps are needed to take into account the multiple scattering effects inside the scattering domain, if the scatterer is of a sufficiently strong power. At the first step, a distribution of large-

Acoustical Imaging, Volume 25, Edited by Halliwell and Wells.
Kluwer Academic/Plenum Publishers, 2000.

scale scattering inhomogeneities is reconstructed by means of a time-of-flight algorithm. These inhomogeneities are areas with a significant deviation of a sound velocity $c(\mathbf{r})$, a density $\rho(\mathbf{r})$ and an amplitude absorption coefficient $\alpha(\mathbf{r},\omega)$ from their mean values c_0, ρ_0, and $\alpha_0(\omega)$ having been previously measured in a homogeneous background medium without the scatterer. At the second step, a fine-scale structure of the scatterer $v(\mathbf{r},\omega)$ is reconstructed. The scatterer function is:

$$v(\mathbf{r},\omega)=\frac{\omega^2}{c_0^2}W(\mathbf{r},\omega);\ W(\mathbf{r},\omega)=g_c(\mathbf{r})+\left(\frac{\omega_2}{\omega}\right)^2\left(\frac{g_\rho(\mathbf{r})}{\omega_2^2}\right)+i\left(\frac{\omega}{\omega_2}\right)^{\nu-1}\left(\omega_2^{\nu-1}\ g_{\alpha(\nu)}(\mathbf{r})\right). \quad (1)$$

Here $g_c(\mathbf{r})=\left(1-c_0^2/c^2(\mathbf{r})\right)$;

$$g_\rho(\mathbf{r})=c_0^2\sqrt{\rho(\mathbf{r})}\nabla^2\rho^{-1/2}(\mathbf{r});\ \ g_{\alpha(\nu)}(\mathbf{r})=-2\frac{\alpha(\mathbf{r},\omega_2)-\alpha_0(\omega_2)}{\omega_2^\nu}[c_0^2/c(\mathbf{r})]. \quad (2)$$

The frequency ω_2 is mean in a frequency working range. An index ν of a frequency dependence degree of the absorption coefficient is proposed:

$$\alpha(\mathbf{r},\omega)=(\omega/\omega_2)^\nu\alpha(\mathbf{r},\omega_2). \quad (3)$$

Three items in (1) correspond to velocity c-, density ρ- and absorption $\alpha(\omega)$- scatterer components, respectively. The second step reconstruction is carried out in the Born (a single scattering) approximation, but against a new inhomogeneous background, what is the large-scale distribution having been already known. At the second step an expected resolution capability is no worse than 0.5 mm in the tomography plane.

In the initial variant of the two-dimensional tomography scheme, transmitter-receiver transducers of the antenna array are set in such a manner that constituents of their cylindrical sensitive surfaces are perpendicular to the tomography plane (x, y). *In the improved variant* the sensitive surfaces of the different transducers are inclined at different angles to the tomography plane. The slope angles allow to obtain an information (which is absent in the initial scheme) about scatterer space spectrum $\tilde{v}(\vec{\xi})$ along the third coordinate, at the least in some low-frequency domain. The space frequency domain range is proportional to the range of the slope angles. Then for obtaining a scatterer estimation $\hat{v}(\mathbf{r},\omega)$ by processing the scattering data at the second step, an additional phase correction is induced into the initial two-dimensional algorithm [5]:

$$\hat{v}(\mathbf{r},\omega)=\frac{1}{2}\frac{\omega^2}{c_0^2}\int_0^{2\pi}d\varphi(x,y)\int_0^{2\pi}d\varphi'(x,y)\ f(\varphi,\varphi',\omega)\left|\sin(\varphi-\varphi')\right|\times$$
$$\times\exp\left\{i\omega\left[x(\cos\varphi'-\cos\varphi)/c_0+y(\sin\varphi'-\sin\varphi)/c_0+\Delta t(\mathbf{r})\right]\right\},\ \ \mathbf{r}=\{x,y,z\}. \quad (4)$$

Here $\varphi=\varphi(x,y)$ and $\varphi'=\varphi'(x,y)$ are polar angles of vector components $(\mathbf{k_0})_{x,y}$ and $(\mathbf{l_0})_{x,y}$ in the plane (x,y) for falling and scattered waves. These waves are considered as being local-plane waves in a neighborhood of a fixed point $\mathbf{r}'=\{x,y,0\}$, i.e. of projection of the point $\mathbf{r}$ (though in reality the waves are transmitted and received by the cylindrical transducers). Values φ and φ' depend on coordinates x,y and on a geometrical position of a current pair "transmitting transducer S - receiving one R". Scattering amplitude $f(\varphi,\varphi,\omega)$ is formed by the scattered wave analogously to the two-dimensional case. A time correction is:

$$\Delta t(\mathbf{r})=\Delta t_S(\mathbf{r})+\Delta t_R(\mathbf{r})\ ,\ \ \mathbf{r}=\{x,y,z\}. \quad (5)$$

An item Δt_S (or Δt_R) is a difference between time invervals of a signal propagation from the points $\mathbf{r}$ and $\mathbf{r}'$ to the transmitter S (or to the receiver R); this difference is caused by the slope of the transmitter axis. For the sake of a simplicity, the relation (4) is given for the homogeneous background without an absorption. However, for the real tomograph the mentioned inhomogeneous large-scale background is taken into consideration.

Some schemes corresponding to different variants of the transducers slopes have been analyzed. To estimate the scheme resolution capability over the third z-coordinate, a point spread (PS) function of the given scheme is reconstructed. For this purpose a very small (point-like) model three-dimensional scatterer $v(\mathbf{r},\omega)\rightarrow\delta(\mathbf{r}-\mathbf{r}_0)$ is disposed to a fixed point $\mathbf{r}_0=\{x,y,0\}$ and its imaging, i.e. the PS function $\hat{v}(\mathbf{r},\omega)\rightarrow\hat{v}_\delta(\mathbf{r},\omega|\mathbf{r}_0)$, is built by the equation (4). In the simplest case $\mathbf{r}_0=0$, the time correction (5) for points $\mathbf{r}$ along the vertical z-axis is: $\Delta t(x=0,y=0,z)\approx -z(\sin\alpha_S+\sin\alpha_R)/c_0$, where α_S, α_R are slope angles of the transmitter and the receiver.

1.1. Unified Transmitter-Receiver Transducers

As mentioned above, each of the 26 transducers can work both as the transmitter and the receiver. The axis slopes of the transducers can be organized by different variants. In each of the variants sampling slopes values may be distributed over the transducers in an accidental manner or by a certain law. However, numerical modeling showed that chosen distribution did not practically influence on a view and characteristics of the PS function. In case of *uni-oriented* slopes, the vertical axes of all transducers are oriented toward the antenna array's center. For example, at sampling step 0.5° a slope range is α_i=0.5°÷13°; $i=\overline{1,\ 26}$ is transducer's number. Then the PS function $\hat{v}_\delta(x=0,y=0,z\ |\mathbf{r}_0=0)$ along z-direction for $\mathbf{r}_0=0$ is a <u>complex</u> function, in principle (fig.1*a*; the PS function is normalized by the maximal value of its real part). Indeed, the scattering data contain an information about the scatterer space spectrum constituents $\tilde{v}(\vec{\xi})$ for z-space components:

$$\xi_Z=k_0(\sin\alpha_R+\sin\alpha_S). \qquad (6)$$

Therefore, in case of the uni-oriented slopes the information (6) for $\xi_Z\geq 0$ (or $\xi_Z\leq 0$) is only available. It results in appearance of an undesirable image part of the PS function and in "intermixing" the velocity and absorption scatterer components [3]. When a purpose of medical investigations does not demand a separation of these components, the scheme with the uni-oriented slopes is convenient. High resolution of *general* picture of the scatterer components may be provided: for fig.1*a* the 0.7-level width of the PS function magnitude $|\hat{v}_\delta|$ along z-direction is $\approx 2.5\lambda_0$. However, if the scatterer has an absorption and a separate reconstruction of the scatterer components is necessary, then the following scheme would be more convenient.

In case of *bi-oriented slopes*, the transducers axes are oriented both toward the antenna array's center and away from it. For example, α_i=-6.5°÷6.5° at the sampling angle step 0.5°; value $\alpha_i=0$ is eliminated because of the <u>even</u> number of the transducers ($i=\overline{1,\ 26}$). Then the space spectrum distribution (6) is sufficiently symmetrical in areas $\xi_Z<0$ and $\xi_Z>0$. As a consequence, the PS function becomes near-real value (fig.1*b*). A small $\mathrm{Im}\,\hat{v}_\delta$ is caused by a spectrum symmetry error because of the irregular arranging of the transducers and their small number (only 26, in reality). The 0.7-level width of the function $\mathrm{Re}\,\hat{v}_\delta$ is $\approx 2.5\lambda_0$. Increasing this resolution capability is possible by increasing the number of the transducers for enlarging the slopes range. However, some technical restrictions may appear. First, if the slopes are too large, then a common beam zone in the vertical direction of a part of the transducers becomes weak. Secondly, an organ being under tomography may also cause a restriction. For example, a mamma tomography near the chest allows the transducers slopes only toward the mamma center, because otherwise, a tissue nearly the chest is not insonified. Thus, in this case only the scheme of the uni-oriented slopes can be realized in practice. At the same time, the tomography of arms and legs allows both variants – with the uni-oriented and bi-oriented slopes.

1.2. Separated Transmitter and Receiver Transducers

All transducers may be separated in two independent groups: one group of 26 (for example) transmitter transducers and other group of 26 receiver ones. In particular, schemes are possible, in which the slope of all transmitters is absent (α_S=0), but receivers are inclined. Such the schemes have an important advantage: all the transmitters have a unified (in the vertical direction) projector beam zone. This zone must belong to the projector beam zones of all receivers, that is obtained due to receivers height being more than that of transmitters.

In case of *bi-oriented receivers slopes*, the slope angles are, for example, $\alpha_R = -6.5° \div 6.5°$ at the angle step 0.5° (value $\alpha_R=0$ is eliminated); $\alpha_S=0$. The PS function is near-real value with the 0.7-level width $\approx 3.3\lambda_0$ (fig.2*a*). An oscillating character of the PS function is a result of a near-rectangular shape of the space spectrum distribution (6).

In case of *uni-oriented receiver slopes*, the slope angles are, for example, $\alpha_R = 0° \div 12.5°$ at the angle step 0.5°; $\alpha_S=0$. The PS function is complex (fig.2*b*); the 0.7-level width of its magnitude $|\hat{v}_\delta|$ is $\approx 3.3\lambda_0$.

2. SEPARATING RECONSTRUCTION OF SCATTERER COMPONENTS UNDER UNKNOWN FREQUENCY DEPENDENCE OF ABSORPTION COEFFICIENT

Thus, the most appropriate scheme can be chosen according requirements on the imaging quality, which depends on the purposes of medical investigations. The scheme should provide, as much as possible, the greatest amount of the reconstructed constituents in the scatterer space spectrum over the third dimension. It results in the possibility of the separate reconstruction of tissue diagnostic characteristics, i.e. the scatterer components in (1), with the high accuracy. However, if the PS function of the used tomography scheme is complex due to technical or anatomical reasons, then the separation of the c-, ρ-, α(ω)-components demands, in general case, the use of the algorithm described in [3] and based on a different frequency dependence of the scatterer components. Moreover, the separation of the components under conditions of the complex PS function presumes that the index ν of the frequency dependence degree of the absorption coefficient (3) is known. However, in practice, this index is different not only for biological tissues of different organs, but for tissues of the same organ and for the same domain of the given organ of different patients as well [6,7,8]. By these reasons, an influence of a deviation of the true value ν from a priori proposed value ν_0 on an accuracy of the separate reconstruction of the c-, ρ-, α(ω)-components is below researched.

The scatterer components are separated on the base of estimations $\hat{W}(\mathbf{r},\omega_j)$ (reconstructed by three-dimensional algorithm [3] or by the quasi-three-dimensional one (4)) of the functions $W(\mathbf{r},\omega_j)$ at three frequencies ω_j, j=1,2,3, where $\omega_1 < \omega_2 < \omega_3$. [3] If $g_\delta^0(\mathbf{r},\omega_{low}) = \mathrm{Re}\,g_\delta^0 + i\ \mathrm{Im}\,g_\delta^0$ is a reference function of the given tomography scheme ($\omega_{low} \le \omega_1$), then $g_{c,\rho,\alpha(\nu)} \otimes g_\delta^0 = S_{c,\rho,\alpha(\nu)} + i\ N_{c,\rho,\alpha(\nu)}$, where ⊗ means a convolution; $S_{c,\rho,\alpha(\nu)}$ are, correspondingly, estimations of the functions $g_{c,\rho,\alpha(\nu)}(\mathbf{r})$ which should be separated; $N_{c,\rho,\alpha(\nu)}$ are uninformative terms. The convolution of the relation (1) by $g_\delta^0(\mathbf{r})$ leads to the linear system to determine 6 unknowns $S_{c,\rho,\alpha(\nu)}$, $N_{c,\rho,\alpha(\nu)}$ at each fixed point **r**:

$$\begin{cases} S_c + S_\rho/\omega_j^2 - N_{\alpha(\nu)}\omega_j^{\nu-1} = \mathrm{Re}\left[\hat{W}_j^{cor}(\mathbf{r})\right] \\ S_{\alpha(\nu)}\omega_j^{\nu-1} + N_c + N_\rho/\omega_j^2 = \mathrm{Im}\left[\hat{W}_j^{cor}(\mathbf{r})\right] \end{cases};\ j = 1,2,3. \qquad (7)$$

Here functions $\hat{W}_j^{cor}(\mathbf{r})$ correspond to the functions $\hat{W}(\mathbf{r},\omega_j)$, space spectra of which are corrected by space spectrum of the reference PS function [3]. It should be noted that the case of linear frequency dependence of the absorption coefficient ($\nu = 1$) is possible, for example, in liver [8]. Then the frequency dependence of the c- and α(ω)-scatterer components in (1) becomes the same.

Consequently, the separating these two components is impossible by means of the complex PS function.

2.1. Errors in Separating c-, ρ-, α(ω)-Components Under Deviation of True Index ν from Proposed Index ν_0

As mentioned, the true index ν of the frequency dependence degree of the absorption coefficient might be unknown. Then, during the separating, the index ν_0 is proposed and estimations $\hat{S}_{c,\rho,\alpha(\nu)}$ by the following equation (8) take place instead of the true estimations $S_{c,\rho,\alpha(\nu)}$ by the equation (7):

$$\begin{cases} \hat{S}_c + \hat{S}_\rho / \omega_j^2 - \hat{N}_{\alpha(\nu_0)}\omega_j^{\nu_0 - 1} = \mathrm{Re}\left[\hat{W}_j^{cor}(\mathbf{r})\right] \\ \hat{S}_{\alpha(\nu_0)}\omega_j^{\nu_0 - 1} + \hat{N}_c + \hat{N}_\rho / \omega_j^2 = \mathrm{Im}\left[\hat{W}_j^{cor}(\mathbf{r})\right] \end{cases} ; \quad j = 1,2,3. \tag{8}$$

Errors of the separation (8) caused by distinction between ν_0 and ν can be estimated by equating the left parts of the equations (7) to the left part of the corresponding equations (8) (because the right parts of the equations (7) and (8) are the same). As a result,

$$\hat{S}_c - S_c = -N_{\alpha(\nu)}\omega_2^{\nu-1}K_c(\nu,\nu_0); \tag{9a}$$

$$\hat{S}_{\alpha(\nu_0)}\omega_2^{\nu_0-1} - S_{\alpha(\nu)}\omega_2^{\nu-1} = S_{\alpha(\nu)}\omega_2^{\nu-1}K_\alpha(\nu,\nu_0); \tag{9b}$$

$$\frac{\hat{S}_\rho}{\omega_2^2} - \frac{S_\rho}{\omega_2^2} = -N_{\alpha(\nu)}\omega_2^{\nu-1}K_\rho(\nu,\nu_0). \tag{9c}$$

Right parts of the equations (9a,b,c) are the separation errors of the c-, ρ-, α(ω)-scatterer components. Numerical values of weighting coefficients $K_c(\nu,\nu_0)$ and $K_\rho(\nu,\nu_0)$ are shown in fig.3 for three fixed values ν_0, as the true index ν belongs to the interval $1 \le \nu \le 2$ which is characteristic for biological tissues. It appears that $K_\alpha(\nu,\nu_0) \approx -K_c(\nu,\nu_0)$, and $|K_\rho|$ is more than ten times less than $|K_c|$ and $|K_\alpha|$. Then, the ρ-component is practically insensitive to the distinction between ν and ν_0. This fact is illustrated by results of separating the c- (fig.4*a*), α(ω)- (fig.4*b*) and ρ- (fig.4*c*) components by using three frequencies 400, 600 and 800 kHz at $\nu_0=\nu=1.75$ (dashed line) and at $\nu_0=2$ (thick line); for the ρ-component the dashed and the thick lines coincide. These are components of a «needle-type» scatterer $W(\mathbf{r},\omega)$ having point-like character in the plane (x,y) and a finite size along z-direction. Because of this, amplitudes and dimensionalities of the scatterer components differ from physical values.

2.2. Iterative Reconstruction of True c-, α(ω)-Components and Degree Index ν

The separating algorithm (8) appears to remain stable, but leads to respective errors up to 30% while the true index changes within the range $1.25 \le \nu \le 2$ at the fixed proposed ν_0. Below, iterative algorithms of determination of the true estimations $S_{c,\alpha(\nu)}$ and the true index ν at any fixed space point **r** are very briefly described. Determined index ν(**r**) is an additional diagnostic parameter for medical investigations.

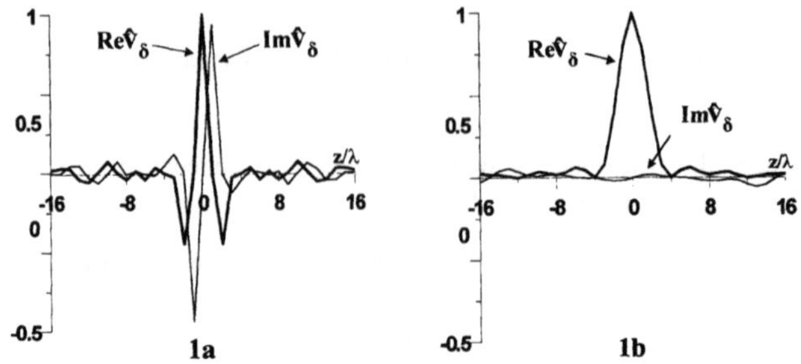

Figure 1. Tomography scheme with *unified* transmitter-receiver transducers.

– Normalized real (thick line) and imaginary (thin line) parts of the scheme's point spread function along vertical direction:

(*a*) - slopes of vertical axes of all transducers are oriented toward array's center (uni-oriented slopes);

(*b*) - slopes of transducer's vertical axes are oriented both toward array's center and away from it (bi-oriented slopes).

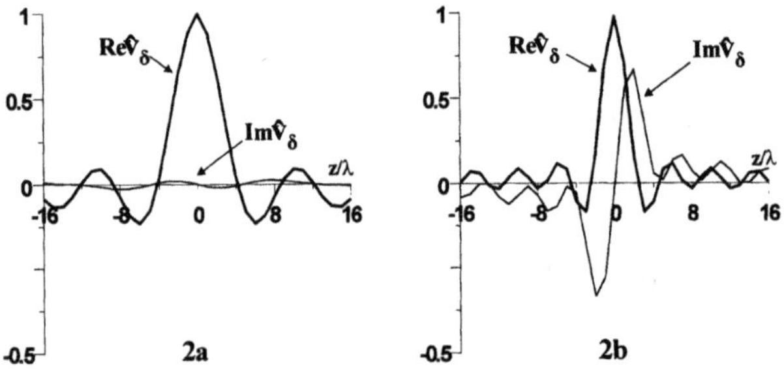

Figure 2. Tomography scheme with *separated* transmitter-receiver transducers. Slope of transmitters is absent; receivers are inclined.

– Normalized real (thick line) and imaginary (thin line) parts of scheme's point spread function along vertical direction:

(*a*) - slopes of receiver's vertical axes are oriented both toward array's center and away from it (bi-oriented slopes);

(*b*) - slopes of vertical axes of all receiver are oriented toward array's center (uni-oriented slopes).

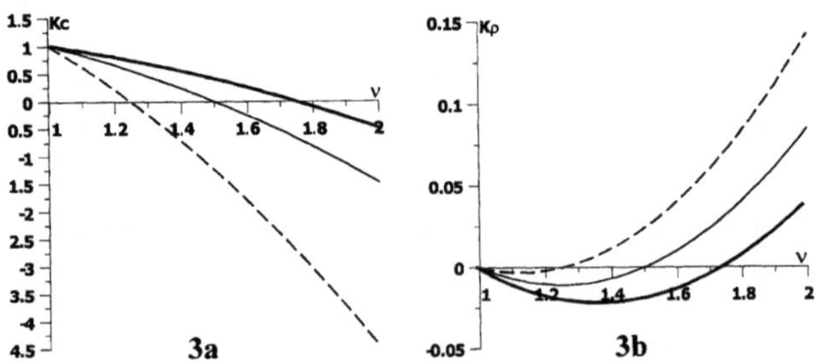

Figure 3. Weighting coefficients at some errors in estimations of scatterer's velocity (*a*) and density (*b*) components.

– These errors arise when true index ν of frequency dependence degree of absorption coefficient is different from proposed index ν_0 : $\nu_0 = 1.25$ (dashed line), $\nu_0 = 1.5$ (thin line), $\nu_0 = 1.75$ (thick line).

Two sets, each of three frequencies, are proposed: $\{\omega_j^{[1]}\}$, $\{\omega_j^{[2]}\}$, j=1,2,3. The reference PS function must be unique for the sets. Mean frequencies of the sets ($\omega_2^{[1]} = \omega_{21}$, $\omega_2^{[2]} = \omega_{22}$) must be different, meanwhile the lowest and the highest frequencies might be the same.

The determination of $S_{\alpha(\nu)}(\mathbf{r})$ and $\nu(\mathbf{r})$ is based on the relation (9b) for α-component, which is valid for any ω_2 and ν_0, if the value ν is true. The initial data for the iterations are estimations $D_1 = \hat{S}^{[1]}_{\alpha(\nu_0)}(\mathbf{r})\omega_{21}^{\nu_0-1}$ and $D_2 = \hat{S}^{[2]}_{\alpha(\nu_0)}(\mathbf{r})\omega_{22}^{\nu_0-1}$ calculated by (8) for two frequency sets at a single ν_0. Further, these data are independent of a number n of an iteration step. An initial approximation $\nu^{(n=0)}$ is given; for example, $\nu^{(n=0)}(\mathbf{r}) = \nu_0 \ \ \forall \mathbf{r}$. Then estimations $Q_1 = S_{\alpha(\nu)}(\mathbf{r})\omega_{21}^{\nu-1}$, $Q_2 = S_{\alpha(\nu)}(\mathbf{r})\omega_{22}^{\nu-1}$ looked for at each fixed step n are evaluated as follows:

$$Q_1^{(n)} = D_1 \,/\, \left\{1 + K_\alpha^{[1]}(\nu^{(n-1)}, \nu_0)\right\}; \quad Q_2^{(n)} = D_2 \,/\, \left\{1 + K_\alpha^{[2]}(\nu^{(n-1)}, \nu_0)\right\}. \tag{10}$$

Then the next relation gives a more accurate estimation $\nu^{(n)}(\mathbf{r})$:

$$(\omega_{21}/\omega_{22})^{\nu^{(n)}(\mathbf{r})-1} = Q_1^{(n)}/Q_2^{(n)}, \text{i.e. } \nu^{(n)}(\mathbf{r}) = 1 + \lg\left(Q_1^{(n)}/Q_2^{(n)}\right) / \lg\left(\omega_{21}/\omega_{22}\right), \tag{11}$$

because the true estimation $S_{\alpha(\nu)}(\mathbf{r})$ is frequency independent.

In a similar manner, the determination of $S_c(\mathbf{r})$ and $\nu(\mathbf{r})$ is carried out on the base (9a) for c-component. However, because the values $P_1 = N_{\alpha(\nu)}(\mathbf{r})\omega_{21}^{\nu-1}$ and $P_2 = N_{\alpha(\nu)}(\mathbf{r})\omega_{22}^{\nu-1}$ are also unknown, the relation of (9a) is simultaneously considered for two values ν_0 at each of the two frequency sets. The estimation $\nu^{(n)}(\mathbf{r})$ is evaluated by (11)-type relation, but the values Q_1, Q_2 are substituted by P_1, P_2. A multitude of points $\mathbf{r}$ of the scattering domain at which the absorption is present, but $N_{\alpha(\nu)}(\mathbf{r}) \cong 0$, represents a "blind " zone. In this zone the determination $\nu(\mathbf{r})$ is possible only by (10), (11) for the α-component.

Results of the numerical iterative reconstruction are shown in fig. 4*a,b* and fig.5. The two frequency sets are {400 kHz, 600 kHz, 1000 kHz} and {400 kHz, 800 kHz, 1000 kHz}. If interference is absent, the reconstructed estimations coincide with their true values. In case of the interference presence, errors appear in space domain, where the corresponding scatterer component is relatively small (thin lines in the figures). The point z=0 in the fig.5*a* corresponds to the "blind" zone of the c-component iterative scheme.

Finally, it should be noted that in relations (9a,b) the same degree index is involved. Therefore an optimal approach to determine the true estimations of the velocity, absorption coefficient and its the frequency dependence degree proposes a simultaneous consideration of (9a,b). The estimation $\nu^{(n)}(\mathbf{r})$ is represented as $\nu^{(n)}=\nu^{(n-1)}+\Delta\nu^{(n)}$. The numerical coefficients in (9a,b) are expanded into a series with $\nu^{(n-1)}$-point center in a linear (up to $\Delta\nu^{(n)}$) approximation. Obtained system of linearized equations is redundant and solved at each iterative step by the least mean square method. This approach provides a higher solution stability to noisy interference. Moreover, it allows to determine $\nu(\mathbf{r})$ at all points $\mathbf{r}$ where the absorption is present, and the "blind" zone is absent.

Acknowledgments. This work was supported by the Russian Foundation for Fundamental Research.

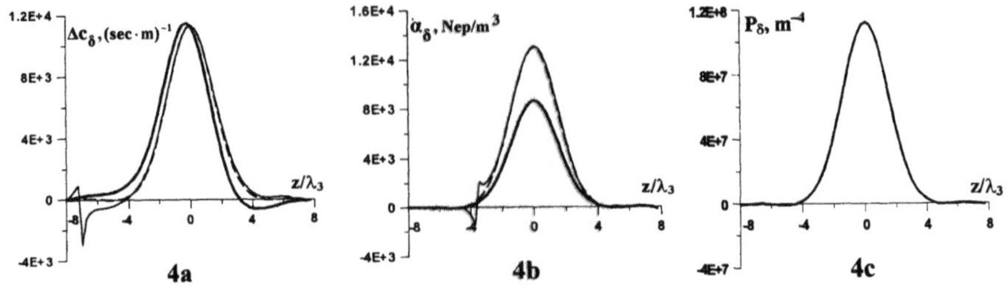

Figure 4. Scatterer's components of velocity (*a*), absorption (*b*) and density (*c*) reconstructed along z-section by incomplete noised scattering data in three-frequencies mode.
– Results of reconstruction:
- Dashed line - true index $\nu = 1.75$ is known.
- Thick line - index $\nu_0 = 2$ is proposed instead of true index $\nu = 1.75$. These noised values of scatterer's components are initial data for iterations (noisy interference leads to a relative error up to 1% in these data).
- Thin line - result of iterative reconstruction.

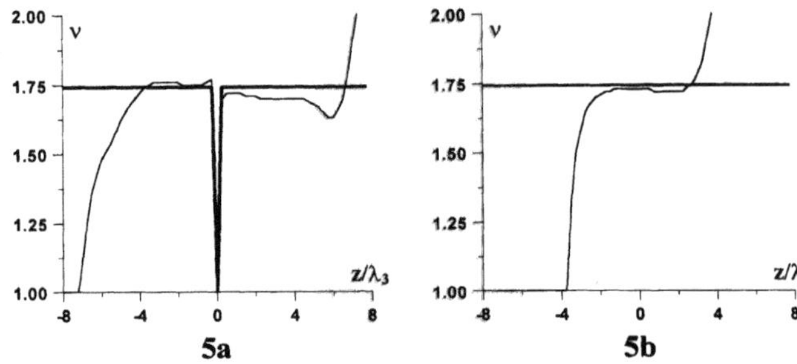

Figure 5. Index of frequency dependence degree of absorption coefficient reconstructed along z-section.
– Reconstruction by iterative scheme for velocity (*a*) and absorption (*b*) components:
- Thick line - noisy interference is absent.
- Thin line - noisy interference leads to a relative error up to 1% in data initial for iterative scheme. Size of scattering domain is $|z| \le 4\lambda_3$.

REFERENCES

1. M. P. André, H. S. Janée, P. J. Martin, G. P. Otto, B. A. Spivey, and D. A. Palmer, High-speed data acquisition in a diffraction tomography system employing large-scale toroidal arrays, *Intl. J. Imaging Systems Technol.* **8**(1), 137-147 (1997).
2. M. P. André, H. S. Janée, G. P. Otto, and P. J. Martin, Reduction of phase aberration in a diffraction tomography system for breast imaging, in: *Acoustical Imaging-22*, edited by P. Tortoli and L. Masotti (Plenum Press, New York, 1996), pp.151-157.
3. V .A. Burov, A. L. Konjushkin, and O. D. Rumiantseva, Multi-dimensional acoustical tomography by incomplete data, in: *Acoustical Imaging-23*, edited by S. Lees and L. A. Ferrari (Plenum Press, New York (1997), pp.589-594.
4. V. A. Burov, et al., *Ultrasonic tomograph and circular antenna array for ultrasonic tomograph. Russia Patent* N 2145797, in Russian (Moscow, 2000).
5. A. J. Devaney and M. L. Oristaglio, Inversion procedure for inverse scattering with the distorted-wave Born approximation, *Phys. Rev. Lett.* **51**(1), 237-240 (1983).
6. S. A. Goss, R. L. Johnson, and F. Dunn, Comprehensive compilation of empirical ultrasonic properties of mammalian tissues, *J. Acoust. Soc. Am.* **64**(2), 423-457 (1978).
7. S. A. Goss, R. L. Johnson, and F. Dunn, Comprehensive compilation of empirical ultrasonic properties of mammalian tissues. II, *J. Acoust. Soc. Am.* **68**(1), 93-108 (1980).
8. *Physical Principles of Medical Ultrasonics*, edited by C. R. Hill (Ellis Horwood Limited, New York, 1986).

PRACTICAL DESIGN OF MULTILAYER PIEZOELECTRIC TRANSDUCERS FOR HARMONIC IMAGING

Rozenn Desmare, Louis-Pascal Tran-Huu-Hue, Franck Levassort, and Marc Lethiecq

G.I.P. Ultrasons/ LUSSI, E.A. 2102, Ecole d'Ingénieur du Val de Loire
rue de la chocolaterie
41034 Blois Cedex
FRANCE

INTRODUCTION

Contrast Agent Imaging (C.A.I.) and Native Harmonic Imaging (N.H.I.) are two recent ultrasonic imaging techniques in which nonlinear effects are used. In these applications, the ultrasound system uses overtone components in the echo signals to form an image. These overtones are generated either from nonlinear propagation in tissue or from non-linear scattering of the pulse from micro-bubbles of the contrast agent. In practice, the ultrasound system must transmit signals at a frequency noted f_a and receive signals at a frequency $f_r=2f_a$. For this, two techniques can be used : (i) band-pass filtering of transmitted and received signals at different center frequencies and / or (ii) pulse inversion of transmitted signals and subtraction of the resulting echoes. Such a dual frequency behaviour can be obtained with a two-layer piezoelectric structure. In this paper, we propose to study the potentiality of multilayer piezoelectric transducers for harmonic imaging.

First, we have developed an optimisation method for ultrasonic transducers, based on the minimisation of a performance index. This method only requires the calculation of the transducer impulse response. In order to study the two-layer transducer, we have developed a new K.L.M. circuit-based model which allows to calculate transfer functions and impulse responses. Using this impulse response, two-layer transducer optimisation is performed using the simplex method. Secondly, we have optimised, with the same method, a single layer broadband transducer for harmonic imaging. Finally, the performances of both transducers are compared.

Acoustical Imaging, Volume 25, Edited by Halliwell and Wells.
Kluwer Academic/Plenum Publishers, 2000.

OPTIMISATION METHOD OF AN ULTRASONIC TRANSDUCER

Optimisation approaches used at present are limited to one configuration and depend on the application. Based on a method and parameters introduced by Thijssen[1], our method is independent of transducer configuration and allows the optimisation of complex transducers. Indeed, it only requires the impulse response to be calculated. The method is based on the minimisation of a performance index. The main difficulty is to determine this performance index corresponding to the chosen application : it must accurately describe transducer performances and must be defined with pertinent parameters. The index used here is composed of three parameters calculated on impulse response and noted d_{20} and d_{40}, which represent the transducer resolution, and *amp*, which represents the sensitivity. d_{20} (resp. d_{40}) is defined as the smallest length corresponding to 96.8% (resp. 99.76%) of the area of impulse response envelope. For a gaussian response, d_{20} (resp. d_{40}) corresponds to the 20 dB (resp. 40 dB) duration expressed in seconds. *amp* is the amplitude of the maximum of the impulse response envelope. This envelope is the inverse Fourier transform of a transfer function without dimension, so, amp is expressed in s^{-1}. The performance index is noted x and is defined as :

$$x = \alpha.10.\log(d_{20}) + \beta.10.\log(d_{40}) - \gamma.10.\log(\text{amp}). \qquad (1)$$

Weighting parameters noted (α, β, γ) allow to take into account the importance of one parameter compared to the others. For medical imaging, we tried several weightings and compared results with those obtained by experimented designers. (3,2,1) weighting gives consistent results.

From an initial configuration of the transducer, using the K.L.M. scheme, the impulse response and consequently the performance index are calculated. Using a simplex minimisation, transducer characteristics are modified to obtain the minimal index. Optimisation is performed for several values of backing acoustic impedance (Z_B), the backing thickness being considered as infinite. Results are shown as variations of performance index x (dB), d_{20} and d_{40} durations (periods), amp (s^{-1}), quarter-wavelength layer impedance (Z_{layer1} in MRa) and thickness (e_{layer1} in $\lambda/4$), active layer thickness (e_{piezo} in $\lambda/2$) and series inductance (L in μH), all versus Z_B.

OPTIMISATION OF TRANSDUCERS FOR HARMONIC IMAGING

Nowadays, single active layer broadband transducers are used in harmonic imaging. In order to evaluate performances of a two-layer transducer for harmonic imaging, both types of transducers are optimised and results are compared.

Optimisation of a two-layer transducer for harmonic imaging

In a recent paper[2], we have developed a multilayer scheme based on the K.L.M. equivalent circuit[3] and a matrix notation[4]. The main advantage of this method is that, for any number of active layers, bonding layers and electrical connections between successive layers as well as classical passive elements of a transducer can be taken into account.

We have fabricated a two-layer structure using soft PZT ceramic discs (Pz27, Ferroperm[5]). We note f_a the resonance frequency of each ceramic. According the nature of electrical connections between both ceramics (series or parallel), the resonance frequency of the stack is f_a or $f_a/2$: a dual frequency behaviour is obtained if a parallel connection in transmission and a series connection in reception are used. We have measured and simulated impulse responses for both transmission and reception configurations (fig.1). The results are in a good agreement. The differences are due to the one-dimensional nature of the model. Radial modes are not taken into account in the simulation and perturb measurement, especially at low frequencies (transmission).

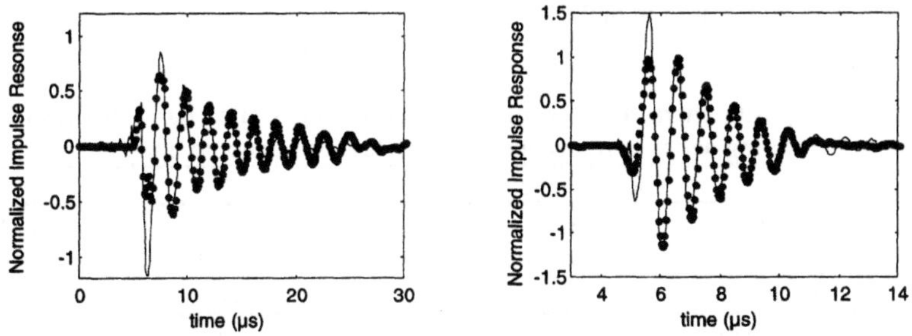

Figure 1 . Normalised impulse response of the two-layer transducer measured (points) and calculated (solid lines) in transmission (a) and in reception (b).

For our optimisation study, we consider a transducer made of two active layers of 1-3 piezocomposite, a quarter-wavelength layer and a backing. Characteristics of the piezocomposite are shown in table 1. Several hypothesis are made. First, the pulse inversion technique is considered. Consequently, in reception, the fundamental signal is eliminated. Secondly, in order to simulate the two-layer transducer behaviour in transmit-receptive mode, the harmonic signal generated within human tissue must be taken into account. We assume that this signal has the same fractional bandwidth as the fundamental signal transmitted by the two-layer transducer while having a double frequency. A 20 dB attenuation is applied on the harmonic signal to take into account the harmonic production rate. This signal and the reception signal, which is centred on f_a, are multiplied to form the transfer function corresponding to harmonic operation. Using this transfer function, our optimisation method is used for the multilayer transducer.

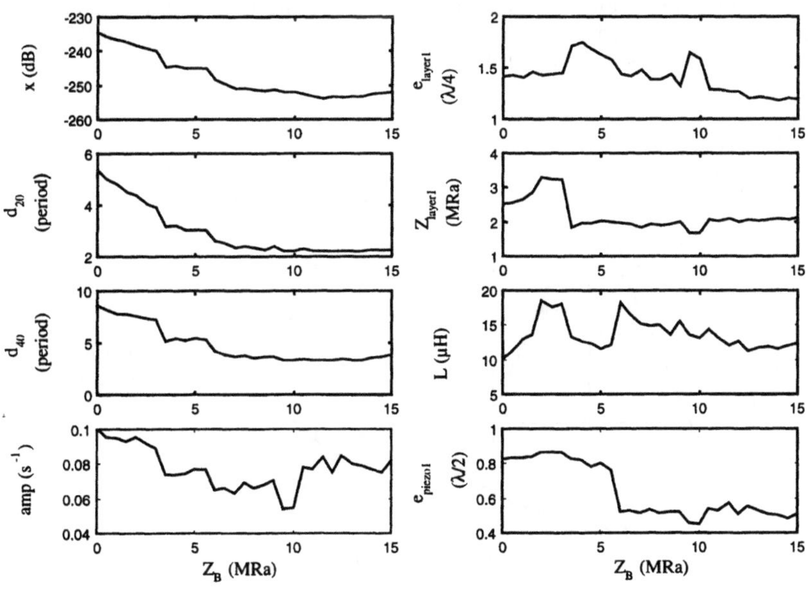

Figure 2 . Optimisation results of a transducer maded of two layers of piezocomposite, a quarter-wavelength layer and a backing for harmonic imaging.

Figure 2 shows optimisation results. Minimal value of performance index is obtained for a backing acoustic impedance around 11 MRa. Optimal performances and characteristics are summarised in table 1.

Table 1. Piezocomposite characteristics, optimal performances and characteristics of two-layer transducer (backing acoustic impedance = 11 MRa) for harmonic imaging

Piezocomposite characteristics					
V_l	k_t	Z	δ_e	δ_m	$\varepsilon^S_{33r}(f_a)$
3000 ms^{-1}	0.67	10 MRa	0.05	0.02	200
Performances			**Optimal characteristics**		
x= -254 dB			$e_{layer1} = 1.3\ \lambda/4$		
$d_{20} = 2.2$ periods			$Z_{layer1} = 2.1$ MRa		
$d_{40} = 3.3$ periods			$L = 13\ \mu H$		
$amp = 0.09\ s^{-1}$			$e_{piezo} = 0.6\ \lambda/2$		

In order to compare these performances to standard technology, we have optimised a single-layer broadband transducer for harmonic imaging.

Optimisation of a single layer broadband transducer

We have considered the same conditions as for the two layer transducer, *i.e.*, an active layer of the same 1-3 piezocomposite, a quarter-wavelength matching layer, a backing and harmonic generation with 20 dB attenuation. However, a new parameter noted f_0, centre frequency of the broadband transducer, must be determined to obtain the optimal dual frequency behaviour between $f_a/2$ and f_a. As shown in figure 3, a gaussian filter is applied on the transfer function of the transducer in order to transmit at $f_a/2$. Filter characteristics are its centre frequency noted nf_0, such as $nf_0 = f_a/2$ with n varying from 0.1 to 0.8, and its bandwidth noted BW, varying from 20 % to 80 %.

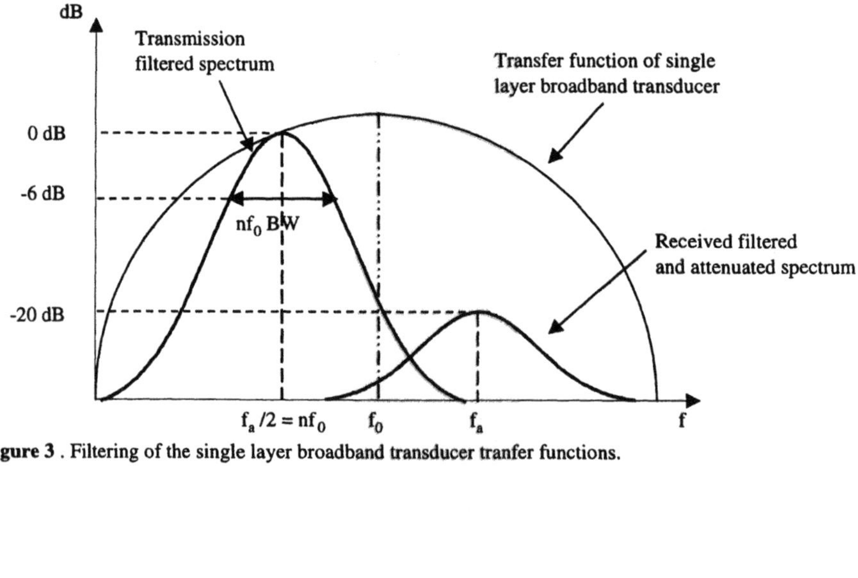

Figure 3. Filtering of the single layer broadband transducer tranfer functions.

In order to determine f_0, we have calculated the transducer transfer function in transmit–receive mode versus n and BW, in the case of a dual frequency behaviour with inverse impulse technique. Single layer broadband performances are evaluated by the performance index.

Results are shown in figure 4. Variations of x, d_{20}, d_{40} and *amp* are represented by contour lines versus nf_0 and BW. We can notice that the minimal value of x is –250 dB and is obtained by a gaussian filter whose centre frequency is $0.7f_0$ and bandwidth is 80 %. Resultant impulse response has d_{20} and d_{40} values respectively around 2.9 and 5.8 periods and an amplitude term in the order of 0.05 s^{-1}.

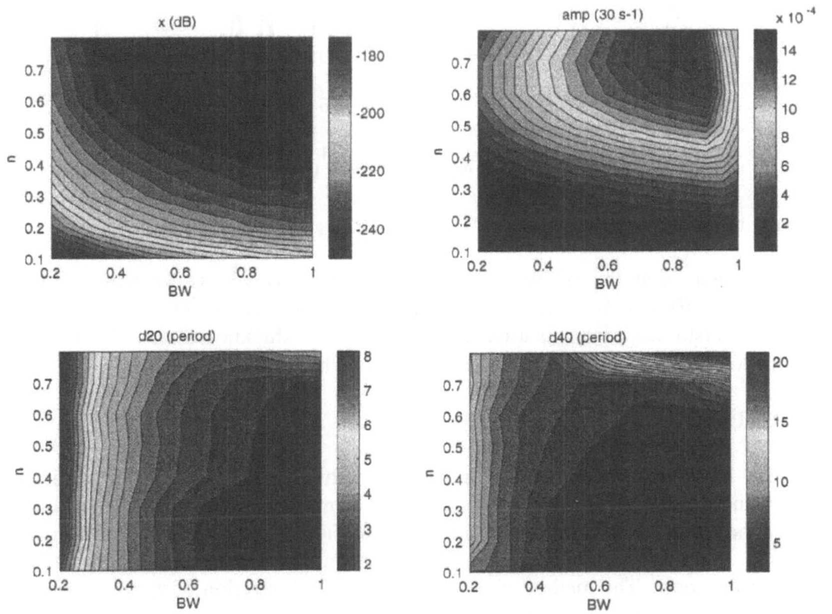

Figure 4 . Contour lines corresponding to variations of x, d_{20}, d_{40} and amp versus characteristics of gaussian filter (n and BW) for the single layer broadband transducer.

DISCUSSION AND COMPARISON

Comparison between two-layer and single layer configurations is summarised in table 2. Performance indexes are respectively about –254 dB for the two-layer transducer and -250 dB for the single layer transducer. Consequently, a performance gain is obtained in favour of the two-layer transducer.

Table 2 . Comparison of performances of two-layer and single layer transducers for harmonic imaging

Two-layer transducer	Single layer transducer
x = -254 dB	x = -250 dB
d_{20} = 2.2 periods	d_{20} = 2.9 periods
d_{40} = 3.3 periods	d_{40} = 5.5 periods
amp = 0.09 s^{-1}	*amp* = 0.05 s^{-1}

The results show that a performance gain is obtained for sensitivity as well as for resolution. Impulse response durations are around 2.2 and 3.3 periods, for the two-layer device, an improvement of 25 % for d_{20} and 40 % for d_{40}. In the same way, the amplitude term is 40% higher for the two-layer transducer.

In order to visualize this performance gain, we have represented impulse responses of both transducers in the case of a dual-frequency behaviour for harmonic imaging. We can notice the two-layer transducer superiority.

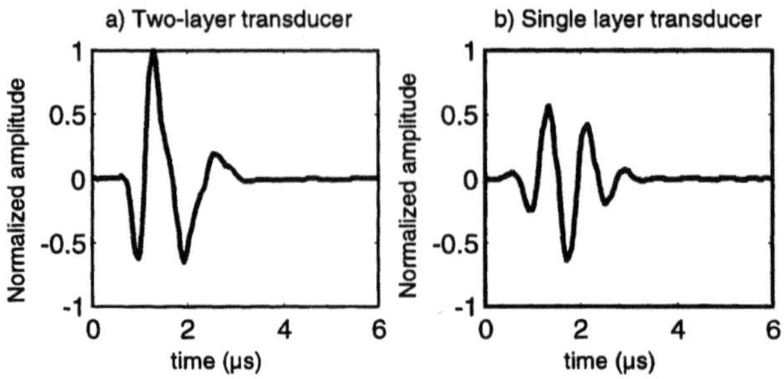

Figure 5 . Normalised impulse responses of two-layer and single layer transducers.

It appears that the optimal characteristics of a two-layer transducer for harmonic imaging are different from those of a transducer designed for classical imaging. Indeed, backing acoustic impedance is around 11 MRa and the thicknesses of active layers are close to λ/4. Moreover, matching layer and active layers have similar thicknesses.

CONCLUSION

We have developed an optimisation method based on the minimisation of a performance index. This index is established from three parameters (d_{20}, d_{40} and *amp*) corresponding to resolution and sensitivity conditions. During transducer optimisation for harmonic imaging, harmonic generation within human tissue and 20 dB attenuation are taken into account. The method gives the optimal configuration of the two-layer transducer (quarter-wavelength acoustic impedances and thicknesses, backing acoustic impedance,...) in the case of a dual frequency behaviour for harmonic imaging.

Finally, we have optimised a single layer broadband transducer for harmonic imaging in order to compare performances of both transducers. We have observed that the two-layer transducer has better performances than the broadband single layer device. We obtain a sensitivity gain around 40 % and a resolution gain around 25 %.

REFERENCES

1. J.M. Thijssen, W.A. Verhoef and M.J. Cloostermans, Optimisation of ultrasonic transducers, *Ultrasonics*, 41 (1985).
2. R. Desmare, L.P. Tran-Huu-Hue, F. Levassort, W. Wolny and M. Lethiecq, Modelling of multilayer piezoelectric structures, *Ferroelectrics*, 224 :195 (1999).
3. R. Krimholtz, D.A. Leedom and G.L. Matthei, New equivalent circuit for elementary piezoelectric transducers, *Electron. Lett.*, 6:398 (1970).
4. S.J.H. Van Kervel and J.M. Thijssen, A calculation scheme for the optimum design of ultrasonic transducers, *Ultrasonics*, 134 (1983).
5. Ferroperm ldt., Piezoelectric Materials Data Book, Kvisgård, Denmark (1996).

A STOCHASTIC APPROACH TO OPTIMISE WIDE-BAND BEAM PATTERNS

Andrea Trucco

Dept. of Biophysical and Electronic Engineering
University of Genoa
Via Opera Pia 11, 16145 Genova, Italy

ABSTRACT

The adoption of wide-band signals in active beamforming is very useful to improve the performances of acoustical imaging systems. Although a great amount of work has been done to develop and optimise weighting windows for an array working in narrow-band, one can verify that these windows do not provide optimal results when applied under wide-band conditions. The synthesis of a weighting window that is well-suited to work on wide-band signals may be considered as an open research problem.

In this paper, after a proper definition of the wide-band beam pattern, a method aimed at determining the values of the weight coefficients to be applied to the elements of a linear array that generates a beam pattern fulfilling some *a priori* fixed constraints is proposed. This method relies on a stochastic scheme based on the simulated annealing algorithm: its high flexibility allows to optimise any wide-band pulse waveform used by the imaging system and to comply with different constraints on the desired beam pattern.

Some results showing a notable improvement in array performances, in terms of main lobe width and side lobes level, with respect to an unitary weighting are hereby presented.

INTRODUCTION

This paper addresses the linear beamforming: a technique aimed at processing the signals sampled by an array of sensors. Although beamforming systems can be used for a lot of disparate problems, here, without loss of generality, their application to the active acoustical imaging is accounted for.

When the spatial aperture of the array is larger than the equivalent spatial length of the pulse used for the insonification, the beam pattern profile depends also on the pulse envelope shape[1-5]. This case is usually referred to as the wide-band condition, since it requires pulses of short time duration (i.e. wide-band signals). In this situation, it is possible to limit speckle effects (i.e., interference effects causing image quality degradation) and to improve the range resolution. Consequently, images obtained by using wide-band signals are generally of better quality and potentially contain more information than those obtained by means of narrow-band signals[4,6]. In addition, wide-band signals permit to overcome a classic problem related to narrow-band conditions: when high spatial resolution and a small number of array elements are required, the spacing between elements becomes wider than $\lambda/2$, (λ being the wavelength related to the carrier frequency) and this causes the presence of grating lobes in the beam pattern. The use of wide-band signals is a very suitable method to reduce the elevation of such grating lobes[3,4,6,7]. The first point to address is the definition of the wide-band beam pattern as the conventional beam pattern is derived assuming narrow-band conditions. Previous papers dealt with two possible criteria to define the wide-band beam pattern[1,2]: here, that based on the measure of the maximum amplitude of every beam signal is preferred as it is more related to imaging applications.

In many practical applications, for the sake of simplicity, conventional weighting windows developed under narrow-band assumptions are applied to arrays working under wide-band conditions. This fact is based on the conjecture that the good performance of a window are naturally extended to the wide-band case. In the recent literature[1,8], it has been shown that such a conjecture often fails: generally, conventional windows provide results of poor quality when applied to wide-band problems.

In this paper, by considering a linear $\lambda/2$-spaced array and using numerical simulations, a stochastic method to synthesize a set of weight coefficients optimised to provide a beam pattern (that is) very close to a desired one is proposed. The weight values are computed by taking into account the adopted pulse waveform, i.e., both the pulse envelope and bandwidth aspects. Owing to an accurate problem definition, the method works over a normalised angular domain, in such a way that the synthesised weighting window keeps its validity for every steering direction. The devised method is based on simulated annealing and it represents a major evolution of the stochastic approach proposed by the author for array synthesis under narrow-band conditions[9,10].

WIDE-BAND BEAM PATTERN

Since, under wide-band conditions, the conventional formulation of the beam pattern is not valid any more, a redefinition of the beam pattern is needed. In literature, two methods assuming the knowledge of the signals received by the sensors when a plane wave reaches the array at a given incidence angle were proposed[1,2]. In the simplest case, such signals can be modelled as a replica of the transmitted pulse, adequately delayed in accordance with the incidence angle and the positions of the array elements[1,6,11,12]. After fixing a steering direction, the knowledge of these signals allows the computation of the beam signal. Let define t the time, α the incidence angle of a plane wave, θ the steering direction, u the arbitrary variable defined as $u = \sin\alpha - \sin\theta$ that ranges in [-2, 2]. Under the far-field hypothesis and with reference to an array composed of M point and omnidirectional elements, the beam signal, $b(t,u)$, can be computed as follows:

$$b(t,u) = \sum_{i=0}^{M-1} w_i \cdot A\left(t + \frac{idu}{c}\right) \cdot \exp\left(j\omega \frac{idu}{c}\right) \tag{1}$$

where w_i is the weight coefficient assigned to the i-th element, $A(t)$ is the envelope of the adopted acoustic pulse (possibly, after pulse compression), d is the array inter-element spacing, ω is the pulsation, and c is the sound velocity in the addressed medium. It is evident that in Eq. (1) a base-band quadrature reception has been hypothesised[13] to represent the signals in terms of envelope and phase information.

The value of the beam pattern, $p(u)$, for a given pair of incidence angle and steering angle can be considered as the maximum of the beam signal over time[1,6,7,11,12,14]:

$$p(u) = \max_t \{ |b(t,u)| \} \tag{2}$$

or as the total energy of the beam signal[1,5,15]:

$$p(u) = \left(\int_{-\infty}^{\infty} |b(t,u)|^2 \, dt \right)^{1/2} \tag{3}$$

It can be noticed that, in both cases, the function $p(u)$ is equal to $p(-u)$, hence it is sufficient to study and visualise the beam pattern over the range $0 \le u \le 2$. Moreover It can be proved that, in both cases, when the wide-band hypothesis is relaxed until a narrow-band condition is achieved, the resulting beam pattern is equal to the conventional narrow-band beam pattern[11,15].

Depending on the application considered, one can choose the more suitable definition of the beam pattern between Eqs. (2) and (3). In acoustical imaging where the amplitude of the whole beam signal is reported along a line of pixel, the definition in Eq. (2) seems the best choice to evaluate the beam pattern of the system, whereas the one in Eq. (3) appears the right selection in applications where the total energy of the beam signal is used to perform further operations/decisions. Thus, in the following of this paper, the wide-band beam pattern will be defined as in Eq. (2). Finally, the conventional visualisation of a beam power pattern, $bpp(u)$, on a logarithmic scale normalised to 0 dB, is obtained as follows:

$$bpp(u) = 10 \log \left[\left(\frac{p(u)}{M} \right)^2 \right] \qquad [\text{dB}] \tag{4}$$

where M is the number of array elements that should be equal to the sum of the (normalised) weight coefficients.

WEIGHT SYNTHESIS USING SIMULATED ANNEALING

The stochastic method for weight coefficients synthesis proposed in this paper is based on the simulated annealing (SA) scheme. More details on the concept and implementation of SA can be found in the Appendix[16,17].

The use of SA requires the choice of an energy function, $f(\mathbf{W})$, that depends on the vector of weights, $\mathbf{W}$, to be optimised. Once a desired normalized BP, $p_d(u)$, has been fixed, the energy function must be able to penalize the array configurations that yield a great difference between the desired BP and the current one. To this end, one can choose:

$$f(\mathbf{W}) = \left[\gamma \int_{u \in S} \left(\frac{p(u)}{M} - p_d(u) \right) du \right]^2 \tag{5}$$

where γ is a constant scale coefficient, S is the set of values of u satisfying the relation $p(u)/M > p_d(u)$. As the integral in Eq. (5) represents the square of the area of the regions in which the obtained beam pattern exceeds the desired one, if the energy function is lowered to zero then the constraints are fulfilled.

Figure 1 shows a flow-chart of the optimization procedure that will be described below. When one starts the iteration j, the element i is randomly chosen (all elements are visited according to a random sequence that does not allow any further visit the same element before all them have been visited once), its weight w_i is perturbed and, on the basis of the temperature T_j, the perturbation can be accepted or not. The initial temperature, T_{start}, is chosen high enough and such that the first configuration perturbations may very often be accepted, even though they lead to a sharp increase in energy.

During the optimization procedure, a constraint is imposed to limit to low values the current taper ratio (CTR), which is the ratio between the maximum and minimum weight coefficients[9,10]. This constraint makes it possible to avoid any consequences of possible unforeseen occurrences regarding the elements with the largest weight coefficients. To limit the CTR, each weight coefficient should be perturbed in a random and continuous way; but one should make sure that the coefficient value is included in an a priori fixed range $[w_{min}, w_{max}]$.

The number of iterations, j, is increased every time all the M elements have been perturbed once. The process terminates when a state of persistent block (freezing) is reached, due to the temperature lowering. Alternatively, according to previous experiences, one can perform a number of iterations that is large enough to ensure that a block state will be reached.

Owing to the probabilistic nature of SA, different temperature scheduling and random initial configurations may lead to different final results. However, if a logarithmic scheduling is chosen, almost all process runnings give slightly different results in terms of energy and weight values. This means that the resulting weighting window is stable and close to the optimal one.

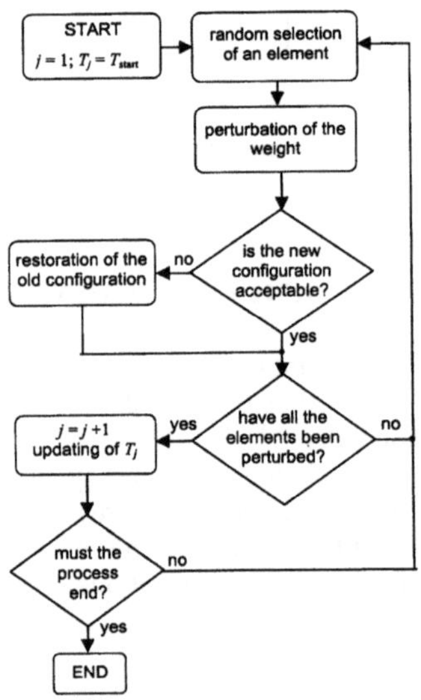

Figure 1. Flow chart of the proposed optimisation scheme.

With respect to narrow-band algorithm presented in[9,10], the computational load of the proposed method, under wide-band hypothesis, is notably higher due to the fact that the evaluation of the beam pattern require the computation of a whole beam signal for each value of u. Moreover, the beam pattern must be computed (or simply updated) at each weight perturbation. Thus, one running of the proposed method may consume some hours of computing time, but once the result is achieved, it is not necessary to run the algorithm any more as the obtained weight values can be uploaded inside the imaging device without requiring any re-computation of them.

RESULTS AND COMPARISONS

To test the efficiency of the proposed method, a linear $\lambda/2$-spaced array composed of 125 elements is considered. The beam pattern is numerically evaluated under the following conditions: carrier frequency equal to 1 MHz, sampling frequency equal to 5 MHz (40 MHz for test and visualisation purposes), and acoustic pulse with a truncated Gaussian envelope lasting 2 μs.

Concerning the simulated annealing scheme, the following parameter have been fixed: $T_{start} = 1$, $w_{min} = 0.2$, $w_{max} = 1.8$, $\gamma = 50$, and number of iterations = 2000. The desired beam pattern, $p_d(u)$ has been fixed as a constant level of −42 dB for any u over the range [0.08, 2].

Figure 2 shows one of the results obtained by the proposed method in comparison with the beam pattern produced by unitary weights. One can notice that the main lobes have a close width in the two cases ($u_{-3dB} = 6.68\cdot10^{-3}$ for unitary weights, $u_{-3dB} = 7.28\cdot10^{-3}$ for optimised weights, i.e., an enlargement of 9%; see Fig. 2b), whereas the side lobe of the optimised case are sharply lower over the region $0.02 \le u \le 0.45$ that has a great practical importance. For other u values the optimised beam pattern is quite constant at −40 dB, as required.

A very important remark is that, contrary to what happens in narrow-band beamforming, the main lobe do not suffer a sharp increase in its width as a consequence of the side lobe reduction and equalisation. This is due to the specific definition of the wide-band beam pattern previously adopted that does not allow to infer about the total area of the beam pattern. In this specific case, one has verified that the area of the optimised beam power pattern (linear scale) is only 6% larger than the one produced by unitary weights.

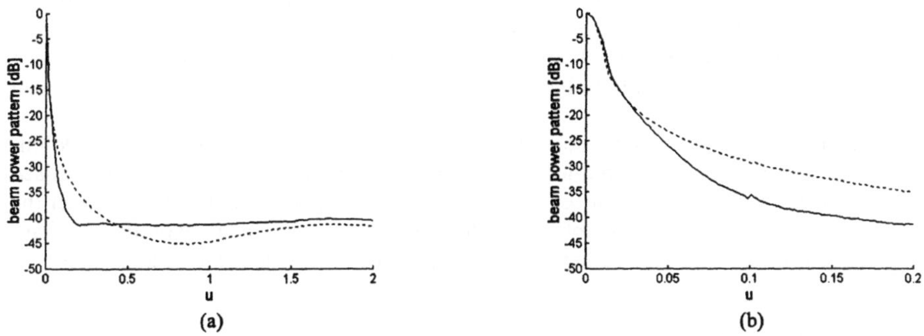

Figure 2. Beam patterns obtained by the proposed method (solid line) and by unitary weights (dashed line) for a 125-element array and Gaussian pulse. (a) Beam patterns comparison over the domain $0 \le u \le 2$. (b) Zoom of such a comparison over the domain $0 \le u \le 0.2$.

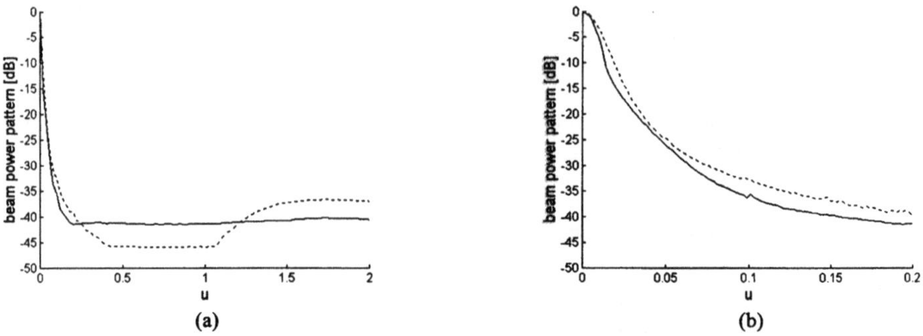

Figure 3. Beam patterns obtained by the proposed method (solid line) and by the Dolph-Chebychev window equalised at –40 dB (dashed line) for a 125-element array and Gaussian pulse. (a) Beam patterns comparison over the domain $0 \le u \le 2$. (b) Zoom of such a comparison over the domain $0 \le u \le 0.2$.

Figure 3 shows the same optimised beam pattern (solid line) compared with the one obtained by the conventional Dolph-Chebychev window (dotted line) computed for an equalisation level of –40 dB. The Dolph-Chebychev window was selected as it offered the best performances over the wide set of conventional windows that were tested. First of all, one can see that the Dolph-Chebychev window does not succeed in keeping constant the side lobe level. In addition, the whole profile of the optimised beam pattern is preferably to that obtained by the Dolph-Chebychev window. The main lobe enlargement produced by the Dolph-Chebychev window is considerable ($u_{-3dB} = 9.25 \cdot 10^{-3}$, i.e., an enlargement of 38% with respect to unitary weights) and also the total beam power pattern area is quite large (36% enlargement with respect to unitary weights).

Finally, the CTR value of the optimised window is 4.6; it is sharply lower than that of addressed Dolph-Chebychev window, that is 10.8.

CONCLUSIONS

In this paper, a proper definition of the wide-band beam pattern has been described and a method aimed at fixing the values of the weight coefficients to be applied to the elements of a linear array that generates a beam pattern fulfilling some *a priori* fixed constraints has been proposed. Such a method relies on a stochastic scheme that has been based on the simulated annealing algorithm. The main implementation details have also been discussed shortly.

Some results showing the poor performances of classical windows (e.g., Dolph-Chebychev) under wide-band conditions have been reported in comparison with optimised results achieved thanks to the proposed method. Finally, it has been remarked that under the wide-band conditions, it is possible to reduce the side lobe level avoiding a sharp enlargement of the main lobe, due to the different physical meaning of the beam pattern.

ACKNOWLEDGEMENTS

The author would like to thank Eng. Emanuele Omodei for his invaluable assistance in developing this work and for the fruitful discussions on the subject. Moreover, a special thank also to Eng. Paolo Repetto for his co-operation in implementing and testing the method.

APPENDIX

Initially, SA aimed to simulate the behaviour of the molecules of a pure substance during the slow cooling that results in the formation of a perfect crystal (minimum-energy state).

The use of this technique to solve other types of problems is based on the analogy between the state of each molecule and the state of each variable that affects an energy function to be minimised. This function is called the energy function, $f(\mathbf{Y})$, $\mathbf{Y}$ being the vector of state variables. The algorithm is iterative: at each iteration, a small random perturbation is induced in the current state configuration $\mathbf{Y}_j$ (j being the iteration). If the new configuration, $\mathbf{Y}_n$, causes the value of the energy function to decrease, it is accepted ($\mathbf{Y}_{j+1} = \mathbf{Y}_n$). Instead, if $\mathbf{Y}_n$ causes the value of the energy function to increase, it is accepted with a probability dependent on the system temperature, in accordance with the Boltzmann distribution. The higher the temperature, the higher the probability that the state configuration that caused the energy function to increase may be accepted. In short, the probability that $\mathbf{Y}_n$ may be accepted as a new configuration, $P\{\mathbf{Y}_{j+1} = \mathbf{Y}_n\}$, can be expressed as:

$$P\{\mathbf{Y}_{j+1} = \mathbf{Y}_n\} = \begin{cases} e^{\frac{f(\mathbf{Y}_j) - f(\mathbf{Y}_n)}{kT_j}} & , \text{if } f(\mathbf{Y}_n) > f(\mathbf{Y}_j) \\ 1 & , \text{otherwise} \end{cases} \tag{6}$$

where k is the Boltzmann constant and T_j is the system temperature. As the iterations go on, the temperature T_j is gradually lowered, following the reciprocal of the logarithm of the number of iterations[16], until the configuration freezes in a certain final state.

Thanks to its probabilistic nature, this method represents a notable improvement over classic methods of local descent, although it is computationally demanding. The repetition of the process, using different initial configurations, increases the reliance on the quasi-optimality of results, even though a full optimality cannot be proved.

REFERENCES

1. X. Lurton and S. Coatelan, "Array Directivity for Wide Band Signals: a Time-Domain Approach," *Proc. Europ. Conf. on Underwater Acoustics*, Luxembourg, September 1992.
2. V. Murino, A. Trucco, and A. Tesei, "Beam Pattern Formulation and Analysis for Wide-Band Beamforming Systems Using Sparse Arrays," *Signal Processing*, vol. 56, pp. 177-183, January 1997.
3. M.H. Bae, I.H. Sohn, and S.B. Park, "Grating Lobes Reduction in Ultrasonic Synthetic Focusing," *Electronics Let.*, vol. 27, pp. 1225-1227, 4th July, 1991.
4. A. Macovski, "Ultrasonic Imaging Using Arrays," *IEEE Proceedings*, vol. 67, pp. 484-495, April 1979.
5. R.F. Follett and J.P. Donohoe, "A Wideband, High-Resolution, Low Probability of Detection FFT Beamformer," *IEEE Jour. on Oceanic Engineering*, vol. 19, pp. 175-182, April 1994.
6. D.H. Turnbull, F.S. Foster, "Beam Steering With Pulsed Two-Dimensional Transducer Arrays," *IEEE Trans. on Ultrasonics, Ferroelectrics, and Frequency Control*, vol. 38, pp. 320-333, July 1991.
7. J. Lu and J.F. Greenleaf, "A Study of Two-Dimensional Array Transducers for Limited Diffraction Beams," *IEEE Trans. Ultrasonics, Ferroelectrics, and Frequency Control*, vol. 41, pp. 724-739, September 1994.
8. A. Trucco, V. Murino, and G. Tacconi, "Beam Pattern Analysis under Generalized Conditions," *3rd Europ. Conf. on Underwater Acoustics*, Heraklion, Crete (GREECE), pp. 525-530, June 1996.
9. V. Murino, A. Trucco, and C.S. Regazzoni, "Synthesis of Unequally Spaced Arrays by Simulated Annealing," *IEEE Tr. Sig. Proc.*, vol. 44, pp. 119-123, January 1996.
10. A. Trucco and V. Murino, "Stochastic Optimization of Linear Sparse Arrays," *IEEE Journal of Oceanic Engineering*, vol. 24, pp. 219-299, July 1999.

11. P. Challande and P. Cervenka, "Theoretical Influence of Various Parameters on the Behaviour of a Linear Antenna," *Journal d'Acoustique*, vol. 3, pp. 17-27, March 1990.
12. G.R. Lockwood and F.S. Foster, "Optimizing Sparse Two-Dimensional Transducer Arrays Using an Effective Aperture Approach," *Proc. 1994 IEEE Ultrasonics Symposium*, pp. 1497-1501, Cannes, November 1994.
13. R.O. Nielsen, *Sonar Signal Processing*, Artech House, Boston, 1991.
14. M.O. O'Donnel, W.E. Engeler, J.T. Pedicone, *et al.* "Real-Time Phased Array Imaging Using Digital Beam Forming and Autonomous Channel Control," *Proc. 1990 IEEE Ultrasonics Symp.*, pp. 1499-1502, 1990.
15. A.P. Goffer, M. Kam, and P.R. Herczfeld, "Design of Phased Arrays in Terms of Random Subarrays," *IEEE Trans. Anten. Propag.*, vol. 42, pp. 820-826, June 1994.
16. S. Kirkpatrick, C.D. Gellatt Jr., and M.P. Vecchi, "Optimization by Simulated Annealing," *Science*, vol. 220, no. 4598, pp. 671-680, 1983.
17. W.A. Kuperman, M.D. Collins, J.S. Perkins, and N.R. Davis, "Optimal Time-Domain Beamforming With Simulated Annealing Including Application of A-Priori Information," *Jour. Acoust. Soc. Am.*, vol. 88, pp. 1802-1810, October 1990.

NON-UNIFORM PIEZOELECTRIC CERAMIC POLARISATION: MINIMISING ULTRASOUND FIELD DIFFRACTION

Vera L. S. Nantes Button [1]; Eduardo T. Costa [1]; Joaquim M. Maia[1] and S. Leeman[2]

[1]Universidade Estadual de Campinas, Depto. de Engenharia Biomédica, Faculdade de Engenharia Elétrica e de Computação and Centro de Engenharia Biomédica, CP6040, 13081-970, Campinas, SP, Brazil
[2]King's College London, Dept of Med. Eng. & Phys., Dulwich Hospital, London, SE22 8PT, UK

INTRODUCTION

Acoustic diffraction can be understood as the interference between two wave components: direct or planar and edge waves propagating outwards from the transducer face[1,2]. Diffraction has been studied for decades and one looks for methods that minimise its effects in the acoustic field produced by ultrasound transducers. With this objective, arrays and single element transducer using piezoelectric ceramic, piezocomposites and PVDF have been described in the literature[3,4,5].

The term "apodisation" is generally used to describe a non-uniformity of the source transducer, which results in a field with reduced diffraction[6,7,8]. In our work we have modified the electric polarisation of a piezoelectric ceramic disc to reduce the edge wave contribution, obtaining a planar single element transducer that generates a less diffractive field.

We have used a non-linear electrical poling field in order to produce intense ferroelectric dipole alignment in the central region of the ceramic disc, and less intense alignment of the dipoles in the edge of the disc. To define the ideal poling profile we have used finite element method (FEM) simulations with ANSYS5.1 software. The ideal electrical field intensity profile was achieved with a spherical poling electrode. We have removed the metallic electrode deposited on one of the disc faces, leaving the other electrode intact to be used as ground electrode. The spherical electrode was then put in contact with the electrode-free disc face. FEM simulations were also performed to study the apodisation effect on the vibrational behaviour of the piezoelectric ceramic disc.

The non-uniform polarisation (apodisation) of the ceramic disc was expected to make it vibrate more intensely in its central region than at its periphery, emphasising the direct wave contribution to the field, while minimising the contribution of the edge wave. Experimental

Acoustical Imaging, Volume 25, Edited by Halliwell and Wells.
Kluwer Academic/Plenum Publishers, 2000.

apodisation of ceramic discs and acoustical field mappings of the ultrasound transducers that we have constructed, confirmed that this goal has been reached.

SIMULATIONS

FEM Electrostatic Analysis was used to simulate the poling electric field produced inside a ceramic disc (1mm thick and 12.7mm diameter) when the surface-deposited electrode was substituted by a spherical electrode placed right in the centre of the top of the disc. Distinct values for the spherical electrode radius were used in the simulations. Figure 1 shows the simulated volume (a) and the resultant poling field (b), which is stronger in the central region of the disc. This simulation was performed with a non-uniform electric field produced by a 2kV voltage applied by a spherical electrode with 5mm radius. The bottom plane face electrode was grounded.

The curve shown in Figure 1(b) was fitted by a polynomial equation and it was used to modify the piezoelectric matrix that describes an apodised ceramic disc during the piezoelectric FEM simulations. Transient, modal and harmonic analyses were performed to study the vibrational behaviour of apodised and non-apodised ceramic discs. In this article we show some results of the transient analysis.

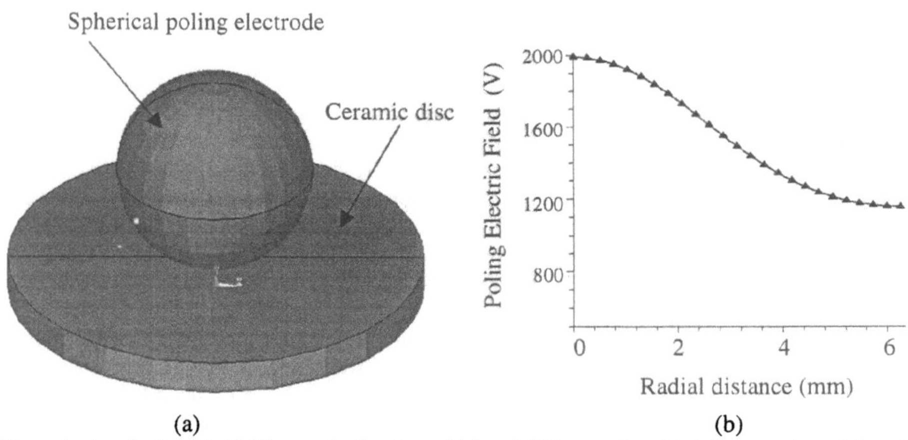

Figure 1. Apodisation of a PZT5 ceramic disc, 1mm thick and 6.35mm radius. (a) Axisymmetric model used in simulations. (b) Non-linear poling electric field as predicted by FEM simulations.

For the transient analysis, the apodised and non-apodised ceramics were excited by a negative narrow rectangular pulse (5V, 400ns). Vertical displacement (UZ) versus time curves were plotted for a node of the ceramic disc mesh located right in the centre of its top face. Figure 2 shows the predictions for the first 5μs after the stimulus application. The ceramic discs initially vibrate at the resonance frequency and the apodised ceramic showed a maximum displacement 13% larger than the ceramic simulated without apodisation.

Transient analysis prediction for the time interval from 0 to 20μs after the stimulus application were used in the PSD (power spectral density) calculations (Figure 3). These show that the vertical displacement energy, around the resonance frequency, was about 30% larger for the apodised ceramic.

These results were used in a preliminary apodisation study to define the format and the size of the poling electrode. In the next section we describe the experimental work accomplished to confirm that the apodised ultrasound transducers generate acoustic fields with less diffraction.

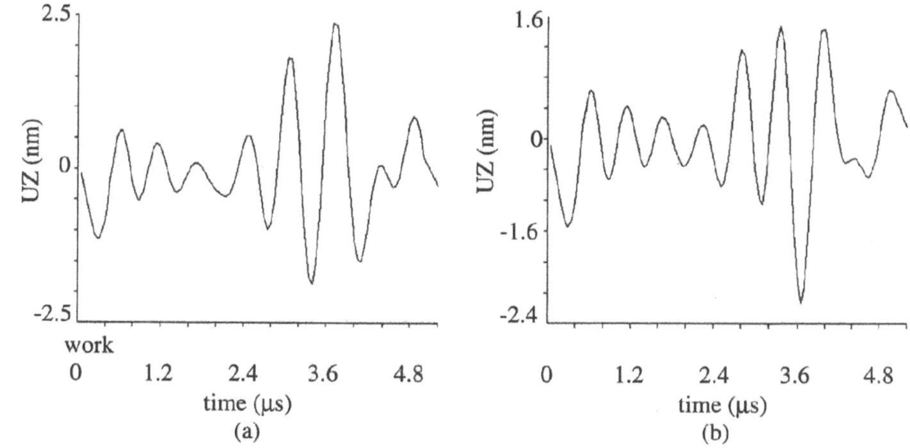

Figure 2. Vertical Displacement (UZ x t) of the central node (top face) from the ceramic disc mesh, for apodised (a) and non-apodised (b) piezoelectric ceramics.

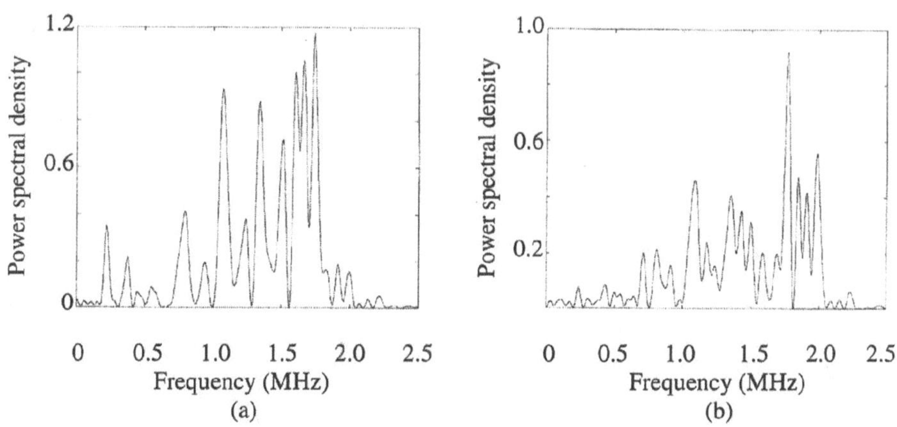

Figure 3. PSD of the vertical displacement of the central node from the mesh of the ceramic disc top face: (a) apodised ceramic and (b) commercially polarised ceramic.

TRANSDUCER DEVELOPMENT

Some piezoelectric ceramic discs were apodised according to the apodisation scheme defined in the FEM simulations. The previously removed face electrodes were reconstructed using a gold sputtering technique.

We have constructed backed ultrasound transducers with non-apodised commercial piezoelectric ceramic discs and with our apodised ceramics, in order to compare their performances.

All the ceramic discs used were PZT5, 12.7mm diameter and 1mm thick. Figure 4 shows the transducer assembly scheme.

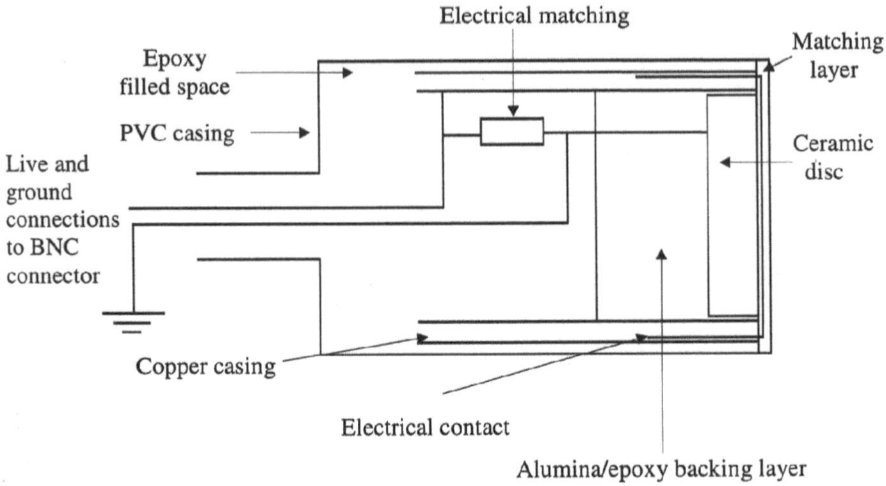

Figure 4. Transducer assembly scheme.

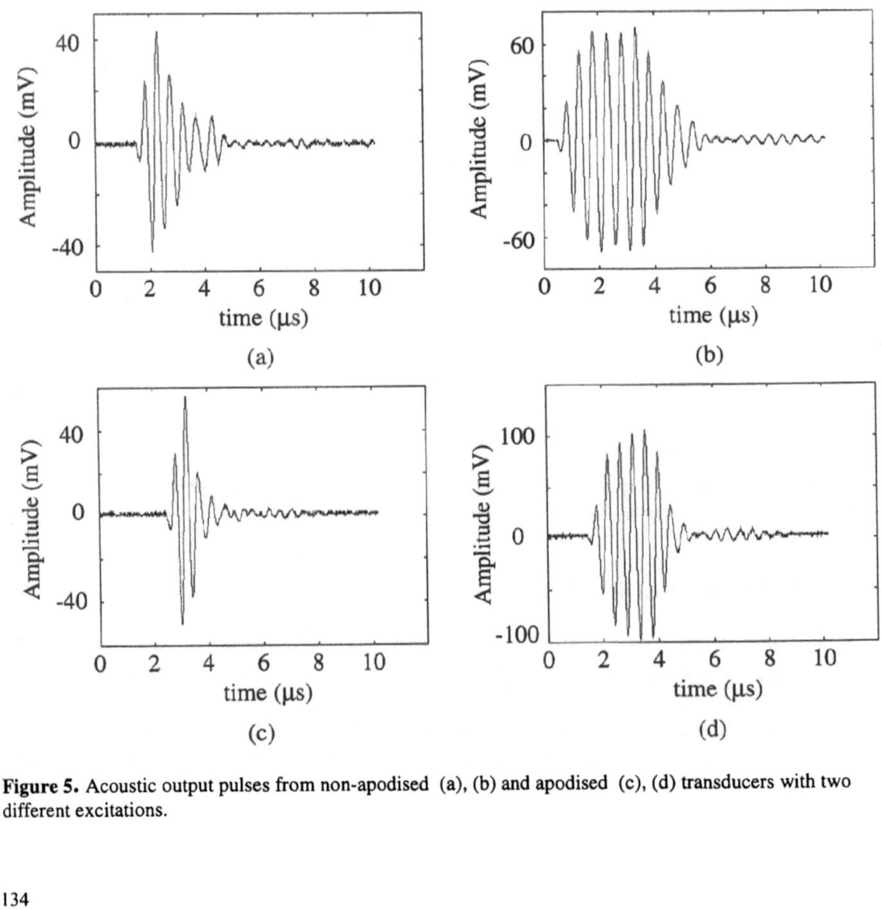

Figure 5. Acoustic output pulses from non-apodised (a), (b) and apodised (c), (d) transducers with two different excitations.

TRANSDUCER TESTING

Pulsed ultrasound fields were produced in water by backed (≅ 2MHz centre frequency) planar transducers constructed with apodised and non-apodised piezoelectric ceramic discs. A point PVDF hydrophone, 1mm diameter, was used to probe these fields.

Figure 5 shows acoustic output pulses from two of these transducers, excited by bursts with 1 and 5 sinusoidal pulses. There was no significant difference between the apodised and non-apodised transducer output amplitudes, but the apodised transducer exhibited output pulses a few cycles shorter than the non-apodised one.

Figure 6 shows the ultrasound field contour plots from apodised and non-apodised transducers excited by a 10-cycle sinusoidal pulse. We noticed that the –3dB and –6dB contour plots from the apodised transducer field extended to larger distances from the transducer face than the same contour plots of the non-apodised transducer field. This result

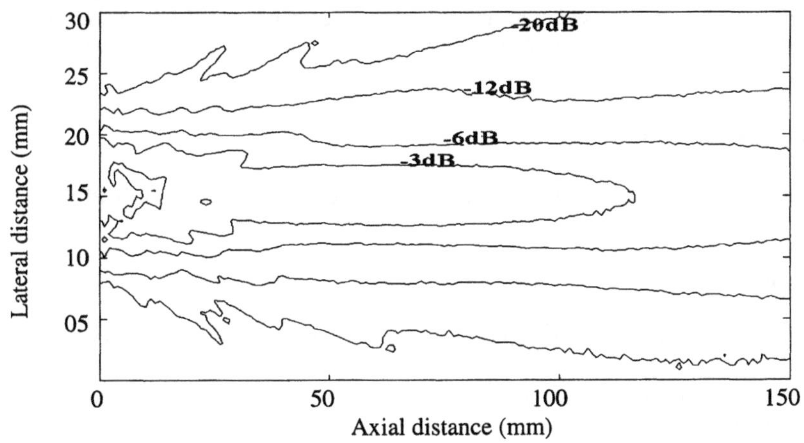

(a)

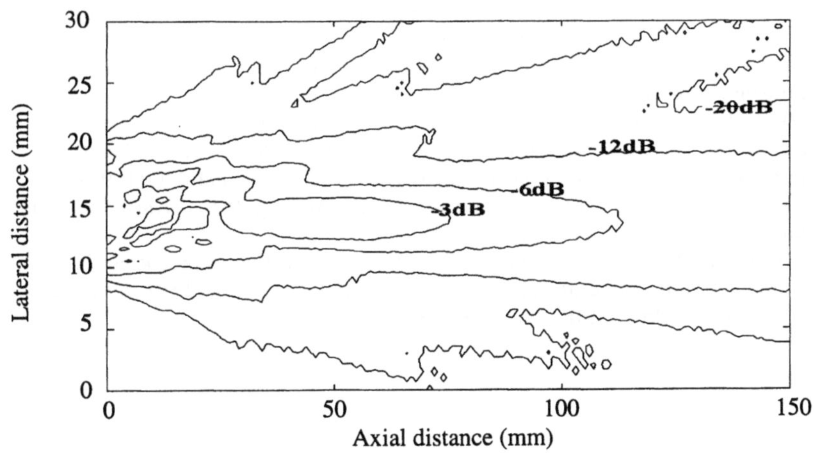

(b)

Figure 6. –3dB, –6dB, -12dB and –20dB contours of the acoustic field of an apodised (a) and a non-apodised transducer (b).

was also observed with the other 4 apodised and 2 non-apodised transducers that we have constructed, and it was understood to be a consequence of the weaker polarisation at the discs periphery that caused a diffraction reduction in the acoustic field.

CONCLUSION

We have manipulated the poling process of piezoelectric ceramic discs in order to reduce the diffraction effects caused by edge waves in the acoustic field. FEM simulation was used to investigate the poling electric field format and the vibrational behaviour of apodised and non-apodised piezoelectric ceramic discs.

Apodised ceramics were used to construct ultrasound transducers and their acoustic fields showed an improved pattern when compared to those of transducers constructed with commercial ceramics poled in a conventional way. The –3dB and –6dB contours of the acoustic field show that an apodised transducer shows a large depth of field, a very desirable result if medical images are to be obtained from deep structures without lowering the ultrasound frequency.

ACKNOWLEDGEMENT

The authors acknowledge the financial support of FINEP (RECOPE) and FAPESP.

REFERENCES

1. J. P. Weight, Ultrasonic beam structures in fluid media, *J. Acoust. Soc. Am.* 76:1184-1191 (1984).
2. P. R. Stepanishen, Transient radiation from pistons in an infinite baffle, *J. Acoust. Soc. Am.* 49:1629-1638 (1971).
3. E. T. Costa and S. Leeman, Diffraction artefacts and their removal, *Acoustical Imaging* 17:403-411, Plenum Press, NY (1988).
4. J-Yu Lu and J. F. Greenleaf, Ultrasonic nondiffracting transducer for medical imaging, *IEEE Trans. UFCC* 37:438-447 (1990).
5. R. H. Brittain and J. P. Weight, Fabrication of non-uniformly excited wide-band ultrasonic transducers, *Ultrasonics* 25:100-106 (1987).
6. D. K. Hsu, F. J. Margetan, M. D. Hasselbuch, S. J. Wormley, M. S. Hughes and D. O Thompson, Technique of non-uniform poling of piezoelectric element and fabrication of gaussian transducers, *IEEE Trans. UFCC* 37:404-410 (1990).
7. D. R. Dietz, Apodized conical focusing for ultrasound imaging, *IEEE Trans. Sonics and Ultrasonics* 29:128-138 (1982).
8. V. L. S. Nantes Button, E. T. Costa and J. M. Maia, Apodization of piezoelectric ceramics for ultrasound transducers, in *Ultrasound Transducer Engineering*, K. K. Shung, Ed., Proc. SPIE 3664:108-118 (1999).

A NEW METHOD FOR MEASURING ULTRASOUND BEAMS PHASE DISTRIBUTION

Gabriele Guidi, Simone Falteri, and Emiliano Maione

Dept. of Electronics and Telecommunications,
University of Florence, Italy

INTRODUCTION

In ultrasound equipment the concept of resolution has different flavors depending on the parameter that is measured. For example with an imaging system two reflecting targets tend to be detected as a single target if their distance is shorter than the ultrasound beamwidth, usually represented by the Full Width at Half Maximum (FWHM[1]). Hence, for the same target-to-target distance, the lower is the FWHM, the better will be the capability of the system to distinguish each reflecting target (i.e., the lateral resolution).

In color flow mapping applications, in addition to the spatial resolution, which is influenced by the same kind of parameters influencing the B-Mode image, the velocity resolution has to be considered, or the capability of the system of detecting Doppler contribution related to different velocities. In the ideal case, where the exploration beam is a plane wave, each flow-line produces a spectral line deviated from the transducer frequency, of a quantity (Doppler shift) proportional to its velocity. If these conditions would be exactly satisfied also for real equipment, a measurement on a real flow, where flow-lines at different velocities are composed together, would give a Doppler signal whose spectrum would be the velocity histogram related to the analyzed Sample Volume (SV).

As a matter of facts, each focused beam gives an acoustic field definitely different by a plane wave, so that for a single velocity component a full spectrum of Doppler frequencies is generated, originating the so-called Intrinsic Spectral Broadening (ISB), whose extent is larger as the focusing is stronger.

In real flow measurements, different flow lines crossing the SV involve the superposition of broadened spectral components, which, as a result, give a Doppler spectrum as distant from an histogram as stronger is the amount of ISB[2]. This deviation might alter significantly the parameters that are usually extracted from the Doppler spectrum (e.g., mean or peak frequency), with consequent velocity estimation errors.

The ISB produced by a specific transducer influence the velocity estimation accuracy of the Doppler flowmeter at which is connected; therefore it is related to the overall velocity resolution of the system. For example, in order to distinguish two parallel flow lines moving at slightly different velocities, a probe producing little ISB is needed, so that the two related spectral components are kept apart and the related spectral indexes are well different each

Acoustical Imaging, Volume 25, Edited by Halliwell and Wells.
Kluwer Academic/Plenum Publishers, 2000.

other. On the other hand little ISB means weak focusing, involving a worst lateral resolution. Hence, if the spatial resolution is increased, also the velocity uncertainty increases, and vice-versa.

Although a compromise focusing can always be found, if ISB is accurately characterized, it may be more convenient to look for the best spatial resolution through a sufficiently strong focusing (with no care about possible Doppler artifacts), and to extract from the "broadened" signals parameters as near as possible to the true ones, by taking into account the intrinsic focusing properties of the transducer at the frequency domain processing stage.

This kind of studies gave some results for single element transducers[3], and for arrays[4], with the limitation that the equation used for estimating the amount of "spurious" Doppler bandwidth on the whole spectrum is an approximate relationship, valid only for a limited range of angles, and for a limited range of distances from the transducer, not too far from the focal zone[5].

Recent researches demonstrated that the ISB analysis can be done by considering the Doppler signal generated by a single target moving at constant velocity, as the result of a double modulation in amplitude and frequency, where the amplitude term is related to the behavior of the acoustic field amplitude met by the target during its passage through the beam, while the frequency modulation term is influenced by the acoustic field phase distribution[6]. According to this point of view the simultaneous knowledge of amplitude and phase field distributions generated by a specific transducer allows to predict precisely its ISB, giving a mean for performing precise Doppler measurement corrections, independently of the beam-axis-to-flow-direction angle and of the position along the beam.

While the techniques available for measuring the acoustic field amplitude are well ascertained, there are few methods as fully established for measuring its phase distribution[7,8]. This paper proposes a new high accuracy method for measuring simultaneously the amplitude and phase spatial fields produced by PW ultrasound transducers, based on a proper post-processing procedure over the field recordings made with an hydrophone.

MATERIALS AND METHODS

The measurement technique is based on time domain analysis of the radio-frequency pressure signals produced in front of the transducer when it is excited with a time windowed sinusoidal signal at frequency f_0.

A field representation in the time domain, as the one detected in such way, is doubtlessly the best for analyzing the effect of a transient excitation. One the other hand, when a beamplot have to be calculated, the most convenient representation is that where the time dependent term is separated by the space varying one.

Indicating with u(x,y,z,t) the behavior in time of the acoustic field at the spatial point {x,y,z}, for a CW excitation or a steady period of a pulsed excitation, the field can be expressed as:

$$u(x,y,z,t) = \mathrm{Re}\{\underline{U}(x,y,z)e^{j\omega_0 t}\}; \quad \omega_0 = 2\pi f_0 \tag{1}$$

where $\underline{U}(x,y,z)$ is the complex function obeying to the Helmoltz equation, which defines the spatial varying part of the field[9], having its own modulus and phase:

$$\underline{U}(x,y,z) = U(x,y,z)e^{j\phi(x,y,z)} \tag{2}$$

Looking at u(x,y,z,t) as the projection on the real axis of a rotating phasor with angular velocity ω_0, if we consider a specific point $\{\tilde{x},\tilde{y},\tilde{z}\}$ at two specific times $\tilde{t}$ and $\tilde{t}-\Delta\tilde{t}$, the field can be written as:

$$u(\tilde{x},\tilde{y},\tilde{z},\tilde{t}) = U(\tilde{x},\tilde{y},\tilde{z})\cos[\omega_0\tilde{t} + \phi(\tilde{x},\tilde{y},\tilde{z})] \tag{3a}$$

$$u(\tilde{x},\tilde{y},\tilde{z},\tilde{t}+\Delta\tilde{t}) = U(\tilde{x},\tilde{y},\tilde{z})\cos[\omega_0\tilde{t} - \omega_0\Delta\tilde{t} + \phi(\tilde{x},\tilde{y},\tilde{z})] \tag{3b}$$

If $\Delta\tilde{t}$ is chosen such as:

$$\omega_0 \Delta\tilde{t} = \frac{\pi}{2}$$

the expression 4b becomes:

$$u(\tilde{x},\tilde{y},\tilde{z},\tilde{t}-\Delta\tilde{t}) =$$

$$= U(\tilde{x},\tilde{y},\tilde{z})\cos\left[\left(\omega_0\tilde{t}+\varphi(\tilde{x},\tilde{y},\tilde{z})\right)-\frac{\pi}{2}\right] = U(\tilde{x},\tilde{y},\tilde{z})\sin\left[\omega_0\tilde{t}+\varphi(\tilde{x},\tilde{y},\tilde{z})\right] \quad (4)$$

Hence combining eq 3a and 4 as the real and imaginary part of a complex number, we have:

$$\begin{aligned} & u(\tilde{x},\tilde{y},\tilde{z},\tilde{t}) + ju(\tilde{x},\tilde{y},\tilde{z},\tilde{t}-\Delta\tilde{t}) = \\ & U(\tilde{x},\tilde{y},\tilde{z})\cos\left[\omega_0\tilde{t}+\phi(\tilde{x},\tilde{y},\tilde{z})\right] + jU(\tilde{x},\tilde{y},\tilde{z})\sin\left[\omega_0\tilde{t}+\phi(\tilde{x},\tilde{y},\tilde{z})\right] = \\ & U(\tilde{x},\tilde{y},\tilde{z})e^{j[\omega_0\tilde{t}+\phi(\tilde{x},\tilde{y},\tilde{z})]} = \\ & \underline{U}(\tilde{x},\tilde{y},\tilde{z})e^{j\omega_0\tilde{t}} \end{aligned} \quad (5)$$

that, as written in eqn (1), is the complex expression whose real part is the quantity measured with the hydrophone. In other terms, eqs. 3a and 4 give the real and the imaginary part of the complex function $\underline{U}(\tilde{x},\tilde{y},\tilde{z})$ at time $\tilde{t}$. Operating in CW this time can be arbitrarily chosen, while for PW excitation the spatial field should remain sufficiently steady around $\tilde{t}$.

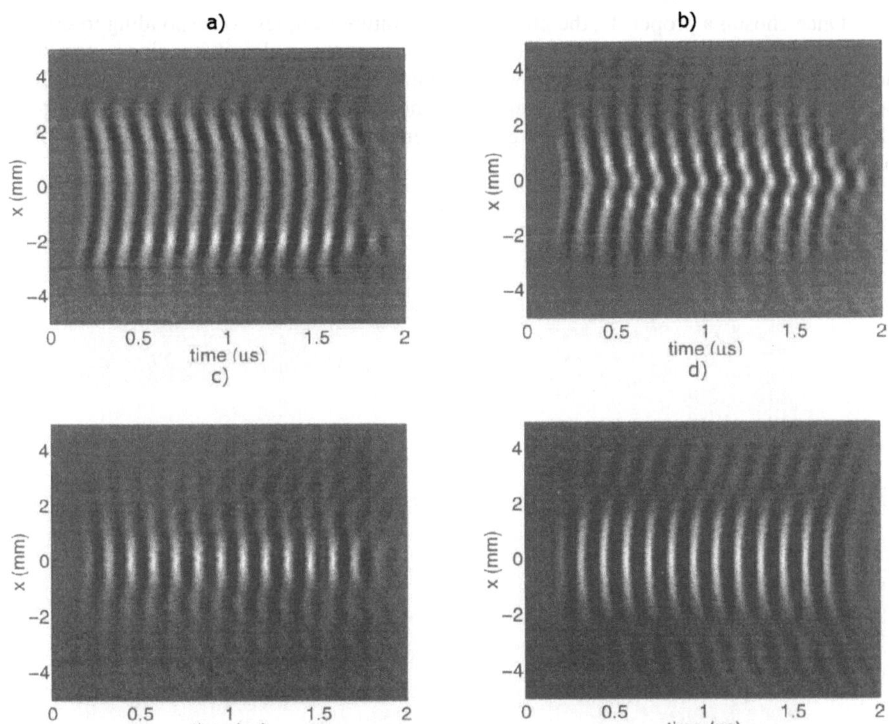

Figure 1 - RF time data acquired at different distances from the aperture: a) z=5 mm (near field); b) z=19 mm; c) z= 33 mm (focal zone); d) z=65 mm (far field)

In the practical technique the transducer is held immersed in degassed water contained in a cubic tank with lateral dimension 50 cm. Pressure signal are detected by a needle hydrophone, moved trough a computer controlled positioning system.

For each transducer-to-hydrophone distance the radio frequency signals are acquired at locations regularly displaced respect to the axial position, in a range symmetrical respect to the beam axis.

The acquisitions reported in this paper were made through a digital oscilloscope LeCroy 9350A (with a bandwidth of 500 MHz and capable of 1Msample/s), at sampling frequency multiple of $4f_0$. In this way the time delay between the I and Q components is covered by an integer number of samples, and the processing described above can be implemented just selecting proper indexes of the acquired data array.

In our experiments a single element spherically focused circular transducer was used, working at 8MHz and manufactured by Krautkramer Branson Inc. (formerly KB Aerotech), Lewistown, PA 17044, USA. The excitation chosen was thus an 8 MHz burst made by 12 pulses.

In figure 1 is shown, for example, the behavior in time of the field generated by the test transducer detected at different distances from the aperture, with lateral displacements ranging from -5 mm to +5 mm. The time axis starts from an offset, set by the operator on the oscilloscope, which represents the global delay of this space-time distribution respect to the application of the transducer excitation. By looking at the figure it can be identified the pulse train evolution starting from the near field (figs. 1a and 1b), proceeding though the focal zone, where the pulse is considerably narrower (fig. 1c), and finishing at the far field (fig. 1d), where the phase fronts appear reverted respect to the near field.

Once chosen a proper $\tilde{t}$, the phase and quadrature samples corresponding to different positions along a direction perpendicular to z, represent the real and imaginary part of the function $\underline{U}$ described by eqn (2), from which module and phase can be easily found. If this processing is applied to a range of z, amplitude and phase fields are measured over a planar region perpendicular to the aperture, giving therefore the full complex behavior of the acoustic field.

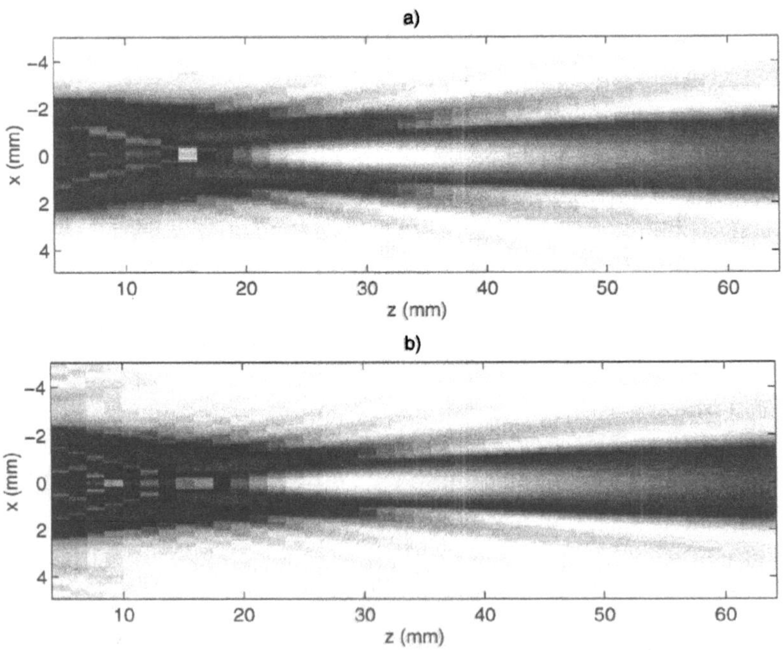

Figure 2 - Comparison between measured (a) and simulated (b) field amplitude on the test transducer

In order to verify the accuracy of the method, especially for the phase measurements, the same signals measured with the hydrophone were simulated.

The tool used for simulating the complex field distribution was the "Field II" program developed by J.A. Jensen[10]. The simulator inputs define the transducer geometry as aperture shape and dimensions, the excitation signal applied to the transducer, and its impulse response. As output it supplies the space-time properties of the pressure field u(x,y,z,t). Such output is analogous to the RF signal displayed by an oscilloscope connected to an hydrophone located at the points of simulation.

RESULTS

The proposed beam characterization technique from time-domain data gave the results summarized by figure 2. A 10 mm wide rectangular area - symmetrical to the beam axis - was explored, spanning in depth from 5 mm to 65 mm from the transducer face. The scan resolution was 1.5 mm along the beam axis and 100 μm on the perpendicular direction.

The simulation parameters of the transducer were measured in laboratory or extrapolated by simulations. The focal distance F (i.e. the point corresponding to the maximum field amplitude) and the FWHM were obtained with direct measurement on the amplitude beamplots. The transducer aperture W was approximately evaluated with a vernier caliper, as the external probe diameter minus two times a roughly approximated value of the plastic shield thickness. The geometrical focus R was initially set equal to the true focal depth. Such aperture and curvature radius values, were used as starting parameters for the beamplot simulations, and were recursively modified to reach a best-fit between measured and simulated field amplitudes. No apodization was considered due to the mechanical assembly of the transducer, made with the active element uniformly glued on a backing layer.

With this procedure the following values of the focusing parameters were found: F=30 mm; FWHM=1 mm; W=6 mm; R=46 mm.

A comparison between measured and simulated amplitude field mapping of the transducer is shown in figure 2. For avoiding discrepancies due to the unknown surface velocity module, v_0, both maps were normalized. For its weak focusing, the measured beamplot of fig. 2a reveals an elongated beam with evident sidelobes and irregularities. Such features are found with extremely good agreement in the simulated beamplot of figure 2b, which slightly differs only in the nearest part of the scanned area (range 5÷20 mm), where the small beamplot features evidenced by the simulation do not appear in fig 2a.

From both the measured and the simulated field data originating such nicely matched beamplots, the phase fields were calculated, according to the method described in the previous section.

The phase behavior at four particular distances from the transducer, corresponding to the RF images of fig. 1, is shown in figure 3. In the very near field (fig 3a) the phase concavity is positive for x in the range [-3, +3] mm, corresponding to the locations directly in front of the 6 mm wide aperture, due to the transducer spherical focusing. While the ultrasonic pulse moves away , this concavity tend to be altered (fig 3b), up to become almost flat around the focus (fig 3c) where all the wave contribution comes simultaneously (i.e. with the same phase), for enhancing the field amplitude in that zone.

At greater distances (fig 3d), the field tend to a spherical wave, so that the corresponding phase has a concavity oriented in opposite way respect to the first picture. In all the four subfigures the dotted lines, representing the simulated results, are near to the measured values, except for large off-axis displacements, where the signal tend to decrease considerably. It has to be reminded though, that the FWHM is 1 mm while the lateral extent of the scanned area covers 10 mm in width.

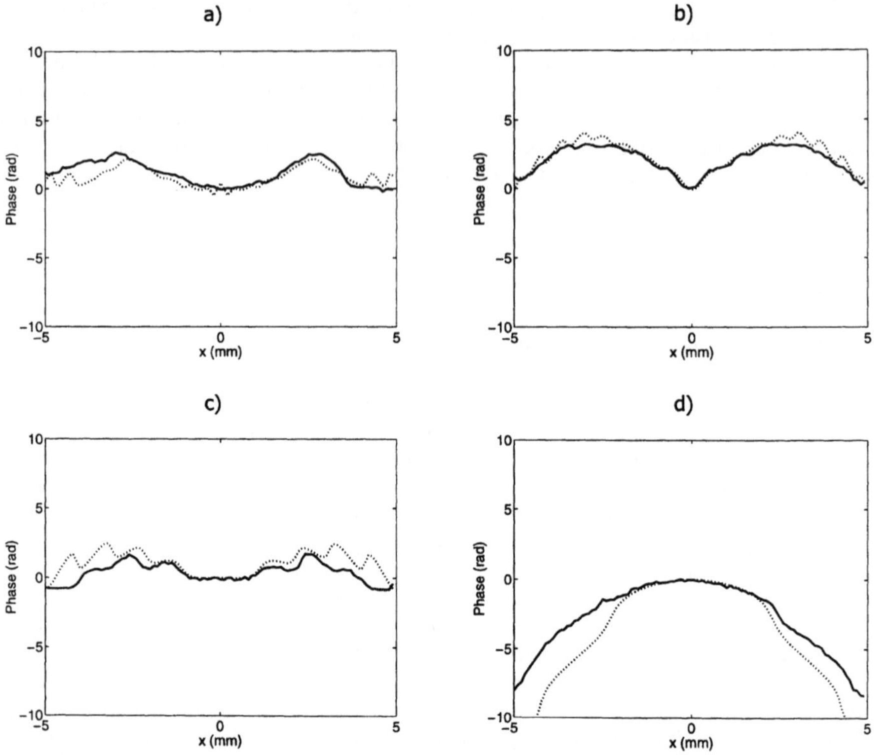

Figure 3 - Phase behavior of the simulated (dotted line) and measured (continuous line) field, at different distances from the aperture: a) z=5 mm (near field); b) z=19 mm; c) z= 33 mm (focal zone); d) z=65 mm (far field)

Finally, in figure 4, an overall comparative view is given, representing as a surface both the measured and the simulated field phases. Although a phase jump, due to a failure of the unwrap algorithm, has to be noticed in the simulated data near 50 mm, also this result seem to confirm qualitatively the correctness of the measurement procedure.

CONCLUSIONS

A new processing method for extracting the complex spatial field, in terms of amplitude and phase, from a proper phase-quadrature sampling of the time field behavior at different spatial points, has been presented. The method has been verified only in the case of spherically focused circular transducers employing time signals acquired by a needle type hydrophone connected a digital oscilloscope, providing an appropriate sampling rate, and a precise (low-jitter) triggering between excitation and field data acquisition.

The same transducers have been properly simulated, and from the time-space field variation, the space dependent amplitude and phase has been extracted through the same method here presented.

Although a thorough validation has still to be done, the measured results seems in good agreement with the expected (simulated) data, suggesting therefore the correctness of the measurement method.

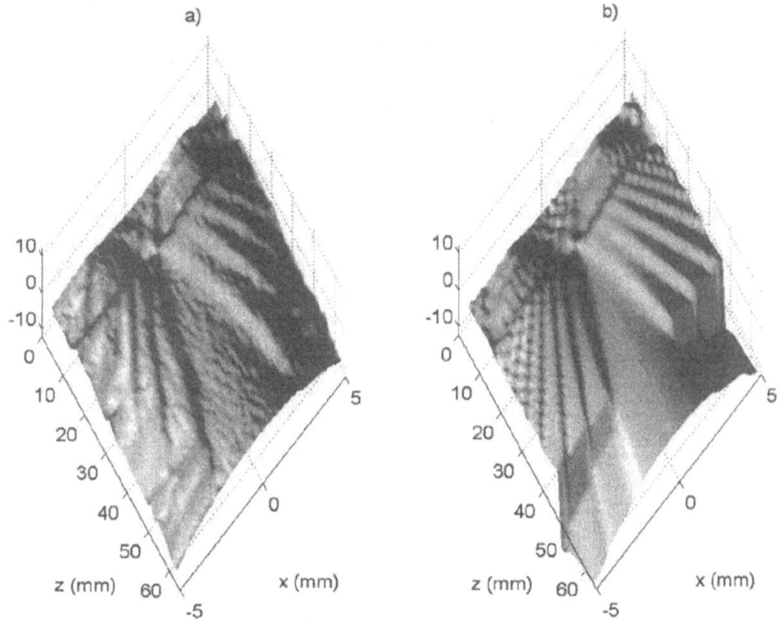

Figure 4 – Overall comparison between measured (a) and simulated (b) field phase distribution on the test transducer

ACKNOWLEDGMENT

The contribution of Dr. Alessandro Gubbini, for giving access to the probe characterization facilities of ESAOTE S.p.a (Florence, Italy) and for setting it up in the best way to implement this method, is gratefully acknowledged.

The authors wish also to thank Professor Piero Tortoli for several useful discussions.

REFERENCES

1. J.W. Hunt, M. Arditi and F. S. Foster, Ultrasound transducers for pulse echo medical imaging. *IEEE Trans Biomed Eng*, 30:453-481 (1983).
2. G. Guidi, V.L. Newhouse, P. Tortoli: A study of Doppler Spectrum shape based on the summation of flow-line spectra, *IEEE Trans Ultrason Ferroelec Freq Cont.* 42-5:907-915 (1995).
3. P.Tortoli, G. Guidi and V.L. Newhouse. Improved blood velocity estimation using the maximum Doppler frequency. *Ultrasound Med Biol.* 21:527-532 (1995).
4. A.J. Winkler and J.Wu. Correction of intrinsic spectral broadening errors in Doppler peak velocity measurements made with phased sector and linear array transducers. *Ultrasound Med Biol.* 21:1029-1035 (1995).
5. V.L. Newhouse, E.S. Furgason, G.F. Johnson and D.A. Wolf. The dependence of ultrasound Doppler bandwidth on beam geometry. *IEEE Trans Sonics Ultrasonics.* 27:50-59 (1980).

6. G. Guidi, C. Licciardello and S. Falteri. Intrinsic spectral broadening (ISB) in ultrasound Doppler as a combination of transit-time and local geometrical broadening, accepted for pubblication on *Ultrasound Med Biol* (2000).
7. J.D. Aindow and R.C. Chivers. Measurement of the phase variation in an ultrasonic field. *J Phys E: Sci Instrum.* 15:83-86 (1982).
8. G. Guidi and S. Falteri. Phase measurements of acoustic fields based on a moving target. *IEEE Trans Ultrason Ferroelec Freq Cont.* 46:679-689 (1999).
9. J. W. Goodman, *Introduction to Fourier Optics*, 33-34, McGraw-Hill, New York (1968).
10. J.A. Jensen, N.B. Svendsen, Calculation of pressure fields from arbitrarily shaped, apodized, and excited ultrasound transducers. *IEEE Trans Ultrason Ferroelec Freq Cont.* 39:262-267 (1992).

A SIMPLE CALCULATION APPROACH FOR THE PARAMETRIC SOUND FIELD GENERATED BY A FOCUSED ANNULAR ARRAY: APPLICATION TO VIBROACOUSTOGRAPHY

Jean-Pierre Remenieras, Olivier Bou Matar, Frédéric Ossant, Frédéric Patat

LUSSI, E.A 2102 / G.I.P ULTRASONS
Faculté de Médecine, 2 bis bd Tonnellé
B.P. 3223, 37032 Tours Cedex
FRANCE

INTRODUCTION

In vibroacoustography, ones tries to form an image of the deformability of a tissue submitted to low frequency excitation forces. These forces can be applied directly onto the surface through an external vibrating system or, as recently proposed by Fatemi and Greenleaf [1], distributed into the volume by using the nonlinear combination of the two interfering primary beams. If the two primary beams are harmonic at frequency f_a and f_b differing by a small difference $f_- = f_a - f_b$, the nonlinear effects will give two types of low frequency forces. Zero frequency forces which are currently named as radiation pressure and a distribution of volume forces, noted F/ϑ, beating at f_- and resulting from the two primary beams intermodulation. The calculation of F/ϑ is the subject of this work. This is important for vibroacoustography since the deformation field of the tissue will result from the convolution of the parametric volume force density by the solid Green's function of the isotropic elasto-dynamic problem. If one ones to have resolution in vibroacoustography imaging, one may use strongly focused primary field so that F/ϑ will develop only in a limited region of space. This is why focused annular arrays have been chosen for calculation.

In general, the combination of diffraction and nonlinearity is not an easy problem to deal with. The chosen approach aims to provide an efficient computation schema. This is done by calculating the source representation as a sum of complex gaussian functions. Then, under the parabolic approximation, the propagation terms and the nonlinear combination of terms keep gaussian. Then, we show how the computation can be performed and used to test different annular arrays configurations.

Acoustical Imaging, Volume 25, Edited by Halliwell and Wells.
Kluwer Academic/Plenum Publishers, 2000.

PRESENTATION OF THE SPHERICAL FOCUSED ANNULAR ARRAY ADAPTED TO PARAMETRIC GENERATION

The parametric focused annular array is shown **figure 1**. The odd elements of the transducer are driven at the primary frequency f_a and the even elements at the frequency f_b. The focusing distance d is determined by our application in the breast and is fixed at 6 cm. The advantage with the annular array, is that the focus is symmetric, i.e. the same both in the scan plane and transverse to the scan plane. With a strong symmetric focusing (focusing gain $G \approx 30$), it is possible to get a very thin scan slice, and important gradient in the transverse direction of the beam. For a uniform amplitude vibration across a focused disc with a diameter $2a$, the focal diameter $D_F(1dB)$ and the depth of focus $L_F(1dB)$ for 1dB definition of the beam width are expressed by the following expression [2]:

$$L_F(1dB) = \pi d/G\,,\ D_F(1dB) = \lambda d/2a\,, \tag{1}$$

with z_0 the Rayleigh distance and G the focusing gain $G = z_0/d = \pi a^2/\lambda d$. If we decrease the frequency f and increase the radius a, with G and d stable, the depth of focus $L_F(1dB)$ and the primaries pressure amplitude are stable. The on axis evolution of the primaries pressure are approximately similar for these two different frequencies. For the design of our annular array, we impose the same focusing gain for the different elements, i.e. $G_a = G_b$. The surfaces are lightly different because the frequencies are different.

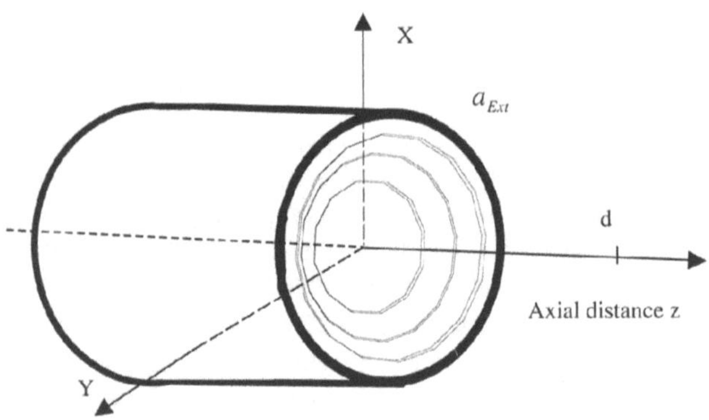

Figure 1. Geometry of the cylindrical parametric annular array

NONLINEAR INTERACTION BETWEEN TWO FINITE AMPLITUDE SOUND BEAMS GENERATED BY A SPHERICAL ANNULAR ARRAY

Our calculation is based on a gaussian beam expansion of the primary fields. In the linear approximation, any axisymetric source can be expressed as a linear combination of gaussian beams with complex parameters. The determination of the latter uses a modified computation method developed by Wen and Breazeale [3]. Following Coulouvrat's work [5], and extending it to the case of multi elements annular arrays, we will see that this gives a convenient way to calculate the parametric field generated by nonlinear interaction of two fields at frequencies f_a and f_b. The analysis is based on the KZK parabolic wave equation, which accounts for nonlinearity, absorption due to viscosity, and diffraction effects:

$$\frac{\partial^2 p}{\partial z \partial \tau} = \frac{c_0}{2}\nabla_\perp^2 p + \frac{\delta}{2c_0^3}\frac{\partial^3 p}{\partial \tau^3} + \frac{\beta}{2\rho_0 c_0^3}\frac{\partial^2 p}{\partial \tau^2} \tag{2}$$

where p is the sound pressure, z is the coordinate along the axis of the beam, $\tau = t - z/c_0$ is the retarded time, c_0 is the small signal sound speed, δ is the diffusivity of sound, ρ_0 is the ambient density and β is the coefficient of nonlinearity. As usual in acoustics, we can assume $p = p_1 + p_2$ where p_1 is the solution of the linearized wave equation and contain two complex amplitudes q_{1a} and q_{1b} at frequencies w_a and w_b:

$$p_1(r,z,\tau) = (1/2j)\left(q_{1a}(r,z)e^{jw_a\tau} + q_{1b}(r,z)e^{jw_b\tau}\right) + cc \qquad (3)$$

q_{1a} for instance satisfies to:

$$\frac{\partial q_{1a}}{\partial z} + \frac{j}{2k_a}\nabla_\perp^2 q_{1a} + \frac{\delta\omega_a^2}{2c_0^3} q_{1a} = 0. \qquad (4)$$

$p_2(r,z,\tau)$ contains the second order fields:

$$p_2(r,z,\tau) = \frac{1}{2j}\left[q_{2a}(r,z)e^{j2w_a\tau} + q_{2b}(r,z)e^{j2w_b\tau} + q_+(r,z)e^{jw_+\tau} + q_-(r,z)e^{jw_-\tau}\right] + cc \qquad (5)$$

and we are interested in the field at the difference frequency $\omega_- = \omega_a - \omega_b$. Combining (2) and (5), we get for q_-:

$$\frac{\partial q_-}{\partial z} + \frac{j}{2k_-}\nabla_\perp^2 q_- + \frac{\delta\,\omega_-^2}{2c_0^3} q_- = -\frac{\beta\omega_-}{2\rho_0 c_0^3} q_{1a} q_{1b}^* \qquad (6)$$

The primary fields are calculated as gaussian decompositions in the transducer plane:

$$q_a(r,0) = \sum_{n=1}^{Na} B_{an}\, e^{-(r/A_{an})^2}, \qquad q_b(r,0) = \sum_{m=1}^{Nb} B_{bm}\, e^{-(r/A_{bm})^2} \qquad (7)$$

Using the Green's functions of equation (4) and (6), the w_- sound field amplitude $q_-(r,z,\tau)$ can be given in an analytical form by:

$$q_-(r,z,\tau) = -\frac{k_-\beta}{2\rho_0 c_0^2} e^{-\alpha_- z} \sum_{n=1}^{Na}\sum_{m=1}^{Nb} B_n B_m^* \int_{z'=0}^{z} \bar{q}_{nm}(z,r,z')dz' \qquad (8)$$

where

$$\bar{q}_{nm}(z,r,z') = \frac{e^{(\alpha_- - \alpha_a - \alpha_b)z'}}{(a_{nm} + b_{nm}z')} e^{-\frac{r^2}{2}\frac{(d_{nm} + e_{nm}z')}{(a_{nm} + b_{nm}z')}}$$

$$a_{nm} = 1 - j\left(\frac{k_a}{z_{an}} + \frac{k_b}{z_{bm}^*}\right)\frac{z}{k_-} \qquad d_{nm} = \frac{1}{z_{an}z_{bm}^*}\left(k_a z_{bm}^* + k_b z_{an}\right)$$

$$b_{nm} = \frac{j}{k_- z_{an} z_{bm}^*}\left(k_a z_{an} + k_b z_{bm}^*\right) + \frac{z}{z_{an} z_{bm}^*} \qquad e_{nm} = j\frac{k_-}{z_{an} z_{bm}^*}$$

Using this method, the parametric sound beam $q_-(r,z,\tau)$ is simply the sum of cross interaction terms produced by the series of Gaussian beams. Each interaction term is the complex integral easily numerically calculable.

NUMERICAL RESULTS

Influence of the curvature of the annular array on parametric beam

In order to study the influence of the curvature, a two elements annular array has been chosen for simplicity. The radius $a_1 = 3cm$, $a_2 = 4.08cm$ as well as the frequencies $f_a = 1MHz$ and $f_- = 10KHz$ are kept fixed. The source pressure at the transducer level is fixed to be 0.5 MPa. The focal distance d is varied between 5 and 50 cm (i.e G_a ranges from 37.7 to 3.7). **Figure 2** shows the on axis pression of the f_- parametric field. The stronger the focusing gain G_a, the sharper the parametric pression distribution. The rapid decay in the post focal region is due to the rapid spherical spreading of the primary beams and is more prononced when G_a is large. However, the maximum parametric pressure keeps around 200 Pa because the focusing effect obtained on the primary beams is counter balanced by the shortening of the interaction region.

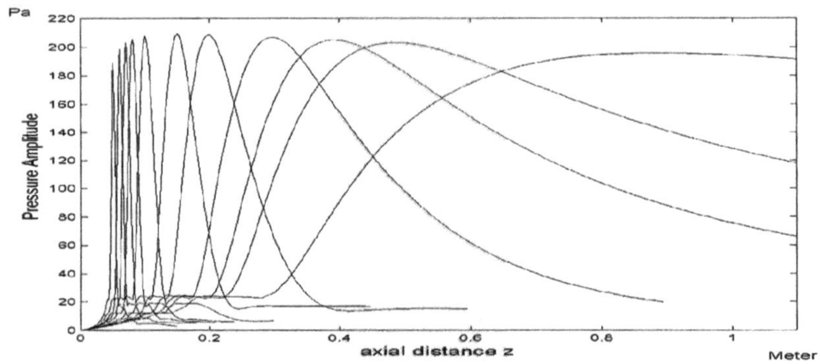

Figure 2. On axis parametric pressure amplitude for two elements focused annular array. $r_a = 1.88m$, $f_a/f_- = 100$. The focusing distance varies from d=5 cm to d=50 cm. $a_1 = 3\ cm$ and $a_2 = 4.08\,cm$.

Influence of the frequency downshift ratio on parametric generation

In order to study the influence of the f_a/f_- ratio alone, we need to keep a constant focusing gain $G_a = 31.4$. So the ratio a_1^2/λ_a is keep constant equal to that obtained with $a_1 = 3cm$, $f_a = 1MHz$. While f_a varies from 1 to 6MHz, the internal radius a_1 goes from 3 to 1.2 cm. In all case, the primary pressures $P_{a\,Max}$ and $P_{b\,Max}$ peaks at focus around 16 MPa. In **figure 3.a** the on axis amplitude of the parametric beam is shown with the downshift ratio being varied from 100 to 600. It can be easily seen that increasing the downshift ratio decreases the difference frequency generation. **Figure 3.b** shows the increase in transverse directivity obtained on the secondary beam.

Influence of the number of rings of the parametric array

We saw that the maximum pressure level for the difference frequency wave is limited by the size of the interaction region of the two primary fields. The latter can be increased if one separates different rings on the transducer, these rings being alternatively excited with frequencies f_a and f_b. This can be compared to the single element configuration which is excited by the sum of excitation sinus at f_a and f_b. Results are plotted on **figure 4** where the calculation has been done with $a_{External} = 4.08\,cm$, $f_a = 1MHz$, $f_- = 10KHz$ and 1,2,4,6 and 8 rings. If we increase the number of rings, we decrease the axial and lateral resolution, but we increase the parametric generation. An array with 4 rings at 1 MHz is an acceptable compromise between sensibility and resolution.

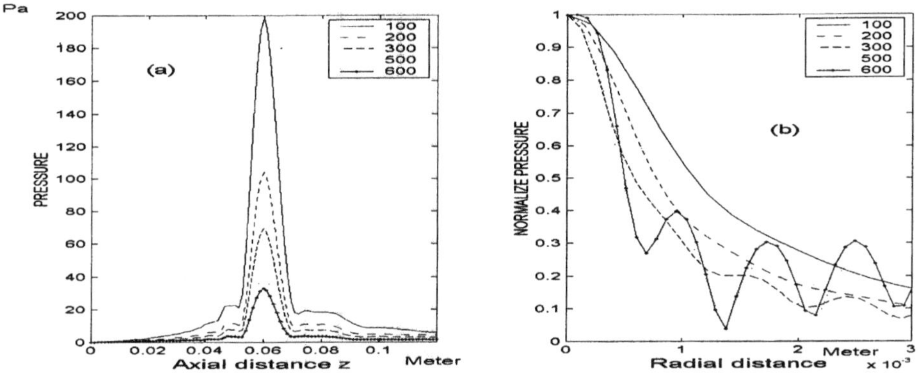

Figure 3. (a) On-axis pressure amplitude, (b) Beam patterns in the focal plane for different downshift ratio

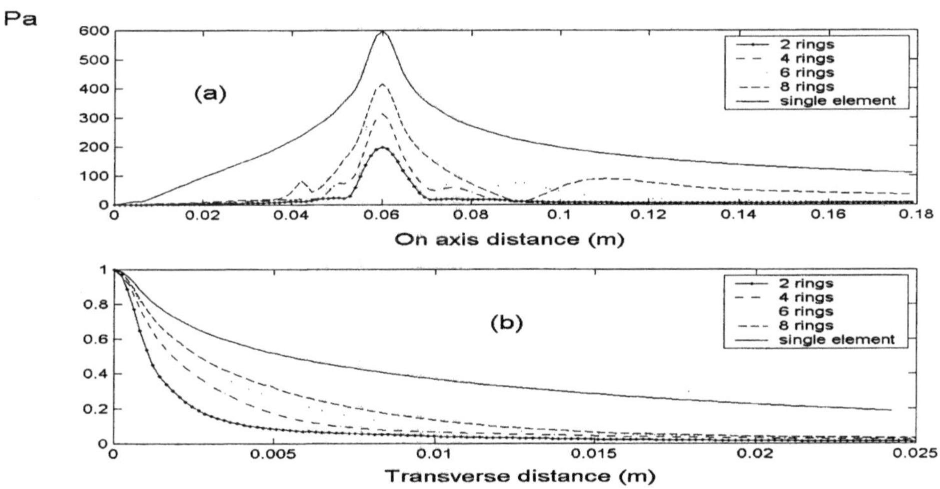

Figure 4. On-axis pressure amplitude (a) and beam patterns (b) of the difference frequency wave

Calculation of the parametric volume force

The parametric pressure field q_- which has been calculated is given by equation (8) as a function of the variable r,z and τ where τ is the retarded time and is modulated by the term $\exp(-j\omega_-\tau) = \exp(j(k_-z - \omega_-t))$ in the (x,y) plane. The real part of the expression $q_- \exp(-j\omega_-\tau)$ has then to be taken in order to draw the real parametric pressure field in this plane. This is represented on **figure 5a** and **b** for two different times opposite in phase which means differing by $50\mu s$ and for a 4 elements annular array. A step further can be then be undergone by calculating the gradient of this field which then gives the distribution of the volume force provided by the nonlinear interaction. This is shown on **figure 6** where we can see that the most important gradients stand in the transverse direction of propagation. This comes from the strong focusing effect of the primary fields.

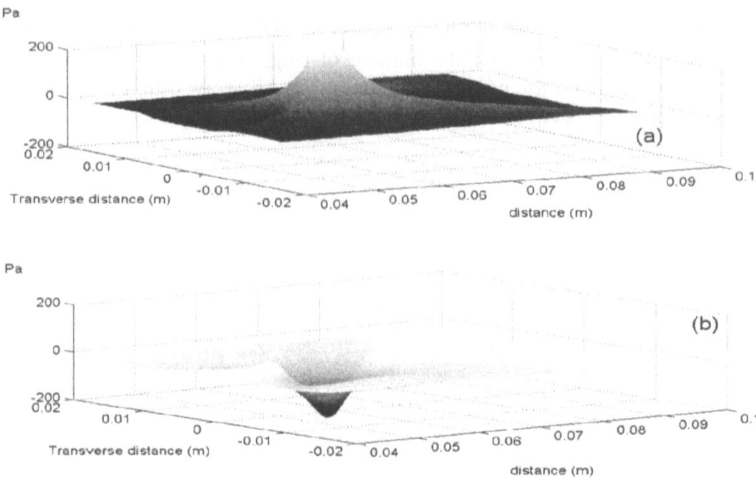

Figure 5. Real part of the parametric pressure in the (x,z) plane for two times separated by $50\mu s$

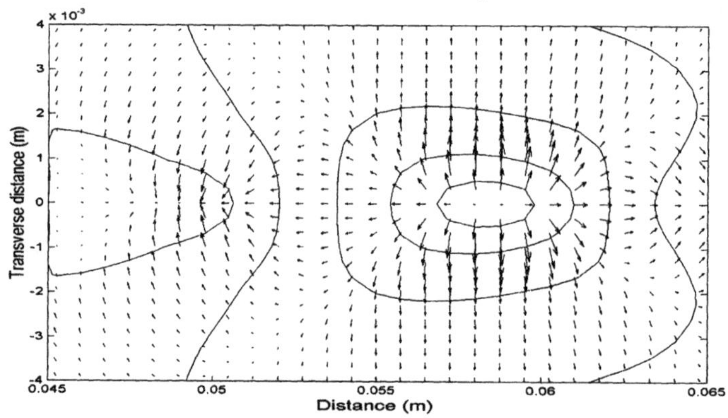

Figure 6. Gradient of the real part of the parametric pressure represented on figure 7.

CONCLUSION

This works shows a computation method for the calculation of the space distribution of the low frequency part of the nonlinear interaction of two strongly focused beams. The approach, based upon the gaussian development of the primary source terms, reduces the calculation to a single integral over a bilinear combination of the source terms. It has been used on multi-element annular arrays in order to predict the volume force distribution that will then act on the propagation media. In our experience, a number of a hundred complex gaussian source terms is enough to give a reasonable precision on the parametric field diagram generated by a four rings transducer.

REFERENCES

1. M.Fatemi and J.F.Greenleaf. Ultrasound-stimulated vibro acoustic spectrography, Science. 280:82 (1998).
2. B.A.Angelsen. Waves, signals and signal processing in medical ultrasonics, .1: (1996).
3. J.J.Wen and M.A.Breazeale. Computer Optimization of the Gaussian Beam Description of an Ultrasonic Field, Computational Acoustics. 2:181 (1990).
4. M.F.Hamilton and D.T.Blackstock, Book Nonlinear Acoustics, Academic Press, San Diego, (1998).
5. F.Coulouvrat, Thesis "Méthodes asymptotiques en acoustique non lineaire", University P&M Curie-Paris 6 (1991).

A CONVEX-CONVEX SHAPED TWO-DIMENSIONAL ARRAY PROBE USING MULTILAYER PZT TRANSDUCERS

Shuzou Sano[1], Takaya Ohsawa[1], Satoshi Tamano[1], Yutaka Sato[1], Syosaku Ishihara[2], Masato Nakamura[2], Yuichi Miwa[3], Takashi Azuma[3]

[1]Hitachi Medical Corp., Research & Development Center
2-1, Shintoyofuta, Kashiwa, Chiba 277-0804, Japan
[2]Hitachi Corp., Production Engineering Research Laboratory
292, Yoshidacyo, Totsuka, Yokohama, Kanagawa 244-0817, Japan
[3]Hitachi Corp., Central Research Laboratory
1-280, Higashikoigakubo, Kokubunji, Tokyo 185-8601, Japan

INTRODUCTION

Three-dimensional ultrasonography is one of the recent technological advances in diagnostic imaging for the easy recognition of organs and blood vessels. This technology uses different devices such as a conventional one-dimentional array probe with a motor drive for scanning and a sonographer with a position sensor for generating the position data of the probe in the wedge, linear, and rotational directions. Another type of probe used for electronic scanning in the volumetric region is the 2-D array probe, and different types of sparse-array geometries have been investigated to improve pulse-echo beam width and reduce sidelobes [1].

The areas of clinical applicability for this technology continue to expand. It has been applied in the field of cardiology for imaging the complex anatomy and dynamics of the heart; to obstetrics for evaluating the fetus which can be difficult due to fatal size and movement; to the study of vascular system for imaging the blood flow and vessel anatomy; and to gynecology for characterizing uterine anomalies [2].

Furthermore, 3-D ultrasonography with conventional 1-D array probe scanning with a motor drive or sonographer has limited image quality and real-time imaging. The 2-D array probe will be essential in diagnostic ultrasound equipment because it improves the clinical image quality and real-time imaging. It achieves this allowing the image thickness to be reduced by focusing in the elevation plane perpendicular to the scanning dimension.

On the other hand, the 2-D array probe with sparse-array requires many beam-forming channels (more than 256 corresponding to the transmit and receive channels), and more and more to improve clinical image quality.

Our purpose is to develop a 2-D array probe with PZT transducer elements arranged with the same pitch for both the short and long axes, and a system to obtain and reconstruct volume data for 3-D ultrasonography.

In this paper, we discuss the 2-D array probe which is structured with multilayer PZT transducer elements to reduce electrical impedance and transmitter voltage, and designed with a convex-convex shape to maintain wide 3-D view.

ARRAY DESIGN

Figure 1 shows an array probe obtaining the volume data for 3-D ultrasonography. To achieve high-imaging quality, we used a focused beam which is equivalent to the beam formed by the circular concave transducers, and we arranged transducer elements in several rings by tying up electrically and focused the ultrasound by setting up the delay time in the order of rings.

By forming the 2-D array probe in a convex-convex shape, volume data can be obtained easily by only shifting the focused beam freely on the 2-D array probe. The beam is shifted by changing the switches connected to the transducer elements.

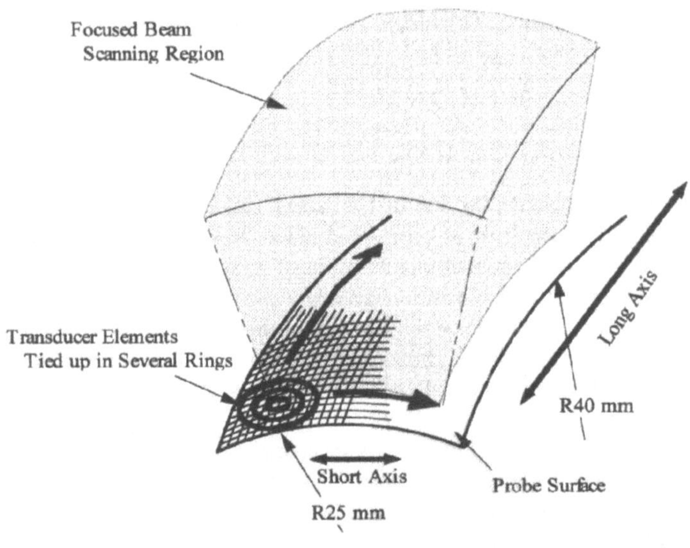

Figure 1. Schematic diagram of obtaining volume data for 3-D ultrasonography by scanning the focused beam freely on the probe surface with a convex-convex shape.

Figure 2 shows the fundamental structure of the 2-D array probe corresponding to a convex-shaped thin-array transducer which consists of multilayer PZT transducer elements with two matching layers, a backing layer, and lead wires. Multilayer PZT transducer elements are connected to lead wires with a fixed pitch and convex shape along long axis between small interconnection pads and inner electrodes on one side of the transducers.

The 2-D array probe has an offset arrangement that is convex in shape because the thin-array transducers form a convex short-axis curvature. This structure is obtained as follows:

1) A multilayer PZT transducer with a structure that is electrically parallel and acoustically serial can obtain a lower driving voltage and suppress the increase of electrical impedance due to the small-size transducers.
2) Manufacturing processes with the offset arrangement of thin-array transducers can secure a high yield compared by directly dicing the PZT transducers into the 2-D structure.
3) Thin-array transducers are useful for connecting and mounting the electronic circuits, such as cross-point switches.

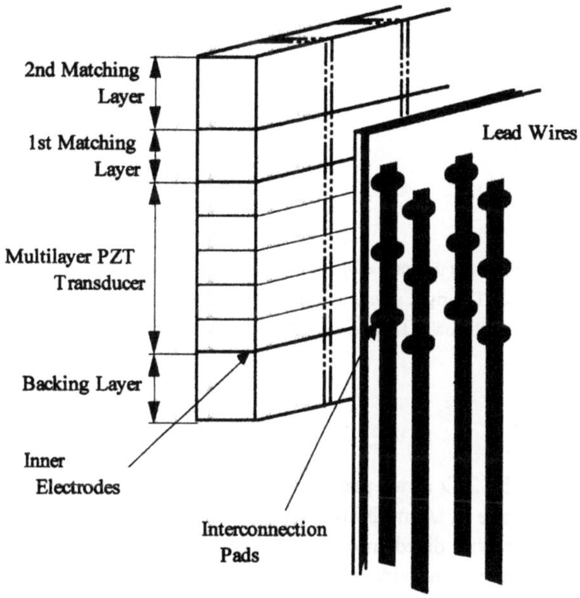

Figure 2. Schematic diagram of the convex-shaped thin-array transducer which mainly consists of multilayer PZT transducer elements with two matching layers, a backing layer, and lead wires. Each transducer element is interconnected to lead wires in a fixed curvature.

The principal specifications of our 2-D array probe are listed below:

1) The arrangement of the transducer elements is 192 ch in the curvature of the 40 mm long axis, and 64 ch in the 25 mm short axis, with 0.38 mm pitch.
2) Element characteristics with a 5-layer transducer are a 2.5 MHz center frequency and a 100 pF capacitance which corresponds to an approximately 600 Ω impedance.
3) The region scanned by the focused beam is expected to be 100 degrees along the long axis, and 64 degrees along the short axis.

In order to investigate some characteristics of the 2-D array probe and system to form the focused beam, some simulations were performed.

Figure 3 shows the result of the acoustic pressure field analysis by using the FEM (Finite Element Method) when the pulsed ultrasound is transmitted from one transducer element which consists of two matching layers, multilayer PZT transducer, backing layer, and lead wires inserted by offset arrangement thin-array transducers. From this result, we can confirm that the ultrasound is continuously transmitted and propagated in front of the transducer element, and is not influenced by the step-shaped structure.

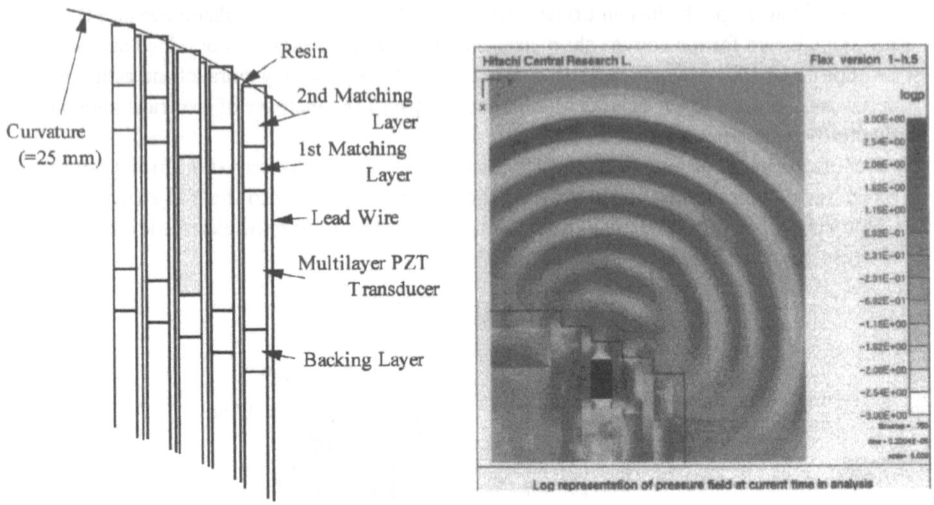

Figure 3. The left side is a schematic diagram of the 2-D array probe model used in the analysis of acoustic pressure field by FEM simulation. The right side is the result when the pulsed ultrasound is transmitted from one transducer element inserted by the offset arrangement of thin-array transducers.

ARRAY FABRICATION

In order to develop the 2-D array probe, we fabricated 64 convex-shaped thin-array transducers, shown in Figure 4. A multilayer PZT transducer sheet with two matching layers and a backing layer was diced into slices. These slices are pretreated in order to interconnect the lead wires on one side of the transducer and diced into 192 transducer elements. These transducer elements are arrayed with a 40 mm curvature.

Next, lead wires are alternately interconnected to the inner electrodes corresponding to each layers at one side of the transducer elements.

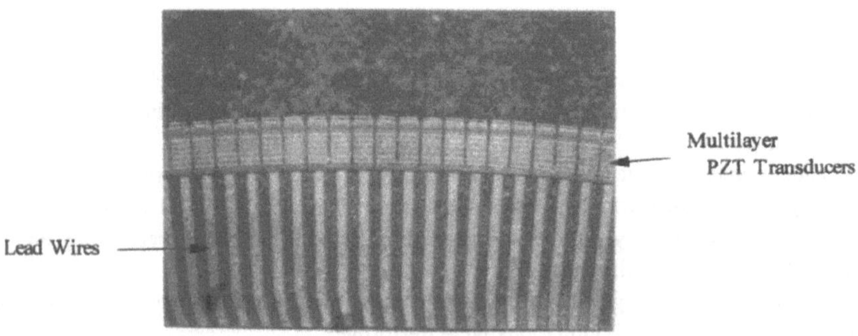

Figure 4. Photograph of convex-shaped thin-array transducer when multilayer PZT transducers are interconnected to lead wires with a 0.38 mm pitch.

In the next stage, 64 convex-shaped thin-array transducers are arranged with an offset structure to achieve a convex short-axis curvature of 25 mm, with a 0.38 mm pitch. They are then fixed by a resin selected as the most suitable for acoustic conditions.

Figure 5 shows a photograph of the 2-D array probe assembled in the probe case. The surface is convex-convex shape, and the interconnection pads in each lead wire sheet are fixed to the electronic circuits in order to transmit and receive ultrasound during beam formation.

Figure 5. Photograph of 2-D array probe with a probe case. Interconnection pads formed on the lead wire sheets are interconnected to electronic circuits.

MEASUREMENT AND RESULTS

To evaluate the pulse and frequency characteristics of the transducer elements, we used an oscilloscope to investigate the impulse response and power spectrum of the echo signal reflected from the metal plate located 2 cm in front of the transducer array. Impedance and phase characteristics were investigated by an impedance analyzer (HP 4194A). Figure 6 shows the results of measurement and simulation using a 3-D model for reference.

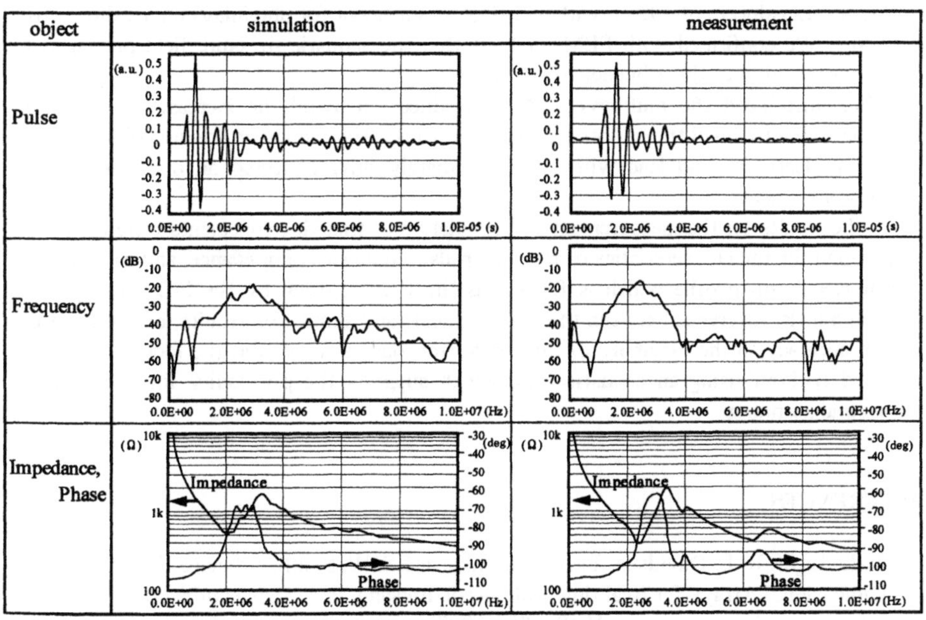

Figure 6. Evaluations of pulse, frequency, impedance, and phase characteristics for transducer elements with the results of simulation using the 3-D model.

In a second evaluation we investigated the beam profile by transmitting as a Fresnel beam with a focal length of 50 mm from the probe surface. The transmitted ultrasound was measured by a hydrophon moved along the long and short axes which were 50 mm from the probe surface. Figure 7 shows the results of measurement and simulation.

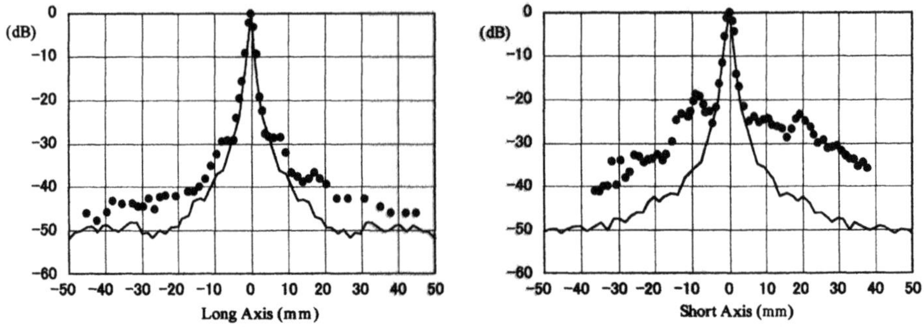

Figure 7. Beam profile of the Fresnel beam transmitted with a focal length of 50 mm from the center surface of convex-convex shaped 2-D array probe. The left corresponds to the long axis and the right corresponds to the short axis (The black circle is measurement and the solid line is simulation).

DISCUSSION AND CONCLUSION

We have developed a 192 ch by 64 ch convex-convex shaped 2-D array probe with curvatures of 40 mm for the long axis and 25 mm for the short axis. It operates at 2.5 MHz for real-time 3-D ultrasonography. This structure was developed by fabricating 64 convex thin-array transducers of 192 ch with an offset arrangement for obtaining convex short-axis curvature. We confirmed with FEM analysis that the acoustic pressure field propagated continuously without the influence of steps.

We introduced the multilayer PZT transducer to avoid the increase of electrical impedance due to small transducer elements and to achieve the lower driving voltage. Two opposing electrodes and lead wires were alternately interconnected on one side of the transducer elements.

The results for the measurement and simulation of transducer elements were found to correspond for the characteristics of received pulse, frequency, impedance, and phase. The transmitted beam profile which is formed as the Fresnel beam by the 2-D array probe showed a little difference between the measurement and simulation results, because of the lack of accuracy in the assembly of thin-array transducers. However, the focused beam is obtained without using an acoustic lens, which suggests the possibility for use in 3-D ultrasound scanning.

REFERENCES

[1] E. D. Light, R. E. Davidsen, T. A. Hruschka, and S. W. Smith, Advances in Two Dimensional Arrays for Real time Volumetric Imaging, *Proceedings of the IEEE Ultrasonics Symposium*, pp.1619–1623(1997).

[2] T. R. Nelson and D. H. Pretorius, Three-Dimensional Ultrasound Imaging, *Ultrasound in Med. & Biol.*, vol. 24, no. 9, pp.1243–1270(1998).

ULTRASONIC IMAGING WITH 2D MATRIX TRANSDUCERS

Roman Gr. Maev, Andrei Ptchelintsev, Aleksei A. Denissov

Department of Physics,
University of Windsor
Windsor, N9B 3P4 Canada

INTRODUCTION

Matrix and array transducers offer several enhanced capabilities in imaging and beam forming compared to a single transducer scanning system. The matrix transducer technology makes fast data acquisition and imaging possible, while using motionless transducers. Advanced technologies such as micro-machining and IC manufacturing can produce matrices of thousand elements of less than 0.1 mm in size, creating an acoustic analog of CCD camera[1]. This size is essential for phased arrays, given the element size must be around half a wavelength for efficient beam-forming[2]. However, for matrices not using phased principles to build the acoustic beam, such small elements represent a disadvantage. Matrix transducers having such a high density can only operate in the pick-up mode due to power dissipation restrictions. Moreover, acoustic wavelengths at megahertz frequencies are at least 10^3 times larger than the light wavelengths, thus, there seems to be no reason to go down to the micron range resolution[3].

Series of experiments with 2D matrices with relatively large elements 1.0×1.0 mm^2 and 0.5×0.5 mm^2 made with a LMN piezoelectric ceramic at 20 MHz have been performed. In the experiments contacting 32- and 52-element matrix transducers have been employed. The transducers have been calibrated on a set of circular reflectors of 2.0 to 7.5 mm diameter. The accuracy of the size prediction from *C*-scan views is within 0.5 mm. Acoustical images of various objects have been also generated.

The tilt of the probe from the normal position to the outer surface is one of the main factors yielding the poor quality of acoustic images. A method to compensate the tilt in a few-degree range and enhance the image quality has been developed. The technique is based on a processing of *A*-scan data. For each *A*-scan the TOF variation due to the tilt has been determined and the tilt angle is evaluated. A correction algorithm taking into account the amplitude vs. the tilt dependence has been incorporated into the software.

Acoustical Imaging, Volume 25, Edited by Halliwell and Wells.
Kluwer Academic/Plenum Publishers, 2000.

EXPERIMENTAL SETUP AND MEASUREMENT

One of the array transducers used in the study has the diameter of 8.0 mm and contained 52 emitting-receiving broadband elements with the central frequency around 20 MHz; the matrix period was 1.0 mm, individual elements being of 0.74×0.74 mm^2 square. A smaller 32-element array with similar characteristics has also been used. The elements were switched one at a time using a custom coaxial relay multiplexer interfaced with the parallel port. Experimental data have been acquired with a USD-15 (ultrasonic pulser Krautkramer) connected to a TDS-520 (Tektronix digitizing oscilloscope), digital data were transferred to the PC via a GPIB interface card. All analog settings were maintained constant through the experimental procedure. The sampling rate (TDS-520) used in the experiments was 500 MSamples/s. For image smoothing and quality enhancing various 2D interpolation algorithms were used[3].

The imaging capability of the system generally depends on the penetration depth of the transducer. The penetration depth drops with the decreasing size of the element due to the beam divergence and the increasing electrical mismatch with 50 Ohms transmitters and receivers. In general the image quality gets worse for smaller and deeper reflectors. Images obtained on ∅1.0 and 2.0 mm flat-bottomed holes drilled inside a 319 aluminum casting sample are given in Figures 1 and 2. The holes were grouped by four for the depth range from 1.0 to 8.5mm separated by 0.5 mm depth step within a group. Featured *C*-scans were obtained to 8.5 mm deep for ∅2.0 mm flat-bottom holes, however reflectors of ∅1.0 mm diameter were only clearly observed to 4.5 mm deep. The experimental diagrams illustrating the reflected amplitude versus the reflector's depth for ∅1.0 mm and ∅2.0 mm flat-bottom reflectors are given in Figure 3.

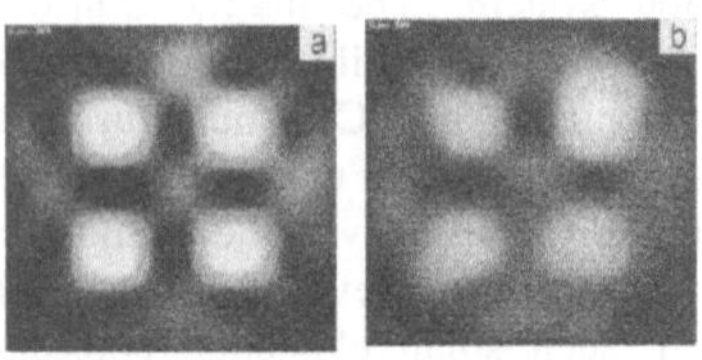

Figure 1. *C*-scans of ∅1 mm flat-bottom reflectors located in the depth range: a) 1.0–2.5 mm; b) 3.0–4.5 mm (0.5 mm step).

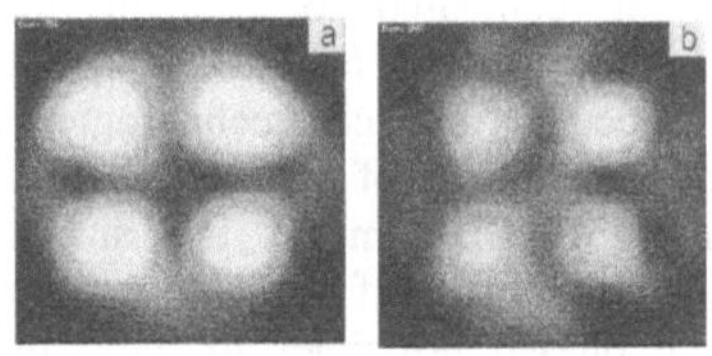

Figure 2. *C*-scans of ∅2 mm flat-bottom reflectors located in the depth range: a) 1.0–2.5 mm; b) 7.0–8.5 mm (0.5 mm step).

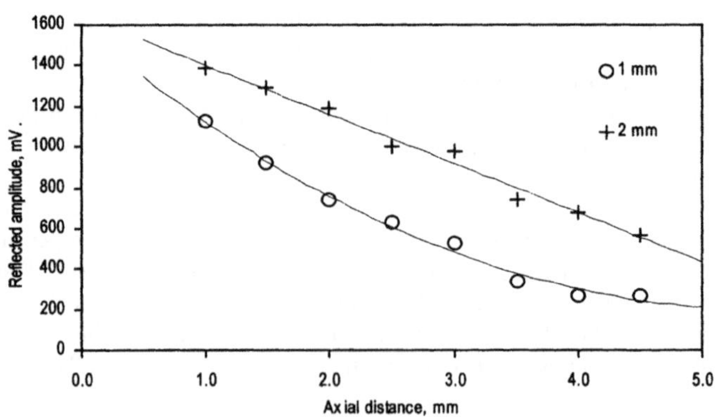

Figure 3. Amplitude of the reflected signal vs. the reflector's depth for ∅1 mm and ∅2 mm flat-bottom reflectors in 319-alloy aluminum casting

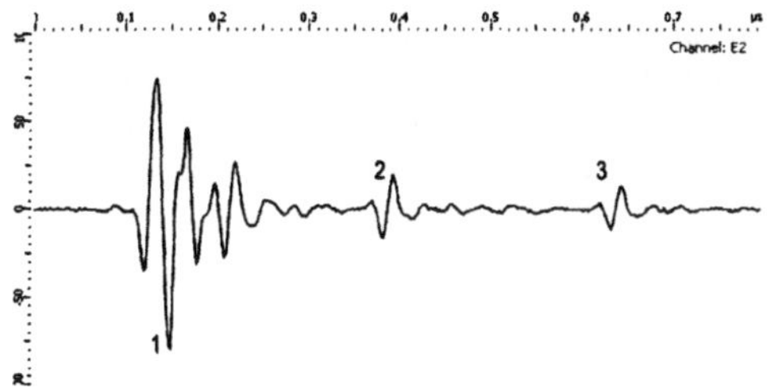

Figure 4. Typical *A*-scan of a single element displaying the interface reflection (1), the flat-bottom hole reflection (2) and the double reflection on the same hole (3)

When producing acoustic images with the matrix contacting transducers one should pay close attention to the condition on the contacting surface. Disturbances on this interface can significantly deteriorate the image quality. In order to compensate for a very common disturbing factor as the tilt, an algorithm taking into account the angular dependency of the refraction coefficient has been implemented. The data acquisition program routinely analyzes sets of *A*-scans data acquired with the matrix transducer. The tilt angle is then determined using the time-of-flight (TOF) delay information. In fact the reflection pulse with the biggest amplitude corresponds to coupling-metal boundary (see Figure 4), and given that TOF inside the acoustic delay are nearly identical, the tilt angle is calculated from the TOF matrix using bilinear LMS regression. Prior to using the algorithm of the amplitude correction, a curve representing the averaged reflected amplitude vs. the tilt angle has been produced. An example of the curve for steel plate of 1.5mm thick is given in Figure 6. S that once the actual tilt of the probe was determined, the amplitude correction curve was used to correct the amplitudes. The accuracy of the tilt angle determination is around 0.2°. As it has been shown experimentally the amplitude-tilt angle dependency is rather sharp for 1.0×1.0 mm^2 transducers, which is determined mostly by the lateral shift of the beam inside the acoustic delay (see Figure 5), where Z_1 is the delay, Z_2 is the coupling gel, Z_3 is the metal.

Thus, at 1.5° tilt the shift inside a 5 mm long delay is around $a = 0.3$ mm, the near field inside the delay being around 2 mm, and the beam spread is $\theta = 3.1°$ at –6 dB level, transmission-reception patterns as can be calculated combined using formula (1):

$$\theta = \arcsin(1.4/ka). \tag{1}$$

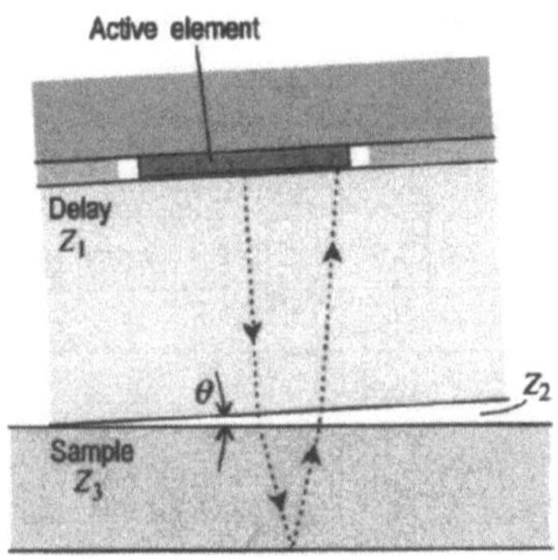

Figure 5. Effect of refraction of the beam on a tilted interface

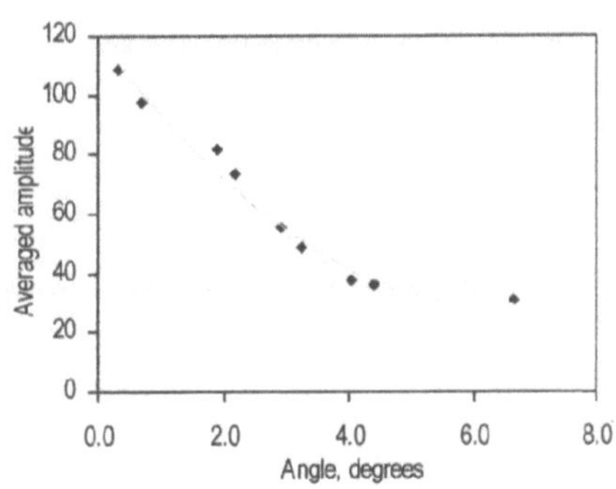

Figure 6. Transmission coefficient vs. tilt: experiment

Due to the variable thickness of the coupling layer between a tilted probe and the surface, modulation of the transmission coefficient is produced. However this effect can be neglected, because the acoustic impedance of the delay (Polystyrene 2.5 MRayl) is close to that of the coupling liquid (ultrasonic gel $Z \approx 2.0$ MRayl). The following simple formula can be used to predict the modulating effect of the layer:

$$D_{123} = \frac{D_{12} D_{23} e^{-jk_2 h}}{1 - R_{21} R_{31} e^{-j2k_2 h}}, \tag{2}$$

where D_{ij} and R_{ij} are the corresponding transmission and reflection coefficients, k_2 is the wave number of the layer, h is the thickness of the layer, indices 1 trough 3 point respectively to the delay, coupling medium and object. With a degree of approximation this formula, however derived for parallel boundaries, can be used for tilted configurations with small tilts. A computation according to formula (2) yields around 5 percent amplitude variations over the transducer surface.

Figure 7 illustrates a number of *C*-scan views of the same object (∅2.5 mm, 1 mm thick calibration sample) taken for various tilt angles and data processing algorithms. The *C*-scans in each column were obtained with the same raw experimental data. The first row

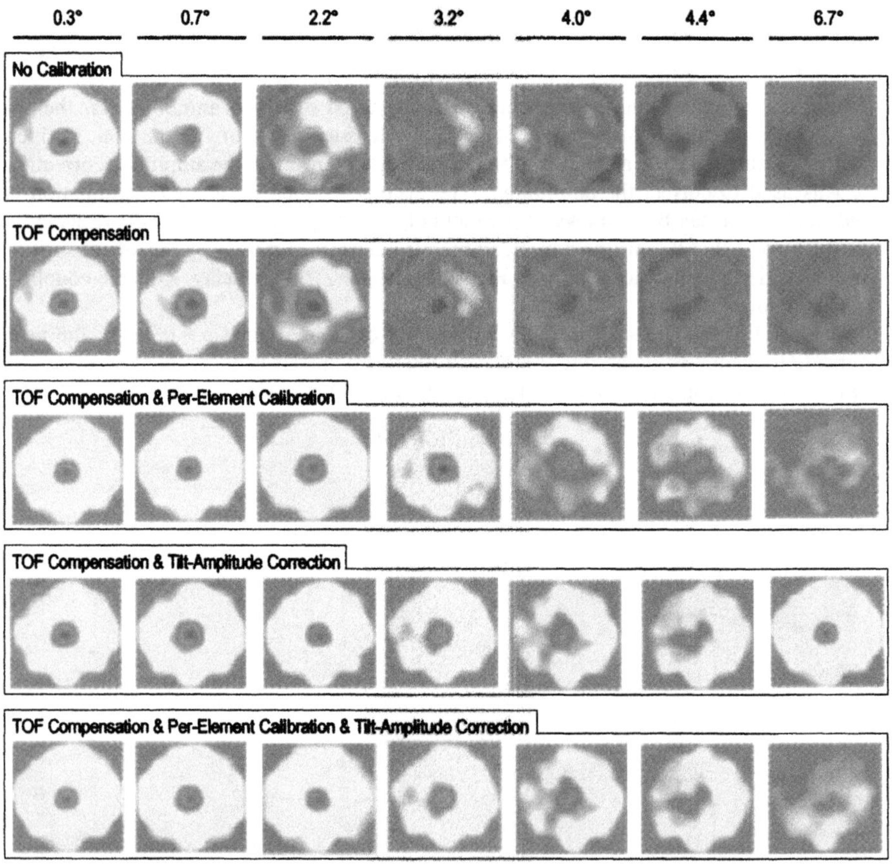

Figure 7. Acoustic images of the 2.5 mm calibration sample obtained from 1 mm depth presenting results of different calibration methods

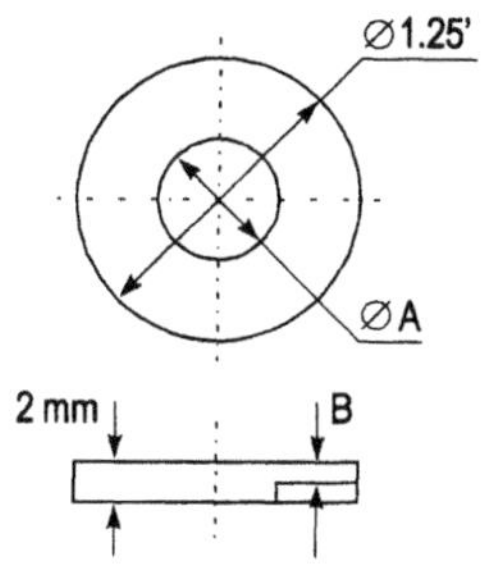

Figure 8. Schematic view of the calibration samples

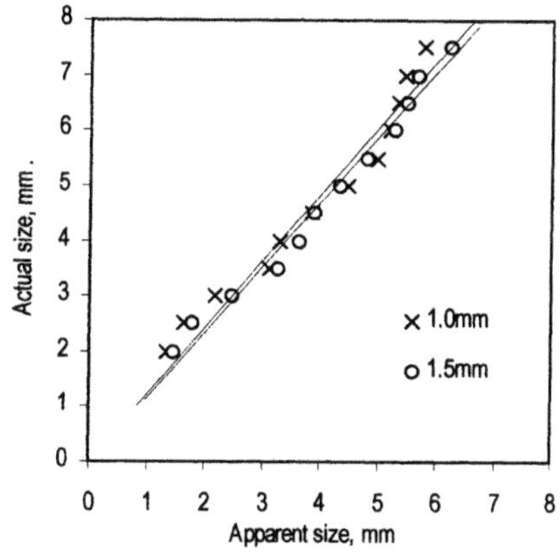

Figure 9. Calibration chart established for 1.0 mm and 1.5 mm thick round calibration samples

represents unprocessed data; the second shows results with the TOF compensation; the third benefits the TOF compensation and the sensitivity calibration for each element of the array. The fourth features both the TOF compensation and the tilt-amplitude correction, and the last row accumulates three of the above methods. As can be seen the images are significantly corrupted with the increasing tilt angle, however the TOF compensation and the tilt-amplitude correction may significantly improve the poor quality at high tilt angles. On the whole these data processing techniques provide a satisfactory quality of the images with tilt angles to 3–4°.

In Figure 10 some *C*-scans obtained on round calibration sets are given. A schematic view of the calibration sample is given in Figure 8. Two sets of the samples of 1.0 and 1.5 mm thick with the diameters ranging from 2.0 to 7.5 mm were used in this experiment. In order to estimate the accuracy of the diameter measurement from the *C*-scan data, calibration charts for two depths 1.0 and 1.5 mm were produced. The *C*-scan image depends on the distribution of reflected signals as well as on the color scale. The largest circle has been inscribed into a *C*-scan and then compared with the actual size of the reflectors. The calibration charts in that way eliminate any influence of color scaling on the measurement. Typical errors of the size determination are within 0.5 mm, and often less than 0.3 mm.

In Figure 11 another example of an imaging application for the presented device is given. The generated acoustic images represent laser spot weld samples of 0.75 and 1.0 mm thick. Surface burn marks were respectively around 3.0 and 7.0 mm.

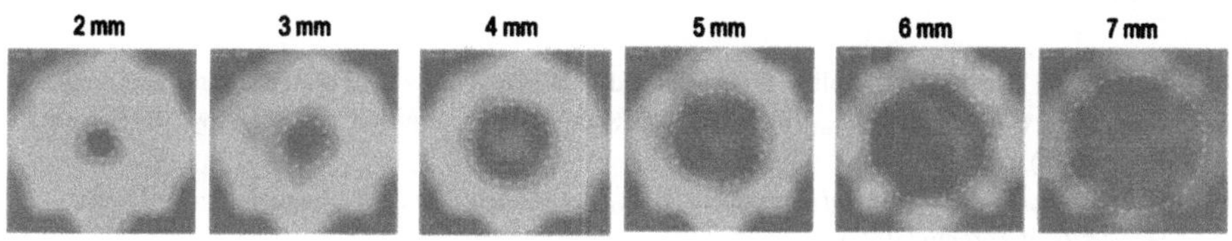

Figure 10. Acoustic images produced with 1.5 mm thick round samples with the 52-element array

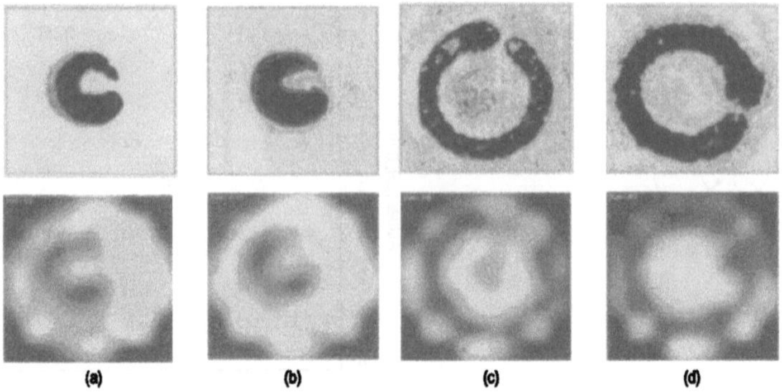

Figure 11. Acoustic images of laser spot welds generated with a desktop scanning acoustic microscope and corresponding images obtained with the 52-element array: (a), (b) 0.75 mm thick steel plate approximately ∅3.0 mm. weld; (c), (d) 1.0 mm thick steel plate around ∅7.0 mm weld.

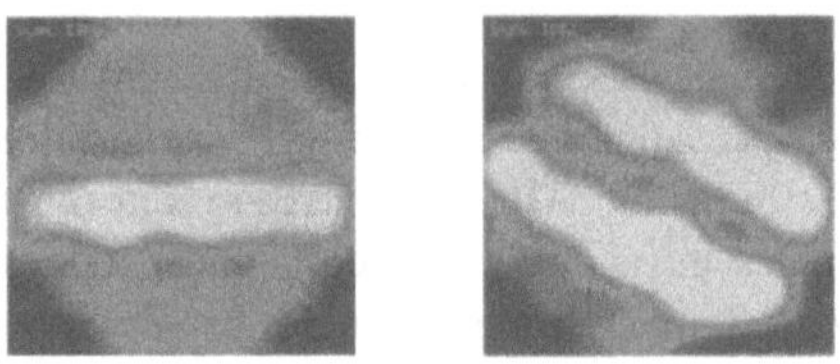

Figure 12. Acoustic images of copper wires inside the flexible printed circuit board, which is incorporated into 1.5 mm thick plastic cover.

Matrix transducer is capable to visualize location and orientation of the flexible printed circuit board (PCB) inside the plastic. In Figure 12, images of 0.2 mm thick and 2 mm wide wires are shown. The separation between the wires in the right image is 1 mm.

CONCLUSIONS

This paper has discussed a broad range of experimental results relating to development of 2D matrix transducer technologies and its potential applications. The major criterion for the imaging capability of the system generally depends on the penetration depth of the transducer and resolution. The influence of the various parameters, including various interface conditions, on the quality of the ultrasonic imaging is an important issue of this research and the goal for the further study. A more detailed analyses and experimental works taking into account those results are underway.

REFERENCES

1. K. Erikson, A. Hairston, A. Nicoli, J. Stockwell, and T. White, A 128×128 ultrasonic transducer array, in: *Proc. IEEE Ultrasonics Symp.*, Vol. 2, eds. S.C. Schneider, M. Levy and B.R. McAvoy (IEEE, New York, 1997), p. 1625.
2. D.H. Turnbull, *Two-Dimensional Transducer Arrays for Medical Ultrasonic Imaging*, Ph.D. Thesis, University of Toronto (1992).
3. R. Gr. Maev, A. Ptchelintsev and A. Denissov, Acoustical imaging using matrix of piezoelectric transducers for nondestructive evaluation of plastic composites, in: *Proc. ICCE*, ed. D. Hui (ICCE Orlando, 1999), p. 665.

NARROW APERTURE PHASED ARRAY COMPUTED TOMOGRAPHY

Iwaki Akiyama,[1] Naoki Takizawa,[1] and Akihisa Ohya[2]

[1]Department of Electrical Engineering
Shonan Institute of Technology,Fujisawa 251-8511, Japan
[2]Institute of Information Science and Electronics
University of Tsukuba, Tsukuba, Japan

INTRODUCTION

Conventional electronic sector scanners for diagnostic ultrasound system steer focusing beam using a phased array. A delay time sequence determines the focal length as well as the beam direction in B-scan plane. Although the focal position is dynamically moved, the focal position for the direction of elevation of the array is fixed because an acoustic lens is used. Thus, it is not good resolution in the region other than the focal region of the acoustic lens. The authors proposed a novel tomographic reconstruction technique using a narrow aperture phased array (NAPA) to improve the spatial resolution. We call it NAPA-CT[1][2][3]. The narrow phased array forms a circular arc-focusing beam; divergent beam along the direction of the elevation and the focusing beam along the lateral direction. When the circular arc-focusing beam is formed by the narrow phased array, the echoes from the scatterers at the circular arc are simultaneously received by the array. Thus the amplitude of the echoes at the specific delay time is expressed as the summation of coefficient of back scattering at the circular arc whose radius is determined by the delay time. Since it is considered as the Radon Transform of the back-scattering coefficient on the hemi-spherical surface, it is possible to map the distribution of scatterers at the hemi-spherical surface by inverse Radon transform. To confirm the feasibility of the proposed NAPA-CT, an experimental system is constructed using a commercially available electronic sector scanner with a 3.5MHz phased array probe. Concave type acoustic lens is used to form the circular arc-focusing beam. Angle of beam spreading is approximately 30 degrees. The system collects sixty projections of 120 sampling within one minute. The distribution of the scatterers is reconstructed at the hemi-spherical surface by the filtered back projection algorithm. The reconstructed images of tips of seven needles and seventeen needles along a spiral are shown. Also we show the reconstruction of dense media like biological tissues. Finally we show the experimental results of phantom made of graphite particles and dextran particles .

NARROW APERTURE PHASED ARRAY COMPUTED TOMOGRAPHY

The proposed technique is based on the circular arc-focusing beam which is formed by a narrow aperture phased array (NAPA). Figure 1 shows schematic illustration which shows the difference of beam pattern between the NAPA and a conventional phased array.

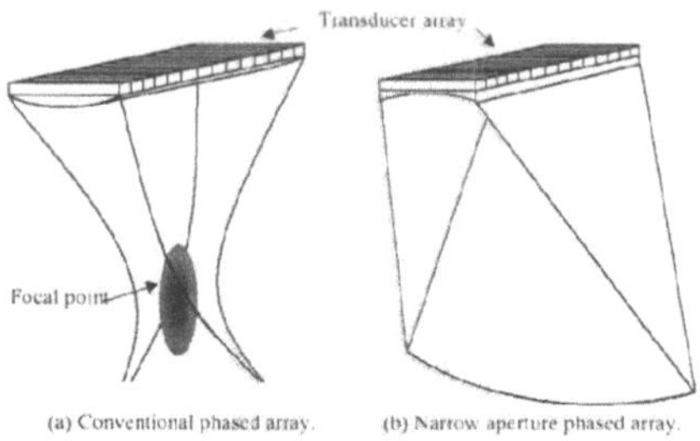

Figure 1. Schematic illustration of beam pattern. (a) is of conventional phased array and (b) is of narrow aperture phased array (NAPA).

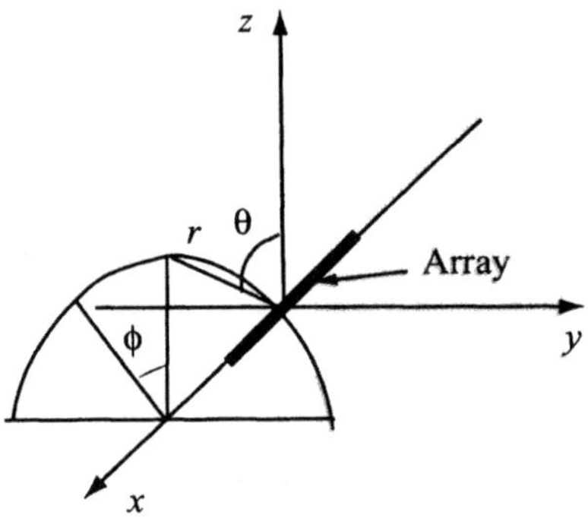

Figure 2. The cordinates system of r, θ, and ϕ. The array is along x-axis.

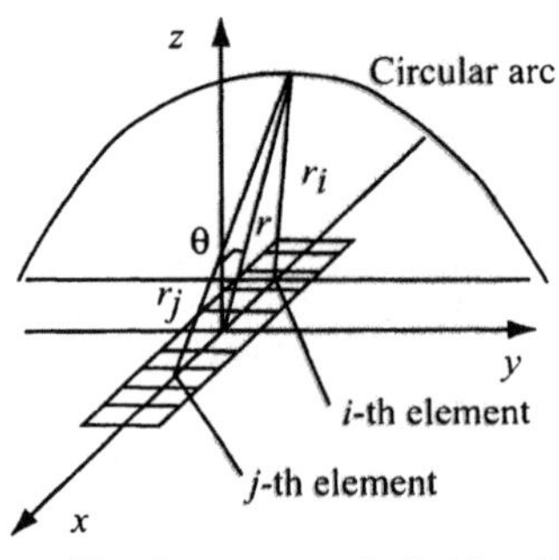

Figure 3. The i-th element insonifies the ultrasound of $p(t)$ and j-th element receives back-scattering waves from (r, θ, ϕ).

First of all we define coordinates system of r, θ, and ϕ as shown Figure 2. Then we cconsider the echo signals received by the NAPA. Let $e_{ij}(t)$ be the signal which is insonified by i-th element and received by j-th element as shown in Figure 3. The Fourier transform of $e_{ij}(t)$ is expressed as[4];

$$E_{ij}(k) = \int\int\int_V k^2 P(k) f(r,\theta,\phi) \frac{\exp\{ik(r_i + r_j)\}}{r_i r_j} r^2 \cos\theta dr d\theta d\phi, \tag{1}$$

where$f(r,\theta,\phi)$ is a distribution of back-scattering coefficient, $P(k)$ is a Fourier transform of transmitting waveform $p(t)$, and k is a wave number. $r_{i,j}$ is a distance between i- and j-th element and a scatterer at (r,θ,ϕ) expressed as;

$$r_{i,j} = \sqrt{r^2 + x_{i,j}^2 - 2rx_{i,j}\sin\theta}, \tag{2}$$

where $x_{i,j}$ is a x-axis value at i-th and j-th element, respectively. Then we consider the focus onto the circular arc of radius of r_f and direction of $\theta = \theta_f$. We apply the time delay due to the distance of l_{ij};

$$l_{ij} = \sqrt{r^2 + x_{i,j}^2 - 2rx_{i,j}\sin\theta} \tag{3}$$

The output signal of the phased array is expressed as a summation of all the received signals with time delay.

$$\sum_i\sum_j E_{ij}(k) = k^2 P(k) \sum_i\sum_j \int\int\int_V f(r,\theta,\phi) \frac{\exp\{ik(r_i + r_j)\}}{r_i r_j} \exp\{-ik(l_i + l_j)\} r^2 \cos\theta dr d\theta d\phi. \tag{4}$$

We consider the summation term. First we define the beam pattern function of $A(k;r,\theta)$.

$$A(k;r,\theta) = \sum_i\sum_j \exp\{ik(\Delta r_i + \Delta r_j)\}, \tag{5}$$

where

$$\Delta r_{i,j} = r_{i,j} - l_{i,j} - (r - r_f). \tag{6}$$

Thus the summation term is expressed as;

$$\sum_i\sum_j \frac{\exp\{ik(r_i + r_j)\}}{r_i r_j} \exp\{-ik(l_i + l_j)\} \simeq \frac{\exp\{i2k(r - r_f)\}}{r^2} A(k;r,\theta). \tag{7}$$

The beam pattern function $A(k;r,\theta)$ is shown in Figure 4 where k=14 rad/mm and array length is 20 mm.

As shown in Figure 4, $A(k;r,\theta)$ is approximately expressed as $\delta(\theta - \theta_f)$. Thus the output of phased array is written by,

$$\sum_i\sum_j E_{ij}(k) = k^2 P(k) \cos\theta_f \int \exp\{ik(r - r_f)\} \int_{\theta=\theta_f} f(r,\theta,\phi) d\phi dr. \tag{8}$$

Then we have output signals by taking the inverse Fourier transform of left hand side of eq.(8) which is divided by k^2.

$$e(t) \simeq cos\theta_f \int\int_{\theta=\theta_f} p(t - \frac{r - r_f}{c}) f(r,\theta,\phi) d\phi dr. \tag{9}$$

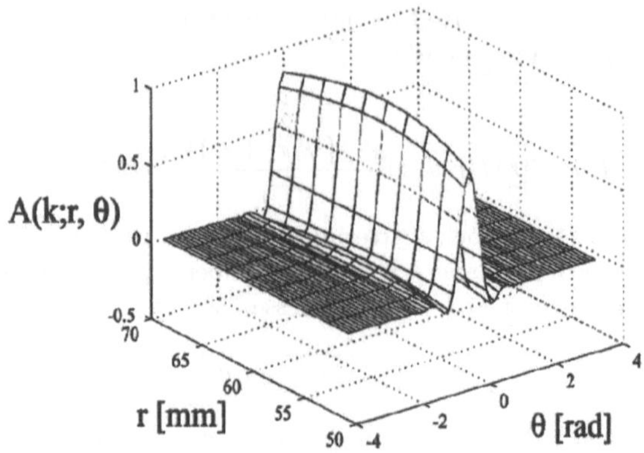

Figure 4. Bird's eye view of the beam pattern function $A(k;r,\theta)$ where k=14 rad/mm and array length is 20 mm.

Sparse media

If the objects are expressed as sparse media, the echoes are expressed as

$$e(t) \simeq cos\theta_f \int_{\theta=\theta_f} f(r_f,\theta,\phi)d\phi. \tag{10}$$

Therefore the echoes are considered as the Radon transform of distribution of back-scattering coefficient on the hemi-spherical surface of radius of r_f.

Dense media

If the objects are considered as dense media such as biological tissues, it is difficult to reconstruct the distribution of back-scattering coefficient. Thus we consider the local average of back-scattering coefficient. We define the weighted average value of back-scattering coefficient of $\hat{f}(r_f,\theta,\phi)$;

$$\hat{f}(r_f,\theta,\phi) = \int p(t - \frac{r - r_f}{c}) f(r,\theta,\phi)dr, \tag{11}$$

where $p(t)$ is the transmitting waveform which is considered as a weighted function. Therefore, the echo is expressed as,

$$e(t) = \cos\theta_f \int_{\theta=\theta_f} \hat{f}(r_f,\theta,\phi)d\phi. \tag{12}$$

This equation shows the Radon transform of $\hat{f}(r_f,\theta,\phi)$ is equal to the echo at $t = r_f/c$.

EXPERIMENTAL SYSTEM AND THE RESULTS

Experimental system

We develop an experimental system as shown in Figure 5. An ultrasonic diagnostic equipment is a commercially available electronic sector scanner(Aloka SSD-2000). The received echoes are digitized by A/D board(Gage Co. CS8012) at 40MHz sampling of 12bits resolution. NAPA is fabricated from the conventional phased array by attaching a concave type acoustic lens to it, as shown in Figure 6. The spread of the circular arc-focusing beam is approximately thirty degrees. This system collects sixty projection of 120 samples within one minute. The NAPA probe is made of 3.5MHz transducers and of 48 elements.

Needles targets

One projection is made of 120 of sampling postion, which covers ninety degrees in sector format. Total number of projections is sixty. The filtered back projection algorithm is used for reconstrution. Figure 7 shows the reconstructed image of the tips of seven needles. The seven hot spots are clearly observed in Figure 7. The intensity of each tips is different, because the direction of each needles is slightly different.

Then we show the experimental results of the tips of seventeen needles which are arranged along a spiral. Figure 8 shows the shaded image of the targets. This image is obtained by thresholding the each reconstructed image.

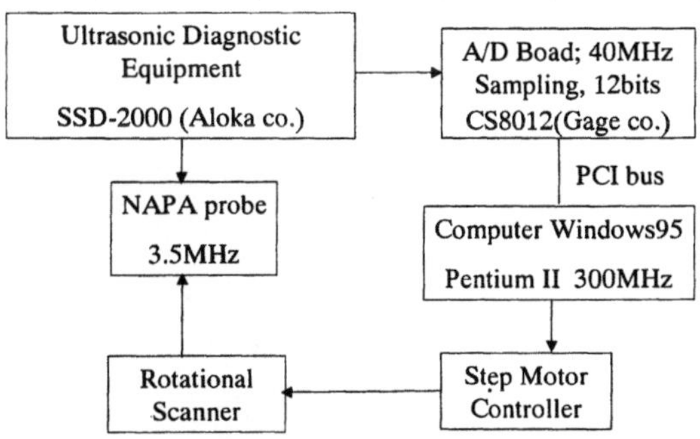

Figure 5. Experimental system. NAPA is equipped with a commercial available electronic sctor scanner.

Figure 6. NAPA probe of 3.5MHz with concave type acoustic lens.

Dense scatterer targets

We use the two types of the tissue mimicking phantom based on agar-gel. The graphite particles and dextran particles are used as scatterers. They are randomly suspended in the agar-gel. Figure 9 shows the experimental results of the phantom with a hole of graphite particles. As shown in figure 9, the hole is observed in the middle of the phantom. Figure 10 shows the experimental results of the phantom made of two blocks of dextran paticles. As shown in figure 10, two blocks are observed in the top view. Both of them are reconstructed by the envelope detection. As a result, the dense scatterer targets are reconstructed.

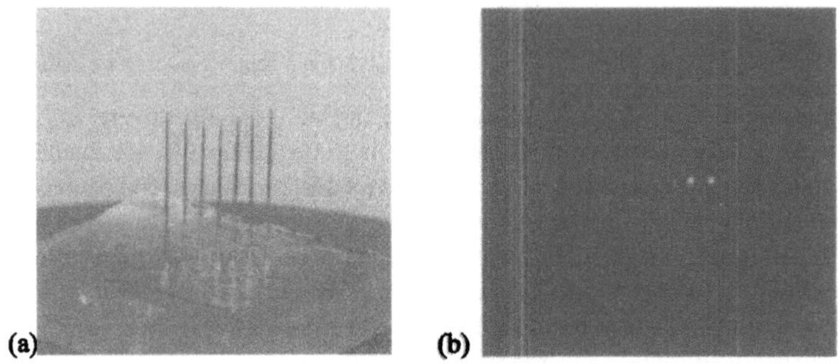

Figure 7. (a) is the imaging targets of seven needles. (b) is the reconstructed image of tips of the seven needles.

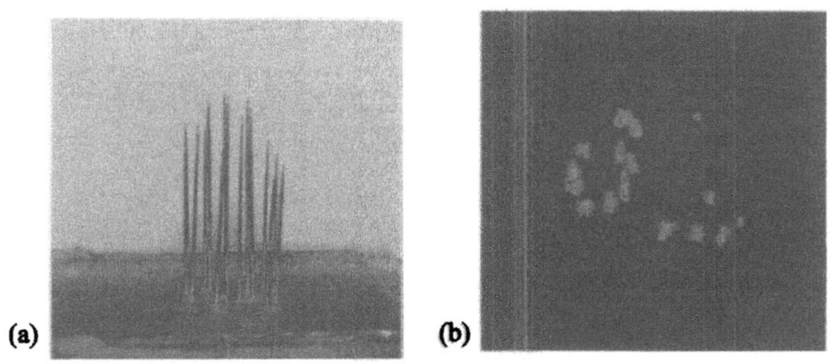

Figure 8. (a) is the imaging targets of seventeen needles whose tips are arranged along a spiral. (b) is the reconsturucted image of tips of seventeen needles along a spiral. This image is obtained by thresholding the each reconstructed image.

CONCLUSION

We shows the theory of computed tomography using narrow aperture phased array. Since this technique is based on pulse-echo method, the application is flexible. Also it is easy to apply the three dimensional imaging by shifting the time delay. We develop the experimental system based on commercially available electronic sector scanner system with the NAPA probe attached with concave type acoustic lens. The NAPA probe is of 3.5MHz and 48 elements. The system collects sixty projections within a minute. We made the experiments using the system to reconstruct the tips of needle targets and the tissue mimicking phantom as dense scatterer targets. Consequently, the results show the feasiblity of this technique.

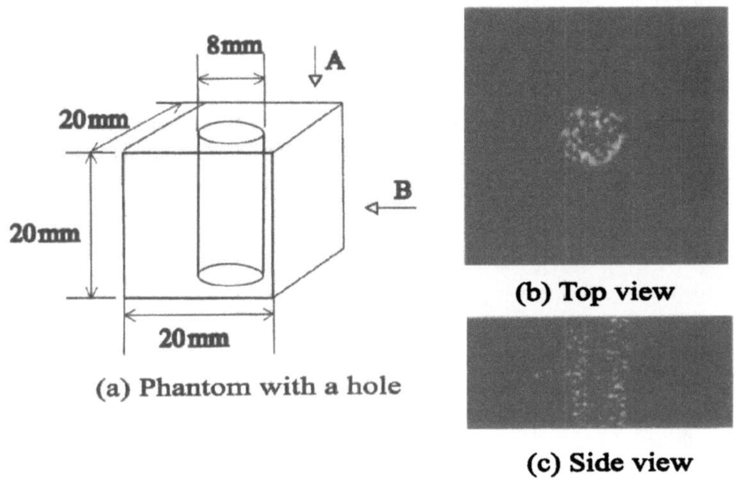

Figure 9. Reconstructed image of the phantom. The graphite particles are used as scatterers.(a) is a top view and (b) is a side view.

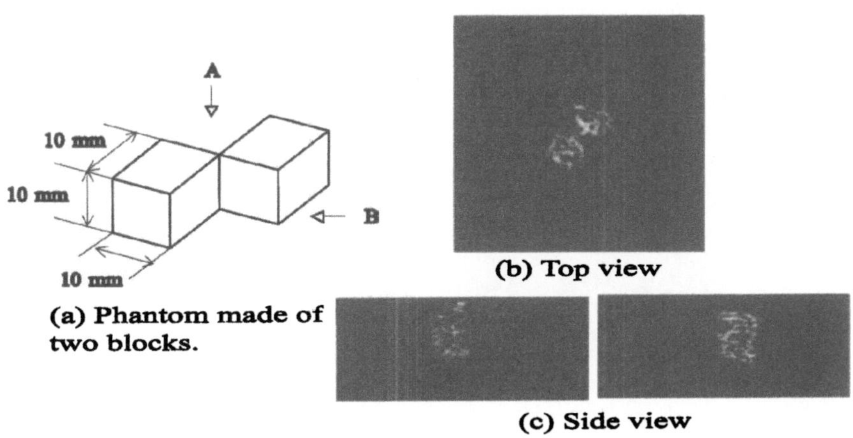

Figure 10. Reconstructed image of the phantom. The dextran particles are used as scatterers. (a) is a top view and (b) is a side view.

ACKNOWLEDGEMENTS

The authors are grateful to Dr.Takashi Itoh and Mr. Nobuo Oshima for their helpful discussion in the system construction.

References

[1] I. Akiyama. Tomographic Reconstruction Technique for Pulse-Echo, International Journal of Imaging System and Technology, 8, 5, 432- 437, 1997.

[2] I. Akiyama and K.Yano, Synthetic Aperture Computed Tomography, Acoustical Imaging, 23, 575- 581, 1997.

[3] N. Takizawa, A. Ohya and I.Akiyama, Ultrasonic To-mography Using Arc Focusing Beam, IEEE Ultrasonics Symposium Proceedings, 1659- 1662, 1998.

[4] S. J. Norton and M. Linzer, Ultrasonic Reflectivity Imaging in Three Dimensions:Exact Inverse Scattering Solutions for Plane, Cylindrical, and Spherical Apertures, IEEE Trans. on Biomedical Engineering, BME-28, 2, 202-220, 1981.

AN ALTERNATIVE TECHNIQUE FOR ACOUSTIC CHARACTERIZATION OF COMPOSITE ULTRASONIC TRANSDUCERS FOR MEDICAL APPLICATION

Pedro Acevedo [a], Dilip Das-Gupta [b]

[a] DISCA-IIMAS-UNAM, Apdo. Postal 20-726 Admon. No. 20, 01000 México D.F., México
[b] UCNW-SEECS, Dean Street, Bangor, Gwynedd LL57 1UT, Great Britain

INTRODUCTION

With the expansion in the use of diagnostic ultrasound that has occurred in recent years, there has been growing concern over the safety of ultrasound exposure. This concern has led to a new emphasis in research into possible ultrasonically induced bioeffects and to new efforts to develop reliable procedures for the measurement of the acoustic output from clinical diagnostic ultrasound equipment. It is important to know how the calibration measurements, performed typically in water are related to the field that actually occurs within the patient, and to be able to predict the field for a variety of exposure conditions.
Both researchers involved in ultrasound bioeffects studies and manufacturers of commercial equipment share the need to characterise adequately the acoustic output emitted by ultrasonic transducers.
The most widely used guide for measuring acoustic output from medical ultrasonic equipment is the FDA guide [1], this document, in turn is based on the American Institute of Ultrasound in Medicine and the National Electrical Manufacturers Association (AIUM/NEMA) Standard [2]. This standard requires four intensities measurements, namely, the spatial-peak pulse-average intensity I_{sppa}, the spatial-peak temporal-average intensity I_{spta}, the spatial-peak temporal-peak intensity I_{sptp} and the spatial-average temporal-average intensity I_{sata}. Additionally, the -6 dB beam area A_6, the total power W, the maximum half cycle intensity I_m and the pulse duration, *PD*, of the waveform must be specified. Definitions of these parameters and the strategy employed in their deduction may be found in references [1], [2] and [3].

INTENSITY MEASUREMENT PROCEDURE

The measurement system used to perform all the experimental part consists of a glass tank containing water, a water deioniser, a hydrphone, a pulse-echo board, a HC11 microcontroller board, a step motor control board, an IBM compatible personal computer, a

Acoustical Imaging, Volume 25, Edited by Halliwell and Wells.
Kluwer Academic/Plenum Publishers, 2000.

digital oscilloscope and a positioning mechanic system. A general diagram of the system is shown in fig. 1.
Commercial composite samples of PZT-5 and VDF-TrFE (ready poled and electroded L-25 PIEZEL™) were provided by Daiken Chemicals of Japan [4] to construct the transducer to be characterised. The active element used in the construction of this transducer was a 250 μm thick and 10 mm diameter disc of this composite. A Schematic representation of the composite transducer is shown in fig. 2.
Before proceeding with the data, it is instructive to make a few comments on the overall procedure adopted and to further clarify terminology used in acoustic beam profiles.
Before a measurement can commence, the location where the peak temporal voltage lies must be found. This is achieved by initially aligning the hydrophone active element in the far-field of the transmitter device. The far-field is considered to be any point that lies beyond the focal length of the transducer.
The hydrophone is subsequently moved in the *x-y* planes (see figure 1) until a point is found where it detects a maximum signal. Upon successfully locating this point, it is moved towards the acoustic source (in the *z* direction) until a new maximum temporal voltage is registered. Minor adjustments are then made in the *x-y* directions to verify this point. As the hydrophone is moved in the direction of the transmitter (i.e. into its near-field) several positions along the beam axis will indicate a series of maximum voltages.
The object of the exercise is to ensure that the maximum located above does not lie in the transmitter near-field. The point where the last signal maximum is observed along the beam axis, should coincide with its focal length, this distance is designated as *g* for an unfocussed transducer.
In practice, it is very difficult to accurately measure *g* directly. Instead, it was estimated here from the reciprocal of the time delay of acoustic echoes observed on the oscilloscope. For an unfocussed transducer,

$$g = \frac{r^2}{\lambda_o} \quad m \tag{1}$$

where r is the radius of the acoustic source and λ_0 the wavelength of the signals propagating in the medium. Equation 1 may be re-written in terms of the acoustic velocity of water as,

$$g = \frac{r^2(mm^2)\, f_0(MHz)}{1.5} \quad mm \tag{2}$$

The sound velocity in water in the above equation is assumed to be 1500 m/s. The distance defined by equation 2 translates into the delay-time observed on the oscilloscope as,

$$\Delta t = \frac{r^2(mm^2)\, f_0(MHz)}{2.25} \quad \mu s \tag{3}$$

where Δt is the delay between firing the pulser and the arrival of the first echo. Thus, by keeping Δt on the oscilloscope in conformance with equation 3, final adjustments to the hydrophone are made to identify precisely the location of the temporal peak voltage.

It is important to keep Δt as small as possible in order to minimise the attenuation of signals in the water path. For an unfocussed transducer, the delay is directly proportional to the square of the active element radius. It is for this reason that the selected transducer radius was 5 mm.
Having establish the location of the temporal peak voltage, the hydrophone is then scanned in the x-y directions along at least four orthogonal radii. The time-voltage waveforms are digitised, averaged, stored and transferred to the microcomputer. The program computes the maximum value of pulse intensity integral and the peak voltage. The whole procedure is subsequently repeated by manually moving the hydrophone to a new position to execute a complete scan along the source radius. It was found more convenient in practice to start at one extreme of the beam (rather than the centre) and continue unidirectionally to the other extreme along the source diameter. The program was consequently modified to permit orthogonal diametric scans.

RESULTS

Data for the transducer under test appear in figures 3 and 4 and table 1. The intensity values for this transducer seem to be higher than those for PVDF transducers, but not higher than the corresponding values for P(VDF-TrFE) [5]. At present, composite materials provide an option for the future advance in the design of transducers for medical diagnosis. Material researchers are aiming to obtain a composite material which possesses a high mechanical coupling coefficient, a high "figure of merit" $g_h d_h$ (where g_h and d_h are the hydrostatic voltage and charge coefficient respectively), a low Q and an acoustic impedance which matches human tissue.

CONCLUSIONS

All intensity calculations are ultimately based on hydrophone measurements of the temporal pressure waveform at the point in the ultrasonic field where the parameter of interest may have its maximum value. All intensity parameters reported in this work were checked and verified using methods described in the FDA's Guide. This was carried out in an attempt to assess the validity of the algorithms employed. The data presented in this work were found to comply with the uncertainty criterion outlined in the FDA's Guide.
The intensity values reported here are appreciably lower than those normally encountered in the literature. This may be attributed to the way in which the transducer was stimulated. Here, a broad-band excitation method was employed throughout. If the tone-burst is used instead, larger voltages are expected to be registered by the hydrophone. Secondly, although, deionised water was used in all the experiments, this needed to be de-gassed. This was not possible here but formation of minute air bubbles in the surfaces of the source and receiver was tried to be kept minimum.
Finally, it can be stated that the present measurement technique is an useful design aid in characterising prototype transducers if results are to be compared on an equal basis with a calibrated probe. If, on the other hand, the absolute transducer intensity values are required then additional equipment needs to be implemented to improve overall accuracy and repeatability.

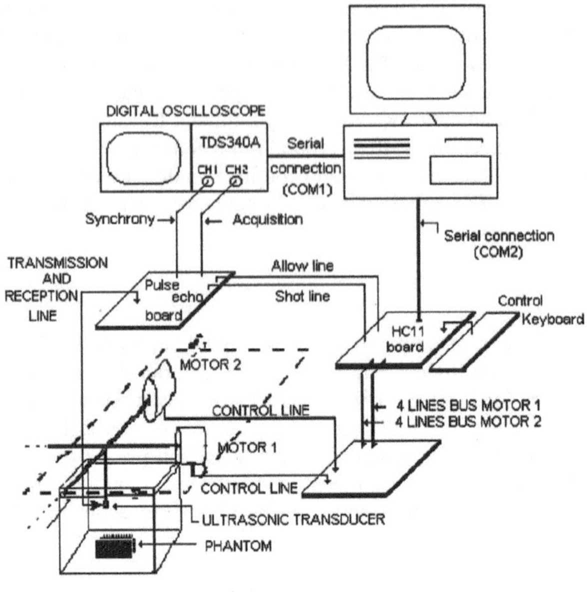

Figure 1. General diagram of the measurement system.

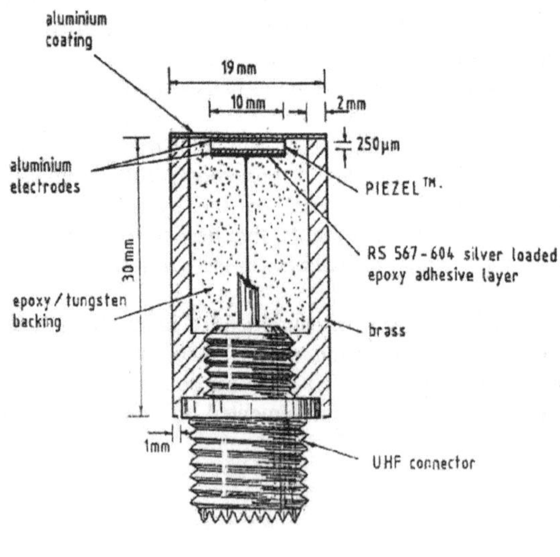

Figure 2. Schematic representation of composite transducer.

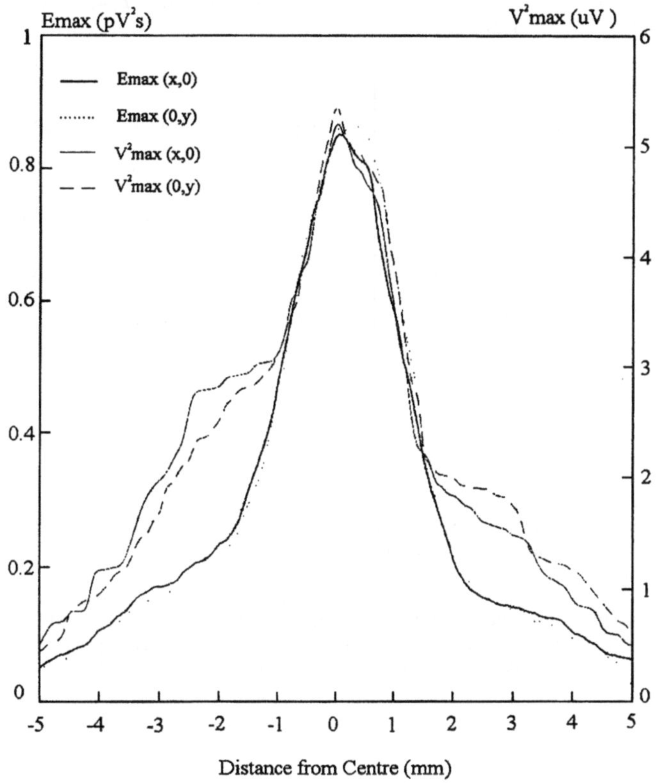

Figure 3. Spatial plot of E_{max} and V^2_{max} along orthogonal diameters for composite transducer.

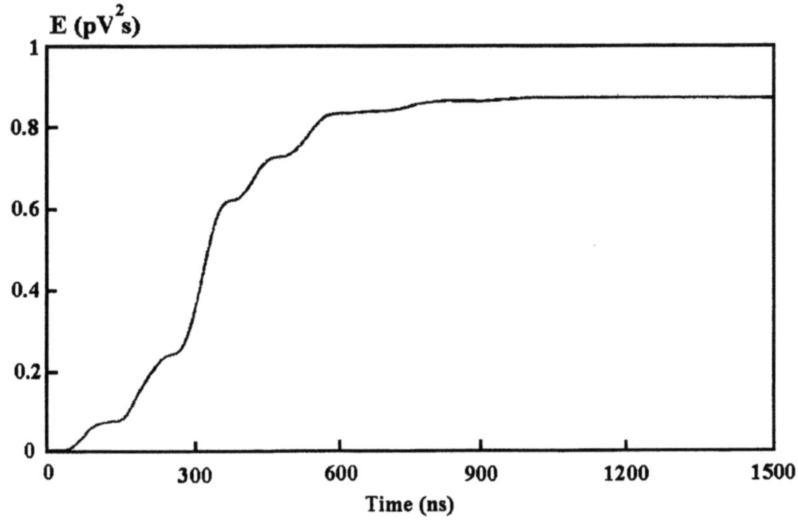

Figure 4. Pulse intensity integral at centre of beam for composite transducer.

Table 1 Intensity Data (with $K_f^2 = 1.98 \times 10^{-11}$ ($V^2cm^2/\mu W$)), $I_{sppa} = 9.6$ (mW/cm^2), $I_m = 18$ (mW/cm^2) and $PD = 456$ (ns).

Radius	I_{sata} ($\mu W/cm^2$)	W (μW)	R_6 (mm)	I_{sptp} (mW/cm^2)	I_{spta} ($\mu W/cm^2$)
1	9.74	3.69	2.39	26.57	21.63
2	10.71	2.95	1.79	26.57	21.63
3	10.57	3.45	2.10	27.38	21.97
4	11.59	3.01	1.83	27.38	21.97

REFERENCES

1. FDA, " Guide for measuring and reporting acoustic output of diagnostic ultrasound medical devices," US Dept. Health and Human services, Food and DrugAdministration, 510(K), 1985.
2. AIUM/NEMA, "Safety Standard for Diagnostic Ultrasound Equipment," Nat. Elect Manuf. Assoc., Washington, DC, Publication UL1-1981, 1981.
3 G. R. Harris, " A discussion of procedures for ultrasonic intensity and power calculations from miniature hydrophone measurements," *Ultrasound ed. Biol.*, Vol.11, No. 6, pp. 803-817, 1985.
4. Data sheet EC-31c, Daiken Fluoro-Chemicals, Daiken Industries Limited, Osaka, Japan.
5. P. Acevedo, *On the Nature of Polymeric and Ceramic Sensors*, Ph.D. Thesis, (1992).

TOWARDS AN INTRA-ARTERIAL PROBE FOR IMAGING AND THERAPY USING LASER-GENERATED ULTRASOUND

Peter A. Payne[1], Chee-Loong Saw[1], Kate A. Roome[3], John V. Hatfield[2] and Richard J. Dewhurst[1]

[1] Department of Instrumentation and Analytical Science
[2] Department of Electrical Engineering and Electronics
University of Manchester Institute of Science and Technology (UMIST)
Manchester M60 1QD, UK
[3] now with Regional Medical Physics Department, Newcastle General Hospital, Newcastle-upon-Tyne NE4 6BE, UK

INTRODUCTION

Coronary heart disease is the scourge of the developed world. In the western hemisphere approximately 50% of all deaths are related to cardiovascular disease, most of these being due to ischaemic heart disease. Some degree of arterial sclerosis (thickening and toughening of artery walls) is a normal age-related phenomenon. However, atherosclerosis which is a build up of deposits in the arterial lumen is much more serious, since the deposits can gradually increase in size and cover practically the entire surface of severely affected arteries. Eventually, calcification of the plaque can develop leading to thrombus formation and weakened sections of the artery known as aneurysms which have a tendency to burst.

Traditionally, arterial disease has been treated surgically[1]. The procedures used include bypass techniques, angioplasty, often balloon angioplasty whereby a guide wire is advanced along the artery to the site of the occlusion and a balloon is slid along the wire and inflated, compressing the stenosis material against the artery wall. There is a 95% success rate initially with this technique, however, there are many problems, including a 30% chance of re-stenosis. Other techniques include stenting (placing a support inside the artery to reduce re-stenosis); atherectomy, where miniature surgical devices are used within the artery to cut away the deposits; ultrasonic angioplasty, although this is difficult due to the high ultrasonic power required; non-laser thermal techniques using plasma sparks, hot tips or heated balloons; laser angioplasty, in which the laser power is used to ablate the tissue forming the blockage. This can involve a number of mechanisms including vaporisation or photochemical mechanisms[2,3]. There are advantages associated with laser angioplasty which usually uses the XeCl excimer laser at 308 nm with pulse durations

Acoustical Imaging, Volume 25, Edited by Halliwell and Wells.
Kluwer Academic/Plenum Publishers, 2000.

ranging from 80 to 250 ns. These include more precise ablation with minimal thermal or mechanical damage to adjacent tissue and less likelihood of particulate debris being released into the blood stream. The laser light is usually guided via optical fibres to the site of occlusion along a guide wire. Several optical fibres are often arranged in an annulus. A significant complication with this technique is the possibility of arterial wall perforation. It is this difficulty that we have been concerned with for some while in our attempts to develop a new form of diagnostic and therapeutic intra-arterial probe.

DIAGNOSTIC CAPABILITIES

The form of probe that we are developing has been described in detail elsewhere[4,5]. It comprises an optical fibre passing concentrically through an ultrasound transducer. This has a forward looking transducer element on the front face and a series of transducer elements arranged around the wall of the probe capable of being used as a phased receive array.

Considerable work has been done aimed at characterising healthy and diseased tissue, originally with healthy porcine aorta[6] and more recently, with post mortem samples of human aorta (buffered, formalin fixed). Typically, ten separate measurements on the post mortem human tissue were conducted by moving the laser spot to different areas on the tissue sample. The laser used was an Nd:YAG with an 8 ns pulse (wavelength 532 nm). In most cases, the waveforms generated from healthy tissue were reasonably smooth in nature and gave rise to frequency spectra that were also of a smooth form (see Figure 1). In contrast, the photoacoustic signals generated from diseased human aorta (also buffered and formalin fixed) had characteristic time domain structure which, it is believed, arise from anomalous absorption sites within the tissue (see Figure 2).

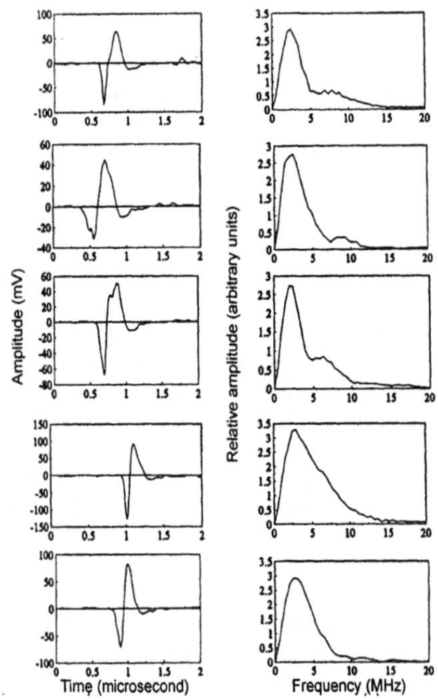

Figure 1. Photoacoustic signals from healthy human aorta and their frequency spectra.

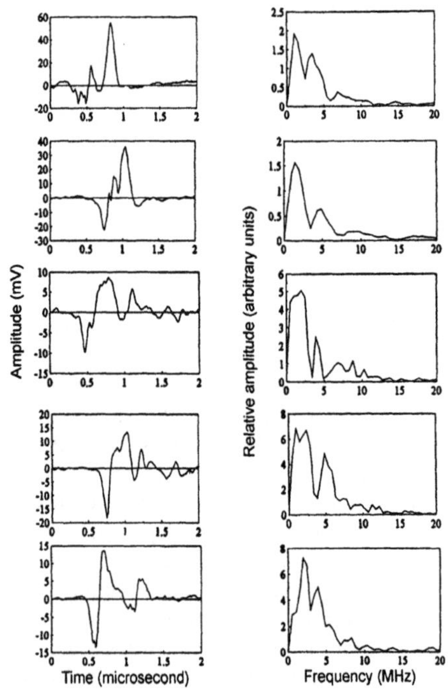

Figure 2. Photoacoustic signals from diseased human aorta and their frequency spectra.

Additional features also appear in the frequency spectrum of the diseased tissue signals, as can be seen from Figure 2. Similar results have also been reported at a wavelength of 532 nm by Beard *et al*[7].

Our experiments on healthy and diseased human tissue were all conducted using the forwards looking mode of the probe. In this mode, the spot size on the tissue to be examined is arranged to be relatively large, compared with the acoustic wavelength in the tissue. This is defined by the cone of laser light emerging from the optical fibre. If the probe, together with the fibre, are brought much nearer the tissue under examination, the spot size can be made smaller and the directivity pattern of the radiated ultrasonic energy will now no longer be directly back towards the probe, but will be at such an angle that it will reflect off the arterial wall and be directed back towards the annular array of receiving transducer elements referred to previously. This is the basis for the sideways imaging mode for the probe being developed.

DIRECTIVITY PATTERN EXPERIMENTS

Initially, we undertook experiments based on tissue mimicking gels with an optical absorption coefficient of 2 mm^{-1}; the laser spot size was 0.19 mm. The same experiments were then conducted on samples of healthy, post mortem human aorta and very similar results were obtained. Examples of both of these experiments are shown in Figures 3 and 4.

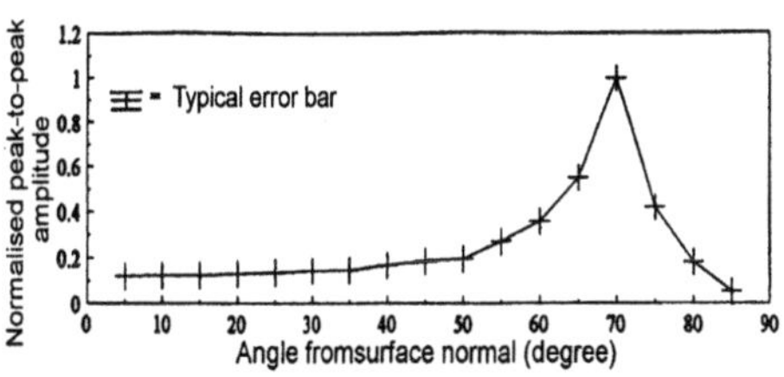

Figure 3(a). Backwards propagating directivity pattern from gelatine (absorption coefficient 2 mm^{-1}); laser spot 0.42 mm; incident energy 7 mJ per pulse.

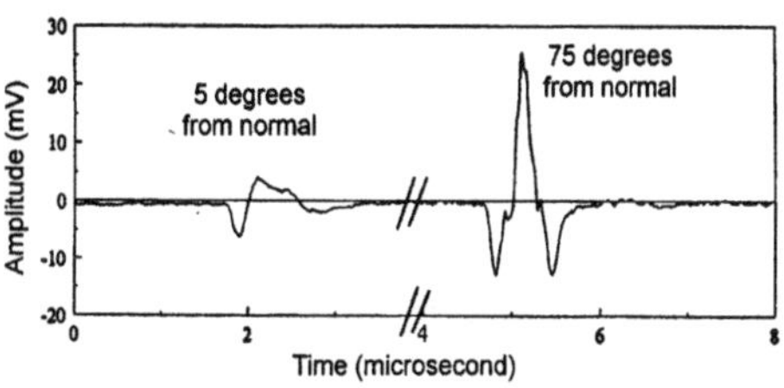

Figure 3(b). Typical received waveforms obtained during the measurements on gelatine

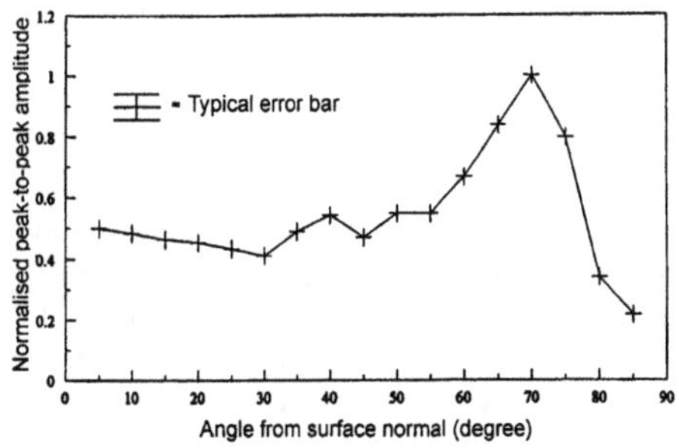

Figure 4(a). Backwards propagating directivity pattern from human aorta; laser spot 0.21 mm; incident energy 2 mJ per pulse.

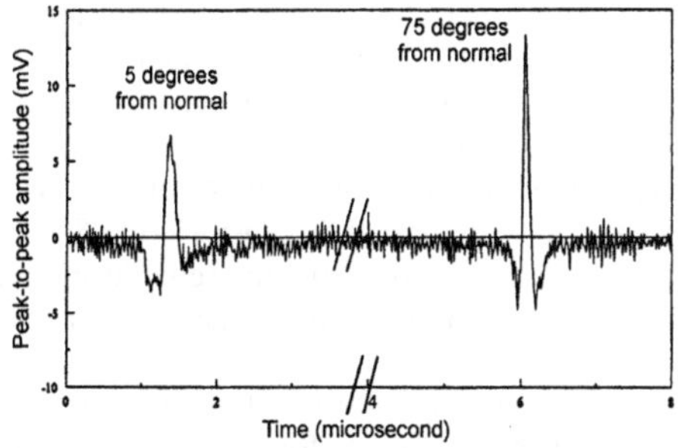

Figure 4(b). Typical received waveforms obtained during the measurements on human aorta.

A further set of experiments were conducted in which the laser light was deliberately introduced at a non-normal angle to the tissue. Little effect was seen on the directivity pattern so produced.

PROBE FABRICATION

Detailed descriptions of the manner in which the forward looking part of the probe is constructed have been published[8]. Our principal concern currently is the best manner of constructing the phased array elements on the side of the probe. A decision to use polyvinylidene fluoride (PVDF) was taken some while ago based primarily on the excellent receive characteristics of this piezoelectric fluoropolymer[9]. Material is commercially available in various thicknesses and our experiments have concentrated on 28 μm film. Ideally, the overall diameter of the probe should be <3 mm and we also regard 32 elements as the minimum from which we can get suitable images to aid in probe guidance. In order to determine the best structure and materials for the probe, it was decided to conduct a set of planar experiments. Photolithography was used to define a set of electrodes with bonding pads at the end. A number of pulse echo experiments were then conducted with the planar form of the probe and from these, decisions made on materials for backing the piezoelectric film. Perspex was found to be a suitable material. A diagram showing the constructional features of the planar probe is shown in Figure 5.

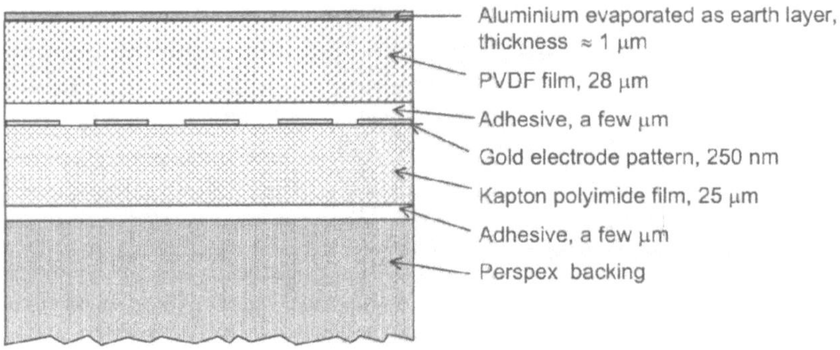

Figure 5. Planar arrangement of transducer micro-film layers (not to scale)

At present, the concept is to form the electrodes on a thin polyimide film, to then glue the piezoelectric polymer onto this with a ground electrode already formed, and then to wrap this structure around a perspex former, again, using a thin glue layer. The overall concept is shown in Figure 6 which also shows that we propose to use signal drivers to cope with the very low impedance of each electrode and suitable devices have been identified with overall dimensions of 1.8 × 1.5 × 0.5 mm*. The greatest hurdle that we have to overcome is that of making reliable and repeatable connections to the electrodes forming the array. Currently, our planar form of the transducer produces less than 50% successful bonds to the pad areas. We are in discussion with a commercial company who claim to be able to conduct this form of bonding and ultimately this may be the only way forward.

Other methods of forming the sideways looking elements have been considered. For example, we could use a solvent to dissolve the PVDF and, with a suitable viscosity, it might be possible to coat the perspex former by a repetitive dipping process. However, initial experiments were not successful. Another possible way forward is to employ the

* National Semiconductors, Quad low-power video buffer CLC114

small degree of shrinkage (3 to 4%) that PVDF possesses when the temperature is raised to 80°C for an extended period (two hours). However, this would require very accurate extrusion of a tube of PVDF which would then have to be slid over the former, raised to 80°C and then possibly, due to the depolarisation that this might produce, the film would need repoling. No experiments based on this concept have yet been performed.

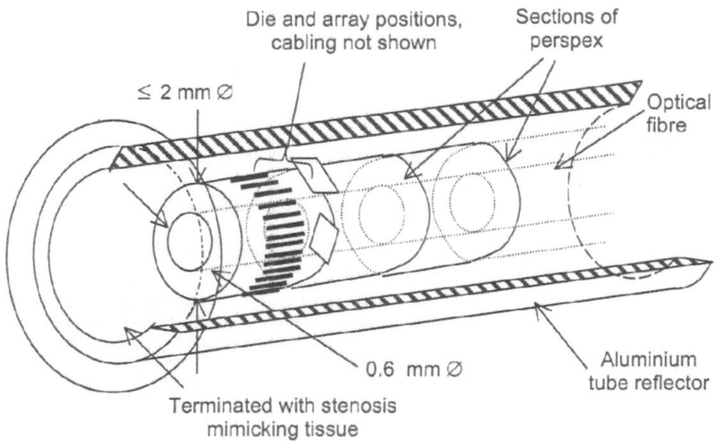

Figure 6. Sideways-looking probe arrangement showing position of buffer chips (die). Also showing layout of forthcoming experiments to further investigate probe performance.

The electrodes are 90 μm wide for the 32 element planar prototype. This is at about the limit for screen printing technology and it is worth investigating whether screen printing electrodes onto a suitable substrate might be a simpler solution capable of being translated to cylindrical topography at a later stage. The substrate material would have to be capable of withstanding high temperatures, since the process would require very high baking temperatures[10].

Referring again to Figure 6, our next set of experiments are to be conducted with the probe inside an aluminium tube to act as a good reflector and with the tube terminated with a piece of stenosis mimicking tissue. This should enable us to investigate the potential performance of the overall concept, despite our problems with making good connections to the 32 or more electrodes. This work has also been delayed due to long delivery times for the signal buffers.

CONCLUSIONS

This paper is in the form of a progress report. Our experience in attempting to build miniature intravascular probes has been mixed. The forward looking version of the probe has performed well, although to date no *in vivo* experiments have been undertaken. The sideways looking device proves extremely difficult to construct and we may have to rely on commercial sources to overcome our interconnect problems.

ACKNOWLEDGEMENTS

We are grateful to the National Heart Research Fund for studentship support for Dr Kate Roome and Chee-Loong Saw. We also wish to thank Mr G. Alban Davies, Cardiac Surgeon, for his continued interest and advice on clinical applications, and Miss Audrey Richardson for her assistance with the production of this paper.

REFERENCES

1. P. Kumar and M. Clark. *Clinical Medicine – A Textbook for Medical Students and Doctors*, 3rd edition, W B Saunders, London (1994).
2. R. Ginsburg and H.J. Geschwind. *Primer on Laser Angioplasty*, 2nd edition, Futura Publishing, Mount Kisco, NY (1992).
3. S.L. Jacques, Laser-tissue interactions – photochemical, photothermal and photomechanical, Surg. *Clin. N. Am.* 72:531 (1992).
4. Q.X. Chen, G.A. Davies, R.J. Dewhurst and P.A. Payne, Photo-acoustic probe for intra-arterial imaging and therapy, *Electron. Lett.* 29:1632 (1993).
5. K.A. Roome, P.A. Payne and R.J. Dewhurst, Towards a sideways looking intravascular laser-ultrasound probe, in: *Eurosensors XII, Vol. 1*, N M White, ed., IOP Publishing, Bristol, (1998).
6. A. Kuhn. *Laser-Ultrasound as a Diagnostic Tool for Medical Applications*, PhD Thesis, Department of Instrumentation and Analytical Science, UMIST, Manchester, UK (1996).
7. P.C. Beard and T.N. Mills, Characterization of post mortem arterial tissue using time-resolved photoacoustic spectroscopy at 436, 461 and 532 nm, *Phys. Med. Biol.* 2:177 (1997).
8. Q. Shan, R.J. Dewhurst, A. Kuhn, K.F. Pang and P.A. Payne, Modelling of a photoacoustic probe designed for medical applications, *Ultrasonics* 34:575 (1996).
9. Q.X. Chen and P.A. Payne, Industrial applications of piezoelectric polymer transducers, *Meas. Sci. Technol.* 6:249 (1995).
10. T. Papakostas, N.R. Harris, S.P. Beeby and N.M. White, Piezoelectric thick-film polymer pastes, in: *Eurosensors XII, Vol. 1*, N M White, ed., IOP Publishing, Bristol, (1998).

PSEUDOSPECTRAL TIME-DOMAIN MODELING OF NONDESTRUCTIVE ULTRASONIC FIELDS

S. A. Nielsen[1,2], J. P. Lynov[2] and W. D. Kristensen[1]

[1] FORCE Institute, Mechatronics and Sensor Technology, Park Allé 345, DK-2650 Brøndby, Denmark
[2] Risø National Laboratory, Optics and Fluid Dynamics Dept., OFD-128, DK-4000 Roskilde, Denmark

INTRODUCTION

The application of steel welds for structural components requires reliable non-destructive evaluation (NDE) models and inspection procedures to determine their integrity and serviceability at the various production stages and in service. However, the material dimensions and its complexity, e.g. the content of large columnar grains in the welding limit the application of traditional ultrasonic NDE. These facts indicate the need for more accurate solutions to the forward problem and associated inspection procedures. In many applications, for example within the nuclear, offshore and subsea fields advanced automated ultrasonic inspection is now available and performed efficiently by the P-scan system (Neergaard et al., 1999). Moreover, Nielsen et al. (1997-98) have proposed inverse ultrasonic procedures for tomographical imaging and elastic constant determination.

The objective of forward modeling for NDE can be posed as "given an assumed material structure, predict what ultrasonic transducers over that structure would record". The modeling is based on a solution to the acoustic/elastic wave equation and one may distinguish between three main types of methods namely, analytical, geometrical and direct methods. Analytical methods are limited to simple material structures because only then closed form solutions exist. Geometrical methods, as ray tracing, do not take into account diffraction phenomena. Direct methods (e.g. pseudospectral, finite-differences, finite-elements) are time consuming and memory expensive, however, unlike the first two methods, direct methods can produce snapshots, which is an important aid in interpreting ultrasonic fields.

In this paper we compare pseudospectral time-domain modeling with finite-difference modeling. The paper is organised in the following way: The first section discusses numerical dispersion. The next section discusses the pseudospectral Chebyshev collocation scheme and its application to non-destructive ultrasonic fields. Then two preliminary examples of pseudospectral modeling are given and our concluding remarks are contained in the final section.

Acoustical Imaging, Volume 25, Edited by Halliwell and Wells.
Kluwer Academic/Plenum Publishers, 2000.

NUMERICAL DISPERSION

Numerical modeling of ultrasonic fields in the time-domain by the finite-difference scheme may handle any kind of waves in complex media but are limited mainly because numerical dispersion introduces large errors when the waves propagate over long distances.

Dispersion curves for the second-order finite-difference scheme may be obtained if a plane harmonic wave is substituted into the wave equation. One obtains, e.g., the normalized P-wave phase velocity

$$\frac{c_p}{c_o} \equiv \frac{\omega}{kc_o} = \frac{1}{\pi\alpha\beta}\sin^{-1}\left\{\alpha\left[\sin^2(\pi\beta\cos\theta)+\sin^2(\pi\beta\sin\theta)\right]^{1/2}\right\} \tag{1}$$

where $\alpha = c_0\Delta t/\Delta x$ is the stability ratio with time step Δt and grid size Δx. $\beta = \Delta x/\lambda$ is the dispersion parameter and θ is the angle between the direction of propagation and the x-axis. Figure 1 shows dispersion curves for P-wave and S-wave velocities versus the dispersion parameter. A dispersion free modeling scheme is characterized by phase and group velocities that do not vary with the dispersion parameter (or wavenumber).

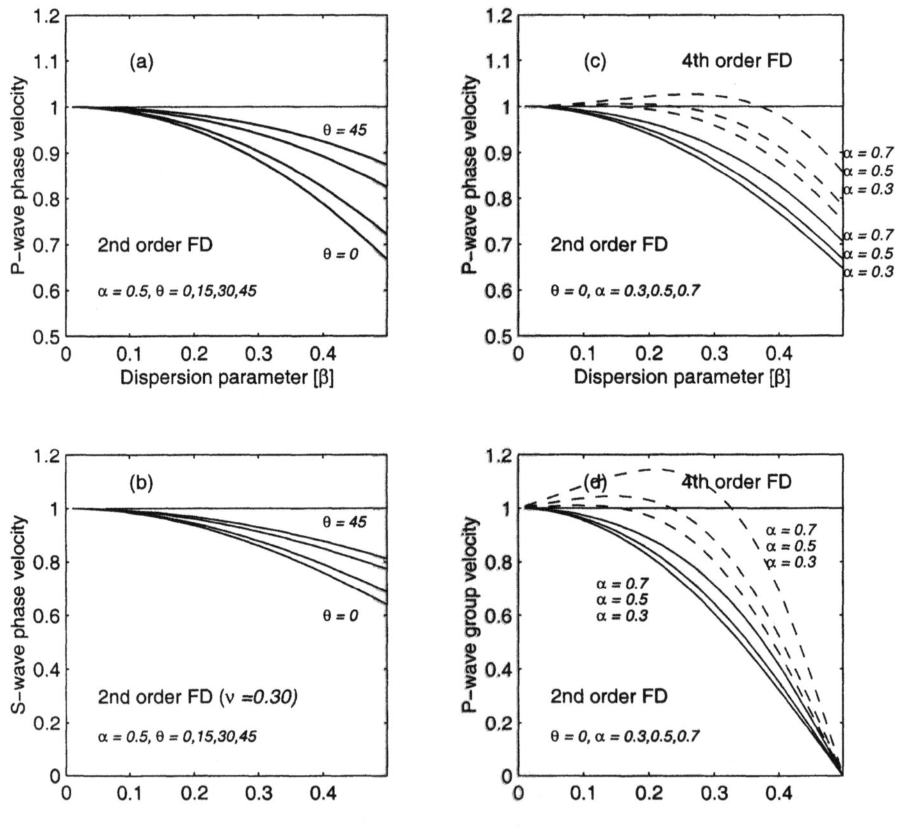

Figure 1. Dispersion curves for normalized velocities. (a) P-wave phase velocity, (b) S-wave phase velocity, (c) P-wave phase velocity and (d) P-wave group velocity. (Left) different propagation angles with respect to grid and (right) different stability ratios.

Figure 1a shows the normalized P-wave phase velocity for different angles of propagation relative to the grid ($\theta = 0$, 15, 30, 45°). Figure 1b shows the corresponding normalized S-wave phase velocity for the same angles and for Poisson's ratio ($\nu = 0.30$). In these figures, phase velocities for the stability ratio ($\alpha = 0.5$) are plotted. It may be seen that the numerical dispersion is largest when the wave propagates along the grid ($\theta = 0$). However, when the P-wave phase velocity is about 2% less than the ultrasonic wave speed the numerical dispersion is often considered sufficiently small. This value corresponds to a wavelength of 10 grid points.

Figure 1c shows the normalized P-wave phase velocities for different stability ratios (α = 0.3, 0.5, 0.7). The corresponding P-wave group velocities $c_g = d\omega/dk$ are shown in Fig. 1d. Dashed curves for a fourth-order finite-difference scheme are shown for comparison. From the figures, it can be seen that the phase and group velocities diminish as the dispersion parameter (or wavenumber) increases. This represents a numerical low-pass filtering effect as higher wavenumber components propagate slower than lower wavenumber components. As a result, finite-difference modeling of ultrasonic pulses with broad bandwidth may result in progressive pulse distortion.

Two important points regarding numerical dispersion may be noticed. Firstly, it may be seen that the stability limit ($\alpha = 0.7$) for the second-order scheme should be used in order to minimize dispersion. Secondly, compared with the second-order scheme less dispersion is predicted for the fourth-order scheme. The latter observation indicates that higher-order finite-difference schemes should be used in numerical modeling to minimize numerical dispersion. Fornberg (1996) has shown that pseudospectral methods may be considered as the limit of the finite-difference schemes, as the order goes to infinity. Hence, pseudospectral methods have theoretically infinite accuracy. Moreover, pseudospectral methods require much fewer grid points to achieve the same accuracy as with finite-difference methods. The pseudospectral method, on the other hand, requires more calculation per grid point. The reduction in number of grid points per wavelength is the main strength of pseudospectral methods. The strength becomes even more significant in 3D modeling.

THE PSEUDOSPECTRAL METHOD

For solving the dispersion problem a global method is proposed in which spatial derivatives are calculated by a Chebyshev collocation scheme. This scheme has been described extensively in the mathematical and electromagnetic literature. See e.g. Canuto et al. (1988) for a review. Pseudospectral methods based on the non-periodic domains have also recently been used to model diffractive optical elements (Dinesen et al., 1999). The Chebyshev polynomial of order $N\text{-}1$ is defined as

$$T_{N-1}(x) = \cos((N-1)\cos^{-1} x) \tag{2}$$

It has extrema on [-1,1] in the so-called collocation grid points (i.e. Chebyshev-Gauss-Lobatto points), as seen in Fig. 2 and given by

$$x_j = \cos\left(\frac{\pi(j-1)}{N-1}\right) \qquad j = 1,\dots,N \tag{3}$$

In the Chebyshev collocation scheme a function *f* is approximated by

$$f(x) \approx \sum_{j=1}^{N} \phi_j(x) f(x_j) \tag{4}$$

where x_j are the collocation points, and interpolation between collocation points is based on Chebyshev - Lagrange polynomials given as

$$\phi_j(x) = \frac{(-1)^j \left(1-x^2\right) T_{N-1}{}'(x)}{c_j (N-1)^2 (x-x_j)} \tag{5}$$

where $c_1 = c_N = 2$ and $c_2 = c_3 = \ldots = c_{N-1} = 1$.
The derivative of *f* at a collocation point is approximated as

$$\frac{df}{dx}(x_j) \approx \sum_{k=1}^{N} D_{jk} f(x_k) \tag{6}$$

where the derivative operator may be represented by a matrix $D_{jk} = \phi_k'(x_j)$, the differentiation matrix, with entries

$$D_{jk} = \begin{cases} \frac{c_k}{c_j} \frac{(-1)^{j+k}}{c_j (x_k - x_j)} & j \neq k \\ -\frac{1}{2} \frac{x_k}{(1-x_k)^2} & j = k \neq 1, N \\ \frac{2(N-1)^2+1}{6} & j = k = 1 \\ -\frac{2(N-1)^2+1}{6} & j = k = N \end{cases} \tag{7}$$

The derivative operator is extended to higher-order derivatives by $D_{jk}{}^{(n)} = \left(D_{jk}\right)^n$.

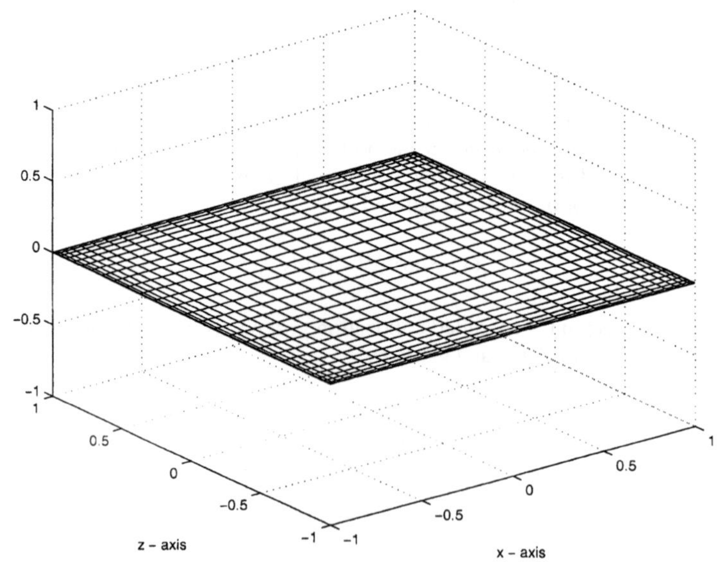

Figure 2. Two-dimensional Chebyshev collocation grid points.

NUMERICAL EXAMPLES

An example of the Chebyshev collocation scheme is shown in Fig. 3. The solid line is the exact differentiation of a damped sinusoidal, while the circles indicate N=30 Chebyshev collocation points, corresponding to about 3-4 grid points per spatial wavelength. This is in contrast to 10 points per wavelength for the second-order finite-difference scheme. The circles indicate the result of the numerical differentiation using Eq. (6) in the Chebyshev collocation points. Figure 4 shows the pseudospectral accuracy of the differentiation. It is seen, that the error between the numerical and exact differentiation is about 10^{-12} for N larger than 45. This error is the limit of the machine accuracy.

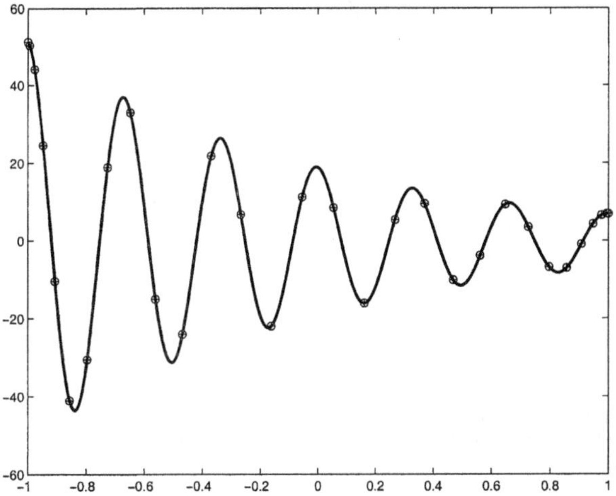

Figure 3. Chebyshev collocation scheme applied on damped sinusoidal.

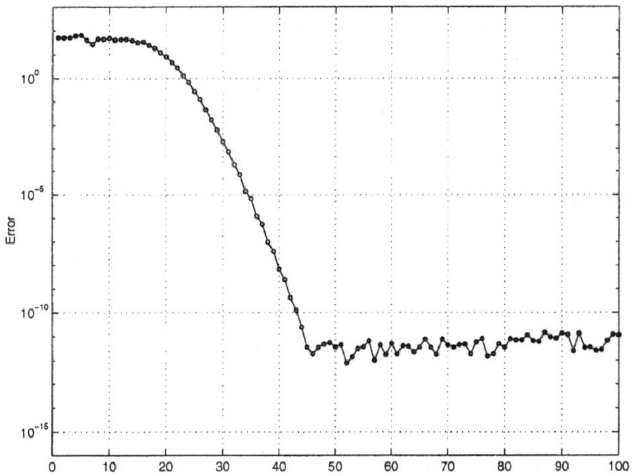

Figure 4. Pseudospectral accuracy versus collocation points.

Figure 5 compares the Chebyshev collocation scheme with a second-order finite-difference scheme for a solution to the scalar hyperbolic equation. The time integration is accomplished using a four-stage, fourth-order, Runge-Kutta method. The solid line represents the theoretical solution to the scalar hyberbolic equation after integration over 2π-periods. The circles indicate the numeric solution based on the Chebyshev collocation scheme, while the dashed line indicates the finite-difference solution. The distorted pulse in the finite-difference scheme is seen clearly. The numerical dispersion causes pulse broadening and high-frequency ringing.

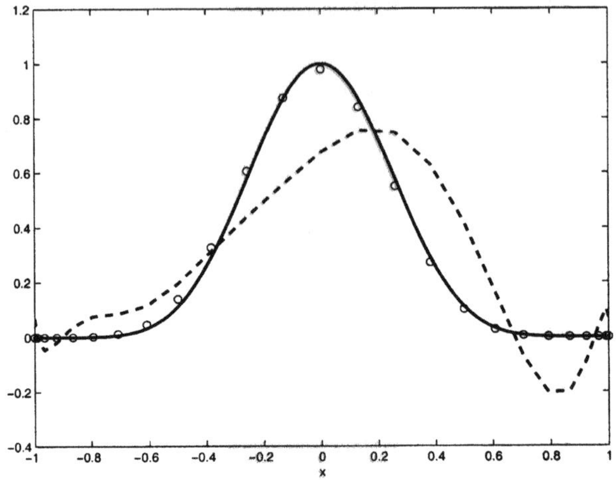

Figure 5. Numerical solution based on the Chebyshev collocation scheme and a second-order finite-difference scheme. In both cases the number of grid-points in x is used.

CONCLUSION

In conclusion, a major limitation of the finite-difference scheme is numerical dispersion. A pseudospectral method is proposed for the accurate solution of spatial derivatives in non-periodic domains. The method is based on a Chebyshev collocation scheme. The method requires much fewer grid points to achieve the same accuracy as with finite-difference methods. The pseudospectral method, on the other hand, requires more calculations per grid point.

ACKNOWLEDGMENTS

This work was supported by the Danish Academy of Technical Sciences.

REFERENCES

Canuto, C.,Hussaini, M.Y., Quarteroni, A. and Zang T.A. *Spectral Methods in Fluid Dynamics,* Springer-Verlag, New-York (1988).

Dinesen, P.G., Hesthaven, J.S, Lynov, J.P and Lading, L. Pseudospectral method for the analysis of diffractive optical elements, *J. Opt. Soc. Am.* A/Vol. 16: 1124-1130 (1999).

Fornberg, B. *A Practical Guide to Pseudospectral Methods,* Cambridge University Press, Cambridge (1996).

Neergaard, S., Kohler, S. and Bogen, E.T. System for advanced automated ultrasonic inspection, *Insight* 41: (11) 692 (1999).

Nielsen, S.A., and Bjørnø, L. Bistatic circular array imaging with gated ultrasonic signals, in: *Acoustical Imaging,* S. Lees and L.A. Ferrari, ed., Plenum Press, New York, 23: 441-446 (1997).

Nielsen, S.A., Andersen, S.I. and Toftegaard, H. An ultrasonic circular aperture technique to predict elastic constants of fiber reinforced composites, in: *Acoustical Imaging,* H. Lee, ed., Plenum Press, New York, 24: (In Press) (1998).

STAMPEDE - "STANDARD TECHNIQUES FOR ACOUSTIC MICROSCOPY FOR PLASTIC ENCAPSULATED DEVICES"

J. Barton[1], T. Randles[1], D Vanderstraeten[2], Gust Schols[2], T Howard[3], P Groeneveld[4]

[1] National Microelectronics Research Centre, Lee Maltings, Prospect Row, Cork, Ireland
[2] Alcatel Microelectronics, Westerring 15, B-9700 Oudenaarde, Belgium
[3] Towcester Technical Services, Towcester Mill, Towcester, Northants NN12 6AD, United Kingdom
[4] Philips Semiconductor, Gerstweg 2, 6534 AE Nijmegen, The Netherlands

INTRODUCTION

The objective of the STAMPEDE project is to develop a comprehensive measurement methodology for the qualification and production monitoring of plastic packages using acoustic microscopy. More specifically it aims at the development of a measurement methodology:

- To evaluate the integrity of interfaces within plastic packages
- To provide an analysis interpretation method which allows us to evaluate the degree of interface degradation or delamination, in order to provide a measure of the robustness of the component to further degradation.
- To provide a simple and objective measurement technique that can be used in the manufacturing environment by everyone with little or no knowledge about failures found by acoustic microscopy.

Indeed, many of the structural defects associated with plastic encapsulated high reliability microelectronics components cannot be detected by traditional inspection methods such as X-rays. Scanning Acoustic Microscopy (SAM) is currently the most effective and widely used technique for the non-destructive examination of plastic packages for structural defects.[1,2] However, there is currently no standard test methodology defined which allows the meaningful, objective comparison of results obtained at a variety of locations employing a range of different acoustic equipments.

This project aims to provide this standard test methodology. It was derived from the work carried out in the Standards, Measurement and Testing (SMT) 4th Framework project STAMPEDE on "Standard techniques for acoustic microscopy for plastic encapsulated

Acoustical Imaging, Volume 25, Edited by Halliwell and Wells.
Kluwer Academic/Plenum Publishers, 2000.

devices" by a consortium comprising four partners from four member states of the EU: Belgium, The Netherlands, Ireland and the United Kingdom. Two partners, Alcatel Microelectronics (B) and Philips Semiconductors (NL), are IC manufacturers (computing, consumer, automotive and telecommunications sectors), one is an acoustic microscopy analysis services company (Towcester Technical Services – UK), and the last one is a microelectronics research centre (National Microelectronics Research Centre – IRE).

The developed test methodology has been thoroughly assessed and validated during the STAMPEDE project by the International Delamination Working Group (IDWG), which acts as an industrial user group, consisting of IC manufacturers, assembly manufacturers, acoustic microscopy equipment manufacturers, research centres, end-users and service providers. Figure 1 illustrates the locations of the various defects found in plastic packaged devices that can be located using acoustic microscopy.

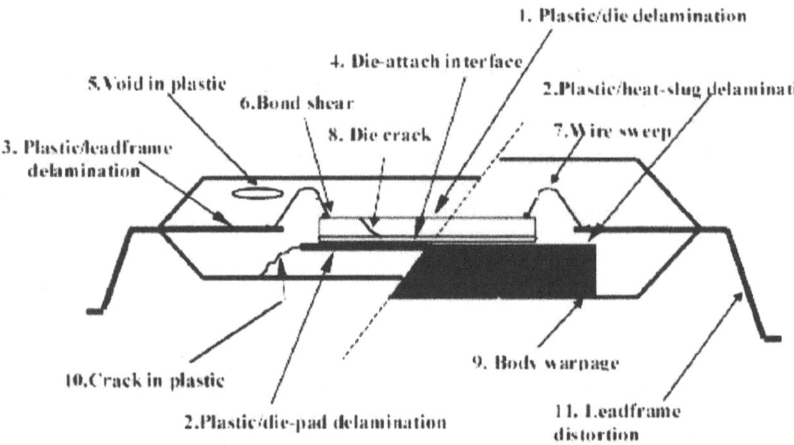

Figure 1. Various defects found in plastic packages

This paper will review the work done in the STAMPEDE project so far with particular emphasis on:

- Model and industrial IC package specification, manufacture, handling and transport protocol
- Calibration test jig design and development
- Acoustic Microscopy analysis
- Intercomparison of results
- Elaboration of a comprehensive measurement methodology

MODEL SAMPLES

The aim of these samples is to investigate the ability of the acoustic lenses to detect and resolve small defects buried in packaging materials. These machined voids also serve to simulate voids and cracks found in real microelectronic packages. Laser micro machining was used to machine the features in the materials selected. This technique was used because it is a rapid and cost effective means of realising test structures in a range of depths, diameters and shapes.

The main conclusion of this work is a design for two sets of void patterns representing cracks and delaminations, to be laser machined into substrates. The group 1 set of patterns contains 100 μm wide and 300 μm long slots, which provide a span of spacing and depths from 10μm to 100 μm respectively. The group 2 model samples include patterns of 300 μm long slots of depth 10 μm but with a large span of widths. Other features were added with various depths from 50 μm to 500 μm. These patterns are for the purpose of both surface and subsurface investigation. The materials selected were PMMA, polyimide (Kapton) and several silica-filled epoxy materials similar to those employed in real electronic packages. Figure 2 shows an acoustic image of model samples machined in PMMA. Attempts at producing tapered samples were discontinued because of the difficulty found in producing features that were representative of cracks. A variety of complementary techniques were used to ensure the quality of the laser machined model samples and calibration tools produced. The techniques used include high frequency acoustic microscopy, SEM, optical microscopy and acoustic profilometry. The complementary techniques used give semi-quantitative information on the depth and width of the machined features. Scanning electron and optical microscopy give qualitative information on the quality of the machined voids, whereas techniques like acoustic profilometry give a profile of the machined voids.

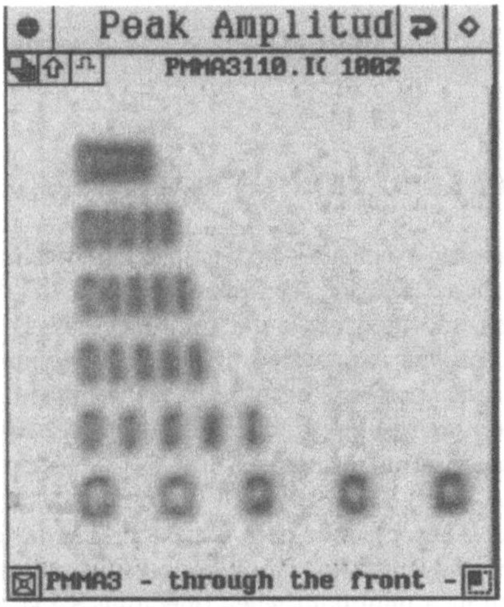

Figure 2. Acoustic image of STAMPEDE model sample machined in PMMA

PLASTIC PACKAGE SAMPLE HANDLING AND TRANSPORT PROTOCOL

A plastic packaged IC is under considerable levels of internal stress after encapsulation. Encapsulation stress is the primary cause of defects such as delamination and cracks. The level of stress within the package is not static and can be altered by handling or changes in the environmental conditions such as ambient and moisture in which the package is stored and tested. As a consequence, defects in the industrial samples, such as delamination or cracks, may increase in size while the samples are being transferred between partners due to handling and changes in environmental conditions. It may be difficult to compare the measurement of defects by the different partners under such circumstances.

The objective, therefore, is to define an easy and cheap handling and transport protocol for both industrial and model samples. This protocol must be designed to minimize the relaxation or aggravation of any process-induced defects that may occur during handling and transport. It is also the aim to incorporate this handling and transport protocol in a later phase in the acoustic microscopy inspection specification.

Because the environmental impact was difficult to tackle it was divided into the following four different items that are more suited for investigating:

- Influence of water absorption
- Influence of dry packing
- Influence of storage in water
- Influence of transport

At the end of the experimental work it was possible to draw the following recommendations for the acoustic microscopy specification:

- Packages may not stay longer than three hours in the water before they need to be dry packed
- Dry packing is certainly recommended when the packaged components need to be transported
- Distilled or DI-water can be used without any problem for the acoustic microscopy inspections
- Several dry packing cycles will not affect the status of the delamination related defects

EQUIPMENT CALIBRATION / CALIBRATION TEST JIGS

The images produced using Scanning Acoustic Microscopes manufactured by various manufacturers are dependent on many factors; some are related to the equipment (including both the microscope itself and the ultrasound transducer used) and some are related to the way in which the microscope is operated. Because the equipment made by one manufacturer does not necessarily perform exactly like that made by any other manufacturer, in order to compare the results produced by various equipment, it is necessary to define standard characterisation principles and test objects which can be transferred from one instrument to another to be used to ensure that there can be concordance of the images produced. The objective here is to define as far as practicable, (i) the scope of the applicability of the characterisation procedures, (ii) the significant parameters which should be recorded as a minimum when operating SAM equipment and (iii) details of the characterisation procedures and test objects.

The philosophy of the calibration/characterisation is to provide the means to control the use of Scanning Acoustic Microscopes for practical applications, not to burden the methodology with unnecessarily precise calibration procedures. It should be noted that because of the variations in equipment performance from manufacturer to manufacturer, it is not possible to specify exactly a set of operating parameters which can be used on all equipments to obtain identical images. The objective of the characterisation procedure is to ensure reproducibility of operation of each individual microscope and to define transferable standards which can be used to make meaningful comparisons between the results obtained on different microscopes.

Limitations Of Scope

1. The operating mode is C-SAM (pulse-echo mode of operation).

2. The preferred coupling medium is de-ionised water with a recommended minimum of 10 MOhm-cm resistivity and maintained within the temperature range 20°C to 25°C during scanning. For static sensitive devices, the water should be re-ionised. As a fall back, distilled water may be used.
3. The applicable operating frequency range is 15 MHz to 100 MHz, and the test objects have been designed accordingly. The basic methodology defined in the equipment calibration procedures may in principle, however, be extended to higher frequencies.
4. The application is the examination of a range of plastic packaged semiconductor devices for the presence of delaminations and cracks in the structures.
5. The analysis is limited to the examination of a single interface depth for each gate setting.

Calibration Test Jigs

As part of this project, two calibration test jigs were devised to facilitate a thorough calibration of acoustic microscopes. The first test jig, while nominally a spot size comparison jig can also be used to calibrate the x-y travel and monitor magnification of the acoustic microscope as well as the gain of the rf receiver. The acoustic velocities of materials can also be measured using this jig. This jig is shown in Figure 3. The second test jig developed is a regular check sample. This regular check sample is recommended to be used to monitor the day to day reproducibility of microscopes in regular usage. An example of the regular check sample in Figure 4 consists of a LQFP80 package on a copper leadframe. The package does not contain any silicon on top of the die-pad. Prior to encapsulation, 4 holes were drilled into the die paddle of the leadframe. The holes measure 0.2mm, 0.4mm, 0.6mm and 0.8mm in diameter and are situated at 2.5mm pitch along the diagonal of the die paddle. The leadframes were then encapsulated resulting in packages 12mm X 12mm X 1.4mm thick.

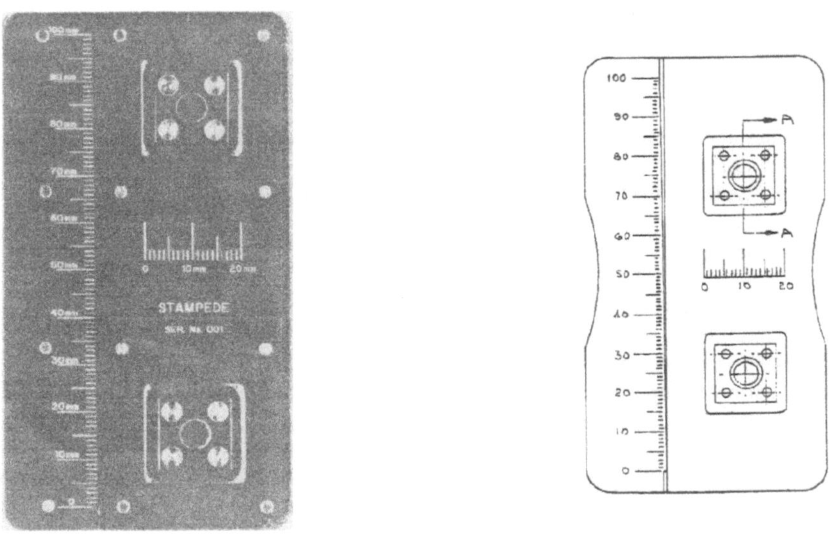

Figure 3. Photograph (a) and schematic (b) of STAMPEDE spotsize calibration jig

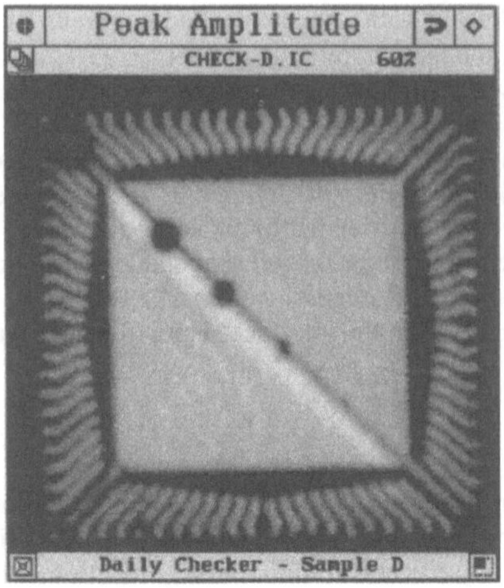

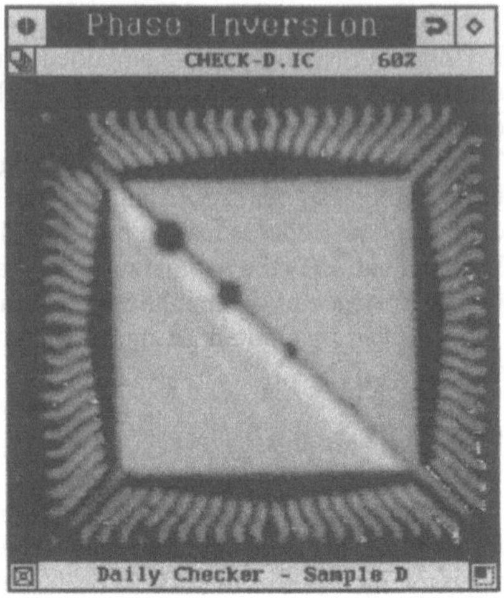

Figure 4. Acoustic images of regular check sample.

ACOUSTIC SIGNAL ACQUISITION ON INDUSTRIAL SAMPLES

A range of industrial samples were circulated to each of the partners in a round robin experiment. This was done to compare the repeatability of the STAMPEDE operating and calibration procedures on different acoustic microscopes at different sites. The selection of the industrial samples was based on two important parameters. The first was the package style and the second the possible defect types. The following selection criteria have been used:

- Package style criteria:

- What types of packages are currently available on the market
- What are the worst case packages for delamination
- Which packages are produced in high volumes
- What kind of material sets are currently the industry standard
- What kind of material sets will be used in the future
- Is there a typical defect type for a specific package style

- Defect type criteria:

- Voids in the plastic, undetectable by X-ray
- Delamination on top of the die pad
- Delamination on top of the leadframe
- Delamination on top of the leadframe bonding lands
- Delamination on top of the silicon

The industrial partners selected the following twelve different package types with different pin counts, ranging from the low cost low pin count packages to the high performance high pin count packages:

- Thin plastic ball grid array (TBGA) 192 pin
- Plastic ball grid array (PBGA) 256 pin
- Thin plastic quad flat package (TQFP) (1.4 mm thick) 176 pin and 100 pin
- Plastic quad flat package (PQFP) 64, 80, 100 and 120 pin
- Plastic leaded chip carrier (PLCC) 68 pin, 44 pin, 28 pin
- Plastic small outline (SOIC) 16, 20, 24 and 28 pin
- Plastic dual-in-line (PDIL) 24 pin
- Plastic shrink small outline package (SSOP) 28 pin
- Low profile quad flat package (LQFP) 80 pin
- SIL9 power package
- SIL9 medium power package
- Wide body small outline (SOIC) 16 pin and 32 pin

The results from all four sites showed excellent repeatability when the calibration, handling and transport protocols were all adhered to and the operating procedure followed. Figure 5 shows a sample of the industrial samples acoustic image and signal acquisition.

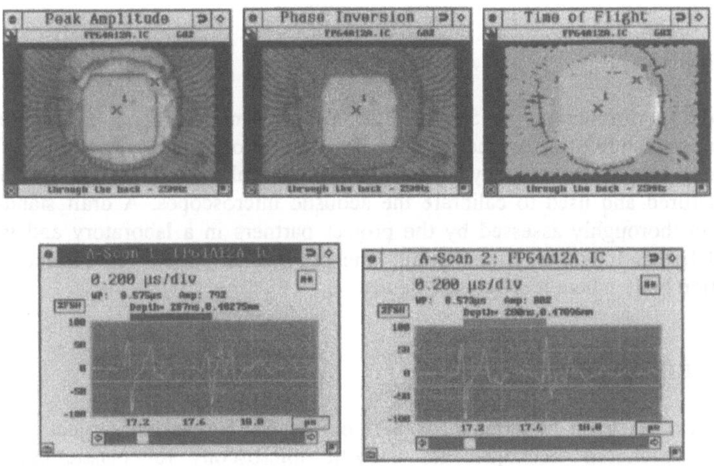

Figure 5. 64 pin plastic quad flat package (PQFP) scanned through the back side with 25MHz transducer showing popcorn crack

ELABORATION OF STANDARD

The Workpackage on the elaboration of the standard was subdivided into 4 tasks :

- Draft of standard
- Industrial assessment of draft standard
- Industrial User Group assessment and feedback
- Submission of pre-standard

Based on the measurements and results of the STAMPEDE project a standard measurement methodology for the imaging and detection of defects in microelectronic plastic packages using acoustic microscopy was drafted. This draft standard included general SAM operating procedure, basic calibration of the Scanning Acoustic Microscope, device handling procedure, presentation, interpretation and processing of results, pro forma for recording scanning parameters and training guidelines

The robustness of the proposed draft standard was assessed through the implementation of the standard in the standard SAM inspection procedures of Alcatel Microelectronics and Philips and at different assembly subcontractors in the Far-East. After the assessment of the proposed draft standard by Alcatel Microelectronics and Philips, the Industrial users in the STAMPEDE project consortium, the draft standard was distributed for comments and suggestions to the members of the Industrial User Group supporting the project since its start, as well as to experts on scanning acoustic microscopy in user companies and institutes. During the first project year the STAMPEDE consortium contacted the International Delamination Working Group (IDWG). The IDWG consists of acoustic microscopy and delamination experts from the major IC companies in Europe. Also American IC manufacturers operational in Europe are members of the IDWG, as well as the major manufacturers of acoustic microscopy equipment, such as Sonix, Sonoscan and Hitachi. The developed test methodology was thoroughly assessed and validated during the STAMPEDE project by the IDWG.

After the draft standard was assessed, the pre-standard was submitted to the European standardisation bodies and is currently being reviewed. Some members of the IDWG are also members of the JEDEC Committee, and have offered their services to help introduce the draft standard into a JEDEC standard.

CONCLUSIONS

The work performed in the STAMPEDE project on developing a comprehensive measurement methodology for the qualification and production monitoring of plastic packages using acoustic microscopy has been presented. Calibration test jigs were designed and manufactured and used to calibrate the acoustic microscopes. A draft standard was produced and thoroughly assessed by the project partners in a laboratory and industrial context and by an Industrial users group. The pre-standard was then submitted to the standardisation body where it is being reviewed.

ACKNOWLEDGEMENTS

This work has been undertaken as part of the SMT research project 97CT-2182 STAMPEDE 'Standard techniques for acoustic microscopy for plastic encapsulated devices' which is supported by the European Commission. The authors gratefully acknowledge their support.

REFERENCES

1. T.M.Moore,.SAM provides critical IC inspection, *Adv. Mat. & Procs*, Oct 1990.
2. T.M.Moore. C-mode acoustic microscopy applied to integrated circuit package inspection, *Solid State Elects*, Vol 3, pages 411-421, 1992

ULTRASONIC NDE USING INSTANTANEOUS PHASE INFORMATION VIA MORLET WAVELET TRANSFORM

Guangming Zhang,[1] Yuwen Wang,[2] Shuyi Zhang,[1] and Bufa Zhang[1]

[1]Lab of Modern Acoustics, Institute of Acoustics
Nanjing University, Nanjing 210093, China
[2]Research Institute for Laser and Infrared Application
Xi'an Jiaotong University, Xi'an 710049, China

1. INTRODUCTION

In ultrasonic nondestructive evaluation (NDE), it is often difficult to distinguish a flaw echo from the background noise. The noise often masks the flaw echo signal, creating a hindrance to detection. Many techniques have been used to analyze the noisy ultrasonic signals. The split spectrum algorithms for the suppression of the interference noise are known to work well when properly tuned. Moreover, the results presented by Gustafsson and Stepinski[1] showed the most split spectrum algorithms such as polarity thresholding and minimization rely on phase characteristics of noise free filter signals and are not using amplitude information. Therefore, it should be considered that the phase information is important in the ultrasonic NDE. The wavelet transform (WT) is the most recent technique for processing signal with time-varying spectra, and has been utilized to improve the ultrasonic flaw detection in the noisy signals. The Morlet WT is a kind of analytic wavelet transform, and can extract the instantaneous phase information of the broadband ultrasonic signals accurately. In this work, we try to apply the instantaneous phase information to the ultrasonic signal processing technique based on the WT.

In the paper, the phase characteristics of the flaw echoes and noise are derived using a geometrical approach. And then according to these characteristics, a new ultrasonic NDE technique is proposed. This technique firstly extracts the instantaneous phases of the broadband ultrasonic signal by using the Morlet WT. Then, The ultrasonic flaw echoes are distinguished from the background noise according to the phase information. Moreover, the pruning[2] and threshold[2] processing are performed for further suppressing the noise. Finally, the flaw signal is reconstructed using the Morlet reconstruction formula.[3]. The proposed technique has been verified both by simulation and ultrasonic flaw detection experiments, yielding significant signal-to-noise ratio enhancement.

2. PHASE CHARACTERISTICS OF FLAW ECHO

In the broadband ultrasonic signal processing, the best-performing mother wavelet[2] is the Gaussian wavelet. The Gaussian wavelet $g(t)$ can be expressed as

Acoustical Imaging, Volume 25, Edited by Halliwell and Wells.
Kluwer Academic/Plenum Publishers, 2000.

$$g(t) = \frac{\exp(-t^2/(4B_{mw}))}{\sqrt{4\pi B_{mw}}}\cos(2\pi f_c t + \varphi) \tag{1}$$

where the parameter B_{mw} determines the bandwidth of the mother wavelet, and f_c is its center frequency. Furthermore, the flaw echo is usually modeled as[1]

$$f(t) = h(t)\cos(2\pi f_{tr} t + \varphi) \tag{2}$$

where the envelop $h(t)$ is a bell-shaped function, and f_{tr} is the center frequency of the received signal. Assuming that $h(t)$ is a Gaussian function, Eq. (2) becomes

$$f(t) = \frac{\exp(-t^2/(4B_{tr}))}{\sqrt{4\pi B_{tr}}}\cos(2\pi f_{tr} t + \varphi) \tag{3}$$

where the parameter B_{tr} determines the bandwidth of the flaw echo.

In addition, according to Eq. (1) the daughter wavelet with scale a can be rewritten as

$$g_a(t) = \frac{\exp(-t^2/(4B_a))}{\sqrt{4\pi B_a}}\cos[(2\pi f_{tr} + 2\pi f_H)t] \tag{4}$$

where B_a is the factor which determines the bandwidth of the wavelet filter, f_H is the frequency offset between the center frequency f_a of the daughter wavelet and the center frequency f_{tr}. Let $\omega_{tr} = 2\pi f_{tr}$, $\omega_H = 2\pi f_H$. Combining Eqs. (3) and (4), the narrow-band flaw signal $f_a(t)$ at the scale a can be obtained (assume the pulse width of flaw is T):

$$f_a(t) = \exp(-\frac{B_{tr}B_a\omega_H^2}{B_{tr}+B_a})\frac{\exp(-\frac{(t-T)^2}{4(B_{tr}+B_a)})}{\sqrt{4\pi(B_{tr}+B_a)}}\cos[(\omega_{tr} + \frac{B_a\omega_H}{B_{tr}+B_a})(t-T)+\varphi] \tag{5}$$

If $B_{tr} << B_a$, then

$$f_a(t) = \exp(-B_{tr}\omega_H^2)\frac{\exp(-\frac{(t-T)^2}{4B_a})}{\sqrt{4\pi B_a}}\cos[(\omega_{tr}+\omega_H)(t-T)+\varphi] \tag{6}$$

In fact, this condition is usually satisfied in the broadband ultrasonic flaw detection. Obviously, for the narrow-band flaw signal at the scale a the instantaneous phase $\phi_a(t)$ at the time t can be written as

$$\phi_a(t) = (\omega_{tr}+\omega_H)(t-T)+\varphi \tag{7}$$

According to the definition of group delay:

$$v(f) = -\frac{1}{2\pi}\frac{d(\phi(f))}{df} \tag{8}$$

where $\phi(f)$ is the instantaneous phase, and substituting Eq. (7) into Eq. (8), then the group delay of the narrow-band flaw signal $f_a(t)$ can be obtained

$$v_s(a,t) = -\frac{1}{2\pi}\frac{d\phi_a(t)}{df_a} = c_1 t + c_2 \tag{9}$$

where c_1 and c_2 are constant, and a is the scale factor. In the practical WT, the shift factor b takes the place of the time t, then Eq. (9) can be rewritten as

$$v_s(a,b_j) = c_1' b_j + c_2\,,\quad b_j \in (b_1, b_n) \tag{10}$$

where c_1' is a constant, b_j is the shift factor, and b_1 and b_n correspond to the time lower limit t_d and upper limit t_u of the flaw echo respectively. Assuming that ultrasonic signals are sampled at the same time interval, obviously the group delay difference of the flaw echo is a constant:

$$v_s(a,b_{j+1}) - v_s(a,b_j) = const \tag{11}$$

where j varies from 1 to n. On the other hand, the noise is a random process, so that the group delay $v_n(a,b_k)$ of the noise signal is a random and the group delay difference is also a random:

$$v_n(a,b_{k+1}) - v_n(a,b_k) = random \tag{12}$$

The randomness of the group delay can be measured by its entropy, and the detailed calculation will be shown in the next section. According to the phase characteristics of the flaw echoes, the flaw echoes can be distinguished from the noisy ultrasonic signal, and enhancement of the detected echoes can be realized.

3. ALGORITHM

The following outlines the procedure of the proposed technique.

3.1. Extraction of the Instantaneous Phases by the Morlet WT

Firstly, perform continuous wavelet transform (CWT) on the received ultrasonic signal by using Morlet wavelet. Then the instantaneous phases of the ultrasonic signals are extracted by utilizing the output of the complex analytic WT. Here, set the central frequency of the mother wavelet to be equal to the central frequency of the ultrasonic signal.

3.2. Calculate the Group Delay and Group Delay Difference

The discrete definition of the group delay in the discrete Fourier transform (DFT) is

$$v(k) = -\frac{N}{2\pi}[\phi(k+1) - \phi(k)],\quad 1 \le k \le N/2 \tag{13}$$

where $\phi(k)$ is the phase component, k the frequency index, and N the total number of data points in the DFT. A similar definition corresponding to the CWT can be given. Let $\phi(a,b)$ be the instantaneous phase of the point (a,b) in the time-frequency plane. According to the relation between scale and frequency: $a = f_c / f$, and using the definition of the group delay, the group delay in the point (a_i, b_j) can be calculated:

$$v_{b_j}(a_i) = -\frac{1}{2\pi f_c}\left(\frac{\phi(a_{i+1},b_j)-\phi(a_i,b_j)}{1/2^{m_{i+1}}-1/2^{m_i}}\right) \quad (14)$$

where $a_i = 2^{m_i}$, $m_{i+1} = m_i + step$. The group delay difference value is defined as the difference of the group delay between the time b_j and the adjacent point b_{j+1} in the same scale a_i, i.e.

$$dv_{b_j}(a_i) = v_{b_{j+1}}(a_i) - v_{b_j}(a_i) \quad (15)$$

3.3. Calculate the Moving Entropy of the Group Delay Difference Values

The following outlines the procedure for obtaining the moving entropy of the group difference values:

1. The group delay difference values, which range from 1 to N in the scale a_i, are separated into many parts using a moving window whose width is M, when the window center is varied from $b_{M/2}$ to $b_{(N-M)/2}$. Calculate the histogram $f_{b_j}(m)$ of the group delay difference values in every window, index m varying from 1 to M. Then normalize the area of the group delay histogram.
2. Calculate the moving entropy I_k of the group delay difference values using formula (16), and let it to be the entropy of the point (a_i, b_k) in the time-frequency plane.

$$I_{b_k}(a_i) = -\sum_{1}^{M} f_{b_j}(m)\log_2 f_{b_j}(m)\,,\; M/2 \le k \le (N-M)/2 \quad (16)$$

Repeating the process, we can obtain an entropy curve of the group delay difference for every scale. If there is no noise and only a flaw in the moving window, the group delay difference entropy becomes zero. Moreover, If there is only noise in the moving window and the phase of the noise is uniformly distributed over [-π,π], the probability density function of the group delay will be a constant over [1,M]. Therefore the group delay difference entropy becomes $\log_2 M$.

3.4. Extraction of Flaw Information

As said above, the flaw has the smaller group delay difference entropy while the noise has the larger group delay difference entropy. Giving an entropy threshold I_p, if $I_{b_k}(a_i) \le I_p$, it can be considered that there is a flaw in the time b_k and scale a_i. Then all the flaws can be found one by one using this method. If a flaw has been found, then remain the wavelet coefficients in the moving window whose center is (a_i, b_k). Repeat this process for all the flaws. Afterward set all the wavelet coefficients in other time points to be zero. At last, perform the pruning[2] and threshold[2] processing on the wavelet coefficients.

3.5. Signal Reconstruction

Finally, the flaw signal is reconstructed from the processed wavelet coefficients by using the Morlet reconstruction formula.[3]

4. EXPERIMENTAL RESULTS

In order to evaluate the performance of the proposed technique, some experiments have been carried out using simulated data and practical detected signals. The experimental results show the proposed technique is efficient in improving the signal strength and reducing the noise. The following shows two examples of these experimental results.

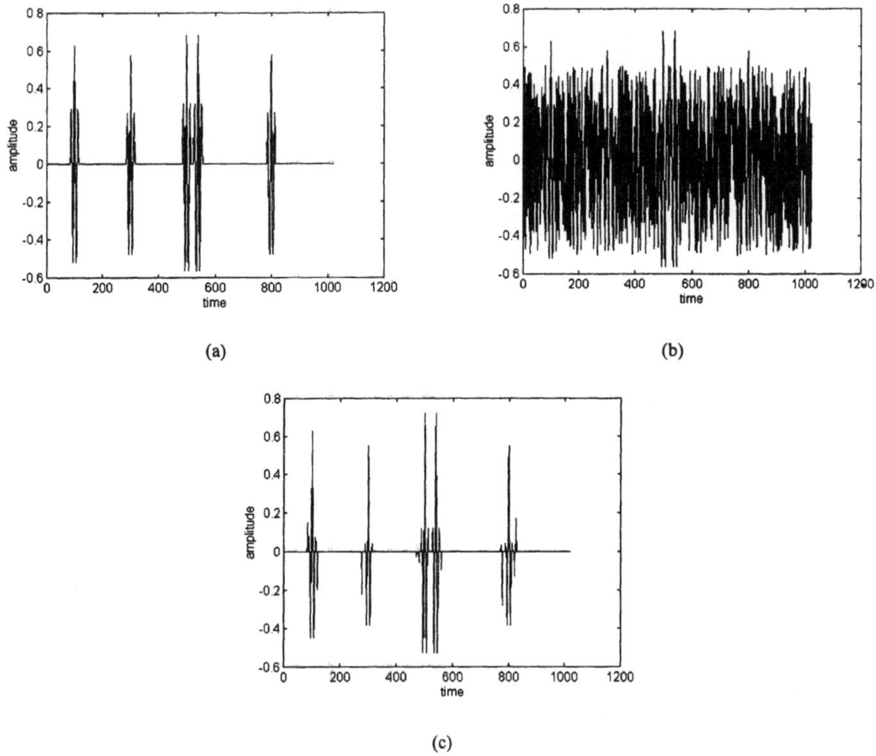

Figure.1. (a) The simulated noise free flaw signal with five echoes. (b) The simulated noisy ultrasonic signal. It was created using the signal in Fig.1 added in noise. (c) The processing output of Fig.1b

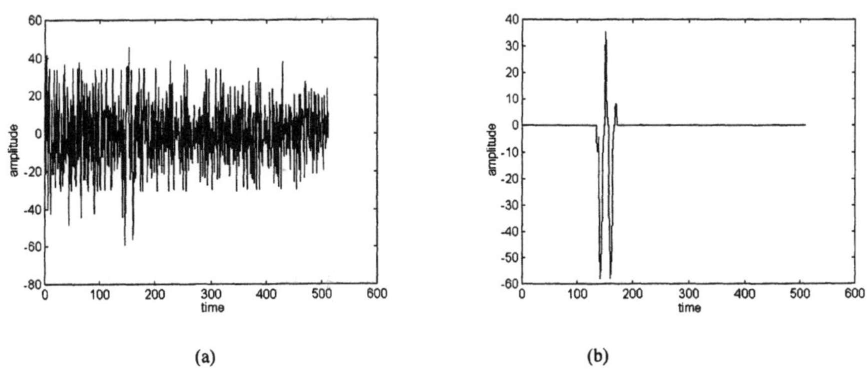

Figure.2 (a) Ultrasonic signal obtained for a cast steel sample. (b) The processing output of Fig.2a.

An example using the simulated data is shown in Fig.1. Fig.1a is a simulated noise free flaw signal, and five flaw echoes are made in the simulated signal. The noisy ultrasonic signal plotted in Fig.1b is the flaw signal added in a random noise. Then, the simulated noisy ultrasonic signal plotted in Fig.1b is processed using the proposed technique, and the output is shown in Fig.1c. The five echoes are clearly detected, and significant signal-to-noise ratio enhancement is achieved.

Another example for the proposed technique applying to the practical ultrasonic flaw detection in highly scattering materials is given in Fig.2. The detected signal plotted in fig.2a is obtained by detecting a block of cast steel using a 2.5 MHz commercial unfocused longitudinal transducer. One transverse hole was drilled in the block of cast steel. Obviously, it is difficult to distinguish the echo of the hole from the background noise duo to the highly scattering of the cast steel. Fig.2b shows the output using the proposed technique. The echo of the artificial hole is clearly detected without any other contribution from the acoustic noise.

5. CONCLUSIONS

In ultrasonic NDE flaw signals may be contaminated by noise, even inundated. Under such circumstances, it is difficult to distinguish the flaw signal from the background noise directly. In this paper, a new ultrasonic NDE technique is presented using the phase characteristics of the noise free flaw echoes by the Morlet WT. These phase characteristics are derived using a geometrical approach. The results presented here show the potential of the proposed technique for detecting flaws in complex environments. Further work is planned to research the determination method of the entropy threshold I_p, and to compare the technique with the other ultrasonic signal processing techniques to further assess the capabilities of the technique.

REFERENCES

1. M. G. Gustafsson, and T. Stepinski, Split spectrum algorithms rely on instantaneous phase information—a geometrical approach, IEEE Trans. Ultras. Ferroe. Freq. Contr. **40(6),** 659-665(1993).
2. A. Abbate, J. Koay, J. Frankel, S. C. Schroeder, and P. Das, Signal detection and noise suppression using a wavelet transform signal processor: application to ultrasonic flaw detection, IEEE Trans. Ultras. Ferroe. Freq. Contr. **44(1),** 14-25(1997).
3. O. Rioul and P. Duhamel, Fast algorithms for discrete and continuous wavelet transforms, IEEE Trans. Inform. Theory **38(2),** 569-586(1992).

RESONANT PHOTOACOUSTIC DETECTION AND IMAGING SCHEME

Tsutomu Hoshimiya

Department of Applied Physics
Tohoku Gakuin University
Tagajyo, 985-8537, JAPAN

INTRODUCTION

Recently, resonant character in modulated photoacoustic (PA) measurement schemes, which utilizes glass capillary resonance or thermal bending, have been proposed and demonstrated. However, resonant characteristic in a rigid solid pipe has never been applied until now. In the present paper, a heat transferred to the environmental surrounding gas for the solid specimen generates a heat source distributed along an inner surface of the rigid solid cylinder, which causes an acoustic resonance.

PRINCIPLE

It is generally well known that a PA effect in a gas medium is governed by a wave equation (strictly speaking, a Navier-Stokes equation). Simply, a pressure wave p(z,t) propagating along a z-direction satisfies an equation,

$$\frac{\partial^2 p}{\partial x^2} - \frac{1}{c^2}\frac{\partial^2 p}{\partial t^2} = \alpha H \quad (1)$$

where H(z ,t) represents a heat generated by a periodically modulated laser beam.

In the case where a laser beam irradiates an open-ended solid pipe with a length L, a sound wave generated by a modulated laser irradiation will reflect at the both ends of an open-ended pipe. As a result, a standing wave with a wavelength of 2L will dominate so that a resonant condition

$$f_0 = \frac{c}{2(L + 2\Delta L)} \quad (2)$$

will be satisfied, where c is a speed of a sound wave, and a parameter ΔL represents the open-end correction and it is given by 0.6r (r denotes a half of an inner diameter of the pipe).

If we expand a pressure wave into a linear combination of time varying amplitudes of normal nodes of an open-pipe resonance, the fundamental one is shown as

Acoustical Imaging, Volume 25, Edited by Halliwell and Wells.
Kluwer Academic/Plenum Publishers, 2000.

$$p = A\cos(\frac{\pi z}{L}) \qquad (3),$$

then the amplitude A(t) satisfies an equation same as a damped harmonic oscillator,

$$\frac{d^2A}{dt^2} + \gamma\frac{dA}{dt} + \omega_0^2 A = \beta \cdot I_0 \qquad (4)$$

where ω_0 represents a resonant frequency, γ denotes a extinction coefficient. Generally, γ corresponds to a viscosity loss of a gas medium. However, if an artificial surface defect is introduced, γ increases correspondingly due to the acoustic power loss caused by excess energy dissipation.[1]

Since an acoustic wave generated by a PA effect behaves as a classical damped harmonic oscillation, an acoustic resonance will take place if an optical modulation frequency is scanned over an acoustic resonant frequency.

The presence of a surface defect on an inner surface of the pipe specimen with both end opened will degrade a Q-factor of the acoustic resonance due to the energy loss of the acoustic wave generated by a PA effect. This detection scheme makes use of the highly sensitive acoustic resonance in the frequency domain, and the spatial resolution will be limited by the laser beam focusing.

The detection scheme based on the degradation in an acoustic resonance can be coupled by a laser beam scanning technique so that an inspection of the inner surface of the cylindrical solid specimen will be available.[2]

EXPERIMENTAL APPARATUS

Basic experimental setup used in the R-PA scheme is shown in Fig. 1, which is similar to that of a photoacoustic microscope (PAM).

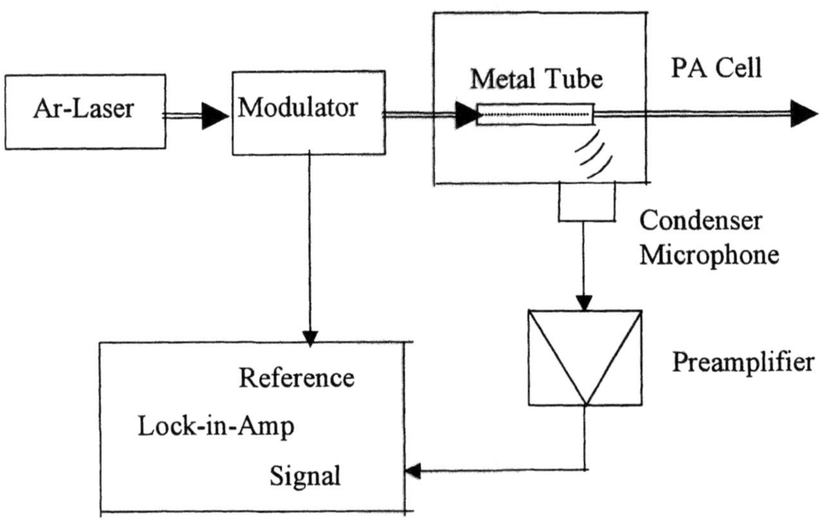

Figure 1. Basic experimental setup

An Ar-ion laser was used as an optical source, the beam of which is modulated by a mechanical chopper (NF Circuit block, CH-353) or an acousto-optic (AO) modulator (Intraction, AOM-125) coupled with a universal function generator. A high-sensitive condenser microphone (Brewer & Kaejer 4166) was attached to a specially designed PA cell in which a metal pipe specimen under inspection was set. A cupper pipe with outer and inner diameters of 4mm and 2mm, respectively and a length of 170mm was used as a metal specimen.

A PA signal is generated by the laser beam irradiation onto an inner surface of a pipe. A signal was detected phase-sensitively by a lock-in amplifier, and both amplitude and phase PA signals were recorded as a function of a modulation frequency of the laser beam. Changing a modulation frequency near the resonance frequency of the open pipe 986.01 Hz calculated by Eq. (2), PA amplitude and phase signals were recorded for a pipe without and with a surface defect on the inner wall of the pipe.

EXPERIMENTAL RESULTS

A typical character of the acoustic resonance was observed in a PA amplitude and phase data in the frequency domain shown in Figs. 2 (a) and (b). The laser power in this case was 351.6 mW. These characteristics are in good agreement with those expected by an idealized damped harmonic oscillator. The obtained central frequency and Q-factor of acoustic resonance were 985Hz and about 10, respectively. The resonant frequency showed an excellent agreement with the value of 986.01Hz calculated by Eq. (2) for an open-ended metal tube with a length and an inner diameter of 170mm and 4 mm, respectively.

In the case when an artificial surface defect was introduced on an inner surface of the pipe, the generated R-PA signal was remarkably deceased so that the laser power was increased up to 710.4mW, about a twice of that of the previous case. The feature of both signal decrease and degradation of the Q-value in tuning characteristic in R-PA amplitude signal is shown apparently in Fig. 3. As a result, resonance Q-value was degraded up to the value of about 4, which apparently showed the dissipation of the acoustic energy due to the presence of the surface defect.

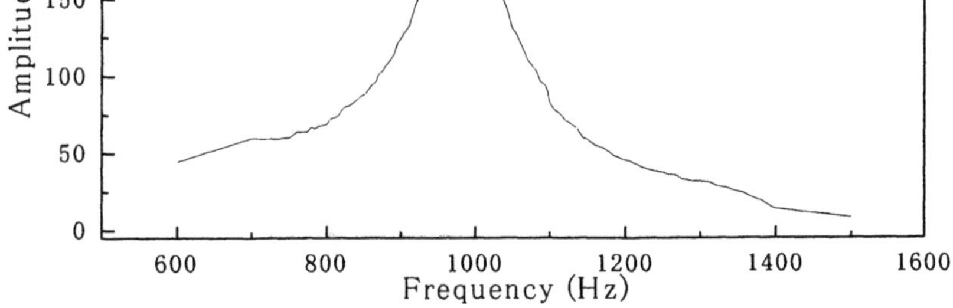

Figure 2. (a) Tuning characteristic of a R-PA amplitude signal.

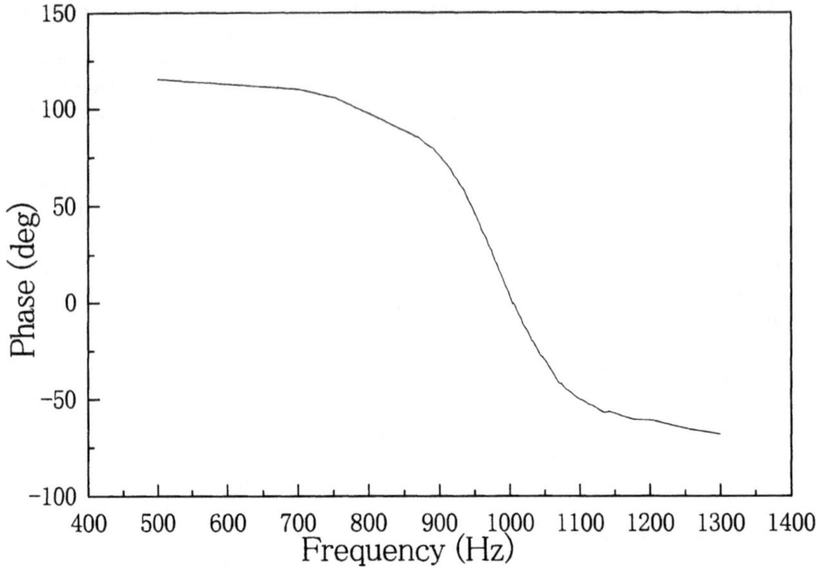

Figure 2. (b) Tuning characteristic of a R-PA phase signal.

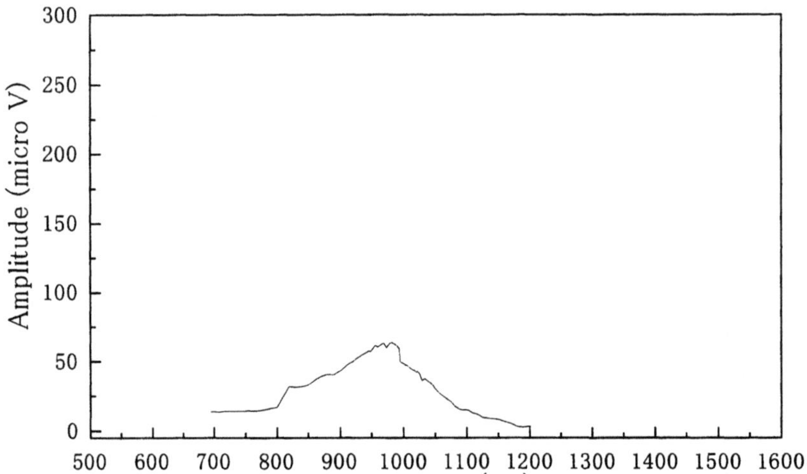

Figure 3. Tuning characteristic of R-PA amplitude signal with defect.

DISCUSSION AND CONCLUSION

In this paper, a new R-PA scheme, which utilizes the combination of a PA effect and an acoustic resonance of a solid pipe, was proposed and demonstrated in an audio-frequency region. NDI of a metal pipe specimen by the use of Q-value discrimination was performed by the R-PA scheme.

The R-PA scheme can be applied both NDI and a nondestructive precise measurement of finished sizes of pipe-shaped materials such as a "variable orifice nozzle (VON)" used for direct injection diesel engines, developed by the Zexel group and Tokyo University, Japan[3].

In a typical VON, a nozzle size is with a 0.18 mm in diameter and a 0.7 mm in length. The resonance frequency calculated for an environment in an atmospheric pressure is 225.5 kHz. An extension of this method beyond audio frequency by the use of ultrasonic air transducer, for example realized by fabrication with a surface micromachine technology[4], will be the next subject of the present detection scheme.

ACKNOWLEDGMENT

The author would like to thank Dr. T. Kobayashi of Zexel for his offer of VON sample to the author. . He is grateful to Prof. B. T. Khuri-Yakub of Stanford University for his courtesy and assistance to collaborated research project of application of micromachine ultrasonic transducer to this study. He also thanks graduated student Mr. T. Saisu and undergraduate student Mr. T. Satoh for their help in the experiment.

REFERENCES

1. C. P. Hsieh and B. T. Khuri-Yakub: Appl. Phys. Lett., vol. 60, 1815(1992).
2. T. Hoshimiya: Japanese patent (submitted, 1999).
3. T. Hasegawa, H. Maehara, R. Tanishige, T. Iwasaki, T. Kobayashi, and Y. Matsumoto: private communication (1999).
4. M. I. Haller and B.T. Khuri-Yakub: IEEE Trans. Ultrason. Ferroelect. Freq. Contr., vol. 43, 1(1996).

ACOUSTOGRAPHY APPLIED TO THE MONITORING OF IMPACT DAMAGE GROWTH IN COMPOSITES UNDER FATIGUE CONDITIONS

A.S. Chen, D.P. Almond and B. Harris

Department of Materials Science and Engineering
University of Bath
Bath, BA2 7AY, UK

1. INTRODUCTION

Acoustography[1,2] is a recently developed ultrasonic technique which offers broad-area, nearly real-time ultrasonic imaging. Acoustography uses an acousto-optic imager formed by a layer of ultrasound-sensitive liquid-crystal which enables large areas to be imaged instantaneously and continuously. A test piece is insonified by a broad beam of ultrasonic waves and the ultrasound which is transmitted through the test piece is detected by the liquid-crystal imaging screen. The system has been adapted to provide a means of continuously monitoring damage-zone growth inside a composite material sample throughout a mechanical test.

Carbon-fibre-reinforced polymer (CFRP) composite materials have pronounced high specific stiffness and strength but they are also highly sensitive to impact damage that may occur in a variety of ways during the service life of a composite component[3,4]. For the safe application of composites in aerospace structures a knowledge of the effect of impact damage on the subsequent load-bearing abilities of composites is essential. Many studies of impact damage in composites have been carried out with the help of a variety of NDT techniques such as X-rays, ultrasonic C-scan and thermography. There has been some success in using these NDT methods to monitor damage development during stressing experiments by interrupting a test and removing the specimen for damage inspection. This stop and restart test procedure, however, is far from satisfactory because of the great uncertainty about, not only the appropriate load level or test stage at which to interrupt a mechanical stressing experiment, but also the effect of the unload-reload process on the true mechanical behaviour of composites under continuous loading.

Acoustical Imaging, Volume 25, Edited by Halliwell and Wells.
Kluwer Academic/Plenum Publishers, 2000.

The objective of the current work has been to utilise the acoustography technique to study impact damage growth under fatigue conditions, uninterruptedly. The project has had two major phases. The first has been a technique development phase: to develop the acoustography technique as a means of monitoring damage in a composite sample subjected to static or cyclic loading in a servo-hydraulic fatigue test machine. The second, the application and data-gathering phase, is to utilise the developed technique to study impact damage growth during fatigue in representative aircraft composite material samples and to provide designers with information that will allow them to select composite systems which are best able to resist damage growth.

2. INTEGRATION OF AN ACOUSTICSCOPE AND A MECHANICAL TESTING MACHINE

The crucial task in mounting an AcousticScope™ (AS), an acoustography system supplied by Santec Systems Inc., USA, onto an Instron mechanical testing machine was the construction of a water tank around the composite sample gripped for loading. The experimental arrangement that has been developed is shown schematically in fig.1. The sample (4 mm thick, 100 mm wide and 200 mm long with 100 mm gauge length) is clamped between two massive testing machine loading grips in an anti-buckling guide with a circular cut-out in the gauge section to allow ultrasound to pass through the sample to facilitate acoustographic imaging. The cut-out, has a diameter of 86 mm and the circular acousto-optic detector has a 75 mm diameter. The composite sample is coated with a thin layer of silicone rubber on both sides to prevent water entering the composite through openings in the damage zone (cracks).

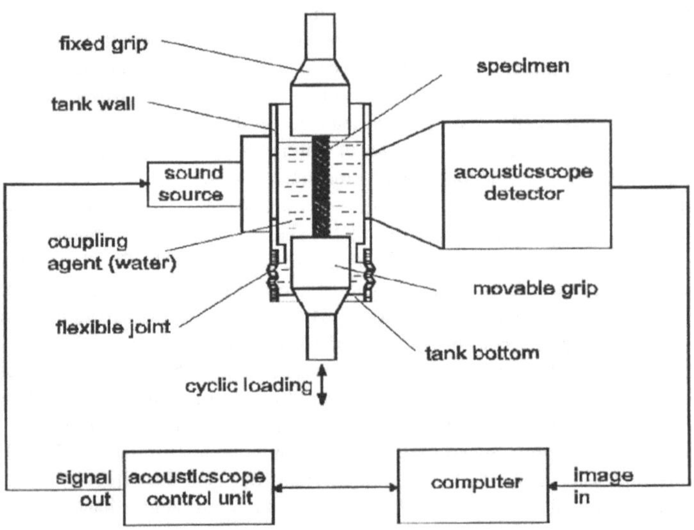

Figure 1. Schematic diagram of experimental set-up.

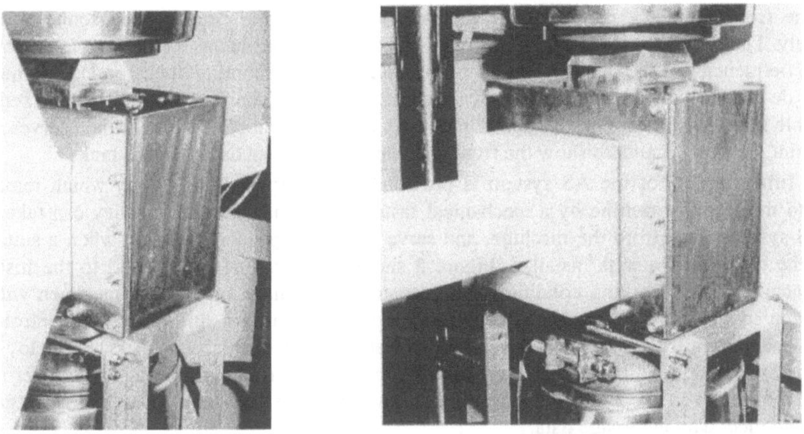

a) front b) back

Figure 2. The water tank for the AcousticScope in operating condition.

Figure 3. The AcousticScope attached to the Instron machine in the working condition.

The front and back panels of the tank are attached to the detector and sound source, respectively. The sound source is a 65 mm square piezoelectric transducer. It is run cw at randomly generated frequencies in the range 3.2 to 3.4 MHz and it is moved continually by an electric motor to further de-phase the sound field to avoid interference effects. The tank bottom plate, fixed on the movable lower grip, is connected to the main tank body by a rubber bellows which serves as a flexible joint. Figures 2a and 2b show the front and rear views, respectively, of the tank.

A firm support for the AS system is provided by a purpose-built trolley which remains attached to the testing machine by a mechanical fastening mechanism. This trolley can take the whole AS system away from the machine, and serve as an independent AS unit when a suitable tank for the coupling water is installed. Figure 3 shows the AS system attached to the Instron testing machine in the working condition. An acoustography image of the test specimen with a damage area in the centre is presented on the computer screen. An automatic AS-controlling computer programme has also been developed at Bath, which enables the user to employ the acoustography technique during either short-term static or long-term fatigue tests. The latter may last as many as eleven days so it is necessary to switch the AS system on and off at regular intervals to collect damage image data.

3. RESULTS AND DISCUSSION

3.1 Calibration

A 4 mm thick CFRP panel sample containing a half-blind flat bottom hole of 12.7mm (0.5 in) diameter packed with an ultrasound absorbing substance was used for the dimensional calibration of the AS. Figure 4 shows three acoustography images of the hole taken with different distances between the sample and the detector. The edge sharpness of the circular hole on the images reduces when the panel is moved away from the detector. The horizontal line-profiles of the intensity across the centre of the hole for the same images are shown in figure 5.

The straight lines drawn under the individual intensity profiles represent the true diameter of the circular hole, *i.e.* the ends of the straight lines correspond to the physical edges of the hole. It is interesting to note that the real edges of the hole correspond well to the bottom corners of all the profile readings where the grey levels start to increase sharply, regardless of the distance of sample to detector. This has been used as a basis for the dimensional measurement of damage monitored by the AS in the following tests.

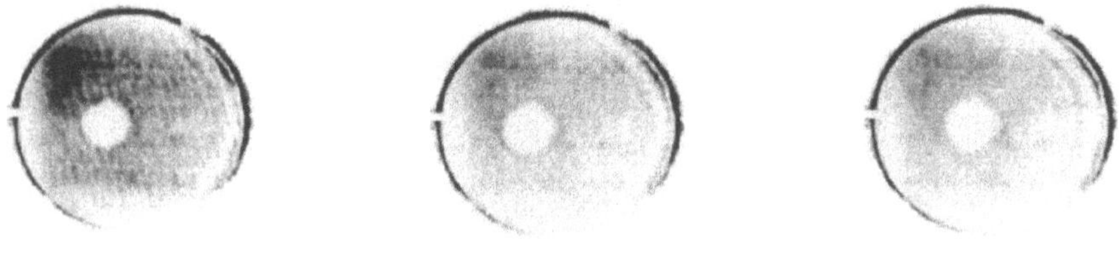

0mm 20mm 40mm

Figure 4. Acoustography images of an 12.7 mm diameter half-blind-flat-bottom hole in a CFRP plate taken with distances indicated between sample and detector.

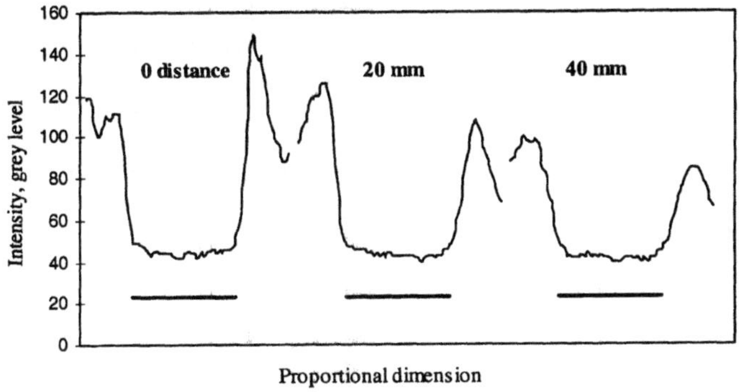

Figure 5. Line profiles of 12.7 mm hole image at different distances between sample and detector.

3.2. Damage growth during fatigue

A CFRP sample (AS4/8552 with $[45/0/-45/90]_{2s}$ lay-up) with damage at its centre equivalent to ~10 J impact was cyclically loaded in compression (R = 10) at 1 Hz and with a maximum compression stress of 150 MPa. After over 10^6 cycles, no significant damage growth was detected by the AS. A further fatigue test with an increased maximum compression stress of 160 MPa was then carried out on the sample. A significant amount of damage zone growth was monitored during the second fatigue test, as shown by selected images in figure 6.

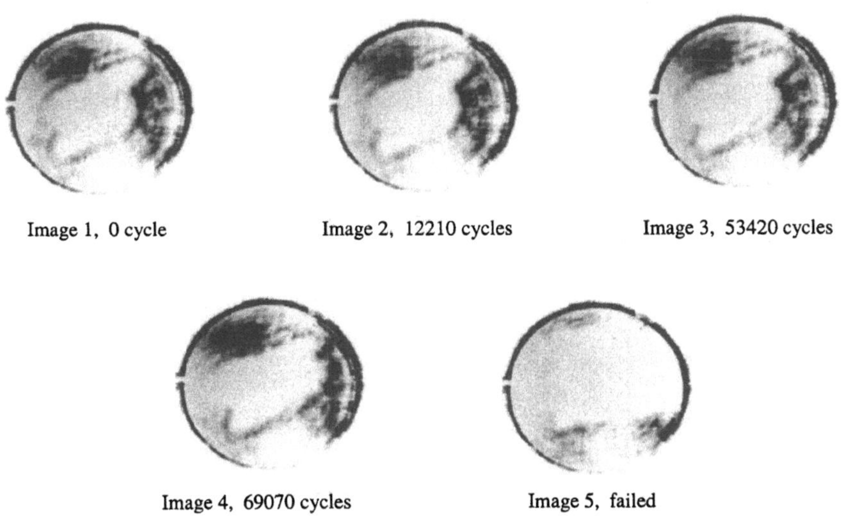

Figure 6. Acoustography images showing damage growth during fatigue.

Image 1 was taken before the fatigue test. The first growth started after ~10000 cycles, which can be seen by comparing images 1 and 2. The central left boundary of the damage zone has extended laterally in image 2. With increasing numbers of cycles, the growth rate increases and damage enlargement becomes more evident, images 3 and 4. Image 5 taken after the sample failed in fatigue(~70000 cycles), shows clearly that the damage area had grown across the whole width of the sample. The sample failed by buckling across the damage weakened central region under the compressive loading. The change of the damage zone can be analysed by the line-profile intensity reading method used in the last section. Figure 7 shows the percentage change of the damage zone width during the fatigue test. The line for the intensity profile reading was located horizontally across the centre of the damage zone. It can be seen that the damage grew with a gradually increasing rate with the increase in the number of cycles. After about 50000 cycles the increase of the growth rate with the number of cycles become more obvious. The damage was then enlarging in a unstable manner and caused the final fatigue failure of the composite sample. To the authors' knowledge, these are the first quantitative experimental data ever reported on damage growth in a carbon-fibre composite during fatigue loading in real-time.

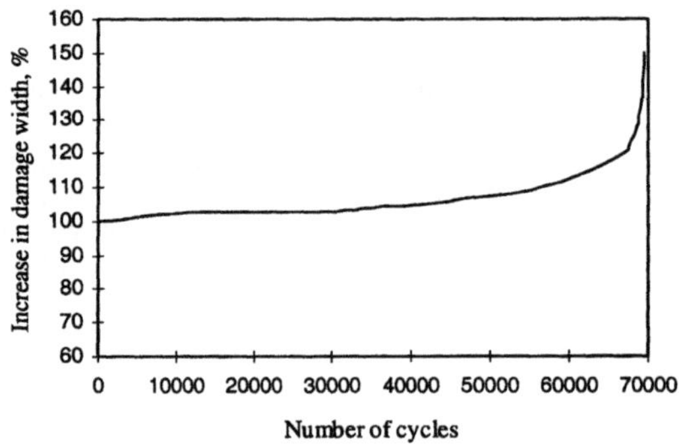

Figure 7. Damage growth during fatigue, measured by the line-profile method.

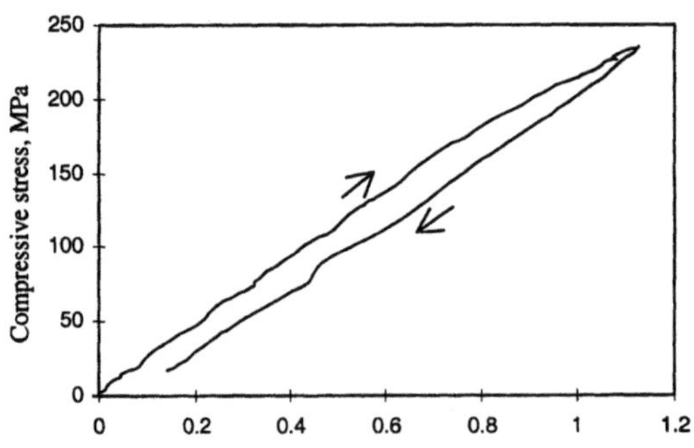

Figure 8. Stress-strain curve of load/unload after fatigue.

3.3. Static test

A compressive loading and unloading test was carried out on a sample (AS4/8552, [45/0/-45/90]$_{2s}$) that had survived over 10^6 cycles of compressive fatigue (R=10) of 150 MPa maximum stress after a ~10 J impact damage. The stress/strain curve for the load/unload process is shown in figure 8. Some images of the damage zones with the stress levels at which they were taken are shown in figure 9.

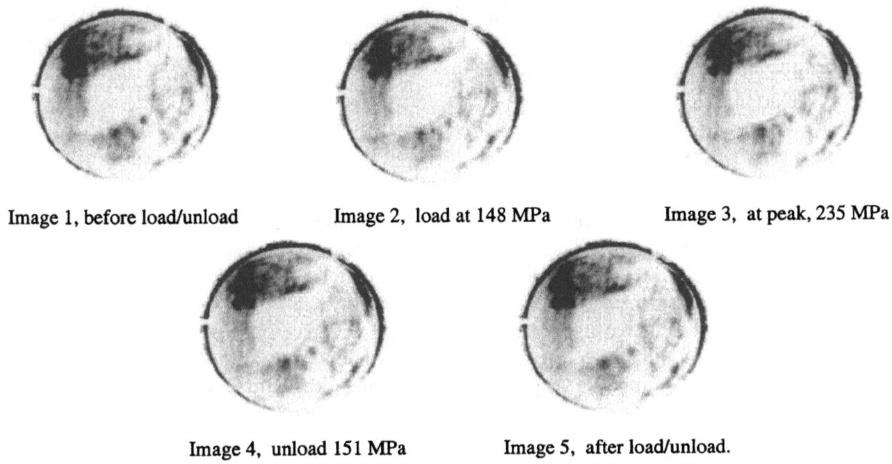

Figure 9. Acoustography images showing damage area changes with the load levels in a load/unload test.

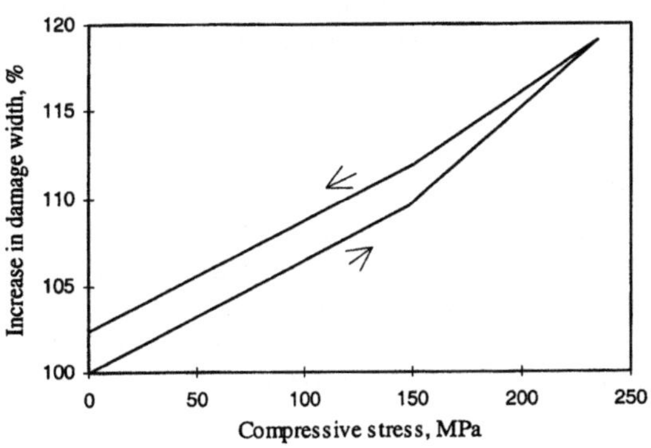

Figure 10. Apparent damage width change during load/unload test, measured by the line-profile method.

The growth of the damage was apparent when the stress reached quite high levels, images 1-3 in figure 9. The important finding of this test, however, is that the monitored damage area appears to reduce as a result of unloading, images 3-5 in figure 9. The result of the line-profile intensity analysis on this group of images is presented in figure 10. The arrows indicate the load/unload sequence. The fact that only a very limited damage area enlargement (a few %) could be measured after the load/unload test (compare images 1 and 5) has raised an important question about the normal NDT practice in using C-scanning to monitor and measure damage in composites without any load applied. It appears that the delamination crack opened and closed as the load was applied and removed. The change in delamination width revealed by loading amounts to about 15 %. The implication is that a conventional ultrasonic C-scan of an unloaded sample would underestimate the true delamination defect size by 15%. Again, real-time acoustography imaging has provided this unique and very important result.

4. SUMMARY REMARKS

Acoustography provides a very powerful means of monitoring damage in composites in real-time during both static loading and long-term dynamic fatigue tests. The successful integration of the AcoustoScope and a mechanical test machine has provided a new analytical technique which significantly enhances the information that can be gained in mechanical testing. The technique shows great potential for increasing our knowledge and understanding of failure processes in composites and other materials. Acoustography is a convenient and versatile ultrasonic imaging system that is superior in a number of respects to conventional ultrasonic C-scan.

ACKNOWLEDGEMENT

This project is supported by the EPSRC, UK. The provisions of materials by BAE SYSTEMS, Airbus is acknowledged. The authors wish to acknowledge J. Ball and M. Gaitonde of BAE SYSTEM, Airbus for their continual interest and help in this work.

REFERENCES

1. Sandhu, JS, Wang, H and Popek, WJ, Proc. SPIE 2944, 1996, 117-120.
2. Sandhu, JS, Wang, H and Popek, WJ, Proc. 2nd Conf. NDT applied to process control of composite fabrication, 1996, St Louis, USA.
3. Cantwell, WJ, Curtis, PT and Morton, J, Int. J. Fatigue, 6(1984) 113-118.
4. Soutis, C and Curtis, PT, Comp. Sci. Tech., 56(1996) 677-684.

TWO UNDERWATER EXAMPLES OF IMAGING BY MULTI-LOOK SYNTHESIS

P. Alais, P. Cervenka, P. Challande, P. Janvrin, J. Marchal

Laboratoire de Mécanique Physique
Université Paris 6, CNRS UPRES A 7068
2 Place de la Gare de Ceinture, F-78210 Saint-Cyr-l'Ecole
France

INTRODUCTION

In airborne radar or satellite imaging, synthetic aperture techniques are successful to obtain high resolution images from the earth surface. The method consists in adding coherently complex amplitude echographic signals obtained from one target for different positions of the platform. The most spectacular feature of the synthetic aperture imaging is that the aperture is limited in principle only by the possible angle of view $\Delta\theta$ and that the cross range resolution becomes independent of the range, much higher than for classical systems. Extension to sonar systems was early proposed by Castella [1]. Preliminary evaluations [2] have revealed limitations due to : the much lower value of the sound celerity and the difficulty of correctly sampling the virtual aperture unless using a slow mapping rate, the necessity to appreciate the platform motion with an accuracy better than one eighth of a wavelength to avoid a loss of coherence within the synthetic aperture length. Different solutions [2, 3, 4, 5, 6, 7] are proposed to improve the spatial sampling, using multiple receiver arrays and to compensate navigation incertitudes using self focusing algorithms. Even if large progresses have been registered these last years, the coherent synthetic aperture technique remains limited to research prototypes.

Another technique which avoids these problems is to build images by addition of the energy instead of the complex amplitude of the echo. This technique was applied by Kossoff [8] in the 70's for obtaining the first good obstetrical images of the fetus inside the womb. But tissue inhomogeneities prevent a good superposition of images due to variations of sound celerity and this technique was not retained in modern operation. The ocean is a much more favorable medium and the multi-look sonar technique has been proposed [7, 9, 10, 11] as a viable alternative as long as problems are not solved with the coherent imaging. At the 24th ISAI [12], we have presented theoretical simulations and first experimental results concerning the non coherent synthetic aperture technique. The simulated motion uncertainty on real data reveals that errors in the estimation of the motion may be accepted

Acoustical Imaging, Volume 25, Edited by Halliwell and Wells.
Kluwer Academic/Plenum Publishers, 2000.

with one or two orders of magnitude larger than in the coherent synthetic aperture process. Moreover, spatial under-sampling of pings caused by the vehicle speed is also well tolerated and allows large coverage rates.

In the present paper, we introduce the results of the experiment at sea of two multibeam sonar systems, built in our laboratory, using this multi-look method. The aim is to reduce the speckle effect, to increase the contrast and to slightly improve the spatial resolution of images. This technique is relatively robust with respect to different sources of uncertainty such as the motion compensation. The first experiment was performed in the bay of Brest (test with a multibeam side-scan sonar). The second one was performed in the coastal zone of Barcelona (test with a front-looking sonar). Both experiments were held on oceanographic ships of IFREMER (Institut Français de Recherche pour l'Exploitation de la Mer). The two sonars have been conceived for the imagery and the bathymetry of the continental shelf. One is a single side-looking sonar, the other one has a front-looking geometry shape. Both have bathymetric capabilities. The later is part of the project *COSMOS* that is partially funded by the European Commission within the *MAST III* program. Images recorded at sea, with an enhanced contrast due to the speckle removing, are presented.

THE SIDE SCAN SONAR SYSTEM

Experiments were carried with a system similar to the one presented at the 24th ISAI. We used the same transmitter i.e. a circular antenna (10 degrees azimuthal aperture). Two receiving arrays 3m-long made of 32 90mm-transducers are separated transversely by 30 mm i.e. 2 wavelengths at the central operating frequency of 100 kHz. The transmitted signal is a chirp (bandwidth B = 3 kHz, pulse duration T = 8 ms) modulated in amplitude by a gaussian shape. The complex amplitude raw data are sampled at 4 kHz. The pulse compression yields a range resolution of the order of 25 cm. A simple theoretical simulation shows that the beamforming should be limited to an angle of view of 5 degrees to avoid grating lobes and a more adapted transmitter will be used in future experiments. In these conditions the angular cross-range resolution obtained with an adequate beamforming is of the order 1/200 radian. It means a metric resolution at 200 m range, a value which will be only slightly improved by the non coherent synthetic aperture technique. In fact, the major improvement concerns the contrast of the images due to the speckle erasing.

This system was experimented in the bay of Brest, fixed to the hull of the oceanographic research vessel "THALIA" belonging to IFREMER. The movement of the vessel was measured by relatively "poor" sensors i.e. a double axis piezoelectric inclinometer and an ultrasonic gyrometer. Happily, the sea state was good. The presented images have been obtained assuming a straight navigation and a constant speed.

The first picture concerns the vicinity of two moorings, the "Ducs d'Albe" (Figure 1) which give infinite shadows and have induced undermining by their interaction with currents. This image may be compared with another one (Figure 2) obtained with a classical 4m-aperture system presented at the 22nd ISAI [14]. Although the navigation track is not exactly the same, the local relief may be recognized. The contrast and evidence of relief details are very much enhanced by the multi-look technique. The resolution of Fig. 2 is about the same but part of the information is destroyed by the speckle. White is for the highest energy level.

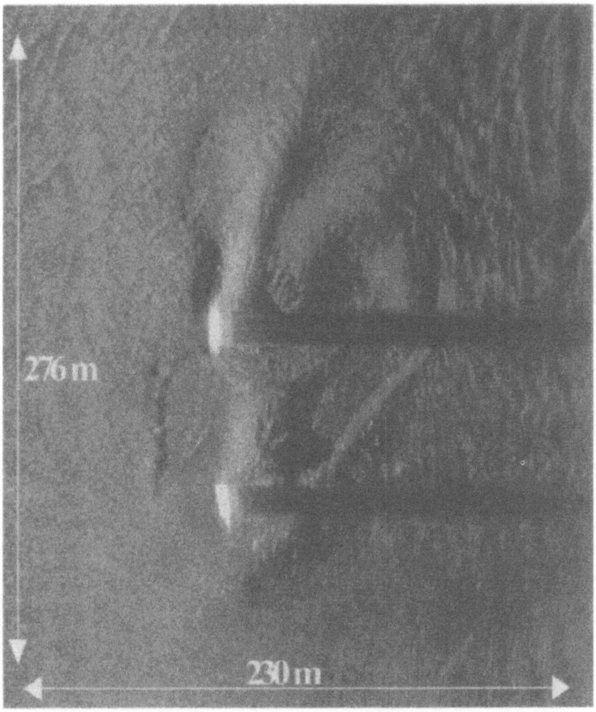

Figure 1 : "Ducs d'Albe" view 1 - multi-look processed image

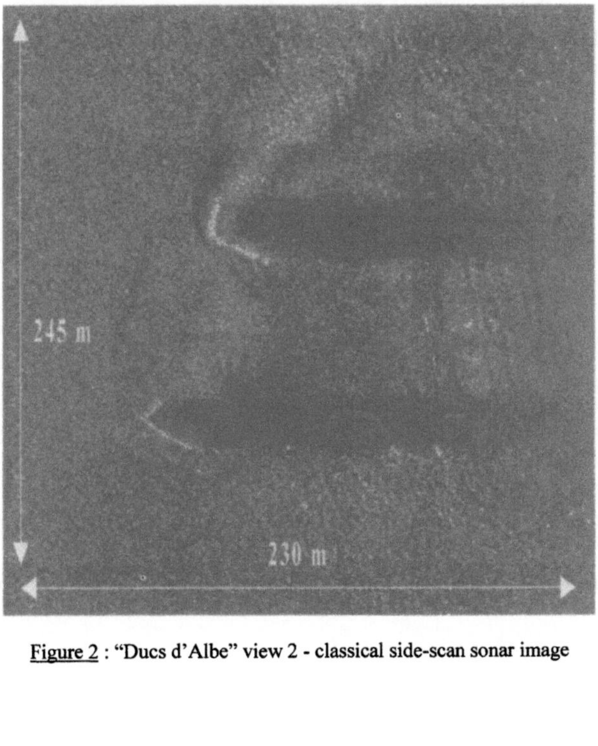

Figure 2 : "Ducs d'Albe" view 2 - classical side-scan sonar image

Figure 3 : "mud area surrounded by rocks"

Figure 3 shows an area of very low reflectivity, probably mud beside sand, and high reflectivity rocks. Figure 4 exhibits a high contrast on the shadows of a ship wreck.

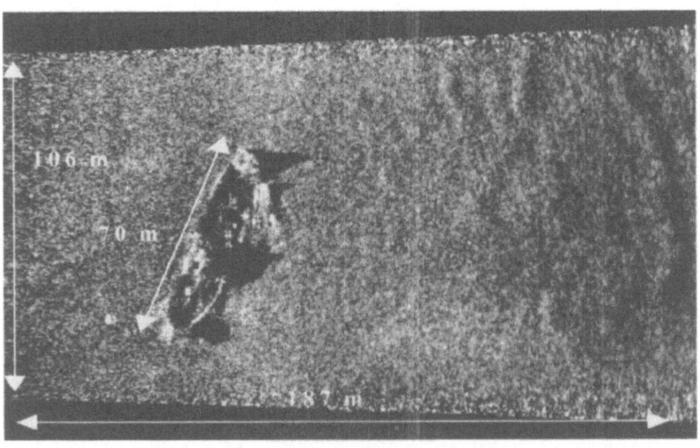

Figure 4 : "The coal cargo *Swansea* ship wreck", depth at nadir around 30m

THE FRONT-SCAN SONAR SYSTEM

The principle of the multi-look technique is to use redundant information obtained of one target from several points of view. This idea can be used also in a forward looking geometry. It concerns the part of the sea bottom which lies near the navigation track and is not imaged by the side scan technique. In this case, the maximal redundancy is obtained for targets which are at the vertical of the navigation axis. These targets may be viewed by a few hundreds pings while the ship is moving from nearly 0° to nearly 80° site angles. With this geometry, a simple reasoning for evaluating the image resolution as presented in [13] is not available. Large range information is affected by poor cross range resolution. Adding information to obtain the final image must be done with an adequate ponderation giving the greatest importance to information obtained under 30-50° site angles to achieve the best resolution. Superposition of one ping images was done over a unique XY map assuming for each one a flat sea bottom at the local vertical depth. The attitude data of the vessel were given by accurate sensors for roll and pitch motion. The navigation data were obtained from the gyrocompass and a DGPS system. In these conditions, a metric resolution was obtained.

The following results have been obtained with a hull-mounted system. We used a 80 cm aperture circular antenna (25° azimuthal aperture) associated with two interferometric receiving arrays made of 32 22mm-transducers operated with the same transmitted chirp (B = 3 kHz, T = 8 ms) centered on 100 kHz. The range resolution and the angular cross range resolution are respectively of the order of 25 cm and 1/50 radian. The experiments were led on an IFREMER vessel "EUROPE" at locations near the Barcelona coast selected by geologists from the Instituto de Ciencias del Mar, partners in the COSMOS project.

Figure 5 shows what was recorded with the COSMOS system during a left turn of the ship. Anthropogenic sediment has been dumped at the end of an underwater pipeline. Several trawling marks can also be noticed. Depth is around 55 m, maximum slant range is around 350 m. Light grey is for low backscattered level.

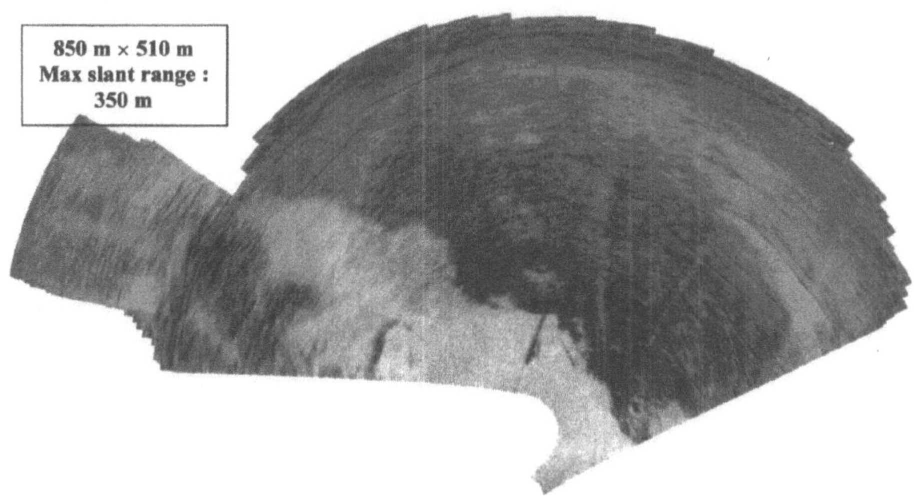

Figure 5 : "Cosmos turning over an anthropogenic sediment area"

The second image was recorded in the area of Barcelona, near the shore of the city of Blanes (figure 6). It represents a whole day of navigation, i.e. about 16000 pings. The size of the image is around 3000 m × 3000 m (pixel pitch 50 cm). Thanks to the speckle cancellation, it lets appear several sedimentary faces. We can see some areas of rocks, sand, mud and some covered by posidonies .

Figure 6 : "Blanes area, one day of survey"

Figure 7 is a part of the Blanes area located at the top north east of the previous image. An interesting sea floor exhibits with bulges and little reliefs.

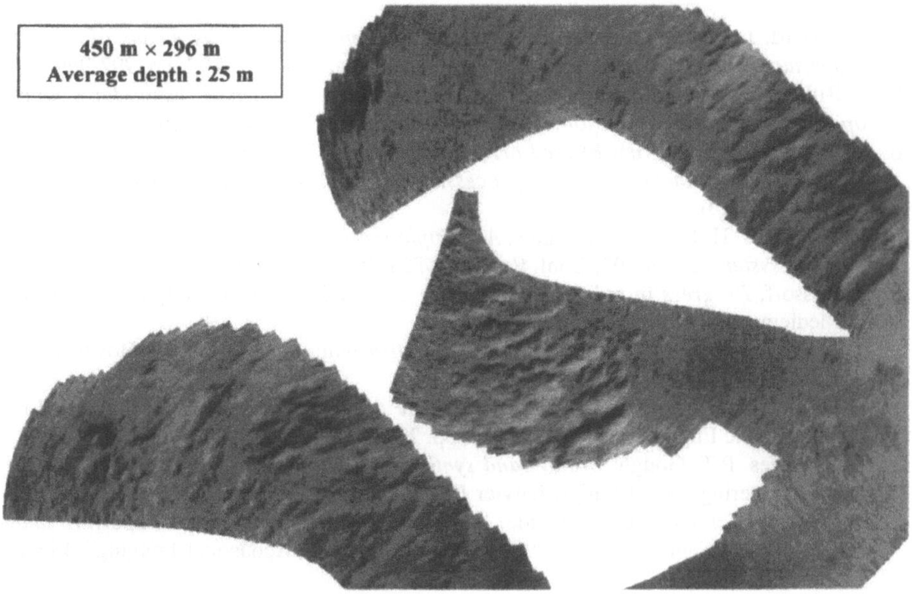

Figure 7 : "Zoom on a sea floor with bulges in the north east of the Blanes area"

CONCLUSION

We point out that images obtained with both systems don't show any speckle patterns in the area where the redundancy is significant. Therefore, contrast and visibility of relatively small targets is enhanced. Non coherent imaging technique seems to be a promising technique for those who are interested in obtained high resolution imaging at low computational expense. It offers also good prospects in term of mapping rates. Our goal is now to implement this multi-look technique in a real time mapping tool.

ACKNOWLEDGMENTS

We thank R. Person, X. Lurton from IFREMER, for their very efficient cooperation when carrying the experiments.

REFERENCES

1. F.R. Castella, *Application of one-dimensional holographic techniques to a mapping sonar system*, Acoustical Holography, Vol. 3, 1970, pp. 247-271.
2. L.J. Cutrona, *Comparison of sonar system performance achievable using synthetic aperture techniques with the performance achievable by more conventional means*, J.A.S.A., Vol. 58, n° 2, August 1975, pp. 336-348.
3. J. Chatillon, M.E. Zakharia, M.E. Bouhier, *Self-focusing of synthetic aperture sonar : validation from sea data*, Proc. 2nd European Conference on Underwater Acoustics, Lyngby, Denmark, July 1994, pp. 727-731.

4. V. Tonard, J. Chatillon, *Acoustical imaging of extended targets by means of synthetic aperture sonar technique*, Acustica, Acta Acoustica, Vol. 83, 1997, pp. 992-999.
5. D. Billon, F. Fohanno, *Theoretical performance and experimental results for synthetic aperture sonar self-calibration*, Oceans'98, September 1998, Nice, France.
6. G. Shippey, T. Nordkvist, *Phased array acoustic imaging in ground coordinates, with extension to synthetic aperture processing*, IEEE Proc. Radar Sonar Navigation, Vol. 143, n° 3, Juin 1996.
7. B.L. Douglas, H. Lee, C.D. Loggins, *A multiple receiver synthetic aperture active sonar imaging system*, Oceans'92, Conf. Records MTS, IEEE, pp. 300-305, 1992.
8. G. Kossoff, *Progress in pulse echo techniques*, Proc. 2nd World Congress Ultrasonics in Medicine, pp. 37-48, Rotterdam 1974.
9. P. D. Hansted, *Simplified digital synthesis of ultrasonic images*, Proc. of Royal Society of London 1981, A Vol. 374, pp.491-502.
10. P. De Heering, *Alternates Schemes in Synthetic Aperture Sonar Processing*, IEEE Journal of Oceanic Engineering, Vol. 9, n° 4, pp. 277-280, October 1984.
11. M.P. Hayes, P.T. Gough, *Broad band synthetic aperture sonar*, IEEE Journal of Oceanic Engineering, Vol. 17, n° 1, Janvier 1992.
12. P. Alais, P. Cervenka, P. Challande, V. Lesec, *Non coherent synthetic aperture imaging*, 24th ISAI, Santa Barbara 1998, to be published in "Acoustical Imaging", Plenum Press.
13. F. Ollivier, P. Alais, P. Cervenka, *A high resolution bathymetric sidescan sonar using dynamic focusing and beam steering*, proc. 22nd ISAI, Florence 1996, "Acoustical Imaging", Plenum Press, pp. 573-581.

RECONSTRUCTION OF EVOLVING SHALLOW WATER BUBBLE CLOUDS BY ACOUSTIC TOMOGRAPHY

Daniel Rouseff,[1] Steven G. Kargl,[1] Frank S. Henyey,[1] and Jerald W. Caruthers[2]

[1]Applied Physics Laboratory, College of Ocean and Fishery Sciences University of Washington, Seattle, WA 98105 U.S.A.

[2]Naval Research Laboratory, Stennis Space Center, MS 39529 U.S.A.

INTRODUCTION

Even in small concentrations, bubbles have a major effect on underwater acoustic propagation. Bubbles change the effective index of refraction of the ocean and can cause sound to be strongly attenuated and dispersed. The dominant effect at a particular acoustic frequency is from bubbles that resonate at that frequency. Consequently, probing a bubble cloud at multiple frequencies gives information about the bubble size distribution.

In the present paper, results from a recent experiment are reported where tomography was used to visualize shallow water bubble clouds. The clouds were probed over a range of frequencies. By examining a sequence of tomographic reconstructions, information about the effects of swell waves and rip tides can be inferred.

OVERVIEW OF THE EXPERIMENT

In the winter of 1997, a shallow water experiment was conducted near the Scripps Pier in La Jolla, California, USA. The goal of the experiment was to determine the dynamics, distribution and acoustic effects of bubbles just offshore from active surf. A major component of the experiment was the "Delta Frame" designed and constructed by the Naval Research Laboratory, Stennis Space Center. Figure 1 shows top and side views of the frame. The outer frame was in the shape of an equilateral triangle, approximately 9.4 meters a side. Two sources, labeled A and B, were positioned at vertices, and eight receivers were distributed around the periphery of the frame. To avoid spurious reflections, extenders boosted the sources and receivers above the frame to 2.8 m above the seabed in water nominally 4.3 m deep. For a more complete description of the experimental geometry, see Caruthers et al.[1]

Acoustical Imaging, Volume 25, Edited by Halliwell and Wells.
Kluwer Academic/Plenum Publishers, 2000.

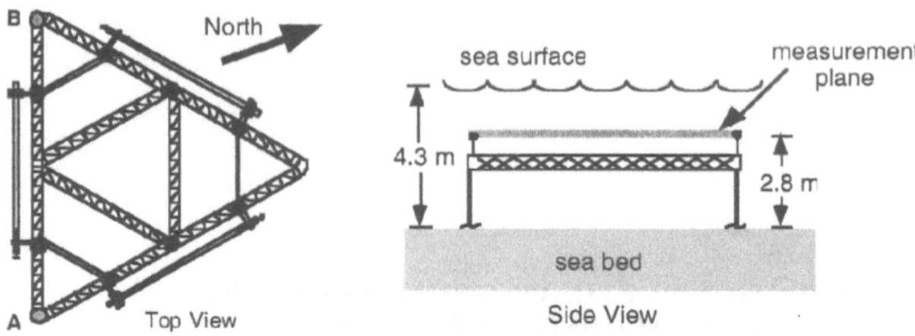

Figure 1. Delta Frame. Triangular frame, 9.4 m a side, supports two acoustic sources and eight receivers.

With two sources and eight receivers, there were a total of 16 acoustic paths ranging in length between 2.5 and 8.6 m. Transmission measurements were made along the paths at eight frequencies: 39, 59, 78, 98, 117, 146, 186, and 244 kHz. Each frequency component was 0.1 ms in duration with a modified Hanning window. The transmission sequence was repeated every second. Over the course of the experiment, the Delta Frame was deployed ten times, each time for approximately 90 minutes.

For the frequencies and path lengths used in the experiment, it is not realistic to look for individual bubbles. Rather, one seeks the effective index of refraction produced by an aggregate of bubbles. This effective index of refraction is complex, having a real part related to the sound speed and an imaginary part related to the attenuation. It is well known that a bubble driven near its resonant frequency will strongly absorb and scatter sound. At resonance, the bubble's scattering and absorption cross-sections are typically a factor of a thousand greater than its geometric cross-section.[2] In the resonant bubble approximation, the scattering and absorption at a particular acoustic frequency is attributed to bubbles of the corresponding resonant size. Consequently, one can think of the Delta Frame experiment as probing for bubbles of eight different sizes. The measurements are made in a horizontal plane and repeated every second allowing the spatial and temporal distributions of different sized bubbles to be tracked. Tomography represents a convenient technique for combining the assorted measurements and visualizing the evolution of bubble clouds.

Ideally, one could directly reconstruct both the sound speed and attenuation associated with the bubble cloud. From measurement of acoustic travel time, one hopes to reconstruct sound speed, and from measurement of acoustic intensity, one hopes to reconstruct attenuation. Without specialized processing, however, only the later inversion is practical for bubble clouds. Because of the strong dispersion, a transmitted pulse is greatly distorted as it propagates. Under such conditions, the very definition of travel time becomes problematic.

As an example, Fig. 2 shows the observed transmission along a particular path at two different times. During the quiescent period, four packets corresponding consecutively to the 244, 186, 146 and 117 kHz transmission packets are observed. During a rip current event 650 seconds later, each packet's amplitude is greatly attenuated; at 117 kHz, it is comparable to the noise floor. Each packet's shape bears little resemblance to what was transmitted. Note that the 244 kHz transmissions are the least attenuated. This is typical and is dictated by the processes that remove bubbles from the water. Large bubbles rise rapidly to the sea surface and small bubbles dissolve. The remaining intermediate size bubbles have the strongest effect. For the frequencies used during the experiment, the greatest attenuation was usually observed at 59 kHz.

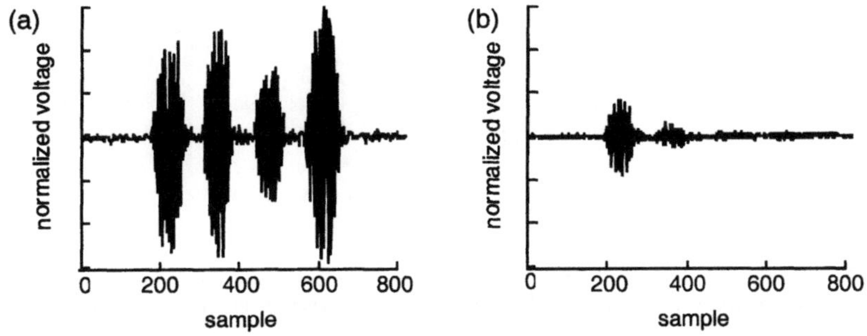

Figure 2. Measured waveforms. Path from source A to receiver on oposite leg, nearest source B (see Fig. 1). Response contains packets transmitted at 244, 186, 146, and 117 kHz. (a) Quiescent period. (b) Active period 650 seconds later

TOMOGRAPHIC RECONSTRUCTION

Let $\alpha(\mathbf{r})$ be the attenuation associated with the bubble field in the horizontal measurement plane (Fig. 1). Taking the negative of the logarithm of the intensity measured along path i yields the observable d_i modeled as

$$d_i = \int \alpha(\mathbf{r}_i) d\eta_i + \varepsilon_i \tag{1}$$

The vector $\mathbf{r}_i$ is constrained to lie along the straight ray path connecting source and receiver. The noise is given by ε_i. Implicitly, each term is a function of transmission frequency and time.

The tomography problem reduces to reconstructing $\alpha(\mathbf{r})$ from measurements of its projections. This is a classic problem, and when there are a sufficient number of ray paths the reconstruction usually proceeds using well-known backprojection techniques.[3] In the present case, however, only 16 paths are available. In such cases, tomographic reconstruction can be cast as an estimation problem. In implementation, often the medium is divided into cells in which the field being reconstructed is assumed constant. The goal of the inversion is to estimate these constants. This is the so-called discrete data/discrete reconstruction approach that is often used in ocean imaging.[4] For the Delta Frame data, a discrete-data/continuous reconstruction approach was used. In this formulation, the reconstruction is a superposition of weighted continuous functions. The experimental geometry and the expected correlation length of the bubble clouds determine the functions. The associated weights are determined by solving the least-squares minimization problem. Numerical simulations suggested that the Delta Frame could resolve details of the bubble cloud down to about 2 m. The recent book by Munk et al.[4] offers a comprehensive overview of ocean acoustic tomography. The details of the inversion algorithm as implemented for the Delta Frame data is available in Rouseff et al.[5]

The tomography algorithm has been applied to interesting sections of the Delta Frame data. Recall that each transmission sequence consisted of eight frequencies; consequently, eight images can be generated from each sequence. Such images have been generated to study the spatial distribution of different sized bubbles at an instant in time. For a particular frequency, a series of images can be generated to track the evolution of the bubble cloud as a function of time. As an example, Fig. 3 shows four 146 kHz images spaced two seconds apart. The images are of a rip current event. Features of the bubble cloud are evident at scales between the size of the frame and the expected 2 m resolution.

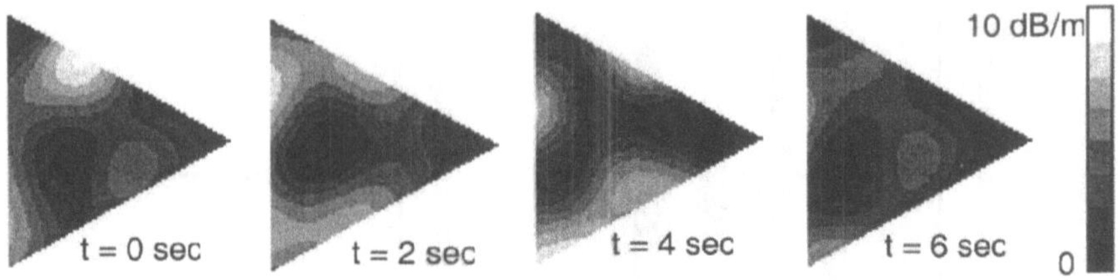

Figure 3. Tomographic images of attenuation. Area reconstructed corresponds to area of Delta Frame. Time relative to Sequence 3250 of the fifth deployment. 146 kHz data used to construct images.

The images in Fig. 3 show significant frame-to-frame variability. To understand this result, the attenuation was averaged over the frame for a sequence of images. The solid line in Fig. 4 shows a two-minute time-series of average attenuation. Also shown is a synoptic measurement of the local water depth.[6] When the attenuation is significant, the two lines are anti-correlated. A physical interpretation is that at a wave trough, the bubble cloud dips into the horizontal tomography plane. At a wave crest, the cloud rises and attenuation decreases. The rapid variation in the images, then, is caused by ocean swell advecting the bubble clouds in the vertical. In part, changes in the horizontal tomography images are cause by the vertical structure of the cloud. Horizontal advection is at longer time scales.

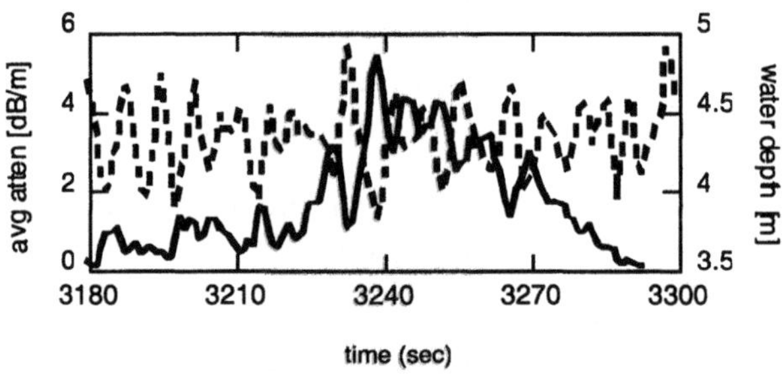

Figure 4. Comparison of average attenuation (solid line) and water depth (dashed line).

ACKNOWLEDGMENT

This work was supported by the United States Office of Naval Research.

REFERENCES

1. J.W. Caruthers, S.J. Stanic, and P.A. Elmore, Acoustic attenuation in very shallow water due to the presence of bubbles in rip currents, J. Acoust. Soc. Am. 106: 617 (1999).
2. C. S. Clay and H. Medwin, *Acoustical Oceanography*, John Wiley: New York (1977).
3. A. C. Kak and M. Slaney, *Principles of Computed Tomographic Imaging*, IEEE Press: New York (1988).
4. W. Munk, P. Worcester and C. Wunch, *Ocean Acoustic Tomography*, Cambridge Press: Cambridge (1995).
5. D. Rouseff, F. S. Henyey, J. W. Caruthers, S .J. Stanic, Tomographic reconstruction of shallow water bubble fields, To appear in IEEE J. Oceanic Eng. (2000).
6. P. Dahl, personal communication.

ACOUSTICAL RAY TRACING SYSTEM FOR VISUALIZING SIMULATED SOUND REFLECTION FROM A PLANE ROCK WALL IN ACOUSTICAL ARCHAEOLOGY

Naofumi Aoki, Yoshinao Aoki, Yohji Kudoh,
Norifumi Tonouchi, and Masafumi Takase

Graduate School of Electronics and Information Engineering,
Hokkaido University,
N13, W8, Kita-ku, Sapporo 060-8628, Japan
aoki@media.eng.hokudai.ac.jp, FAX +81-11-700-2651

INTRODUCTION

It is conceivable that some ancient structures were designed for exploiting their acoustical effects, with a particular purpose such as amplification of speech for an audience. Since no acoustical marks remain, above speculation can only be verified by measurement of the acoustical characteristics of those structures (Jhan, 1996). The results of investigation into how ancient people acoustically used structures may provide insights into the lifestyle of ancient people from an acoustical point of view. Therefore, such a study could be treated acoustical archaeology (Aoki and Kudoh, 1996).

In order to develop a methodology for acoustical archaeology, we propose an acoustical ray tracing method to visualize the acoustical reflection characteristics of ancient structures. Since this was a pilot study, we have focused our attention on the case of a plane rock wall that reflects acoustical rays like an optical mirror. In such a condition, the speech of someone standing in front of the wall may be amplified by the reflection of the sound off the wall. In this paper, we also describe the development of a graphical user interface (GUI) -based simulation system written in Java computer language for the convenience of visualizing how the acoustical ray tracing method focuses the locations of sound sources.

THEORY OF ACOUSTICAL RAY TRACING

Excavated Rock Wall and Model for Developing Theory

An ancient pyramid excavated in Cholula, Mexico (see Fig.1) has a rock wall

Acoustical Imaging, Volume 25, Edited by Halliwell and Wells.
Kluwer Academic/Plenum Publishers, 2000.

(indicated by an arrow in Fig. 1) that is thought to have been a sound reflection wall for amplifying the speech of monks. We consider the case shown in Fig. 2, in which a speaker and audience, represented by a loudspeaker and microphone, respectively, are in front of a plane rock wall. If the wall has an acoustically perfect mirror surface, then the mirror image of the sound source appears at the symmetrical position with respect to the wall. Therefore, it is possible to determine the location of the sound source by finding the mirror image, which can be calculated from the curvature of the recorded spherical wave that is expressed as a collection of acoustical rays.

Fig.1 Rock wall in a excavated pyramid in Cholula, Mexico.

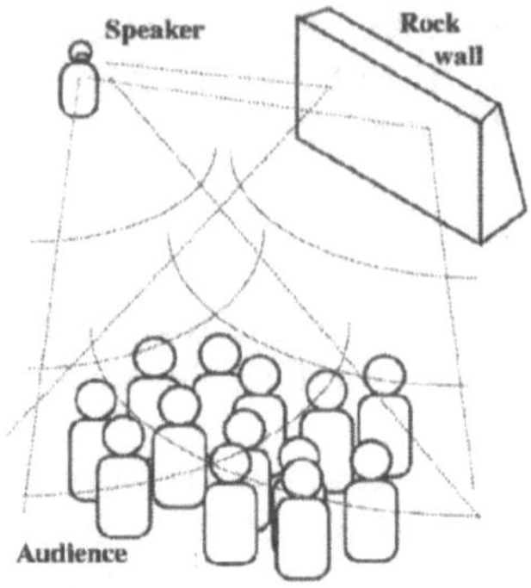

Fig.2 A model of a sound-reflecting wall.

Figure 2 shows an acoustical model of the wall shown in Fig. 1. The model is based on the assumption that the rock wall has an acoustical reflecting surface. In order to use an acoustical ray tracing method to analyze the model, we assume that the model is a two-dimensional one. Let z be the distance between the sound source and the plane to record the propagating sound wave and x denotes a coordinate of this recording plane. For simplicity of analysis, we assume that there is a single frequency $f\,(=\omega/2\pi)$ of the sound wave. In order to extract acoustical rays from the wave fronts of propagating sound waves, we introduce a plane wave expressed as $\exp(j\omega t)\exp(j2\pi kx)$. The reflected plane wave from the wall has a retardation time Δt and is superposed with the direct wave at the recording plane. In this condition, the sound field intensity $s(x)$ of the superposed acoustical waves at the recording plane is expressed as follows:

$$s(x) = |\exp(j(\omega t + 2\pi kx)) + \exp(j\omega(t - \Delta t))|^2 = |\exp(j\omega t)|^2 \cdot |\exp(j2\pi kx) + \exp(-j\omega\Delta t)|^2, \quad (1)$$

where k is a wave-number. If the propagation of sound waves satisfies the condition of Fresnel diffraction, the following approximation is valid:

$$\Delta t = \frac{\sqrt{z^2 + (x - x_0)^2}}{v} \cong \frac{z}{v} + \frac{(x - x_0)^2}{2zv}, \quad (2)$$

where x_0 is the location of the sound source with respect to the x-axis, and v is the propagation speed of an acoustical wave. Therefore, since $\omega = 2\pi v/\lambda$,

$$s(x) = |\exp(j2\pi kx) + \exp(-j\omega\frac{z}{v})\exp(-j\frac{\pi(x - x_0)^2}{\lambda z})|^2 = 2 + 2\cos(k + \omega\frac{z}{v} + \frac{\pi(x - x_0)^2}{\lambda z}), \quad (3)$$

where λ is the wavelength. This represents a chirp signal for which the frequency depends on the x position.

Relation between Spatial Frequency and the Direction of an Acoustical Ray

A plane wave q propagating in the three-dimensional space with propagation directions α, β, γ can be written as follows:

$$q(x,y,z) = \exp[j\frac{2\pi}{\lambda}(\alpha x+\beta y+\gamma z)], \quad \gamma=\sqrt{1-\alpha^2-\beta^2}. \tag{4}$$

Using the relation expressed in Eq. (5), the above equation can be rewritten as the kernel of a two-dimensional Fourier transform, namely $\exp[j2\pi(\mu x + \nu y)]$, where μ and ν represent the spatial frequencies expressed as,

$$\mu = \alpha/\lambda, \quad \text{and} \quad \nu = \beta/\lambda. \tag{5}$$

Suppose that an acoustical ray can be treated under the condition of paraxial approximation in geometrical optics. As illustrated in Fig.3, the following approximation then becomes possible:

$$\alpha = \cos\theta \cong x_2/z. \tag{6}$$

Therefore, from Eq. (5),

$$x_2 = \alpha z = \lambda\mu z. \tag{7}$$

As shown in Fig. 4, the position of an acoustical ray on the x-axis can be expressed in a two-dimensional case as follows (Aoki 1986):

$$\xi_1 + x_2 = \xi_1 + \lambda\mu z. \tag{8}$$

Based on these relations, the propagation direction of an acoustical ray, namely λ, μ, can be estimated by calculating the spatial frequency μ.

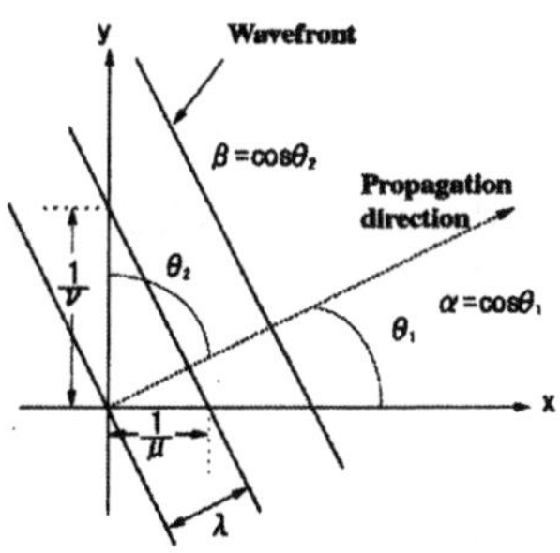

Fig. 3 Spatial frequencies λ, μ associated with a plane wave of wavelength λ.

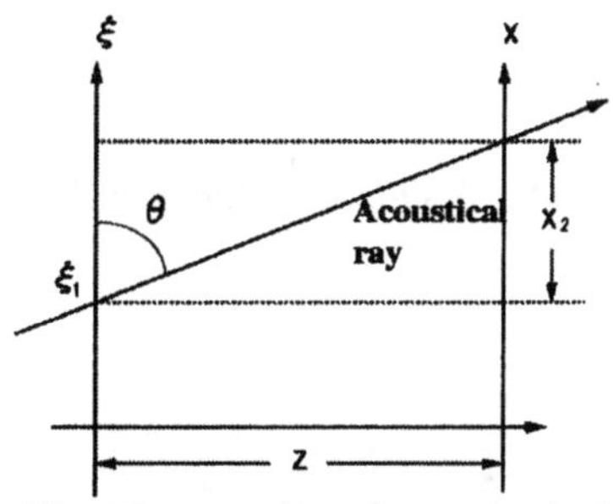

Fig. 4 Propagation of an acoustical ray.

Local Spatial Frequency Estimated by either Differential or Integral Calculation

Theoretically, a local spatial frequency can be estimated as the differential of the phase information of a signal with respect to the spatial variable x;

$$\mu(x) = \frac{1}{2\pi}\frac{\partial}{\partial x}[2\pi kx + \omega\frac{\pi}{\upsilon} + \frac{\pi}{\lambda z}(x-x_0)^2] = k + \frac{1}{\lambda z}(x-x_0). \tag{9}$$

In general it is difficult to deal with local spatial frequency, which, as shown in Eq. (9), a function of x, is to determine the direction of an acoustical ray. Therefore a peak value of the power spectrum distribution estimated by the Fourier transform for a short range, in which $(x - x_0)/\lambda z$ is assumed to be constant, is defined as a local spatial frequency instead of Eq. (9). An acoustical ray can be drawn with the direction that is estimated from the calculated local spatial frequency.

EXPERIMENT

Computer Simulation

Based on the above-described theory, some cases were analyzed by computer simulation. According to Eq. (3), an output signal, which is a pure tone generated by a loudspeaker, can be represented as a chirp signal on the observation axis, as shown in Fig. 5. From this chirp signal, the original location of the sound source can be estimated by the proposed technique. As shown in Fig. 6, computer simulation was used to examine the case in which the x values used for the calculation range from 0 to 400. The source is located at $(z, x) = (300, 100)$, and the frequency of the pure tone is set at 5 kHz. It appeared that the inverse propagation of acoustical ray tracing results in focus of acoustical rays at the original source location. The theory can also be applied for the case of multiple sound sources. Figure 7 shows the recorded output signal under the condition of three sound sources for which the x values were set at –1000, 2000, and 3000, respectively. As can be seen in Fig. 8, the locations of these three sound sources could be focused by computer simulation.

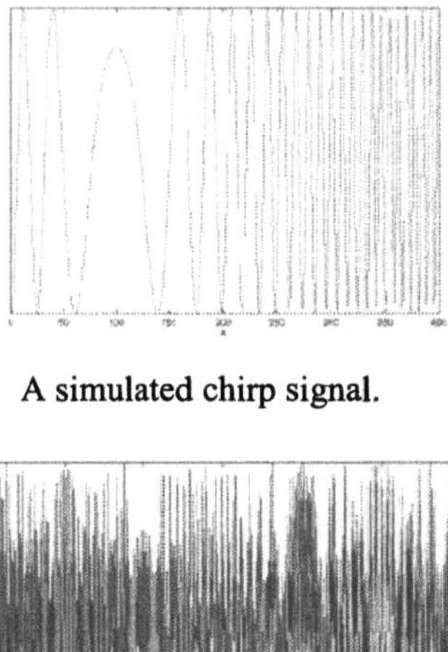

Fig. 5 A simulated chirp signal.

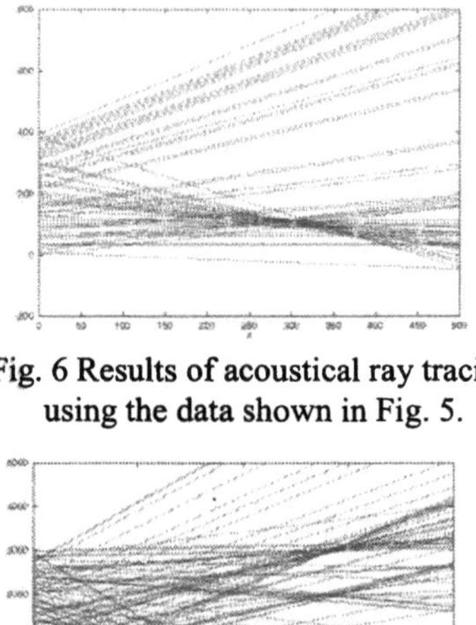

Fig. 6 Results of acoustical ray tracing using the data shown in Fig. 5.

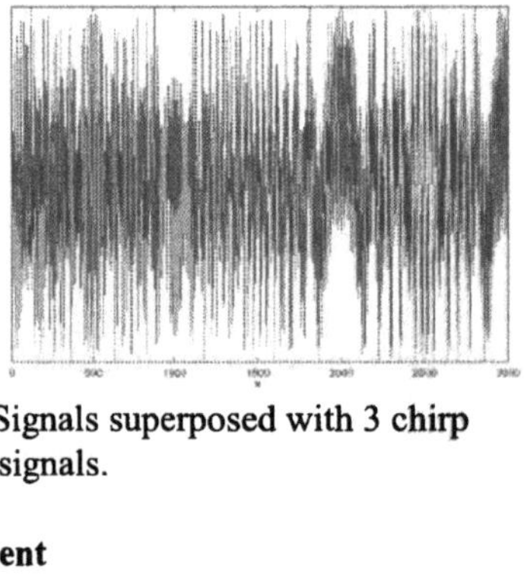

Fig. 7 Signals superposed with 3 chirp signals.

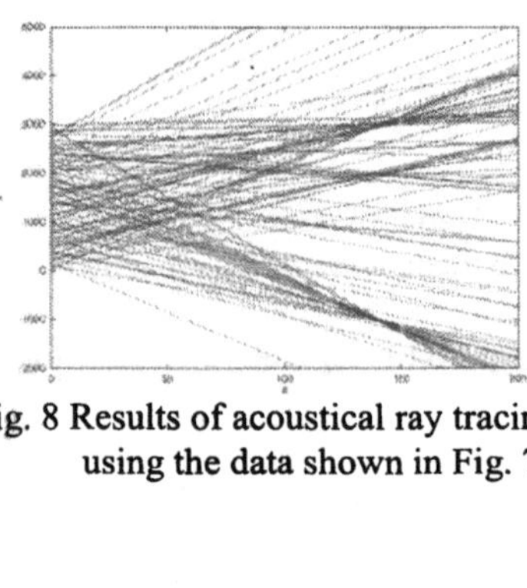

Fig. 8 Results of acoustical ray tracing. using the data shown in Fig. 7.

Measurement

To evaluate the validity of computer simulation results, measurements were carried out using the experimental setup shown in Fig. 9. A computer-generated pure tone was radiated by a loudspeaker, and was also used as a reference signal that was superposed onto a signal recorded by a probe microphone. The distance between the microphone and the speaker with respect to the x-axis was set at 300 cm. The microphone was located at $x = 200$ cm. The frequency of the pure tone was set at 8 kHz. The sampling frequency and quantization precision was set at 44.1 kHz and 8 bits, respectively. The resolution for the sampling along the x-axis was 1 cm. In order to avoid the influence of reflection by walls, the measurements were conducted outdoors. Figure 10 shows one of the recorded signals.

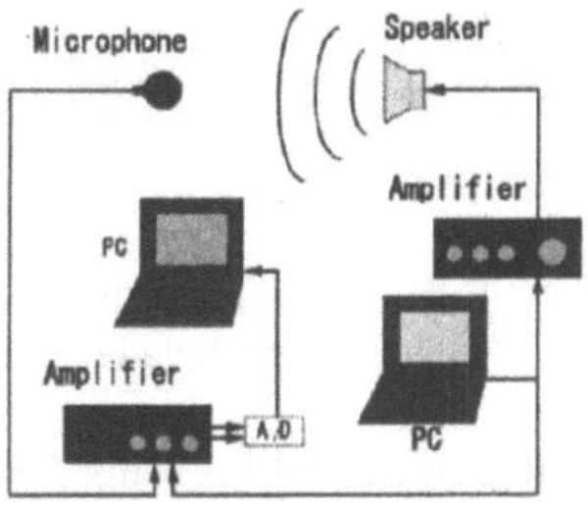

Fig. 9 Block diagram of the experiment.

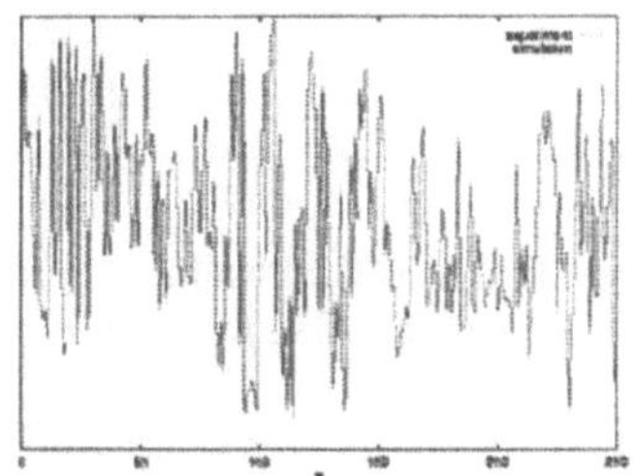

Fig. 10 Recorded acoustical signals.

Comparison Computer Simulation and the Measurement Results

The results of computer simulation were compared with those of measurement. Acoustical rays were calculated from local spatial frequencies and were plotted as shown in Figure 11 for the computer simulation. Figure 12 shows the result obtained by using the data shown in Fig.10. In the case of computer simulation, although the original source location with respect to the z-axis was overestimated, the acoustical rays focus at about $x = 200$ cm. The measurement results also showed that acoustical rays focus at about $x = 200$ cm. However, the measurement results somewhat differs from the original source location with respect to both the x and z-axes.

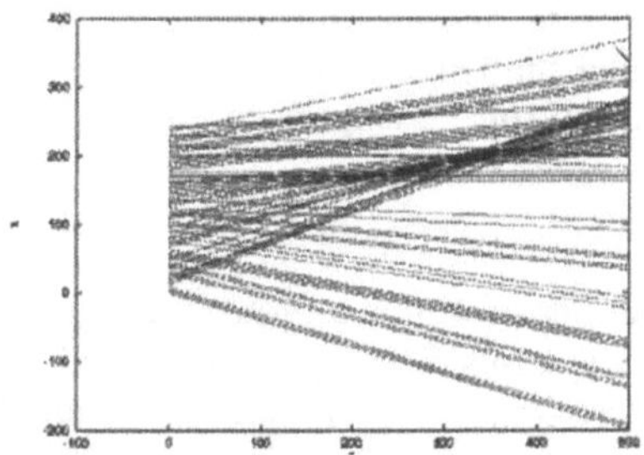

Fig. 11 Acoustical ray tracing with simulated data.

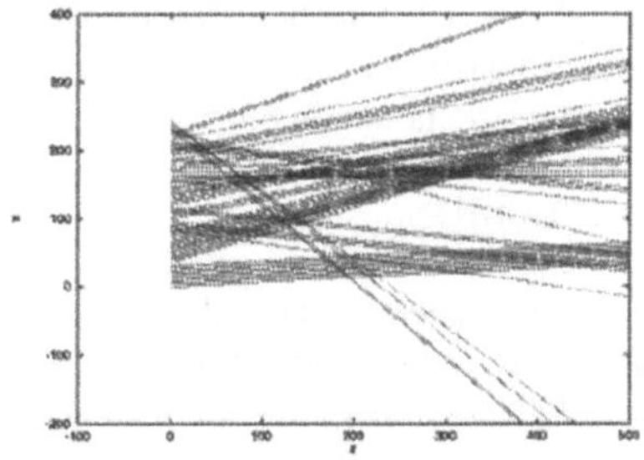

Fig. 12 Acoustical ray tracing with experimental data.

Development of a GUI-based Simulation System using Java Computer Language

In order to create an environment that enables efficient computer simulation, a GUI-based system using Java computer language was developed (Ammeraal, 1998). This system allows users to locate sound sources at arbitrary positions with a simple mouse-pointing operation. Figure 13 shows a model displayed on the Java-controlled screen, where the acoustical wall corresponding to the rock wall mentioned before and a sound source is located.

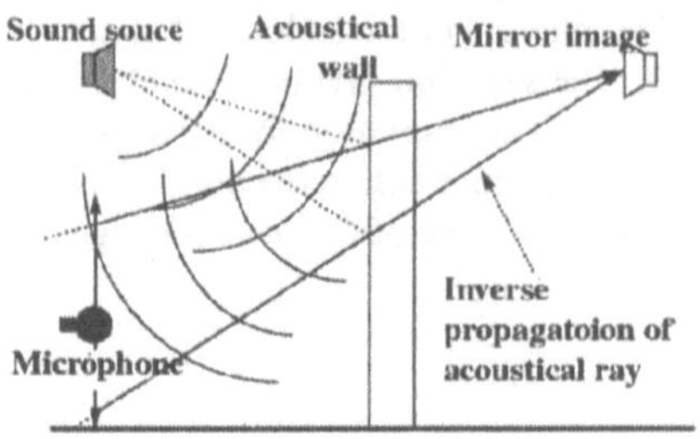

Fig. 13 Acoustical wall and mirror image of sound source.

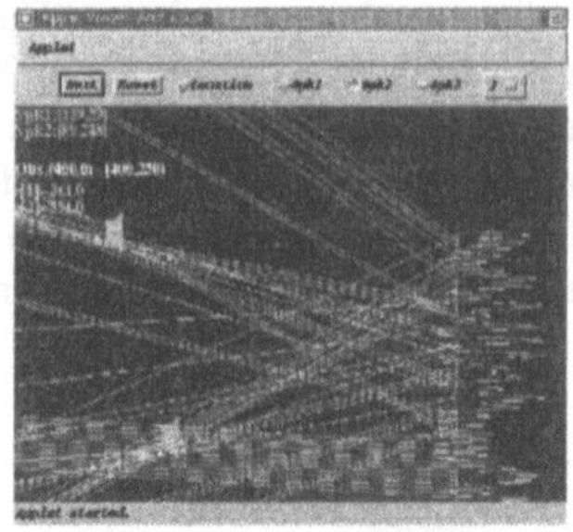

Fig. 14 Example of a GUI-based system using Java language.

Figure 14 shows an example of acoustical ray tracing calculated by using the GUI-based system, in which the locations of two sound sources are shown.

DISCUSSION

Until now, a computer simulation has been used to examine several cases in which more than three sound sources are located. When the number of acoustical rays drawn on the screen increases due to an increase in the number of sound sources, it becomes difficult to visually determine the location of sound sources. A more efficient visualizing technique appears to be needed to overcome this problem.

A single frequency, which is a coherent sound source, is employed in computer simulation. However, this condition does not necessary reflect the acoustical characteristics of the real world. In order to evaluate the validity of the proposed technique, we have to therefore perform computer simulation by taking into account more practical conditions. It is known that a human voice contains several dominant spectra called formants. Thus, signals consisting of many frequency components like chirp signals must be employed in computer simulation, in order to examine how the human voice is amplified by a wall.

CONCLUSION

In this paper, we described a technique that can be used to investigate the acoustical characteristics of ancient structures. As a pilot shtudy, a model of a plane rock wall was used. Both computer simulation and measurements were carried out to evaluate the validity of the proposed technique. A GUI-based simulation system was developed for visualizing the results of acoustical ray tracing.

REFERENCES

R.G. Jhan, P. Devereux, and M. Ibison, Acoustical resonances of assorted ancient structures, *J. Acoust. Soc. Am.* 99, pp.649-658, (1996).

Y. Aoki and Y. Kudoh, Application of complex MEM processing for measuring reflective architecture in acoustical archaeology, *Proc. ASA and ASJ 3rd Joint Meeting*, pp.1359-1364, (1996).

Y. Aoki, *Wave Signal Processing*, Morikita Pub. Co., Tokyo, (1986).

L.Ammeraal, *Computer Graphics for Java Programmers*, John Willey & Sons Ltd, (1998).

METHODS OF MULTIBEAM ECHOSOUNDERS

RESOLUTION INCREASING

Oleg V.Dmitriev, Kirill V.Kolosov, Inessa P.Prudnikova

Reacont-Ltd company
Russia, Moscow, 115201
Kashirskoye shosse, 13A

INTRODUCTION

Multibeam echosounders are widely used in many branches of industry and technique, such as search of the fish shoals and bottom mapping. Resolution is a main value which determines efficiency of these systems. This value is considered to be determined by receiving system aperture. Nevertheless, scanning pulse duration, pulse type and other parameters should be taken into account too. Besides, input signal/noise ratio of multibeam echosounders is usually high, there is a priori information concerning scattering acoustic signal in many cases. All of this allows to use resolution increasing methods of spectral estimation during signal processing.

This paper is devoted to the applications of these methods to the multibeam input samples processing in order to get an optimal result. Generalized Prony method[1] and modified Capon method[2] are used in this work. Multiflash model of the scattering acoustic field is used to generate model patterns of signal. We have also processed the experimental data (Norwegian Sea, 1996) to reconstruct a real bottom profile.

RESOLUTION OF MULTIBEAM ECHOSOUNDERS

We supposed that echosounder has a wide radiation pattern in one plane (XOY for example) and a narrow pattern in other (perpendicular, YOZ) plane. Hence, two-dimensional (in XOY plane) problem can be considered. Resolution is supposed to depend on receiving system aperture as a general rule. If receiving array is placed in (0,0) point along the OX axis then angle resolution of such system $\Delta\theta$ can be estimated as

$$\Delta\theta \approx A/\lambda \,, \tag{1}$$

where A is the system aperture, λ — main wavelength.

On the other hand, if the radiating acoustic pulse τ is short enough (matched processing can be used for pulse compressing), bottom line is described by smooth function $b(\theta)$, then angle range $\delta\theta$ of bottom profile region is insonified can be evaluated as

$$\delta\theta = \frac{c \cdot \tau}{2 \cdot [db(\theta)/d\theta]} . \tag{2}$$

Depending on real conditions situation when $\delta\theta$ is less then $\Delta\theta$ may occur. This case happens, for example, for model profile (Fig.1) for $\tau = 0.8$ ms when vertical angle θ is less then -14° and more then 34°. Thus, resolution can be easily increased for more then half of the model bottom surface. Generalized Prony method must be effective for this situation.

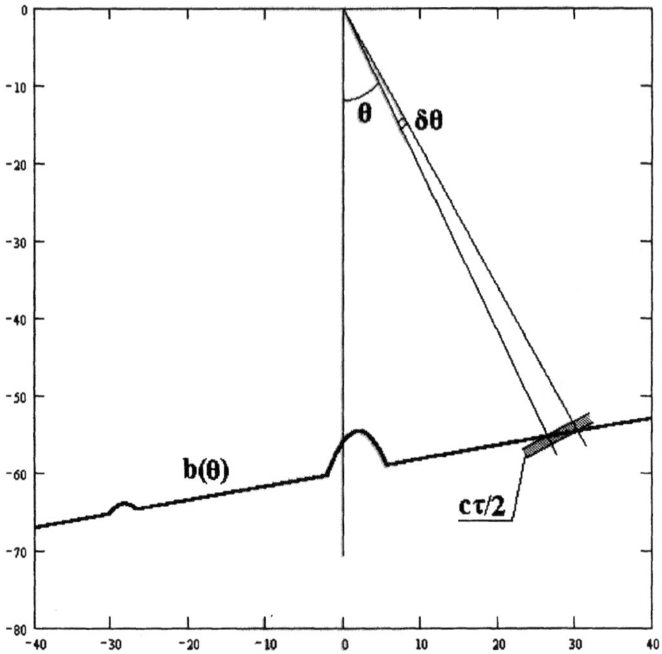

Figure 1. Model profile.

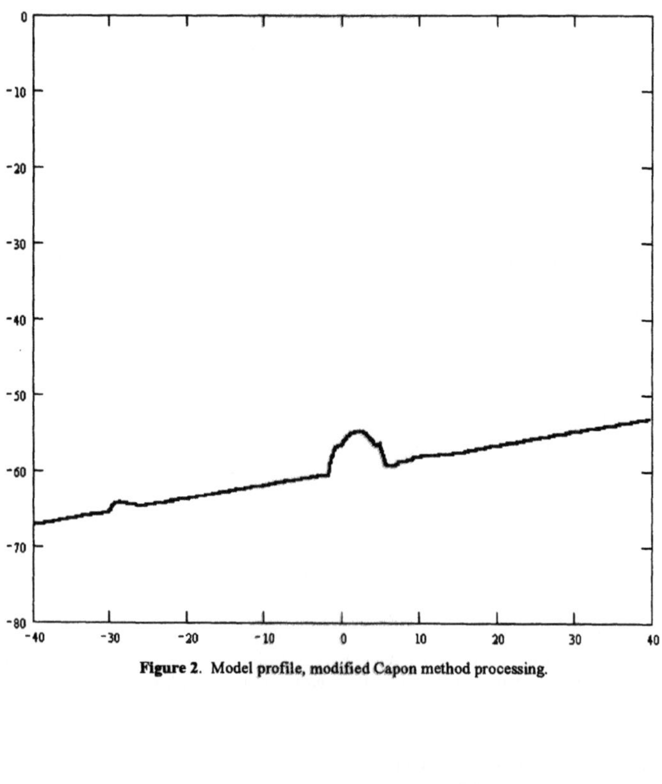

Figure 2. Model profile, modified Capon method processing.

More complicated case occurs when acoustic wave falls on a bottom surface almost normally (θ is near to 10° for the model profile). Angle range of reflecting spot $\delta\theta$ is significantly more then beam width $\Delta\theta$. We considered modified Capon method like the method of interbeam processing for this case.

These methods will perform properly while signal/noise ratio is considerable, and it is true almost for all of situations concerned with multibeam echosounders.

We have generated the model echo signal scattered by the model profile to verify our algorithms and compare them with standard processing. Figure 1 illustrates the model profile line, elementary scatterers (500÷700) is uniformly placed along the line. We also used real echo data was recorded in spring 1996 year in Norwegian Sea. The main radiating-receiving parameters are: radiating pulse duration — 0.8 ms, time discrete — 0.27 ms, receiving array has 32 equidistant elements (λ/2 step).

STANDARD PROCESSING

Standard processing consists in discrete Fourier transform (DFT) of input samples. This method does not take into account the effects concerned with real dimensions the reflecting spot $\delta\theta$. Result of this processing is mean beam power vs time. Beam width $\Delta\theta$ is determined by (1) for this situation.

Figure 3 illustrates the result of 64-points DFT processing of the model echo-signal. The small hillock (left part of the model profile) is not reconstructed satisfactory. Figure 5 illustrates the result of real echo data processing. These results can be compared with the results of high-resolution methods processing for the same echo signals

MODIFIED CAPON METHOD

The following algorithm is based on modified Capon method. This method[2] allows to determine a surface profile using interbeam processing. To solve the problem a great number of directions (with step less then beam width) is taken into account. Signals $A(t,k_1)$, $A(t,k_2)$, ...,$A(t,k_M)$ of beams are located close to selected beam are used to signal power estimation of the beam.

Power for k-direction can be estimated as

$$P(t,k) = \frac{1}{\mathbf{D}(k)\mathbf{R}^{-1}(t)\mathbf{D}^{\mathrm{T}}(k)} \tag{3}$$

where

$\mathbf{R(t)} = \mathbf{A}^{\mathrm{T}}\mathbf{(t)A(t)} + \gamma\mathbf{E}$ — normalized correlation matrix;

$A(t)=\{A(t,k_1), A(t,k_2),\dots,A(t,k_M)\}$ — signal row-vector, M = 3÷9;

E — unitary matrix;

γ — normalization coefficient , we used $(10^{-4}\div10^{-1})\cdot\max\|A\|$ range;

$D(k)=\{DN(k,k_1), DN(k,k_2),\dots,DN(k,k_M)\}$ — space response row-vector;

$DN(k,k_n)$ — k_n-beam response on identity k- direction signal.

Profile point distance for k-direction is found from $P(t,k)$ maximum estimation. Fig. 2 shows the result of model profile (Fig.1) reconstruction. Initial step of directions was 2°. Profile curve is defined more precisely by means of interbeam processing according to the algorithm being mentioned above in the range ±1° near selected beam. Power estimation is calculated using signals of 9 beams. Secondary step was 0.25°.

Thus, the method allows to restore accurate form of dissimilarities with angle dimensions less then aperture resolution (~2° in our case), i.e. really augments a resolution. This method appears to perform efficiently for full angle range of received signal.

GENERALIZED PRONY METHOD

If input sample $\{Z_n\}$ ($n=1,2..N$) is a sum of several exponents (real or complex), generalized Prony method allows to determine the parameters of these exponents. Prony algorithm consists of following steps.

The first one is linear equations system calculation

$$\mathbf{Pm}\cdot\mathbf{Zv} = \mathbf{Av} \tag{4}$$

where

$$Pm_{i,j} = \sum_{k=1}^{N-P} Z_{P+k-i} \cdot Z^{*}_{P+k-j}$$

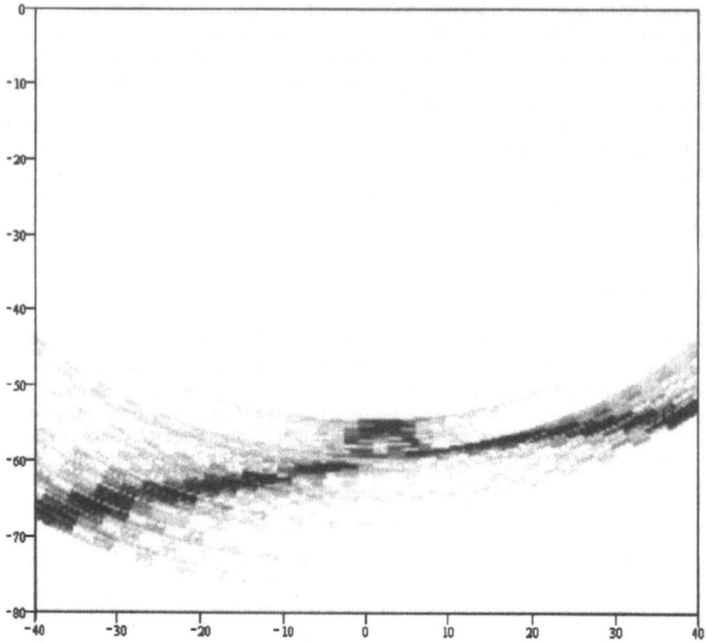

Figure 3. Model profile, standard DFT processing.

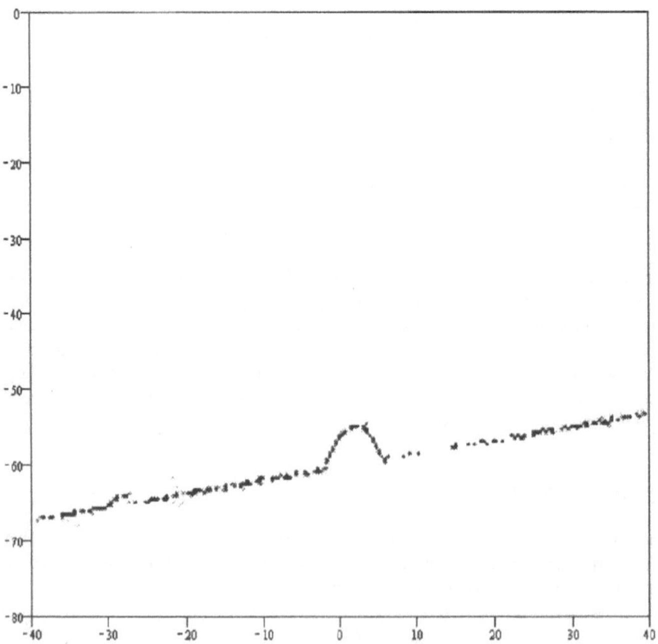

Figure 4. Model profile, generalized Prony method processing.

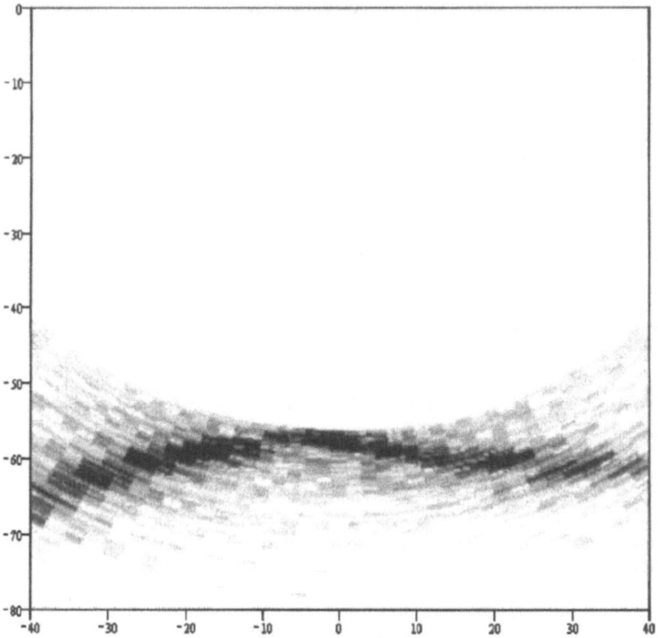

Figure 5. Real profile, standard DFT processing

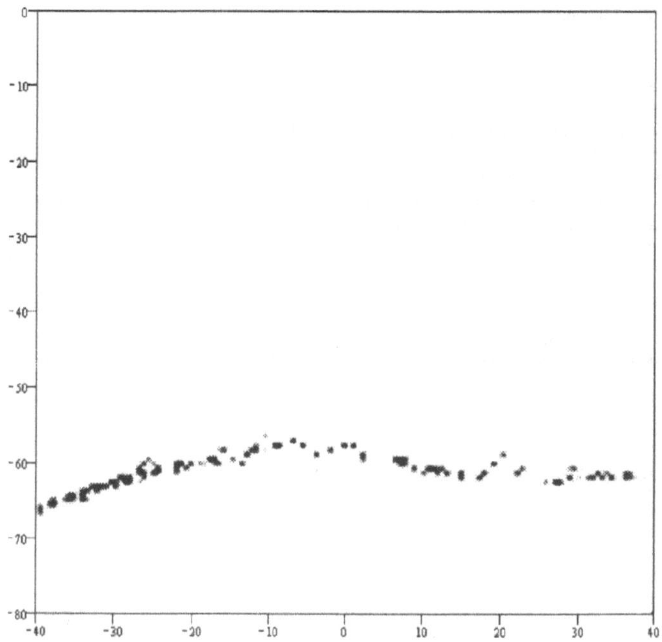

Figure 6. Real profile, generalized Prony method processing.

$$Av_j = \sum_{k=1}^{N-P} Z_{P+k} \cdot Z^{*}_{P+k-j}$$

$i,j = 1,2...P$, P is Prony method order

Zv vector is determined at this step.

The second step consists in polynomial factorization

$$\Phi(z) = \sum_{j=0}^{P} Zv_j \cdot z^{P-j} , \qquad Zv_o = 1.0 \tag{5}$$

This step allows to determine root vector ***Zr*** for $\Phi(z)$ polynomial.

The purpose of the third step is to calculate Prony magnitude vector ***Ar*** . This vector can be found from the matrix equation

$$(\mathbf{Mv}^{+} \cdot \mathbf{Mv}) \cdot \mathbf{Ar} = \mathbf{Mv}^{+} \cdot \mathbf{Z} \tag{6}$$

where

$$Mv_{n,j} = (Zr_j)^{n-1} \ , \ j = 1,2..P,$$

"+" means hermitian conjugation

Each component of ***Ar*** is a complex magnitude of point scatterer, and direction (vertical angle, as shown on Fig.1) of this scatterer Dr_j can be found as

$$Dr_j = \arcsin\left[\frac{1}{\pi} \cdot \arg(Zr_j)\right] \tag{7}$$

Input sample $\{Z_n\}$ is supposed to be obtained from equidistant receiving array with $\lambda/2$ discrete. On applying the algorithm (4)-(7) to each sample for all time discretes parameters all of scatterers are located within the field of view can be defined. In the case concerned such scatterers reconstructs a bottom profile (Figures 4,6). We did not use any additional treatment for these figures.

CONCLUSIONS

The methods of resolution increasing were considered in this work. These methods are applied to the samples of experimental data as well as model data. The results obtained can be compared with ones are found by the standard DFT processing.

Modified Capon method and generalized Prony method allow to increase a resolution of echosounder, especially for small insonified region. These methods can be used in a real equipment.

REFERENCES

1. S. Lawrense Marple, Jr., Digital Spectral Analysis with Applications., Prentice-Hall, Inc., 1987
2. V.A.Burov, E.E.Kasatkina, Maximum Likehood Processing of Signals from a Graded Profile Receiver., Acoustical Physics, 1997, v.43, № 1

A SATELLITE LABORATORY FOR ACOUSTICAL IMAGING USING A COMMUNICATIONS SATELLITE

Yoshinao AOKI[1], Kohji IIDA[2], Tohru MUKAI[2], Miho KATAGIRI[1], and Shinji OHGAKI[1]

[1]Division of Information Media Engineering
Graduate School of Engineering, Hokkaido University
Kita-ku, N13 W8, Sapporo, Japan
Tel +81-11-706-6502, Fax +81-11-700-2651
[2]Division of Marine Environment and Resources
Graduate School of Fisheries Sciences, Hokkaido University
Minato-cho 3-1-1, Hakodate, Japan
Tel +81-138-40-8852, Fax +81-138-40-8854

INTRODUCTION

Acoustical imaging is a relatively new and sophisticated technique for visualizing various kinds of targets by the use of acoustical waves. For the purpose of teaching acoustical imaging techniques, lectures on acoustical imaging are given in classrooms by referring to the wave propagation theory and imaging techniques. However, it is difficult to demonstrate the actual imaging techniques used in real fields. The purpose of this study was therefore to construct a satellite laboratory of acoustical imaging by connecting computers in remotely located laboratories and laboratories or classrooms in a university through a communications satellite.

DATA ACQUISITION USING A FISH FINDER

A marine vessel used for research purposes is a good example of a field laboratory. Such a research vessel is equipped with many underwater acoustical imaging systems such as a scientific echo-sounder, scanning sonar, acoustic Doppler current profiler (ADCP), and seismic profiler. This equipment allows observation of fish resource distribution, sea bottom features, ocean currents, and under sea bottom resources.

Usually, acoustical data obtained in a survey are analyzed on the vessel using simple and preliminary data-processing techniques, because the infrastructure of the computer network on board is not sufficient for high-degree complicated analysis. After the vessel returns to port,

Acoustical Imaging, Volume 25, Edited by Halliwell and Wells.
Kluwer Academic/Plenum Publishers, 2000.

advanced post-processing is performed on the pre-processed acoustical data using a computer network in the laboratory of a university connected to powerful computer resources, and three-dimensional maps of fish resource distribution, which provide useful information for both students and researchers, are produced.

Fig. 1. Laboratory on the T/S Oshoro-Maru used for oceanographic investigation.

Figure 1 shows a picture of the T/S Oshoro-Maru (G.T. 1,383), which belongs to Hokkaido University. This vessel is used for investigation of fisheries and oceanography, and it is equipped with many kinds of acoustical imaging instruments such as an echo sounder and sonar, enabling various acoustical images to be obtained.

Figure 2 shows an echogram obtained for the purpose of estimating fish abundance by a scientific echo sounder operated at frequencies of 28 kHz and 96 kHz. This echo sounder is a sophisticated system with a time-varied-gain (TVG) control circuit and a digital echo integrator, which enable images of high quality and fish abundance indices to be obtained.

Figure 3 shows a sonar image obtained by using a horizontal scanning sonar. Generally, a scanning sonar is used to scan an acoustical beam rapidly in a forward half space to visualize a fish school distribution and sea bottom features in a plan position indication (PPI) manner. Both of these acoustical imaging systems are useful tools for scientific investigation in the ocean.

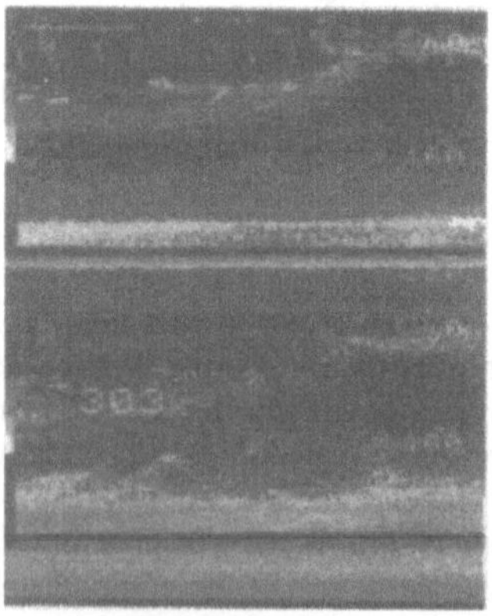

Fig. 2. Echogram obtained by a scientific echo sounder.

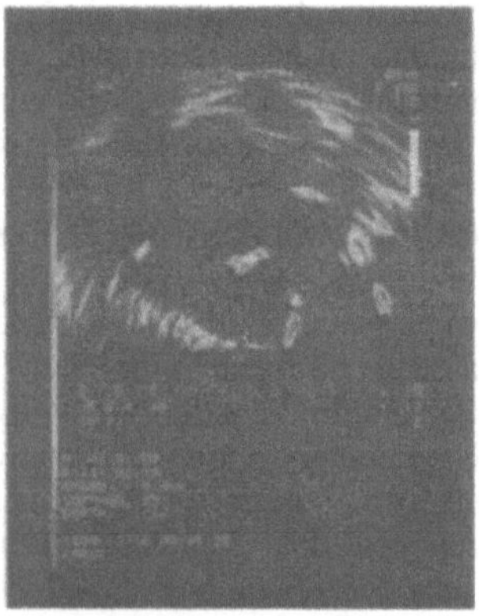

Fig. 3. Sonar image obtained by a horizontal scanning sonar.

One of our purposes is to send such scenes and images to laboratories and classroom on land where high-degree investigations and lectures are carried on as shown in Fig.4. In the laboratory on land, we can not only observe the acoustical data in remote place on the sea in real time, but also analyze using computer network connected to powerful computer resources and software. Then we can send back the high-degree processed data to the field laboratory and to the classroom for lectures.

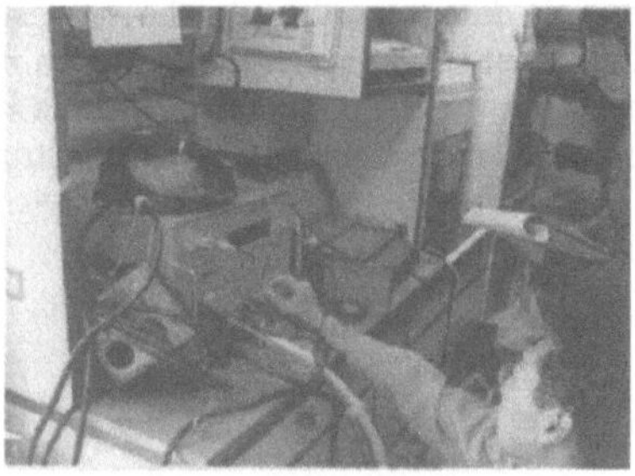

Fig. 4. Operation of acoustical imaging and satellite communication systems.

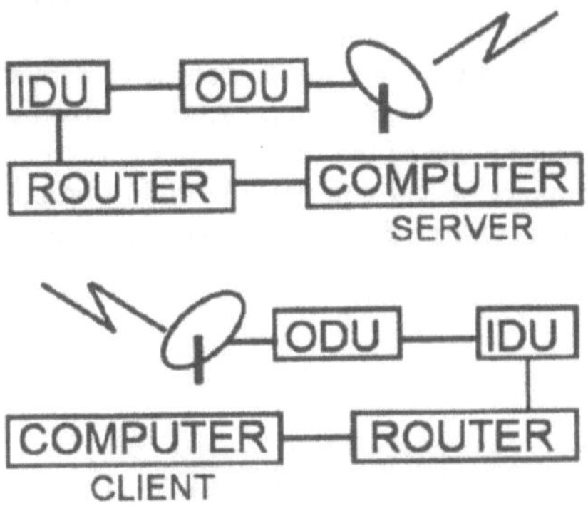

Fig. 5. Block diagram of satellite laboratories.

COMMUNICATION SATELLITE SYSTEM FOR SATELLITE LABORATORIES

For constructing a satellite laboratory, we developed the infrastructure of a satellite communication system connected to personal computers in laboratories, as shown in Fig. 5, by using a Japanese communications satellite, JCSAT-1 [1]. Figure 6 shows a maritime mobile earth station, which has an automatic direction-controlled antenna attached to the upper deck of the ship.

Figure 7 shows a photograph of a VSAT attached to the roof of the School of Engineering building in the campus of Hokkaido University. We tried to send data of acoustical images obtained by the echo sounder and sonar shown in Fig. 2 and Fig. 3, respectively, via the communications satellite, and we were able to transmit and receive in turnback transmission. In the near future, we plan to try direct transmissions of image data between the School of Engineering and the School of Fisheries of Hokkaido University, which are located about 150 km apart, and between the School of Engineering and the T/S Oshoro-Maru at sea, about 1000 km away.

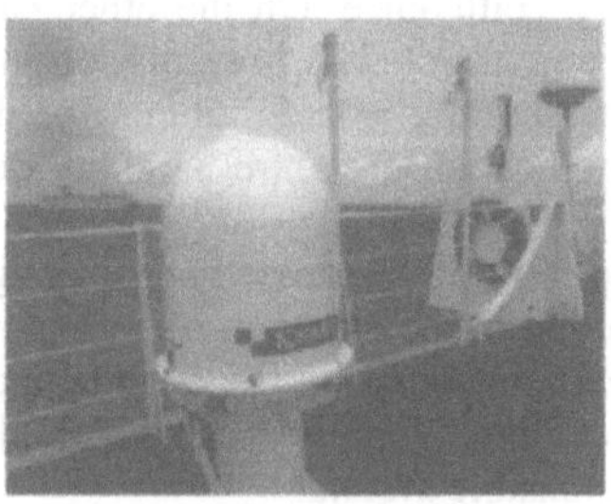

Fig. 6. Maritime mobile earth station.

Fig. 7. A VSAT for satellite communication.

CLIENT-SERVER SYSTEM FOR SATELLITE LAORATORIES

In order to simplify the transmission operation of satellite communication by users, we have developed a client-server system for data transmission. In this system, we can process the fundamental data transmission operation according to indications on the computer screen, as shown in Fig. 8. A TCP/IP protocol is used in this system for communications between the server and client computers. Since many clients can access the server computer by using the TCP/IP protocol, the server can send acoustical imaging data according to the requests of clients like an e-mail system. Another advantage of the client-server system is that even if the computer power of the client is not sufficient for processing the huge amount of data, a high-performance server can provide highly processed acoustical images such as 3-D reconstructed images, as will be described in the next section.

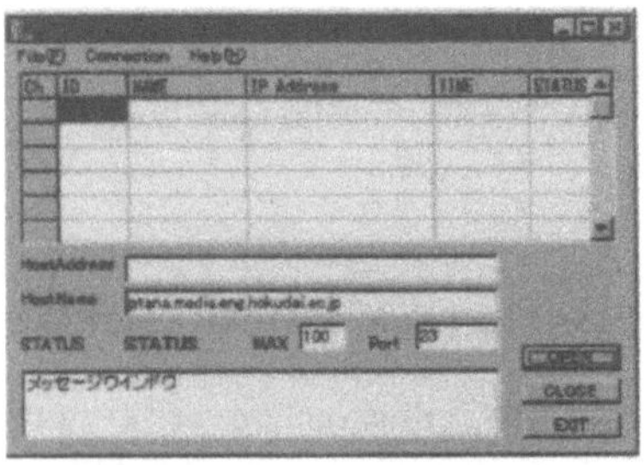

Fig. 8. Screen of the executing server program where ID number, name, IP address are the beginning of data transmission.

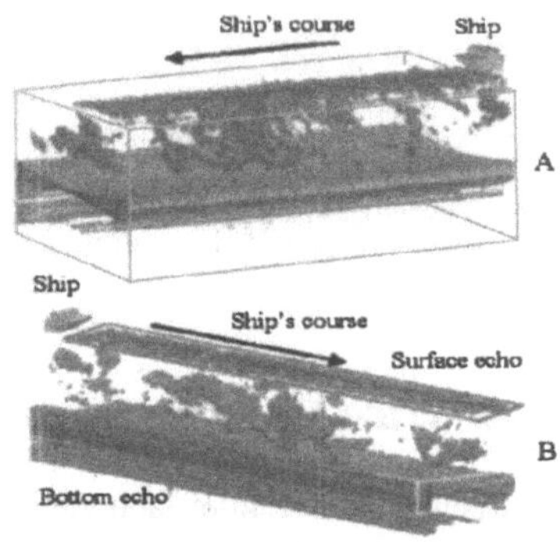

Fig. 9. Three-dimensional acoustical image of a fish school reconstructed from hundreds of 2-D sonar images.

TRANSMISSION AND PROCESSING OF FISH FINDER DATA

The transmission speed of the communication satellite is 64 k bps, and compressed image data (for example, by JPEG compression) obtained by the echo sounder is transmitted. Usually, a time delay is troublesome in satellite communication when explaining the process of an experiment using video, as is the case of a satellite TV conference. On the other hand, the delay-time is reduced in data transmission, but a large amount of data transmission increases the communication cost. Particularly in the case of transmitting a large amount of data for reconstructing 3-D images after processing, low communication costs are naturally desirable [2].

The image shown in Fig. 9 was constructed from about 600 pre-processed sonar images that were obtained by sampling for a time interval of 1.0 second in the direction of ship movement [3]. In the field laboratory, most of the acoustical images obtained by the echo sounder and sonar in real time are 2-D images. A merit of a satellite laboratory using a high-speed communication satellite exists in the case of a partial responsible system.

In a partial responsible system, pre-processing of raw data can be completed in a remotely located field laboratory, and high order-processing, such as 3-D processing, of the pre-processed data transmitted from the field laboratory is performed in a laboratory on land. The highly processed data is then sent back to the remotely located field laboratory or to a lecture room through the communication satellite. Displaying 3-D acoustical images such as the image shown in Fig. 9 helps not only researchers but also students in the classroom to understand the spatial distribution and shape of a fish school.

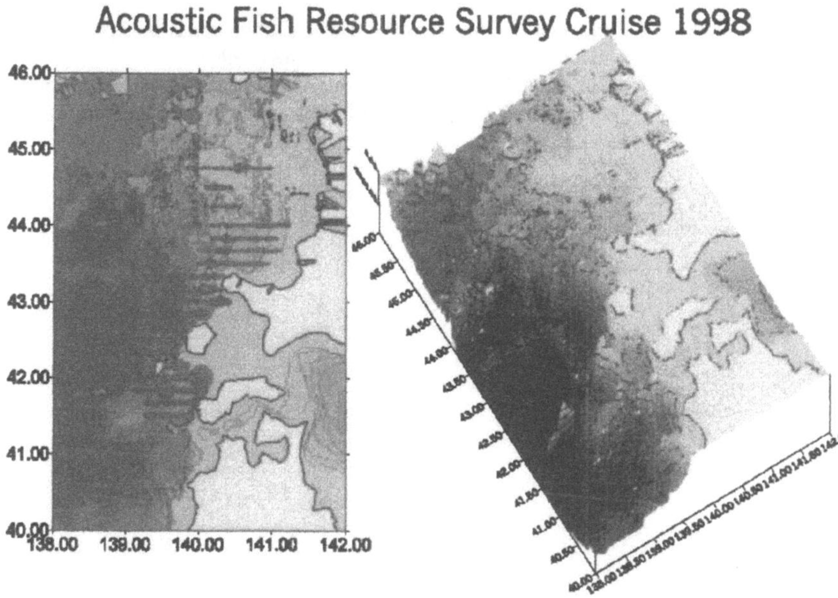

Fig. 10. Three-dimensional distribution map of walleye pollock around Hokkaido obtained by an acoustic survey using a scientific echo sounder.

Also, when a huge amount of data to be processed is obtained during a long survey cruise, researchers on board want to obtain highly processed data, such as a 3-D distribution map of fish resources overlaid on oceanographic data. In such a case, the satellite laboratory discussed here is useful. Figure 10 shows a 3-D distribution of walleye pollock resources around Hokkaido that was analyzed by the geographical information analyzing system (GIS) in a laboratory on land.

In order to produce such highly processed 3-D data, data on the location of the ship, depth, echograms, sonar images, ADCP, and CTD profiles are sent from the field laboratory to a laboratory on land through the communication satellite.

There is currently a limitation of the transmission rate from the field laboratory to the land laboratory (2400 bps) compared to transmission in the opposite direction (64k bps). In the near future, we hope to realize bi-directional high-speed data transmission through a communication satellite.

ACKNOWLEDGEMENT

This research was partially supported by grants from the Ministry of Education, Culture and Science of Japan (grant number 11792011). We wish to express our appreciation to this grant.

REFERENCES

[1] Y. Aoki, R. Mitsuhasi and S. Tanahashi, "Transmission of Holographic Imaging Data of Micro- and Acoustic Waves by Communications Satellite," Proc. of the 20th International Symposium on Acoustical Imaging, Vol. 20, pp. 57-64 (1993).

[2] K. Iida, T. Mukai, Y. Aoki and T. Hayakawa, " Three-Dimensional Interpretation of Sonar Image for Fisheries," Acoustical Imaging Vol. 22 pp. 583-588 (1996).

[3] K. Iida, T. Mukai, N. Horiuchi, "Three-Dimensional Visualization of Fish Schools Using Sector Scanning Sonar," International Archives of Photogrammetry and Remote Sensing. Vol. 32, Part 5, pp. 729-734 (1998).

ANGULAR SPECTRUM APPROACH TO TIME–RESOLVED ACOUSTIC MICROSCOPY

S. A. Titov and R. Gr. Maev

Centre for Imaging Research and Advanced Material Characterization,
School of Physical Science, University of Windsor, Windsor,
N9B 3P4, Canada

INTRODUCTION

A time–resolved acoustic microscopy has been recently developed for quantitative materials characterization [1,2]. In this method transducer output waveform v(z,t) is recorded as a function of time t and specimen displacement z along acoustic axis. Usually wide–band low–frequency large aperture cylindrical transducers are used in the measurements [3]. Surface leaky waves and directly reflected wave can be observed separately in the output waveform at defocused position of the transducer. Relative time delays of these pulses are measured as a function of defocusing distance z for determining wave velocities [4] or elastic constants of the materials [2]. However the received pulse form can be inconstant at various distance z because of, for example, the specimen dispersion or complexity of the transducer impulse response. In this paper an angular spectrum approach to the v(z,t) data processing is described.

THEORY

Typical experimental setup for time–resolved measurements is shown in Fig. 1. To determine the output waveform of the system we should analyze spatial and temporal structure of the acoustic field. Let $h_1(x,y,t)$ be an acoustic pressure generated by the probing wave in the focal plane of the transducer (x,y). By $H_1(k_x, k_y, \omega)$ denote the three–dimensional Fourier transform of field $h_1(x,y,t)$, where k_x, k_y are components of wave number $k_0 = \omega / c$, $\omega/2\pi$ is frequency, c is the wave velocity in the fluid. Then the angular spectrum of the reflected wave can be expressed in the focal plane as:

$$U(k_x, k_y, \omega) = H_1(k_x, k_y, \omega) \cdot R(k_x, k_y, \omega) \cdot \exp(+2ik_z z) \tag{1}$$

Here $R(k_x, k_y, \omega)$ is reflection function, $k_z = k_0 \cos\theta$, θ is incident angle, and phase factor $\exp(2ik_z z)$ is propagation function for space between focal and specimen planes [5].

Acoustical Imaging, Volume 25, Edited by Halliwell and Wells.
Kluwer Academic/Plenum Publishers, 2000.

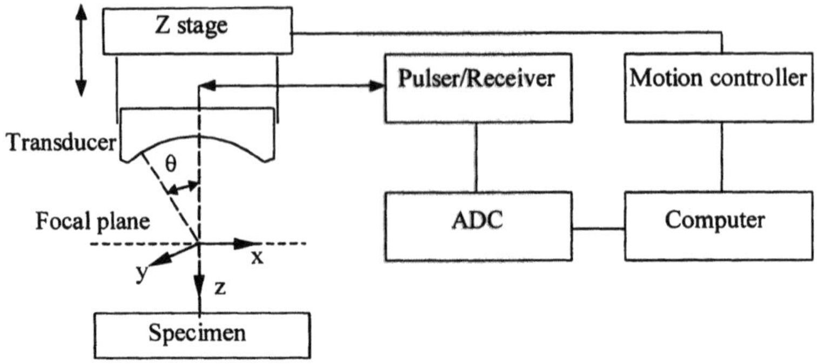

Figure 1. Experimental setup.

Now we define the sensitivity function $H_2(k_x,k_y,\omega)$, which gives the voltage response of the transducer to a plane wave of unit amplitude at the focus. Therefore the output voltage of the transducer is

$$v(z,t) = \int\int\int_{-\infty}^{\infty} H_1 H_2 R(k_x,k_y,\omega)\exp(2ik_z z - i\omega t)d\omega dk_x dk_y \quad (2)$$

This formula is valid for any form of an electrical exciting signal as well as for any type of ultrasonic transducer. For narrow–band tone burst mode time t and frequency ω may be omitted in this expression and it can be reduced to the well–known formula for v(z) curve [6]. In continuous wave Doppler microscope transducer is moved at constant velocity $z=v_t t$ [7]. Because of this relation between spatial and temporal variables the output signal can be expressed as a function of one variable v(t). In these two special cases the inverse problem can be solved for a particular frequency only. The time–resolved microscope has wide–band excitation and obviously, its output waveform should be considered as two–dimentional function.

To reconstruct the reflectance function from the equation (2) it is necessary to make some assumptions. Suppose the specimen is isotropic and uniform; then the reflectance function is axial symmetric $R(k_x, k_y, \omega) = R(k_z, \omega)$. By changing of variables we obtain

$$v(z,t) = \int\int_{-\infty}^{\infty} H_0(k_z,\omega)R(k_z,\omega)\exp\{i(2k_z z - \omega t)\}dk_z d\omega = F^{-1}{}_{k,\omega}[H_0(k_z,\omega)R(k_z,\omega)] \quad (3)$$

where $F^{-1}[\,]$ is two–dimentional inverse Fourier transformation and general pupil function $H_0(k_z, \omega)$ is found by integrating over azimuthal angle φ

$$H_0(k_z,\omega) = \int_0^{2\pi} H_1(k_z,\varphi,\omega)H_2(k_z,\varphi,\omega)k_z d\varphi \quad (4)$$

For anisotropic material a cylindrical transducer is used for the measurements. Let φ_0 be an orientation angle and assume that the pupil function is sensitive only in this direction

$$H_0(k_z,\omega,\varphi)=H_1(k_z,\omega)H_2(k_z,\omega)\cdot k_z\cdot\delta(\varphi-\varphi_0). \tag{5}$$

where δ( ) is delta–function. Then the formula (3) can be also used in this case for reconstruction of the reflection function $R(k_z, \omega) = R(k_z, \omega, \varphi_0)$ at given specimen orientation.

To analyze the time–resolved microscopy let us define new variable

$$\tau=2k_z z/\omega=2z\cos\theta/c. \tag{6}$$

This value τ is just time of the plane wave propagation from the focal plane of the transducer to the specimen and back. By $\tau_0=2z/c$ denote time–of–flight of the directly reflected wave at θ=0. Since $\tau/\tau_0=\cos\theta$, by change of variable we obtain from (3)

$$v(z,t)=\frac{1}{2z}\int_{-\infty}^{\infty}\int H_0\left(\frac{\tau}{\tau_0},\omega\right)R\left(\frac{\tau}{\tau_0},\omega\right)\omega\exp\{-i\omega(t-\tau)\}d\tau d\omega. \tag{7}$$

Suppose $H_0(\tau/\tau_0, \omega) = P(\tau/\tau_0)\, G(\omega)$, i.e. general pupil function can be represented as a product of the of function of aperture angle and a function of frequency. This is a good approximation for lensless line focus microscope. Then we have in the case of dispersion free reflectance function

$$v(z,t)==\frac{1}{2z}\int_{-\infty}^{\infty}R\left(\frac{\tau}{\tau_0}\right)P\left(\frac{\tau}{\tau_0}\right)g(t-\tau)d\tau=\frac{1}{2z}\left[R\left(\frac{t}{\tau_0}\right)P\left(\frac{t}{\tau_0}\right)\right]*g(t) \tag{8}$$

where $g(t)=F^{-1}[G(\omega)]$ is impulse temporal response of the transducer and electronics units and * is a convolution operation. If $g(t)=\delta(t)$, then

$$v(z,t)==\frac{1}{2z}\left[R\left(\frac{t}{\tau_0}\right)P\left(\frac{t}{\tau_0}\right)\right]. \tag{9}$$

Thus for frequency unlimited system the complex output signal at given z can be expressed as a product of the scaled reflectance and pupil functions. If pupil function P=1 within aperture angles $[\theta_m, 0]$ and P=0 outside this range, then at given z the output waveform v is nonzero for t belonging to $[\tau_0 \cos\theta_m, \tau_0]=[2z\cos\theta_m/c, 2z/c]$ only. The amplitude of the signal has a sharp steps at points $\tau_0\cos\theta_m$, τ_0 and phase undergoes approximately 2π jump at the moment $\tau_0\cos\theta_R$, where θ_R is critical surface wave angle. Moreover, phase jumps occur at negative as well as at positive values of z.

Experimentally surface waves pulses are observed only at negative z. Let us consider now an actual ultrasonic system with limited frequency bandwidth. Its temporal impulse response g(t) has nonzero duration and zero mean. Therefore convolution according to (8) gives approximately zero result where product R(t)P(t) is smooth slowly varying function. Output pulses appear at the points where the reflectance or pupil functions have phase or amplitude singularities. Suppose impulse response can be expressed as

$$g(t)=A(t)\exp(-i\omega_0 t) \tag{10}$$

where ω_0 is central frequency and A(t) is complex envelope. Assume also that the phase is approximately a linear function in the vicinity of the critical angle

$$R\left(\frac{t}{\tau_0}\right)P\left(\frac{t}{\tau_0}\right)=C\exp(\frac{i\alpha(t-\tau_R)}{z}) \tag{11}$$

Here coefficient $\alpha>0$ defines the slope of the phase, and C is constant. Combining (8), (10), and (11) we obtain formula for Rayleigh pulse

$$v(z,t)\approx C\exp(-\frac{i\alpha\tau_R}{z})\exp(-i\omega_0 t)\int_{-\infty}^{\infty}A(t-\xi)\exp(i\xi(\alpha/z+\omega_0))d\xi \tag{12}$$

For z>0 phase factor $\exp(i\xi\ (\alpha/z+\omega_0))$ changes rapidly and function A is smooth therefore the output signal is approximately equal to zero. For z<0 this phase factor is slowly varying function and we have pulse at $t=\tau_R$.

Numerical simulation according formula (8) has been carried out to confirm these theoretical results. It was assumed that the central frequency of the probing signal is equal to 10 MHz, the waveform duration equals 0.1 μs, and pupil function $H_0=1$. Waveforms calculated for fused quartz – water interface are shown in the figure 2.

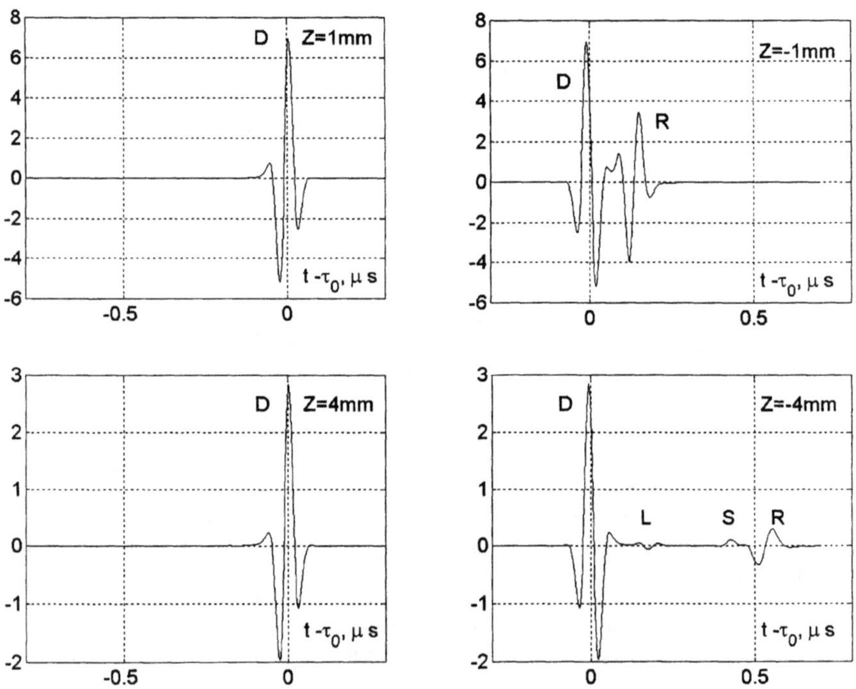

Figure 2. Output waveforms at several defocusing distances z.

For z>0 only directly reflected pulse D is observed. On the other hand, for z<0 a large leaky surface wave R appears in addition to pulse D. Furthermore, leaky surface skimming longitudinal and shear waves L and S are visible at a sufficient defocusing distance.

Generally, for an arbitrary measuring system $H_0(\tau/\tau_0, \omega) \neq P(\tau/\tau_0)\,G(\omega)$, and formula (8) isn't valid any more. In this case we can rewrite (7) as superposition integral:

$$v(z,t)=\frac{1}{2z}\int_{-\infty}^{\infty}R\left(\frac{\tau}{\tau_0}\right)\int_{-\infty}^{\infty}H_0\left(\frac{\tau}{\tau_0},\omega\right)\exp\{-i\omega(t-\tau)\}d\omega d\tau=\frac{1}{2z}\int_{-\infty}^{\infty}R\left(\frac{\tau}{\tau_0}\right)h\left(\frac{\tau}{\tau_0},t-\tau\right)d\tau \quad (13)$$

Impulse response h is now a function of z and τ. Also, the waveform R depends on defocusing distance because of reflectance function is scaled inversely to z. Therefore, the wave shape at critical angles is variable. Time position of pulse peaks or zeroes relative to actual time of flight might be a nonlinear function of z. Thus in the case of complicated pupil function as well as in the case of the reflectance function dispersion we should use (3) for acoustic properties measurements.

EXPERIMENTAL

Configuration of the experimental setup is shown in the Fig. 1. A point focus acoustic lens was fabricated at the end of fused quartz buffer rod. Lens focus distance is 3.4 mm and its half angular aperture is 37°. 30 V magnitude and 20 ns duration negative pulse was used for the piezoelectric transducer excitation. The output waveform was digitized by 8 bit 140 MHz bandwidth ADC at 400 MHz sampling rate. The acoustic system was translated by step motor actuator and spatial sampling interval is 8 μm. The output waveform v(z,t) measured for hardened steel is shown as gray–scale image in the Fig. 3.

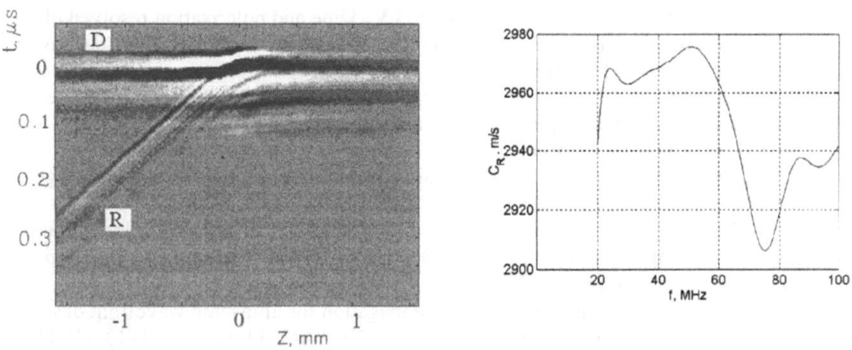

Figure 3. Waveform v(z,t) for steel (left); estimated leaky Rayleigh wave velocity (right).

The waveforms v(t) are translated along vertical axis by delays of direct reflected pulses $\tau_0=2z/c$, thus the direct reflected wave D has horizontal alignment. Clearly visible at negative z line R is generated by leaky Rayleigh wave. Its velocity c_{R0} = 2952 m/s was calculated from the slope of this line similarly to works[1,4]. To estimate function H_0R spectral analysis of the acquired data v(z,t) was carried out according to (3) (Fig. 4).

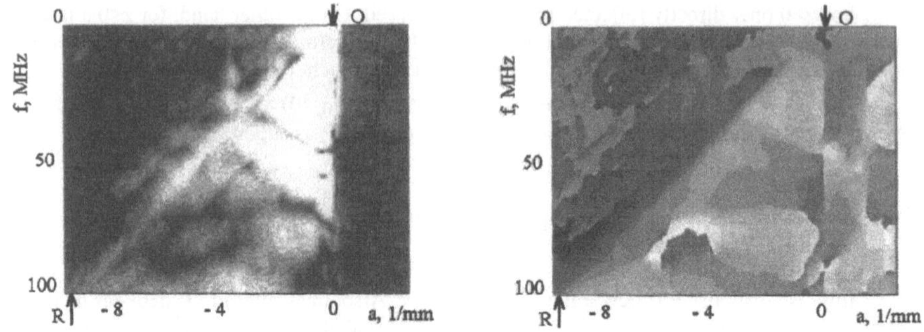

Figure 4. Amplitude (left) and phase (right) of the output waveform spectrum.

The spectrum H_0R calculated by two–dimentional fast Fourier transform is expressed as a function of frequency ω and spatial frequency $a=(k_z-k_0)/2\pi = k_0(\cos\theta-1)/2\pi$. Thus an angular spectrum component at incident angle θ_0 corresponds in this plot to line (ω, $k_0(\cos\theta_0-1)/2\pi$). The direct reflected wave is represented by vertical line (ω, $a=0$). Rayleigh wave line (ω, $k_0(\cos\theta_R-1)/2\pi$) is marked by points O and R. Position of 2π phase jump was used to detected critical angle and to calculate leaky surface wave velocity c_R . Result of c_R measurement is shown in the Fig. 3 as a function of frequency. Deviation of c_R from value c_{R0} described above can be estimated within 1% approximately for frequency range of 20–100 MHz.

REFERENCES

1. Hsu N.N., Xiang D., Fick S.E., and Blessing G.V. Time and polarization resolved ultrasonic measurements using a lensless line–focus transducer, in: *Proc. IEEE Ultrason. Symp.*, pp. 867–871, (1995).
2. Li W., and Achenbach J.D. Determination of elastic constants by time–resolved line–focus acoustic microscopy, *IEEE Trans. on Ultrason., Ferroelect., Freq. Contr.*, vol. 44, n. 3, pp. 681–687, (1997).
3. Xiang D., Hsu N.N., Blessing G.V. The design, construction an application of a large aperture lens–less line–focus PVDF transducer, *Ultrasonics*, 34:641, (1996).
4. Xiang D., Hsu N.N., Blessing G.V. Statistical error analysis of time and polarization resolved ultrasonic measurements, *IEEE Trans. on Ultrason., Ferroelect., Freq. Contr.*, vol. 45, n. 4, pp. 1006–1015, (1998).
5. Liu D., and Waag C. Propagation and backpropagation for ultrasonic wavefront design, *IEEE Trans. on Ultrason., Ferroelect., Freq. Contr.*,vol. 44, n. 1, pp. 1–13, (1997).
6. Somekh M.G., Bertoni H.L., Briggs G.A.D., and Burton N.J. A two–dimentional imaging theory of surface discontinuities with the scanning acoustic microscope, Proc. R. Soc. Lond., A401, pp. 29–51, (1985).
7. Titov S.A., and Maev R.G. Doppler continuous wave scanning acoustic microscope, *Proc. IEEE Ultrason. Simp.*, pp. 713-718, (1997).

IMAGING OF FERROELECTRIC DOMAINS BY ATOMIC FORCE ACOUSTIC MICROSCOPY

Ute Rabe, Sandrine Amelio, Sigrun Hirsekorn, and Walter Arnold

Fraunhofer-Institute for Nondestructive Testing
University, Bldg. 37
D-66123 Saarbrücken, Germany

INTRODUCTION

Atomic Force Microscopy (AFM) is a near-field technique to generate high-resolution images of surfaces. A microfabricated elastic beam with an integrated sharp sensor tip at its end is scanned over the sample surface. With a variety of dynamic modes, leading to microscopies such as Force Modulation Microscopy [1, 2], Ultrasonic Force Microscopy [3], Atomic Force Acoustic Microscopy [4, 5], Microdeformation Microscopy [6], Scanning Local Acceleration Microscopy [7] or Pulsed Force Microscopy [8], images can be obtained in which the contrast depends on the elasticity of the sample surface. In our Atomic Force Acoustic Microscopy (AFAM) setup [4, 5], we evaluate the cantilever vibration spectra at ultrasonic frequencies in order to discern local elastic data quantitatively. Either the sample is insonified or the cantilever suspension is excited at ultrasonic frequencies. The vibration spectra of the cantilever depend on the local sample stiffness and hence on the local elasticity of the constituents of the sample. Acoustical images can be obtained by measurement of the cantilever vibration amplitude at a fixed frequency close to a contact resonance with the help of a lock-in amplifier. In this contribution we present AFAM results for contact spectroscopy and imaging of the domain structure of ferroelectric ceramics.

Below their Curie temperature, ferroelectric single crystals have a permanent electric dipole moment. A mechanical stress applied to such a crystal changes the electrical polarization, and vice versa, an external electric field causes a mechanical strain [9]. Within one single crystal there are different domains of homogeneous polarization. The possible domain orientations depend on the crystal structure of the material. Technical polycrystalline piezoelectric materials like Lead Zirconate Titanate (PZT) have substituents in the $Pb(Zr,Ti)O_3$ perovskite structure at the morphotropic phase boundary which enables the coexistence of phases with different crystal structures [10]. This results in a very complex domain structure. These materials are of particular scientific and technological interest. As the

Acoustical Imaging, Volume 25, Edited by Halliwell and Wells.
Kluwer Academic/Plenum Publishers, 2000.

grain size of commercial PZT ceramics is in the μm range and the domain size consequently in the sub-μm range, it is appropriate to examine these materials by AFM with regard to a variety of scientific questions, for example, domain switching or fracture. To this end, different techniques have already been applied. The domain structure of ferroelectric materials can be imaged indirectly by AFM [11] by imaging the surface topography. The surface of such a sample is polished and etched such that domains of different orientations have different etch rates. This results in height differences reflecting the domain structure. The domains at the surface can also be imaged directly using the so-called piezo-mode [12, 13]. An ac electric field is applied between the AFM tip and a counter electrode below the sample. This leads to an induced vibration, the phase of which depends on the orientation of the domains.

MATERIALS

Pb(Zr, Ti)O_3 ceramics were examined in the experiments described in this paper. They are solid solutions. $PbTiO_3$ is tetragonal in its ferroelectric phase. Zr^{+4} substitution for Ti^{+4} causes the appearance of an additional ferroelectric phase of rhomboedral symmetry. The Pb content of technical ceramics varies between 48 and 51 mole % PbO, close to the morphotropic phase boundary (MPB) between tetragonal and rhomboedral forms. Though attempts have been made, large single crystals of PZT at the MPB which would allow the determination of elastic and electric properties were not obtained so far [14]. The PZT ceramic examined here is PIC 151 which is a soft ceramic doped with antimony and nickel. The linear macroscopic elastic and electric constants of the polycrystal are known [15]. Poled plates with a thickness of 7 mm were cut with a diamond saw into stripes 2 mm wide. The resulting samples were glued sidewise to a steel plate and polished (using a Wirtz Phoenix 4000 polishing machine). Diamond paste with grain sizes of 12, 6, 3, 1, and 0.25 μm was used subsequently. Finally, the samples were polished with alumina (grain size 0.25 μm) to remove remaining grooves from the diamond polishing procedure. In the specimens manufactured in this way, the polarization vector was then parallel to the surface examined.

THEORY

The tip-sample contact stiffness in an AFM experiment depends on the static forces applied on the sensor tip adjusted by the cantilever deflection Δz and by the attractive tip-sample forces such as electrostatic forces and adhesion forces. The contact stiffness k* also depends on the sensor tip radius, the Young's modulus and the Poisson's ratios of the tip and the sample, respectively. When the sensor tip is in repulsive contact with a sample, the system is stiffened and the resonance frequencies of all modes increase relative to the free resonances [5]. By using cantilevers with high spring constants of the order of 40-70 N/m, it is ensured that the main contribution to the contact stiffness is given by elastic interaction forces between tip and sample, and not by other forces such as adhesion forces (Fig. 1). According to the Hertzian contact model [16] a contact area with a radius of

$$r_H = \sqrt[3]{\frac{3F_H R}{4E^*}} \, . \tag{1}$$

is formed when an isotropic sphere of radius R contacts a mechanically isotropic flat surface. F_H is the sum of the attractive and the static cantilever forces, and E^* is the reduced elastic

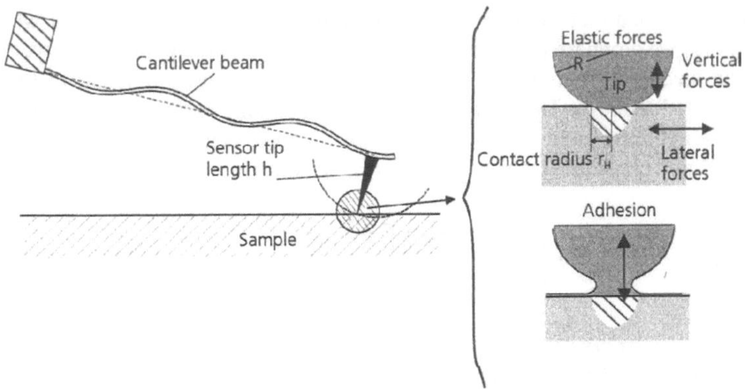

Fig. 1. Principle of contact spectroscopy to measure local elastic constants on a material. From the vibrational spectra of a cantilever in an AFM, the contact stiffness k* is measured. The contribution of adhesion forces to the contact stiffness is made small by applying a sufficiently high static force on the cantilever so that the Hertzian contact forces are dominant. The lateral forces to k* are neglected. Details of data evaluation are discussed elsewhere [17]. The indented volume is about $\pi r_H^2 \times 3r_H \approx 500$ nm^2.

modulus $1/E^* = (1-\nu_S^2)/E_S + (1-\nu_T^2)/E_T$, where E_S, E_T, ν_S, and ν_T are the Young's moduli and the Poisson ratios of the surface and the tip, respectively. The mechanical stress field of the contact extends about $3r_H$ normal to the surface and gives an estimate of the surface volume which is probed by the tip. The maximum pressure in the contact area is given by $P_C = F_H/(2\pi r_H^2)$. The radius of the sensor tips can vary by almost two orders of magnitude. While new sensor tips can have radii as small as 2 nm, the radius of a tip can increase dramatically due to wear when it is used on hard surfaces and a radius of 200 nm or more is obtained after a few contact experiments [18]. As a consequence, the contact stiffness can be extremely different even when the tip material, the sample, and the static cantilever force are the same. To estimate this, let us take a cantilever with a spring constant of 50 N/m and with a silicon cantilever tip oriented in <100> direction (E_T = 130 GPa, ν_T = 0.181) and which is pressed with a static deflection of 20 nm into an isotropic sample having a Young's modulus of 100 GPa and a Poisson's ratio of 0.3. The theoretical contact data according to the Hertzian model are listed in Tab. 1. It is noticeable that despite large tip radii R, the contact radius r_H remains small. The values listed in Tab. 1 for k* and P_C were calculated with a constant Poisson's ratio despite the fact, that the Si tip is not isotropic in the (100) plane. Hence, ν depends on the direction in the (100) plane making the indentation profile and the ensuing stress-field non-spherical. However, these effects change the contact stiffness only by a few percent for an anisotropy factor A less than 2 (Si: A = 1.56) [19].

Table 1. Calculated contact radius r_H, contact stiffness k*, and maximum pressure P_C for a silicon tip with different radii R on a sample having a Young's modulus of 100 GPa and a Poisson's ratio of 0.3

	R = 2 nm	R = 20 nm	R = 200 nm
r_H [nm]	0.8	1.7	3.7
k* [N/m]	96	206	445
P_C [GPa]	10	2.2	0.47

In order to obtain an estimate of the variations in local contact stiffness which can be expected in PZT, we compared the stress/strain relationships of the polarized polycrystals parallel to the electrical polarization (in 3-direction) to those perpendicular to the electrical polarization (in 1-direction). The linear piezoelectric equations in the case of uniaxial stress in 1- and in 3-direction yield the corresponding Young's moduli for constant electric displacement D and constant electrical field E, respectively:

$$Y_1^{D,E} = \left(c_{11}^{D,E} - c_{12}^{D,E}\right)\cdot\left(1 + \frac{c_{12}^{D,E}c_{33}^{D,E} - \left(c_{13}^{D,E}\right)^2}{c_{11}^{D,E}c_{33}^{D,E} - \left(c_{13}^{D,E}\right)^2}\right) = \frac{1}{s_{11}^{D,E}}, \tag{2}$$

$$Y_3^{D,E} = c_{33}^{D,E} - \frac{2\left(c_{13}^{D,E}\right)^2}{c_{11}^{D,E} + c_{12}^{D,E}} = \frac{1}{s_{33}^{D,E}}. \tag{3}$$

Here, c_{ij}^{D} and c_{ij}^{E} are the elastic constants and s_{ij}^{D} and s_{ij}^{E} are the compliance constants for constant electric displacement (superscript D) and constant electric field (superscript E), respectively. The macroscopic elastic data for PIC 151 given in [15] yield the numerical results Y_1^E = 60.0 GPa, Y_3^E = 52.6 GPa, Y_1^D = 70.0 GPa, Y_3^D = 118.7 GPa. The values for constant electric field hold for the case that the two surfaces between which the mechanical stress is applied are electrically short-circuited. In this case, there exists only the mechanical anisotropy between the directions parallel and perpendicular to the electrical polarization direction [20]. When the two surfaces are not connected by electrodes (superscript D), no electrical charges can flow to compensate the surface charges. Hence the change in the polarization field due to the mechanical strain of the lattice persists. This polarization causes an additional restoring force which leads to a higher Young's modulus and to a considerable anisotropy between Y_1^D and Y_3^D.

EXPERIMENTAL RESULTS

Contact Resonances

For the contact resonance experiments the options available in a commercial AFM (AFM/LFM II and Dimension 3000, Digital Instruments/Veeco, Santa Barbara, CA).are used to control the static cantilever forces. An external frequency generator provides a stable sinusoidal excitation of a piezoelectric transducer coupled to the back side of the sample. It emits longitudinal acoustic waves into the sample which cause out-of-plane vibrations of the sample surface. These surface vibrations are transmitted into the cantilever via the sensor tip. The cantilever vibration signal is measured by the optical beam-deflection sensor of the instrument, and is then evaluated using a lock-in amplifier. The set-up can be used either to acquire cantilever vibration spectra at a selected location or to take acoustic images. In the spectroscopy mode the excitation frequency is changed stepwise, the digitized lock-in output at a given frequency is read, and the vibration amplitude of the cantilever as a function of frequency is stored. In the imaging mode, a frequency close to a contact resonance is selected for excitation and while the sample surface is scanned, the lock-in output is fed into an auxiliary channel of the commercial instrument, digitized, and displayed as a color-coded image. The frequency range employed covers the flexural modes of the cantilever from 10 kHz up to about 3 MHz. With the type of cantilevers used, we are usually able to evaluate the first three flexural resonance frequencies. For the quantitative evaluation of the local Young's modulus, data of several modes must be used allowing to obtain information on all pertinent parameters, in particular the contact radius. Details of the AFAM technique are published elsewhere [5, 17].

The surfaces prepared with the polishing process described above had a mean roughness R_a of less than 2 nm within one grain. Height differences from grain to grain were larger and ranged from several nm up to 20 nm. Therefore, the grains could be imaged in the AFM topography mode without further etching. The average grain size of the PIC ceramic sample was 2 μm. Examples of resonance spectra on a piezoelectric ceramic sample are shown in Fig. 2. Apart from exceptions, the contact resonance frequencies are generally lower on PIC 151 than on silicon <100>, which means that the corresponding elastic constants of silicon are higher than the one of PIC 151. When all results from the various contact spectra are averaged, one obtains a Young's modulus of 106.5 ± 14 GPa on location (1) and 99.5 ± 16 GPa on location (2) on the PIC 151 sample.

Though our tips and cantilevers are doped and therefore electrically conductive, the range of Young's moduli which we obtained from our contact resonance experiments corresponds to the anisotropy expected in the case of constant D. We found no influence on the contact resonance frequencies when we grounded the cantilever. Contrary to macroscopic experiments where a sample is stressed and strained as a whole, the mechanical stress field of the AFM tip is inhomogeneous and decays within a thin surface layer. Therefore the local change in the polarization field within the indented volume cannot be compensated by charge transfer between the tip and a counter electrode on the opposite side of the sample. Hence, our experimental data must be compared to the theoretical values $Y_1^D = 70$ GPa and $Y_3^D = 118.7$ GPa which are derived from a polarized ensemble of grains, see above. We actually measure within a given domain of a grain and hence obtain single crystal data.

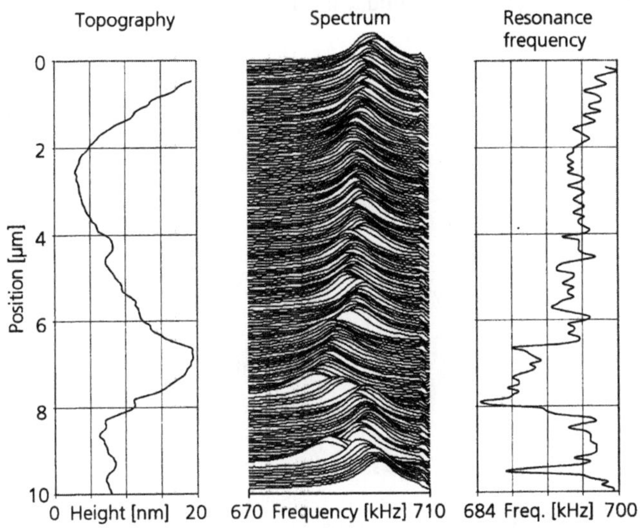

Fig. 2. Left: Line-scan of a length of 10 μm showing the topography obtained on PIC 151. Center: Corresponding contact resonance spectra. The variation of the contact resonance frequency as a function of position becomes very evident. Using the inversion procedure described earlier [17], this corresponds to a change of the elastic modulus from 87 GPa to 120 GPa. Right: Line-scan displaying the resonance frequency.

Imaging

Ferroelectric domains can also be imaged directly by AFAM. In contact resonance imaging the sensor tip is vibrated in contact with the sample surface with a preset static load. An acoustical excitation frequency close to a contact resonance frequency on the material under examination is chosen. The vibration amplitude of the cantilever at this frequency is detected with the help of a lock-in amplifier and displayed as a color-coded image. A local difference in contact stiffness causes a shift of the contact resonance frequency and hence a change of cantilever vibration amplitude at the preset frequency. Fig. 3 shows the topography and a contact resonance image obtained on the surface of a PZT ceramic. The main corrugations in the topography image are the grain boundaries and scratches from polishing. Within individual grains, the surface is relatively flat. In the acoustic image, substructures in the grains which represent the domain structure are clearly resolved.

The contrast depends on whether the set-point frequency is chosen above or below the average contact resonance maximum. If the set-point is below the contact resonance, regions with higher contact stiffness lead to a lower cantilever vibration amplitude and hence darker areas in the image. If the setpoint is above the contact resonance, regions with higher contact stiffness appear as bright areas. Fig. 4 shows a section of the surface area shown in Fig. 3 imaged at higher magnification. While the left image was taken at an excitation frequency above the contact resonance, the right image was taken at a frequency below the contact resonance. As can be seen, contrast inversion occurs which leads to the conclusion that the contrast between the domains is caused by differences in contact stiffness. The effect of contrast inversion has been noticed earlier [7, 21]. However, at that time, it was not yet possible to obtain quantitative data in AFAM images. Most features in the images, for example the triangular area with high contact stiffness in the middle, appear inverted in contrast. Some dark areas, however, remain dark in both images. It is likely that these are areas where the contact resonance frequencies are far away from the excitation frequency, so that there is only a small vibration amplitude of the cantilever at both excitation frequencies.

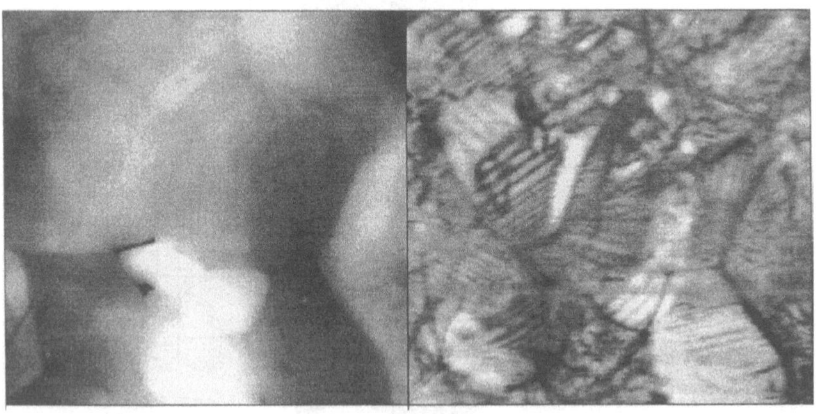

Figure 3: Topography (left) and contact resonance image of a polished PZT ceramic. The size of the images is 10 × 10 μm^2. The gray scale covers 20 nm in the height image. The first, second, and third free resonance frequencies of the cantilever are 162 kHz, 1004 kHz, and 2624 kHz, respectively. The spring constant of the cantilever is 41 N/m. The first contact resonance on PZT is obtained at 697.5 kHz. The image is taken at 698 kHz, i.e. above the first contact resonance.

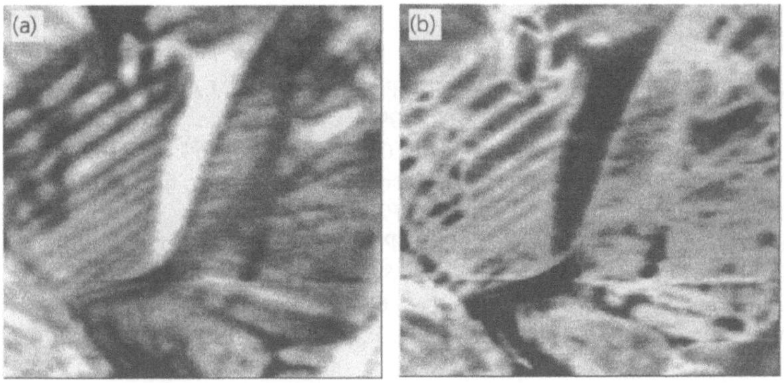

Figure 4: Contact resonance images taken on PIC 151 at two different frequencies: Left 699 kHz (above contact resonance) and right 695 kHz (below contact resonance). The image size is $4.3 \times 4.3\ \mu m^2$.

Such large frequency shifts can be caused by surface topography. If the tip is for example over a deep and steep pore, it might touch the side-walls of the hole instead of touching a flat area with its apex. This changes the tip-sample contact stiffness dramatically. The behavior of the line-scan spectra (Fig. 2) and the contrast inversion let us conclude that the contrast of the substructures in the grains is caused by differences in contact stiffness from domain to domain. Furthermore, since the sample surface is locally strained by a contacting and vibrating AFM tip, polarization charges are generated due to the piezoelectric effect. It is well-known that the bulk electric polarization due to a preferred macroscopic orientation of the domains influences the elastic response of a ferroelectric material when stresses are applied [22]. The contrast in AFAM is caused by the elasticity variations from domain to domain due to their different electrical polarization and hence ferroelastic behavior.

SUMMARY

We were able to image domains with AFAM on PIC 151. The changes in the resonance frequencies as shown in the line scan (Fig. 2) demonstrate clearly that the contrast is due to variations in contact stiffness caused by changes of the local Young's modulus. Until now, no elastic data of the single crystals which constitute PIC 151 exist. The technique described here opens the way to obtain such data and can also be applied to other multiphase materials. Such measurements were undertaken not only in piezoelectric ceramics but also in other types of materials [23] but were not reported here. The AFAM technique opens the way to calibrate the gray scale of the images in units of contact stiffness by using a feedback loop in which the values of the local resonance frequency are used for imaging. If the sample is flat, the calibration may also be obtained in local elastic moduli.

ACKNOWLEDGEMENT

We thank the German Science Foundation for financial support within the research program "Multifunctional Materials". This research was also partially supported by the Volkswagen Foundation. We thank D. C. Lupascu and J. Rödel from the TH Darmstadt, Department of Materials, for close collaboration.

REFERENCES

[1] P. Maivald, H.T. Butt, S.A. Gould, C.B. Prater, B. Drake, J.A. Gurley, V.B. Elings, and P.K. Hansma, *Nanotechnology* 2: 103 (1991)

[2] M. Radmacher, R.X. Tillmann, and H.E. Gaub, *Biophys. J.* 64:735 (1993)

[3] K. Yamanaka, H. Ogiso, and O. V. Kolosov, *Appl. Phys. Lett.* 64:178 (1994)

[4] U. Rabe, K. Janser, and W. Arnold, in: *Proc. 22nd Int. Symp. Acoustical Imaging*, Ed. P. Tortoli, and L. Masotti, Plenum Press Plenum Press, New York (1996), pp. 669

[5] U. Rabe, K. Janser, and W. Arnold, *Rev. Sci. Instrum.* 67:3281 (1996)

[6] P. Vairac and B. Cretin, *Appl. Phys. Lett.* 68:461 (1996)

[7] N.A. Burnham, G. Gremaud, A.J. Kulik, P.J. Gallo, and F. Oulevy, *J. Vac. Sci. Tech.* B14:1308 (1996)

[8] A. Rosa, E. Weilandt, S. Hild, and O. Marti, *Meas. Sci. Technol.* 8:1 (1997)

[9] Ch. Kittel, *Introduction to solid state physics*, John Wiley, New York (1971)

[10] M. Hammer and M. J. Hoffmann, *J. Am. Ceram. Soc.* 81:3277 (1998)

[11] O. Auciello, A. Gruverman, H. Tokumoto, S.A. Prakash, S. Aggarwal, and R. Ramesh, *MRS Bulletin* 23:33 (1998)

[12] U. Rabe, E. Kester, V. Scherer, and W. Arnold, to be published in: *Proc. 24th Int. Symp. Acoustical Imaging*, edited by H. Lee, Plenum Press, New York (2000)

[13] L. M. Eng, H.-J. Güntherodt, G. A. Schneider, U. Köpke, and J. Munoz Saldana, *Appl. Phys. Lett.* 74:233 (1999)

[14] B. Jaffe, W. Cook, and H. Jaffe, *Piezoelectric Ceramics*, Academic, London (1971)

[15] Data sheet of PI Ceramic, Lindenstraße, D-07589 Lederhose, Germany

[16] K. Johnson, *Contact Mechanics*, Cambridge University Press (1995)

[17] U. Rabe, S. Amelio, E. Kester, V. Scherer, S. Hirsekorn, and W. Arnold, to be published in *Ultrasonics*, 2000.

[18] U. Rabe, E. Kester, and W. Arnold, *Surf. Interface Anal.* 27:386 (1999)

[19] J.J. Vlassek and W.D. Nix, *Phil. Mag. A*, 67, 1045 (1993)

[20] T. Ikaeda, *Fundamentals of Piezoelectricity*, Oxford Science Publications (1990)

[21] U. Rabe, PhD-thesis, Technical Faculty, University of Saarbrücken, and Fraunhofer-Institut for Nondestructive Testing, Saarbrücken, Germany, (1996), unpublished.

[22] see for example A. Moulson and J. M. Herbert, *Electroceramics*, Chapman and Hall, London (1990)

[23] E. Kester, U. Rabe, L. Presmanes, Ph. Tailhades, and W. Arnold, J. Phys Chem Solids, in press

SCANNING ACOUSTIC MICROSCOPY CASE STUDIES OF MICROELECTRONIC PACKAGING

John Barton, Tony Compagno and John Barrett, *Senior Member, IEEE*

National Microelectronics Research Centre
Lee Maltings, Prospect Row
Cork, Ireland

INTRODUCTION

In recent years, scanning acoustic microscopy(SAM) has been found to be a very successful technique when used in the microelectronics industry to evaluate, from a reliability perspective, standard plastic packaging technologies such as PQFP's, PLCC's, DIP's and SOP's. The recent explosion of advanced packaging techniques such as Chip-on-Board, Flip-chip and BGA and the recent proliferation of Microsystems has further widened the arena of what constitutes microelectronics. With such a wide breadth of devices from standard plastic packages to state-of-the-art microsystems, it is difficult to find a failure analysis technique which can cope competently with that scope. Scanning Acoustic Microscopy is one such technique. This paper will show that with the right equipment and range of transducers, an acoustic microscope can reliably be used across the whole spectrum of microelectronics.

SCANNING ACOUSTIC MICROSCOPY (SAM)

Scanning Acoustic Microscopy (SAM) is based on the interaction between ultrasonic waves and matter. The presence of inhomogeneities and discontinuities along their propagation paths inside the matter causes modifications in the amplitude and polarity of ultrasonic waves. Acoustic microscopy works through the transmission of a collimated pulse of sound along an acoustic lens. A piezoelectric transducer generates the pulse, which is focused onto the sample via a liquid coupling medium. Pulses within the sample are collected by the lens and returned to the transducer, where they are reconverted to an electrical signal. Figure 1 shows a schematic of an acoustic microscopy system. The amplitude of the returned pulses, and the exact time elapsed from the original signal, indicate the magnitude and the depth of physical changes in the sample itself. The microscope facilitates rapid interpretation of the depth information, and the lens is mechanically scanned over the surface of the sample to produce a two-dimensional image.

Acoustical Imaging, Volume 25, Edited by Halliwell and Wells.
Kluwer Academic/Plenum Publishers, 2000.

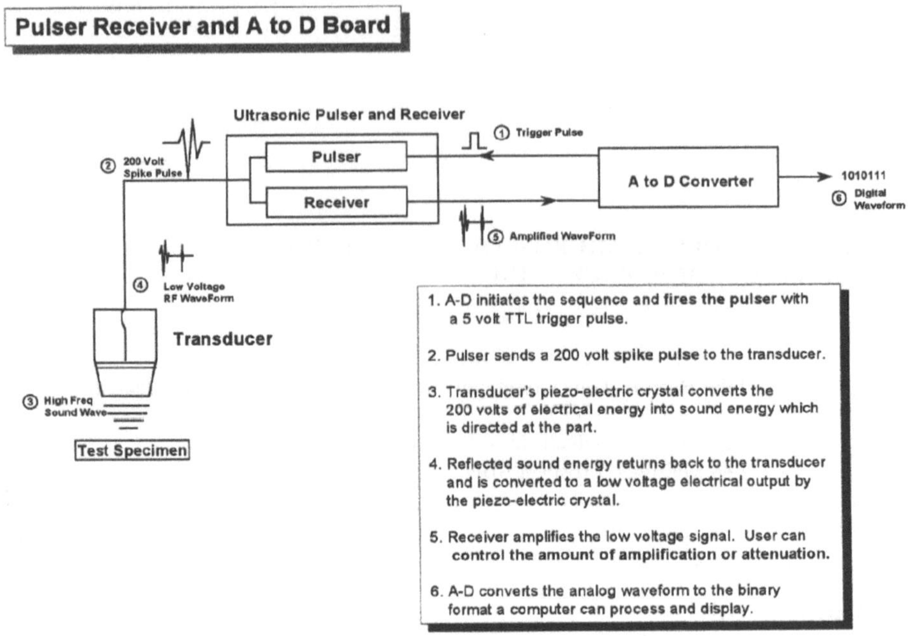

Figure 1. Schematic of Acoustic Microscopy system

Two basic operation modes can be used: the reflection mode and the through transmission mode. In the reflection mode, the scanning ultrasonic beam is focussed on a plane inside the object. The reflected portion of the ultrasonic waves is received by the same transducer that works also as a detector, the amplitude and polarity of the detected signal are analysed and processed for each point in a colour map fashion and displayed on a cathode ray tube monitor. The correspondent image shows the internal features of the object at the chosen focal plane. In transmission mode, the transmitted portion of the ultrasonic beam is focussed on another detector, situated on the other side of the object with respect to the source. Only the amplitude of the correspondent image is analysed and gives a low resolution black and white image of the internal features of the object through its entire thickness. Typical ultrasound frequencies used for analysis of microelectronic components range from 10MHz to 200MHz. However a trade-off exists in that the higher the frequency used the greater the lateral resolution obtainable but the depth of penetration into the sample is reduced and vice versa. The acoustic microscope used in all these case studies was a SONIX HS1000 with software IC5.98d.

CASE STUDY 1: PLASTIC PACKAGES

Standard plastic packages come in a wide variety of shapes and forms but most resemble the cross-section shown in Figure 2. Because of the many clearly defined interfaces in such a package, acoustic microscopy is ideal for detecting a multitude of sub-surface defects as seen in Figure 2. This case study shall deal with die attach voiding/delamination.

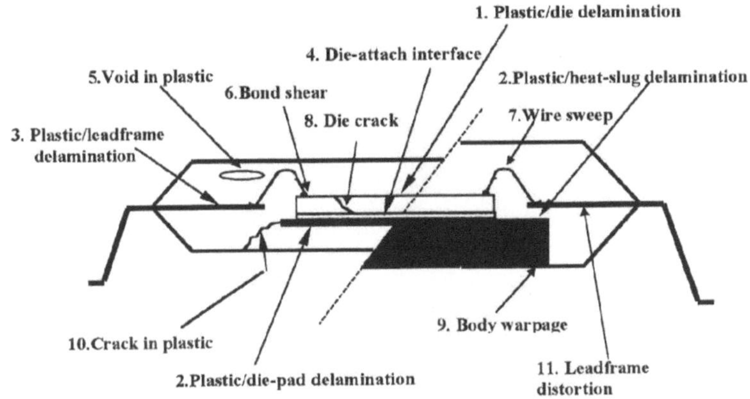

Figure 2. Anatomy of standard/power plastic package

A quantity of 128 lead TQFP packages were received for failure analysis. This analysis consisted of acoustic microscopy, X-ray and microsectioning. The sample was analysed by acoustic microscopy using a 15MHz transducer. One of the devices showed voiding/delamination in the die attach as seen in Figure 3 where the delamination is shown in black. The sample had been previously x-rayed and nothing was detected so the sample was microsectioned for verification. A cross –section through the centre of the device revealed the die-attach delamination exactly as detected with the SAM. Figure 4 shows the x-ray of the device while Figure 5 shows a microsection of the delamination. The pattern of voiding/delamination observed in the microsection matched exactly the pattern seen by the SAM.

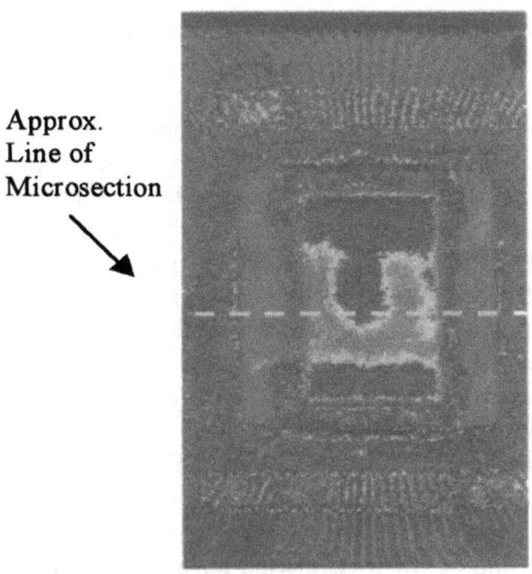

Figure 3. Die attach delamination

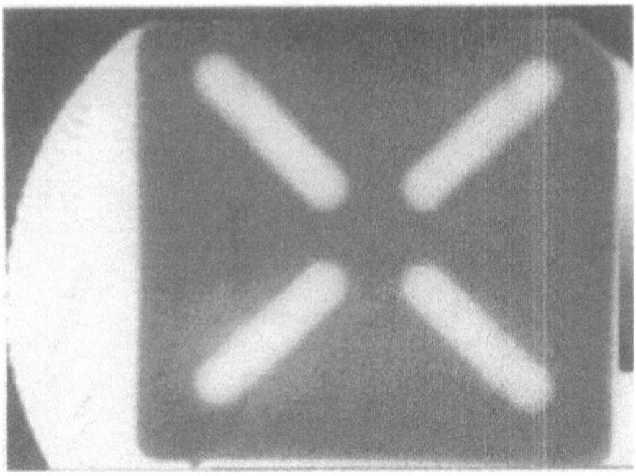

Figure 4. X-ray of package

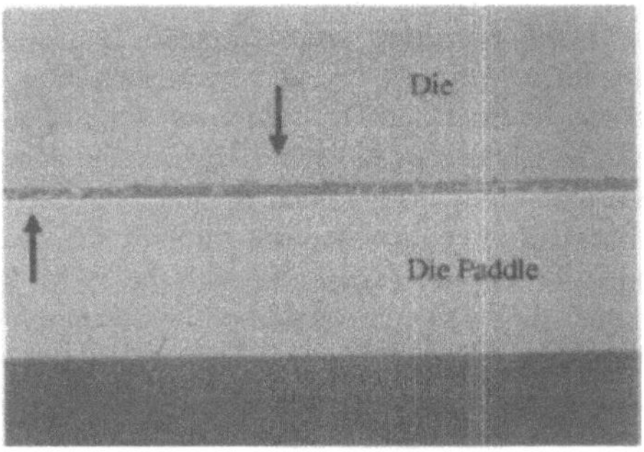

Figure 5. View highlights voided areas and non-voided areas in the die attach. Mag x200

CASE STUDY II: LOW TEMPERATURE CO-FIRED CERAMIC (LTCC) INTEGRATED CAPACITORS

LTCC technology uses a glass-ceramic dielectric and thick-film conductor materials for a multilayer fabrication strategy that uses only one sintering step. The ability to combine such process-related advantages with high performance means that it is among the most promising ceramic technologies available. Recently, the scope of the technology has been further increased by the successful development of methods and materials for integration of passive components into the co-fired substrate. Fully realised, the new technology includes integral capacitors, resistors, and inductors, which have each been fully electrically characterised and successfully tested for reliability in a high stress automotive environment. To achieve integral capacitors in LTCC, one of the most important issues is to maintain the structural integrity of the finished substrate. This is achieved by using 'inserts' containing high permittivity material ($BaTiO_3$). These are cylinders, 500μm in diameter and 100μm thick, which are slotted into the substrate layers. To develop large capacitors, a regularised array of these inserts is introduced into the LTCC layer. Metal plates are printed above and below to sandwich the array. The effect is equivalent to placing a number of small-value capacitors in parallel. Figure 6 shows an exploded diagram of an integrated capacitor.

These capacitors were found to be susceptible to voiding due to their insert format and the inherent performance of the LTCC materials. In order to investigate this effect, acoustic microscopy was employed to firstly locate these voids and verify the primary influencing factors in the LTCC process, including in particular the lamination pressure level. Subsequently, electrical and SAM analyses of a number of capacitor formats was completed to quantify the level of voiding present, to define optimum process parameters, and to resolve the impact of varying the capacitor size and geometry (i.e. number of layers) upon the parameters.

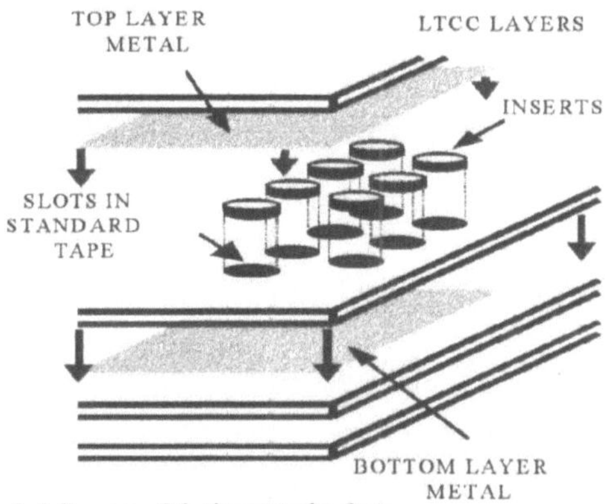

Figure 6. Exploded diagram of the insert technology.

Evaluation Methodology

A total of 20 test samples were analysed. Capacitors on three different test vehicles were analysed. Test vehicles TV1, TV2, and TV3 contain six [1.2nF], nine [140pF to 9.6nF], and eight [1.38nF to 4.14nF] capacitor structures, respectively. The lamination pressures used during fabrication processing ranged between 100 bar and 300 bar. Both single and two-layer buried capacitor structures were analysed. Lower lamination pressure samples consisted primarily of single-layer structures, while the two-layer samples tended to have higher lamination pressures applied.

Scanning acoustic microscopy is an ideal technique for locating voids in the buried inserts of LTCC material. A non-destructive technique, ultrasound is particularly sensitive to sub-surface voids and delaminations. These scans were taken using a 75MHz transducer. With careful positioning of the electronic gates, only the first sub-surface interface is analysed. All of the inserts are clearly distinguishable and appear greyish in colour. Any voids in the inserts are shown in white. Figure 7 shows scans taken from the front and back of the same 104-insert array laminated at 300 bar. Note that early every insert contains a void. A microsection confirmed the presence of the voids as shown in Figure 8.

The reason for the existence of these voids may be traced back to the shrinkage rates of the LTCC materials. The fact that the insert material shrinks at a lower rate means that it must be placed in a slot that is larger than its own volume in order to achieve a common volume after sintering. Thus, there is the potential for a built-in void, unless appropriate processing steps are taken This specifically means increasing the lamination pressure to remove the air in the gap. Process development completed by the manufacturer clearly demonstrated that this adjustment works effectively.

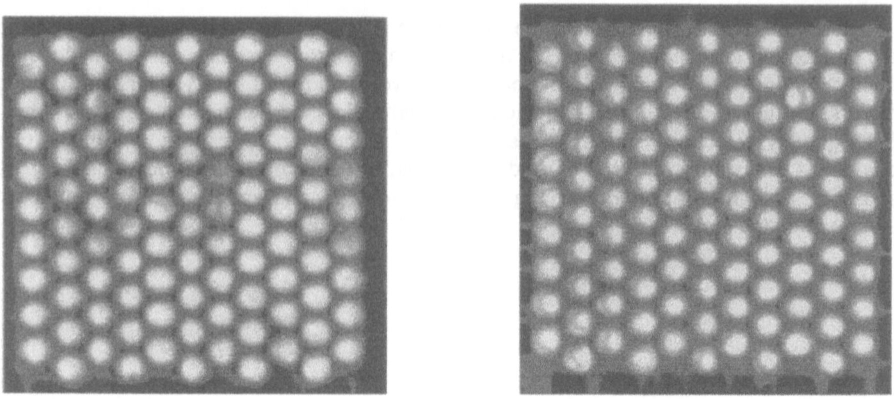

[A] [B]

Figure 7. The images show a uniform level of voiding between [A] the Top plate and the inserts, and [B] the bottom plate and the inserts.

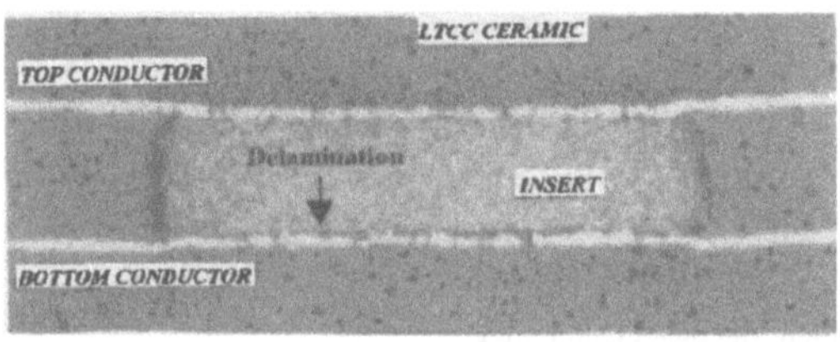

Figure 8. This microsection of TV2 (100 bar) clearly shows the presence of voids at the insert interfaces.

Effect of Varying Lamination Pressure

An acoustic microscopic analysis of a number of test vehicles, over a range of lamination pressures and other parameters, was completed which confirmed this. Increasing the lamination pressure from 100 bar to 150 bar resulted in a reduction in the severity of the voiding, particularly towards the centre of the capacitor plate insert pattern. However, this was non-uniform, with some segments containing inserts with quite low levels of voiding while other, usually larger, segments contained voiding levels which were very similar to 100 bar test vehicle. Increasing to 180 bar results in a further reduction, though the non-uniformity remains. A trend is emerging in which the larger voids are now visible primarily on inserts at the periphery of the capacitor matrix patterns. However, voiding at central inserts was also seen. At 200 bar, the larger voids are now limited to the capacitor's peripheral inserts. Voiding on central inserts appears much lower. The highest pressure applied was 300 bar. Analysis of the single-layer capacitors indicates a continuing decrease in the presence of voiding at the insert-metal interface. Voiding is now exclusively seen on the peripheral inserts, and at a much lower level than was seen on samples fabricated using 200 bar or less. Figure 9 shows a scan of a complete test vehicle laminated at 300 bar.

Impact of the Number of Capacitor Layers

When two-layer samples were analysed, the SAM results indicated a similar trend to that seen for the one-layer samples. Notably, the higher relative level of the lamination pressures applied and analysed is indicative of a need for a general increase when more than one capacitor layer is used. This transfers to the levels of voiding seen in these samples. For instance, at 200 bar, the level of voiding is higher for a two-layer capacitor than for a one-layer structure with the same surface plate area. It can be seen that the level of voiding is higher in both cases for the two-layer structure. However, the relative difference at 300 bar is significantly smaller.

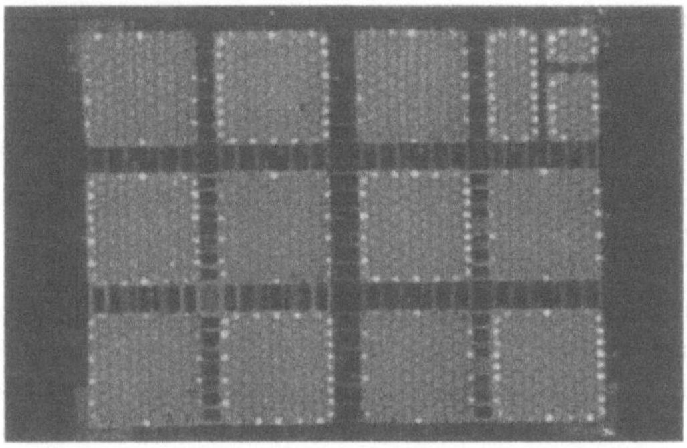

Figure 9. The lowest level of voiding appears when 300 bar was applied. In this case, the voiding is vastly reduced, concentrated only at the periphery of the capacitor patterns, and, in some cases, eliminated totally.

CASE STUDY III: FLIP-CHIP UNDERFILL

The term flip-chip is used for an assembly method where the chip is mounted directly on the board by means of soldering or thermocompression. Electrical contact between board and chips is accomplished by metal bumps located on the chip surface. A cross-section of a typical flip-chip is given in Figure 10.

Figure 10. Cross-sectional view of Flip-Chip package

The basic flip-chip process is as follows:

1. The most important task in wafer bumping is to put down an underbump metal layer (UBM) on the aluminium pads. The UBM defines the region of terminal metallurgy on the top surface of the chip.
2. Solder is deposited by either electroplating or evaporation methods.
3. The chip is fluxed,

4. Placed and,
5. Reflowed.
6. The underfill is deposited and cured.

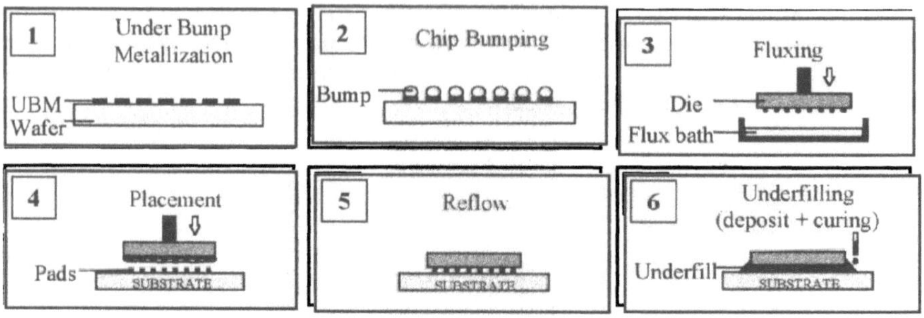

Figure 11. Flip Chip assembly process

The liquid underfill is applied after flip chip attachment. By capillary force, the gap between chip and substrate is filled. After curing, the underfill becomes solid. The primary purpose of underfilling is to guarantee interconnection reliability. The underfill is used to reduce the stress on the solder bumps during temperature cycling. Underfill absorbs stress caused by the mismatch in the coefficient of thermal expansion (CTE) between the chip and the substrate. As temperature increases, the substrate and die start to expand at different rates, but the solder bumps attempt to keep the die and substrate together. Therefore, a large shear stress is developed in the bumps. The introduction of a rigid underfill layer reduces that stress. However, during the filling action, there is the danger that air can be trapped resulting in voids or delaminations in the underfill after curing. These air-gaps can cause serious reliability hazards to the chip. For example, an initially harmless underfill delamination may grow until it encounters and shears a solder bump. These air-gaps in the underfill are easily detectable by acoustic microscopy. With a high resolution system and high frequency transducers, defects within the solder bumps and the bump/chip or bump/substrate integrity can be analysed.

A number of flip-chip Assembly Test Vehicles (ATV's) were analysed using the acoustic microscope as part of an intensive characterisation process. Each ATV had four flip-chipped die on it:

- 2 large area flip-chip die : 10 x 5 mm sq., 200 µm pitch, 116 I/Os UBM : Electroless Ni/Au , bump : Eutectic Sn/Pb printing
- 2 small area flip-chip die: 2.8 x 3.7 mm sq., 223 µm pitch, 48 I/Os, C4 technology with eutectic Sn/Pb coating

A picture of an ATV is given in Figure 12. A typical acoustic image of a large area flip-chip die taken using a 150MHz transducer is given in Figure 13. The defects that can be imaged include voids and delaminations (white) in the underfill, density variations in the underfill and defects within the solder bumps including cracks and voids. For the purpose of this case study, the underfill voids and delamination will be focussed on. Each die on the ATV's was scanned with a 150MHz transducer at the start of the characterisation process. A total of 9 ATV's were scanned making a total of 36 flip-chip die. By utilising a template together with peak amplitude thresholds, the underfill delamination as a percentage of area can be accurately calculated. For the small area flip-chips, the baseline delaminations were very small ranging from 0% to 0.7%. For the larger flip-chip die the base-line delaminations ranged from 0% to 26.39%.

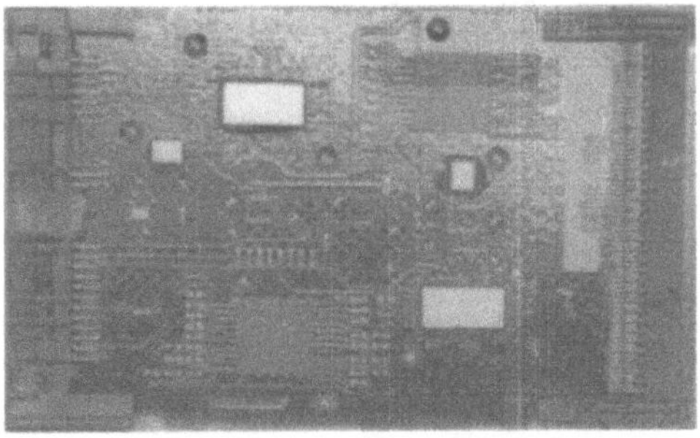

Figure 12. Flip-chip ATV

Figure 13. Large area flip-chip acoustic image

A number of the ATV's were then environmentally stressed, either through Humidity testing or Temperature Cycling. The tests were stopped at regular intervals and one ATV in each test was scanned at every interval while the others were only analysed at the beginning and end of the test. Table 1 shows the environmental tests and intervals.

Table 1. Reliability Tests and environmental levels.

Test	Description	Intervals
Accelerated Thermal Cycling Temperature cycling with dwell time = 15mins	Temperature cycling of 0°C to +100°C for up to 1000 hours. Followed by vibration + mechanical shock.	100,250,500 and 1000 cycles
High T/H: High Temperature/Humidity	Humidity and temp levels @ 85°C/85%RH for up to 1000hrs.	100,250,500 and 1000 hours

The following two figures show representative acoustic images from one large area flip-chip and one small area flip-chip during either Humidity testing and Temperature Cycling. The underfill delamination in the large area die starts to increase after 100 hours 85/85 and continues to increase up to the end of the test ending with between 31-46% delamination on each die. The small area flip-chip underfill delamination showed no signs of increasing until 500 hours had passed. Two die then showed increases in delamination of up to 25% and 50%. The Temperature Cycling had a greater effect on the underfill delamination of the larger area flip-chips than the Humidity test. At the end of the 1000 cycles, the delaminations range from 55% to 75% with on exceptionally low value of

3.37%. Temperature Cycling had less effect on the smaller area flip-chips with delaminations at the end of the test ranging from 0.15-4.77%.

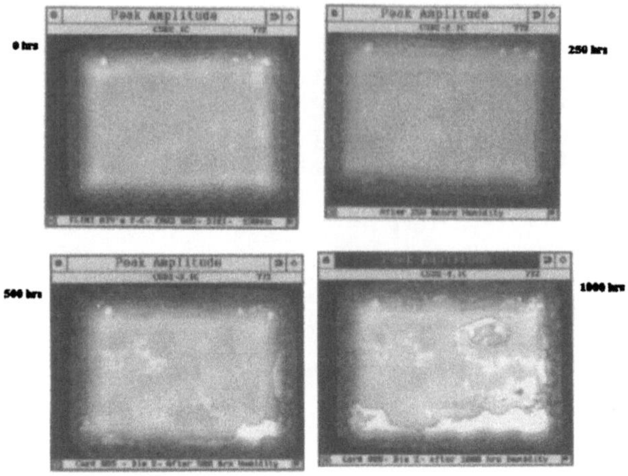

Figure 14. Underfill delamination for IBM flip-chip during Humidity Testing

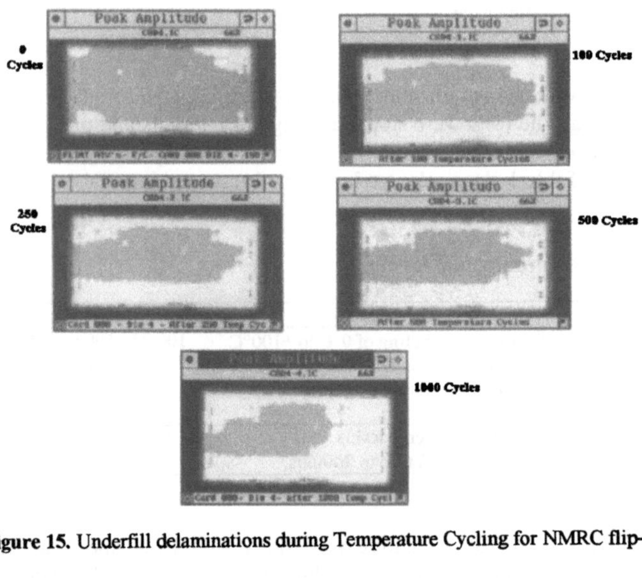

Figure 15. Underfill delaminations during Temperature Cycling for NMRC flip-chip

CASE STUDY IV: BONDED SILICON WAFERS

One bonded silicon wafer test sample was received for trial SAM analysis. This sample consisted of two silicon wafers bonded together using anodic bonding. The two wafers were 380μm and 30μm thick respectively. 300μm wide channels were etched in the thicker wafer with the thinner 30μm wafer acting as a thin membrane over the channels. Figure 16 shows a schematic of the bonded Silicon wafers including channels.

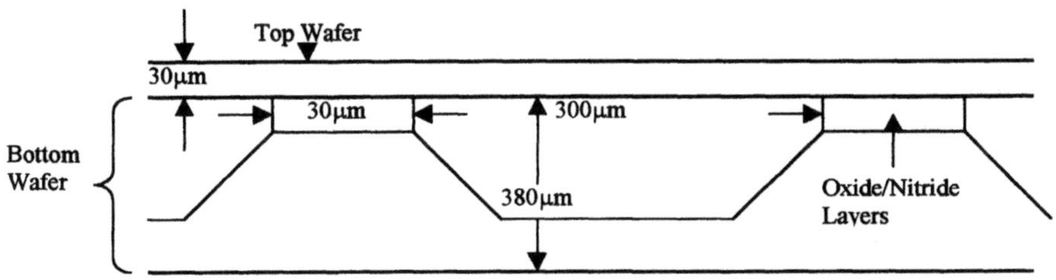

Figure 16. Schematic of Bonded wafers.

There were two main objectives of this analysis. (1) To check the general adhesion quality between the two bonded silicon wafers. and (2) To see if the Scanning Acoustic Microscope could detect small disbonded spots in the 30μm area between the channels cut in the 30μm thick wafer. Silicon bonded wafers are particularly challenging because the samples are usually very thin (<1mm) and sound propagates through silicon quickly. Because the sample was only 410μm thick in total, it was decided to use the 180 MHz Ultra High Frequency lens using a 2 GHz sampling frequency to scan the silicon bonded wafers.

After initial set-up time, very good images were obtained of the channels and the wafer interface. Image time and definition is dependent on picture quality required. Generally a four inch wafer can be scanned in about 5/6 minutes depending on picture quality and the number of data points picked. This type of scan will only give a general view of the wafer interfaces. Figure 17 shows a scan of an entire wafer focussing on the silicon/silicon interface. Poor adhesion is indicated by the brighter spots on the wafer. In order to see specific defects it is necessary to focus in at a particular area on the wafer. To pick up the defects/disbonded spots in the 30μm area between the channels it is necessary to do a 5mm X 5mm or 2mm X 2mm scan, which takes between one to seven minutes depending on the resolution.

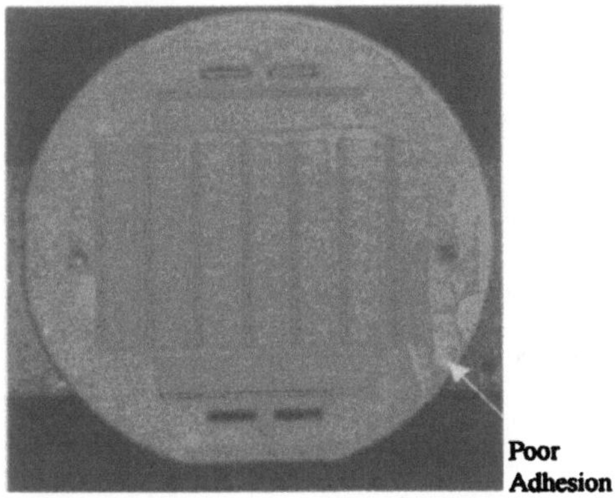

Figure 17. Acoustic image of 4 inch Silicon bonded wafers

By increasing the resolution and reducing the scan size to 10mm X 10mm, disbonded areas in the microchannels begin to be observed while the poor adhesion is also seen in Figure 18.

Figure 18. 10mm X 10mm acoustic image of microchannels

Figure 19 shows the highest resolution and a 2mm X 2mm scan size clearly shows the disbonded regions on the 30μm channels. Scanning Acoustic Microscopy has proved to be a very good technique to determine the bonding quality between the wafers. The overall general adhesion quality of the bonding can be observed with a very quick low resolution scan. Defects such as the small disbonded spots in the 30μm area between the channels are readily detected with high resolution smaller scans. The only drawback is that a high resolution scan of an entire wafer would be extremely time-consuming.

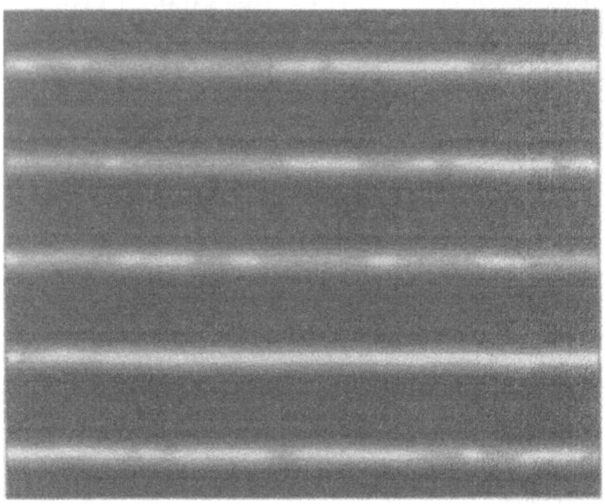

Figure 19. 2mm X 2mm acoustic image of microchannels

CONCLUSIONS

This paper has demonstrated the usefulness of the Scanning Acoustic Microscope as a microelectronics failure analysis tool through a number of very different case studies. With state of the art equipment, technical proficiency and a wide range of acoustic transducers, the entire gamut of microelectronics from plastic packaging to microsystems can be analysed quickly and accurately for a variety of defects.

RESIDUAL STRESS CHARACTERIZATION BY SCANNING ELECTRON ACOUSTIC MICROSCOPY

Y. Hong, Z. N. Zhang, S. Y. Zhang, Z. Q. Li and X. J. Shui

Lab of Modern Acoustics and Institute of Acoustics
Nanjing University, Nanjing, 210093, P. R. China
Tel.: +86-25-3593310, Fax: +86-25-3315557
E-mail: paslab@nju.edu.cn

1. INTRODUCTION

The Scanning electron-acoustic microscopy (SEAM), also referred to thermal wave microscopy, has been widely used in characterizations of materials since it was firstly developed.[1,2] It has been applied to study subsurface defects, doping regions in semiconductors,[3] domain structures in ferromagnetic and ferroelectric materials[4-6] and other materials[7] due to its special abilities of nondestructive evaluation for subsurface.

Characterization of the residual stress has been proved to be useful to evaluate elastic modulus, yield strength and hardness of materials. Although many experimental methods[8,9] have been developed to show the distribution of the residual stress around the indentation, they cannot exhibit nondestructively the subsurface characteristics in micro-scales. The present work is to propose a nondestructive method for detection the distribution of the residual stress, which not only reveals the outline of the subsurface distribution but also gives much detailed information in different depths by SEAM.

2.METHODS

The experimental system is modified from a scanning electron microscope, in which the electron beam is intensity-modulated at frequency ω and incident on the specimen. The local periodic heating on the specimen surface gives rise to the generation of thermal wave and then acoustic wave. But the thermal wave is so highly

Acoustical Imaging, Volume 25, Edited by Halliwell and Wells.
Kluwer Academic/Plenum Publishers, 2000.

damped that typically propagates no more than a thermal diffusion length. Solution of the equation of heat diffusion shows the thermal diffusion length (thermal wave length) μ_t is expressed as

$$\mu_t = \sqrt{2K / \omega\rho C} \tag{1}$$

where K, ρ, C are the thermal conductivity, density and specific heat of the specimen, respectively. The surface and subsurface features within μ_t in the specimen may serve as a thermal wave scattering objects during it propagates, thus the acoustic signal will be generated as a result of thermal-to-acoustic mode conversion. Then it is reasonable to consider the output of a PZT transducer as a sum of electron-acoustic (EA) signals produced by different acoustic sources at different depths. Considering that the output of PZT is mixed with a reference signal, the final imaging signal is[10, 11]:

$$\overline{V} = \int_0^{x_m} |V(x')| \cos[\psi(x') - \Psi_0] dx' \tag{2}$$

where x_m is maximum penetration depth of the thermal wave, Ψ_0 is the phase shift of the reference signal, $V(x')$ and $\psi(x')$ are amplitude and phase lag of the EA signal caused by the acoustic source at depth x', respectively. The output signal $\overline{V}$ reflects the structures of the specimen in the thermal-wave penetration range. Therefore it is possible to get different EA amplitude images by varying frequency to display the properties and structures in different depth range because of μ_t inversely proportional to square root of ω. On the other hand, the phase lag, as well as the amplitude, contains the information about the thermoelastic features of the specimen at a depth x'. Since the structures at different depths induce different phase lags, the EA signal resulting from different subsurface depths can be enhanced selectively by adjusting the phase shift of the reference signal, which makes it possible to get much detailed information at a certain depth.

In SEAM, the electron beam is modulated at the frequency in the range of 10*KHz*-1*MHz* with an accelerating voltage ~30*KV*. The EA signal is detected by a PZT transducer contacted with the bottom of the specimen and fed to the preamplifier and then a lock-in amplifier. Both amplitude A and phase ψ of the output of the PZT can be acquired and processed by a computer. Finally, the alternative images corresponding to A, ψ, $A\cos\psi$ (X component) and $A\sin\psi$ (Y component) of the EA signal can be obtained and displayed in the screen. The secondary electron (SE) image and the EA image can be obtained in the same situation simultaneously.

3. EXPERIMENTAL RESULTS AND DISCUSSION

3.1. EA images at different frequencies

For an aluminum plate indented with 98*N*, we get the EA amplitude images at different frequencies of 12*KHz*, 39*KHz* and 187*KHz* as shown in Figs. 1(a)-1(c), respectively. For a comparison, the SE image of the same area is shown in Fig. 1(d). From Figs. 1(a)-1(c), several characteristics can be observed:

(1) The stress distribution expands from the cross to rectangle shape with the decrease of the modulation frequency. It is shown that the stress is diffusive with the depth increasing because the images obtained in lower frequency include information in deeper depth.

(2) The residual stress distribution is asymmetric, which may indicate the asymmetry of the loading.

(3) There is a defect marked with "A" in the SE image [Fig. 1(d)]. It is interesting to note that the defect "A" can be seen in all the other EA images except Fig. 1(b). But the corresponding frequency of Fig. 1(b) is not the lowest, which means that the frequency decrease is not the only reason for the resolution decay in this case. Considering the other two EA images are taken at the radius or thickness resonance frequencies of the PZT other than Fig. 1(b), it clearly shows that the frequency response of the PZT also plays an important role in determination of the spatial resolution. Those images at the resonance frequencies, where the EA signal is more sensitive to thermoelastic properties, are sure to reveal much more details. Therefore, the spatial resolution is optimized as a consequence.

3.2. EA images of different reference phase shifts at a selected frequency

For the same specimen, we get a set of EA vector images at a selected modulation frequency of 187*KHz* as shown in Figs. 2(a)-2(d) by changing the reference phase shift. A careful study of these images shows that the stress concentration area changes from one diagonal to the other. Unlike being visible in the Figs. 2(c)-(d), defect "A" disappears gradually from Fig. 2(b) to Fig. 2(a), indicating that these images reveal much deeper features. Specially, the distribution of residual stress along the diagonals in Fig. 2(a) is much different from the others, which illustrates that the residual stress does not decrease monotonically beneath the indentation as usually considered.[7, 8]

In general, the laminated imaging can be implemented in SEAM either by changing the modulation frequency or the phase shift of reference signal. The former is based on the variation of the thermal diffusion length μ_t with modulation frequency. Thus both surface topography and the subsurface features in the range of μ_t are all included in an EA image indiscriminately if no signal procession is taken. The latter, on the other hand, is extracted the contribution of the acoustic sources at a depth x' within μ_t. Therefore, the laminated features at different depth x' can be displayed directly. This is the difference between the EA amplitude images and vector images.

3.3. Dependence of EA signal on experimental parameters

The following figures illustrate the dependencies of the EA amplitude signal of Al on experimental parameters, such as modulation frequency, accelerating voltage

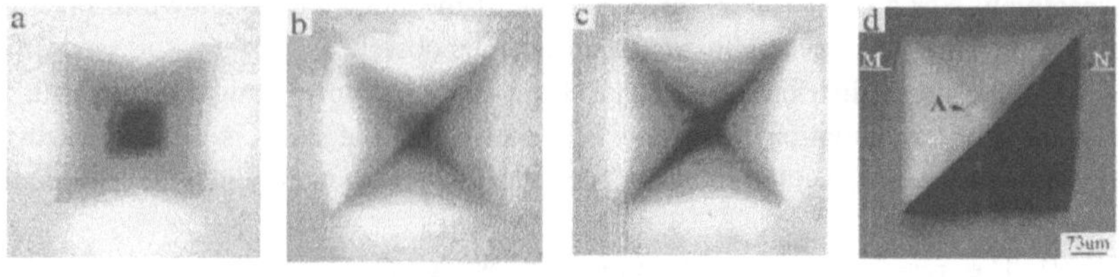

Figures 1. Vicker's indentation in aluminum: (a), (b) and (c) EA amplitude images at modulation frequency of 12*KHz*, 39*KHz* and 187*KHz*, respectively, (d) SE image.

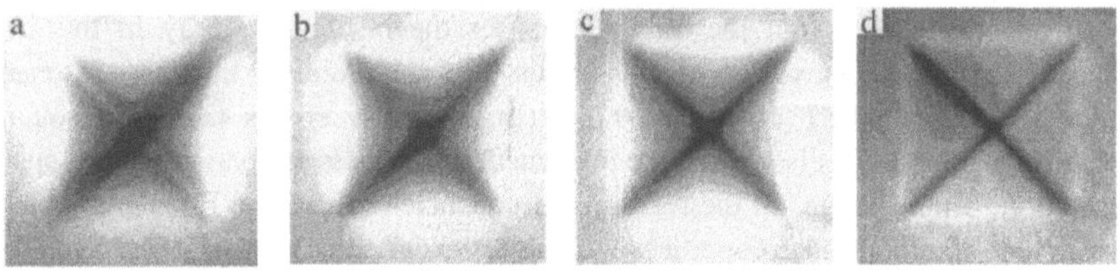

Figures 2. Vicker's indentation in aluminum: (a), (b), (c) and (d) are EA vector images with the reference phase shift of 11°, 127°, -37° and -160°, respectively at 187*KHz*.

and electron beam size. All line profiles are taken by scanning in the same position between "M" and "N" shown in Fig. 1(d).

Fig. 3 illustrates the lateral distribution of the EA signal of Al plate at 39*KHz*, 190*KHz* and 234*KHz*, respectively. The EA signal is normalized to eliminate the effect of frequency response of the PZT. By a comparison of these line profiles, it is shown that the detection depth of EA images changes with the modulation frequency. With the increase of detection depth, the gradient of the stress variation tends to slow down, corresponding with what is observed from EA images. Moreover, these line profiles show the EA signal is in direct proportion to the magnitude of residual stress. Qian et al.[12] has calculated the dependencies of the EA signal on Lame constant λ_s, mass density ρ_s and thermal conductivity K_s. According to Ref. [12], the presence of residual stress fields around an indentation may either increase λ_s or decreases K_s or both, which is the signal generation mechanism of EA images and needed to be further studied.

The variation of EA signals with electron beam accelerating voltage is shown in Fig. 4. With the accelerating voltage increasing, the magnitude of EA signals increases clearly, and the detection depth seemly increases at the same time, which is due to the reliance on the incident energy of the electron beam and penetration depth given by

$$z_0 = V^{1.43} / 20\rho_s \tag{3}$$

where V is the accelerating voltage, ρ_s is the mass density of the specimen.

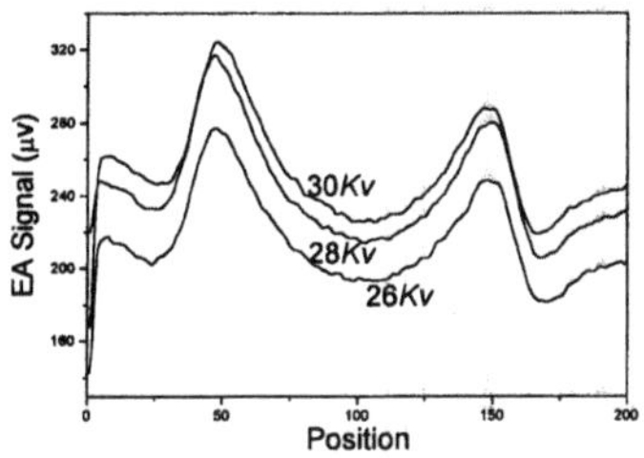

Figure 3. Lateral distribution of EA signal at the modulation frequencies, 39*KHz*, 190*KHz* and 234*KHz*.

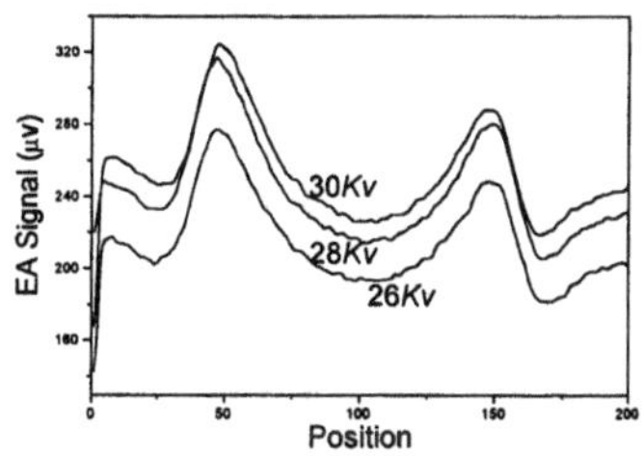

Figure 4. Lateral distribution of EA signal at accelerating voltages, 30*KV*, 28*KV* and 26*KV*.

The relationship between the EA signal and the spot size of electron beam has also been studied. As shown in Fig. 5, the detection depth of the EA signal, as well as its magnitude, increases with the spot size decreasing. However, it is impractical to get infinitely high resolution by decreasing the spot size because it is limited by the sensitivity of the PZT transducer.

3.4. EA images of different materials

In addition, EA images of brass and shape memory alloy (SMA) Nitnol indented with a load of 306*N* are also obtained as shown in Fig. 6. Although the outline of the EA images of brass and SMA are similar to those of Al, some differences are apparent. The EA images of the alloys (brass and SMA) are much more complex than that of Al. This complexity can be explained by the inhomogeneity of alloys in which different compositions respond differently to the indentation and then give rise to the complexity of stress distribution. Such differences can be related with material properties (especially elastic modulus, yield strength and hardness etc.).[13, 14]

4.CONCLUSIONS

The experimental observations in this work provide much information about the residual stress distribution in metals left by an indentation. The unique nondestructive capabilities of the SEAM, such as laminated imaging with either frequency or phase varied, signify a potential application of SEAM as a nondestructive tool for the characterization of residual stress in micro-scales. A further work on the mechanism and theoretical explanation of the residual stress distribution in EA images is in progress.

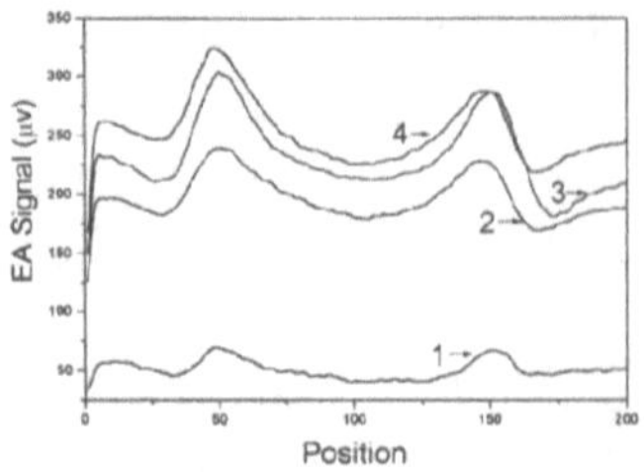

Figure 5. Lateral distribution of EA signal with the spot size decreasing from 1 to 4.

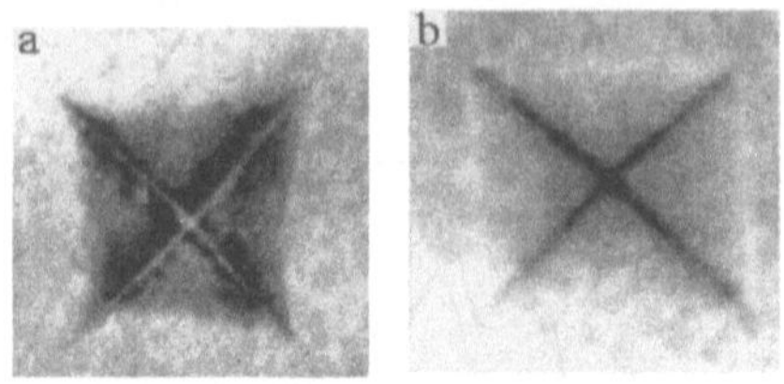

Figure 6. EA images of Vickers's indentation in (a) SMA and (b) brass at modulation frequency of 145*KHz*

REFERENCES

1. E. Brandis and A. Rosencwaig, Thermal-wave microscopy with electron beams, *Appl. Phys. Lett.*, **37**(1), 98-100 (1980).
2. G. S. Cargill III, Ultrosonic imaging in scanning microscopy, *Nature*, **286**, 691-693 (1980).
3. A. Rosencwaig, Thermal-wave imaging in a scanning electron microscope, *Ann. Rev. Mater. Sci.*, **15**, 103-118 (1985).
4. L. J. Balk, D. G. Davies and N. Kultscher, Investigation of Si – Fe transform sheets by scanning electron acoustic microscopy (SEAM), *IEEE Trans. Magn.*, Mag-**20**(5), 1466-1468 (1984).
5. Q. R. Yin, B. Y. Zhang, Y. Yang, F. M. Jiang and M. L. Qian, Electron-acoustic imaging of advanced ceramics and related materials, *Prog. Nat. Sci.*, **6**, S115-118 (1996).
6. M. L. Qian, X. M. Wu, Q. R. Yin, B. Y. Zhang and J. H. Cantrell, Scanning electron acoustic microscopy of electric domains in ferroelectric materials, *J. Mater. Res.*, **14**(7), 3096-3101 (1999).
7. G. S. Cargill III, Electron beam-acoustic imaging, in: *Physical Acoustics*, edited by W. P. Mason and R. N. Thurston (Academic, New York, 1988), Vol **XVIII**, pp. 125-165.
8. T. O. Mulhearn, The deformation of metals by vickers-type pyramidal indenters, *J. Mech. Phys. Solids*, **7**(2), 85-96 (1958).
9. A. G. Atkins and D. Tabor, Plastic indentation in metals with cones, *J. Mech. Phys. Solids*, **13**(3), 149-164 (1965).
10. S. Y. Zhang and L. Chen, Photoacoustic microscopy and detection of subsurface features of semiconductor devices, in: *Photoacoustic and Thermal Wave Phenomena in Semiconductors*, edited by A. Mandelis, (Elsvier, New York, 1987) pp.29-52.
11. Y. C. Shen and S. Y. Zhang, Piezoelectric photoacoustic evaluation of Si wafers with buried structures, *IEEE Trans. UFFC*, **39**(2), 227-231, (1992).
12. S. G. Qian, S. Y. Zhang, J. C. Cheng and Z. Q. Wang, Three-dimensional theory of scanning electron-acoustic imaging, *Prog. Nat. Sci.*, **6**, S115-118 (1996).
13. S. S Chiang, D. B. Marshall and A. G. Evans, The response of solids to elastic/plastic indentation. 1. stresses and residual stresses, *J. Appl. Phys.* **53**(1), 298-311 (1982).
14. D. B. Marshall, B. R. Lawn and A. G. Evans, Elastic/plastic indention damage in ceramics: the lateral crack system, *J. Am. Ceram. Soc.*, **65**(11), 561-566 (1982).

DOPPLER ULTRASONIC IMAGING

Hai_Dong Liang, Michael Halliwell, Peter N T Wells

Centre for Physics Engineering Research in Medicine
University of Bristol
Bristol General Hospital
Guinea Street Bristol BS1 6SY UK

Introduction

The Doppler effect is the phenomenon by which the frequency of a wave received after reflection by a moving target is shifted from that of the source. When the directions of movement of the reflector and of the ultrasonic beam are at some angle θ, the Doppler frequency shift is given by:

$$f_d = \frac{2 f_0 v_r \cos\theta}{c} \tag{1}$$

where c and v_r are the speed of ultrasound in the medium and the velocity of the reflector toward the transducer respectively, f_0 is the source frequency, and $c >> v_r$.

Continuous wave (CW) Doppler ultrasound is very sensitive to the detection of scattered ultrasound because it is a narrow band technique, i.e., the ultrasonic and electrical signals are confined to a very small range of frequencies. This permits the use of low-noise electronics and transducers that have a low insertion loss, i.e., little energy is lost during the conversion from electrical signal to mechanical vibration and back again.

X-ray computed tomography (CT) images are slices through the tissue object reconstructed from a large number of measurements of X-ray transmission through the patient. CT employs a number of beams of X-rays incident upon the patient and an array of detectors positioned on the far side of the patient to measure profiles of the transmitted radiation. The X-ray source is rotated around the patient and the detector array is simultaneously moved physically or electronically. The received intensity depends on the line integral of the X-ray attenuation of the patient tissues traversed by the beam. The individual beams are called rays. After an initial set of measurements has been made the source and detector are rotated by φ (angular interval) to a new position in the plane of the scan and a new projection is obtained. The number of angular intervals (K) is equal to the number of views or projections, and is determined by:

$$K = \frac{\pi}{\varphi}. \tag{2}$$

This assumes that the measurements are made for a total rotation of 180°. An image can be reconstructed by a process of backprojecting from each angle at which the source-detector system is aligned. It is the process of producing an image of a two-dimensional distribution from estimates of its line integrals along a finite number of lines with known locations. The quality of the image depends on the number of rays in a projection, the number of projections, and the deblurring function. [1]

Previous investigators have examined the use of the principles of CT for processing images made from ultrasonic transmission[2,3] and reflection data.[4] CW ultrasound is generally considered to be unable to discriminate range and distance between scattering targets. [5] However, when targets are in motion with respect to an ultrasonic beam, scattering targets return echoes that are correspondingly Doppler-shifted in frequency. The idea of using this aspect of the Doppler effect for spatial imaging was first proposed more than twenty years ago by Wade et al.[4]. Greenleaf and Ylitalo[6] used a tomographic method to detect the existence of flowing fluid in a phantom; however, they did not rotate the phantom to cause the Doppler effect. Ultrasonic imaging using the Doppler effect caused by a linearly moving transducer was carried out by Nagai and Greenleaf. [7] The production of ultrasonic images by artificially moving the object while keeping the transducer static does not seem previously to have been realised.

In this paper, the theory, construction and performance of a continuous wave Doppler ultrasonic microscanner (DUMS) are described. A backprojection algorithm, similar to those used in X-ray CT, was applied to process the Doppler shift frequencies from rotating objects containing scattering targets as a preliminary to investigating tissue specimens.

Methods

An ultrasound transducer was used to insonify a rotating object at a single frequency. Another similar transducer in the same transducer housing was used to detect the field scattered from the object. The set-up is illustrated in Fig. 1. As the object was rotated through 180°, the amplitude of the scattered field at each aspect angle φ was recorded. The angular speed of the rotation of the sample was ω radian/s.

As an object rotates with respect to a stationary ultrasonic transducer, the scattering centres within the object return echoes that are Doppler-shifted in frequency by amounts depending on the velocities of the individual scatterers. The scattering centres lying on a line of constant cross-range all have the same effective velocity in the direction pointing toward the transducer, so the amplitude of any particular frequency value is the line integral of the scattered radiation at the cross-range corresponding to that frequency (Fig. 1). The amplitudes of the returned signals at other frequencies give the line integrals for the scatterers at the corresponding cross-ranges. The amplitude as a function of frequency can be interpreted as a tomographic projection. A continuum of the projections at different angular positions is generated while the object is rotating. A CT reconstruction algorithm can produce an image of the distribution of scatterers in the insonified object from these projections.

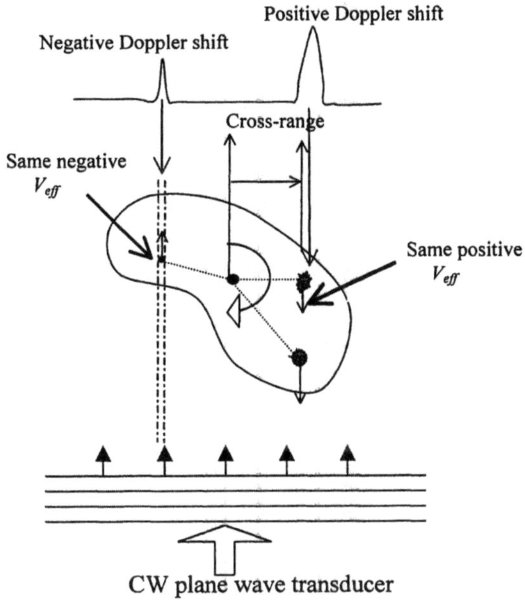

Fig. 1. Geometry of Doppler tomographic imaging.

Determination of Projections and Rays

To determine the rays and projections corresponding to the Doppler measurements, consider an object that can be enclosed in a circle of radius R divided into N cross-ranges in each of which the Doppler shift is to be measured. As the scatterers move through a cross-range strip of width $2R/N$, they radiate the Doppler shifted frequency for that cross-range. In order for the scatterers to remain in the same cross-range strip during the measurement interval Δt, the shortest allowable crossing time must be greater than the measuring time. Thus,

$$\Delta t < (2R/N)/(\omega R) = 2/(N\omega). \tag{3}$$

The shortest allowable time is inversely proportional to the number of cross-ranges (i.e., rays) and to the angular velocity. Because one side of the object gives positive Doppler shifts while the other side gives negative shifts, the overall span of the Doppler shift frequencies is $4\omega R/\lambda$. Dividing this span into N values gives the required discrimination in frequency:

$$\Delta f = 4\omega R/(N\lambda). \tag{4}$$

To measure to this precision in the interval Δt, we require

$$\Delta t > 1/\Delta f = N\lambda/(4\omega R). \tag{5}$$

The upper limit of (3) can be combined with the lower limit of (5) to obtain the possible number of cross-ranges:

$$N < (8R/\lambda)^{1/2} \tag{6}$$

and the spatial resolution, δ (width of each cross-range slice), is given by:

$$\delta = 2R/N > \sqrt{R\lambda/2}\,. \tag{7}$$

The spatial resolution δ has a numerical value greater than $\sqrt{R\lambda/2}$. It is determined only by the ultrasonic wavelength and the object size. If we assume that the object to be scanned is 10 mm in diameter, let λ=0.153 mm (for the 9.79 MHz ultrasound that we used in water, in which the speed is 1500 m/s), then: $N < (8\times 5.0/0.153)^{1/2} = 16.2$.

Thus, the maximum number of rays is about 16. The resolution is calculated to be 0.63 mm, i.e., 10.0/16 mm.

Next, we shall determine the number of projections K. From (2) and (3):

$$\Delta t = \frac{T}{K} < \frac{2}{N\omega} = \frac{2}{N\frac{\pi}{T}} = \frac{2T}{N\pi} \tag{8}$$

and

$$K > N\pi/2\,. \tag{9}$$

From (2) and (5)

$$\Delta t = \frac{T}{K} > \frac{1}{\Delta f} = \frac{N\lambda}{4\omega R} = \frac{N\lambda}{4\frac{\pi}{T}R} = \frac{N\lambda T}{4\pi R} \tag{10}$$

and

$$K < \frac{4\pi R}{N\lambda}\,. \tag{11}$$

From (9) and (11), the number of projections is about 25.

The angular interval: $\varphi = \pi/K = 7.2°$.

If sampling rate is 100 MHz, and the sample size is 1 Mbyte, the total sampling time is:

$$T = \frac{KM}{f_s} = \frac{25\times 1048576}{100\times 10^6} = 0.26 \text{ s.} \tag{12}$$

The optimum rotation speed is: $f_r = 1/(2T) = 1.92$ Hz = 115.4 rpm.

There is a factor of 2 here, as the rotation during T is half a revolution. It needs to be converted to a full revolution. The 115.4 rpm rotation speed is easily achievable in practice with a 10 mm diameter object.

Finally, to ensure that all scatterers in a cross-range band add together in the receiver over the maximum duration of the allotted measurement interval Δt, the travel time of all reflections in that band must be much less than this interval (say, 1%). Thus:

$$\Delta t > 100(2R/C) \quad (13)$$

and

$$T > 100K(2R/C). \quad (14)$$

If we let $K = 25$, $R = 5$ mm and $c = 1500$ m/s, then $T > 0.017$ s. From the above analysis, $T = 0.26$ s, so (14) is satisfied. If we increase the number of samples M, e.g., from 1 M to 2 M, the frequency resolution is better, from about 100 Hz to 50 Hz, but the rotation speed must be reduced, resulting in a lower Doppler shift. If we keep the same rotation speed, the angular interval is greater, and the frequency shift is obtained from a 14.4° angular interval. This blurs the image. If we keep on increasing the rotation speed so that the rotation speed is 1 revolution per 10 ms, then the frequency shift is from the whole revolution and K=1, which is not suitable for image reconstruction. This means that there is a trade-off between the numbers of the projections and of the rays.

Image Reconstruction

The ultrasonic signal from each projection was digitised and Fourier-transformed to find the frequency distribution. The Doppler frequency shift was extracted by finding the frequency spectrum difference of the received signal and the transmitted signal. Then the backprojection method was used to find the spatial positions of the scattering centres in the object.[1] The backprojection algorithm for reconstruction estimated the scatterer density at each spatial point by summing all the rays through that point. We just used the backprojection without filtering to demonstrate the use of the CT method to reconstruct an image from continuous wave ultrasound.

System Design

The rotation of the object containing the scattering targets was controlled by a servomotor. The motor controller was type DMC-1040 (Galil Motion Control Inc., Mountain View, California, USA), which was a 4-axis PC-based servo-control system, of which only 1 axis was used for rotation; the other 3 axes were used to align the transducer at the beginning of the scan.

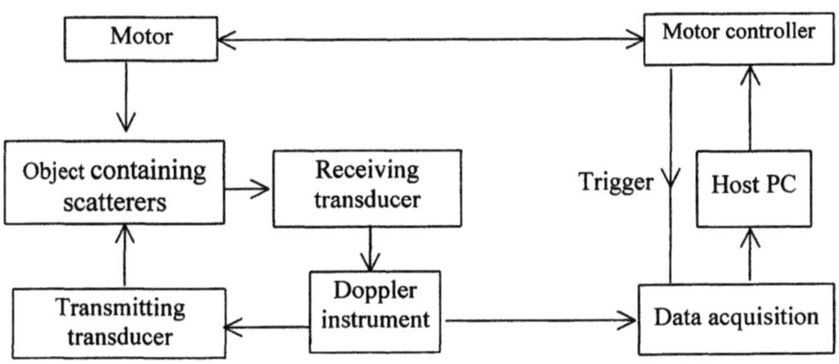

Fig. 2. Schematic of the DUMS hardware.

The host was a PC (Fig. 2). The Doppler instrument (made in our laboratory) sent an electrical sinusoidal waveform to drive the transmitting transducer. A separate transducer received the ultrasound signal which was then amplified by the Doppler instrument. The RF signal was bandpass-filtered and then digitised by an 8-bit CS8500 data acquisition board

(Gage Applied Sciences Inc., Montreal, Quebec, Canada). The digitisation rate for the ultrasound signal was 100 MHz.

As the ultrasonic signal frequency was 9.79 MHz, the Doppler shift frequency was always less than 21 kHz if the target was less than 10 mm away from centre of rotation and the rotating speed was less than 1500 rpm. Consequently, signals other than those in the range 9.76-9.82 MHz were superfluous and were filtered out. A bandpass filter having Q=13 was used to remove these signals.

While the object was rotating, the data acquisition was triggered as it reached a fixed angular position and the Doppler shift signal was sampled. This procedure was repeated until the desired number of samples (projections) had been acquired. Each set of signals was then Fourier-transformed to find its frequency spectrum, and the resulting set of spectra was used as the line projection data to reconstruct the image. The final image displayed the scattering centres of the targets.

Phantoms

Stainless steel needles were used as the material for the phantom, because of their small diameters and strong angle-independent scattering characteristics.

A single stainless steel needle phantom with a separation of 4.5 mm from the needle to the centre of rotation was constructed. The diameter of the needle was 0.81 mm. The phantom shown in Fig. 3 was also constructed. It comprised four parallel stainless steel needles. Two were 0.81 mm in diameter and the other two were 0.65 mm in diameter. All were 20 mm in length. The separation between the needles was 5.8 mm and 6.0 mm respectively.

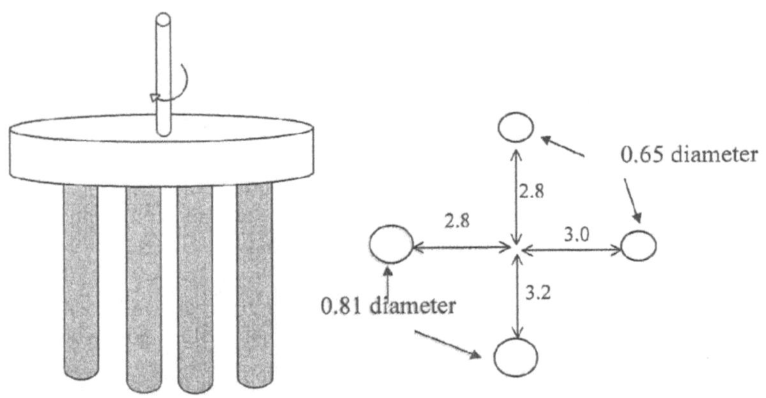

Fig. 3. Diagram of phantom

Results

Single Target Imaging

The single needle phantom was rotated at 115 rpm. Measurements were taken every 7.2°. The amplitude spectra are shown in Fig. 4. The Doppler frequency shift was proportional to the target cross-range. The amplitude was not constant even though the target was a uniform cylinder, due to the shape of the beam profile.

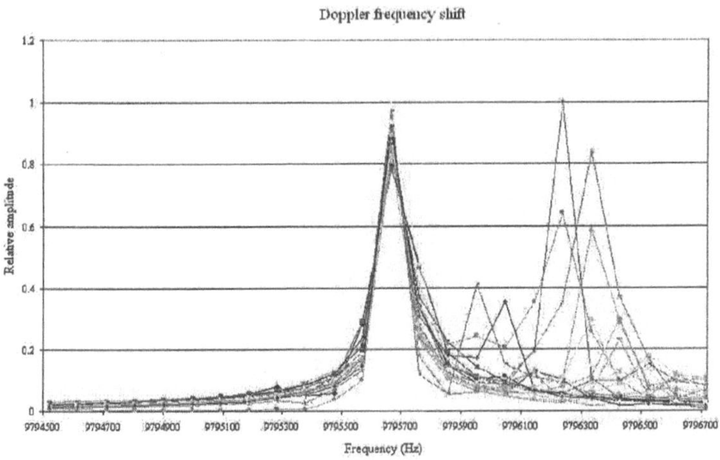

Fig. 4. Frequency shift from needle which was 4.5 mm away from centre of rotation, measured at 26 angular intervals of 7.2° in the range from 0° to 180°; rotation speed was 115 rpm.

Images were reconstructed from the phantom with two different rotation speeds, 115 rpm and 300 rpm respectively (Fig. 5). Measurements were taken every 7.2°. The image display grey scale was linear from 0 to 255. When the image was processed so that only the pixels with amplitudes greater than half of the maximum scale were displayed, it was found the radial resolution was better at 300 rpm (0.7 mm) than that at 115 rpm (1.2 mm).

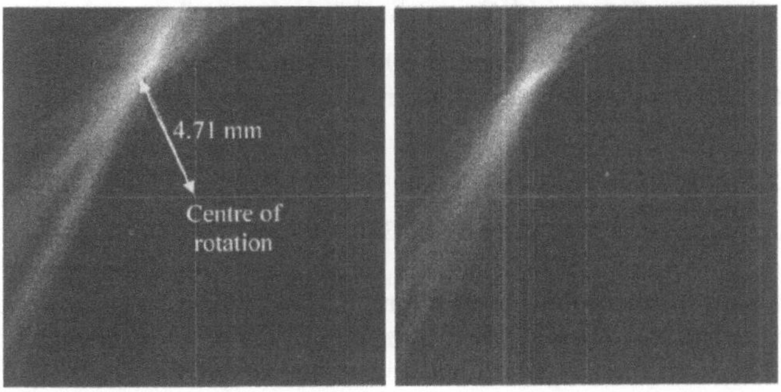

Fig. 5. Left panel is an image reconstructed from rotating needle 4.5 mm away from the centre of rotation; rotation speed was 115 rpm. Right panel is same target but at 300 rpm.

When the rotation speed was faster, the object travelled through a larger angular range as the sample time was fixed. There was a wider spread of frequency content, i.e., the circumferential resolution was worse (Fig. 6). But the radial resolution was better.

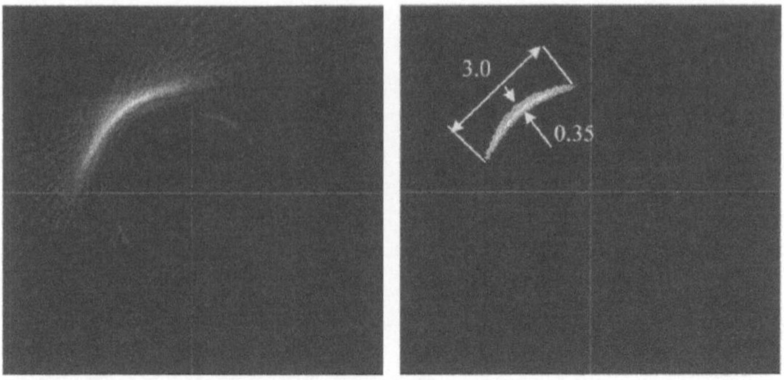

Fig. 6. Left panel is an image reconstructed from rotating needle 4.5 mm away from centre of rotation; rotation speed was 1000 rpm. Right panel only displays pixels with amplitudes greater than half the maximum grey scale.

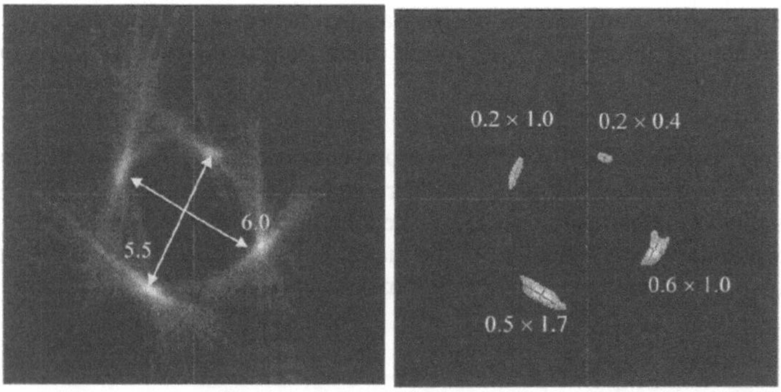

Fig. 7. Left panel is an image of 4 targets scanned at 300 rpm and every 5°. Right panel only displays pixels with amplitudes greater than half the maximum grey scale.

Multiple Target Imaging

The phantom with multiple targets was scanned at 300 rpm and measurements were taken every 5°. The result is shown in Fig. 7. The needles can be clearly identified from the image. There is about 10% error in determining the separation of the needles. The displayed diameters of the needles were less than their actual diameters, i.e., the physical diameter of the needle was 0.81 mm but, from the full width half maximum image, the width of the needle appeared to be about 0.6 mm.

Small Plant Branch and Porcine Intervertebral Disc Imaging

A small botanical plant branch was scanned at 300 rpm and measurements were taken every 2°. The result is shown in Fig. 8. From the image, the ring structure can be seen. But the ring structure might not be the real structure of the plant branch. It might be an artifact due to moving through the resolution cell, as can be seen with single target phantom. If the rings are not concentric to the centre of rotation, it is sure that they are the real structures. The brightest point in the bottom left part of the image is from the needle which attached the branch to the rotation shaft. There is a dark hole in the centre of rotation, as there is no Doppler shift from there.

The anatomy of porcine intervertebral disc indicates that it has a nucleus in the centre and is surrounded by annular structures of fibres. This can be seen in Fig. 8. The disc was scanned at 300 rpm and measurements were taken every 2°. The image is brighter in the bottom left part because the ultrasonic probe was not positioned to point to the centre of the disc. The other part of the disc was not insonified with strong ultrasonic field intensity.

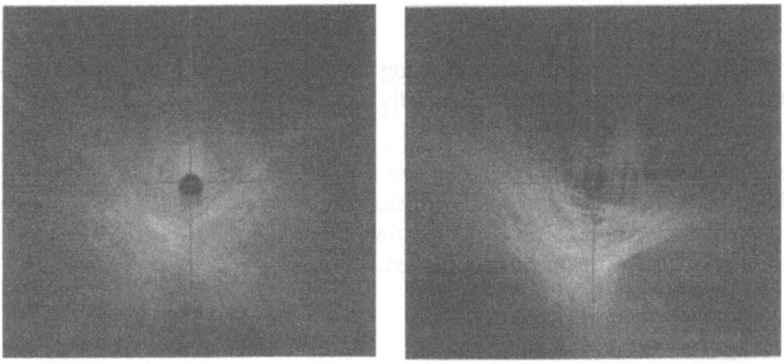

Fig. 8. Left panel is the image of small plant branch. Right panel is the image of porcine intervertebral disc.

Discussion

The resolution of the microscanner is determined by the compromise between the frequency resolution and the rotation speed. Although increasing the rotation speed increases the Doppler frequency shift, the information gathered is in a wider angle and this decreases the circumferential resolution.

Only the backprojection method was used to reconstruct the image, mainly because we wanted simply to demonstrate the feasibility using the CT method to reconstruct an image from Doppler ultrasound. Further research is needed to include the differentiation and Hilbert transformation and to find the optimum deblurring methods.

Targets situated on the axis around which the transducer rotates do not give rise to Doppler signals and therefore are undetectable. In principle, this limitation could be overcome by mounting the object eccentrically with respect to the axis of rotation, with an alignment rod and suitable bearing to maintain the beam direction so that it always enveloped the object.

An ultrasonic probe could be rotated around the object while the object was stationary. The Doppler shift would be caused by the relative movement between the probe and the object. A scatterer image could be reconstructed in a similar way. This scanning mode might be of use for scanning large organs or soft tissue specimens.

The object could also be rotated in one plane, tilted and rotated again in another plane. This could be repeated until all the scatterers had been detected and filled in the 3D space to form a 3D image. This scanning mode would require the ultrasonic beam profile to be of constant amplitude in all directions.

If the object was rotated at a variety of speeds, the scatterers would experience different centrifugal forces and consequential displacements within the object. As the centrifugal acceleration is related to the rotation speed and the distance between the scatterers could be found from the reconstructed image, the stress could be derived if the density of the object was known. A strain image or a Young's modulus image could thus be constructed to map the elasticity of the object.

From Fig. 4, it can be seen that the amplitude of the frequency spectrum depends on the intensity profile of beam. This indicates that DUMS could also be used to determine the ultrasonic beam profile: this might be the basis of a quality assurance procedure.

Conclusion

The method of image reconstruction using the backprojection technique employed to create X-ray CT scans has been successfully applied to reconstruct images of targets rotating in a continuous wave ultrasonic beam.

The resolution is limited by the target size and the ultrasound wavelength. Theoretical analysis shows that the resolution of the microscanner is determined by the compromise between the frequency resolution and the rotation speed. At a carrier frequency of 10 MHz, a resolution of about 0.3 mm has been achieved.

References

1. G T Herman, "Image reconstruction from projections" New York: Academic Press, 1980.
2. J. F. Greenleaf, S.A. Johnson, S. L Lee, G. T. Herman and E. H. Wood, "Algebraic reconstruction of spatial distribution of acoustic absorption within tissue from their two dimensional acoustic projections," in Acoustical Holography, Vol. 5, P. S. Green Ed. New York: Plenum Press, 1974, pp. 591-603.
3. J. F. Greenleaf, S. A. Johnson, W. F. Samayoa and F. A. Duck, "Algebraic reconstruction of spatial distribution of acoustic velocities in tissue from their time of flight profiles," in Acoustical Holography, Vol. 6, N. Booth Ed. New York: Plenum Press, 1975, pp. 71-90.
4. G. Wade, S. Elliott, I. Khogeer, G. Flesher, J. Eisler, D. Mensa, N. S. Ramesh and G. Heidbreder, "Acoustic echo computer tomography," in Acoustical Imaging, Vol. 8, A. F. Metherell Ed. New York: Plenum Press, 1978, pp. 565-576.
5. P. N. T. Wells, "Basic principles and Doppler physics" in Clinical Applications of Doppler Ultrasound, K. J. W. Taylor, P. N. Burns, P. N. T. Wells Eds. New York: Raven Press, 1995, pp. 1-17.
6. J. F. Greenleaf and J. Ylitalo, "Doppler tomography" in IEEE Ultrasonics Symposium, 1986, pp. 837-841.
7. K. Nagai and J. F. Greenleaf, "Ultrasonic imaging using the Doppler effect caused by a moving transducer" Optical Engineering, 1990, Vol. 29, pp. 1249-1254.

MODELING OF THE ACOUSTIC SIGNAL BACKSCATTERED BY A BIPHASIC SUSPENSION : APPLICATION TO THE CHARACTERIZATION OF RED BLOOD CELL AGGREGATION

David Savéry and Guy Cloutier

Laboratory of Biomedical Engineering, Clinical Research Institute of Montreal, and Laboratory of Biorheology and Medical Ultrasonics, University of Montreal Hospital, Montreal, Québec, Canada

INTRODUCTION

The purpose of this study was to assess the effect of the microstructural characteristics of a medium composed of weak scattering particles on its ultrasonic backscattering properties. The relationship between the geometrical features of the scattering medium (size and orientation of particles) and the power of the backscattered signal was specifically studied.

Blood is a typical case of such medium and its echogenicity depends on the red blood cell (RBC) aggregation level. Determining the aggregation state of RBCs in blood is of clinical relevance[1] for the diagnosis of several cardiovascular pathologies such as hypertension. The pathophysiological consequences of RBC hyperaggregation can be explained by the hemodynamical perturbations that RBC clusters cause on the blood flow. The variability of the scattering properties of blood with the RBC aggregation level could lay the basis of an acoustical characterization of blood rheology.

In particular, experimental measurements of the backscattering coefficient (*BSC*) in Rayleigh conditions and at a low particle concentration have shown that the *BSC* increases linearly with the effective scatterer volume weighted by the volumic fraction[2]. Experiments also revealed that, for frequencies above 30 MHz, the frequency dependence is reduced in the presence of RBC aggregation[3]. These results suggest that the ultrasonic study of RBC aggregation is possible.

To theoretically clarify the relationship existing between the RBC aggregation level and the echogenicity of blood, an analytical model is proposed to simulate the signal backscattered by a dilute suspension of particles. RBC aggregates are modeled as Gaussian shaped inhomogeneities in the bulk acoustic impedance. The effect of the particle growth on the backscattered wave is particularly studied to explain experimental observations concerning the behavior of the *BSC* at high frequencies.

Acoustical Imaging, Volume 25, Edited by Halliwell and Wells.
Kluwer Academic/Plenum Publishers, 2000.

METHODS

When an acoustic wave encounters spatial fluctuations in density and compressibility, a scattered wave results from the radiation in all directions of the incident wave. For a pulse-echo measurement, the backscattered component is measured by the emitting transducer. In blood, RBCs create impedance fluctuations, and consequently their spatial arrangement and shape affect the macroscopic acoustical properties of blood.

Let us consider a blood volume insonified by a plane pressure wave of amplitude *P* and wave vector *k*. The Born hypothesis is valid because of the low contrast between red blood cells and plasma acoustical properties. The amplitude of the backscattered wave measured in *r*, located far from the blood volume and in the opposite direction of the wave propagation (as seen when the scattering volume is in the far field of the transducer), is given by[4,5]:

$$p_{backscattered}(\vec{r}) = P\frac{e^{i\vec{k}.\vec{r}}}{\|\vec{r}\|}\frac{\vec{k}^2}{4\pi}\int \gamma_z(\vec{r_0})e^{2i\vec{k}.\vec{r_0}}d^3\vec{r_0} \quad (1)$$

with:

$$\gamma_z(\vec{r_0}) = 2\frac{z(\vec{r_0}) - z_0}{z_0} \quad (2)$$

This last quantity describes the local relative deviation in r_0 of the acoustical impedance $z(r_0)$ from the spatial mean impedance z_0.

This formulation allows to calculate the backscattering coefficient (*BSC*) defined as the acoustic intensity backscattered by a unit volume *V* insonified by an incident plane wave with unit intensity and received per steradian:

$$BSC = \frac{k^4}{16\pi^2}\frac{1}{V}\left|\int_V \gamma_z(\vec{r_0})e^{2i\vec{k}.\vec{r_0}}d^3\vec{r_0}\right|^2 \quad (3)$$

If we assume that all spatial inhomogeneities have the same shape, such as individual red blood cells or their aggregates, one can then dissociate the random process of the *M* particle center positions $\{r_i\}_{i=1...M}$ from their common shape $C(r)$ and write[6,7]:

$$\gamma_z(\vec{r}) = C(\vec{r}) * \sum_{i=1}^{M}\delta(\vec{r} - \vec{r_i}) \quad (4)$$

In this expression, $C(r)$ represents the perturbation in acoustic impedance due to a single scatterer centered at the origin, and δ is the Dirac delta function. Using Eq. 3, *BSC* can be computed in a factorized form involving two new terms: the form factor $F(k)$, which describes the effect of the particle shape on the backscattered intensity, and the structure factor $W(k)$ that takes into account interference between echoes scattered by different clusters.

$$BSC = \frac{k^4}{16\pi^2}HV_s\, F(\vec{k})\, W(\vec{k}) \quad (5)$$

In this equation, H and V_s represent respectively the volumic fraction of the scatterers (hematocrit) and the volume of a single scatterer. The following two equations define the form factor and the structure factor:

$$F(\vec{k}) = \frac{1}{V_s^2}\left|\int_{scatterer} C(\vec{r})e^{2i\vec{k}.\vec{r}}d^3\vec{r}\right|^2 \quad (6)$$

$$W(\vec{k}) = \frac{1}{V}\frac{V_s}{H} < \sum_{i=1}^{M}\sum_{j=1}^{M} e^{2i\vec{k}.(\vec{x_i}-\vec{x_j})} > \tag{7}$$

In fact, the form and structure factors are not strictly independent. In very dense media for instance, the positions $\{r_i\}$ are regularly spaced out by intervals that depend on the scatterer morphology as in a crystal. Thus, the shape of the scatterer will affect the structure factor $W(k)$. In the current study, H is supposed small enough to consider $W(k) \approx 1$ *i.e.* the medium is assumed *diluted*. This entails that *BSC* will increase linearly with H in the case of non-interacting particles.

Very small particles compared to the wavelength (their maximal dimension a is such that $ka < \pi/10$) could be considered as Rayleigh scatterers. In this case, $F(k) \approx \gamma_0^2$, where γ_0 is the average relative impedance inhomogeneity inside the scatterer (as defined in Eq. 2). *BSC* is then proportional to $k^4 V_s$. But for higher frequencies (non-Rayleigh scattering), the exact form and orientation of particles can also alter the form factor.

Consequently, to assess the influence of the non-punctual shape of scatterers on *BSC*, one assumes that they have a Gaussian tridimensional shape with an axial symmetry. Therefore, the form factor will depend on a set of three geometric parameters: its inertia axis σ_1, σ_2 and the angle θ between the major axis and the wave vector (Fig. 1). These geometrical features were chosen because of the different possible configurations that RBC aggregates can have. Alternatively, RBC clusters can either be long and lined up along stream lines, or they can form compact clumps in the case of pathological enhanced aggregation. This set of geometrical parameters can simulate these different types of cluster shape. The Gaussian assumptions finally give the following expression for the prototype function $C(r)$, where the propagation axis is Oz (with unit vector e_z):

$$C(\vec{r}) = \gamma_0 \ \exp(-\frac{1}{2}\left\|\mathbf{S}^{-1}\mathbf{R}_{-\theta}\vec{r}\ \right\|^2) \tag{8}$$

This equation involves γ_0, the average relative impedance inhomogeneity inside the scatterer, the inertia matrix $\mathbf{S}$ and the rotation matrix $\mathbf{R}_{-\theta}$ around Ox defined by:

$$\mathbf{S} = \begin{pmatrix} \sigma_1 & 0 & 0 \\ 0 & \sigma_1 & 0 \\ 0 & 0 & \sigma_2 \end{pmatrix} \quad \mathbf{R}_{-\theta} = \begin{pmatrix} 1 & 0 & 0 \\ 0 & \cos\theta & \sin\theta \\ 0 & -\sin\theta & \cos\theta \end{pmatrix} \qquad (9)\ (10)$$

Applying basic formulas for Fourier transforms to Eq. 8, the form factor can be analytically derived from Eq. 6:

$$F(k\vec{e_z}) = \gamma_0^2 e^{-4(k\sigma_1 \sin\theta)^2 - 4(k\sigma_2\cos\theta)^2} \tag{11}$$

Inserting Eq. 11 into Eq. 5 and reminding that the dilution of the medium allows to write $W(k) \approx 1$, the backscattering coefficient can be found. We assumed that V_s was $(2\pi)^{3/2}\sigma_1^2\sigma_2$ (the integral of the 3D unit Gaussian) and introduced the anisotropy index $\varepsilon = \sigma_2/\sigma_1$ to obtain:

$$BSC = \frac{1}{2^{5/2}\pi^{1/2}}\varepsilon k^4 \sigma_1^3 \gamma_0^2 H e^{-4k^2\sigma_1^2(1+(\varepsilon^2-1)\cos^2\theta)} \tag{12}$$

In Rayleigh conditions, when $k\sigma_1 \rightarrow 0$, *BSC* increases linearly with $\varepsilon\sigma_1^3 k^4$, which means that *BSC* varies linearly with the volume of scatterers and with the fourth power of the ultrasound frequency. In non-Rayleigh scattering, for a fixed shape of scatterer, the spectral slope defined as the slope of the log-log curve *BSC* vs. frequency can be extracted and is given by the quadratic relation:

$$slope = \frac{\partial \log BSC}{\partial \log k} = 4 - 8k^2\sigma_1^2(1 + (\varepsilon^2 - 1)\cos^2\theta) \tag{13}$$

The assessment of this slope is of particular interest for tissue characterization. For a fixed frequency, this slope decreases with an increasing size of scatterers. The spectral slope approaches 4 for Rayleigh conditions.

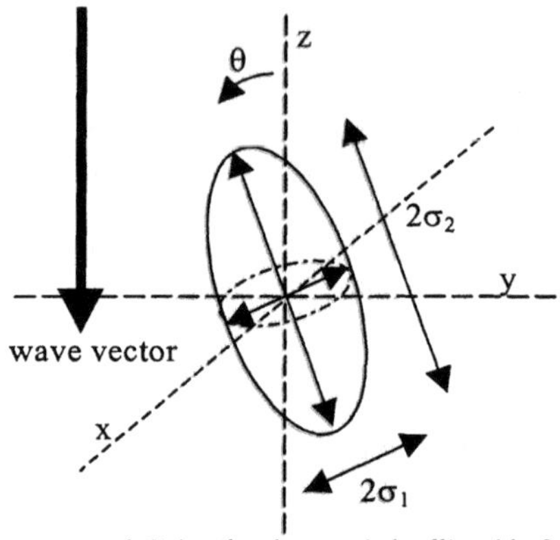

Figure 1. Geometrical parameters defining the characteristic ellipsoid of the Gaussian scatterer.

RESULTS

Fig. 2 illustrates the relationship between the backscattering coefficient and the frequency of the acoustic wave for different sizes of isotropic scatterers ($\varepsilon = 1$). Results were obtained with the acoustic properties ($\gamma_0 = 0.26$ and sound celerity $c = 1540$ m/s) of a red blood cells/plasma suspension. Experimental data obtained at low frequencies and at two different hematocrits by Shung et al.[8] for human suspended red blood cells are also shown for comparison. Fig. 3 shows the effect of an increase in the isotropic scatterer' sizes on the power, at a fixed frequency. Fig. 4 shows the evolution of the spectral slope with the scatterers' size at various frequencies. Finally, Fig. 5 presents the influence of the anisotropy ε and of the orientation angle θ on the backscattered power.

DISCUSSION

The model intends to simulate the backscattering of a plane acoustic wave by a medium composed of discrete scatterers immersed in a dilute medium. The geometrical assumption of their Gaussian shape allowed to determine analytically the effect of several microstructural parameters such as the particles' size, elongation, and orientation on the backscattering coefficient. This model gives relevant results that may allow to solve the inverse problem of inferring quantitative geometrical features of the backscattering medium from acoustical data.

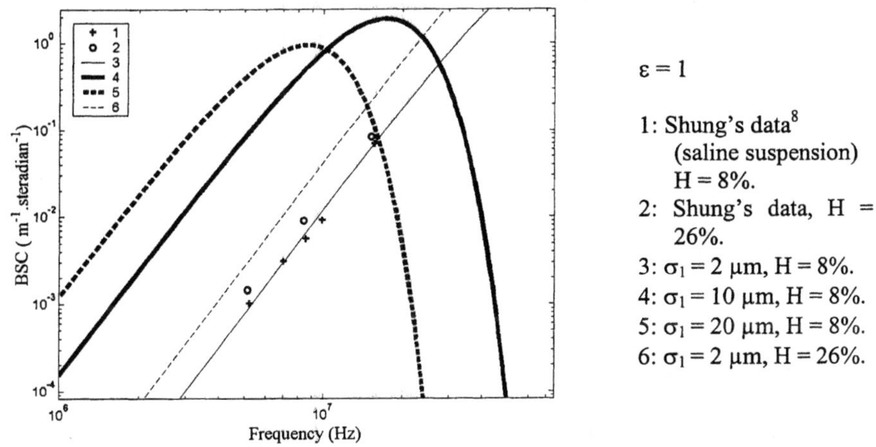

Figure 2. Backscattering coefficient as a function of frequency for isotropic scatterers.

In the hypothesis of a low hematocrit, *BSC* increases proportionally to *H*. This allows to fit data obtained by Shung et al.[8] for $H \approx 8\%$, and to estimate for a dilute non aggregating and isotropic suspension of red blood cells, the characteristic size $\sigma_1 \approx 2$ μm, agreeing with the average thickness of an individual erythrocyte. However, this linear assumption is no longer valid for higher hematocrits as shown at 26% by the discrepancies between curves 2 and 6 of Fig. 2. Random correlations in positions of particles appear at this volumic fraction because of the previously mentioned crystallographic effect. The model cannot take into account the decrease in backscattered power observed[8] as hematocrit exceeds 13%.

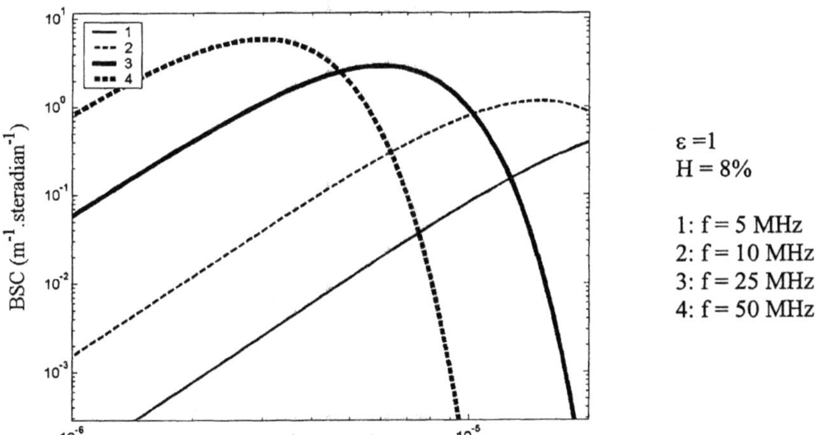

Figure 3. Backscattering coefficient as a function of the isotropic scatterer's size.

Figs. 2 and 3 show that non-aggregating erythrocytes ($\sigma_1 \approx 2$ μm) can be considered as Rayleigh scatterers for frequencies lower than 50 MHz. For Rayleigh scatterers, the power increases linearly with the particle volume (Fig. 3) and with the 4th power of frequency (Fig. 2). However, for bigger scatterers, *BSC* reaches a maximum and then decreases when

$\sigma_1 > 0.11\lambda$. This entails that the 4th power frequency dependence for Rayleigh scattering is no longer valid for large particles, such as large RBC aggregates. This decrease can be explained by the internal destructive interference occurring between echoes produced by different regions of the same big scatterer.

For a given particle size in Fig. 2, the slope of the log-log curve *BSC* vs. frequency decreases from 4 to 0, for frequencies ranging from 0 to the critical value $f_c = 0.11c/\sigma_1$. Fig. 4 shows the evolution of this slope at a fixed frequency when isotropic scatterers grow ($\varepsilon = 1$). It reveals that the growth of particle results in a decrease of the spectral slope. This diminished frequency dependence has already been experimentally observed[3] for aggregating red blood cells circulating at low shear rates. The assessment of this slope for Gaussian scatterers can directly give an effective scatterer size $\sigma_1 (1+(\varepsilon^2-1)\cos^2\theta)^{1/2}$ by use of Eq. 13.

The power strongly depends on the orientation θ for anisotropic particles (Fig. 5). This angular dependence of the backscattered power was previously observed by our group[9] in the presence of RBC aggregation. Fig. 5 reveals that the backscattered power is maximal when scatterers expose their longest section to the ultrasound beam. The peak-to-peak amplitude of the cosine-shaped curve *log BSC* vs. θ depends on ε and equals $8k^2\sigma_1^2|\varepsilon^2-1|$. Thus, for a fixed σ_1, deviation of ε from 1 increases the sensitivity of *BSC* to the orientation angle θ.

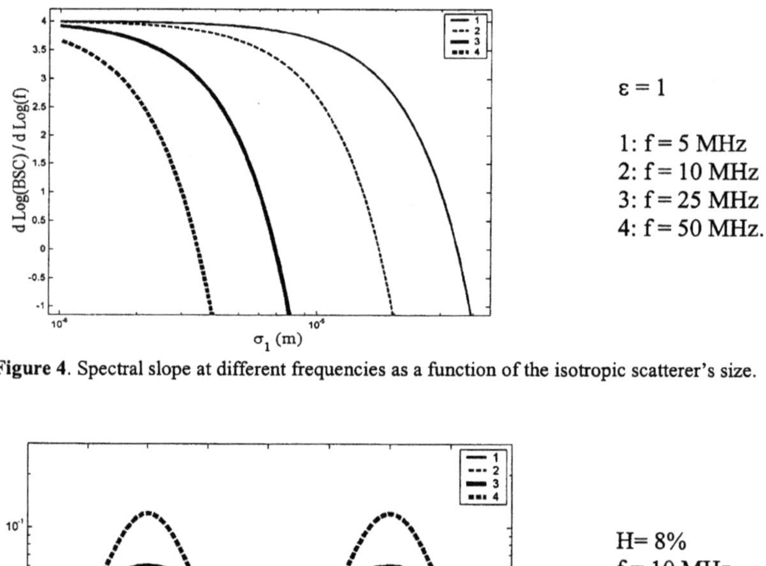

Figure 4. Spectral slope at different frequencies as a function of the isotropic scatterer's size.

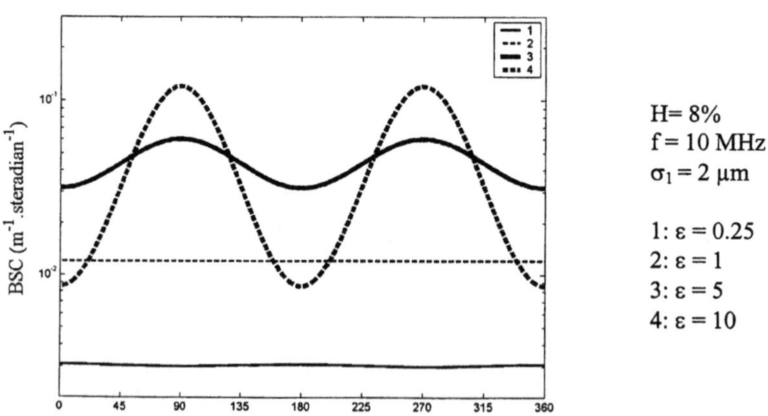

Figure 5. Backscattering coefficient as a function of the orientation angle θ.

CONCLUSION

The model explained experimental observations on the power backscattered by flowing blood, namely the increase in power at low frequencies and the decrease of the spectral slope at higher frequencies both due to the growing size of scatterers. The angular dependence of *BSC* for anisotropic particles could also be reproduced by the simulations. However, some improvements would be necessary to get more insight into the understanding of ultrasound backscattering by blood. Firstly, the model is valid for diluted suspensions but cannot be easily generalized for a dense medium, where the packing of particles plays an important role. Another limitation is the hypothesis that all scatterers are identical and have Gaussian shapes. Polydispersity and variations in orientation are not currently taken into account. It remains yet that blood characterization by the backscattering method has a good potential to provide accurate information about the RBC aggregation level.

ACKNOWLEDGEMENTS

Scholarship from the Fonds de la Recherche en Santé du Québec, and grants from the Medical Research Council of Canada (# MOP-36467) and the Heart and Stroke Foundation of Québec. The first author was supported by a studentship from the Group of Biomedical Modeling of the Institute of Biomedical Engineering of the University of Montreal.

REFERENCES

1. F. Zannad, J.F. Stoltz, Blood rheology in arterial hypertension, *J Hypertens* 10:69 (1992).
2. M. Boynard, J.C. Lelièvre., and R. Guillet, Aggregation of red blood cells studied by ultrasound backscattering, *Biorheology* 24:451 (1987).
3. F.S. Foster, H. Obara, T. Bloomfield, L.K. Ryan, and G.R. Lockwood, Ultrasound backscatter from blood in the 30 to 70 MHz frequency range, *IEEE Ultrasonics Symposium Proceedings* 1599 (1994).
4. M.F. Insana and D.G. Brown, Acoustic scattering theory applied to soft biological tissues, in: *Ultrasonic Scattering in Biological Tissues*, K.K. Shung., G.A. Thieme, CRC Press, Boca Raton, Ann Arbor, London, Tokyo (1993).
5. B.A.J. Angelsen, A theoretical study of the scattering of ultrasound from blood, *IEEE Trans Biomed Eng* 27:61 (1980).
6. J. Meunier and M. Bertrand, Echographic image mean gray level changes with tissue dynamics: A system-based model study, *IEEE Trans Biomed Eng* 42:403 (1995).
7. I. Fontaine M. Bertrand, and G. Cloutier, A system-based approach to modeling the ultrasound signal backscattered by red blood cells, *Biophysical J* 77:2387 (1999).
8. K.K. Shung, R.A. Sigelmann, and J.M. Reid, Scattering of ultrasound by blood, *IEEE Trans Biomed Eng* 23:460 (1976).
9. L. Allard, G. Cloutier, and L.G. Durand, Effect of the insonification angle on the Doppler backscattered power under red blood cell aggregation conditions, *IEEE Trans Ultrason Ferroelec Freq Cont* 43:211 (1996).

FREQUENCY DEPENDENCE OF SIMULATED ULTRASOUND SIGNALS BACKSCATTERED BY AGGREGATING RED BLOOD CELLS

Isabelle Fontaine and Guy Cloutier

Laboratory of Biomedical Engineering, Clinical Research Institute of Montreal, and Laboratory of Biorheology and Medical Ultrasonics, University of Montreal Hospital, Québec, Canada.

ABSTRACT

The frequency dependence of the ultrasound signal backscattered by blood was studied with a simulation model. The ultrasound backscattered signal was simulated with a linear model that considers the characteristics of the ultrasound system and the tissue scattering properties. The tissue function was described by the position and the shape of the red blood cells (RBCs). The effect of the flow on the spatial organization of aggregating RBCs was simulated with a 2D model. It is an iterative model that considers the effect of the flow and the adhesive and repulsive forces acting on each RBC. RBC aggregation was simulated at 40 % hematocrit and shear rates of 0.05 s^{-1}, 0.08 s^{-1}, 0.1 s^{-1}, 0.5 s^{-1}, and 5 s^{-1}. Ultrasound frequencies between 5 and 500 MHz were studied. Results suggest that in the presence of RBC aggregation, the frequency dependence may vary depending on the shear rate and the range of frequencies studied. The enhancement of the backscattered power with RBC aggregation may be more important at frequencies below 30 MHz. At higher frequencies, changes in the frequency dependence of the ultrasound backscattered power were noted in the presence of RBC aggregation. It is concluded that further studies may be necessary to understand the relationship between the ultrasound backscattered power and the spatial organization of aggregating RBCs.

INTRODUCTION

Rayleigh scattering occurs when the acoustical inhomogeneities encountered by the ultrasonic wave are much smaller than its wavelength. In such conditions, it is well recognized that the ultrasound backscattered power is proportional to the fourth power of the incident wave frequency. For non aggregating red blood cells (RBCs), the fourth power frequency dependence was observed experimentally at least up to 30 MHz[1]. Several groups observed that the frequency dependence of the backscattered power decreases in the

Acoustical Imaging, Volume 25, Edited by Halliwell and Wells.
Kluwer Academic/Plenum Publishers, 2000.

presence of RBC aggregation[1-3]. Yuan and Shung[1] measured a fourth power frequency dependence of the backscattering coefficient between 3.5 and 12.5 MHz at shear rates of 10 s^{-1} and 22 s^{-1}. They observed a smaller frequency dependence at 2 s^{-1} for the same frequencies. In the absence of significant aggregation, at 32 s^{-1}, Foster et al.[2] measured a 3.5 frequency dependence between 40 and 60 MHz. At a lower shear rate of 0.16 s^{-1}, which results in a higher level of aggregation, they found a much smaller slope of 0.4. Van der Heiden et al.[3] performed several measurements at 30 MHz for shear rates of 0 to 1000 s^{-1}. At low shear rates, the spectral slope of the power backscattered by whole blood was 1 compared to 3.3 for non aggregating RBCs.

Recently, a system-based model was proposed by our group to simulate the ultrasound signal backscattered by blood[4]. The backscattered power was modeled for the case of non-aggregating[4] and aggregating[5] RBCs. The backscattering coefficient was studied as a function of the hematocrit, frequency, and volume of the scatterers for non aggregating RBCs. The power backscattered by aggregating RBCs was studied at a fixed frequency of 7.5 MHz as a function of the hematocrit and shear rate. We now propose to study the effect of the incident wave frequency for the case of RBC aggregation.

METHODS

The simulation model assumes a linear wave propagation, and is valid for a small region of interest in the far field of the transducer. The backscattered radio-frequency *(RF)* signal is defined as the convolution of the ultrasound system characteristics (function *T*), the function characterizing the shape of the RBCs *(C)*, and the function representing their organization in space *(N)*:

$$RF(x,y)=\frac{\partial^2}{\partial y^2}T(x,y)\otimes C(x,y)\otimes N(x,y), \tag{1}$$

where y is the ultrasonic wave direction of propagation. The complete mathematical development leading to Eq. (1) can be found in Fontaine et al.[4]

A rheological simulation model was developed to predict the spatial organization of aggregating RBCs under different flow conditions. As detailed previously by Fontaine et al.[5], the model considers globally the effect of the flow, and the adhesive and repulsive forces between RBCs. Initially, the particles are positioned in a random configuration. They are then moved iteratively considering the forces acting on each RBC. This is repeated until the mean size of the aggregates reaches a stable state. The rheological model was used to simulate the spatial organization of the RBCs in a Couette flow. This type of flow is characterized by a constant shear rate in the gap between two parallel plates or cylinders. The relative displacement speed of the plates or the rotation speed of the cylinders determines the flow shear rate. The simulations were performed on a 2D surface area of 256 μm by 256 μm. In this study, the following shear rates were simulated: 0.05 s^{-1}, 0.08 s^{-1}, 0.1 s^{-1}, 0.5 s^{-1}, and 5 s^{-1}. All simulations were performed at 40% hematocrit.

The backscattered ultrasound RF signal was computed with Eq. (1). The positions of the RBCs at the steady state of aggregation were used to determine the function N. The blood characteristic function was obtained by convoluting this function N with the scatterer prototype function C. For all simulations, the scatterer prototype function was a sphere of 5.5 μm in diameter. The transducer transfer function T was modeled as a Gaussian envelope modulated by a cosine function. In this study, the standard deviations of the

Gaussian envelope were 0.05 mm in the axial direction y, and 0.005 mm in the lateral direction x. Ultrasound frequencies between 5 and 500 MHz were simulated. For all simulations, the direction of propagation of the ultrasonic wave was modeled perpendicular to the flow.

RESULTS

Examples of tissue samples obtained with the simulation model are illustrated in Fig. 1. Fig. 1a was obtained at a constant shear rate of $0.05\ s^{-1}$, whereas Fig. 1b corresponds to a constant shear rate of $5\ s^{-1}$. At $0.05\ s^{-1}$, the simulation results are characterized by a higher level of aggregation than at $5\ s^{-1}$. The mean spectra of the functions N characterizing the position of each RBC in these conditions are illustrated in Fig. 2. The spectrum at $0.05\ s^{-1}$ shows regular oscillations in both directions. At $5\ s^{-1}$, the flow introduces more randomness in the RBC positions, and the spectrum becomes similar to a white noise.

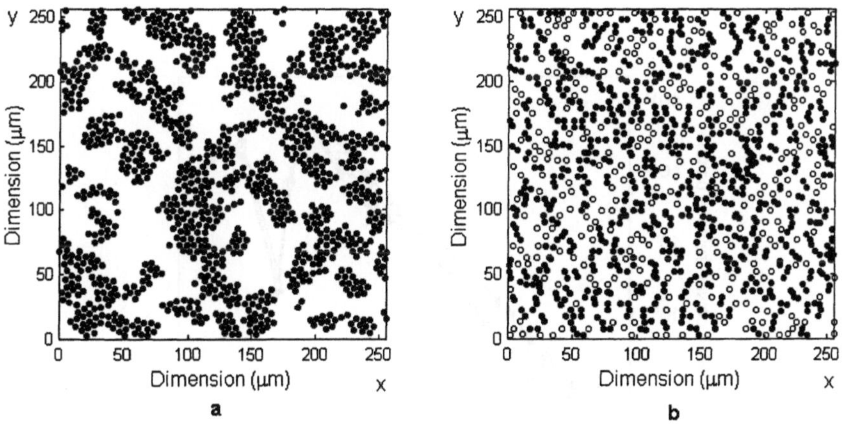

Figure 1. Simulation results obtained in Couette flow by using the rheological model at 40% hematocrit. A) corresponds to a shear rate of $0.05\ s^{-1}$ and b) to a shear rate of $5\ s^{-1}$. Filled circles represent aggregated RBCs.

The backscattered power at different frequencies is shown in Fig. 3 for shear rates of $0.05\ s^{-1}$, $0.1\ s^{-1}$ and $5\ s^{-1}$. The frequency dependence of the backscattered power was quantified by the slope of the log-log plots. For a minimal amount of RBC aggregation, at $5\ s^{-1}$, the backscattered power increased linearly (log-log scale) as a function of the frequency up to 120 MHz. In the presence of significant RBC aggregation, at shear rates of $0.05\ s^{-1}$ and $0.1\ s^{-1}$, the frequency dependence was not constant between 5 and 120 MHz. These plots show an inflexion point above 35 MHz. In the low frequency range (< 30 MHz), the power was higher for lower shear rates (higher level of aggregation). Above 50 MHz, the backscattered power was lower at $0.05\ s^{-1}$ than at $0.1\ s^{-1}$ and $5\ s^{-1}$. All 3 curves become superposed and oscillate at very high frequencies (above 120 MHz).

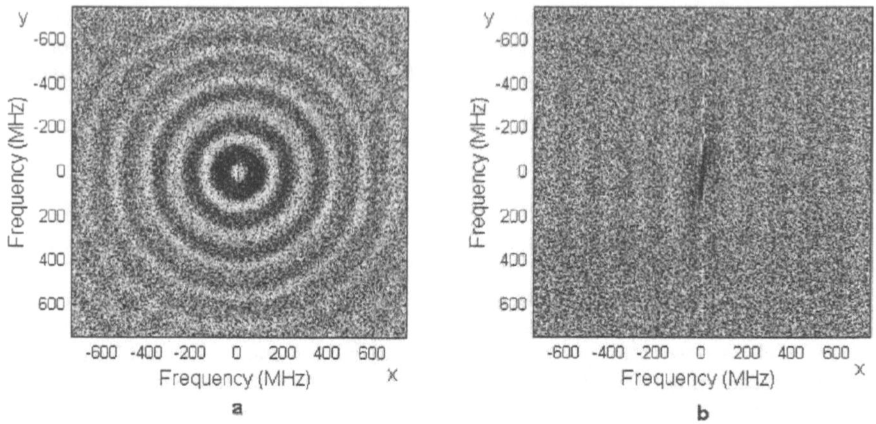

Figure 2. Mean spectra of the functions *N* characterizing the spatial organization of RBCs in Couette flow at 40% hematocrit. A) corresponds to a shear rate of 0.05 s^{-1} and b) to a shear rate of 5 s^{-1}. Each mean spectrum was averaged over four simulations performed under similar conditions.

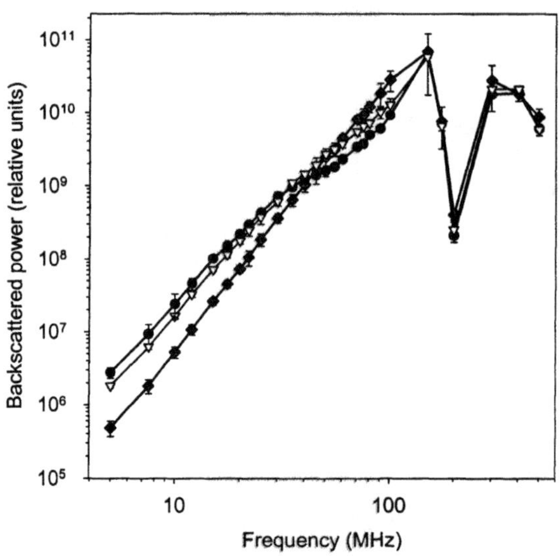

Figure 3. Backscattered power as a function of the frequency at 40% hematocrit. Circles, ●, correspond to 0.05 s^{-1}; triangles, ▼, to 0.1 s^{-1}; and diamonds, ◆, to 5 s^{-1}. Results are expressed in terms of mean ± one standard deviation (n = 4 simulations).

For all simulated shear rates, the slopes of the backscattered power versus frequency were evaluated between 5 and 30 MHz, and 35 and 55 MHz (Table 1). For those frequency ranges, the slopes vary between 1.34 (0.05 s^{-1}) and 3.74 (5 s^{-1}). A two-way analysis of variance was performed (Sigma Stat®, SPSS Science, Chicago, IL, version 2.03 for

Windows®) to confirm the differences observed in Table 1. The results demonstrate that the frequency dependence of the simulated backscattered power depends both on the shear rate ($p < 0.001$) and the frequency ($p < 0.001$). No significant variation in the frequency dependence as a function of the shear rate was noted in the low range of frequencies. At 5 s^{-1} and 0.5 s^{-1}, there was no significant difference between the slopes estimated for each range of frequencies. The slopes of the backscattered power evaluated between 35 and 55 MHz were significantly different from those estimated between 5 and 30 MHz for the simulations at 0.05 s^{-1} (1.34), 0.08 s^{-1} (1.90) and 0.1 s^{-1} (2.39) ($p < 0.001$). Each of these values was also significantly different ($p < 0.001$) compared to the slope estimated at 0.5 s^{-1} (3.49) and 5 s^{-1} (3.55) for the same range of frequencies (35 to 55 MHz).

Table 1. Frequency dependence of the backscattered power over the range of frequencies specified. Results are expressed in terms of mean ± one standard deviation (n = 4 simulations).

	0.05 s^{-1}	0.08 s^{-1}	0.1 s^{-1}	0.5 s^{-1}	5 s^{-1}
5 to 30 MHz	3.16 ± 0.17	3.24 ± 0.11	3.31 ± 0.12	3.50 ± 0.05	3.74 ± 0.11
35 to 55 MHz	**1.34 ± 0.38 ***	**1.90 ± 0.28 ***	**2.39 ± 0.46 ***	3.49 ± 0.41	3.55 ± 0.36

* $p < 0.001$

DISCUSSION

The effect of the flow on the spatial organization of the RBCs was studied by using a 2D model. Various configurations of the spatial organization of the RBCs were obtained as a function of the flow shear rate (Fig. 1). For all simulations, the function C (shape of the RBC) was constant. The second derivative of the function T (system) strictly follows the fourth power frequency dependence. Based on Eq. (1), changes in the backscattered power as a function of the shear rate and frequency can thus be associated to the variations of the spectra of N. This function contains information related to the regularity in the spatial organization of the RBCs. It can be observed that the spectral characteristics of this function are significantly modified by the level of aggregation (Fig. 2).

The computation of the backscattered power as a function of the frequency leads to several observations (Fig. 3). At a given frequency, between 5 and 30 MHz, the simulated backscattered power was higher at low shear rates. This can be explained by the level of aggregation, which is more important when the shear rate is decreased. The backscattered power was higher in the presence of large aggregates, which is in agreement with previous experimental results[2,3]. The inversion of this trend at frequencies higher than 30 MHz is surprising. Van Der Heiden et al.[3] and Foster et al.[2] measured higher backscatter at low shear rates for frequencies of 30 and 40 MHz, respectively. This contradiction may be due to the underestimation of the level of aggregation in our study because we simplified the computation to a 2D simulation model.

In the literature, it is often hypothesized that the frequency dependence of the power backscattered by aggregating RBCs deviates from the fourth power relationship characterizing Rayleigh scattering[1-3]. Our results suggest that the deviation from the fourth power dependence occurs in an intermediate frequency range, between 35 and 100 MHz. A significant decrease in the frequency dependence was observed in the presence of RBC aggregation at shear rates of 0.05 s^{-1}, 0.08 s^{-1} and 0.1 s^{-1}. The superposition of the 3 curves above 120 MHz suggests that the shape of the RBC (function *C*) becomes predominant at very high frequencies.

CONCLUSION

The effect of the flow on the spatial organization of the RBCs was modeled with a 2D simulation model. The results shown in this paper suggest that the frequency of the incident wave may be an important parameter to consider in the study of RBC aggregation using ultrasound. The slope of the backscattered power as a function of the frequency seemed to be more sensitive to the level of aggregation in the range of 35-55 MHz. At a fixed frequency, the sensitivity of the backscattered power to detect RBC aggregation is emphasized in the low frequency range of 5-30 MHz.

The simulation results allow to better understand the effect of the spatial organization of the RBCs on the backscattered power. In the future, it would be important to improve the modeling to better predict the level of aggregation for other flow conditions. These results could be of great importance for the development of an ultrasonic technique to characterize the level of aggregation.

ACKNOWLEDGEMENT

Scholarship from the Fonds de la Recherche en Santé du Québec, and grants from the Medical Research Council of Canada (#MOP-36467) and the Heart and Stroke Foundation of Quebec.

REFERENCES

1. Y.W. Yuan and K.K. Shung, Ultrasonic backscatter from flowing whole blood. II: Dependence on frequency and fibrinogen concentration. *J Acoust Soc Am.* 84:1195 (1988).
2. F.S. Foster, H. Obara, T. Bloomfield, L.K. Ryan, and G.R. Lockwood, Ultrasound backscatter from blood in the 30 to 70 MHz frequency range. *IEEE Ultrasonics Symposium Proceedings.* 1599 (1994).
3. M.S. Van Der Heiden, M.G.M. De Kroon, N. Bom, and C. Borst, Ultrasound backscatter at 30 MHz from human blood: influence of rouleau size affected by blood modification and shear rate. *Ultrasound Med Biol.* 21:817 (1995).
4. I. Fontaine, M. Bertrand, and G. Cloutier, A system-based approach to modeling the ultrasound signal backscattered by red blood cells. *Biophysical J.* 77:2387 (1999).
5. I. Fontaine, M. Bertrand, and G. Cloutier, A system-based simulation model of the ultrasound signal backscattered by blood. *IEEE Ultrasonics Symposium Proceedings.* In press (1999).

20-MHZ US BACKSCATTER MEASUREMENTS AND REPRODUCIBILITY : NORMAL HUMAN DERMIS *IN VIVO*

Céline Fournier, S. Lori Bridal, Geneviève Berger, and Pascal Laugier

Laboratoire d'Imagerie Paramétrique
UMR CNRS 7623
Université Paris VI
Paris, FRANCE 75006

INTRODUCTION

B-scan ultrasound images obtained in the 20-MHz range are currently used in clinical dermatology for morphological studies such as preoperative assessment of tumor extent[1] and the detection of psoriasis and follow-up of its treatment by measurement of the dermis and epidermis thicknesses.[2] As useful as these images are, they do not provide sufficient specificity for characterization of the composition of structures within the skin leaving biopsy the only recourse for tissue identification.

Ultrasonic tissue characterization aims to more fully exploit the information contained in the radio frequency (RF) signal after its interaction with the tissue. Signal processing techniques are applied to estimate acoustical properties of the tissue (attenuation and backscatter coefficients and the variation of these parameters as a function of frequency) that can be linked to the physical properties such as tissue composition, average size of its scattering structures and their number per unit volume.[3]

Very few quantitative ultrasonic measurements have been reported for skin. Most of the existing data have been measured *in vitro* on animal samples (canine, ovine and from mice).[4,5,6] As far as human skin is concerned, one can cite the work of Moran et al.[7] who measured the acoustical properties of ultrasound velocity, attenuation and backscatter coefficients *in vitro* across transverse sections of normal human dermis and epidermis. The recent study of Guittet et al.,[8] *in vivo*, including a hundred subjects, showed that the slope of the attenuation coefficient (measured in a 45% bandwidth about 40 MHz) decreases with age.

The goal of this work is to develop techniques for quantitative ultrasonic characterization of the skin first, by implementing methods for the *in vivo* measurement of integrated backscatter (IBS) and frequency dependence (n) of backscatter (12-25 MHz), and second, by evaluating the reproducibility of these measurements.

Acoustical Imaging, Volume 25, Edited by Halliwell and Wells.
Kluwer Academic/Plenum Publishers, 2000.

MATERIALS AND METHODS

Acquisition System

Images and RF signals backscattered from the skin were acquired with a clinical dermatologic imaging system, the Dermcup 2020 (Ultrasons Technologies, Tours, France), equipped with a 20-MHz center-frequency mechanically scanning transducer and a computer for real-time image visualization, acquisition and storage. The Dermcup 2020 was modified to allow digitization of the RF signals. Each image acquired displays a 6mm-wide and 5mm-deep region and is composed of 296 RF lines with 512 points per line sampled at 100 MHz. The tip of the probe consists of a small cavity in which the 20-MHz transducer is mechanically displaced along a linear scan axis. The cavity is filled with an antiseptical solution (Septeal®, Laboratoires SINBIO, Boulogne, France) which serves as a coupling medium and is sealed with a membrane having an acoustical impedance close to that of the skin. When the probe is held against the skin, its focal zone is generally centered in the dermis. The -6 dB bandwidth of the Fourier transform of the specular echo from a planar reflector placed at the focal length ranged from 14 to 26 MHz.

Acquisition Protocol

Prior to probe placement, the skin-surface was wiped with Septeal® in order to remove sebum or cream. The probe was first placed very lightly against the skin (without pressure), an image of data was acquired and then the probe was pressed against the skin (with pressure) with an approximate pressure of 8 kPa (corresponding to the weight of the 200 g of the probe distributed across its surface) and a second image was acquired. In this manner, two B-scan images and corresponding sets of RF lines were acquired at 10 independent and approximately parallel sites (at least 200 μm apart) of the left anterior mid-forearm of each subject. Room temperature, skin-surface temperature, the age and sex of each subject were recorded.

Subjects

This acquisition protocol was followed for 29 subjects (13 women, 24 to 76 years; 16 men, 20 to 76 years) presenting no skin pathology or lesion at the site of interest. In order to study measurement reproducibility, the acquisition protocol was repeated on three different days for each subject and on 28 different days over a two-month period for a single subject (24 year-old woman).

Ultrasound Parameter Estimation

Beginning 30 points after the skin-surface echo, each set of 296 RF lines was segmented using a 32-point (~ 0.256 mm) sliding Hamming window with 50% overlap (total studied segment of 96 points corresponding to five windows). The center of window number one was 368 μm below the skin surface and window centers were separated by 128 μm. Typically, all windows were in the dermis layer of the skin. The FFT of each windowed RF segment was calculated, and all power spectra from the same depth were averaged to yield the average apparent (not compensated for attenuation between skin surface and region of measurement) backscattered power spectrum as a function of frequency. Backscattered power spectra were compensated for the system response and diffraction using the reference spectrum of the echo from a planar reflector placed at a matched distance from the transducer. For each average backscattered power spectrum, the average or integrated backscatter (IBS) from 12 to 25 MHz was calculated and, within the same bandwidth, the frequency dependence of backscatter was determined by calculating the slope (n) of the least-squares linear fit of the logarithm of apparent backscatter versus the logarithm of frequency.

Estimation of the Reproducibility

Assuming that the random variations of the N repeated measurements on an individual j are normally distributed, the reproducibility or coefficient of variation,[9] is typically expressed as a percentage:

$$CV_j = \left(SD_j / \overline{x}_j\right) \cdot 100\% \qquad (1)$$

where SD_j is the standard deviation and $\overline{x}_j$ the mean of all the measurements from subject j. To calculate the global reproducibility[9] of the measurement technique (CV_{global}), the following relationship is used:

$$CV_{global} = SD_{global} \Big/ \sum_{i=1}^{M} \left(\overline{x}_j / M\right) \qquad (2)$$

where SD_{global} is the global standard deviation of the measurements made on M subjects :

$$SD^2_{global} = \sum_{j=1}^{M} SD_j^2 / M \qquad (3)$$

RESULTS

Single-Subject

Results from the single subject studied on multiple days showed (figure 1) that the intra-individual coefficient of variation (CV) for IBS was less than 10% for all depths and that the CV of the frequency dependence of backscatter n was less than 10% for measurements from the two most superficial windows. CV increased with depth for both parameters and pressures.

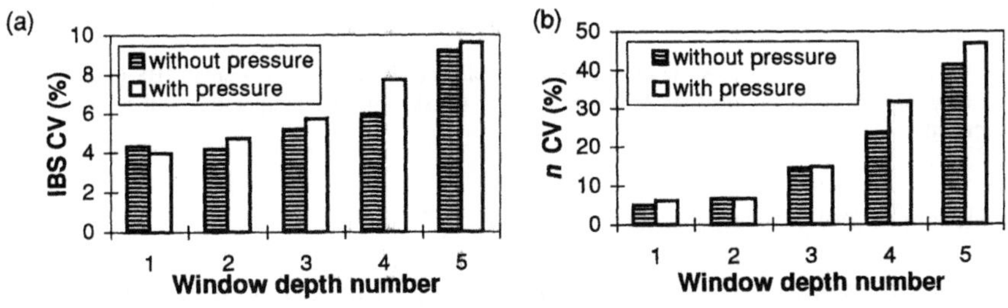

Figure 1. Coefficient of variation of (a) IBS and (b) n for the subject measured on 28 different days.

Average parameter values from the 28 different measurements are summarized in figure 2. The average IBS increased as a function of depth (figure 2.a). IBS values at each depth were significantly greater for measurements with pressure than without (statistical significance: $p \leq 0.05$). The frequency dependence of backscatter n (figure 2.b) decreased with depth and was significantly greater for measurements with pressure than without, in the first two windowed regions.

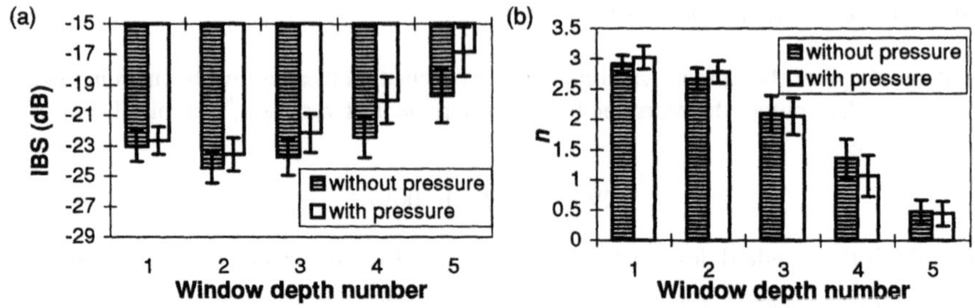

Figure 2. Mean IBS (a) and mean n (b) for a subject measured on 28 different days (± SD).

At each depth, the integrated backscatter increased significantly as a function of both room and skin-surface temperature. For all depths, the Pearson correlation coefficient (R) between IBS and room temperature ranged from 0.67 to 0.82 and between IBS and skin temperature ranged from 0.39 to 0.60 (figure 3.a). Room and skin-surface temperature for all subjects and measurements were in turn correlated ($p \leq 0.001$, $R = 0.45$) (figure 3.b).

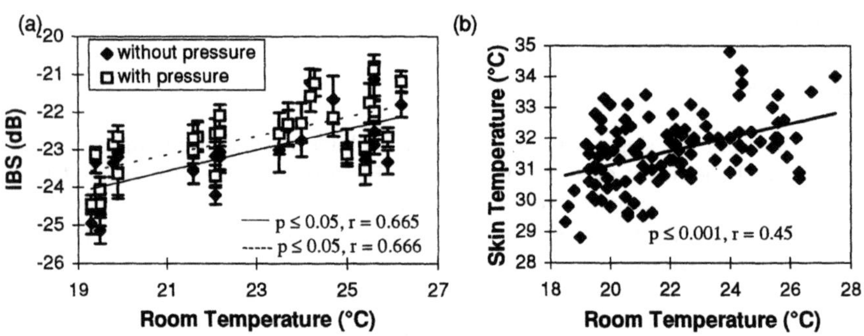

Figure 3. Correlation between room temperature and measured parameters: (a) mean IBS (± SD) in the first window for one subject studied on 28 different days ; (b) skin-surface temperature (all subjects and all measurements).

Comparison of 29 Different Subjects

As shown on figure 4, the CV_{global} for the 29 subjects measured on three different days was less than 7% for both IBS (all depths) and n (first two depths). It was again observed that the CV_{global} for both parameters and pressures increased with depth.

Each set of three measured values for a given subject was averaged then all resulting values were averaged among subjects. Resulting group averages and their standard deviations are summarized in figure 5. The average group IBS increased with depth (figure 5.a) while n decreased (figure 5.b). For each window, both parameters were significantly greater with pressure than without. When group averages were separated according to sex, the IBS was significantly greater for men than for women in the first window both with and without pressure.

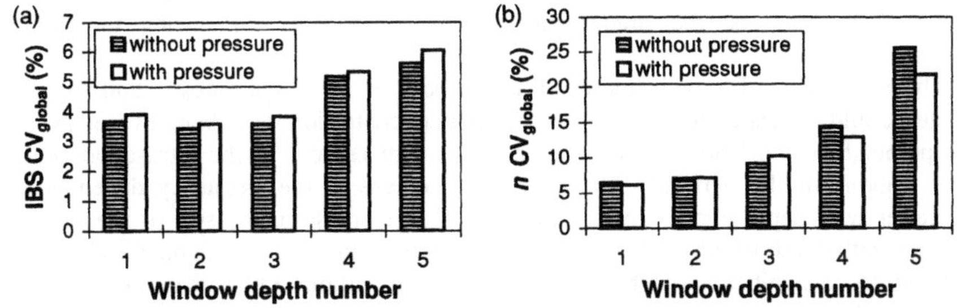

Figure 4. Global coefficient of variation of (a) IBS and (b) n for 29 subjects, each measured on three days.

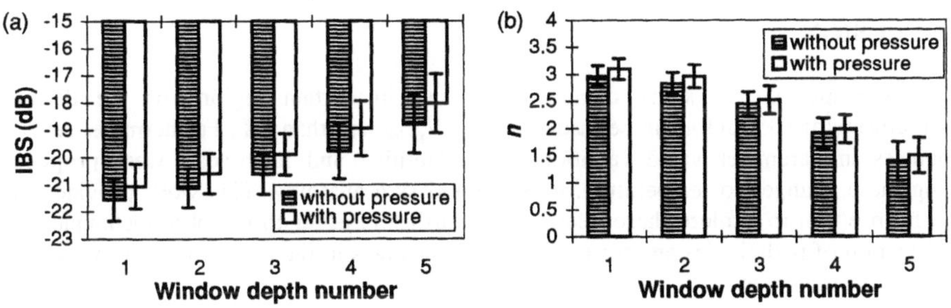

Figure 5. Mean IBS (a) and mean n (b) for 29 subjects, each measured on three days, (± SD).

DISCUSSION

A number of factors can be identified that influence measured values of parameters and measurement reproducibility. For example, the correlation of IBS with temperature suggests that correction for this effect could further improve the reproducibility of measurements. This relationship identified for 28 measurements on the same subject was not confirmed in the multiple-subject study. This could imply either that the temperature effects on backscatter measurements are subject-dependent or that they are hidden by other types of inter-subject variations. Significant differences between parameters with and without pressure demonstrate the importance of controlling this experimental factor as well.

The reproducibility (CV_{global}) of the measurement technique remains less than 7% for IBS (all window) and for n (first two windows) even without special systems for precise control of the pressure applied to the skin, the angle of the probe with the skin surface, the exact position of the measurement site or temperature. IBS appeared to be the more robust of the two parameters although measurements of n could be exploitable in superficial regions of the dermis.

As the goal of measurement of ultrasound parameters is to detect differences in skin tissues, it is useful to compare variations observed in ultrasound parameters with information about the general composition of skin. Histological studies of human skin reveal that the dermis is composed of cells, mostly fibroblasts, and extracellular material such as fibrous proteins (collagen and elastin). Collagen fibers represent 98% of the total dermis dry weight.[10] Collagen is also one of the constituents of normal soft tissue known for having an important influence on ultrasound backscatter.[11] These two facts suggest that collagen fibers may be the predominant tissue components influencing the acoustic properties of skin.

The increase in IBS with depth, observed both in the single subject and for group averages, could be related to increases in collagen concentration with depth as well as to the more perpendicular orientation of collagen fibers with respect to the ultrasound beam in deeper regions of the dermis.[12] The observed decrease in the frequency dependence of backscatter with depth implies there may also be an increase in the size of the scattering structures with depth which could correspond to a thickening of collagen bundles.[12] The IBS was found to be significantly higher with pressure than without. Other than the implications this finding holds for measurement reproducibility, this increase in backscatter could be explained by an increase in the effective number of scatterers (collagen bundles) per unit volume induced by compression of the skin.

CONCLUSION

In conclusion, this study demonstrates that, in certain regions of the dermis, measurements of IBS and n can be made with a CV_{global} less than 7%. Furthermore, several tendencies in parameter value variation were identified and hypothesis were proposed linking these changes to temperature, pressure and skin anatomy. This work provides the basis from which to explore these relationships further in order to establish techniques for the detection of pathologies and for monitoring of changes in the skin produced by medical and cosmetic therapies.

REFERENCES

1. N. Lassau, *et al.*, Value of high-frequency US for preoperative assessment of skin tumors, *RadioGraphics.* 17:1559 (1997).
2. S. Seidenari, Biophysical methods for disease monitoring in dermatology, *Curr Probl Dermatol.* 26:108 (1998).
3. V. Roberjot, S.L. Bridal, P. Laugier and G. Berger, Absolute backscatter coefficient over a wide range of frequencies in a tissue-mimicking phantom containing two populations of scatterers, *IEEE Trans UFFC.* 43:970 (1996).
4. F.K. Forster, J.E. Olerud, M.A. Riederer-Henderson and A.W. Holmes, Ultrasonic assessment of skin and surgical wounds utilizing backscatter acoustic techniques to estimate attenuation, *Ultras Med Biol.* 16:43 (1990).
5. J.H. Cantrell, R.E. Goans and R.L. Roswell, Acoustic impedance variations at burn-nonburn interfaces in porcine skin, *J Acoust Soc Amer.* 64:731 (1978).
6. P.K. Bhagat and W. Kerrick, Ultrasonic characterization of aging in skin tissue, *Ultras Med Biol.* 6:369 (1980).
7. C.M. Moran, N.L. Bush and J.C. Bamber, Ultrasonic propagation properties of excised human skin, *Ultras Med Biol.* 21:1177 (1995).
8. C. Guittet, *et al.*, High-frequency estimation of the ultrasonic attenuation coefficient slope obtained in human skin: simulation and in vivo results, *Ultras Med Biol.* 25:421 (1999).
9. C.C. Glüer, *et al.*, Accurate assessment of precision errors : how to measure the reproducibility of bone densitometry techniques, *Osteoporos Int.* 5:262 (1995).
10. J. Kanitakis. Structure histologique de la peau humaine, in: *Biologie de la Peau Humaine.* D. Scmitt, ed., Les Editions INSERM, Paris (1997).
11. M. O'Donnell, J.W. Mimbs and J.G. Miller, Relationship between collagen and ultrasonic backscatter in myocardial tissue, *J Acoust Soc Amer.* 69:580 (1981).
12. A. Mélissopoulos and C. Levacher. *La Peau : Structure et Physiologie*, Tec & Doc Lavoisier, Paris (1998).

RECENT EXPERIENCE WITH ACOUSTICAL HOLOGRAPHY FOR BREAST IMAGING

Michael P. André,[1] Constance D. Lehman,[2] Victor Neely,[3] Barbara A. Fecht,[3] Jerod O. Shelby,[3] and Tuan Nguyen[1]

[1]Department of Radiology, University of California School of Medicine and Veterans Affairs Medical Center, San Diego, CA 92093
[2]Department of Radiology, University of Washington, Seattle, WA 98195
[3]Advanced Diagnostics, Inc., Richland WA 99352

INTRODUCTION

Acoustic holography is a method of ultrasound imaging that utilizes a coherent laser beam to measure the transmitted acoustic wave through an object. The image is produced from the perturbations in both phase and amplitude of the acoustic beam (dependent on the acoustical properties of the object) at the surface of a fluid-filled cell in the form of an interference pattern which is measured by reflected light from a collimated laser beam. Since reconstruction is performed optically, image formation is real time. This method was the subject of intense research in the 1970's and 1980's for both medical and industrial applications.[1-7] Many in attendance at the 25th International Acoustical Imaging Symposium recall that the original name of this well-known conference is "Acoustical Holography." Some important applications have been found for this technology, but with few exceptions medical imaging research has not seen continued development. The widespread availability in medicine of general-purpose hand-held real-time ultrasound systems diminished interest in the more cumbersome dedicated systems. Furthermore, the quality and cost of appropriate optical systems and source transducers available at the time substantially reduced the practicality of acoustical holography.

Although a number of potential medical applications were explored, one promising area identified for acoustical holography was diagnostic breast imaging for breast cancer management.[4,7] With the sharp increase in incidence of breast cancer, the potential importance of this application may be even better appreciated today than it was in the 1970's. X-ray mammography is the accepted screening tool for diagnosis of breast cancer. It has superb resolution and clarity, is capable of detecting microcalcifications and delineates the borders of many—but not all--masses. However, 60-80% of biopsies recommended from mammography result in benign diagnoses. It is widely recognized that improvements are needed to increase the positive predictive value of conventional breast imaging.[8] In addition, mammography is

Acoustical Imaging, Volume 25, Edited by Halliwell and Wells.
Kluwer Academic/Plenum Publishers, 2000.

less sensitive in detecting cancers in women with mammographically dense breast tissue. In the U.S.A. alone, each year approximately 700,000 women undergo breast biopsies (surgical or needle). Approximately 80% of the tumors biopsied are benign, 20% are malignant. Surgical biopsies--the most common--cost between US$2,500 and US$5,000 while needle biopsies cost from US$750 to US$1,000. Patients experience both physical and emotional effects when undergoing biopsy procedures and resultant internal scarring may be problematic since it complicates interpretation of future mammograms. In the U.S. ultrasound is used primarily only to distinguish cystic from solid breast masses and to guide needle biopsies. In most U.S. institutions, if a lesion is suspected to be solid a biopsy is performed. A number of studies have suggested that improved ultrasound accuracy could help reduce the number of unnecessary biopsies by 40% with a cost savings of as much as US$1 billion per year in the U.S.A. alone.

Conventional pulse-echo ultrasound is an attractive modality for adjunctive characterization of certain breast lesions, but it is not considered specific for cancer and it is not recommended for screening. Furthermore, it is limited by several technical and practical issues: 1) quality of the examination is highly operator dependent, 2) results are not easily reproducible, 3) the field of view and range are small especially with high resolution transducers, 4) tissue behind solid masses is obscured, 5) the features of a suspicious mass are difficult to image in all dimensions, and 6) near-field limitations make many tumors difficult to assess. A number of sophisticated ultrasonic imaging techniques were explored in the past that directly addressed these limitations.[9-11] Many of these methods measure transmission sound and/or a larger portion of the acoustic scatter field than 180° backscatter. These studies suggest that a diagnostic imaging device that is sensitive to the subtle differences in speed of sound, attenuation and/or compressibility of tissues may be uniquely capable of enhancing the detection and characterization of disease. With the exception of acoustical holography these methods such as diffraction tomography are very difficult to implement in a rapid or real-time imaging system necessary for most medical procedures.

The purpose of this study was to review the merits of acoustical holography by assembling a research system incorporating several new developments in design including ones not available in to early investigators. This paper describes results of laboratory studies and a clinical investigation in 54 patients conducted over the past two years with this system (Optical Sonography™) developed by Advanced Diagnostics, Inc. (Richland, WA) for non-destructive evaluation.[12] A number of additional improvements to several key components have been identified and are being implemented.

MATERIALS AND METHODS

The experimental acoustical holography system in this study is based on a design described in detail elsewhere with several modifications (Figure 1).[12,13] The system uses three matched large area transducers (77.42 cm^2), two rotating at a slight angle to fully illuminate the object and one for the reference beam. A sequence of 12 discrete frequencies from 2.45-3.0 MHz is transmitted with equalized output at a PRF of 120 sec^{-1}. The interference pattern generated at the detector is illuminated with a laser and viewed in real time using a 12-bit digital CCD camera. Patient imaging is usually performed with real-time averaging of 4 or more video frames. The acoustic lenses have equivalent diameter of 20 cm and a focal length of 38 cm providing an f-number of 1.875. Laboratory measurements were made of resolution, modulation transfer function, dynamic range, distortion, effective focal plane thickness, acoustic power output, and artifacts.

Three types of patient interfaces were developed for this study, which are interchangeable. In the first, the object of interest is immersed in a water bath that allows dynamic positioning

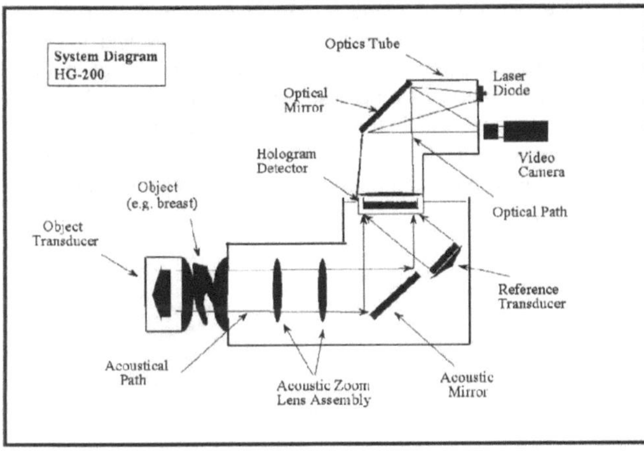

Figure 1. System diagram.

and simultaneous palpation and manipulation by the clinician or operator in conjunction with two acoustically transparent compression plates. A second inter-face uses two liquid-filled bolus bags that couple to and stabilize the body segment. To accomplish the objectives of this preliminary breast imaging study, a third patient interface was developed. The interface consisted of two shaped water-filled bladders that were maintained in contact with the medial and lateral aspects of the breast during imaging. Compression of the bladders against the breast was sufficient to hold the breast stable while contacting as much of the skin of the breast as possible to increase coupling area and thus the area viewed. All imaging modes allow for scanning through layers of a body segment by progressively moving the focal plane and for acoustical zooming of the field of view via acoustic lens adjustments.

During human studies subjects lie prone on a padded table, similar to a biopsy table, with one breast lowered into the acoustic beam through a large hole. Baby oil was used as a skin coupling agent when the water-filled pillows were used and compression against the skin completed the coupling. Transmit intensity and lens magnification were adjusted, and scanning through the breast was completed via focusing from the lateral aspect of the breast to the medial aspect via acoustic lens adjustments. Images were acquired and viewed in real-time and stored as video sequences on tape and as still images in the system computer.

A pilot study was completed in a group of 54 normal and symptomatic female volunteers ranging in age from 18 to 83 years. The study population consisted of approximately twice as many women with lesions as without. Included in the study were women with dense parenchyma, palpable ductal carcinoma in situ with negative mammography, superficial and deep fibroadenomas, and calcifications. Of the women with no lesions, two had breast implants and two had ductal ectasia. Corresponding mammograms and sonograms were obtained for each volunteer with confirmed benign and malignant lesions. Overall the mammographic breast density of the subjects was more dense than fatty: 81% demonstrated heterogeneously or extremely dense breast tissue, while 19% had scattered fibroglandular densities or almost entirely fatty breast tissue. Categorization by tissue composition, lesion location, lesion type, and lesion size was completed using medical records for this group. For a subset of benign and malignant lesions, detailed comparative analyses of lesion characteristics on mammography, pulse-echo ultrasound and OS were made. The following variables were measured: maximum diameter, shape, margins and internal architecture. Mammographic breast density for all patients was rated using the American College of Radiology's BI-RADS™ 4-point breast density categorization. For malignant lesions, specific histology and maximum diameter of the lesions were recorded from the pathology report. For benign masses, histology from core needle biopsies was recorded.

Other modifications to the original design include: a) single-element large area transducers which produce increased uniformity in acoustic output, image bandwidth and aperture, b) computer-controlled manipulation of transducer position and pulse timing to increase image information storage and help eliminate out-of-focal plane interference, c) re-design of the liquid-surface detector to improve image bandwidth, sensitivity, repeatability and general reliability of the system, d) improved optical reconstruction system with a new high output laser and lens with automated level control both of to increase system reliability and image repeatability, e) new acoustic lens technology which reduces reflections, aberrations, and attenuation in the acoustic path, f) acoustic zoom feature which allows for magnification of the field of view, and g) computer-controlled precision alignment of the system even during movement of the patient interface.

RESULTS

System Performance

Acoustic intensity output was measured with an NBS-traceable hydrophone at the entrance point to the patient with maximum transducer power for the swept-frequency beam (2.45-3.0 MHz) with the following results: I_{spta} 140 mW/cm^2, I_{sppa} 14 W/cm^2, I_m 14 W/cm^2, $MI_{2.0}$ 0.46 MPa. The transducer beam patterns for the source transducers are plotted in Figure 2.

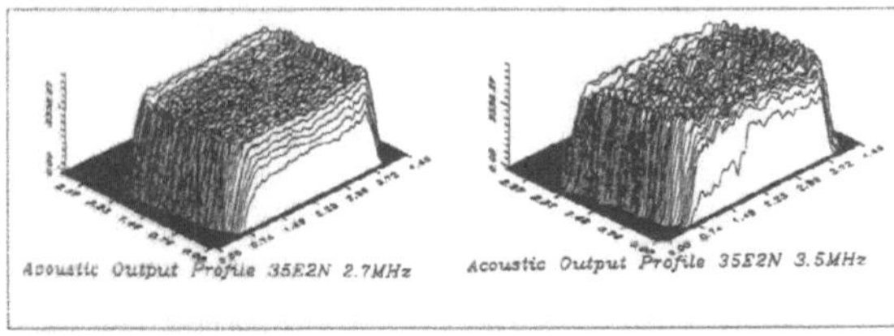

Figure 2. Source transducer beam patterns.

System limiting resolution was estimated using a line-pair target with decreasing line thickness and equal separation (Figure 3). For a swept-frequency beam from 2.6-3.0 MHz, the minimum resolution was 0.5 lp/mm in water. The line spread and modulation transfer functions for a small line target in water are shown in Figure 4. These MTF approximates a sinc function at 2.45 MHz, the low end of the transmit bandwidth. The focal plane sensitivity in the z-axis is shown in Figure 5 that indicates an effective thickness of about 40 mm. Total system dynamic range was computed and found to be 1500:1 with the 12-bit camera in a linear mode.

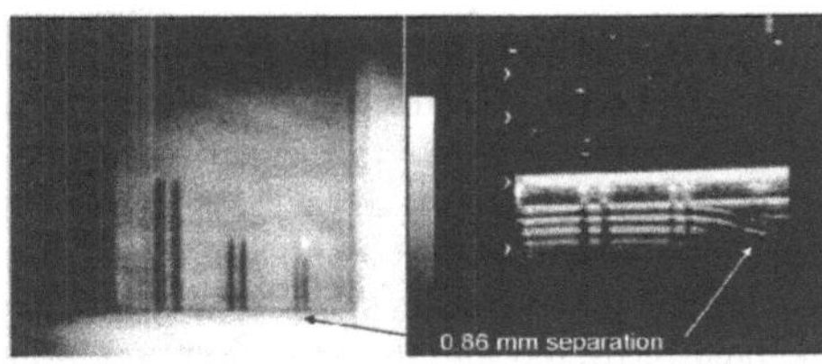

Figure 3. Resolution block by OS (left), reflective (right).

Apatite crystals were suspended in castor oil to approximate breast microcalcifications. Crystals from 0.425-1.18 mm in diameter were well resolved in the acoustic zoom mode. Needle visibility was examined with both a 14-gauge biopsy needle and a 0.6-mm needle. The needle tip was clearly visible throughout the dynamic

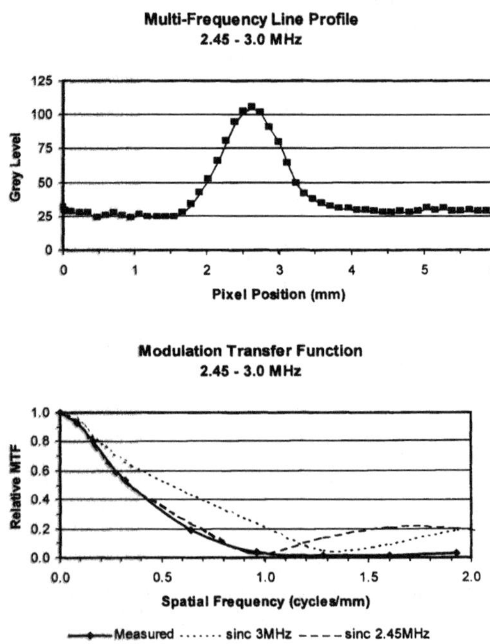

Figure 4. Line spread function and MTF for line target in water.

imaging sequence as it was slowly inserted into a RMI tissue-equivalent breast biopsy phantom. Two physicians were consistently successful engaging the target masses with each pass.

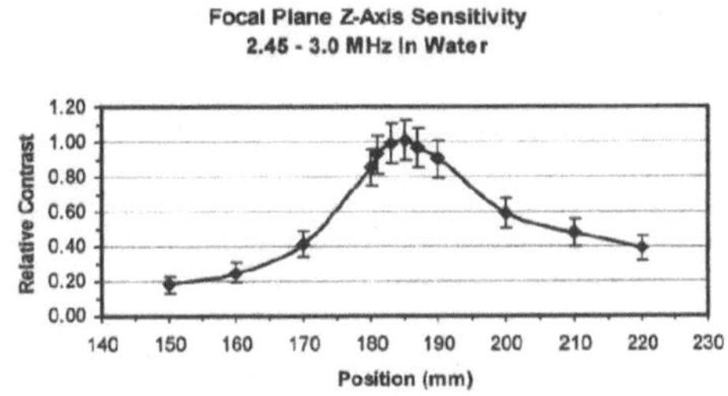

Figure 5. Effective focal plane thickness is ~40 mm.

Human Subjects

In the normal subjects, OS produced breast images that revealed structural features within the breast tissue and a consistent grey scale in soft tissues. Normal tissue structures such as the nipple, skin line, subcutaneous fat, fibrous connective tissue and glandular tissue were well visualized in all breast types. In particular, clarity of the ducts within the retroareolar complex and clear definition of the skin and subcutaneous tissue was noted. The compression needed was significantly less than that applied with traditional mammography. All patients tolerated the imaging session well. No sessions were terminated based on patient discomfort or anxiety. Two representative images of normal women with markedly different breast tissue types (dense vs. fatty replaced) are shown in Figure 6.

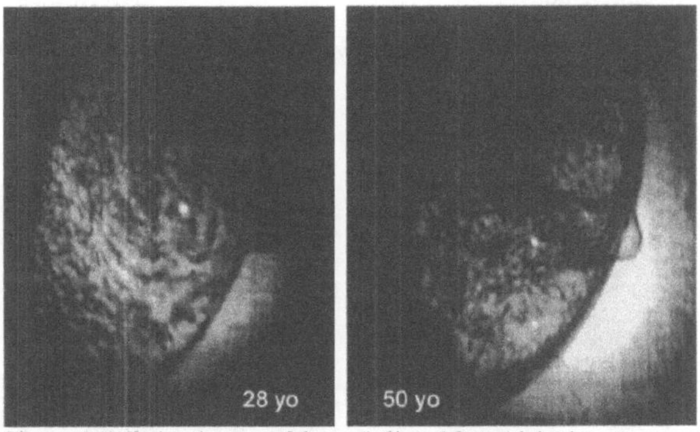

Figure 6. Full view images of dense (left) and fatty (right) breasts.

The breast masses imaged in the study population ranged in size from 4 mm to 5 cm, with a mean of 1.8 cm. Of the 10 cysts most were more acoustically transparent (whiter gray level on this display representing lower acoustical scattering) compared to adjacent breast tissue. Three representative cases are presented in Figures 7-9. Corresponding mammograms, pulse-echo sonograms and optical sonograms are shown. These masses included one infiltrating ductal carcinoma (Figure 7), a benign fibroadenoma (Figure 8) and a cluster of benign cysts (Figure 9). Pathology of surgically excised tumors documented invasive ductal carcinomas for the first case and revealed a second patient lesion. Benign fibroadenoma were diagnosed by histology from core needle biopsy samples in the second case. Fine needle aspiration under ultrasound guidance resulted in complete collapse of simple cysts in the third case.

Two masses (one cancer, one benign fibroadenoma) were mammographically occult, even with retrospective review. In both cases, dense overlying fibroglandular tissue completely obscured the masses on mammography. Both lesions were detected with traditional ultrasound and with OS. In the mammographically occult cancer (Figure 7), determination of the extent of tumor with traditional ultrasound was difficult. A 9-mm mass was detected adjacent to a 1.6-cm ill-defined hypoechoic region. On OS, the lesion was seen as a central mass close to the chest wall with fingerlike extensions protruding into the central breast tissue. The total abnormal area covered at least 2.3 cm, with the most posterior edge of the mass not included in the field of view of the optical sonogram. On pathology, a 2.7-cm invasive carcinoma was confirmed.

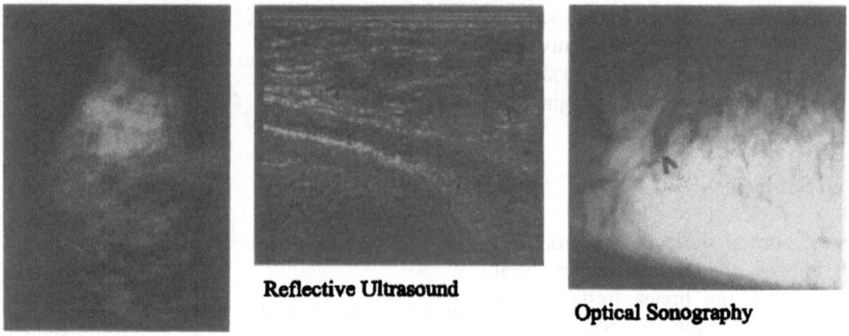

Figure 7. Invasive ductal carcinoma in a 38 year-old female.

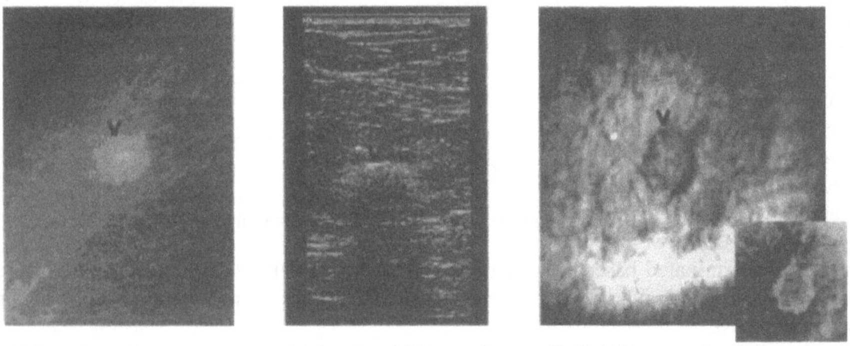

Figure 8. Fibroadenoma in a 47 year-old female.

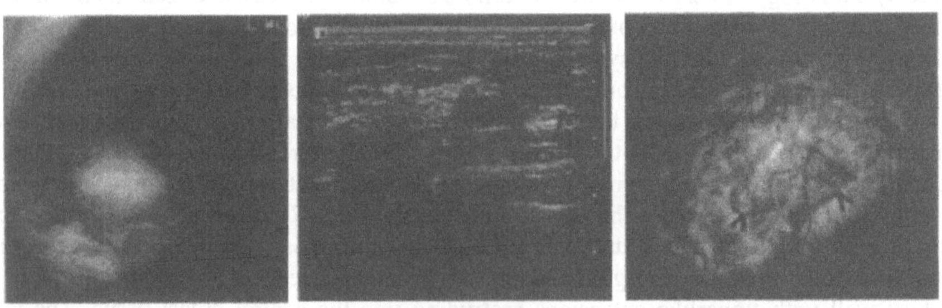

Figure 9. Benign cysts in a 38 year-old female.

For all benign lesions, maximum diameter measurements correlated well among all three modalities. However, it is interesting to note that the two malignant masses appeared significantly larger on OS compared to mammography and pulse-echo ultrasound. In these two cases, pathology reports after surgical removal of the tumors confirmed tumor sizes that were more closely correlated with measurements from OS compared to mammography or ultrasound. In the second malignant mass, the maximum diameter was 3 cm by mammography and 2.6 cm by ultrasound. The lesion appeared to be closer to 5 cm by OS. At pathology, a 3.5-cm invasive carcinoma was found with multi-centric foci of tumor extending 1.7 cm beyond this primary lesion.

The internal architecture of the malignant masses tended to be heterogeneous with darker gray level (higher acoustical scatter) overall. The benign fibroadenomas tended to be homogeneously light centrally with dark rims (perhaps edge enhancement). Some of these dark rims were thin and well defined, while others were thick and microlobulated or indistinct. In most cases, cysts appeared relatively light internally (low scattering) with dark margins. However, there were cysts that appeared homogeneously darker compared to surrounding tissue. Larger microcalcifications were apparently detected on the optical sonograms corresponding to the same region in the breast on mammography.

Resulting Optical Sonography images indicate unique distinction of breast structures, particularly of lesion margins and internal texture, with substantial contrast relative to neighboring fatty and dense parenchyma. Several examples with comparative sonograms and mammograms will be presented. Margins of lesions and internal textures were particularly well defined and provided substantial contrast to fatty and dense parenchyma. In two malignant lesions, OS appeared to approximate more closely tumor extent compared to mammography than pulse-echo sonography.

CONCLUSIONS

These preliminary studies indicate the method has unique potential for detecting, differentiating, and guiding the biopsy of breast lesions using real-time acoustical holography. Promising applications were identified although much more work is needed to establish its utility. In addition, several attributes of the system design will require modification to enhance its suitability for clinical use. Although it greatly simplifies coupling the large source beam to the patient and facilitates manipulation by the clinician, the water bath will likely be unacceptable for many examinations.

OS produces images of breast tissue in women with a variety of breast tissue types and can detect and characterize a variety of lesions, some of which are mammographically occult. Comparisons between OS, mammography and traditional ultrasound showed high correlation in many lesion characteristics. In two malignancies, OS appeared more accurate in assessing tumor extent compared to conventional imaging. It is well known that extent of tumor cannot be accurately, consistently predicted from the mammogram or ultrasound. Thus, for many patients opting for lumpectomy rather than mastectomy, repeat surgeries are all too often needed in order to obtain surgical margins clear of tumor. For this reason, it is particularly important to determine if this new technique may be able to more accurately assess tumor extent and thus more effectively guide surgical planning or biopsy.

This report represents preliminary clinical data. The study was not blinded. The emphasis in this clinical series was more on the diagnostic contribution rather than screening abilities of OS. Thus, the process of locating the lesions was performed with knowledge of the mammographic and sonographic findings.

While data obtained from the prototype OS system must be considered preliminary, it should be noted that this technology provides high resolution images while overcoming some of the limitations attributed to more conventional imaging modalities. This low cost, non-ionizing imaging modality differentiates the gross as well as subtler variations among tissues, and delineates well the edges of breast structures, including cysts, ducts, fibroadenomas, and cancers. It provides a large field-of-view which is necessary for screening and has an acoustic zoom feature to enlarge areas of interest without loss of resolution. Its real-time features are advantageous for possible applications to image-guided interventions. More work is needed to improve the breast-imaging interface to allow for a field of view that includes the entire breast. In addition, large clinical trials with blinded comparative image analyses are needed to further evaluate this promising new breast-imaging tool. This work will also include a retrospective review of all findings to help establish guide to interpretation of normal structures and lesions in these unusual images.

Laboratory results suggest that there are several refinements to the data acquisition subsystems that will significantly improve image quality by enhancing detection of small features and increasing edge delineation. Future design changes in progress are: a) modifications to the acoustic lenses, b) adjustments in transducer center frequencies (investigations are planned for both 2.4 MHz and 3.9 MHz), c) refinement of sweep frequencies (steps ranging from 50 kHz to 200 kHz), d) image post-processing, e) use of unique contrast agents, and f) 3D visualization.

ACKNOWLEDGEMENTS

This work is supported in part by grants from the US Army Medical Research & Materiel Command Breast Cancer Research Program and the Washington Technology Center.

REFERENCES

1. R.K. Mueller, et al, "Sound holograms and optical reconstruction," *Appl. Physics Lett.* 9(1), pp. 328-329, 1966.
2. B.B. Brenden, "A comparison of acoustical holography methods," *Acoustical Holography* 1, Plenum Press, New York, NY, 1969.
3. D.R Holbrooke, E.M. McCurry, V. Richards, R. Shibata, "Acoustical holography for surgical diagnosis," *Annals Surgery* 178(4), pp. 547-559, 1973.
4. A.F. Metherell, K.R. Erickson, J.E. Wreede, et al., A medical imaging acoustical holography system using linearized subfringe holographic interferometry, *Acoustical Holography* 5:453-470 (1973).
5. P. Pille and B.P. Hildebrand, Rigorous analysis of the liquid-surface acoustical holography system, *Acoustical Holography* 5:335-371 (1973).
6. B.P. Hildebrand and B.B. Brenden. *An Introduction to Acoustical Holography*, Plenum Press, New York (1972).
7. L. Weiss, "Some patho-biological considerations of detection of breast cancer by ultrasonic holography," *Ultrasonic Imaging and Holography,* G.W. Stroke, W.E. Kock, Y. Kikuchi, J. Tsujiuchi, Eds., Plenum Press, New York, NY, 1973.
8. F.M. Hall, "Sonography of the Breast: Controversies and Opinions," *AJR* 169, pp. 1635-1636, 1997.
9. J. F. Greenleaf, S. A. Johnson, W. F. Samayoa, F. A. Duck, "Algebraic reconstruction of spatial distributions of acoustic velocities in tissue from their time-of-flight profiles," *Acoustical Holography* 7, Plenum, New York, 1975.
10. G. H. Glover, J. C. Sharp, "Reconstruction of ultrasound propagation speed distributions in soft tissue: Time-of-flight tomography," *IEEE. Trans. Sonics Ultrasonics* SU-24, pp. 229-234,1977.
11. M.P. André, H.S. Janée, G.P. Otto, P.J. Martin, "High speed data acquisition in a diffraction tomography system with large-scale toroidal arrays." *Intl. J. Imaging Systems & Technol.*, 8(1), pp. 137-147, 1997.
12. G.F. Garlick, G.C. Dunham, "Ultrasonic holography imaging method and apparatus," U.S. Patent No. 5,329,817, July 19, 1994.
13. B. A. Fecht, M.P. André, G.F. Garlick, R.L. Shelby, J.O. Shelby, C.D. Lehman, "Investigation of an acoustical holography system for real-time imaging, *Physics of Medical Imaging SPIE* 3336, pp. 106-112, 1998.

A THEORY TO ACCOUNT FOR PHASE VARIATION AND DIFFRACTION IN ULTRASOUND IMAGE FORMATION

Paul E. Barbone

Department of Mechanical Engineering
Boston University
Boston, MA 02215

ABSTRACT

Traditional ultrasonic imaging utilizes the assumption that the image is formed by distributions of point reflectors in an otherwise homogeneous background. We derive relations between the backscattered signal and the medium acoustic properties that are extensions of the traditional assumptions. Our theory is valid for short wavelength fields generated by suitably apodized transducers in weakly inhomogeneous media. In this parameter regime, Born expansion is not valid. We find, however, that *both* the forward propagating and backscattered fields are well modeled to leading order by one-way wave equations. The one-way equations depend on the sound speed variations along the direction of beam propagation. By back-propagating the scattered field, it is possible to reconstruct the sound speed variation within the medium. The reconstruction relies on an assumed correlation between density and sound speed fluctuations.

INTRODUCTION

Ultrasonic imaging utilizes highly collimated beams which propagate into a slightly inhomogeneous medium. This simple statement exposes two key facts related to ultrasound imaging: it utilizes a short wavelength (to obtain a collimated beam) and the medium variation about some mean value is small.

The fact that soft tissue has nearly constant acoustic properties implies that the scattered field amplitude is typically small. In this case, the Born scattering assumption can usually give accurate results, provided the wavelength isn't too small. In the short wavelength limit, however, the range of validity of the Born expansion goes to zero. Thus in mathematical models of ultrasound imaging, there is a competition between two small parameters: the magnitude of the medium variation, and the wavelength.

In this paper, we discuss a model for ultrasound propagation and backscatter which balances the short wavelength and weak medium variation assumptions. The model results in a pair of coupled one-way wave equations. One of the equations models the incident field, while the other models the backscattered field. The one-way equations, if applied to homogeneous media, are consistent with the classical Fresnel approximations. They have the advantage, however, of accounting for the leading order cumulative effect of multiple scattering over a large propagation distance.

Acoustical Imaging, Volume 25, Edited by Halliwell and Wells.
Kluwer Academic/Plenum Publishers, 2000.

As the incident beam propagates into the body, part of its energy is scattered by the medium inhomogeneities. Not all the scattered field, however, propagates forward or backward. Indeed, a small compressibility perturbation will scatter energy uniformly in all directions. Thus, an assumption that all scattered energy is either forward or backward propagating is inappropriate. Nevertheless, the *interesting* part of the scattered field is only that which propagates back toward the transducer.

We present a model equations to compute the forward and back-scattered fields. The scattered field model is based on an approximate representation of the Green's function for the inhomogeneous medium. Our technique accounts for radiation (scattering) in all directions by solving the full wave equation in the vicinity of an isolated source (scatterer). Away from the source but along the beam axis, however, the scattered field is again nearly unidirectional. Thus, a one-way equation may be used. We use matched asymptotic expansions to make these arguments precise and to fit the two approximations together into one theory.

The primary accomplishment here is the derivation of a single equation for the radiated (scattered) amplitude. It can be solved by itself without any reference to an inner solution valid near the source location. The pair of equations modeling the forward and backscattered fields can be solved simultaneously, and the sound speed variations in the beam can be reconstructed.

FORMULATION

We model soft tissue as a weakly inhomogeneous acoustic fluid with density $\rho(\boldsymbol{x})$ and compressibility $\kappa(\boldsymbol{x})$. For the purposes of this contribution, we shall neglect damping, noting that it can be included in a natural way in the forward and backward propagation equations to be derived.

We use $a \approx 1\,cm$ as a typical transducer dimension, and scale all lengths with respect to a. This leads to the following problem formulation for the acoustic pressure $p(\boldsymbol{x}, t)$:

$$\nabla \frac{1}{\rho} \nabla p - \kappa \partial_{tt} p = 0 \tag{1}$$

$$p(0, y, z, t) = \begin{cases} p_o(k^{1/2} y, k^{1/2} z, kt) & (y, z) \in s_o(y, z) \\ 0 & \text{otherwise} \end{cases} \tag{2}$$

$$p(x, y, z, t) = 0 \quad t \leq 0; x > 0. \tag{3}$$

The problem above models high frequency waves radiated by a transducer of unit size. The frequency of the transducer is determined by the nondimensional parameter k. When k is large, then $\partial_t p_o$ is large. (k can be thought of as the wavenumber for the carrier frequency, nondimensionalized by the transducer size, a.) The factors of $k^{1/2}$ in the arguments of p_o allow for a weak focusing of the beam through phase variation over the transducer the surface.

In ultrasonic imaging, the following assumptions apply

$$\rho(\boldsymbol{x}) = \rho_o(1 + \epsilon \rho_1(\boldsymbol{x})), \tag{4}$$

$$\kappa(\boldsymbol{x}) = \kappa_o(1 + \epsilon \kappa_1(\boldsymbol{x})), \tag{5}$$

$$\epsilon \ll 1, \tag{6}$$

$$k \gg 1, \tag{7}$$

$$L \gg 1. \tag{8}$$

Here L is a measure of the propagation distance of the beam. Typical values for the parameters defined above are $1/\epsilon \approx 20$, $k \approx 50$ and $L \approx 20$. Thus, none of the parameters "dominates" the others. They are all of the same order of magnitude.

Exploiting (6) in a regular perturbation scheme gives rise to the Born approximation. This is shown below to be invalid in the regime $\epsilon \sim k^{-1} \sim L^{-1}$. Exploiting (7) puts us in the realm of geometrical optics. The transducer is usually focused, however, and so geometrical optics would lead to a singularity at the focus. Further, geometrical optics predicts a discontinuous field along the edges of the beam. To the author's knowledge, (8) cannot be exploited, but rather puts a limitation on the range of validity of other would-be approximation strategies.

INVALIDITY OF BORN EXPANSION: EXAMPLE

Here we show an example which demonstrates the limitations of the Born approximation in the short wavelength regime. We consider one dimensional propagation in (1-3), with a small step in compressibility:

$$p_o = p_o(kt) \tag{9}$$

$$\kappa_1(x) = H(x - x_o) \tag{10}$$

$$\rho_1(x) = 0. \tag{11}$$

We then assume a solution for $p(x,t)$ in the form:

$$p(x,t) = p_o(t - x/c_o) + \epsilon p_1(t - x/c_o) + \epsilon^2 p_2(t - x/c_o) + O(\epsilon^3) \tag{12}$$

Here, $c_o^2 = 1/\kappa_o \rho_o$. Substituting (12) into (1) with (9) and (10) then gives

$$c_o^2 \partial_{xx} p_1 - \partial_{tt} p_1 = H(x - x_o)\partial_{tt}[p_o(k(t - x/c_o)) + \epsilon p_1(x,t) + O(\epsilon^2)] \tag{13}$$

The approximation proceeds on the assumption that

$$\epsilon p_1(x,t) \ll p_o(k(t - x/c_o)). \tag{14}$$

Thus p_1 is determined by the solution of (13) with $\epsilon = 0$. This yields for $x > x_o$:

$$p_1(x,t) = \frac{-k}{2c_o}(x - x_o) p_o'(k(t - x/c_o)) \qquad x > x_o. \tag{15}$$

With the solution (15), we can examine the validity of the assumption (14). We recall that L is a characteristic propagation distance of interest. Thus we see that assumption (14) is satisfied provided:

$$k\epsilon L \ll 1, \tag{16}$$

which is typically not true. While ϵ may indeed be small, k and L are both large, which makes (16) rather dubious.

We can learn how to obtain a useful approximation by studying where this one went wrong. The exact solution for $x > x_o$ is of the form:

$$p(x,t) = A p_o(t - x/c(x)). \tag{17}$$

Substituting $c(x) = c_o(1 + \epsilon c_1)$ into (17) and expanding to $O(\epsilon)$ gives (15) for p_1. This calculation shows that origin of the nonuniformity in the Born expansion is the accumulated phase variation with propagation distance. This conclusion is confirmed in the analysis in the next section.

EQUATION FOR INCIDENT FIELD

In order to obtain a result valid for small ϵ and large k, we must treat these two parameters simultaneously. We shall assume that $\epsilon \sim k^{-1}$. Consistent with the form of the transducer field at $x = 0$, we make the following scaling assumptions on the behavior of $p(x,t)$:

$$p(x,y,z,t) = u(x, k^{1/2}y, k^{1/2}z, k\tau) \tag{18}$$

$$\tau = t - x/c_o. \tag{19}$$

The scalings introduced in (18) permit us to introduce the large parameter k directly into the wave equation, and so eliminate terms that are small. To see this, we note that the chain rule gives the relations:

$$\partial_x p = \partial_x u - \frac{k}{c_o}\partial_\tau u \tag{20}$$

$$\partial_{yy}p + \partial_{zz}p = k\partial_{yy}u + k\partial_{zz}u \equiv k\nabla_\perp^2 u. \tag{21}$$

Using $\frac{1}{\rho} = \frac{1}{\rho_o}(1 - \epsilon\rho_1 + O(\epsilon^2))$ and the subsitution (18) in (1) gives us

$$\begin{aligned} k^2 & [\frac{1}{\rho_o c_o} - \kappa_o]\partial_{\tau\tau}u + k[\frac{-2}{\rho_o c_o}\partial_{x\tau}u + \frac{1}{\rho_o}\nabla_\perp^2 u] \\ & + k^2\epsilon[\frac{-\rho_1}{\rho_o c_o} - \kappa_o\kappa_1]\partial_{\tau\tau}u = O(k^2\epsilon^2, k\epsilon, k^o). \end{aligned} \tag{22}$$

We see in (22) that the leading order term (k^2) is automatically eliminated by the definition of c_o. This indicates that we chose the retarded time (19) correctly. The next term in (22) requires us to balance k and $k^2\epsilon$. If $\epsilon \ll k^{-1}$, then we could neglect $k^2\epsilon$ relative to k. On the other hand, if $k \gg \epsilon^{-1}$, then we should neglect k relative to $k^2\epsilon$; this would give us a correction to the retarded time that we defined in (19). Since neither of these two conditions in satisfied in general, we must keep *both* terms, treating them as of equal importance. This leads to the leading order equation for u:

$$-2\partial_\tau[\partial_x u - k\epsilon\frac{c_1}{c_o}\partial_\tau u] + c_o\nabla_\perp^2 u = O(k\epsilon^2, \epsilon, k^{-1}). \tag{23}$$

In simplifying (23), we have introduced $c_1(x) = -(\rho_1 + \kappa_1)/2$. It is defined so that the sound speed $c(x)$ is to first order:

$$c(x) = c_o(1 + \epsilon c_1(x) + O(\epsilon^2)). \tag{24}$$

Equation (23) is the leading order equation that can be used to predict the incident field. It allows only "one-way" propagation (in the $+x$ direction and at small angles with respect to x), but accounts for multiple scattering effects while it propagates through the variation in the sound speed. Equation (23) is simpler than the original equation (1) in that it has only one derivative with respect to x. This makes it possible to "march" the solution in the x direction. Solving (23) in the special case of $c_1 = 0$ gives an approximation for the incident field that is tantamount to using the Fresnel approximation in a homogeneous medium. We note that equation (23) is independent of ρ_1 and κ_1 except in their relation to c_1. That is, in the parameter range considered here, the leading order effect of medium inhomogeneities on a propagating wave is the sound speed variations.

The interpretation of the various terms in (23) is perhaps more evident when (23) is rewritten in terms of the pressure, p_{inc}. Doing so gives

$$-2c_o\partial_t[\partial_x p_{inc} + \frac{1}{c_o}(1 - \epsilon c_1)\partial_t p_{inc}] + c_o^2\nabla_\perp^2 p_{inc} = 0. \tag{25}$$

We see that the first term in (25) is responsible for propagating the field in the $+x$ direction with speed $c = c_o(1 + \epsilon c_1)$. The last term in (25) controls the diffraction of the beam. Despite (25) being written in original coordinates, equation (23) is preferable for numerical implementation.

FORMULATION FOR SCATTERED FIELD

Since p_{inc} does not exactly satisfy (1), it will give rise to a scattered wave field. To determine the equation for the scattered field, we first define the total acoustic pressure, p_{tot} as:

$$p_{tot} = p_{inc} + p_{scat}. \tag{26}$$

For notational convenience, we will denote equation (1) by $\mathcal{L}p_{tot} = 0$, and equation (25) as $\mathcal{L}_R p_{inc} = 0$ (the 'R' is mnemonic for "right-going".) Thus equation (1) can be written

$$\mathcal{L}p_{scat} + \mathcal{L}p_{inc} = \mathcal{L}p_{scat} + [\mathcal{L} - \mathcal{L}_R]p_{inc} = 0. \tag{27}$$

In solving (27) for p_{scat}, we may consider p_{inc} as known as the solution of equation (25). Evaluating the known terms in (27) gives to leading order

$$\begin{aligned}
[\mathcal{L} - \mathcal{L}_R]p_{inc} &= c_o^2\partial_{xx}u + k\epsilon c_o^2 c_1 \nabla_\perp^2 u + k\epsilon c_o^2[\partial_x\rho_1(\frac{1}{c_o}\partial_\tau u - k^{-1}\partial_x u) + k^{-1/2}\nabla_\perp\rho_1 \cdot \nabla_\perp(28) \\
&= c_o^2\partial_{xx}u + k\epsilon c_o^2 c_1 \nabla_\perp^2 u + k\epsilon c_o^2 \partial_x\rho_1 \frac{1}{c_o}\partial_\tau u(1 + O(k^{-1/2}). \qquad (29) \\
&= \mathcal{F}u(x, y, z, \tau) \qquad (30)
\end{aligned}$$

The scattered field, therefore, is the result of the equivalent "sources" in (29). These sources depend on both the sound speed and density variations, independently. Further, except for the diffraction terms, the sources depend on *derivatives* of ρ_1. This dependence assures that our leading order solution accurately captures the effects of overall changes of level in the medium parameters. Because of the diffraction terms, however, this is not true for waves propagating at significant angle to the $x - axis$.

In order to solve (27) for p_{scat}, we will need a Green's function for the inhomogeneous medium.

SCATTERED FIELD GREEN'S FUNCTION

The Green's function that we will use to solve (27) satisfies

$$\mathcal{L}g(\boldsymbol{x}, t; \boldsymbol{x}_o) = f(kt)\delta(\boldsymbol{x} - \boldsymbol{x}_o). \tag{31}$$

We can use the same approximations ($k \sim \epsilon^{-1} \to \infty$) to find an approximate representation for g, but doing so yields different approximations in different regions of space.

Inner asymptotic expansion

In finding a simplified equation for p_{inc}, we were able to take advantage of the fact that the incident field propagates in essentially the $+x$ direction. Near a source, however, the radiated field propagates in all directions. Thus, g does not have a preferred propagation direction (except away from the source) as did p_{inc}. To find a leading order approximation for g valid near the source, we assume the following scalings:

$$g(\boldsymbol{x}, t; \boldsymbol{x}_o) = G(k(\boldsymbol{x} - \boldsymbol{x}_o), kt; \boldsymbol{x}_o). \tag{32}$$

Substituting (32) into (31) leads to the following equation for G:

$$k^2[\frac{1}{\rho}\nabla^2 G - \kappa_o\partial_{tt}G] + k^2\epsilon[-\frac{\rho_1}{\rho_o}\nabla^2 G - \kappa_o\kappa_1\partial_{tt}G] - k\epsilon\frac{1}{\rho_o}\nabla\rho_1 \cdot \nabla G = f(kt)\delta(\boldsymbol{x} - \boldsymbol{x}_o). \tag{33}$$

Thus, provided $k|\boldsymbol{x} - \boldsymbol{x}_o| \ll \epsilon^{-1}$ (i.e. $\boldsymbol{x}$ is within a few wavelengths of the source point, $\boldsymbol{x}_o$) we can expand G in a regular perturbation series:

$$\begin{aligned}
G &= G_o + \epsilon G_1 + O(k^{-1}\epsilon, \epsilon^2) &(34) \\
G_o &= \frac{-1}{4p_{inc} r \rho_o} f(k(t - r/c_o)) &(35) \\
\frac{1}{\rho}\nabla^2 G_1 &- \kappa_o\partial_{tt}G_1 = 2\kappa_o c_1 \partial_{tt}G_o. &(36)
\end{aligned}$$

Here, $r = |\boldsymbol{x} - \boldsymbol{x}_o|$.

Outer asymptotic expansion

To obtain an approximation valid in the entire domain, we can use geometrical acoustics. We assume that g is of the form of a carrier wave times a slowly varying amplitude function:

$$g \sim A(\boldsymbol{x}; \boldsymbol{x}_o)\, w(k(t - s(\boldsymbol{x}; \boldsymbol{x}_o)). \tag{37}$$

Further, we expand the phase in terms of ϵ

$$s = s_o + \epsilon s_1. \tag{38}$$

Substituting (37) into (31) and noting (38) and $k \sim \epsilon^{-1}$ leads to

$$A \sim A_o/r \tag{39}$$

$$s_o = r/c_o \tag{40}$$

$$s_1 = -\int_{\boldsymbol{x}_o}^{\boldsymbol{x}} (c_1/c_o) dr. \tag{41}$$

Equations (40) and (41) may be combined to give

$$s \sim \int_{\boldsymbol{x}_o}^{\boldsymbol{x}} (1/c) dr. \tag{42}$$

The constant A_o and the as yet undetermined function w can be found by requiring (37) to match (34) in their common range of validity. Performing this calculation gives

$$g \sim \frac{-1}{4 p_{inc} \rho(\boldsymbol{x}_o) r} f(k(t - s))\; (1 + O(\epsilon, k^{-1})). \tag{43}$$

Equation (43) is valid throughout the domain.

Simplification for backscattered field

In terms of g, the scattered pressure p_{scat} can be found through integration:

$$p_{scat}(\boldsymbol{x}, t) = \int_{V_o} g(\boldsymbol{x}, t; \boldsymbol{x}_o)\, \mathcal{F}u(x_o, y_o, z_o, \tau_o - s)\, dV_o. \tag{44}$$

We remark that (44) is not a convolution since g is not a function of r alone; the medium inhomogeneity does not permit such a representation g.

Equation (44) can be evaluated to leading order by substituting the approximations for g and $\mathcal{F}$ in equations (43) and (29), respectively. The representation of p_{scat} so obtained is analogous to that typically used for the scattered field (c.f. Macovsky, 1983). In evaluating the scattered field at the transducer, i.e. the *backscattered field*, one may use the approximation that the field is primarily a left-traveling wave along the beam axis. In the case of propagation through a homogeneous medium (consistent with the Born approximation), such an approximation is achieved by using Fresnel approximation to the phase in the Green's function. A similar approximation may be made in (44), but it is rather inconvenient. This is because the phase $s(\boldsymbol{x})$ is defined as an integral for each point. Thus, evaluating G on the transducer face and at one source point requires an integration from the source point to each point on the transducer face. Repeating such calculations for each source point in order to evaluate the integral indicated in (44) is obviously a laborious task.

Instead of the Fresnel approximation, we may can use the fact that the *backscattered* field is primarily unidirectional. Barbone (1999) has shown in the time-harmonic case that the full Green's function can be replaced by its one-way (parabolic) approximation when evaluating the backscattered field along the beam axis. In the present context, this gives the following equation for the *backscattered* field

$$\begin{aligned} 2\partial_\eta(\partial_x - \frac{\epsilon c_1}{c_o}\partial_\eta)v + c_o\nabla_\perp^2 v &= k^{-1}\rho_o c_o \mathcal{F}u && (45) \\ p_{scat} &= v(x, k^{1/2}y, k^{1/2}z, k\eta) && (46) \\ \eta &= t + x/c_o. && (47) \end{aligned}$$

We emphasize that v represents the backscattered field, only, along the axis of the beam. The rest of the field in the rest of the domain is not represented by the solution of (45). The apparent "extra" factor of k^{-1} that appears on the right hand side of (45) accounts for the fact that only a small fraction of the total scattered energy propagates backward.

CORRELATION FOR RECONSTRUCTING $c(\boldsymbol{x})$

The right hand side of (45) is expressed in terms of the medium variations, c_1 and ρ_1, and the incident field u. If we knew both u and v for all space and time, then we could simply substitute those functions into (45) to find an equation for the medium variations, c_1 and ρ_1. Such a procedure, however, yields just a single equation for the two unknown functions. Therefore, in order to reconstruct the sound speed, we need to assume something about the variation of the density. In the Born approximation, one assumes that the sound speed variation is negligible. Gray and colleagues (Gray, 1980; Gray and Bleistein, 1986) assumed that the density variation is negligible. Shuman & Barbone (1997) and Nassiri and Hill (1986) both take advantage of the assumption that variations in sound speed are correlated with variations in density. While this correlation is not perfect in soft tissue, it is more accurate than either of the other two extremes.

Therefore, we assume that

$$\rho \equiv \rho(c) = \rho(c_o) + \frac{d\rho}{dc}(c - c_o) + O(\epsilon^2). \tag{48}$$

For soft tissue applications, we may use (Shuman & Barbone, 1997) $\frac{d\rho}{dc} \approx 2\rho_o/c_o$. This leads to the conclusion that

$$\rho_1(\boldsymbol{x}) \approx 2\,c_1(\boldsymbol{x}). \tag{49}$$

With (49), equations (45) and (23) can be simultaneously marched in the $+x$ direction. In so doing, the unknown sound speed, $c_1(\boldsymbol{x})$ can be evaluated. Similar splitting and marching ideas have been explored by Fishman, *et al.* (1997). We have not yet studied the issues of such a procedure, but we expect that some variation of layer-stripping or other profile inversion method (e.g. Bruckstein & Kailath (1987); Hagin (1981); Raz (1981)) should be stable and accurate when applied to this system of equations.

SUMMARY

We have derived a pair of one-way wave equations for the forward incident and backscattered fields. The derivation was based on simultaneously treating the medium variations as weak and the wavelength as short. The one-way equations account to first order for the small but cumulative effect of multiple scattering over a large propagation distance.

We also derived an expression for the scattered field valid everywhere in the domain. We describe a simplification of the scattered field equations valid in the backward direction along the beam axis. This simplification results in a one-way wave equation for waves propagating in the $-x$ direction.

Further work is required to use the derived equations to solve for the sound speed variation in an imaged medium.

REFERENCES

A.M. Bruckstein and T. Kailath, "Inverse scattering for discrete transmission-line models," *SIAM Review*, **29**(3): 359–389 (1987).

L. Fishman, A.K. Gautesen, Z. Sun, "Uniform high-frequency approximations of the square root Helmholtz operator symbol," *Wave Motion*, **26**:127–161 (1997).

S.H. Gray, "A second-order procedure for one-dimensional velocity inversion," *SIAM J. Appl. Math.,* vol. 39, pp. 456-462 (1980).

S.H. Gray and N. Bleistein, "Imaging and Inversion of Zero-Offset Seismic Data," *Proc. IEEE,* vol. 74, no. 3, pp. 440-456 (1986).

F.G. Hagin, "A stable approach to one-dimensional inverse problems," *SIAM J. Appl. Math.,* vol. 40, no.3, pp. 439-453 (1981).

E.J. Hinch, *Perturbation Methods.* Cambridge University Press, New York (1991).

R.A. Lerski, ed., *Practical Ultrasound.* IRL Press, Washington, D.C. (1988).

D.K. Nassiri and C.R. Hill, "The differential and total bulk acoustic scattering cross sections of some human and animal tissues," *J. Acoust. Soc. Am.*, **79**:2034–47 (1986).

S. Raz, "A direct profile inversion: Beyond the Born Model," *Radio Sci.,* vol. 16, pp. 347-353 (1981).

R. Shuman and P.E. Barbone, "Renormalized Born Inversion," *Acoustical Imaging*, Vol. 23, S. Lees and L.A. Ferrari, eds. Plenum Press, New York (1997).

HIGH FREQUENCY REFLEX TRANSMISSION IMAGING: FEASIBILITY FOR EVENTUAL APPLICATION TO THE DIAGNOSIS OF SKIN TUMOURS

Christelle M. Guittet, Jeffrey C. Bamber, Nigel L. Bush, David S. Bell, Peter S. Mortimer*

Joint Department of Physics and Department of Medicine*, Institute of Cancer Research and Royal Marsden NHS Trust, Downs Road, Sutton, Surrey, SM2 5PT, U.K.

INTRODUCTION

Melanoma is currently the eighth most common cancer and as it continues to remain resistant to drugs and radiotherapy, successful treatment depends upon early diagnosis. The eventual aim of the present work is to develop an ultrasonic instrument to aid skin tumour diagnosis at primary care level and thus reduce the number of benign skin lesions unnecessarily excised.

Having shown that ultrasonic attenuation can discriminate malignant melanoma from basal cell papilloma[1], we are investigating the application of a pulse echo technique called "Reflex Transmission Imaging" (RTI) to improve this discrimination.

The specific objectives of the work in progress that is reported in this paper are a) to demonstrate the feasibility of implementing Reflex Transmission Imaging using an existing high-frequency ultrasound skin scanner and b) to develop objective methods for the assessment of RTI system performance by imaging phantoms.

MATERIALS AND METHOD

Method

Reflex Transmission Imaging is a pulse echo technique proposed for imaging attenuation, that requires B-mode equipment and the use of a highly focused transducer (f-number < 3) [2]. A two-dimensional image of relative ultrasonic attenuation at constant depth in the focal plane of a strongly focused transducer may be calculated from the averaged energy backscattered from the broad-beam region beyond the focus. Each pixel is obtained in a plane normal to the beam at the transducer focus, by setting a gate on the backscattered signal beyond the focus where reflections are amplified, amplitude detected and integrated. The focal point is scanned in a two-dimensional pattern to obtain a complete image. The use of a very strongly focused transducer not only improves the spatial resolution in the focal plane but also provides averaging over the broad beam of spatial fluctuations in attenuation and backscattering outside the focal zone thus emphasising fluctuations of attenuation from the focal plane. This technique assumes a homogeneous region in the integration zone behind the focus so that any change in detected signal is dominated by the transmission properties at the focus. The intention is to place the focus at the skin surface, where most of the attenuation in benign lesions occurs.

Acquisition system

RTI was studied by adapting a commercial B-mode skin scanner (Dermascan C, Cortex Aps, Denmark) to acquire 3D radio frequency echo data. Transmission ultrasound images for the plane at the transducer focus, positioned at the skin surface, were reconstructed by integrating the energy backscattered from beyond the focus. Diffraction phenomena that would have caused errors in the relative attenuation estimate, and hence artifacts in the RTI images were eliminated by scanning with a constant skin surface distance.

Images were obtained with two transducers (Table 1). One was not ideal, being designed for B-scan imaging and therefore insufficiently focused for RTI. The other was designed for RTI in terms of focus but was discovered to have poor sensitivity.

Table 1. Nominal values of transducer characteristics as stated by the manufacturers

	1st transducer	2nd transducer
Manufacturer	Panametrics	TMS, U.K.(custom made)
Frequency	20 MHz	15 MHz
Bandwidth	14 MHz (70%)	7.8 MHz (52%)
λ	79 μm	98 μm
F-number	4.8	1.3
Lateral resolution	250-300 μm	158 μm
Axial resolution	50-60 μm	143 μm
Beam diameter	0.38 mm	0.13 mm
Focus depth	12.9 mm	1.2 mm

To produce the 3-D data set necessary for RTI image reconstruction, each transducer was positioned so that its focus was at the surface of the object and a radio frequency B-scan was acquired using water as a coupling medium between the transducer and the object. Translation of the sample beneath the transducer was performed automatically in order to acquire a series of B-scans separated by 200 μm.

Performance assessment

Relative image quality was assessed with a technique equivalent to that employed in X-ray computed tomography using "Rose phantoms"[3]. These were constructed from a homogeneous backscattering medium containing discs of attenuating zones (lesions) at the surface, whose diameters varied from 1 to 3.5 mm. These permitted measurement of contrast-to-noise ratio and the plotting of contrast-detail curves.

The phantoms[4] were made of gelatin containing diffusely scattering suspensions of spherical glass beads. Flat thin attenuating discs of paraffin wax of varying thickness (0.4 to 1.5 mm), and thus varying attenuation, were set at the surface of the gelatin in order to provide a range of details and attenuation contrast for the plotting of contrast-detail curves from observer opinion of the relative visibility of the discs.

The contrast to speckle ratio[3] (CSR) was also evaluated in order to have a physical index of system performance, related to the visibility of the objects, that takes into account the contrast fluctuations due to speckle.

$$\mathrm{CSR} = \frac{S_1 - S_0}{\left(\sigma_0^2 + \sigma_1^2\right)^{\frac{1}{2}}} \tag{1}$$

where S_0 and S_1, and σ_0 and σ_1, are the means and standard deviations of the RTI signal levels respectively in a disc and an area of RTI background speckle.

RESULTS

Contrast assessment of the disc phantoms

Reflex transmission images of the disc phantoms, obtained using both transducers, are shown in Figure 1. All images were acquired using a backscattering integration zone of about 1 mm in depth (130 samples) located about 0.6 mm below the surface, these values having been previously determined as those that would maximize the CSR.

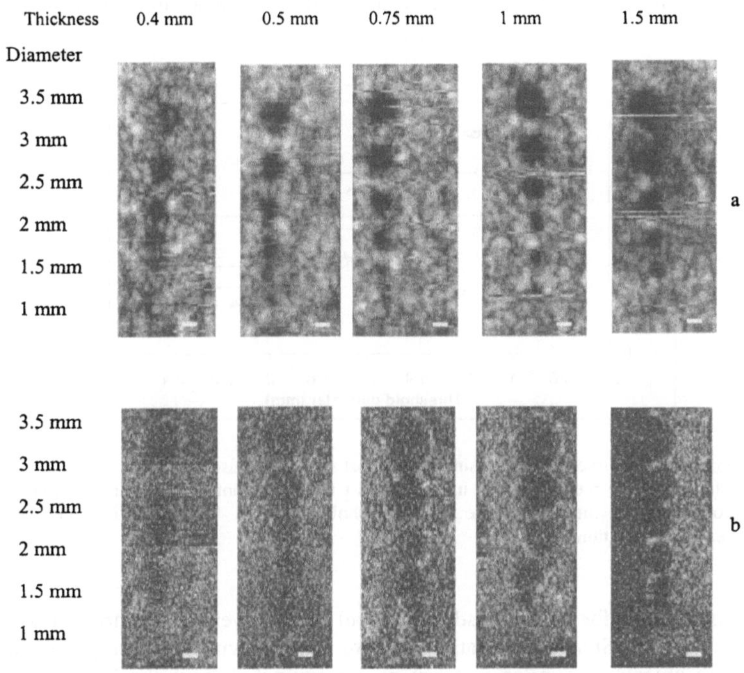

Figure 1. Reflex transmission images of six wax surface discs of increasing thickness (attenuation) and diameters, obtained with a) the first transducer, b) the second transducer. Scale bar: 1.4 mm.

Table 2 presents the CSR related to the RTI images of the discs displayed in Figure 1. For the first transducer, the CSR is higher than for the second one, which is due to its better sensitivity. No significant CSR difference with increasing paraffin thickness was apparent from the images obtained using the second transducer due to its poor signal to noise ratio. In the case of the first transducer, the CSR increased with attenuation due to increasing paraffin thickness.

Table 2. Contrast to speckle ratio (CSR) measured on RTI images of paraffin discs of varying thickness obtained with both transducers.

Paraffin thickness (mm)	0.4	0.5	0.75	1	1.5
CSR 1st transducer	2.1	3.1	3.5	3.9	4.0
CSR 2nd transducer	1.7	1.3	1.8	2.3	2.2

Contrast-Detail curves

Ten observers were asked to state for each disc thickness the smallest disc that they could see with a 50% certainty. The observers knew beforehand the position of the discs, which results in a measure of threshold "perception" and not "location". In Figure 2 the measured image contrast is plotted as a function of the mean of the disc diameters at the threshold perceptibility, from the 10 observations of each disc thickness for both transducers.

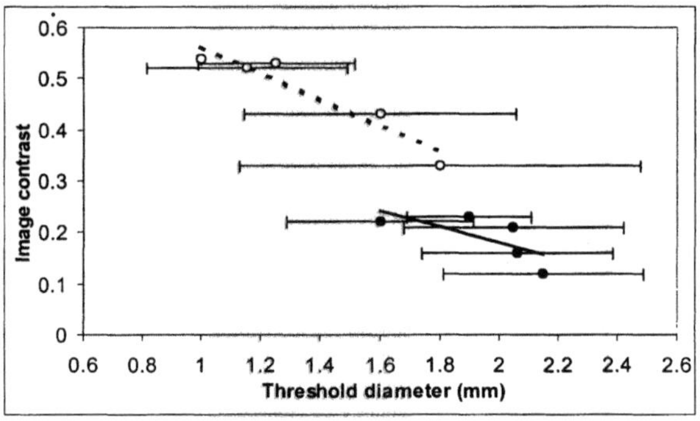

Figure 2: Contrast-detail curves obtained using **O** the most sensitive transducer and ● the most focused transducer, which plot the measured disc image contrast (that may not be a linear function of the attenuation contrast) as a function of the mean threshold diameter. Error bars indicate standard deviations over ten observations.

Figure 2 shows that for both transducers small details were easier to detect with increasing image contrast. Standard deviations tend to increase with decreasing contrast. It appears that some objects of similar size can be discriminated from the background at lower contrast using the second transducer than when using the first transducer, consistent with its better spatial resolution. However, although this is a real perceptual result it provides little useful information about the relative quality of the RTI images from the two transducers. The measured image contrast for the second transducer is lower than for the first due to its poorer sensitivity, resulting in the display of noise that limits the image contrast that can be achieved for a given object contrast. If disc thickness, which varies linearly with paraffin attenuation, is used as a measure of object contrast it becomes clear that the first transducer allows smaller discs to be seen despite its poorer spatial resolution (Figure 3). This is because the second transducer provided much better transfer of object contrast to image contrast, as may be seen by comparing Figure 3 with Figure 2.

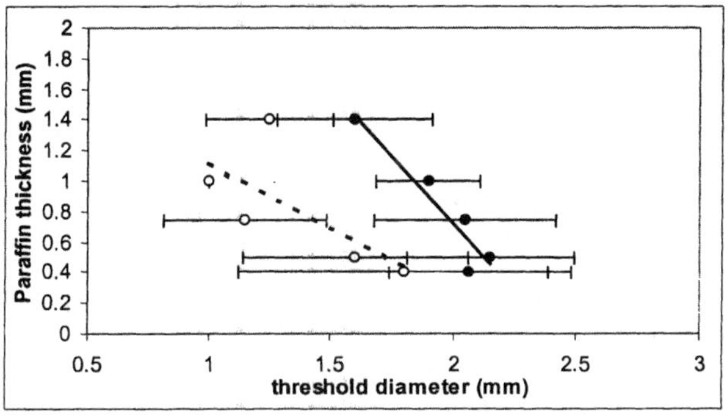

Figure 3. Contrast-detail curves obtained for **O** the most sensitive transducer and ● the most focused transducer, which plot the paraffin disc thickness (related to attenuation contrast) as a function of the mean threshold diameters for visible discs.

PRELIMINARY *IN VITRO* SKIN LESION IMAGE

A preliminary assessment was made of the system's capability to image the attenuation produced by a benign skin tumour. A benign lesion constituted of organising granulation tissue (probably traumatic in aetiology) was placed in formalin immediately after excision and transported to the laboratory. It was scanned in water 2 hours after

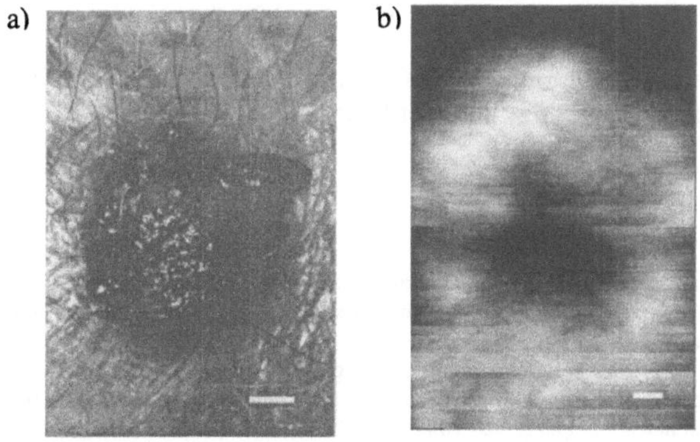

Figure 4: a) Photograph of a benign skin lesion *in vivo*, b) RT image of the same lesion *in vitro* (step: 50 µm, Integration zone: 200 samples). Scale bar: 1 mm.

excision, following the same procedure as described for the phantoms with a 50 µm step using the first (most sensitive) transducer. The RTI image displayed in Figure 4 (right) was obtained with an integration zone of about 1.5 mm (200 samples) placed in the dermis.

Sufficient contrast was obtained to discriminate the lesion from the surrounding healthy tissue. The boundary between the lesion and the healthy tissue is clearly visible, although it is blurred by the 300 μm lateral resolution of this transducer. The visible horizontal banding is an artifact due to movement of the specimen during the acquisition. The image on the left of Figure 4 is a skin surface photograph of the lesion, obtained *in vivo* and is not to the same scale as the RT image.

SUMMARY AND DISCUSSION

Reflex Transmission Imaging (RTI) has been implemented in order to map the relative attenuation characteristics of the skin surface. Phantoms made of a homogeneous backscattering gelatin material, containing surface attenuating discs of varying thickness, were used to evaluate image quality by measuring contrast-to-speckle ratio and plotting contrast-detail curves. Using these phantoms it was shown that of two transducers evaluated for RTI, a 20 MHz poorly focused transducer provided images of better quality than a 15 MHz well-focused transducer, due to its better sensitivity.

Attenuating discs as small as 1 mm in diameter in the phantom were clearly visible with this experimental RTI skin imaging system and a preliminary RTI skin image that was obtained *in vitro* clearly visualised an attenuating benign tumour against surrounding healthy tissue. Further work will aim to optimise high frequency RTI and assess its usefulness for skin tumour diagnosis.

A preliminary contrast detail phantom for RTI, constructed from surface attenuating discs, proved to provide a successful method for assessing relative image quality. However, the limits of spatial resolution for both transducers could not be explored using attenuating discs because diameters smaller than 1 mm were not available due to construction difficulties. This restriction has since been circumvented using a phantom constructed from surface attenuating triangles that can provide linear object dimensions of hundreds of microns, which will be reported in a future paper.

Acknowledgements

The authors are grateful to the EPSRC, British Council and INSERM for supporting this work, and to Cortex Technology ApS for their help in modifying the Dermascan C T.M.

REFERENCES

1. J.C. Bamber, P. Frederiksen, C.C. Harland, S.G. Kale, P.S. Mortimer, Differentiation of benign lesions from melanoma by quantitative analysis of high resolution ultrasound images, *Skin Res Tech*. 3:187 (1997). (Abstract)
2. P.S. Green, M. Arditi, Ultrasonic reflex transmission imaging, *Ultrasonic imaging*, 7, 201:214 (1985).
3. G. Cohen, F.A. Dibianca, The use of Contrast-Detail-Dose evaluation of image quality in a computed tomographic scanner, *J Comput Assist Tomogr*, 3(2), 189:195 (1979).
4. N.L. Bush, C.R. Hill, Gelatine-Alginate Complex Gel: A new acoustically tissue-equivalent material, *Ultrasound Med. Biol.*. 9(5), 479:484 (1983).
5. M.S. Patterson, F.S. Foster, The improvement and quantitative assessment of B-mode images produced by an annular array/cone hybrid. *Ultrasonic Imaging*, 5, 195:213 (1983).
6. P. Schattner, T.K. Whitehurst, J.F. Jensen , A.S. Shah, P.S. Green, Three-dimensional ultrasonic reflection and attenuation imaging, *IEEE Trans. Ultrason. Ferroelec. Freq. Control*, 21, 102:111 (1992).

REAL-TIME SPATIAL COMPOUND IMAGING: TECHNICAL PERFORMANCE IN VASCULAR APPLICATIONS

Robert R. Entrekin[1], James R. Jago[1], and Steen C. Kofoed[2]

[1]ATL Ultrasound, Inc.
P.O. Box 3003
Bothell, WA 98041-3003 USA

[2]Department of Vascular Surgery
Gentofte Hospital
University of Copenhagen
DK-2900 Hellerup, Denmark

ABSTRACT

Spatial compound imaging is a technique in which a number of co-planar, tomographic ultrasound images of an object are obtained from different directions, then combined into a single compound image. Real-time spatial compounding uses electronic beam steering to rapidly acquire component frames from different view angles. The component frames are combined at real-time frame rates to produce compound images with reduced speckle and improved continuity of specular reflectors. Real-time spatial compounding was first reported over 30 years ago, but it has only recently become available on commercial ultrasound systems. This paper describes the technical performance characteristics of real-time spatial compound imaging on the ATL HDI 5000 system, and considers the clinical relevance of these characteristics in vascular sonography.

BACKGROUND

Spatial compounding can be traced back to the earliest days of medical ultrasound imaging. Researchers in the 1950's [1] quickly recognized that the specular nature of reflections from tissue interfaces limited the amount and continuity of information available in early ultrasound images, and used compound mechanical scanning to remedy the problem. Later, when static B-scanners were adopted into clinical practice in the 1960's, systematic manual rocking of the transducer, i.e. "compounding," became an accepted technique [2] for the same reason.

With the introduction of grayscale ultrasound imaging in the 1970's, a second motivation for spatial compounding became evident. Acoustic speckle, caused by wave interference between unresolved scatterers, produced an undesirable random mottle pattern in ultrasound images. Speckle degraded the appearance of tissue texture and made it more difficult to detect small, low contrast lesions in a tissue. Spatial compounding of independent view directions was recognized as a way to reduce speckle noise and improve contrast resolution [3]. Automated water-path B-scanners, such as the "Octoson" [4], utilized spatial compounding with eight transducers as a strategy for both reducing speckle and improving the continuity of specular reflectors.

With the advent of real-time scanners in the middle 1970's through the early 1980's, ultrasound examination became faster and easier to perform, and was widely adopted into many areas of clinical practice. Spatial compounding, which required multiple acquisitions of the same target, was sacrificed in the quest for real-time imaging at increasingly higher frame rates. A number of researchers extended spatial compounding to real-time scanning [5, 6, 7, 8, 9] but none of these systems were developed substantially beyond the technical proof-of-concept stage. By the late 1980's, ultrasound researchers had apparently lost interest in real-time spatial compound imaging.

Renewed interest in real-time spatial compound imaging began in the mid-1990's when at least three groups, working independently, re-implemented the idea with digital ultrasound scanners. These efforts resulted in the published description of a proof-of-concept laboratory system [10] and with the commercial product introduction of real-time spatial compound imaging by two different ultrasound companies [11, 12]. Commercial availability allowed physicians and sonographers to evaluate real-time spatial compound imaging in a variety of medical applications, and the initial clinical impressions have been positive [13, 14]. However, very little technical information has been available to date on the performance characteristics of these products. The purpose of this article is to quantify the performance characteristics of real time spatial compound imaging for one such system, and to illustrate the clinical relevance of these characteristics.

PRINCIPLE OF OPERATION

System Description

The ATL HDI 5000 is a digital ultrasound imaging system that employs a family of broadband transducers to address a wide range of clinical applications. "SonoCT™ Real-time Spatial Compound Imaging" was introduced commercially by ATL in September of 1999 on two transducers, the L12-5 38mm and L12-5 50mm broadband linear arrays. These two transducers have similar performance characteristics, with the major difference being the number of elements (192 vs. 256) and the subsequent fields of view (38 mm vs. 50 mm). The primary clinical applications of these transducers include imaging the breast, thyroid, testicle, peripheral blood vessels, and the musculo-skeletal anatomy of the shoulder, knee, and other joints, tendons, and ligaments. The examples in this paper will focus on the impact of spatial compounding in vascular ultrasound imaging.

A simplified signal path block diagram for real time spatial compounding is shown in Figure 1. As in conventional imaging with a linear array transducer, the beamformer is programmed to transmit a set of parallel ultrasound beams (a "frame") where each beam is perpendicular to the face of the transducer array. For the L12-5 38mm transducer operating with 192 beams per frame, each beam is laterally offset by the array element pitch of 0.2 millimeters. Returning echoes from each beam are amplified, digitized, delayed, summed, and processed in the digital domain to detect the envelope of the beam-formed radio frequency signal. The detected envelope samples are then geometrically warped and spatially re-sampled ("scan converted") into an accurately scaled pixel space and a video signal is generated for display on a video monitor.

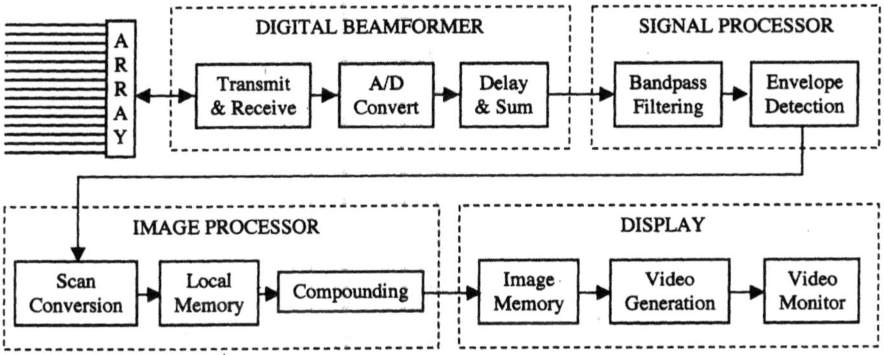

Figure 1. Signal path block diagram of the HDI 5000 ultrasound system, showing the steps involved in generating real-time compound images.

In real-time spatial compound imaging on the HDI 5000, a number N of "component" frames is acquired from different view angles, where N can range from 3 to 9. For each frame in the acquisition sequence, the beamformer acquires a series of parallel ultrasound beams that are steered in the scan plane by a predetermined angle, typically ranging from 0 to 20 degrees in either direction. The maximum useful steering angle is limited primarily by grating lobe artifacts, which are in turn determined by the characteristics of the transducer array, such as element pitch, operating frequency, and element directivity.

Prior to compounding, each of the steered component frames is scan converted onto a common sampling grid and saved in a local memory in the image processor. The scan conversion process insures that a target in one component frame is spatially registered at the same location in the other N-1 component frames. The compound image is formed by averaging data from corresponding locations in all of the component frames. Regions of the steered frames that lie outside the rectangular boundaries of the non-steered (0°) frame are not rendered in this implementation. The acquisition process is repeated continuously.

PERFORMANCE CHARACTERISTICS

Frame Rate and Persistence

Figure 2 illustrates how a compound image is formed in real time from a moving average of the N most recently acquired component frames, and updated with each new frame. The compound image frame rate is therefore identical to the acquisition frame rate. The default acquisition frame rate for the L12-5 38mm is 61 Hz when operating with 192 beams per frame at an image depth of 3 centimeters. Updating the compound image at this rate prevents the "jerkiness" that would be perceived if only one compound frame was produced after acquiring all N component frames, especially for large values of N.

Nevertheless, averaging N component frames constitutes a temporal low pass filter that can cause "persistence" or image blurring artifacts if either the transducer or the target move too rapidly. Jespersen [10] quantified this persistence for compound imaging in terms of the "maximum temporal separation" Δt_{max}. This is the time needed to scan a given point in the tissue from all steering directions, and represents the maximum window of time over which motion blurring can occur. It is well approximated by (N-1)/R, where N is the number of component frames being compounded and R is the acquisition frame rate. Compounding more frames increases the persistence Δt_{max}. However, more compounding

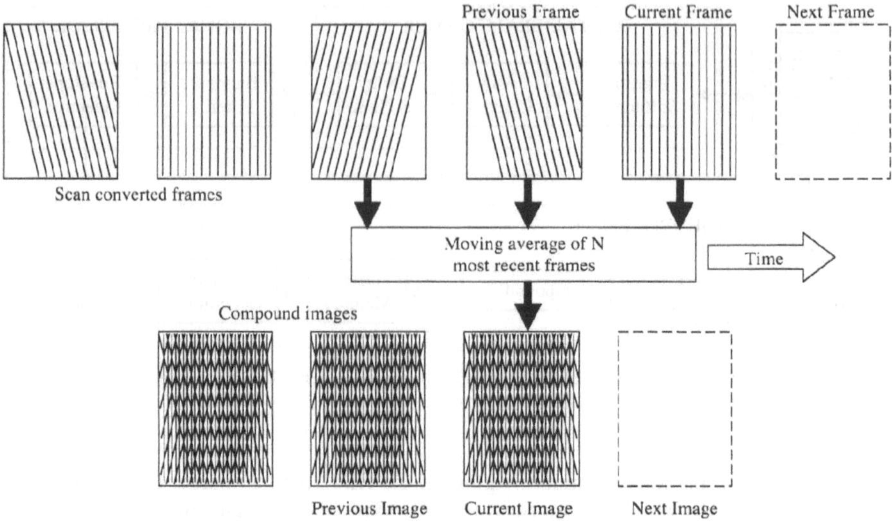

Figure 2. Real time spatial compounding. Incoming samples of the detected echo envelope are geometrically warped and re-sampled onto a common geometry and stored in a local memory. The scan-converted frames are then combined with a temporal moving average filter to form compound images.

usually results in better image quality, creating a tradeoff between image quality and persistence.

A solution to this tradeoff is to provide different degrees of compounding for different clinical applications. In addition, the HDI 5000 provides users with different compounding modes where: (a) spatial compounding is turned off to minimize frame averaging ("conventional" mode); (b) only a few scan angles are compounded ("survey" mode); or (c) spatial compounding is maximized by acquiring frames from many scan angles ("target" mode). As shown in Table 1, when the transducer is selected for examination of the carotid artery, a structure that exhibits significant pulsatility, the "survey" mode of compound imaging only acquires 3 component frames, while the "target" mode acquires 7 frames. However, when the same transducer is selected for imaging the breast, 5 frames are compounded in the survey mode and 9 frames are compounded in the target mode. By providing users with application specific optimization of compounding, as well as allowing them to select the compounding mode (conventional, survey, target), good results can be obtained in virtually all clinical situations.

Table 1. Temporal parameters for various compounding modes and clinical applications

Clinical Application and Compounding Mode for L12-5 38mm transducer	Frame Rate 30 mm depth, 1 Tx zone	N Number of component frames	Δt_{max} Maximum temporal separation per [10]
Cerebrovascular/Carotid Survey mode	61 Hz	3	33 ms
Cerebrovascular/Carotid Target mode	61 Hz	7	98 ms
Small Parts/Breast Survey mode	61 Hz	5	66 ms
Small Parts/Breast Target mode	61 Hz	9	131 ms

Speckle Reduction

The speckle signal-to-noise ratio (SNR_0) is an objective figure of merit derived from the first order echo amplitude statistics of speckle [15]. SNR_0 is easily measured from the echo amplitude histogram in a phantom containing only random scatterers. SNR_0 is simply the mean divided by the standard deviation, and has an expected value of 1.91 for fully developed speckle. The effect of spatial compounding on SNR_0 has a well-founded basis in theory [16]. If N frames are acquired from sufficiently different view angles, the frame-to-frame speckle patterns will be uncorrelated, and averaging these frames will increase the SNR_0 by a factor of $\sqrt{N}$. If the view angles are not sufficiently different, the speckle patterns will be partially correlated and the improvement in SNR_0 will be less than $\sqrt{N}$. It is possible to predict the correlation between view angles based on the overlap of the apertures [10, 16]. However, it is more convenient to simply calculate N_{eff}, the "effective" number of independent frames from the measured SNR_0 of the compound image as prescribed by Jespersen [10]. The measured ratio $(SNR_0/1.91)^2$ of the compound image indicates the effective number of independent frames for a specific implementation.

For these measurements, the L12-5 38mm transducer was used in the "Cerebrovascular: Carotid" application specific setup, with a single transmit focus at 3 centimeters. The phantom consisted of agar gel containing a uniformly random suspension of polystyrene micro-spheres. Raw echo envelope data were captured from the HDI 5000 and analyzed on a PC using proprietary analysis software. Twenty images, each taken from an independent slice through the phantom, were acquired for the three compounding modes available to the user in this application setting. A square region of interest was located near the center of each image, extending from a depth of 20 millimeters to a depth of 30 millimeters. The first order statistics were calculated after the logarithmic compression which is normally applied was "backed out" of the envelope data to allow for comparison with theory. Each histogram is an ensemble of all 20 regions of interest.

The results of these measurements are shown graphically in Figure 3 and numerically in Table 2. In Figure 3, the histogram for conventional imaging (N=1) closely resembles the expected Rayleigh probability density distribution for fully developed speckle and the measured SNR_0 is in good agreement with the expected value of 1.91. The histogram for the survey mode shows a taller, narrower peak, and the histogram for the target mode demonstrates the highest peak with the least variance about the mean. Table 2 shows quantitatively that spatial compounding reduces speckle noise, increasing the speckle signal-to-noise ratio by 37% and 64% for the survey and target modes, respectively.

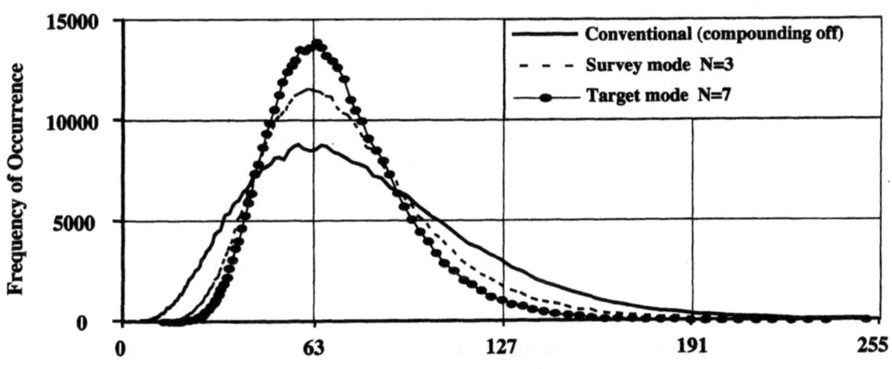

Figure 3. Echo amplitude histograms for conventional and compound imaging.

However, these results also show that the frames acquired from adjacent view directions were partially correlated, since even for survey mode N_{eff} was less than N. Partial correlation of component frames can be explained by the overlap of the apertures used to acquire the frames from adjacent steering angles. These overlaps result from the relatively small incremental steering angles and the relatively large transmit and receive apertures used in the "Cerebrovascular/Carotid" application setting (Tx≈8.4mm; Rx≈25.6mm). Since the maximum steering angle is limited by grating lobe artifacts, the alternative to increase SNR_0 is to decrease the aperture overlap by decreasing the size of the apertures used for transmit and receive beamforming. However, aperture reduction degrades lateral spatial resolution. Experience suggests that sonographers are not willing to substantially sacrifice spatial resolution in order to gain better contrast resolution (e.g. speckle reduction). For this reason, transmit and receive apertures of the HDI 5000 were left unchanged to maintain good lateral resolution, and the actual gains in speckle reduction were modest compared to the theoretical maximum.

Table 2. Speckle SNR results for compound imaging in "Carotid" application setup. Increases in SNR_0 (conventional to survey, survey to target) are statistically significant ($p>0.001$).

N Number of component frames acquired	$\Delta\theta$ Incremental steering angle	SNR_0 Speckle signal to noise ratio (mean ± std deviation)	$SNR_0/1.91$ Improvement ratio	N_{eff} Number of independent views
N=1 Conventional mode (compounding off)	0°	1.929 ± 0.017	N/A	1
N=3 Survey mode 0°, ±12.75°	12.75°	2.619 ± 0.059	1.37	1.88
N=7 Target mode 0°, ±4.25°, ±8.5°, ±12.75°	4.25°	3.128 ± 0.060	1.64	2.68

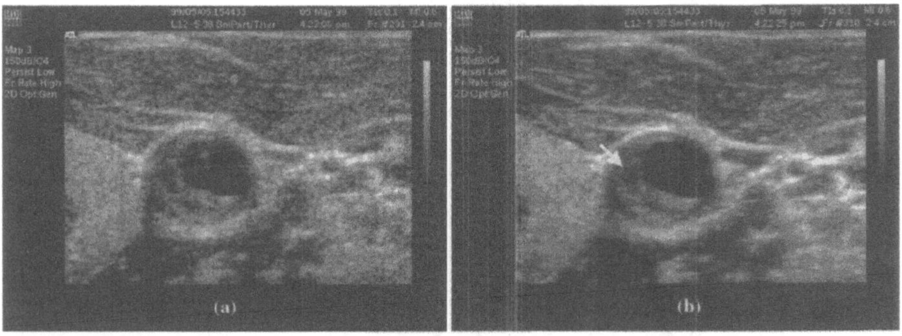

Figure 4. Conventional (a) and compounded (b) views of an atherosclerotic plaque in the carotid artery.

Nonetheless, the clinical significance of even a modest amount of speckle reduction can be appreciated in Figure 4, which shows conventional and compound short axis views of an atherosclerotic plaque in the right common carotid artery. The reduction of speckle in the compound image in Figure 4b results in improved tissue contrast between the plaque, the vessel walls, and blood, as well as between the surrounding thyroid and muscle tissues. In another example, Figure 5 shows conventional and compound longitudinal views of an atherosclerotic plaque at the carotid bifurcation. The inset in the lower left corner of Figure 5b shows that speckle reduction from spatial compounding more clearly defines areas of high echogenicity, which may correspond to calcification. Spatial compounding clearly improves subjective contrast resolution and tissue differentiation in these clinical images.

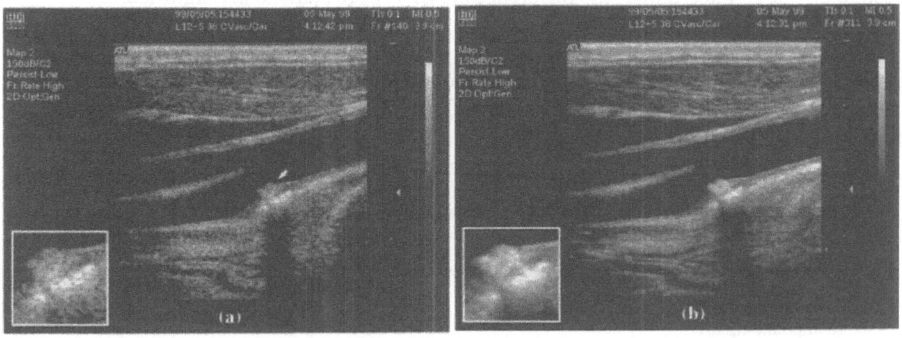

Figure 5. Conventional (a) and compound (b) views of an echogenic plaque in the carotid artery.

Detectability of Low Contrast Regions

The detectability index (denoted d', also known as the "lesion signal-to-noise ratio") is a figure of merit for the detection of low contrast lesions against a speckle background by an ideal observer [17, 18]. Lesion detectability is a function of spatial resolution, statistical properties of the speckle pattern in the lesion and background, as well as lesion contrast and size. However, the effect of spatial compounding on lesion detectability has been treated theoretically and - like SNR_0 - compounding with N independent views should increase the detectability index by a factor of $\sqrt{N}$ [17].

Measurements of the detectability index were made using a random micro-sphere/agar gel phantom with a background concentration of 400 spheres per cubic millimeter. A 4-millimeter diameter cylindrical lesion contained 200 spheres per cubic millimeter, resulting in a backscatter intensity ratio of 0.5 (-3 decibels). This lesion size and contrast were chosen to simulate the task of detecting a small, hypo-echoic, lipid-rich region within a carotid artery plaque. Twenty images were acquired from independent slices through the cylindrical lesion and background for each of the three compounding modes. Raw echo envelope data were captured, the logarithmic compression was "backed out," and the intensity was calculated by squaring the linear envelope as prescribed by Smith [17]. For each image, a 3.5-millimeter circular region of interest was centered inside the lesion (to avoid lesion boundary reflections), and a similarly sized region was located at the same depth in the background. The mean intensity within each region was calculated, and the "mean of the means" μ and the "variance of the means" σ^2 for the 20 lesion/background pairs were used to determine the detectability index from equation (1).

$$d' = \frac{\mu_{lesion} - \mu_{background}}{\sqrt{\left(\sigma^2_{lesion} + \sigma^2_{background}\right)/2}} \qquad (1)$$

The results of these measurements are shown in Table 3. As expected, spatial compounding increased the detectability index by a factor approximately equal to $\sqrt{N_{eff}}$. However, there was a difficulty in determining the uncertainty associated with these measurements, since both means (μ) and variances (σ^2) for the twenty lesion/background pairs were used to form a single estimate of d'. Wear and colleagues [19] have examined this problem, and have offered estimates of the uncertainty - not of d' - but of $(d')^2$ for various conditions of lesion contrast, size, and number of lesion/background pairs. For the conditions in this experiment, the coefficient of variation of $(d')^2$ had a value of about 0.33. Despite this large uncertainty, the differences in $(d')^2$ were statistically significant.

Table 3. Detectability index results for compound imaging in "Carotid" application setup. The increases in $(d')^2$ (conventional to survey, survey to target) are statistically significant ($p>0.001$).

N Number of component frames acquired	$\Delta\theta$ Incremental steering angle	d' Detectability index 4 mm lesion –3dB contrast	d'_{comp} / d'_{conv} Improvement ratio	$(d')^2$ ± std deviation per [19]
N=1 Conventional mode (compounding off)	0°	13.6	N/A	184 ± 106
N=3 Survey mode 0°, ±12.75°	12.75°	18.7	1.38	350 ± 201
N=7 Target mode 0°, ±4.25°, ±8.5°, ±12.75°	4.25°	24.7	1.82	611 ± 351

Clinically, the importance of improving the detectability of low contrast regions can be appreciated by closely examining the plaque seen previously in Figure 4. Spatial compounding subjectively increases the conspicuity of a hypo-echoic region (arrow) within the plaque. This may be significant because lipid-rich plaques appear hypo-echoic, and have been associated with a higher risk of intra-plaque hemorrhage [20]. Improvements in both speckle reduction and in lesion detectability suggest that real-time spatial compounding may improve the ability of ultrasound to characterize plaque composition [21] and subsequent risk of stroke.

Visualization of Specular Reflectors

The amplitude of the echo received from a planar tissue boundary is strongly dependent on the angle of incidence of the ultrasound beam with respect to the boundary. Specular reflection is illustrated in Figure 6, which shows three different views of a smooth, polypropylene cylinder suspended in a water/glycol solution (1540 meters per second). The ultrasound beams in the three views were steered at angles of 17 degrees, 0 degrees, and –17 degrees in Figure 6a, 6b, and 6c, respectively. The pair of dashed white lines represents the front and rear boundaries of the cylinder wall. Note that the bright reflection or "glint" appears at a position along the circumference where the surface of the cylinder is perpendicular to the incident beam. This is shown graphically in Figure 6d, which plots the relative brightness of the reflection as a function of angular position along the circumference of the cylinder. The peak brightness occurs at the nominal steering angle for each of the three frames.

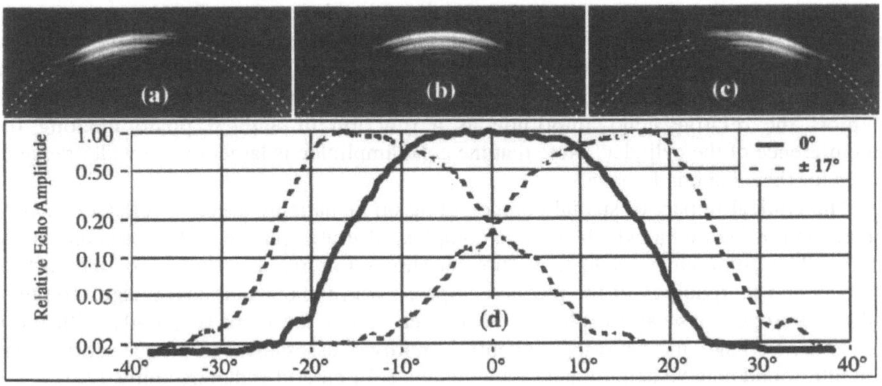

Figure 6. Specular reflection from a cylindrical reflector for steering angles of (a) 17 degrees, (b) 0 degrees, and (c) –17 degrees, respectively. Echo amplitude is plotted as a function of angular position in (d).

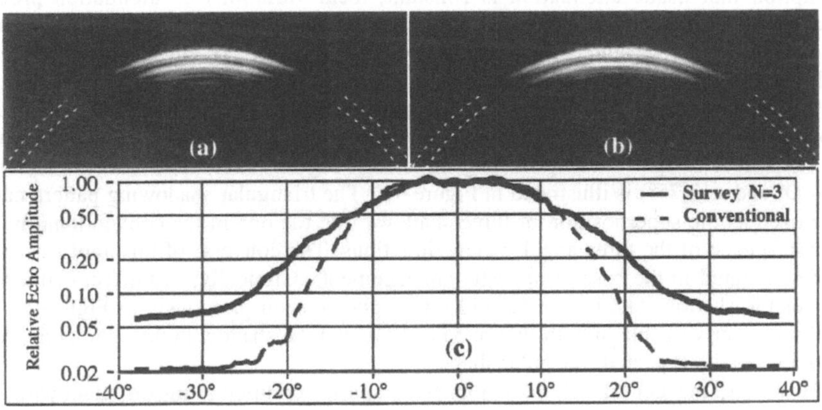

Figure 7. Specular reflection from a cylindrical reflector for (a) conventional and (b) compound imaging for steering angles of 17, 0, and –17 degrees. Echo amplitude is plotted as a function of angular position in (c).

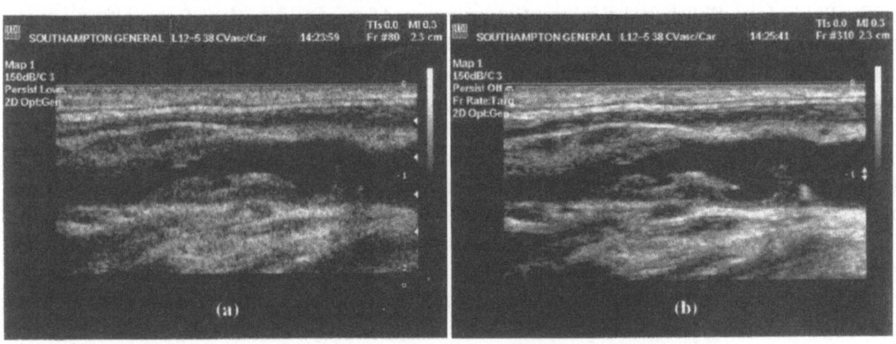

Figure 8. Conventional (a) and compound (b) views of an ulcerated carotid artery plaque.

Spatial compounding improves the visualization of specular reflectors by averaging views from multiple steering angles, increasing the length and consistency of the reflection from curved or angled surfaces. This is illustrated in Figure 7a, which shows the conventional view, and Figure 7b, which shows the compound view of the cylinder. Figure 7c plots the relative echo amplitude as a function of angular position along the circumference of the cylinder. Note that the echo amplitude is larger over a wider range of angles for the compound image.

The clinical impact of spatial compounding on visualizing specular reflectors can be appreciated in Figure 8a, which shows a complex, ulcerated plaque in the common carotid artery. The compound image in Figure 8b shows improved definition of the surface contours and morphology of the plaque compared with the conventional image. Irregular and/or heterogeneous carotid plaques have been associated more frequently with intra-plaque hemorrhage and neurologic events. Improved assessment of plaque morphology with ultrasound may be helpful in selecting patients for carotid endarterectomy. [22].

Shadowing and Enhancement

Acoustic shadowing, and its converse, enhancement, can be generated by inhomogeneous attenuation within tissues. Since ultrasound systems operate on the assumption that tissue attenuation is constant, local variations in attenuation produce characteristic artifacts in the image. A strongly attenuating structure casts a dark shadow, while a region of low attenuation casts a "negative shadow" or enhancement behind it. Shadowing in conventional ultrasound imaging is illustrated in Figure 9a, where a thin slab of absorbing material was placed on the surface of a speckle phantom. This slab introduced approximately 20 dB of attenuation, and produced a stark rectangular shadow.

The effect of spatial compounding on shadowing for three view directions (survey mode, 0° and ±12.75°) is illustrated in Figure 9b. The triangular shadowing pattern can be recognized as the superposition of three shadows like the one in the conventional image, but cast in each of the three angular view directions. The converse of this appearance can be demonstrated in the case of acoustic enhancement. Clinically, this effect can also be seen for the shadow cast by the carotid artery plaque seen previously in Figure 5. The shadow is clearly visible in both Figure 5a and 5b, and the shadow in the compound image approximates the expected triangular shape.

However, the general clinical impact of spatial compounding on shadowing and enhancement is not easy to predict, for several reasons. First, the magnitude and shape of shadows in conventional imaging of real tissues are frequently more complicated than for this simple example. Second, beam refraction effects due to variations in tissue sound speed may also cause shadowing. Refractive shadows have an entirely different etiology, since they only occur when the ultrasound beam encounters a tissue interface at certain "critical" angles. Since real-time spatial compound imaging acquires frames from many angles, refractive shadows may be suppressed because they are not likely to be located at the same position in each component frame [13]. Third, depending on the clinical situation, a shadow can be considered a useful diagnostic cue, or an annoying distraction. In breast imaging, for example, increased attenuation has been associated with cancer, and central shadowing is one of several diagnostic findings suggestive of malignancy [23]. However, refractive shadows from Cooper's ligaments – normal anatomic structures – can mimic the shadows from small occult malignancies, and produce false positive findings. Therefore, the diagnostic relevance of shadowing depends strongly on the specific clinical application. A more complete assessment of the impact of spatial compound imaging on shadowing and enhancement will require comparative clinical studies in specific areas of application.

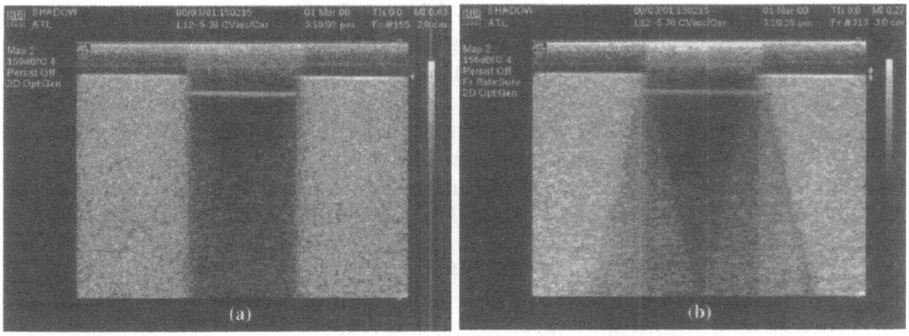

Figure 9. Conventional (a) and compound (b) views of strong acoustic shadowing.

CONCLUSIONS

The ATL HDI 5000 ultrasound system forms spatial compound images in real time from a moving average of the N most recent frames. The compound image is updated at the acquisition frame rate, which prevents the perception of "jerkiness" for large values of N. By providing users with clinical application specific optimization of compounding, as well as allowing them to select the compounding mode, good results without blurring artifacts can be obtained in all clinical situations.

Real-time spatial compound imaging was shown to significantly reduce speckle compared to conventional imaging, with an improvement in speckle signal-to-noise ratio of up to 64%. This amount of speckle reduction, although modest compared to the theoretical maximum, was achieved without reductions of transmit or receive apertures that would otherwise degrade lateral spatial resolution.

Spatial compounding was shown to increase lesion detectability in a contrast phantom. Compounding also improved apparent tissue contrast in images of the carotid artery, and could lead to improvement in the ability of ultrasound to characterize plaque composition.

Real-time spatial compound imaging was shown to improve the continuity of an artificial specular reflector. Clinically, compounding appeared to enhance the visualization of surface ulcerations in a carotid artery plaque, which have been associated with an increased risk of stroke.

Acoustic shadowing from inhomogeneous attenuation appeared in compound images as the superposition of the shadows cast in the various view directions. For the simple case of an absorbing surface layer, compounding preserves the shadow but changes it's shape from rectangular to triangular. Clinically, the impact of compounding on shadowing will require comparative clinical studies in specific areas of application.

In summary, real-time spatial compounding appears to be a promising tool for improving image quality and enhancing the clinical value of vascular sonography.

ACKNOWLEDGEMENTS

The authors are indebted to Susan Juarez for help with manuscript preparation.

REFERENCES

1. D.H. Howry, Techniques used in ultrasonic visualization of soft tissue structures of the body, *I.R.E. Conv. Rec.* 9: 75-81 (1955).

2. I. Donald and T.G. Brown, Demonstration of tissue interfaces within the body by ultrasonic echo sounding, *Br. J. Radiol.* 34: 539-546 (1961).
3. C. Burckhardt, Speckle in ultrasound B-mode scans, *IEEE Trans. Son. Ultrason.* 25: 1-6 (1978).
4. D. Carpenter, G. Kossoff, W.J. Garrett, K. Daniel, and P. Boele, The UI octoson--a new class of ultrasonic echoscope, *Australas Radiol* 21(1): 85-89 (1977)
5. G.F. Tomey and J.M. Reid, Rotational compound scan, *8th International Conference on Medical and Biological Engineering*, Chicago, IL, July 24, 1969.
6. D.A. Carpenter, M.J. Dadd, and G. Kossoff, A multimode real time scanner, *Ultrasound in Med. & Biol.* 6: 279-284 (1980).
7. M. Berson, A. Roncin, and L. Pourcelot, Compound scanning with an electrically steered beam, *Ultrasonic Imaging* 3: 303-308 (1981).
8. D. Shattuck and O.T. von Ramm, Compound scanning with a phased array, *Ultrasonic Imaging* 4: 93-107 (1982).
9. R. Fehr, P. Grandchamp, and P. Krummenacher, Real time display of an ultrasonic compound image, United States Patent No. 4,649,927 (1987).
10. S.K. Jespersen, J.E. Wilhjelm, and H. Sillesen, Multi-angle compound imaging, *Ultrasonic Imaging* 20: 81-102, (1998).
11. Vingmed Sound, Inc., Santa Clara, CA, 1997 (product brochure).
12. ATL Ultrasound, Inc., Bothell, WA, September 27, 1999 (press release).
13. R. Entrekin, P. Jackson, J. Jago, and B. Porter, Real time spatial compound imaging in breast ultrasound: technology and early clinical experience, *Medica Mundi* 43(3): 35-43 (1999).
14. C.R.B. Merritt, Real-time compound ultrasound takes whole new angle, Diagnostic Imaging, published by Miller Freeman, Inc., December 1999; pp. 95-96.
15. R.F. Wagner, S.W. Smith, J.M. Sandrik, and H. Lopez, Statistics of speckle in ultrasound B-scans, *IEEE Trans. Son. Ultrason.* 30: 156-163 (1983).
16. G.E. Trahey, S.W. Smith, O.T. von Ramm, Speckle pattern correlation with lateral aperture translation: experimental results and implications for spatial compounding, *IEEE Trans. Ultrason. Ferroelect. Freq. Contr.* 33: 257-264 (1986).
17. S.W. Smith, R.F. Wagner, J.M. Sandrik, and H. Lopez, Low contrast detectability and contrast/detail analysis in medical ultrasound, *IEEE Trans. Son. Ultrason.* 30: 164-173 (1983).
18. H. Lopez, M.H. Loew, and D.J. Goodenough, Objective analysis of ultrasound images by use of a computational observer, *IEEE Trans. Med. Imaging* 11: 496-506 (1992).
19. K.A. Wear, R.M. Gagne, and R.F. Wagner, Uncertainties in estimates of lesion detectability in diagnostic ultrasound, *J. Acoust. Soc. Am.* 106: 1161-1173 (1999).
20. M.L. Gronholdt, B.M. Wiebe, H. Laursen, T.G. Nielsen, T.V. Schroeder, and H. Sillesen. Lipid-rich carotid artery plaques appear echolucent on ultrasound B-mode images and may be associated with intraplaque haemorrhage, *Eur J Vasc Endovasc Surg* 14(6): 439-45 (1997).
21. J.E. Wilhjelm, M.L. Gronholdt, B. Wiebe, S.K. Jespersen, L.K. Hansen, and H. Sillesen, Quantitative analysis of ultrasound B-mode images of carotid atherosclerotic plaque: correlation with visual classification and histological examination, *IEEE Trans Med Imaging* 17(6):910-22 (1998).
22. A.F. AbuRahma, P.D. Kyer 3rd, P.A. Robinson, and R.S. Hannay, The correlation of ultrasonic carotid plaque morphology and carotid plaque hemorrhage: clinical implications, *Surgery* 124(4):721-6 (1998).
23. A.T. Stavros, D. Thickman, C. Rapp, M. Dennis, S.H. Parker, and G.A. Sisney, Solid breast nodules: use of sonography to distinguish between benign and malignant lesions, *Radiology* 196:123-134 (1995).

MEDICAL APPLICATION OF SOLITON IMAGING

Woon S.Gan

Acoustical Technologies Singapore Pte Ltd,
209-212 Innovation Centre, NTU,
16 Nanyang Drive, Singapore 637722
Republic of Singapore

INTRODUCTION

Solitary waves do not deform or change in shape as they propagate to infinite distance, while maintaining a high localization. In theory, these beams have an infinite depth of field and high resolution which will be of great use for medical imaging. The velocity of a longitudinal wave in solids is much more than on the liquid surface. This is a reason for unsucessful soliton observation in experiments in solids. In recent years, a new field of theoretical physics-soliton theory- is being rapidly developed. Great success has been attained in the elaboration of novel approaches to soliton theory. In particular, a classical inverse scattering transform (IST) has been developed, several problems have been solved on the basis of the quantum IST, complete integrability of some multidimensional nonlinear wave equations has been proved. In this paper, we propose the use of ultrasonic solitary waves for medical imaging. The experimental scheme of diffraction tomography will be used and the design of the transducer will be described in the final section of this paper.

THE FORWARD SCATTERING PROBLEM

We use the unpertubed nonlinear Schrodinger equation (NLSE) to describe the propagation of ultrasound through the human tissue. Our problem here is an inverse scattering problem to reconstruct the object function from the asymptotic data which results from the scattering data.

First we have the Schrödinger equation

$$Ly = \frac{-\partial^2 y}{\partial x^2} + Qy = k^2 y \tag{1}$$

where Q = object function, k = wave vector and L = Schrödinger operator.

For the class of function Q which we are considering, the eigenvalue problem (1) is called a scattering problem. This is because eqn(1) can be interpreted as an incident plane wave interacting with an object defined by the function Q(x,t). The incident and scattered waves can be expressed in terms of certain fundamental solutions which asymptotically either as $x \to +\infty$ or $x \to -\infty$ have the plane wave character of solutions for the problem $Q(x,t_0) \equiv o$. The family of solutions parametrised by k are usually called the Jost solutions for (1).

The nonlinear Schrödinger equation is the first one to, which the IST has been applied[1] and its complete integrability has been proved. The NLS in the canonical form is

$$i q_t + q_{xx} + 2 |q|^2 q = 0 \tag{2}$$

where q = scattered wave amplitude.

A solution of this equation is sought in the form

$$q(x,t) = A\,(x\text{-}vt)e^{\,i\varphi t + i\psi x} \tag{3}$$

where v = wave velocity

Substitution of (3) into (2) yields

$$A_{\xi\xi} + i(2\psi - v)A_\xi - A(\psi + \varphi^2) + 2A^3 = 0,\ \xi = x - vt \tag{4}$$

Let the complex coefficient before A_ξ vanish. Then

$$\psi = \frac{v}{2}$$

Further ,from the equality

$$\varphi = -\psi^2 + B$$

We find that

$$\varphi = -\frac{v^2}{4} + B$$

Finally , from (4) we obtain

$$A_{\xi\xi} - BA + 2\,A^3 = 0 \tag{5}$$

The last equality is integrated, to result in

$$(A_\xi)^2 = C + BA^2 - A^4$$

Then the solution for A is found by inverting an elliptic integral

$$\int_{A_0}^{A} \frac{dA}{\sqrt{(C + BA^2 - A^4)}} = \pm\xi$$

(6)

Let the quadratic polynomial occurring here be represented as

$$P(A) = (\alpha_1 - A^2)(A^2 - \alpha_2)$$

Then we have, with F(A,K) the incomplete elliptic integral

$$\frac{1}{\sqrt{\alpha_1}}\{K(k) - F(A,k)\} = \pm\xi, k = \sqrt{\left(\frac{\alpha_1 - \alpha_2}{\alpha_1}\right)},$$

$$\alpha_{1,2} = \frac{B}{2} \pm \sqrt{\left(\frac{B^2}{4} + C\right)}, \alpha_{1,2} = A_{1,2}^2$$

From this, we have

$$A(\xi) = A_1\left\{1 - \left[\left(1 - \frac{A_1^2}{A_2^2}\right) sn^2(\xi,k)\right]\right\}^{1/2}$$

(7)

For $C \to 0, A_1 \to A_0, k \to 1$, $A \to A_0 \operatorname{sech} A_0\xi$, $A_0 = \sqrt{B}$

Ultimately the soliton solution is

$$q_s(x,t) = \frac{\sqrt{B}}{\cosh\sqrt{B(x - vt)}} \exp\left\{-i\frac{v^2}{4}t + \frac{ivt}{2} + iBt\right\}$$

Insertion of $\sqrt{B} = 2\eta, v = -4\xi$ into the above expression leads us to a canonical form for the one-soliton solution

$$q_s(x,t) = 2\eta \operatorname{sech} 2\eta(x + 4\xi t - x_0)\exp\{-2i\xi x + 4i(\eta^2 - \xi^2)t + i\delta_0\}$$

(8)

The NLS soliton amplitude is independent of its velocity. The NLS soliton describes a modulated wave packet propagating in a nonlinear dispersive medium with a constant velocity. (see Fig 1.)

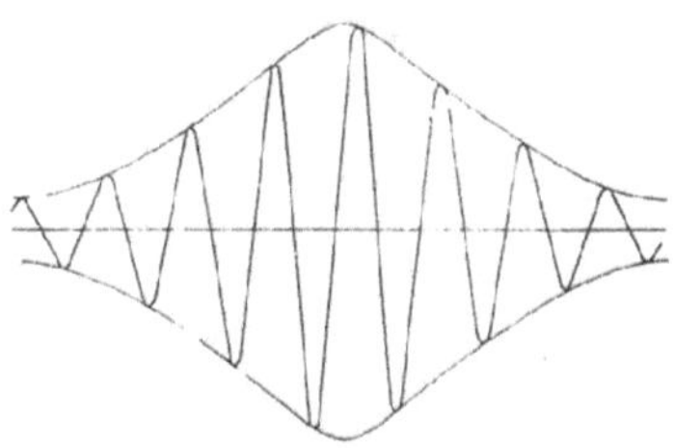

Fig 1. Envelope soliton of NLS equation.

THE INVERSE SCATTERING PROBLEM

The inverse – scattering problem for the Schrödinger equation is to reconstruct the object function from the asymptotic data which results from the scattering process. The scattering data completely determines the spectral distribution function which can be associated with the Schrödinger equation for given --- boundary conditions. Our principal interest is that Q can be recovered from solving a linear integral equation which involves the scattering data so that if $S_{\pm}(t)$ is determined from the initial data $S_{\pm}(0)$ at t =0 then we can in principle construct Q(x,t) at arbitrary later times from the initial Q(x,0). This linear integral equation is called the Marchenko equation[2]. The main purpose of this work is to derive the Marchenko equation for the Schrödinger equation subject to the condition

$$\int_{-\infty}^{\infty}(1+x^2)|Q(x)|dx<\infty$$

The most direct derivation starts from the fundamental relation

$$T_k(k)\phi(x,k)=\psi^*(x,k)+R_+(k)\varphi(x,k) \tag{9}$$

where $T_+=a^{-1}$ and $R_+=ba^{-1}$ where T and R continuous bounded functions

$$R_+(k)=0(1) \text{ as } |k|\to\infty \text{ and } T_+(k)=0(|k|^{-1}) \text{as } k\to\infty \tag{10}$$

There properties of the scattering matrix entries, together with the boundary conditions defining the Jost solutions ensure the existence of the integrals in the Fourier transform of (9)

$$\begin{aligned}&\frac{1}{2\pi}\int_{-\infty}^{\infty}(T_k(k)-1)\phi(x,k)e^{iky}dk\\&=\frac{1}{2\pi}\int_{-\infty}^{\infty}R_+(k)(\varphi(x,k)-e^{ikx})e^{iky}dk+\\&\frac{1}{2\pi_{-}}\int_{-\infty}^{\infty}R_+(k)e^{ik(x+y)}k+\frac{1}{2\pi}\int_{-\infty}^{\infty}(\psi^+(x,k)-e^{-ikx})\\&e^{iky}dk+\frac{1}{2\pi}\int_{-\infty}^{\infty}(e^{-ikx}-\phi(x,k)e^{iky}dk\end{aligned} \tag{11}$$

The time-independent Marchenko equation can be obtained as

$$\begin{aligned}&K_+(x,y,t)+\Omega_+(x+y,t)+\int_x^{\infty}K_+(x,u,t)\Omega_+(u+y,t)du=0, y\geq x\\&\Omega_+(x,t)=\hat{R}_+(x,t)+\sum_{J=1}^{M}D_{+j}(t)e^{-\eta_j x}\end{aligned} \tag{12}$$

$$where\,\Omega_+(x,t) = R_+^{\wedge}(x,t) + \sum_{J=1}^{M} D_{+j}(t) e^{-\eta_j x}$$

and η_j= imaginary part of the wave vector. Solving the Marchenko eqn under condition

$$P_0(\infty,t) << 1 \tag{13}$$

we obtain

$$Q(x,t) = \frac{2i}{\pi} \int_{-\infty}^{\infty} KR_+(k) e^{2ik(x - C(k^2)t)} dk \tag{14}$$

which gives the reconstructed object function.

DIFFRACTION TOMOGRAPHY

The diffraction tomography method will be used to implement the medical imaging using the solitary wave. The integral (14) represents the object function in terms of its Fourier transform in a rectangular coordinate system representing the frequency domain. A diffraction tomography experiment measures the Fourier transform of the object along circular arcs, thus it will be easier to perform the integration if we modify it slightly to use the projection data more naturally. We will use two coordinate transformations to do this: the first one will exchange the rectangular grid for a set of semicircular arcs and the second will map the arcs into their plane wave decomposition. We first exchange the rectangular grid for semicircular arcs. To do this, we represent $\vec{k}$=(kx,ky) in (14) by the vector sum

$$\vec{k} = k_0(\vec{s} - \vec{s}_0) \tag{15}$$

where $\vec{s}_0 = (\cos\phi_0, \sin\phi_0)$ and $\vec{s} = (\cos\chi, \sin\chi)$

are unit vectors representing the direction of the wave vector for the transmitted and the received plane waves respectively. To find the Jacobian of this transformation write

$$k_x = k_0(\cos\chi - \cos\phi_0) \tag{16}$$

$$k_y = k_0(\sin\chi - \sin\phi_0) \tag{17}$$

and

$$\begin{aligned} dk_x dky &= \left|k_0^2 \sin(\chi - \phi_0)\right| dxd\phi_0 \\ &= k_0\sqrt{1 - \cos^2(\chi - \phi_0)}\, dxd\phi_0 \\ &= k_0\sqrt{1 - (\vec{s}\,.\,\vec{s}_0)^2}\, d\chi\phi_0 \end{aligned} \tag{18}$$

and then (14) becomes

$$Q(x,t)=\frac{2t}{\pi}\left(\frac{1}{2}\right)\int_0^{2\pi}\int_0^{2\pi}k_0^3(\vec{s}-\vec{s}_0)R_+(k)e^{2ik_0(\vec{s}-\vec{s})(r-c(k^2)t)}$$

$$\sqrt{1-(\vec{s}.\vec{s}_0)^2}\,d\chi d\phi_0 \tag{19}$$

This integral gives an expression for the scattered field as a function of the (χ,ϕ_0) coordinate system. The data that are collected will actually be a function of ϕ_0, the projection angle, and k the one-D frequency of the scattered field along the receiver line

REALIZATION OF SOLITARY WAVES WITH FINITE APERTURE RADIATORS

The extension and therefore the aperture requirement of solitary waves in space is infinite. But solitary waves can be realized with finite aperture radiators over large axial distance. To calculate solitary waves from the finite aperture radiation a spectrum of solitary waves is used. If we assume that the radiator is circular and has a diameter of D, the Rayleigh -Sommerfeld formulation of diffraction by a plane screen can be written as

$$\Phi_{Rn}(\vec{r},k)=\frac{1}{i\lambda}\int_0^{2\pi}d\phi'\int_0^{D/2}r'dr'q(r,k)\frac{e^{ikr_{01}}}{r_{01}^{\;2}}z$$

$$+\frac{1}{2\pi}\int_0^{2\pi}d\phi'\int_0^{D/2}r'dr'q(r,k)\frac{e^{ikr_{01}}}{r_{01}^{\;3}}z(n=0,1,2,\text{---}) \tag{20}$$

and

$$q(\vec{r},t)=F^{-1}\left[q(\vec{r},\frac{\omega}{c})\right],(n=0,1,2,---\_$$

where λ is wavelength r_{01} is the distance between the observation point, $\vec{r}$, and a point on the source plane (r',ϕ') and F^{-1} represents the inverse Fourier transform. The first and second terms in (20) represent the contribution from high and low frequency components, repectively.

CONCLUSIONS

We have approached the inverse scattering or reconstruction problem using a different reconstruction algorithm based on the Marchenko equation. We also show that it is possible to apply solitary wave to medical imaging using the diffraction tomography scheme.

REFERENCES

1.M.J. Ablowitz, and H.Segur in : Solitons and Inverse Spectral Transform (SIAM, Philadelphia,1980)
2.A.K.Agranovich and V.A.Marchenko, in: The Inverse Problem of Scattering Theory (Gordon and Breach,London,1963).

ULTRASONIC TISSUE CHARACTERIZATION USING SINGULAR SPECTRUM ANALYSIS AND ENTROPY: SIMULATED RESULTS

Adriana Kauati[1], Carlos Dias Maciel[1] and Wagner Coelho de Albuquerque Pereira[1]
[1] Biomedical Engineering Program - COPPE/UFRJ – PO Box. 68510
Zip 21945-970 Rio de Janeiro - RJ, Brasil - kauati@peb.ufrj.br wagner@peb.ufrj.br

INTRODUCTION

During the last decades ultrasound (US) has been established as a powerful tool to investigate internal structures from different kinds of media. In particular, for biomedical applications the basic properties that made it possible are that ultrasound is a nonionizing radiation and is capable to generate (low cost) real time images. Nevertheless, the most attractive approach from the scientific point of view is to apply ultrasound for quantitative tissue characterization with the so-called noble objective of differentiating normal from diseased specimens.

Classical parameters have been used to estimate tissue properties, like attenuation of US waves, backscattering and wave speed, including *in vitro* experiments and the generation of parametric images of tissues like eye, bone, cartilage, liver and skin[2,5].

More recently, the spatial periodic structure of tissues has also been estimated using different methodologies[7] to obtain the parameter Mean-Scatterer-Spacing (MSS). Such tissues can be modelled as having regular and diffuse structures from which backscattered echoes are obtained (liver tissue, e.g.). In our approach, periodicity is investigated by the application of the Singular Spectrum Analysis – SSA[6,8] where it is possible to create two individual subspaces for the signal and noise components.

The objective of this paper is to study an optimum way to separate these two subspaces from the entropy point of view.

THEORY

The Singular Spectrum Analysis is a data-analysis method that has been used in nonlinear dynamics[8] and applied in several signals processing areas[6]. SSA is based on the classical Principal Component Analysis (PCA) where a short and noisy time series is projected in a new optimized orthogonal basis by taking the eigenvalues and eigenvectors of the data autocorrelation matrix. Vautard & Ghil[8] have shown that if a periodic component is present in a time series there will necessarily be an eigenvalue pair (with near-equality values) related to it. The frequency of the periodicity can be found from the power spectrum peak of the associated eigenvectors.

With this approach it is possible to choose a "correct" set of eigenvalue pairs and reconstruct only the periodic part of the time series. Than it can be understood as a signature of that tissue. There has to be a concern in the analysis of eigenvalue pairs from the covariance matrix: A random process can also generate pairs of eigenvalues that have no relation to the structure periodicity we are interested in. The standard subspace separation using Wax e Kailath algorithms[6] is not applied to this class of problems because it requires white noise and due to physical properties of ceramic ultrasound transducer (band limited in frequency), the noise is "colored".

Given this scenario, the next step to be pursued would be finding a quantitative criterion to define what a "correct set of eigenvalues" is. The concept of entropy will be used with this purpose here by quantifying the reconstruction of the signal components from the eigenvectors' linear combination. The classical definition of entropy is associated to the disorder level present in the medium[4]. Entropy may also be interpreted as the level of information necessary to describe a given process[1,3]. One basic definition of it includes a function that is not dependent on values of X but on its probability distribution, as shown in eq (1).

$$H(X) = -\sum_{x\in X} p(x)\log p(x) \qquad (1)$$

where: H (X) = entropy function;

X = discrete variable;

p(x) = Pr{X = x} = probability function;

log is to the base 2.

Another useful concept is the relative entropy, that is a measure of the distance between two distributions, or it is a measure of the inefficiency of assuming that the distribution is q when the true distribution is p[3]:

$$D(p\,||\,q) = \sum_{x\in X} p(x)\log\frac{p(x)}{q(x)} \qquad (2)$$

where: *D(p||q)*: relative entropy;

p(x): probability distribution;

q(x): probability distribution.

Both concepts will be used in this work.

METHODOLOGY

For simulation of the tissue backscattered rf signals it was used a unidimensional model with discrete particles having its position and scattering amplitude as parameters. This approach was obtained from the literature[6,7].

Four types of backscattered signals were simulated with different Signal-to-Noise Ratios (SNR) and jitter (for the periodic part), according to Table 1:

Table 1. Simulated signals in the US frequency of 2.5MHz

FreqMSS is the inverse of the MSS (kHz)

	SNR	Jitter	FreqMSS	FreqMSS* estimated by SSA
Signal 1	3 dB	1 %	616kHz	611kHz
Signal 2	3 dB	20 %	616kHz	706kHz
signal 3	0 dB	1 %	616kHz	611kHz
signal 4	0 dB	20 %	616kHz	755kHz

* MSS – Mean Scatterer Spacing

For each signal it was obtained the set of eigenvalues (and respective eigenvectors) in a descending order of magnitude. Then the reconstruction of the signal was made adding sequentially, two by two eigenvectors. For instance, the first reconstruction is made with eigenvectors 1 and 2. The second reconstruction is with eigenvectors 1,2 and 3,4 and so on. As an example, in Figure 1A (up) one can find the envelope signal of the US echo (simulation of Signal 1 of Table 1) and its power spectrum (below). The first sharp peak is in frequency 616kHz, which is compatible to the simulated signal. In Figure 1B (superior-left) there is the first reconstruction with eigenvectors 1 and 2. Below is its power spectrum with only one sharp peak in the frequency 611 kHz. In Figure 1B (superior-right) it can be seen the signal obtained by subtracting the first reconstruction from the original signal (so called residual subspace) and its power spectrum just below. The sharp peaks in the power spectrum of this residual signal indicate that there is still periodic activity in the signal which means we should look for other pairs of eigenvalues.

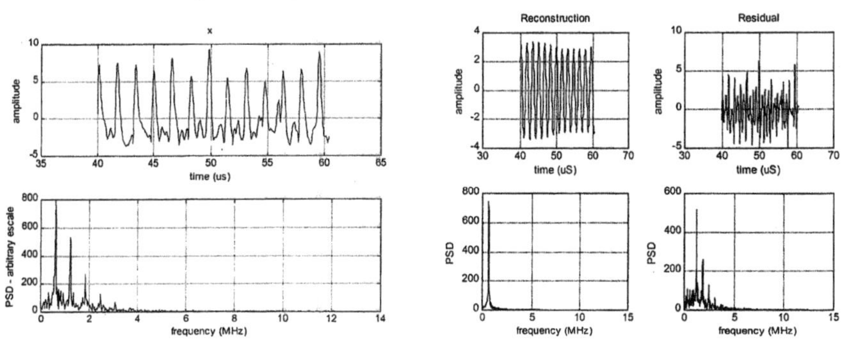

(A) (B)

Figure 1 – (A) Envelope of the simulated RF Signal 1 (Table 1) and its power spectrum (below). (B) Signal reconstructed with the first 2 eigenvectors (superior-left), its power spectrum (inferior-left), residual signal (superior-right) and its power spectrum (inferior-right).

In figure 2A it is shown the eigenvalue set in descending order of magnitude (Singular Spectrum). In figure 2B there are the first 8 eigenvector pairs. The arrows in 2A suggest the pairs of eigenvalues that could be taken as being responsible for the periodicity of the signal. In 2B the arrows show the corresponding eigenvector pairs.

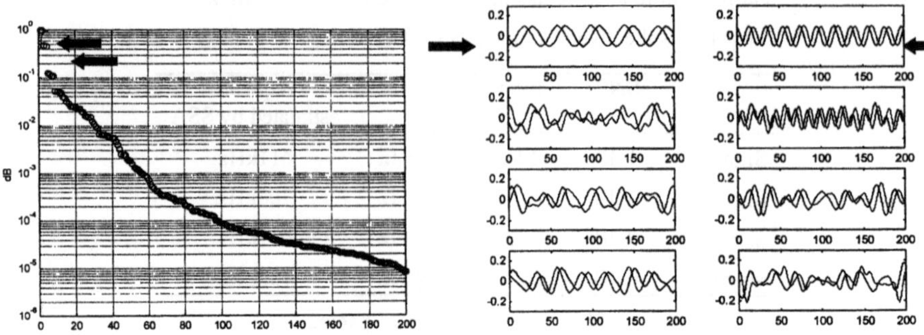

(A) (B)

Figure 2 – (A) Singular Spectrum showing the 200 eigenvalues relative to signal 1. The arrows point to the two first eigenvalue pairs related to periodic activity into the signal. (B) First 8 eigenvector pairs corresponding to the 8 eigenvalue pairs. Arrows point to the first two pairs.

For the simulated signals 2, 3 and 4, the MSS Frequency estimated by the method are, respectively, 706kHz, 611kHz and 755kHz (Table 1).

To try to understand the underlining process the absolute entropy was calculated for every reconstructed signal and respective residual subspace, after each inclusion of a new eigenvector pair. Figure 3 (superior) presents the absolute entropy profile of the reconstructed subspace (RC). Below it can be found the entropy profile for the residual subspace (RD).

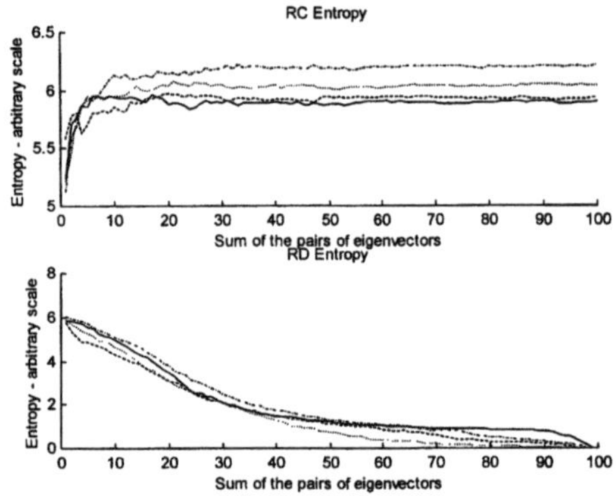

Figure 3 – Absolute entropy for the reconstructed (RC) and residual (RD) subspaces. Horizontal axis represents the number of pairs used for each reconstruction. Signal 1 – dashed line, signal 2 – solid line, signal 3 – dotted line, signal 4 – dashdotted line.

It was calculated the absolute entropy for the RC Fourier Transform (Figure 4 – superior) and RD Fourier Transform (Figure 4 – inferior).

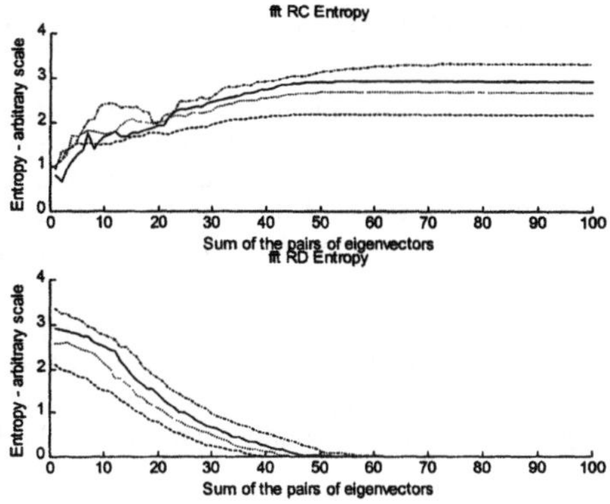

Figure 4 – Absolute entropy for the fft reconstructed (RC) and fft residual (RD) subspaces. Horizontal axis represents the number of pairs used for each reconstruction. Signal 1 – dashed line, signal 2 – solid line, signal 3 – dotted line, signal 4 – dashdotted line.

Figure 5 shows the relative entropy for the RC Fourier Transform (superior) and the relative entropy for the RD Fourier Transform (inferior). Both have the Fourier Transform of the original envelope signal as the $q(x)$ distribution.

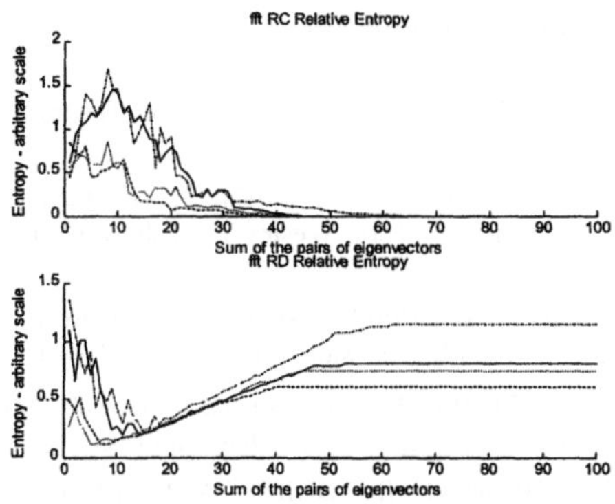

Figure 5 – Relative entropy for the fft reconstructed (RC) and fft residual (RD) subspaces per fft envelope signal. Horizontal axis represents the number of pairs used for each reconstruction. Signal 1 – dashed line, signal 2 – solid line, signal 3 – dotted line, signal 4 – dashdotted line. The Fourier Transform of the original envelope signal was used as the reference.

DISCUSSION AND CONCLUSION

The periodicity estimated by SSA was close to the simulated one for the cases of low jitter (Signals 1 and 3), even for de 0dB of SNR. For jitter of 20% there is a small discrepancy (<15%) for the estimated value (Signals 2 and 4). These results give an idea of the potential of the SSA method.

As the first attempt to understand the SSA method from the entropy preliminary results, in figure 3(superior) it is possible to see that when the reconstruction is made with around 10 eigenvalue pairs the absolute entropy reaches a plateau. The maximum peak in figure 5 (superior) and the minimum peak in figure 5(inferior) also happen around the reconstruction with 10 eigenvector pairs, for the relative entropy.

To try to clarify this behavior a simple simulation was performed using a signal composed only by a sum of 10 sinusoidal waves and colored noise. After the application of the SSA method, the entropies were calculated in the same way as before. The curves obtained were similar to the ones in figures 3, 4 and 5, pointing to the importance of the first 10 pairs of eigenvectors (one at each sinusoidal frequency) as the ones containing the formation present in the signal. This was the expected result, once the simulated signal contained only 10 sinusoidal waves in the signal subspace.

So guided by this result, it seems to be reasonable to suggest that, for the case of simulated backscattered US echoes presented here, the first 10 eigenvector pairs have most of the information of the signal. Therefore, quantification of any parameter should take into account the signal reconstructed with these 10 pairs. It is being presently investigated the limits of validity of this conclusion.

The next step is to look for other forms of calculating entropy to study different aspects of reconstruction.

REFERENCES

1. L. ALSEDA and J. LLIBRE, "Combinatorial Dynamics and Entropy in Dimension One", World Scientific, Advanced Series in Nonlinear Dynamics, vol.5 (1993).
2. S.L. BRIDAL, A. SAIED, E. CHERIN, F. LEFEBVRE, P. LAUGIER and G. BERGER, "High Resolution Quantitative Echography with Tissue Parameters", J.E.M.U., N.2/3, PP.204-211 (1998).
3. T. M. COVER and J. A. THOMAS Elements of Information Theory, Wiley-Interscience, USA, Chapter 2 (1991).
4. R. P. FEYNMAN, R. B. LEIGHTON and M. SANDS, The Feynman Lectures on Physics, California Institute of Technology, Addison-Wesley Publishing Company, U.S.A. Chapter 46-Ratchet and Pawl (1977).
5. F. LIZZI, M. ASTOR and E. FELEPPA, "Statistical Framework for Ultrasonic Spectral Parameter Imaging", Ultrasound in Med. & Biol., vol.23, no.9, (1997).
6. C. D. MACIEL and W. C. A. PEREIRA, "Rf Ultrasound Echo Decomposition Using Singular-Spectrum Analysis", Acoustical Imaging, vol. 24 (1998), in Press.
7. C. SIMON, J. SHEN, R. SEIP and E. S. EBBINI, "A Robust and Computationally Efficient Algorithm for Mean Scatterer Spacing Estimation", IEEE Trans. on UFFC vol. 44 n. 4 (1997).
8. R. VAUTARD and M. GHIL, "Singular Spectrum Analysis in Nonlinear Dynamics, With Applications to Paleoclimatic Time Series", Physica D 35, North (1989).

THE MAXWELL MODEL AS A SOFT TISSUE DESCRIPTOR

Sidney Leeman[1] and Azadeh Peyman[2]

[1]Department of Medical Engineering and Physics
[2]Department of Electrical Engineering
King's College London
London SE22 8PT, U.K.

INTRODUCTION

The use of the linear Maxwell model as a descriptor of the acoustical properties of human soft tissues is presented. It is explained how it may be incorporated into the derivation of a wave equation for the propagation of ultrasound pressure pulses in homogeneous soft tissues. Use of the model automatically results in absorption and velocity dispersion, and the frequency-dependence of these is calculated.

In vivo transmission measurements of the attenuation in human skeletal muscle have been performed on approx. 20 volunteers, and it shown that the predictions of the Maxwell model are in moderate agreement with the data. A reasonable fit to published data from other workers has also been found.

However, the velocity dispersion predicted by the model is not, at first sight, reasonable. It is emphasised that the model exhibits rather different dispersive behaviours for the phase, group, and signal velocities, which can be explicitly calculated. There is some ambiguity in the literature as to which velocity is measured in any particular experimental configuration, and a possible resolution of the observed conflict with observation is presented.

An appealing feature of the Maxwell model is that it allows an analytic solution for the propagation of a 3-D wideband pressure pulse. This is shown, for the case of a circular planar disk transducer radiating into a homogeneous linear Maxwell medium. It is shown how the model may be extended to encompass scattering from density and compressibility fluctuations, and even how scattering from absorption fluctuations may be incorporated.

The Maxwell model provides a reasonable way to characterise attenuation in certain soft tissue organs. However, at this stage it does not provide a complete approach towards describing the 'ideal' soft tissue, *viz* exhibiting linear-with-frequency attenuation and minimal dispersion.

Acoustical Imaging, Volume 25, Edited by Halliwell and Wells.
Kluwer Academic/Plenum Publishers, 2000.

THE MAXWELL MODEL

The Maxwell model is successful for describing the mechanical properties of bouncing putty and warm tar, and it is of some interest to examine its suitability for modelling soft tissue. For simplicity, only homogeneous tissues will be considered. To fulfill this task, the model must be capable of predicting an absorption that varies as $a(\omega) \approx a_0\omega^n$, with $n \approx 1$, at least in the diagnostic frequency range, where a_0 is a constant, and ω is the frequency expressed in radians/second. Also, for diagnostc frequencies, the predicted phase velocity would be expected to show virtually no dispersion -- but it should be borne in mind that experimental evidence for this is poor, and ambiguous. The model is a viscoelastic one, with the stress/strain relationship symbolically represented by a spring (elastic element) and a dashpot (viscous element) arranged in series.

A straightforward approach towards deriving an appropriate one dimensional wave equation for pressure waves in a Maxwell medium may be based on an extension of the usual simplified linear theory for elastic media. The continuity equation (mass conservation) and force law (assuming only pressure forces operative) may be combined with a modified constitutive equation

$$\varrho(z,t) = \varrho_0\kappa_0 p(z,t) + A\int^{t} p(z,t')dt' \tag{1}$$

where ϱ_0 and κ_0 denote the constant values for the density and compressibility of the undisturbed medium, A is a model parameter, p is the acoustic wave pressure, and z and t are the distance and time variables. The first term on the right hand side of Equation (1) represents the elastic component of the model, and the second term the viscous element. This leads to a wave equation

$$\frac{\partial^2 p}{\partial z^2} - \frac{1}{C^2}\frac{\partial^2 p}{\partial t^2} - A\frac{\partial p}{\partial t} = 0 \tag{2}$$

with $C = 1/\sqrt{\varrho_0\kappa_0}$. This equation implies loss, and it may be shown that the absorption predicted is

$$a(\omega) = \frac{\omega}{\sqrt{2}\,C}[(1 + \frac{A^2C^4}{\omega^2})^{1/2} - 1]^{1/2} \tag{3}$$

with an associated phase velocity

$$V(\omega) \equiv \frac{\omega}{k} = \frac{2a(\omega)}{A} \tag{4}$$

with $k = 2\pi/wavelength$. It is informative to examine the low and high frquency limits of the absorption:

$$\begin{aligned} a(\omega) &\sim \sqrt{\tfrac{A}{2}}\,\sqrt{\omega} && \text{for low } \omega \quad (\omega \ll AC^2) \\ &\sim \tfrac{1}{2}AC && \text{for high } \omega \quad (\omega \gg AC^2) \end{aligned} \tag{5}$$

Thus, the Maxwell model predicts frequency dependent absorption and dispersion. It may be regarded that the predicted absorption rises too slowly with frequency, but it should be pointed out that few direct measurements with mammalian tissues have actually been carried out, although attenuation measurements abound. For some mammalian tissues, however, the model can be seen to be in quite good agreement with the data[1,2]. The predicted velocity dispersion is strong, and it could be argued that observations do not support such a behaviour. Again, it may be pointed out that few direct measurements of *phase* velocity, to which Equation (4) refers, have actually been carried out. Speed of sound measurements in tissues are generally carried out via pulse techniques, and it can be argued that it is the *group* velocity, $\partial\omega/\partial k$, which generally shows a slower frequency dependence in this model, that is being measured. Moreover, the model can be shown to have a constant *signal* velocity, and it cannot be ruled out that some pulse techniques actually measure this entity. The simple Maxwell model may also be extended, by including a velocity modifying term in Equation (2), but the physical justification for the form of such terms is obscure.

PULSE PROPAGATION IN MAXWELL MEDIA

A wave equation identical to Equation (2) was conjectured, on phenomenological grounds, to be applicable to ultrasound propagation in tissues, and some features analysed[2]. The wave equation may be solved analytically, by resorting to Laplace transforms, for pulse emission from a circular, planar, unapodised transducer (actually: aperture). Reference 2 may be consulted to obtain the full expression for the pulse, whose front propagates at a constant speed, C -- the signal velocity. The expression for the pulse in cylindrical polar coordinates (h,z) is :

$$p(h,z;t) = P(h,z;t) \circledast f(t) = B(h)e^{-\beta z}\delta(t - z/C) + \ldots\ldots$$

$$\ldots\ldots Cze^{-\beta\tau}H(t - z/C)\frac{1}{\xi}\frac{\partial}{\partial\xi}\int_0^\infty ds.sB(s)\int_0^\infty dy.yJ_0(sy)J_0(hy)J_0(\xi\sqrt{y^2-\beta^2} \qquad (6)$$

where $\circledast$ denotes convolution, H is the Heaviside step function, $B(h)$ represents the apodisation function, $\tau = Ct$, $\beta = AC/2$, and $\xi = \sqrt{\tau^2 - z^2}$. It may be shown that this rather cumbersome expression actually contains the usual direct/edge wave decomposition, albeit modified by the effects of absorption. It is to be noted that the pulse contains a component which is damped according to β, the (constant) high frequency limit of the absorption, and does not exhibit the frequency dependent loss implied by the model.

The velocity dispersion of the model may be modified by phenomenologically adding the term $b\dfrac{\partial^2 p}{\partial z\partial t}$ to the wave equation. This has the effect of modifying the velocity dispersion without altering the absorption characteristics, and produces a near constant-with-frequency speed of sound except at very low frequencies.

LOSS MEASUREMENTS IN SKELETAL MUSCLE

An attempt was made to verify the applicability of the Maxwell model by measuring loss in the calf muscle in the leg of 20 normal volunteers, using a pulse transmission technique. Volunteers were of both sexes, aged from about 9 to 50 years old. Measurements were made at roughly the same height in the leg, approximately at right angles to the muscle fibre direction. The arrangement is schematically shown in Figure (1). Data were digitised at up to 100MHz sampling rate, to a nominal 8 bit accuracy. Attenuation values were expressed relative to water (absorption considered negligible), via

$$a(\omega) - a_W(\omega) = (\ln V_W - \ln V)/d + const \qquad (7)$$

where the subscript 'W' denotes the entity as measured for a water path, d is the propagation path length, and V denotes the receiver voltage as measured. No diffraction corrections were applied. The constant in Equation (7) was not estimated, so that the a-values are not

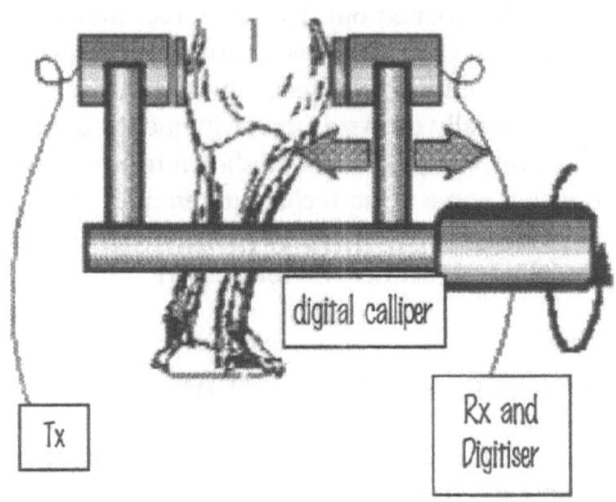

Figure 1 Schematic depiction of muscle measurement arrangement. The Tx/Rx distance may be adjusted, and is recorded digitally. A similar measurement is made in water.

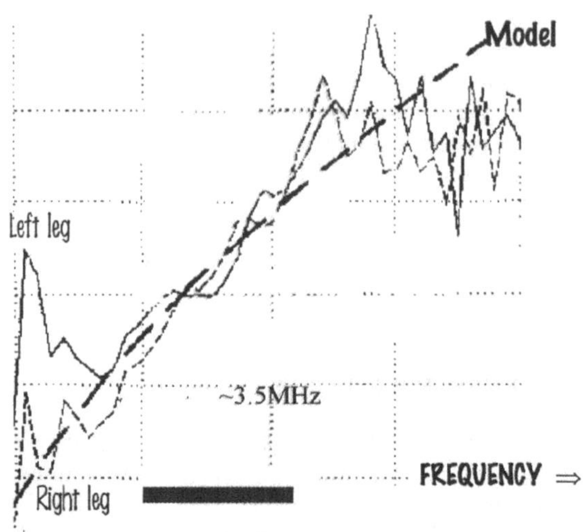

Figure 2 Typical fit of loss data to Maxwell model. Both legs (same volunteer) give very comparable results. Only the data in the shaded frequency range are significant. Values of the loss are not on an absolute scale.

quantitative. However, a fit to the distinctive (decreasing) curvature of the Maxwell model absorption curve was interpreted as being indicative, to some degree, of the essential validity of the model.

Some typical results are shown in Figure (2). The frequency range over which a model fit is sought corresponds to frequencies over which both the incident and transmitted pulses have substantial spectral amplitudes. Outside this range, noise starts to dominate the results. The degree of fit is satisfying, considering the many approximations made, but can not be considered as being conclusive, as no other models were considered. It is, of course, to be noted that the experimental arrangement actually estimates attenuation, rather than absorption, but, at this stage, no attempt has been made to correct for the effects of scattering.

The Maxwell model predicts velocity dispersion, and some time-of-flight measurements were carried out to reveal this. A qualitative review of the measurements does not reveal any compelling reason to suspect a marked effect of this nature. However, the experimental arrangement indicates a parameter more closely related to the group velocity, and the errors of measurement do not allow for an accurate assessment to be made.

EXTENSION TO INHOMOGENEOUS MEDIA

The extension to inhomogeneous media is relatively straightforward, and is effected by assuming that the density and compressibility of the undisturbed medium are spatially varying functions. This results in the conventional scattering terms appearing in the wave equation. One advantage in resorting to the Maxwell model for the medium, is that it introduces loss into the equation. If the parameter A is then also allowed to be location dependent, then scattering from absorption fluctuations is introduced. Such a scattering source has been conjectured to be present[3], for ultrasound propagation in tissues, but has never been experimentally verified.

CONCLUSIONS

The Maxwell model has some interesting properties, only some of which were presented here. It predicts velocity dispersion and a frequency dependent absorption. The latter may be modified by fairly simple patches to the model. Three different speeds of sound are predicted: phase, group, and signal (which turns out to be a constant). Pulse propagation in such a (homogeneous) medium may be analytically calculated.

Direct measurements of skeletal muscle in normal volunteers have shown no more than that the data are consistent with the model, but not conclusively so.

REFERENCES

1 J.K. Brady, S.A. Goss, R.L. Johnston, W.D. O'Brien and F. Dunn, *J. Acoust. Soc. Am.* 60:1407, 1976

2 S. Leeman, L. Hutchins and J.P. Jones, Bounded pulse propagation, *Acoustical Imaging*, P. Alais and A.F. Metherell (eds), 10:427, 1982

3 S. Leeman, R. Leeks and P. Sutton Analysis of pulse-echo ultrasonic images, In: *Information Processing in Medical Imaging*, R. Di Paola, and E. Kahn (eds.), 88:35, 1979 (INSERM, Paris)

CHARACTERIZATION OF COLLAGEN FIBERS IN ATHEROSCLEROTIC PLAQUES IN MICE

Yoshifumi Saijo[1,2], Claus S. Jørgensen[2], Erling Falk[2]

[1]Department of Medical Engineering and Cardiology,
Institute of Development, Aging and Cancer, Tohoku University
4-1 Seiryomachi, Aoba-ku, Sendai 980-8575, JAPAN
[2]Institute of Experimental Clinical Research, Århus University Hospital
Brendstrupgaardsvej, Århus N 8200, DENMARK

INTRODUCTION

Coronary plaque disruption with superimposed thrombus is the main cause of the acute coronary syndrome such as unstable angina or acute myocardial infarction[1]. Pathological investigations have shown that plaque size or degree of the lumen narrowing is less important than the composition of the plaque; lipid-rich and soft plaques are more vulnerable than the collagen-rich hard plaques[2-4]. In clinical settings, intravascular ultrasound (IVUS) imaging has been applied for detecting a vulnerable plaque, which is sometimes found in an angiographically normal coronary artery. The composition of atherosclerosis is defined as the collagen fibers, lipid and calcification in the intima, the smooth muscle and elastic fiber in the media, and the collagen fibers in the adventitia, respectively. However, variety of the degree of staining for optical microscopy indicates that the collagen content or the chemical property is not homogeneous in fibrotic lesions in the intima.

In the present study, a scanning acoustic microscope (SAM) system was employed to characterize the collagen type in the fibrous caps of Apo-E deficient mice, which are well known to develop atherosclerotic plaques. Polarized microscopy and picrosirius red staining were also equipped for qualitative and quantitative analysis of characterization of collagen fibers.

MATERIALS AND METHODS

Twelve Apo-E deficient mice, 6 female and 6 male, were fed with high-fat diet, containing 21% fat and 0.15% cholesterol by weight. Three C57BL mice were used for control. At the end of study, the mice were anesthetized and the arterial tree was perfusion-

Acoustical Imaging, Volume 25, Edited by Halliwell and Wells.
Kluwer Academic/Plenum Publishers, 2000.

fixed at 100 mmHg with phosphate-buffered 4% formaldehyde (PH 7.2) via the left ventricle and immersed in fixative overnight. The aortic root was sectioned serially at 4-ì m interval and embedded in paraffin. The paraffin was removed by coconut oil and graded alcohol process. The sections were unstained and used for the SAM investigations. After the SAM investigation, the sample was stained with Sirius red and covered by a cover glass.

A scanning acoustic microscope (SAM2000; KSI, Herborn, Germany) was equipped for semi-quantitative ultrasonic measurements. In the present study, an acoustic lens with the optimal frequency range between 800 MHz and 1.3 GHz was used. The SAM frequency in the experiments was 1.1 GHz because the image quality was best at 1.1 GHz. Distilled water was used for the coupling medium between the transducer and the sample. The temperature of the transducer, coupling medium, sample and the glass substrate was kept at 37 C° by a heating plate. The Rf output gain was fixed for whole experiments. In the SAM system, 80 dB difference of ultrasonic intensity corresponded to 256 gray scales (0-255). The reflection from the glass surface was set as level 255 (white). The average thickness of the specimen was determined by laser micrometer. Ultrasonic attenuation at each pixel was calculated from the number of the gray level and the thickness. All of the SAM images were stored in a personal computer for statistical analysis.

The optical and polarized images of atherosclerotic plaque were captured to make the same field taken by SAM. The images were analyzed using computer-assisted image analysis equipment (Leitz DMRBE light microscope; Leica Wetzlar, Germany, JVC KY-F55 color video camera; Victor Co.Ltd., Tokyo, Japan, Scion CG-7 RGB color frame grabber, and ScionImage; Scion Corporation, Frederick, MD, USA, and Photoshop; Adobe Systems Inc., Seattle, WA, USA). Although the picrosirius red staining was not a specific staining technique, the area with collagen was specifically visualized by polarized microscopy. The areas were differentiated according to the polarized color; longer wavelength (red or yellowish color) or shorter wavelength (green).

RESULTS

Figure 1 shows the normal light (a), polarized light (b) and acoustic (c) microscopy images of a normal aortic wall observed in a C57BL mouse. Endothelial cells aligning in a single layer comprise the intima. The smooth muscle cells and elastic fibers consisting of the media are clearly shown by the acoustic image. Collagen fibers in the adventitia are highlighted by the polarized image. No pathological atherosclerotic lesions were found in the C57BL mice in the present study. Figure 2 shows an atherosclerotic plaque found in an aortic root of an Apo-E deficient mouse. The structure in the lower left is a part of one coronary cusp. The collagen network in the plaque, the each elastic fiber in the media, and a single macrophage cell near the fibrous cap are shown in the acoustic image. Although the lipid in the plaque was completely removed by alcohol dehydration, the remained outer shape of the cholesterol crystal indicates that the cholesterol had existed in the spaces. The fibrosis in the plaque would be classified into two categories; the inner part with low ultrasonic attenuation and weak polarization, and the center part with high ultrasonic attenuation and strong polarization.

Whole results of the study showed that the fibrous caps in the advanced atherosclerotic plaques were consisted of thick collagen fibrils. In these regions, the polarized color was orange and the polarized intensity was the highest. In acoustic images, these regions showed high attenuation. In contrast, thin fibrous caps exhibited green polarized color showed low polarized intensity. The ultrasonic attenuation in the thin cap was relatively low.

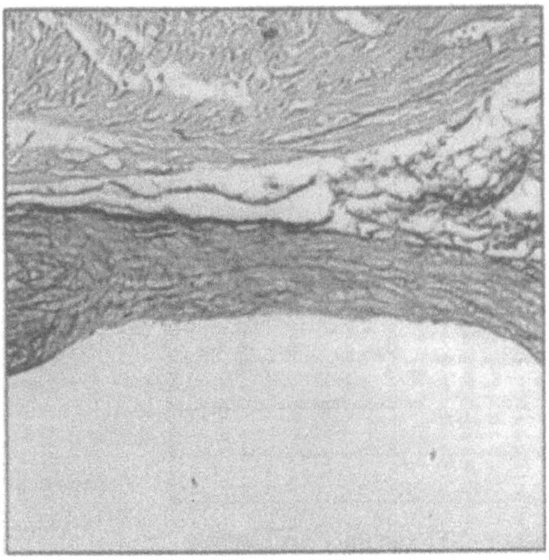

Figure 1(a) Normal light microscopic image of the normal aortic wall.

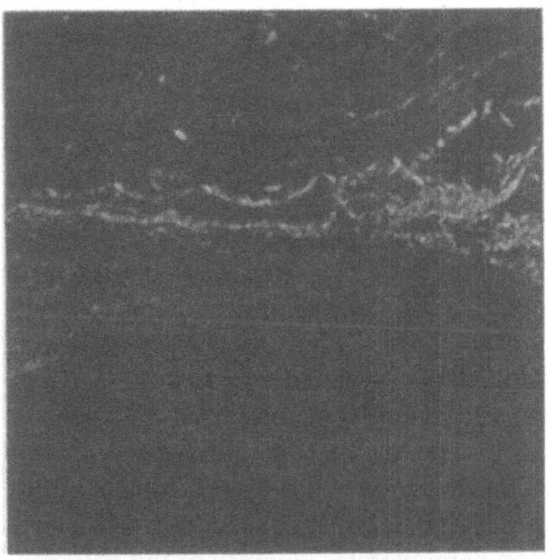

Figure 1(b) Polarized light microscopic image of the normal aortic wall.

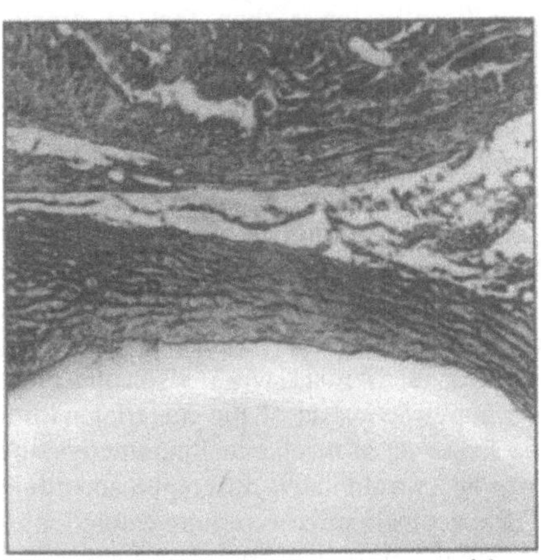

Figure 1(c) Acoustic microscopic image of the normal aortic wall.

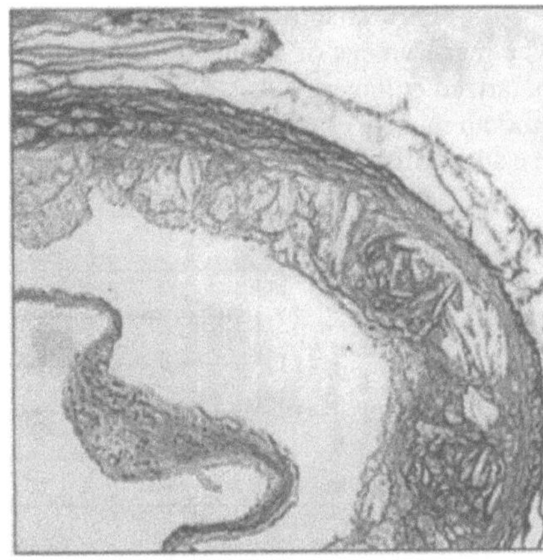

Figure 2(a) Normal light microscopic image of the atherosclerotic plaque.

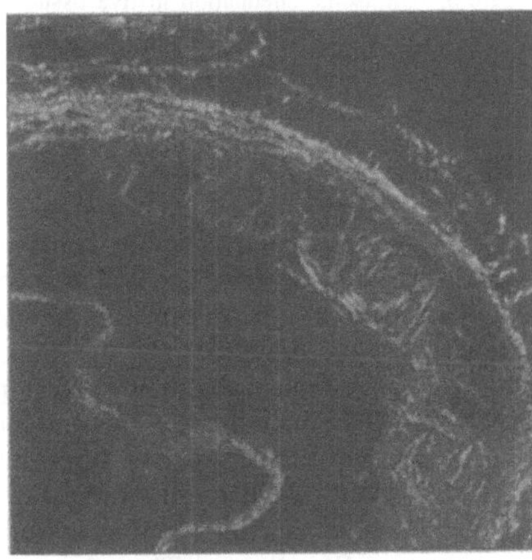

Figure 2(b) Polarized light microscopic image of the atherosclerotic plaque.

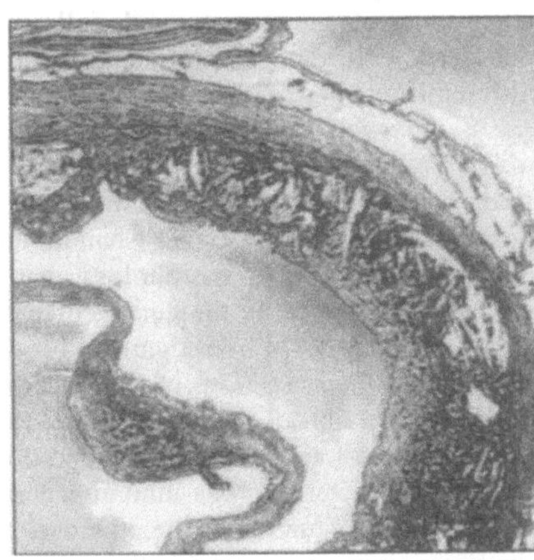

Figure 2(c) Acoustic microscopic image of the normal aortic wall.

Figure 3 shows the results from all the regions from twelve Apo-E deficient mice. The mean values of attenuation in tissue components were 17.2×10^3 dB/mm in the orange polarized collagen, 6.6×10^3 dB/mm in the green polarized collagen, 4.7×10^3 dB/mm in the smooth muscle of media, 15.5×10^3 dB/mm in the elastic fiber of media, 16.1×10^3 dB/mm in the adventitia, respectively.

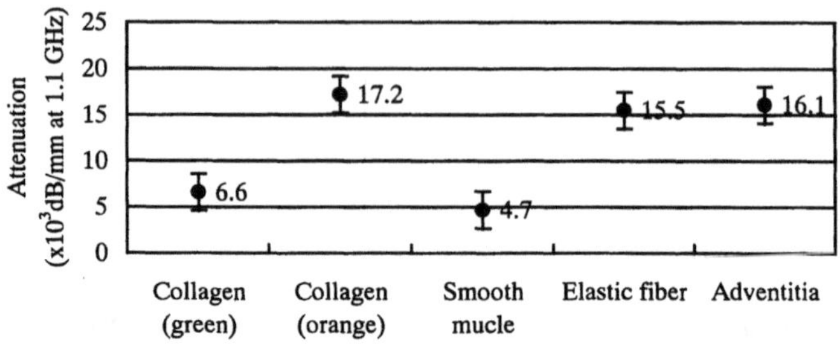

Figure 3. Ultrasonic attenuation in five tissue components found in the atherosclerotic plaques in Apo-E deficient mice.

DISCUSSION

Under polarized light, the collagen fiber stained with picrosirius red is highlighted by induced birefringence. The collagen content is known to be proportional to the intensity of polarization. In other words, the intensity of polarization demonstrates "quantity of the collagen". In the present study, the thick fibrous caps showed high intensity polarization while the thin fibrous caps showed weak polarization. The results indicate that the collagen content is rich in thick fibrous caps and that thin caps contained low amount of collagen.

The wavelength of the birefringence light depends on the collagen type and thickness of the fibril; orange indicates thick collagen fibrils while green indicates thin fibrils[5]. Usually, the thick collagen fibrils compose type I collagen and the thin fibrils compose type III collagen[6]. The color of polarization demonstrates "quality of the collagen". Since type III collagen is often associated with macrophage infiltration[7], green polarized collagen may correlate with vulnerable plaques. The acoustic microscopy revealed that the attenuation of green polarized collagen was significantly lower than that of orange polarized collagen. Consequently, high frequency ultrasound has a possibility to differentiate collagen types I and III in the atherosclerosis.

It is well recognized that the collagen plays a key role in determining attenuation of many tissues because it has particularly high values of elastic modulus. However, recent studies using high frequency ultrasound revealed that the collagen has different acoustic properties in some situations. Chandraratna et.al studied the relation between the intensity and the collagen fiber morphology using 600 MHz acoustic microscopy. The ultrasonic intensity was higher in the dense collagen fiber than in thin collagen fiber in their study[8]. In our previous SAM investigations for the myocardium[9] and kidney[10] using 100-200 MHz ultrasound, the variation of the acoustic properties in fibrosis were also observed. Although the fibrous cap of the atherosclerosis is mainly consists of the material with a single term "collagen", and was supposed to have a single set of mechanical parameters, the acoustic characteristic was inhomogeneous in a cap and would show difference according to the stage or situation of the atherosclerosis. Thus, acoustic microscopy investigation of the fibrous cap would provide the useful information in the mechanical point of view.

Fixation process is known to increase the ultrasonic attenuation. However, we used the specimens fixed by formaldehyde and deparaffinized by coconut oil and ethanol in the present study. In order to make a high quality image by acoustic microscopy, we tried several sets of histological preparations such as frozen sections, alcohol-fixed sections and formalin-fixed, paraffin-embedded sections. The best quality image was made by the formalin-fixed and paraffin-embedded sections. We succeeded to identify each elastic fiber or a single macrophage in atherosclerotic plaques by acoustic microscopy. The formalin-fixed material would be still one of the best choices for high quality acoustic microphotographs unless the fixation influences the absolute acoustical parameters,

Although the discussion of the absolute value of attenuation is not meaningful, this study may demonstrate that the fixation process would preserve the structure-related acoustic properties of the tissue. Both type I and III collagens were formalin-fixed by the standard pathological preparation technique. After the process, the polarization color for type I and III collagens still remained different. Since the polarized color indicates the difference of collagen fibril structure, the different values of attenuation may be also from the micro-structural difference of collagen fiber.

We have reported the attenuation and sound speed of the tissue elements in human aorta by using a 100-200 MHz range acoustic microscope[11]. The attenuation had strong correlation to the sound speed in that study. For example, both the attenuation and sound speed were highest in the calcified lesion and both values were lowest in the lipid. The SAM studies on other kinds of biological material also showed the similar correlation between the attenuation and sound speed. In the present study, the sound speed was not measured, but considering the relation between attenuation and sound speed, the high attenuation regions such as orange polarized area would exhibit high sound speed. Consequently, the attenuation measured in the present study should have a direct correlation to viscosity and an indirect relation to the elasticity of the tissues.

In the previous studies assessing stress distribution on the fibrous caps by using finite element analysis[12], the circumferential stress was concentrated to the thin fibrous cap. In those studies, a single set of values of mechanical properties was given to either a thick or a thin fibrous cap. However, not only the thickness but the mechanical properties also changed associating with the development of atherosclerosis in our study. One of the reasons should be the adaptive change of the elasticity of the thin fibrous cap for preventing plaque rupture. Further evaluation is needed to prove the mechanical property change and the stress on these thin caps from both actual mechanical testing and computer simulation.

CONCLUSIONS

Scanning acoustic microscopy (SAM) was equipped for collagen characterization in experimental animals. Comparing the polarized and acoustic images of atherosclerosis in Apo-E deficient mice, the collagen with orange polarized color showed higher ultrasonic attenuation than that of the collagen polarized with green polarization. The results indicated the possibility of characterizing types I and III collagen by high frequency ultrasound. The data can be applied to the assessment of intravascular ultrasound imaging, which is widely used in the clinical settings today. If a vulnerable plaque is detected before its rupture, the interventional strategies would be not only widening the already narrowed lumen, but also prevention of the plaque rupture and the development of atherosclerosis.

REFERENCES

1. E. Falk, A. Fernandez-Ortiz, Role of thrombosis in atherosclerosis and its complications, *Am J Cardiol.* 23:3B(1995).
2. E. Falk, Why do plaques rupture? *Circulation.* 86:III30(1992).
3. P.K. Shah, Pathophysiology of plaque rupture and the concept of plaque stabilization, *Cardiol Clin.* 14:17(1996).
4. A.P. Schroeder, E. Falk, Vulnerable and dangerous coronary plaques, *Atherosclerosis.* 118:S141(1995).
5. L.C. Junqueira, G.S. Montes, E.M. Sanchez, The influence of tissue thickness on the study of collagen by the Picrosirius-polarization method, *Histochemistry.* 74:153(1982).
6. D. Dayan, Y. Hiss, A. Hirshberg, J.J. Bubis, M. Wolman, Are the polarization colors of Picrosirius red-stained collagen determined only by the diameter of the fibers? *Histochemistry.* 93: 27(1989).
7. G.K. Sukhova, U. Schonbeck, E. Rabkin, F.J. Schoen, A.R. Poole, R.C. Billinghurst, P. Libby, Evidence for increased collagenolysis by interstitial collagenases-1 and -3 in vulnerable human atheromatous plaques, *Circulation.* 99:2503(1999).
8. P.A.N. Chandraratna, P. Whittaker, P.M. Chandraratna, J. Gallet, R.A. Kloner, A. Hla, Characterization of collagen by high-frequency ultrasound: Evidence for different acoustic properties based on collagen fiber morohologic characteristics, *Am Heart J.* 133:364(1997).
9. Y. Saijo, M. Tanaka, H. Okawai, H. Sasaki, S. Nitta, F. Dunn, Ultrasonic tissue characterization of infarcted myocardium by scanning acoustic microscopy, *Ultrasound Med Biol.* 23:77(1997).
10. H. Sasaki Y. Saijo, M. Tanaka, S. Nitta, Y. Terasawa, T. Yambe, Y. Taguma, Acoustic properties of dialysed kidney by scanning acoustic microscopy. *Nephrol Dial Transplant.* 12:2151(1997).
11. Y. Saijo, H. Sasaki, H. Okawai, S. Nitta, M. Tanaka, Acoustic properties of atherosclerosis of human aorta obtained with high-frequency ultrasound, *Ultrasound Med Biol.* 24:1061(1998).

QUANTITATIVE IMAGES BASED ON US BACKSCATTER AND ATTENUATION (5 TO 28 MHz) FOR CHARACTERIZATION OF CAROTID PLAQUE CONTENT

S. Lori Bridal,[1] Bernard Beyssen,[2] Paul Fornès,[3] Pierre Julia,[4] and Geneviève Berger[1]

[1]Laboratoire d'Imagerie Paramétrique
UMR 7623 CNRS-University Paris VI 75006

[2] Radiologie Cardiovasculaire
[3] Laboratoire d'Anatomie Pathologique
[4] Service de Chirurgie Cardiovasculaire
Hôpital Broussais 75014
Paris, FRANCE

INTRODUCTION

Highly stenotic carotid plaques are associated with increased clinical risk. Currently, patients with stenosis > 70 % are treated by surgical endarterectomy.[1] Percutaneous endoluminal carotid angioplasty with stenting is being investigated as an alternative to surgical carotid endarterectomy. However, during the procedure, the risk of emboli depends, not only on plaque extent, but also on its biochemical composition and structure.[2] A noninvasive method for the characterization of plaque nature could add information for evaluation of the risk of producing emboli and thus help choose the most appropriate revascularization technique.

Clinical ultrasound is routinely used to evaluate carotid stenosis but provides only a qualitative description of plaque echogenicity and regularity.[3] A more quantitative analysis of plaque composition could be obtained by analyzing the ultrasonic radio frequency (RF) signals to estimate quantitative material parameters such as the attenuation and backscatter coefficients.[5-8]

The goal of this study was to develop methods for quantitative ultrasound imaging suitable for noninvasive assessment of carotid plaque composition prior to revascularization. This work describes a method to construct quantitative images from local measurements of integrated backscatter, frequency dependence of backscatter, integrated attenuation and slope of attenuation using intermediate-frequency (5-12 MHz and 12-28 MHz) RF signals backscattered from carotid plaque. Initially, average values from regions of interest in these individual parameter images are correlated to plaque type

Acoustical Imaging, Volume 25, Edited by Halliwell and Wells.
Kluwer Academic/Plenum Publishers, 2000.

as assessed by histology. Using the information thus obtained about the relationship between parameter values and plaque composition, discriminant analysis is made to find a set of predictive equations to classify sets of parameter measurements as calcification, intraplaque hemorrhage or lipid. Finally, each set of pixels from different parameter images is classified using the predictive equations, and a single multi-parametric image displaying gray values linked to tissue composition is constructed.

MATERIALS AND METHODS

Data Acquisition

Closed, fresh carotid plaques (N=15) removed from patients selected for endarterectomy (stenosis diameter > 70 %) were pinned to a support in a 0.9% saline bath at 37°C. Two different broadband transducers, in turn, having nominal center frequencies of 10 MHz (0.25" diameter, 1" focal length) and 20 MHz (0.125" diameter, 0.75" focal length) were scanned in a plane above the specimen (scan-line spacing: 500 µm along and 200 µm perpendicular to the longitudinal axis of the plaque). Backscattered RF signals collected along the entire length of each plaque were digitized at 100 MHz (Lecroy 9350 AL Oscilloscope) and stored on computer for off-line analysis.

Parameter Estimation

The general techniques previously described[8] for constructing high resolution (30-50 MHz) quantitative attenuation and backscatter images of atherosclerotic plaque using backscattered RF signals were adapted for lower frequencies. Data were divided into columns of 5 adjacent RF lines (corresponding to a 1 mm width perpendicular to the long axis of the plaque) and 4 consecutive B-scans (corresponding to a 2 mm length along the long axis of the plaque) for a total of 20 RF lines. RF signals were further divided axially with Hamming windows having a length of 64 samples (480 µm) for 10 MHz data and 32 samples (240 µm) for 20 MHz data. These time windowed estimation voxels were displaced along the length of the RF lines with 50 % overlap. For a given voxel, each windowed segment of RF was Fourier transformed, and squared in magnitude. The resulting power spectra from different RF lines within the estimation voxel were averaged and normalized by the power spectrum of the specular echo from a polished stainless steel reflecting plate placed at the same distance from the transducer. For each average power spectrum (not compensated for attenuation in tissue between the transducer and the measurement region), the average or Integrated Backscatter (IB) was calculated and the frequency dependence of backscatter was determined by calculating the slope (n) of the least-squares linear fit of the log of backscatter versus the log of frequency within a 5 to 12 MHz bandwidth for 10 MHz data and a 12 to 28 MHz bandwidth for the 20 MHz data. The attenuation was estimated as a function of frequency from the power decay of six consecutive spectra using a multi-narrowband technique.[8] From the resulting frequency-dependent attenuation and within the same bandwidths, the Integrated Attenuation (IA) (average attenuation) and slope of attenuation β values were calculated.

Histological Evaluation

After data acquisition was completed, each plaque was fixed in 4 % formalin and decalcified if necessary. Beginning at the bifurcation (marked with India ink), the specimen was cut perpendicular to the longitudinal axis of the plaque at 2 mm spacings such that histological slices were obtained with the same spacing as parametric images. Histologic sections were stained with orcein for elastic fibers and hematoxylin-eosin-saffron for nuclei, cytoplasm and collagen, respectively. Structural features in digitized sequential images of histologic sections were compared to those in sequences of ultrasonic B-scans of the same plaque to identify corresponding views. At segments of high stenosis (N = 59), plaque composition was assessed on original histologic sections using an optical microscopy scoring of the level of collagen, lipids, calcification, and intraplaque hemorrhage (0 = absence, 3 = high concentration). Plaque sections were classified as lipid (N = 14) , intraplaque hemorrhage (N = 8) or calcified (N = 15) if these plaque constituents were clearly present in highest concentration (score of 3 for dominant constituent, lower scores for all other constituents). Mixed plaques (N = 19) contained a variety of plaque constituents, none of which were identified as predominant (scores smaller than or equal to 2 for several different constituents). Some sections were identified as being complicated by the presence of thrombus (N = 3).

Multi-Parameter Analysis and Images

To combine information from the different parameter images, a two-step approach was followed. First, in 59 independent quantitative images of highly stenotic plaque, the average parameter values in a 5 by 3 pixel ROI, centered laterally and beginning approximately 800 µm beneath the plaque surface of each parametric image were correlated with plaque composition as assessed by histology (lipid, intraplaque hemorrhage, calcification, mixed and thrombus). Using the information thus obtained about the relationship between parameter values and plaque composition, discriminant analysis was made using NCSS software (NCSS 2000, Kaysville, Utah USA) and a set of predictive equations was found that can be used to classify sets of parameter measurements by plaque type. Secondly, each set of pixels from different parameter images was classified by tissue-type using the predictive equations, and a single, local tissue composition image was constructed. Since the 20 MHz parametric images have an axial resolution 2 times better than the 10 MHz parametric images, each set of pixels on the 10 MHz images was compared to two sets of pixels on the 20 MHz images. If the tissue identified by the two combinations of pixels differed, the pixel on the multi-parametric tissue identification images was marked as unidentified. Multi-parametric tissue identification images are displayed with different levels of gray for masked (low signal) region, unidentified, hemorrhage, lipid, and calcification.

RESULTS

Average parameter values were calculated for each tissue group. To facilitate comparison of parameters measured across the two bandwidths, sets of average values (slope and amplitude) were used to plot the lines shown in figure 1. The attenuation increased with frequency in all cases. This increase in attenuation led to the apparent decrease in backscatter and negative backscatter slopes at higher frequencies. The attenuation is highest for calcified plaques followed by lipid plaques (particularly at higher frequencies) and is generally lower in other groups. At 10 MHz, the slope of attenuation was higher in calcified plaques than hemorrhage or mixed plaques. Slope of backscatter in the lower frequency bandwidth is negative only for calcified plaques but is negative for all groups at higher frequencies. Backscatter is highest in calcified plaques. Parameter values from thrombus were eliminated from further analysis due to the small number of sections.

The ability of each parameter to discriminate groups by pairs ($p < 0.05$) was evaluated with a multiple Kruskal-Wallis Z comparison. All parameters except the β from 12-28 MHz were found to be able to discriminate at least one group from the others. No single parameter was able to discriminate between all groups. For separation of plaques in

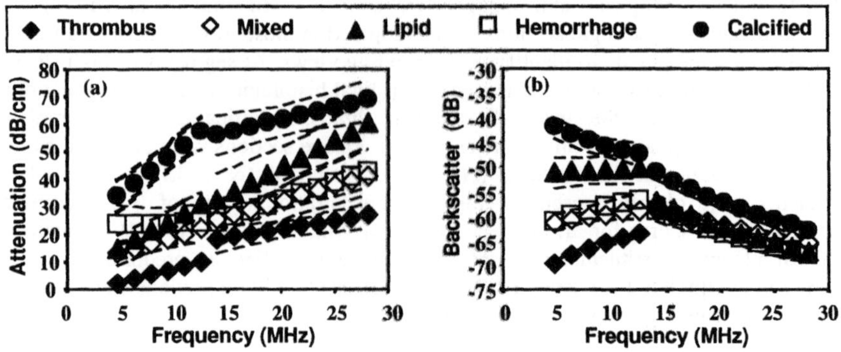

Figure 1. Linear representations of average parameter values for each plaque type. Panel (a): plotted based on IA and β average values. Panel (b): plotted based on IB and n average values. Dashed lines represent approximate range of variation based on standard errors of the means.

four groups (calcification, lipid, intraplaque hemorrhage or mixed), a stepwise variable-selection retained the IB, n, IA, and β from 5-12 MHz and the IB and n from 12-28 MHz parameters. Discriminant analysis classified individuals as summarized in table 1. If only the three groups of calcification, lipid, and intraplaque hemorrhage are considered, the parameters selected for discriminant analysis are IB, n, and β from 5-12 MHz along with IB from 12-28 MHz. The classification of individuals is shown in table 2.

Table 1. Classification performance for separation of plaques into four groups

Type	Correct Classifications	False Classifications	
Calcification	100 %	-	-
Hemorrhage	62.5 %	Mixed: 37.5 %	-
Lipid	64.3 %	Calc.: 7.1 %	Mixed: 28.6 %
Mixed	84.2 %	Calc.: 5.3 %	Hem.: 10.5 %

Table 2. Classification performance for separation of plaques into three groups

Type	Correct Classifications	False Classifications	
Calcification	100 %	-	-
Hemorrhage	75 %	Lipid: 25 %	-
Lipid	71.43%	Calc.: 7.14 %	Hem: 21.43 %

Discriminant analysis also provided a set of predictive equations that allow to classify new sets of parameter measurements by tissue type (calcification, lipid, and intraplaque hemorrhage). Each set of pixels from different parameter images was classified using these predictive equations, and a single, local tissue composition image was constructed. Multi-parametric tissue identification images are shown in figures 2 and 3. Images are presented corresponding to sections that were assessed by histology to demonstrate homogeneous regions of intrapalque hemorrhage (2.a) , lipidic (2.b), calcified (3.a), and a more mixed distribution of plaque constituents (3.b).

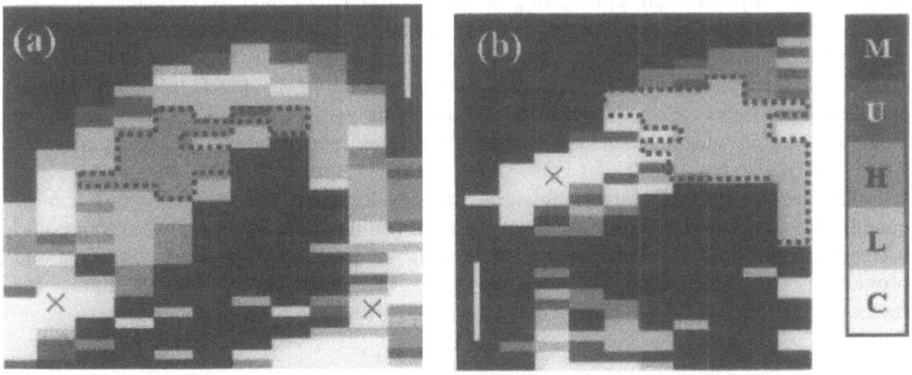

Figure 2. Multi-parametric images. Composition of corresponding histologic slices was assessed as (a) intraplaque hemorrhage containing some lipids surrounded by a thin collagen cap (b) lipidic core surronded by a thin collagen layer with scattered microcalcifications. Good identification (outlined) and poor (x). Grayscale (Masked M, Unidentified U, hemorrhage H, lipid L and calcification C). White bar = 2·mm.

Images in figure 2.a indicate intraplaque hemorrhage content within the outlined central region of the plaque. The predominately lipidic content is well identified within the outlined central region of the plaque in figure 2.b. A fibro-lipidic plaque with small calcifications throughout (fig. 3.a) is strongly identified as calcified. The multi-parametric image of a segment containing more focally distributed calcification in the collagen surrounding a lipidic core (fig. 3.b) displays several regions of calcification with some indication of lipid content towards the center of the plaque intermixed with quite a few pixels indicating intraplaque hemorrhage and unidentified content. Poor correspondence can be identified at pixels indicating calcification in deeper regions of figure 2.a and 2.b (marked with crosses). In the first case, poor identification could potentially be due to distortion of backscatter parameters by signal attenuation in the intervening tissue. The falsely calcified region identified in figure 2.b could be associated with attenuation estimation artifacts at the plaque/saline interface. In figure 3.a the region around the lumen is identified as collagen on the histologic section while pixels in this region indicate intraplaque hemorrhage and lipid. In fact, identification of collagen plaque is impossible with the current model, since the discriminant analysis did not include parameter measurements from this plaque type.

DISCUSSION

By combining IB, n, IA and β quantitative parameters, correct classification of 10:14 lipidic, 15:15 calcified and 6:8 intraplaque hemorrhage regions was obtained. When mixed plaques were included in the discriminant analysis 16 of the 19 regions were identified as mixed, however, 3:8 intraplaque hemorrhage and 4:14 lipidic regions were also identified as mixed. Predictive equations generated from multi-variate discriminant analysis of these results were applied to parametric images pixel-by-pixel to form single tissue-type parametric images. Regions of good correspondence could be identified between local content predicted by these images and the tissue composition indicated on corresponding histologic sections.

One limitation of this study is that measurements were obtained from only a small number of specimens. More specimens are needed to ensure that measurements are obtained from a population representing all types of plaque of interest. Another limitation is in the purity of the plaque types considered. Separation between plaque components by ultrasound quantitative parameters may have been hampered by the fact that plaques classified as predominantly one plaque type contained lower levels of other plaque constituents. Effects of the attenuation in overlying tissues and differences in tissue condition *in vivo* and *in vitro* may change the precise realationships between parameter values and plaque type.

The results lead us further towards the development and clinical validation of quantitative ultrasound imaging for carotid plaque composition. Such a clinical tool could increase insight into the relationship of plaque composition to patient condition and provide useful information for the choice of the most appropriate technique for revascularization.

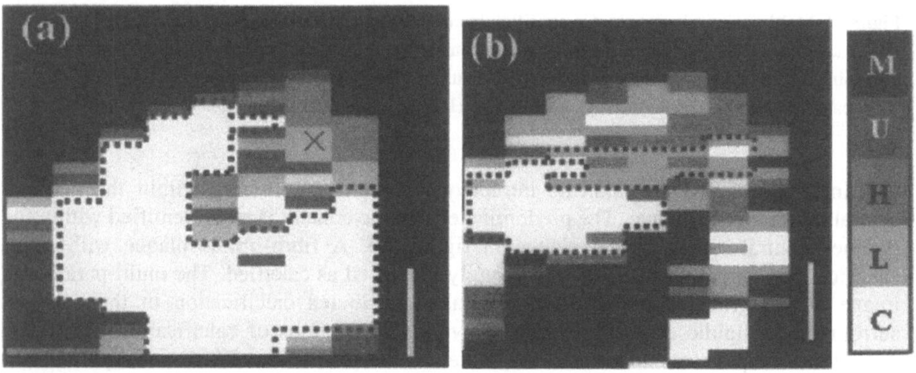

Figure 3 Multi-parametric images. Composition of corresponding histologic slices was assessed as (a) fibro-lipidic plaque with numerous small calcifications (b) large lipidic core surrounded by focally calcified collagen. Good identification (outlined) and poor (x). Grayscale (Masked M, Unidentified U, hemorrhage H, lipid L and calcification C). White bar = 2 mm.

REFERENCES

1. North American Symptomatic Carotid Endarterectomy Trial Collaborators, Beneficial effect of carotid endarterectomy in symptomatic patients with high grade carotid stenosis, *N. Engl. J. Med.* 325:445 (1991).
2. J.G. Theron, G.G. Payelle, O. Coskun, H.F. Huet and L. Guimaraens, Carotid artery stenosis treatment with protected balloon angioplasty and stent placement, *Radiology* 201:627 (1996).
3. European Carotid Plaque Study Group, Carotid artery plaque composition – Relationship to clinical presentation and ultrasound B-mode imaging, *Eur. J. Vasc. Endovasc. Surg.* 10:23 (1995).
5. B. Barzilai, J.E. Saffitz, J.G. Miller and B.E. Sobel, Quantitative ultrasonic characterization of the nature of atherosclerotic plaques in human aorta, *Circ. Res.* 60:459 (1987).
6. T. Noritomi, B. Sigel, V. Swami, J. Justin, V. Gahtan, X. Chen, E. Feleppa, A. B. Roberts and K. Shirouzu, Carotid plaque typing by multiple-parameter ultrasonic tissue characterization, *Ultra. Med. Biol.* 23:643 (1997).
7. L. Wilson, M. Neale, H. Talhami and M. Appleberg, Preliminary results from attenuation-slope mapping of plaque using intravascular ultrasound, *Ultra. Med. Biol.* 20:529 (1994).
8. S.L. Bridal, P. Fornès, P. Bruneval G. and Berger, Correlation of ultrasonic attenuation (30 to 50 MHz) and constituents of atherosclerotic plaque, *Ultra. Med. Biol.* 23:691 (1997).

ASSESSMENT OF SKELETAL STATUS USING AN AXIAL TRANSMISSION TECHNIQUE: A THEORETICAL AND EXPERIMENTAL INVESTIGATION

E. Camus[1,2], M. Talmant[1], J. Guyonvarch[1], G. Berger[1], and P. Laugier[1]

[1]Laboratoire d'Imagerie Paramétrique CNRS UMR 7623
Université Paris 6
Paris, France
[2]Diagnostic Medical Systems
Montpellier, France

INTRODUCTION

In recent years quantitative ultrasound measurements of bone has played a growing role in the assessment and management of osteoporosis and is gradually becoming an integrated part of the diagnosis of osteoporosis. Ultrasound investigations of bone properties include transverse transmission techniques in which the ultrasound wave passes through bone, e.g. the heel bone and the phalanges, and the so-called axial transmission techniques in which the ultrasound wave propagates along the long axis of bone, such as the tibia (Foldes et al., 1995) or radius (Hans et al., 1999). This latter technique was first developed in 1958 to study cortical bone status during fracture healing (Siegel et al., 1958) and has been used subsequently to investigate bone status in osteoporosis. However, to the best of our knowledge, the nature of the propagating waves has not been documented. The present paper is a contribution to the understanding of the propagation mode involved in this axial transmission technique. Theory predicts the existence of a longitudinal lateral wave propagating along the interface in the case of a spatially confined source (Brekhovskikh and Godin, 1992; Officer, 1958). We have compared the experimental arrival times and propagation velocity in test materials immersed in water with the predicted values for the longitudinal lateral wave, and then applied this technique to cortical bone specimens (immersed in water) to determine the longitudinal velocity along the long axis of the bone.

Acoustical Imaging, Volume 25, Edited by Halliwell and Wells.
Kluwer Academic/Plenum Publishers, 2000.

THEORY AND SIMULATIONS

Let us consider a plane interface separating two elastic half-spaces, medium 1 and medium 2, and an acoustic wave source located in medium 1 incident upon the interface (Figure 1, left). The reflected pressure field at an observation point in medium 1 is the sum of the geometric contribution, i.e. the incident and the specularly-reflected waves, the contribution of the so-called lateral or head waves and that of the well-known surface or leaky waves (Brekhovskikh and Godin, 1992; Officer, 1958). Lateral waves arise when the incident field emanates from a spatially confined source and when the bulk wave velocity in medium 2 is greater than that in medium 1. The lateral wave is excited at the longitudinal critical angle, propagates along the interface at the bulk velocity in medium 2 and re-radiates back into medium 1 at the same critical angle, which results in a minimum propagation time. The contribution of the lateral wave to the total pressure field is negligible except when the associated signal is separated in time from the rest of the signal.

In our work the upper medium is water and we assume that the lower medium, either a test material or a bone sample, is an isotropic homogeneous solid and that the interface is plane. It is important to note that for such a plane fluid-solid interface, the above description of the longitudinal lateral wave is still valid. Only the longitudinal lateral wave is considered because the shear wave if any arrives at a later time. Similarly, surface wave contributions which have an analogous ray representation do not arrive before the longitudinal lateral wave. Theoretical predictions of the arrival times of the direct, reflected and lateral waves at the receiver (observation point P) as a function of range *r* (distance from transmitter to receiver along the interface) for a water-solid interface with c_1= 1490 m.s^{-1} and c_2= 3000 m.s^{-1} (Figure 1, right) shows that the lateral wave arrives before the reflected wave at any range *r*, and for large enough *r* values it arrives before the direct wave. This is the basis of our experimental set-up: for given transmitter- and receiver- sample distances (*d* and *z*), the range *r* is chosen such that the lateral wave corresponds to the first arriving signal.

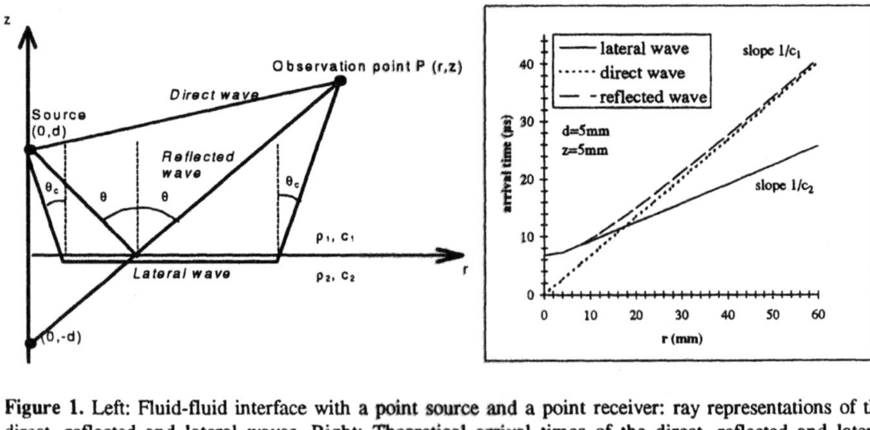

Figure 1. Left: Fluid-fluid interface with a point source and a point receiver: ray representations of the direct, reflected and lateral waves. Right: Theoretical arrival times of the direct, reflected and lateral waves for a fluid-solid interface: c_1 = 1490 m.s^{-1}, c_2 = 3000 m.s^{-1}.

EXPERIMENTS

Experimental set-up

A quasi-point source broadband piezoelectric element (transmitter) was used to generate a wide ultrasonic beam incident upon the water-sample interface and the signals radiated from the interface were received at different positions by a similar element (receiver) moving stepwise along the sample surface (Figure 2). Both piezoelectric elements operate at a central frequency of 0.5 MHz. In our experiments, two parameters were used to identify the first arriving signal, separated in time from the other signals, and compare it to the lateral wave: 1) the difference in arrival times of the first arriving signal between two measurements at consecutive receiver positions, from which the longitudinal velocity in the sample can be calculated; 2) the absolute time-of-flight measurements as a function of range *r*.

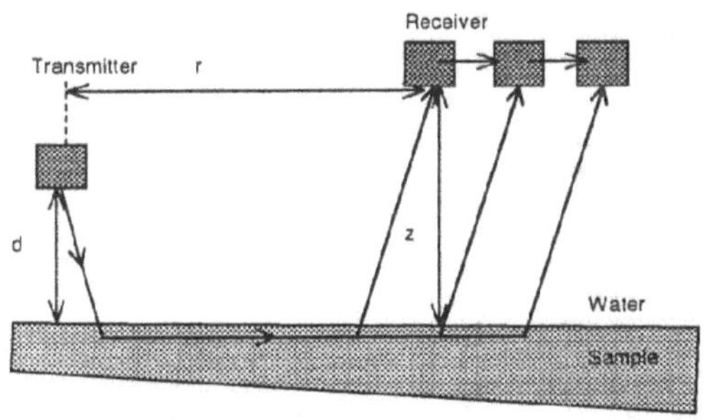

Figure 2. Experimental set-up.

Specimens/Materials

A preliminary study included ultrasonic measurements on solid materials of known acoustical properties: a Plexiglas plate with a longitudinal ultrasound velocity of 2680 m.s^{-1} and an aluminum plate with a longitudinal ultrasound velocity of 6300 m.s^{-1} (Ensminger, 1973). The experimental method was then applied to two human femur specimens with soft tissue and marrow removed. The longitudinal ultrasound velocity in the cortical part of the femur reported in the literature ranges from 3480 to 4200 m.s^{-1} along the axial direction (Siegel et al., 1958; Lowet and Van der Perre, 1996; Lee et al., 1997).

Results

Typical recordings (RF signals and associated *(r,t)* diagram) obtained for the Plexiglas sample are shown in Figure 3. The arrow indicates the first arriving signal which is analyzed. In the *(r,t)* diagram, the amplitude of the radio-frequency signals has been converted into gray level and plotted versus the range *r* (vertical axis). The different waves received at range *r* from the source are shown with their associated trajectories in the *(r,t)* plane. Figure 4 is the corresponding plot of the absolute time-of-flight versus range *r*. The velocity (mean ± standard deviation) measured in the Plexiglas sample is 2650±150 m.s^{-1}; in the aluminum sample, we have obtained a velocity of 6100±300 m.s^{-1}. Both results are in good agreement with the reported values for these materials.

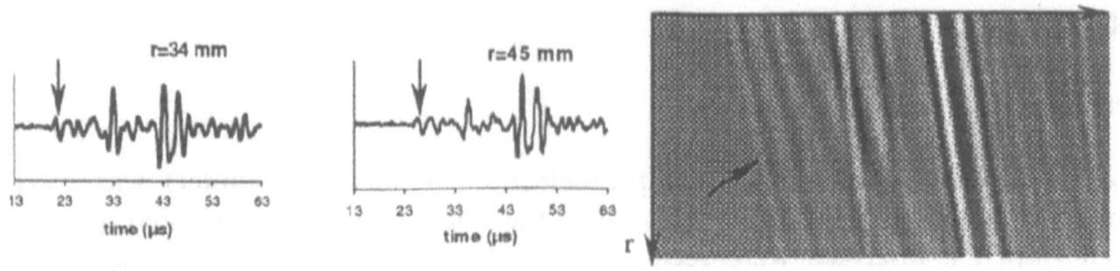

Figure 3. RF signal and associated (r,t) diagram for a water-Plexiglas interface.

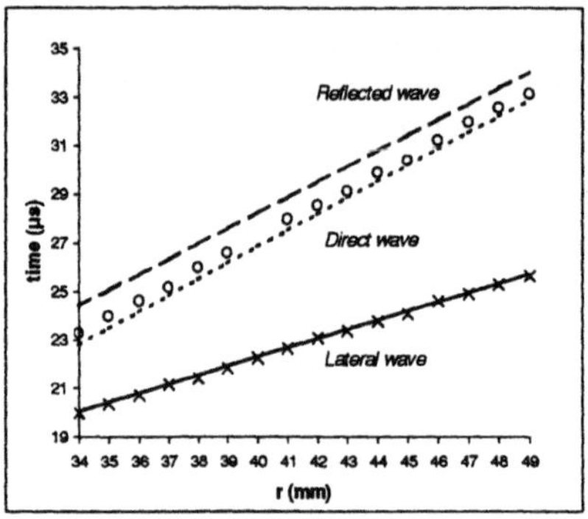

Figure 4. Time-of-flight versus range r for a water-Plexiglas interface. Solid, dotted and dashed lines: predicted arrival times of the lateral wave, direct wave and reflected wave; dots: experimental arrival times of the first arriving signal and direct wave.

Figure 5 displays the RF recordings obtained in one human femur sample. The arrow indicates the first arriving signal which is analyzed. Figure 5 is an interesting case in that several waves seem to arrive at the same time and interfere with each other at small values of range r. In this experiment, we suspected that the requirements for observation of a lateral wave as the first arriving signal were not met, i.e. that the direct wave arrived first, shortly followed by the lateral wave or both waves arrived at the same time and interfered with each other. Therefore the first arriving signal was studied after digital subtraction of a separate data set containing only the direct wave (experimentally obtained after removing the sample) from the first data set containing all waves. The corresponding plot of the absolute time-of-flight versus range r for this femur sample is displayed in Figure 6. The velocities (mean ± standard deviation) measured in this femur sample are 3750±150 $m.s^{-1}$ and in another femur sample 3600±100 $m.s^{-1}$ and 4050±100 $m.s^{-1}$ at two different sites, all in agreement with reported values.

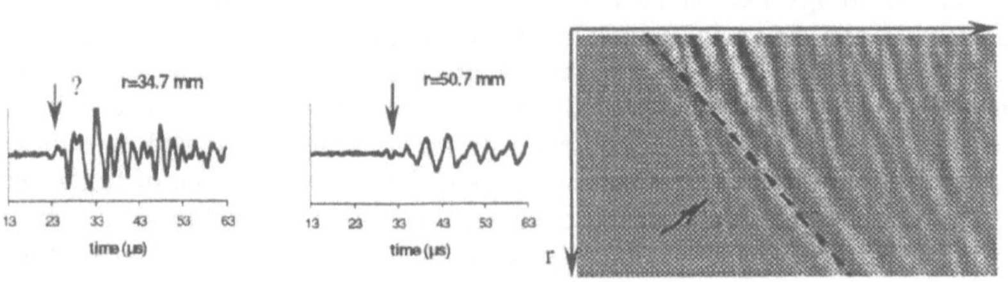

Figure 5. RF signal and associated (r,t) diagram for a water-femur interface. The dashed line indicates the direct wave.

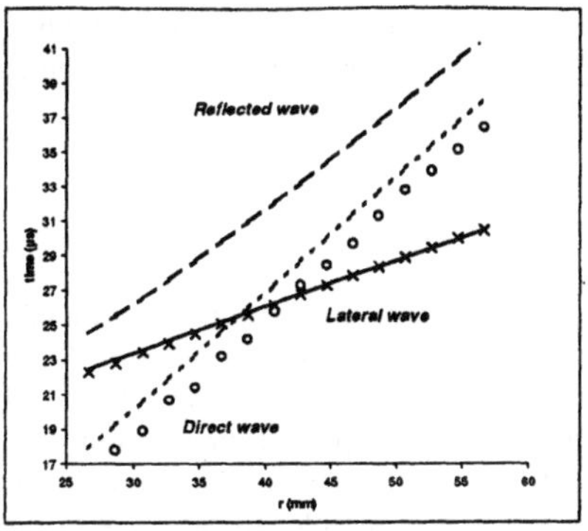

Figure 6. Time-of-flight versus range r for the water-femur interface. Same legend as in Figure 5.

DISCUSSION

The time-of-flight determinations of the first arriving signal for the test materials are in good agreement with the theoretical predictions of the lateral wave. This validates the ultrasonic path followed by the first arriving signal. As the transmitter-receiver separation r along the interface increases, the lateral wave separates more and more in the time-domain from the other received signals and therefore it is easier to track it on the RF recordings. However, its amplitude drops swiftly with increasing range r and the performance of the data acquisition system may limit the detection of the lateral wave for large r values. For both test materials, the propagation velocities of the first arriving signal are close to the longitudinal velocities found in the literature.

The results obtained in one of the human femur specimens (Figure 5) are very interesting in that for small r values the direct wave arrives slightly before the lateral wave. This is predicted by the theoretical analysis and confirmed by the experimental (r,t) diagram and measurements of the time-of-flight as a function of range r. The measured propagation velocities range from 3600 to 4050 for the two femur specimens and are in agreement with reported values.

These *in vitro* results in test materials and human bone specimens strongly suggest that the longitudinal lateral wave was successfully excited and detected as the first arriving signal under specific geometrical arrangements of the transducers relative to the sample surface.

Error sources on the velocity measurements include the determination of the accurate time-of-flight of the first arriving signal and the variation of the receiver-sample distance z with range r along the interface. Indeed, improvements on the signal-to-noise ratio could be made by optimizing the sensitivity of the transducers and would result in better time-of-flight determinations. In our experiments, the receiver-sample distance varied only slightly and no correction was performed in the velocity calculations. If necessary, correction for this variation, which we have already described (Camus et al., 1998), may be done.

Finally, it is important to note that we have considered the model of a plane interface between a homogeneous fluid and an isotropic homogeneous non-attenuating solid. These conditions might not be met when experimenting with biological tissues (soft tissue and bone). Therefore, further work should include the influence of heterogeneity, density, anisotropy and attenuation on the propagation of a longitudinal lateral wave. Effects of the sample thickness in relation to the extension in depth of the lateral wave in the sample should also be investigated.

CONCLUSION

We have investigated the nature of the first arriving signal in the axial transmission technique in terms of arrival time and propagation velocity for test materials and cortical bone specimens. For all samples, good agreement was found between the experimental data and the theoretical arrival time and propagation velocity of the lateral wave, which strongly suggests that the first arriving signal corresponds to the propagation of the longitudinal lateral wave along the sample surface. With the understanding of the propagation along bone surface using the axial transmission technique, further development of this technique and evaluation of the mechanical properties of human cortical bone may be obtained at various skeletal sites, and in particular at peripheral sites, in relation to the diagnosis of osteoporosis.

ACKNOWLEDGMENTS

We acknowledge the support of GIP Ultrasons, Tours, France for providing the transducers.

REFERENCES

Brekhovskikh, L. and Godin, O., 1992, *Acoustics of Layered Media II: Point Sources and Bounded Beams. Wave Phenomena*, Vol. 10, H. Lotsch.

Camus, E., Berger, G., and Laugier, P., 1998, Cortical bone velocity mapping using leaky surface acoustic waves, *1998 IEEE International Ultrasonics Symposium*, Sendai Japan.

Ensminger, D., 1973, *Ultrasonics - The Low- and High-intensity Applications*, Marcel Dekker Inc.

Foldes, A.J. *et al.*, 1995, Quantitative ultrasound of the tibia: a novel approach for assessment of bone status, *Bone*, 17(4):363-367.

Hans, D. *et al*, 1999, Does combining the results from multiple bone sites measured by a new quantitative ultrasound device improve discrimination of hip fracture?, *Journal of Bone and Mineral Research*, 14(4):644-651.

Lee, S.C., Coan, B.S., and Bouxsein, M.L., 1997, Tibial ultrasound velocity measured in situ predicts the material properties of tibial cortical bone, *Bone*, 21(1):119-125.

Lowet, G. and Van der Perre, G., 1996, Ultrasound velocity measurement in long bones: measurement method and simulation of ultrasound wave propagation, *J Biomechanics*, 29:1255-1262.

Officer, C.B., 1958, *Introduction to the Theory of Sound Transmission with Application to the Ocean*, McGraw-Hill Book Company, Inc.

Siegel, I.M., Anast, G.T., and Fields, T., 1958, The determination of fracture healing by measurement of sound velocity across the fracture site, *Surg. Gynec. Obstet.*, 107:327-332.

WIDEBAND ULTRASOUND TO DETERMINE LIPID CONCENTRATION IN FISH

Penny Probert and Russ Shannon

Department of Engineering Science,
University of Oxford,
Parks Road, Oxford, UK OX1 3PJ
pjp@robots.ox.ac.uk, ras@robots.ox.ac.uk

INTRODUCTION

Ultrasound is used commonly in medical imaging for determining the boundaries between different tissues. In this paper we investigate its use in the aquaculture industry, specifically to measure the lipid content of salmonoid fish. Emphasis is on an automated low cost method with rapid throughput. This paper describes the use of low intensity ultrasound to determine fish composition, particularly the fat content.

The "condition" of fish, especially their oil or fat content, is particularly important in fisheries biology because of the overwhelming importance of fat or oil as an energy reserve for reproduction (see e.g. Henderson et al. 1984; Henderson and Almatar, 1989, Marshall 1999). Many fish naturally accumulate very large body reserves of energy – generating oil in relation to their breeding cycle. Molecular genetic breeding programmes are being developed to select for salmon with more uniform and more controlled flesh oil levels, whether high or low. Such programmes depend critically on a rapid, accurate and above all non-invasive method of determining oil levels in live fish.

The only reliable techniques for fat measurement are destructive (e.g. the Soxhlet test). Techniques to image fat developed medically use primarily nuclear magnetic resonance (Jorgenson et al, 1995) but these are unsuitable because of the cost and size of the equipment required. For fish, other methods under investigation include contact methods such as near infra-red diffusion spectroscopy (Isaksson et al 1995, Mayes 1992) and impedance measurement (the Torry Fish Fat Meter - Kent 1990). However, there are problems with reliability since repeatable contact between transducer and live fish is very difficult. In addition, an inspection system that involves disturbing and handling the fish may be expected to cause high levels of stress and some physical damage (Pickering 1993, Lines 1999). These methods also are biased towards the area close to the surface.

Ultrasound is of interest for fish for two reasons. First, it can be used remotely, as water is an excellent coupling medium between transducer and fish. Second, it provides a genuine bulk measurement. A significant proportion of the lipids in fish is distributed as droplets in muscle and other tissue, so bulk measurement is important.

This paper describes preliminary work which investigates the interaction of ultrasound with fish tissue with a view to the application of reflection and transmission sonar in a low cost, time-of-flight sonar system. Such a system would consist of a single transmitter and receiver taking spot meaurements at particular cross sections of a the fish.

Acoustical Imaging, Volume 25, Edited by Halliwell and Wells.
Kluwer Academic/Plenum Publishers, 2000.

ULTRASOUND PROPERTIES IN TISSUE

Basic properties

A number of properties of ultrasound vary with different tissues. Two of the main ones are velocity and attenuation. Velocity has been used as the basis of fat measurement in a number of applications, including dairy products. The velocity of ultrasound is related to the elasticity and the density of tissue. To a first approximation, for tissues of similar density, it is possible to assume that velocity sums linearly according to the volume fraction [Ghaedian, 1998]:

$$\frac{1}{c^2} = \sum_j \frac{x_j}{c_j^{\,2}}$$

where c is the average velocity, x_j is the volume fraction of a particular tissue type and c_j the velocity of that type of tissue. This relationship gives a good description of the ultrasonic properties of food materials in which the density of the different phases are similar and scattering is not appreciable [Pinfield, 1997].

Velocity depends strongly on temperature. The velocities in fat and muscle vary in opposite senses with temperature with linear dependence: whereas the velocity in fat increases, the velocity in muscle decreases. This means that although they are separated quite well at room temperature they converge at between 5 °C and 11°C before diverging again. This must be borne in mind when taking measurements of fish in non-controlled environments.

Attenuation is less reliable as an indicator of tissue composition. Attenuation is measured by monitoring the intensity of ultrasound received. It is affected significantly by scattering and absorption. In addition the intensity at the receiver depends strongly on conditions at the interface where ultrasound is coupled into the medium. In particular, its orientation must be controlled very closely if attenuation is to be used as the prime indicator. It is probably not viable to provide such close control for fish.

Other properties of ultrasound also vary with tissue type. The non-linearity coefficient, normally referred to by the ratio of the first two harmonic components B/A, has been investigated to delineate fat tissue from muscle by Fatemi and Greenleaf (Fatemi-96). Measurement of the non-linearity coefficient requires measurement of the second harmonic component of the signal, and wide bandwidth transducers. Although little work has been done in this area, it is an interesting area to pursue since non-linearity is significant even at low acoustic powers.

Properties of ultrasound in fish

There is significant published literature on values of velocity, attenuation and non-linearity coefficient, summarised by Duck [Duck, 1990]. At room temperature, for example, some typical values of velocity from measurements on beef, pigs and humans are summarised in Table 1. Exact values of velocities vary over about 100msec-1 between species and the organs in which the tissue is situated. Note that, as we might expect, ultrasound in bone behaves significantly differently than in other tissue. Note also the large temperature coefficient of the velocity in fat. However few measurements are available for fish, and then only for whitefish (which do not include salmon).

Fish contain primarily proteins (muscle – typically 15%-30% by weight), fat (typically, in salmon, 5% to 20% by weight), water (50% - 80%) and minerals. Fish protein varies from meat primarily in containing less connective tissue and more water. In general it contains around 80 to 85% water, higher than in birds and mammals. In muscle, there are also differences in the ways in which muscle is segmented [Hoar, 1978]. We would therefore expect some difference from animal protein in ultrasonic properties. The skin also has a higher water content.

Table 1. Typical values of velocity through various tissues. First set of entries (no asterisk) are taken from mammals [from Duck, 1990], second set (asterisked) are from whitefish [Freese and Macow 1968]. The large range of velocities results from different mammal species of different muscle within a single species

	Typical range of temp coefficients	Typical range of velocity values
Muscle	+1 msec^{-1} °C^{-1}	1550-1590 msec^{-1}
Fat	–3 to -7 msec^{-1} °C^{-1}	1412-1487 msec^{-1}
Tendons	Unknown	1630-2640 msec^{-1}
Bone	-6.6 msec^{-1} °C^{-1}	2800-3300 msec^{-1}
*White flesh	Unknown	1540 msec^{-1}
*Dermis	Unknown	1530 msec^{-1}
*Scales	Unknown	1870 msec^{-1}

Most fats have similar ultrasonic properties [Ghaedian, 1998]. Studies on Atlantic salmon have identified sources of lipid in various parts of the body including white and dark muscle, the belly flap, skin, and various other organs [Katikou, 1998] [Reinitz, 1981]. The quantity of lipids is strongly dependent on the cross section taken. Lipid content along the body ranged from 4% to 13% in one set of specimens [Ackman, 1995]. The distribution is dissimilar to that in mammals, in which most fat is deposited in well defined layers. In fish, the lipids in white muscle, which constitutes the main area of the fish that is eaten, are in the form of droplets dispersed in the muscle segment and, to a lesser extent, in the connective tissue surrounding the muscle fibres. A small amount of fat is layered beneath the skin.

Complete fish of course contain other tissue too – in particular bone and organs, skin and scales. For a swimming fish its internal organs are near the bottom, with the swimbladder immediately above them. The swim bladder is an air-filled sack which enables the fish to alter its depth in water as it swims. All of these features cause scattering, absorption and internal reflection. For teleostean fish such as salmon, all these are situated below the backbone; above this is largely muscle and ribs. There is a layer of scales on the outside of the dermis but this is very thin in telesoteans.

The internal structure of the fish presents an obvious difficulty for measurements. In live fish, the air sack provides a huge mismatch to the surrounding tissue (effectively an air/water interface) and effectively reflects any incident ultrasound. The effects of internal organs too may entirely mask both reflection and transmission effects from changes in tissue composition. The choice of a suitable region for measurements in crucial. Currently, to avoid these organs, we are working on a portion below the dorsal fin, and above the backbone. This is a region in which fat is distributed in the muscle. In a final system, a camera would be used to locate suitable sections for measurement.

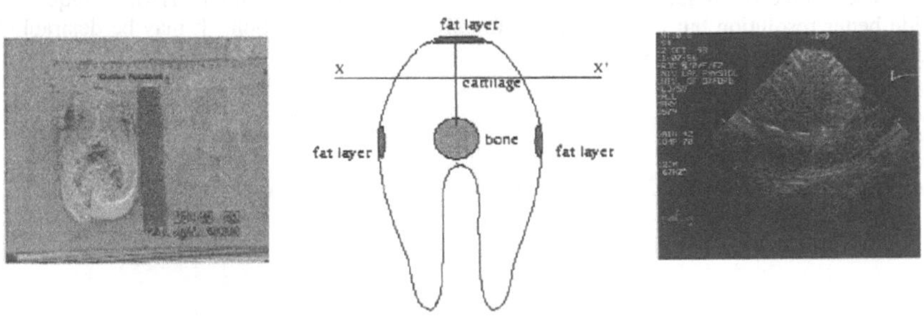

Figure 1. On the left, a photograph of the portion of fish used typically in the experiments, with a schematic (central) showing the various tissues present (tendons are omitted for clarity). On the right an ultrasound B-mode scan, using 2MHz probe, looking down from the top.

A typical cross section of fish used for experiments is shown in Figure 1. The photograph shows the central bone, the skin and the pattern of tendons in the flesh; the region which includes the organs has been removed. A schematic is shown alongside. The structure and orientation of the tendons is likely to affect the velocity significantly. A region of cartilage or bone runs up from the backbone to the top of the back dorsal fin. The ultrasound B-mode image of the volume images clearly the tendon pattern.

Measurements so far have all been performed on dead fish. A freshly-killed fish has been imaged as well to ensure that changes in the skin, scales and muscle structure after death did not alter too significantly the coupling of ultrasound into the fish. The overall characteristics of ultrasound behaviour were similar. Although we would expect some differences in velocity and other ultrasound properties between dead and living fish, these could be handled through calibration.

Information in the radio frequency (RF) signal

The system under development would work directly from the RF signals received, both from transmission through the fish and from reflection. By contrast, medical ultrasound and NDT equipment is optimised for presenting visual images, and masks the raw properties of the RF signal through significant internal processing. The image intensities in the visual images result from differences in ultrasound properties, especially velocity, between different tissues but effectively suppress them. In addition, the eye is very insensitive to small discontinuities, so, for example, there is no need to estimate velocity accurately to present depth information from time-of-flight measurements. By contrast, in this work it is necessary to assess the properties of the rf signal accurately.

The RF signal contains significant information about the material insonified. First, it contains information on average velocity. Published data suggests that velocity is a much surer indication of tissue composition fat content than reflection amplitude [Ghaedian, 1998]. Second, frequency analysis of the signal provides a measure of non-linearity. Finally, it provides information on the quality of the reading. Different pulse distortion results from different paths through a sample and the pulse shape at the receiver provides indication on the homogeneity of the path. If the ultrasound energy is shared between tissue with different velocities pulse spreading results as we see below.

LABORATORY MEASUREMENTS

Figure 2 shows a schematic of the laboratory rig used for measurements. Two 2.25MHz transducers are mounted with one each side of the specimen of fish. Each may operate either in transmit or receive mode. They are operated slightly below specification (in fact at 2MHz) as this improves impedance matching.

The choice of frequency must balance attenuation with resolution. Higher frequencies provide better resolution but more attenuation, and thus less penetration. It may be desirable to increase the frequency at some stage.

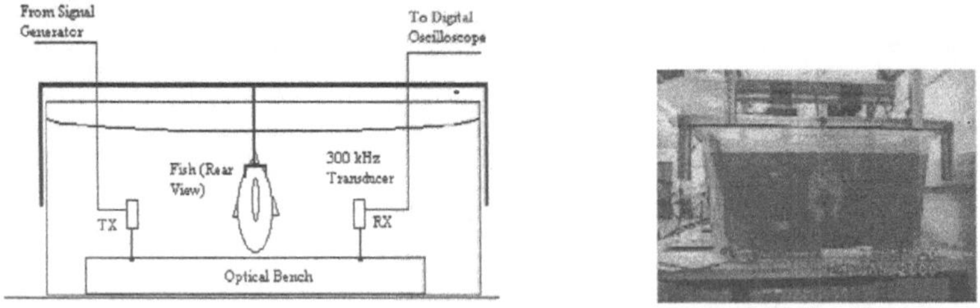

Figure 2. Test rig. The experiments are performed in a water bath, of dimensions approximately 0.7x0.4m. The lining of the bath is non-reflective to sonar. The two sonar transducers were mounted on an optical bench to ensure collinearity. The fish is held in a clamp between the transducers. 2.25MHz transducers were used in the final experiments.

The transducers are typical of those used in non-destructive testing. The bandwidth is over an octave. The transducers operate in the near field with a beam diameter of less than 1cm; hence a selected portion of the fish can be investigated. Focussed and non-focussed transducers have been used. Focussed transducers (optimised for 4cm distance) gave slightly higher power at the receiver up to a distance of about 8cm, but at larger distances the power fell off sharply. In addition they received less power in reflection because scattering is over a wide angle.

Currently the transmitter is driven with a low voltage signal, directly from a 10V signal generator. It may operate either in continuous wave (CW) mode or pulsed mode. In CW mode there is significant feed-through between the transducers through the structure which holds them and this obscures the time of arrival of the received wave. Using short pulses allows this feed-through to die away and provides better resolution. However, measurement of the non-linearity coefficient would require pseudo-CW operation (so harmonics of the signal could be extracted). The signal at transmitter and receiver are monitored using an HP Infinium oscilloscope, sampling at 100MHz.

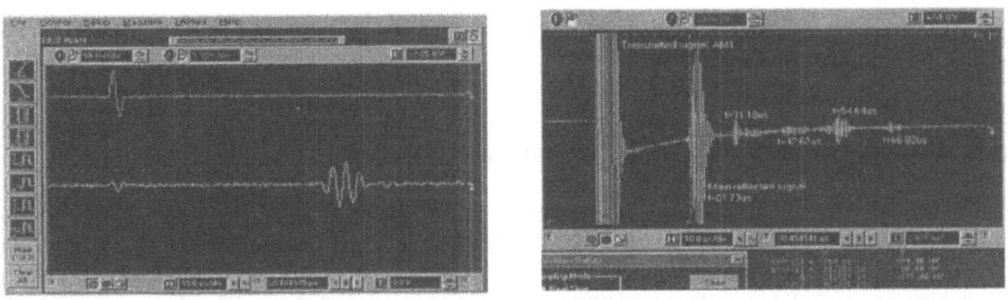

Figure 3. On the left, the top trace shows the pulse transmitted and the lower trace the pulse received. The time distance between these is a measure of mean velocity. On the right, the results from reflection sonar, looking at a complete *trout*. Trout have similar anatomy to salmon. Note that, in this case, the same transducer is used to transmit and receive, and there is a residual quasi-DC level (recovering approximately exponentially) over the entire trace. The signal is sampled at 100MHz. The approximate calibration from time to distance is about 1.5mm/μsec.

Figure 3 shows typical transmission and reflection measurements using this equipment. In transmission there is a small amount of feed-through on the receiver (lower trace) and then only a very low level until the pulse reaches the receiver. In reflection, trout rather than salmon was chosen so the effects of internal structure could be seen clearly (the test tank is too small for a complete salmon to be inserted). Trout are similar in anatomy. The reflection measurements were taken looking at the side, at about the level of the spine (using a different type of waveform). Approximate elapsed times for each reflection are marked. The first portion of the signal is the pulse transmitted. The next portion, at around 22μsec, is from the skin, and weaker multiple reflections from this interface occur at around 44μsec and 66μsec. This leaves us with internal reflections at around 31μsec and 54μsec, corresponding to distance in the fish of around 6.75mm and 24mm). The former may be from ribs, the latter from the air sack. Although there is significant transmission through the spine as fish bone is so soft, at this scale we do not see any reflections from this area. (It is shown by Langton [Langton, 1991] that ultrasound penetrates the right into bone; it does not just take paths around it).

In a typical cross section of salmon (for example along line XX' in figure 1) the acoustic signal passes through a number of soft tissues: in particular skin, the region of muscle, tendon and distributed fat, and a thin region of bone. In addition there is a narrow layer of fat under the skin, with more significant deposits (a few mm thick) at the sides and at the top.

The pulse envelope at the receiver gives information on the homogeneity of the region traversed by the ultrasound. We can see this most clearly by taking measurements with the fish turned round 90 degrees, with the axis of the sonar orthogonal to the plane of the specimen shown in Figure 1. Figure 4 shows the signals received through the fish, expanded in time to show a region of about 10μsec. The first shows the signal received when no fish if present. The second shows the signal received in a region above the spine (for example passing through axis XX'), with the sensor placed centrally on the cross section. The final one was taken with the sensor positioned at the bone. Because propagation is longitudinal along the bone, the ultrasound travels primarily through marrow and the attenuation is low. As the transducer was moved around these regions, the time of arrival of the first part of the pulse remained constant, but the overall shape altered, with the peaks effectively moving within the overall pulse envelope. The centre picture shows significant distortion: the ultrasound propagates through paths in the fatty muscle, tendon and bone.

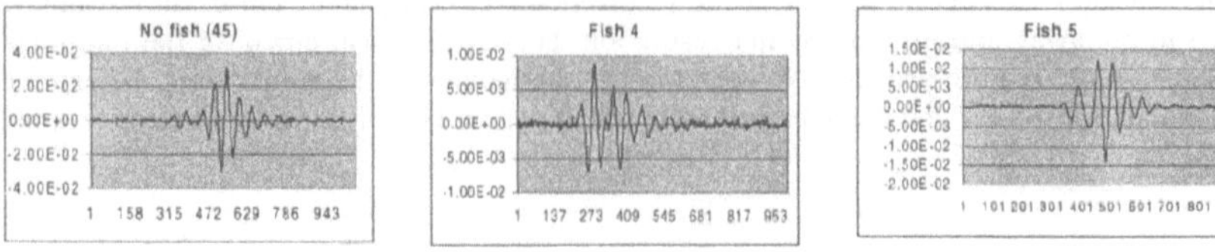

Figure 4: Expanding the time axis over the region in which the pulse is received in transmission sonar. Samples are numbered in units of 10nsec.

Transmission Measurements – bulk fat content

Preliminary measurements have compared an estimate of velocity with 'ground truth' values of fat content obtained through the Soxhlet test. It must be emphasised that these results are approximate only. Measurement of time difference is accurate to around 100nsec, of dimensions to around 1mm. These errors lead to a total error of about 4%. Measurements were taken as follows:

Table 2. measurements taken during tests

d_o:	distance between transducers
d(i):	width of fish along transmitter-receiver line of sight
t_o:	elapsed time between transducers with no fish present
t(i):	elapsed time for fish I

The fat content was determined by the Soxhlet test in a portion of the flesh. The effective velocity was calculated using the expressions:

$$t_{fish}(i) = t(i) - \frac{(d_o - d(i))}{v_w};$$

$$v(i) = \frac{d(i)}{t_{fish}(i)}$$

Where v_w, the velocity of sound in water, was determined from t_o and d_o, and t_{fish} is the time delay through each fish.

Table 3. Measurements on salmon pieces. t_o was 74.9818μsec

	Fish 1	Fish 2	Fish 3	Fish 4	Fish 5
Fish type	A	A	A	B	B
Fat content x % by weight	15.1	13.3	7.5	5.9	3.9
Width of fish d(I) mm	71.8	79.6	67	41	50.4
Time Δt(i):μsec	70.635	70.831	73.071	72.130	73.486
Effective velocity: v(i) msec^{-1}	*1714.2*	*1692.2*	*1627.2*	*1743.6*	*1630.2*

Two types of specimen were used. Type A was a salmon steak, as shown in figure 1a, presented so the ultrasound was orientated along XX'. The steaks of type B were prepared and prepackaged with skin removed. The steaks were positioned so that measurements were made in a direction orthogonal to the plane of the view shown in figure 1.

Discussion of results

Table 3 shows a trend of velocity increasing significantly with fat content. This is reasonable at the temperature used in the tests (around 10°C), in which the velocity of fat is higher than the velocity of muscle. The time of flight was taken as the distance between the start of the pulses transmitted and received. The velocity trend is encouraging but the difference in velocity is greater than we would expect just from the difference in fat content if the fat is all in muscle. In addition since the pulses received were distorted in a similar fashion to Figure 4b it was clear that multiple paths are involved in the propagation.

The results suggest that there may be correlation between other high velocity tissue and overall fat content. Further understanding of fish physiology should provide more information in this area. Experimentally the structure of the tissue is being investigated using B-mode medical imaging. It may be possible to identify certain peaks on the waveform received with particular paths.

Reflection images provide information too. If sufficient resolution could be obtained bone and skin thickness might be measured. This would be significant progress towards simplification.

Other parameters may also be available to provide further information, in particular the non-linearity coefficient which is related to tissue type.

Fat layers

There is of course an easier target we can look at – the regions of fat deposit under the skin which reflection sonar may detect. Ultrasound sensors to measure the depth of subcutaneous fat in animals are already available, but higher resolution is required. In farmed salmon the fat deposits are a few mm thick. High precision reflection sonar should provide an estimate of this depth.

PRECISION SONAR

Precision is crucial in this work. We need to measure time differences to an accuracy of tens of nanoseconds. One way of achieving this is to increase the frequency. For example, at 10MHz we would be aiming for precision of about a quarter of a wavelength, which is reasonable. However increasing the frequency increases attenuation and hence lowers the signal-to-noise ratio at the receiver.

Ideally we would like too to pick out weak reflections, for example to determine the skin depth. We would expect a large reflection at the skin-water interface, then a smaller one about a microsecond later. High signal-to-noise ratio is required, together with discrimination between reflections about wavelength apart.

In medical instruments, harmonic imaging provides improvement in resolution without too much degradation of the signal to noise. However, other methods are available. In non-destructive testing, different frequencies may be used to reach different depths. High power, ultra-short pulses also improve resolution, e.g., 200V pulses are transmitted in some applications. It is feared that very large powers may damage fish [Frenkel, 1999].

However, there is a way to improve resolution without significantly altering the attenuation characteristic or using huge amounts of power: through using wideband techniques. Instead of the ultrasound being transmitted at a single frequency it is 'chirped'; a frequency modulation is produced across the pulse. This technique is used widely in radar and enables us to use lower frequency transducers but still maintain sufficient resolution [Altes, 1995].

The success of the technique lies in frequency modulation of the ultrasound. By putting a characteristic frequency chirp on the transmitter pulse, it can be picked out with precision at the receiver. In air-borne applications we have achieved sub-millimetre resolution from sonar operating with a sweep of about 15kHz at a centre frequency of 45kHz (wavelength about 7.2mm) [Gilkerson 1997].

This method offers high resolution and good noise immunity. In addition it can differentiate between reflections from surfaces which are too close to provide distinct independent reflections.

CONCLUSIONS

This work is still in early stages. It is providing fascinating insights into possibilities of using ultrasound to measure properties of tissues in fish. Improvements in the resolution of ultrasound through broadband operation have offered a solution to better location of interface boundaries. However a number of issues to do with the interaction of ultrasound with various tissue types in fish, and with the physiology of fish remain to be solved before the technique can be considered proven.

ACKNOWLEDGMENTS

We are grateful to Fares Mayia of the Oxford Health Trust for many helpful discussions, and to James Lee for work on reflection measurements. The work has been partly funded by Silsoe Research Institute.

REFERENCES

Richard Altes (1995), Signal Processing for Target recognition in biosonar, *Neural Networks*, 1995, 8 7/8, 1275-1295

Francis Duck: chapter 4 of *Physical Properties of Tissue: A Comprehensive Reference*. Academic Press 1990

Fisher, A. V. (1997), A review of the technique of estimating the composition of livestock using the velocity of ultrasound. *Computing and Electronics in Agriculture* **17**, 217-231

M. Fatemi and J.F. Greenleaf, Real time assessment of the parameter of nonlinearity in tissue using non-linear shadowing *Ultrasound in Medicine and Biology* 22 9 (1996) 1215-1228

M. Freese and D. Makow, High frequency Ultrasonic Properties of Fresh Water Fish Tissue *J. Acoustical Society of America* 44 5 1282-1289 (1968)

R. Ghaedian, J.N. Coupland, E.A. Decker and D.J. MacClements, Ultrasonic Determination of Fish Composition *J Food Engineering* 35 323-327 (1998)

V. Frenkel, E. Kimmel and Y. Iger, Ultrasound-induced cavitation damage to external epithelia of fish skin *Ultrasound in Medicine and Biology*, 25 8 1295-1303 (1999)

P. Gilkerson, Digital Signal Processing for Time of Flight Sonar. Submitted for DPhil, University of Oxford 2000

W.S. Hoar and D.J. Randall., *Fish Physiology Volume 7 (Locomotion),* Academic Press Inc, 1978.

C.M. Langton, C.M. Riggs and G.P. Evans, G.P., Pathway of ultrasound waves in the equine third metacarpal bone, *J. Biomed. Eng.,* 13 113-118 (1991)

V.J. Pinfield and M.J.W. Povey: Thermal scattering must be accounted for in the determination of adiabatic compressibility *J. Phys Chem* **B** 101 1110-1112, 1997

D.J. McClements and M.J.W. Povey: Ultrasonic analysis of edible fats and oils *Ultrasonics* 30 6 383-388, 1992

NON-CONTACT ULTRASONIC IMAGING FOR THE QUANTITATIVE AND NONINVASIVE EVALUATION OF THERMAL INJURY

Joie P. Jones[1] , David Lee[1] ,Saeed Iraniha[2], Mahesh Bhardwaj[3], Victoria VanderKam[2], and Bruce Achauer[2]

[1] Department of Radiological Sciences
University of California Irvine
Irvine, CA 92697-5000 USA
[2] Division of Plastic Surgery
University of California Irvine Medical Center
Orange, CA 92868 USA
[3] Ultran Laboratories Inc.
Boalsburg, PA 16827 USA

INTRODUCTION

At the 23rd International Symposium on Acoustical Imaging, in Boston, Massachusetts, we introduced the concept of non-contact (i.e., through the air) ultrasonic imaging for the evaluation of thermal injury as well as a host of other biomedical applications, and presented some preliminary data in the form of ultrasonic A-lines (or pulse-echo time traces) from burned skin which demonstrated the ability of the technique to measure burn depth[1]. Here we present the results of a double blind clinical study conducted over the past several years and involving over 500 burn sites in some 100 patients. The conclusions, based on computer generated images made from sequential A-lines, clearly demonstrate the accuracy and sensitivity of non-contact imaging for burn evaluation and its superior performance when compared to all other methods currently available.

PROBLEM DEFINITION

We began this project some eight years ago trying to address the following problem: develop a quantitative but noninvasive method for burn evaluation and treatment specification. In addition, the method should be applicable to large scale triage and should be able to be applied by minimally trained personnel. Current methods for burn evaluation generally involve visual inspection of the burn which has proven to be non-predictive. That is, the burn physician only has a fifty-fifty chance in correctly predicting whether or not any given burn will require surgical intervention. Conventional imaging methods give much better results but are either painful for the patient (as is ultrasound) or time consuming and cumbersome (as is CT and MRI). Here the clinical objective is to determine the depth of the burn and, in particular, to determine whether or not the capillary bed at the dermal / fat interface is intact. Such determination defines the clinical course followed. For example, if the capillary bed is intact then there is a variable thickness of viable dermis beneath a layer of coagulated necrosis which requires only the application of appropriate creams and bandages for survival and for burn healing. The above is known as a partial thickness burn. On the other hand, if no residual dermis survives and the underlying capillary bed is destroyed, then survival requires that the dead skin be removed and replaced by a transplant. Such a full thickness burn demands immediate surgical intervention. The quantitative measurement of burn depth would allow the differentiation between a partial thickness and a full thickness burn and would permit a more aggressive patient management which has been shown to reduce infection and to shorten the duration of the illness. Improved function and appearance have also been noted.

Since conventional ultrasound has been shown capable of measuring burn depth[2] (but is not practical because of patient pain and the time required for a study), we propose non-contact ultrasound as a possible means for burn evaluation. Although conventional wisdom suggests that ultrasonic imaging of the body cannot be accomplished without direct contact (or at least via water coupling), we have shown that non-contact imaging is possible, certainly for superficial body regions, provided judicious choices of piezoelectric materials and matching layers are made[1]. Here we present a continuation of our earlier work.

NON-CONTACT IMAGING

The challenges (as well as the utility) of non-contact imaging are easily appreciated from the results of a simple calculation of the insertion loss experienced by an ultrasonic pulse when reflected from skin. For example, the signal reflected from the dermal/fat interface using a 5 MHz water or direct coupled PZT transducer (i.e., a transducer with a quarter-wavelength matching layer which matches the PZT to water) is about 30 dB down from the original pulse. If this same transducer were operating in air, the signal loss would be about 200 dB, which explains why non-contact imaging is considered impossible. However, if the PZT transducer is now matched to air, the loss is about 90 dB -- still significant but certainly not impossible to overcome, especially with superficial tissue structures. In our experiments, described below and using far from optimized materials, we observed losses on the order of 100 dB, clearly demonstrating the feasibility of using this technology for burn evaluation.

Clearly the key to any non-contact system is the air-matched transducer. For some years Ultran Laboratories has offered commercially air-matched transducers operating in the kHz frequency range. Such transducers normally used PZT as the active element with a special ceramic coating which matched to air. Unfortunately, such coatings proved to be no longer feasible when the frequency was increased to the MHz range. Working together, UCI and Ultran have developed proprietary materials which together serve as a matching layer from PZT to air. Although these materials, consisting of layers of a polymer and a pressed fiber, do not provide an exact match to air, they have proven adequate for the preliminary studies described here. Here the basic idea is to replace a single quarter-wavelength matching layer with several layers that progressively match the impedance of each succeeding layer. The method is similar to one in optics that matches an index of refraction through a series of layers of different refractivity. Ultran provided a series of such air matched transducers for the experiments described here. The transducers had center frequencies of 1.0, 2.25, and 5.0 MHz, respectively, and each was an unfocused, broad-banded, plane-piston source with an active diameter of about 1 inch. All of the clinical studies were done at 5 MHz.

Initial experiments at UCI used a simple A-mode scanner consisting of these non-contact transducers with a Panametrics pulser/receiver and a Tektronix digital oscilloscope. This A-mode system was applied to a number of specially constructed test phantoms which verified the capabilities of such a system to make accurate spatial measurements with only air between the transducer and the test object. Applying this non-contact scanner to a series of human volunteers, reflections from the dermal/fat interface were clearly seen using both the 2.25 and the 5.0 MHz transducers.

In order to test the feasibility of non-contact imaging for burn evaluation, we constructed a special self-contained data acquisition system for use in the clinic. This system, shown in Figure 1 consists of a simple pulser/receiver, a 50 MHz ADC, and 32 MB of storage which can retain over 1000 A-line records. The unit is powered by a self contained battery pack and was designed to operate with the 5.0 MHz air-coupled transducer. The operation of the device is quite simple: the on/off switch is engaged, the transducer is held about one inch above the region of interest, and the data collect button is pushed. Engaging the data collect button causes a single ultrasonic pulse to be generated and the backscattered signal to be recorded for a spatial extent of about 2 cm. A front panel dial allows a specific time delay to be set before the ADC begins operation, thereby insuring that only relevant data are recorded. The RF A-line data recorded by the clinical data acquisition system are off-loaded to a laboratory computer (Power Macintosh 9500) for evaluation and analysis.

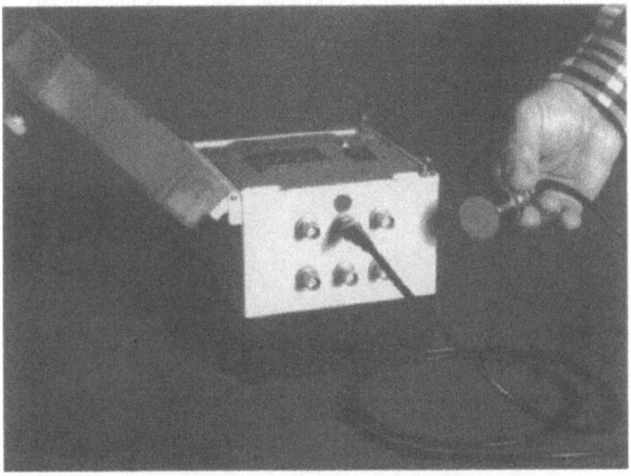

Figure 1. The self contained data acquisition system used in the clinic. The dark bottom is an integrated battery pack. Controls on the top panel consist of an on/off switch, a dial to select the time delay before data acquisition, and a button which engages the generation of a single ultrasonic pulse and the recording of the backscattered signal. The hand held transducer is a 5 MHz plane piston source.

CLINICAL STUDIES

Over the past several years we have collected data on over 500 burn sites in some 100 patients. For each patient enrolled in our clinical study, the burned areas as well as normal skin sites were scanned by non-contact ultrasound on day one and on day three following the burn event. Selected patients were also scanned following a period of several weeks. Two experienced physicians, blind to the results of ultrasound, made independent clinical assessments of the burned areas on days one and three and on later days when feasible. An investigator blind to the clinical findings interpreted the ultrasound results and predicted the type of burn visualized and whether or not the burn could heal without surgical intervention. The effectiveness of non-contact imaging is demonstrated in the series of images presented in Figures 2 and 3. Here the figure captions describe the details shown in each of the computer generated ultrasound images.

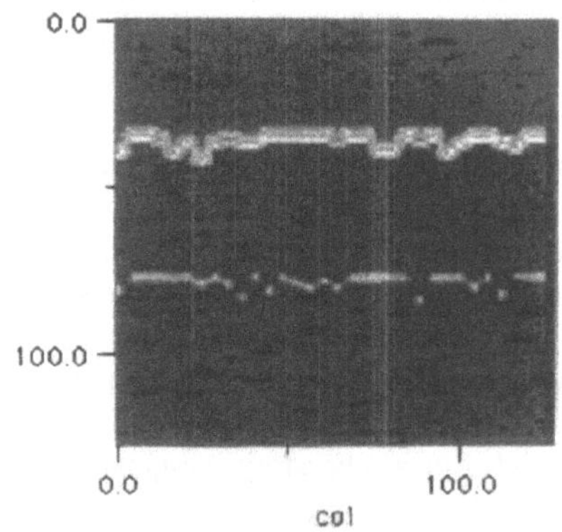

Figure 2a. Computer generated image of normal skin produced from a sequence of A-lines recorded by the data acquisition system. The image is of normal skin from the upper right arm of subject #A. The top solid white line is the surface of the skin while the white broken line below it is the capillary bed at the dermal/fat interface.

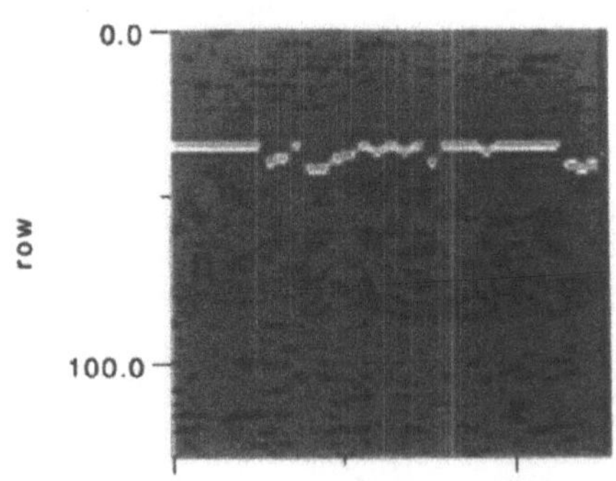

Figure 2b. Image of upper left arm on subject #A showing burn. Note the burned regions in the dermas and the intact capillary bed.

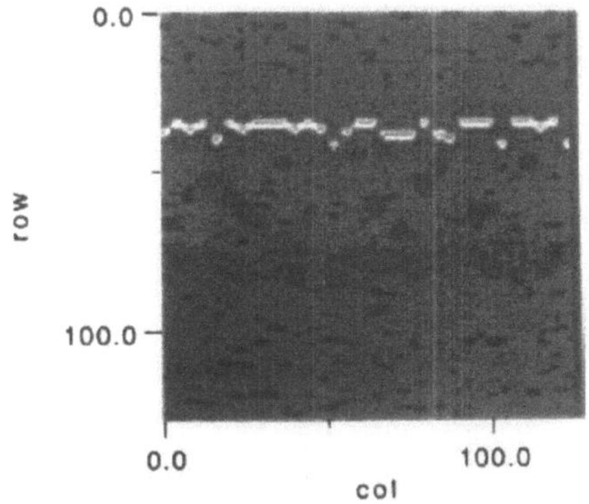

Figure 2c. Same as Figure 2b but three days later. Note healing in the dermas.

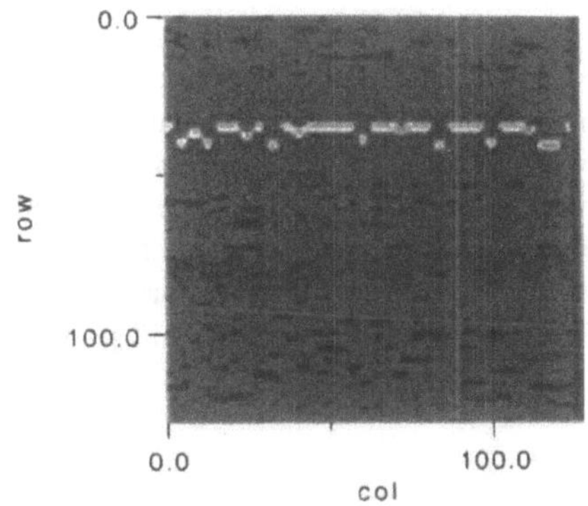

Figure 2d. Same as Figure 2b but three weeks later. Note that the healing is complete.

In Figure 2b we show an example of a partial thickness burn taken only a few hours after the event. Coagulated necrosis is much in evidence in the dermas but the underlying capillary bed is intact. The burn was treated with appropriate topical agents and covered. Three days later the healing process is seen to be much in progress (see Figure 2c) while three weeks later the healing process is complete (see Figure 2d).

In Figure 3, the left half of the image is a partial thickness burn (note the presence of the capillary bed) while the right half is a full thickness burn (here the capillary bed is missing). This differentiation between partial thickness and full thickness burns is precisely our clinical objective. In the first case, the application of appropriate creams and bandages is sufficient for full recovery; however, the second case requires surgical intervention and, in fact, recovery is not possible without it.

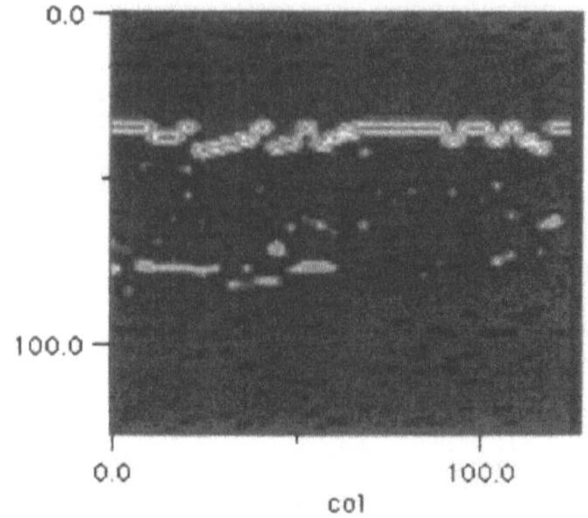

Figure 3. An example, on the left, of a partial thickness burn with intact capillary bed and, on the right, a full thickness burn with the capillary bed burned away. These two cases require totally different clinical management and their differentiation is essential for proper burn care.

Using our clinical data acquisition system, we recorded data from over 500 burn sites in some 100 patients. Following a double blind study, non-contact ultrasound predicted the type of burn visualized and comparisons were made with conventional clinical assessments made by two experienced physicians. Non-contact ultrasound showed an accuracy of 96% while standard clinical assessment showed an accuracy of 80%. The sensitivity for non-contact ultrasound was 100% compared to a sensitivity for standard clinical assessment of 65%. Thus, our study clearly demonstrates the efficacy of non-contact ultrasonic imaging for the assessment of thermal injury. We believe that our method will prove particularly effective for large scale medical triage.

REFERENCES

1. J. P. Jones, D. Lee, M. Bhardwaj, V. VanderKam, and B. Achauer, Non-contact ultrasonic imaging for the evaluation of burn-depth and other biomedical applications, *Acoustical Imaging , Vol. 23*, 89-93, Plenum Press (1997).
2. A. M. Kalus, J. Aindow, and M. R. Caufield, Application of ultrasound in assessing burn depth, *Lancet, 1*, 188-189 (1979).
3. S. Iraniha, M. Cinat, V. VanderKam, A. Boyko, D. Lee, J. Jones, and B. Achauer, Determination of burn depth with noncontact ultrasonography, *Journal of Burn Care and Rehabilitation* (in press).

ULTRASOUND TEMPERATURE AND ATTENUATION MONITORING FOR CONTROLLING THE LASER INDUCED THERMO THERAPY

Robert M. Lemor, Bernhard V. Kleffner, Steffen Tretbar, Rainer M. Schmitt

Fraunhofer Institut für Biomedizinische Technik
Dept. Ultrasound
Ensheimer Str, 48
66386 St. Ingbert
Germany

Introduction

For treatment of tumors and metastases in the human liver the laser induced thermo therapy (LITT) offers a new alternative for minimally invasive surgery[1]. During this treatment a laser applicator is positioned in the liver and malignant tissue is being heated and destroyed. The level of destruction depends on the induced energy and duration of the treatment.

Up to now only experience of the surgeon and preoperative simulation[2,3] can give criteria for controlling the therapy. On-line monitoring could be performed by NMR but is connected with the high costs and high material expenditure of NMR imaging[4,5].

In this paper we present an alternative possibility for on-line monitoring the LITT through the use of diagnostic ultrasound. Two different parameters were chosen for evaluation. 1. The dependence of the speed of sound on temperature. 2. The attenuation of sound. With these two parameters 2-dimensional mapping of the temperature distribution in the tissue as well as a 2-dimensional mapping of structural changes of the tissue is possible.

Methods

1. Temperature Mapping

For determining the temperature distribution in the tissue a method was used, which allows spatially resolved measurement of changes in time of flight. The method analyses RF signals within time gates and determines the displacement of echo parts through autocorrelation of two consecutive A-scans in the time domain[6,7]. With the known local displacement of parts of the returned signal it is possible to calculate the local variations in speed of sound. With an estimated initial overall speed of sound the mapping of speed of sound throughout the treatment is possible. Since the speed of sound depends on temperature, it is possible to determine the temperature distribution in the tissue during the treatment[6,7,8].

The temperature map is resolved in three steps.

a.) Measurement of time of flight displacement

$$D_{t_n}(\alpha,r) = TOF_{t_n}(\alpha,r) - TOF_{t_0}(\alpha,r)$$
$$= \int_0^r \delta D_{t_n}(\alpha,r')dr'$$

b.) Speed of sound calculation

$$\frac{1}{c_{t_n}(\alpha,r)} = \frac{dD_{t_n}(\alpha,r)}{dr}$$

c.) Temperature mapping

$$T_{t_n}(\alpha,r) = T_{t_0}(\alpha,r) + (c_{t_n}(\alpha,r) - c_{t_0}(\alpha,r)) \cdot \frac{\partial T(c)}{\partial c}$$

2. Structural Mapping

The structural variation of the tissue during the treatment can be estimated through the variation in the attenuation coefficient[9]. For estimating the attenuation we used a method in the frequency domain[10]. Through determination of the spectral downshift in two signal parts from different depth of one single A-scan the attenuation coefficient slope can be estimated. During the coagulation process a variation in spectral shifts can be observed and therewith the variation in frequency dependence of the attenuation coefficient can be measured. For measuring the shift in the frequency domain, a center frequency method was used[11].

The structural map is resolved in three steps

a.) Measured center frequency

$$F_{m,t_n}(\alpha,r) = \frac{\int_F F \cdot P_{\alpha,r,t_n}(F)dF}{\int_F P_{\alpha,r,t_n}(F)dF}$$

b.) Attenuation coefficient

$$\beta_{t_n}(\alpha, r, r') = \frac{2\pi\tau^2(F_{m,t_n}(\alpha, r) - F_{m,t_n}(\alpha, r'))}{r - r'}$$

c.) Structural mapping

$$\Delta\beta_{t_n}(\alpha, \bar{r}) = \beta_{t_n}(\alpha, r, r') - \beta_{t_0}(\alpha, r, r')$$

Whereby the variables represent the following definitions:

α,r - spatial coordinates, c - speed of sound, δD - local displacement, T - temperature, TOF - time of flight, t_0 - initial starting point, F - frequency, P - power spectrum, D - measured displacement, t_n - time of measurement, τ - time constant.

Experiments

The experiments were conducted using a diagnostic ultrasound system (Kretz, Combison 310 A) with a 5 MHz scanning probe. The rf-data was digitized using a 50 MHz 8 bit analog digital converter. With this system it is possible to acquire 12 B-Scans (4000 samples, 200 lines) per second. For signal processing and data displaying a Pentium II 450 MHz personal computer with 1 Gbyte of memory was used. The heating source was a ND:YAG laser operating at a wavelength of 1064 nm,. (Dornier, mediLas 4060 fibertom). For the experiments cooled and uncooled laser applicators were used. (Dornier ITT Lightguide H-6190-I, Hüttinger micro dome applicator A13-0540/ Somatex) For calibration purposes the temperature was measured invasively using miniaturized PT-100 temperature probes. The in vitro samples were porcine and bovine liver. In vivo measurements were conducted on pigs.

Results

Figure 1 shows the measured displacement at different distances from the heating source during a heating and cooling cycle. The temperature was measured invasively.

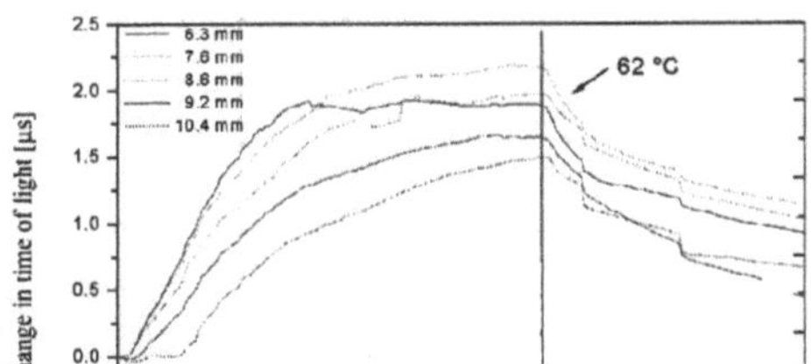

Figure 1.

Figure 2 shows two dimensional temperature maps over treatment time.

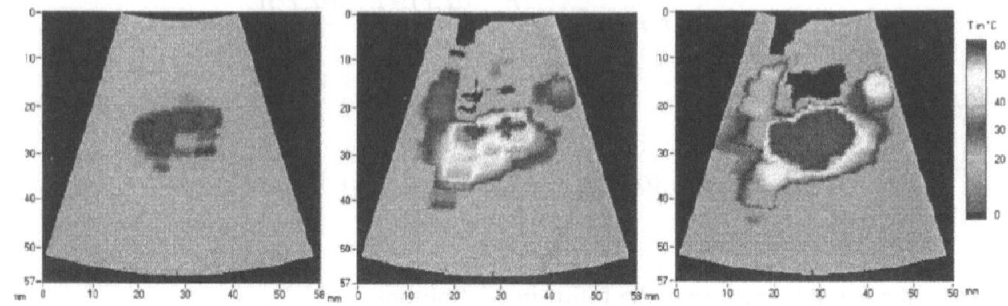

Figure 2.

Figure 3 shows two dimensional maps of the variation of the attenuation coefficient in percent over treatment time. The graphs show the percentage of change extracted at a horizontal line through the map at the center position. As well from the temperature map as from the structural map a demarcation of the coagulated zone (figure 3, porcine liver) is possible.

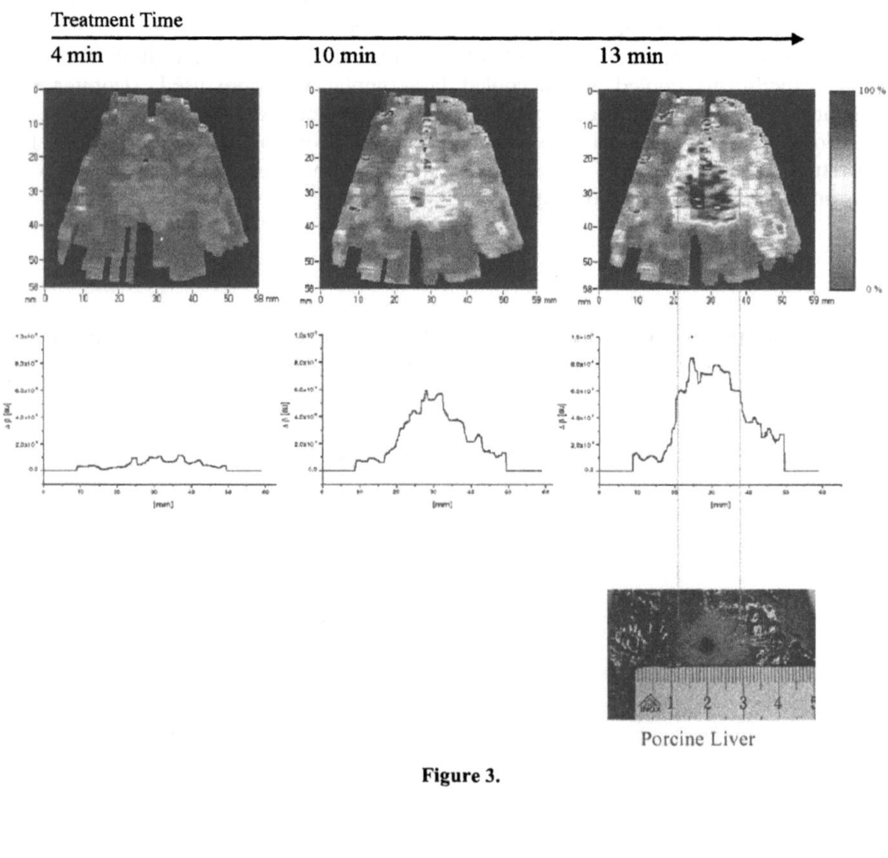

Figure 3.

Conclusion

The results obtained during this work clearly show the value of ultrasound RF data analysis for monitoring the LITT. Two ultrasound parameters could be identified with which it is possible to demarcate the coagulated zone during the treatment under real time conditions. The study so far employed only two-dimensional maps of the region of interest. However, the algorithms derived for this work can be modified for three dimensional data processing.

In vivo measurements gave explanation on artifacts due to patient movement. Effects like breathing and pulse movements could be compensated using image transformation methods.

Outlook

The use of diagnostic ultrasound as a tool for controlling a therapy is not restricted to laser induced thermo therapy. Other therapies as RF-Application or Cryotherapy can also be monitored. In this project an experimental setup for clinical evaluation was developed, which is currently beeing tested.

Acknowledgements

We would like to thank Prof. Hildebrandt and Prof. Menger from the University of Saarland, Medical Center for supporting the in vivo measurements. We also would like to thank Dr. Schwerdtfeger, University of Saarland, Neurology, for giving us access to the laser system. This project was supported by a national grant (BMBF - FKZ 13 N 7063/9).

References

1. Müller G, Roggan A.: Laser-induced interstitial thermotherapy; SPIE Print Monograph; Bellingham, 1-549 (1995)

2. Wheelan WM, Wyman DR: Dynamic Modeling of Interstitial Laser Photocoagulation: Implications for Lesion formation in Liver In Vivo Lasers in Surgery and Medicine 24:202-208 (1999)

3. Olsrund J, Wirestam R, Peterson BRR, Tranberg K-G: Simplified Treatment Planning for Interstitial Laser Thermotherapy by Disregarding Light Transport: A Numerical Study, Lasers in Surgery and Medicine 25:304-314 (1999)

4. Gewiese B, Beuthan J, Fobbe F, Stiller D, Müller G, Böse-Landgraf J, Wolf KJ, Deimling M: Magnetic Resonance Imaging – Controlled Laser Induced Interstitial Thermotherapy, Invest Radiol 29:345-351 (1994)

5. Busse H, Rassek M, Goust A, Kahn T, Schwarzmaier HJ: Improved quantitative MR thermometry using a 1.5 T scanner to monitor cooled applicator systems during laser-induced interstitial thermotherapy (LITT), SPIE Vol. 3594: 196-203 (1999)

6. Seip R, Ebbini ES: Noninvasive Estimation of Tissue Temperature Response to Heating Fields Using Diagnostic Ultrasound, IEEE Trans Biomed Eng 42:8 (1995)

7. Seip R, VanBaren P, Cain CA, Ebbini ES: Noninvasive Real-Time Multipoint Temperature Control for Ultrasound Phased Array Treatments, IEEE Trans Ultrason Vol 43, No. 6, 1996

8. Simon C, VanBaren P, Ebbini ES: Two Dimensional Temperature Estimation Using Diagnostic Ultrasound, IEEE Trans Ultrason Vol 45, No 4, 1998

9. Ribault M, Chapelon JY: Cathignol D, Gelet A: Differential Attenuation Imaging for the Characterization of High Intensity Focused Ultrasound Lesions, Ultrasonic Imaging 20: 160-177 (1998)

10. Kuc R, Schwartz M: Estimating the acoustic attenuation coefficient slope for liver from reflected ultrasound signals, IEEE Trans Ultrason Vol 26, No 5, 1979

11. Kuc R, Schwartz M: Estimating acoustic attenuation from reflected ultrasound signals: Comparison of spectral-shift and spectral-difference approaches, IEEE Trans Ultrason Vol 32, No 1, 1984

SECOND HARMONIC COEFFICIENT OF NONLINEARITY FOR ULTRASOUND CONTRAST AGENT

Valérie Labat, Jean-Pierre Remenieras, Olivier Bou Matar, Abdeldjalil Ouahabi and Frédéric Patat

LUSSI / G.I.P. Ultrasons, Faculté de Médecine
2 bis bd Tonnellé, B.P. 3223
37032 Tours Cedex
FRANCE

INTRODUCTION

This work proposes a theoretical model for second harmonic propagation in ultrasound contrast agents. Ultrasound contrast agents consist in gas bubbles suspended in a liquid. In order to improve their life-time, they are often encapsulated, which modifies their acoustic behaviour.

In a first stage, we derive the nonlinear propagation equation for a two-phasic medium after homogenisation. This expression links up pressure to small volume variations of the bubbles. In order to solve it, it is combined with the Rayleigh-Plesset equation in which encapsulation is introduced together with damping. The Rayleigh-Plesset equation describes the dynamic behaviour of bubbles. These two-coupled equations are then solved simultaneously with the successive approximations method in quasilinear theory to provide Helmholtz-type equations. Complex wave number, synonymous with dispersion, and a second harmonic coefficient of nonlinearity are introduced. Evolutions with frequency of phase speed, attenuation and nonlinear parameter are calculated. Influences of encapsulation and bubble distribution are also performed. Finally, second harmonic generation is analysed in detail, and the contributions of both dispersion and nonlinearity are specially discussed.

PROPAGATION IN BUBBLY LIQUID

Let a liquid contain spherical encapsulated gas bubbles in suspension. In a first stage, the bubble population is supposed monodisperse (*i.e.* with a same radius R_0) and uniformly distributed throughout the volume of the liquid.

Propagation equation

First, we consider the nonlinear state equation for a two-phase medium after homogenisation[1]. It is combined with the continuity and momentum equations to obtain the following nonlinear propagation equation where local effects are neglected:

$$\nabla^2 p - \frac{1}{c_0^2}\frac{\partial^2 p}{\partial t^2} + \rho_0 N \frac{\partial^2 v}{\partial t^2} = -\frac{\beta}{\rho_0 c_0^4}\frac{\partial^2 p^2}{\partial t^2}, \tag{1}$$

where c_0, ρ_0 and β are respectively the sound velocity, the density and the nonlinear parameter of the liquid, and N the number of bubbles per unit volume. We have there an expression linking up pressure p to small volume variations v of the bubbles. For solving the system, Eq.(1) is not sufficient, on its own: it must be combined with another one.

The Rayleigh-Plesset equation

The Rayleigh-Plesset equation describes the dynamic behaviour of spherical bubbles under a time-varying pressure field in an incompressible liquid. As comparison with free bubbles, encapsulation creates an increased stiffness of the system and an additional loss term due to internal friction inside the bubbles[2]. For encapsulated bubbles, this equation can then be written[2]:

$$R\ddot{R} + \frac{3}{2}\dot{R}^2 = \frac{1}{\rho}\left\{\left(P_0 - P_v + \frac{2\sigma}{R_0}\right)\left(\frac{R_0}{R}\right)^{3K} + P_v - P_0 - p(t) - \frac{2\sigma}{R_0} - S_p\left(\frac{1}{R_0} - \frac{1}{R}\right) - \rho\delta\omega R\dot{R}\right\}, \tag{2}$$

where R and R_0 represent respectively the instantaneous and initial bubble equilibrium radii, ρ the density of the surrounding medium, σ the surface tension coefficient, K the polytropic exponent coefficient of the gas and S_p the shell elasticity parameter. P_0, P_v and $p(t)$ are hydrostatic, vapour and time-varying pressures. The total damping coefficient $\delta = \delta_{vis} + \delta_{rad} + \delta_{th} + \delta_{elas}$ includes viscous, re-radiation, thermal and elastic contributions. Its expression has been formulated by Devin[3] for free bubbles at resonance frequency, generalised by Eller[4] at other frequencies, and completed by De Jong[2] for encapsulated bubbles.

As bubbles are spherical, the relation $v = \frac{4\pi}{3}R^3$ is then introduced into Eq.(2) to give:

$$\ddot{v} + (\omega_0^2 + \omega_1^2)v + \delta\omega\dot{v} = -\varepsilon p + (\alpha + d)v^2 + \beta'(2\ddot{v}v + \dot{v}^2 + 2\delta\omega v\dot{v}), \tag{3}$$

with
$$\begin{cases} \omega_0^2 = \dfrac{3K}{\rho R_0^2}\left(P_0 - P_g + \dfrac{2\sigma}{R_0}\right) \quad \omega_1^2 = \dfrac{2}{\rho R_0^3}(S_p - \sigma) \quad \varepsilon = \dfrac{4\pi R_0}{\rho_0} \\ \alpha = 3\beta'(K+1)\omega_0^2 \quad d = 4\beta'\omega_1^2 \quad \beta' = \dfrac{1}{8\pi R_0^3}. \end{cases}$$

Resolution of the system

In order to solve simultaneously these two-coupled equations, we use a method of successive approximations [5,6] in quasilinear theory:

$$p = p_1 + p_2 = \tfrac{1}{2}\{p_1 e^{j\omega t} + p_2 e^{2j\omega t}\} + cc, \text{ and } v = v_1 + v_2 = \tfrac{1}{2}\{v_1 e^{j\omega t} + v_2 e^{2j\omega t}\} + cc, \quad (4)$$

where p_1, v_1 are linear solutions, p_2, v_2 second order correction terms ($|p_2| << |p_1|$ and $|v_2| << |v_1|$), and cc denotes complex conjugate term. Substitution of Eq.(4) into Eq.(1) and Eq.(3) yields for

fundamental : $$\left[\nabla^2 + \frac{\omega^2}{\tilde{C}_1^2}\right] p_1 = 0, \quad (5)$$

second harmonic: $$\left[\nabla^2 + \frac{4\omega^2}{\tilde{C}_2^2}\right] p_2 = \beta_2(\omega).\frac{2\omega^2}{\rho_0 c_0^4} p_1^2 , \quad (6)$$

in which we have introduced complex celerities $\tilde{C}_n$ (Eq.(7)) at order n, for taking into account dispersion, and a second harmonic coefficient of nonlinearity β_2 (Eq.(8)) by identification with classical nondispersive nonlinear acoustics:

$$\frac{1}{\tilde{C}_n^2} = \frac{1}{c_0^2} + \frac{\rho_0 N \varepsilon}{\omega_0^2 + \omega_1^2 - n^2\omega^2 + nj\omega^2\delta}, \quad (7)$$

$$\beta_2 = \beta + \frac{\rho_0^2 c_0^4 N(\alpha + d - 3\omega^2\beta' + 2j\delta\omega^2\beta')}{\left(\omega_0^2 + \omega_1^2 - \omega^2 + j\omega^2\delta\right)^2\left(\omega_0^2 + \omega_1^2 - 4\omega^2 + 2j\omega^2\delta\right)} . \quad (8)$$

Phase speeds c_n and attenuation coefficients α_n can be deduced from Eq.(7) setting:

$$c_n = Re^{-1}(\tilde{C}_n^{-1}), \text{ and } \alpha_n = -n\omega \operatorname{Im}(\tilde{C}_n^{-1}). \quad (9)$$

If we suppose now a polydisperse distribution[7,8], (*i.e.* with $N(R_0)dR_0$ bubbles per unit volume having equilibrium radius between R_0 and $R_0 + dR_0$), the number of bubbles N is replaced in Eq.(7) and Eq.(8) by $\int_0^\infty N(R_0)dR_0$.

RESULTS

Phase speeds, attenuations and second harmonic coefficient of nonlinearity have been calculated for Albunex® with values[2] : $\mu = 10^{-5}$, $S_p = 8N.m^{-1}$, $S_f = 4.10^{-6} kg.s^{-1}$.

For low frequencies ($\omega << \sqrt{\omega_0^2 + \omega_1^2}$), phase speeds correspond to the sound speed in a liquid where the compressibility has been increased by the presence of bubbles[6] (Figure 1(b)). For high frequencies, phase speeds approach the one of the liquid without bubbles. As expected, celerity (Figure 1(b)) and attenuation (Figure 1(c)) show the most important variations around the bubbles resonance.

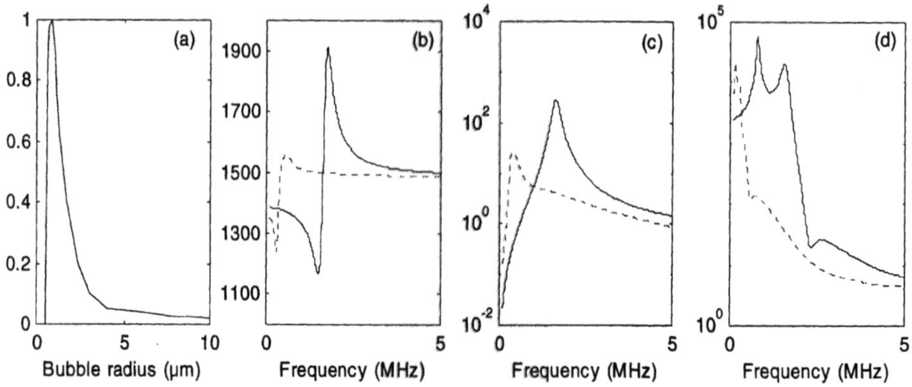

Figure 1. Effect of bubble distribution (a) on : (b) celerity (m/s), (c) attenuation (dB/cm) and (d) nonlinear parameter for free bubbles. On (b),(c) and (d), solid line represents monodisperse distribution with a radius of *2.1 μm*, while dots represent polydisperse distribution with a mean radius of *2.1 μm*.

The nonlinear parameter β_2 presents two resonances (Figure 1(d)). The higher one occurs when the frequency of incident acoustic wave coincides with the bubbles resonance frequency ($\omega=\sqrt{\omega_0^2+\omega_1^2}$). The lower resonance appears when the second harmonic frequency matches the bubbles resonance one ($\omega >> \sqrt{\omega_0^2+\omega_1^2}$).

Influence of bubble distribution

Due to the presence of greater bubbles, the polydisperse bubble distribution (with a mean radius of $2.1\mu m$) shifts the resonance for phase speed, attenuation and nonlinear parameter compared with a monodisperse distribution with the same radius $2.1\mu m$ (Figure 1).

Effects of encapsulation

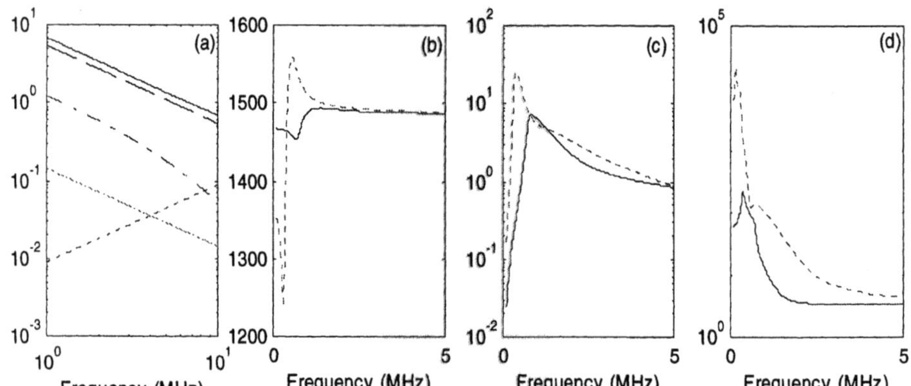

Figure 2. Effect of encapsulation on : (a) damping coefficient δ, (b) celerity (m/s), (c) attenuation (dB/cm) and (d) nonlinear parameter for polydisperse bubble distribution. (a) represents contributions of acoustic (dots), thermal (dashed-dot line), viscous (thin solid line), and elastic (dashed line) damping on total damping (thick solid line) calculated for bubbles with a radius of *2.1 μm*. On (b), (c) and (d) solid line represents encapsulated bubbles as a comparison with free ones (dots).

First, the shell elasticity creates an additional force and produces indeed a higher frequency resonance for encapsulated bubbles than for free ones. Moreover, internal friction inside the shell due to viscosity is responsible for additional losses (Figure 2(a)) and reduces pulsation amplitude of bubbles[2,8]. As shell increases the stiffness of bubbles, their compressibility decreases, which explains smaller dispersion amplitude for encapsulated than for free bubbles[7] (Figure 2(c)). For the same reasons, encapsulated bubbles present lower values for attenuation and nonlinear parameter (Figure 2(c) and (d)). For volume fraction of only 10^{-5}, it attains values as high as 10^4 for free bubbles, but falls with encapsulation down to 250, which is even greater than for pure water ($\beta = 3.5$).

HARMONIC PROPAGATION

Let follow with the detailed study of second harmonic propagation in bubbly liquid. We assume plane waves propagation in the $+x$ direction: $p_n(x) = p_{n0}\exp(-jk_n x)$, where $k_n = n\omega / \tilde{C}_n$ is the wave number at order n. We impose that no second harmonic component is radiated at the transducer surface: $p_1(0) = p_{ac}$, and $p_2(0) = 0$, where p_{ac} is the acoustic emission pressure. Eq.(5) has an obvious solution $p_1(x) = p_{ac}\exp(-jk_1 x)$. It is injected together with the previous boundary condition in Eq.(6) to give:

$$p_2 = \frac{j2\pi f \beta_2 p_{ac}^2}{\rho_0 c_0^4 (\tilde{C}_2^{-1} + \tilde{C}_1^{-1})} x \,\mathrm{sinc}\left[2f(\tilde{C}_2^{-1} - \tilde{C}_1^{-1})x\right] \exp\left[-j(k_2 + 2k_1)x/2\right]. \tag{10}$$

Second harmonic amplitude is directly proportional to β_2. Following the evolution for the nonlinear parameter (Figure 2(d)), p_2 grows and reaches its maximum value at $0.3\,MHz$, and then decays with frequency.

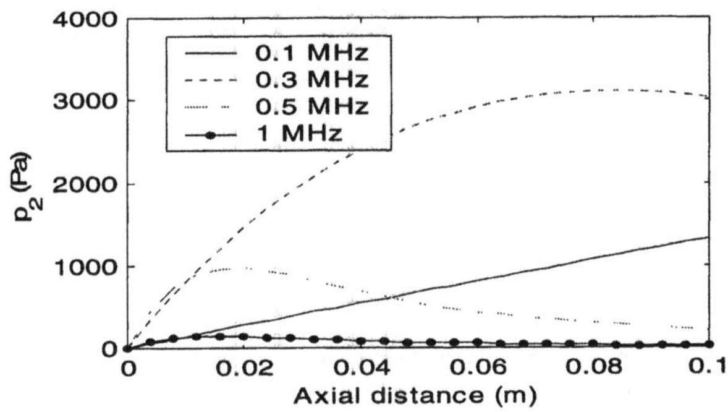

Figure 3. Amplitude of second harmonic p_2 versus axial distance from the transducer surface from various driving frequencies for an initial pressure $p_{ac} = 0.5\,bar$ at the transducer surface.

But, nonlinearity is not the only one contribution to second harmonic generation: dispersion also operates on p_2. To illustrate this point, we derive the nonlinear contribution for distance shorter than the dispersion length ($l_d = 2\pi / |k_2 - 2k_1|$). In this case and for μ sufficiently small (Figure 4), Eq.(10) reduces to Eq.(11), with $\tilde{C}_1 = \tilde{C}_2 = c_0$, which corresponds to the classical nonlinear case in the absence of dispersion[6]:

$$p_{2NL} = \frac{j\omega\beta_2 p_{ac}^2}{\rho_0 c_0^3}\frac{x}{2}. \quad (11)$$

As a consequence, dispersion effect on second harmonic generation can be written:

$$p_{2disp} = \frac{2}{c_0\left(\tilde{C}_2^{-1} + \tilde{C}_1^{-1}\right)}\operatorname{sinc}\left[2f(\tilde{C}_2^{-1} - \tilde{C}_1^{-1})x\right]\exp[-j(k_2 + 2k_1)x/2]. \quad (12)$$

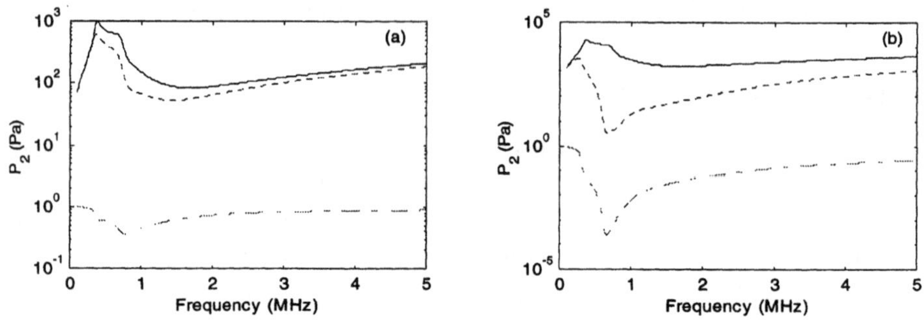

Figure 4. Nonlinear (solid line) and dispersion (dashed dot line) contributions to second harmonic component (dots) for an initial pressure $p_{ac} = 0.5\ bar$ at (a) $x = 1\,mm$ and (b) $x = 1\,cm$ distance from the source.

Close to the source, nonlinear effect is predominant, while dispersion predominates with the distance (Figure 4).

CONCLUSION

This works deals with harmonic propagation in ultrasound contrast agents. Parameters as bubble distribution, encapsulation and damping have been introduced to obtain an accurate model. Evolutions versus frequency of phase speed, attenuation and nonlinearity parameter have been investigated for the previous parameters. Finally, second harmonic generation in ultrasound contrast agents has been discussed. It has been shown, for plane waves, that the increase in second harmonic, due to nonlinearity (*i.e.* β_2) is counter-balanced by a strong decrease imposed by dispersion.

REFERENCES

1. E.F. Kozyaev and K.A. Naugol'nykh, Parametric sound radiation in a two-phase medium, Sov. Phys. Acoust., 26: 48, 1980.
2. N. de Jong. *Acoustic properties of ultrasound contrast agents,* PhD Thesis, Erasmus University (1993).
3. C. Devin, Survey of thermal, radiation and viscous damping of pulsating air bubbles in water, *J. Acoust. Soc. Am.,* 31:1654 (1967).
4. A.I. Eller, Damping constants of pulsating bubbles, *J. Acoust. Soc. Am.*, 47:1469 (1970).
5. E.A. Zabolotskaya and S.I. Soluyan, Emission of harmonic and combination-frequency waves by bubbles in liquids, Soviet Physics, 18: 396, 1973.
6. M.F. Hamilton and D.T. Blackstock. *Nonlinear Acoustics*, Academic Press, San Diego, (1998).
7. K.W. Commander and A. Prosperetti, Linear pressure waves in bubbly liquids: comparison between theory and experiment., *J. Acoust. Soc. Am.,* 85:732 (1989).
8. C.C. Church, The effects of an elastic solid surface layer on the radial pulsations of gas bubbles, *J. Acoust. Soc. Am.*,97:1510 (1995).

THE APPLICATION OF WAVELET TRANSFORMS TO TISSUE CHARACTERIZATION AND FLOW IMAGING

Joie P. Jones[1] , David Lee[1] ,and Sidney Leeman[2]

[1] Department of Radiological Sciences
University of California Irvine
Irvine, CA 92697-5000 USA
[2] Department of Medical Engineering and Physics
King's College School of Medicine and Dentistry
London SE22-8PT, UK

INTRODUCTION

Fourier methods have long been used in acoustical imaging, and almost universally in other areas of science and engineering,to relate and interchange time domain signals with those in the frequency domain. Although Fourier methods have proven to be a powerful means for analysis, they are not generally appropriate for transient signals. To overcome this fundamental problem, time blocks of short duration have been used and the short-time Fourier transform has been utilized. However, a short time window results in low spectral resolution and a longer time window (to increase spectral resolution) compromises the assumption of stationarity and gives poor spatial resolution. Wavelet methods. on the other hand, are said to provide both high temporal resolution as well as high spectral resolution. In addition, they are robust for even relatively small data sets. Here we develop an effective but relatively simple wavelet formalism which, when applied to pulse echo ultrasound data, can yield estimates of parameters such as attenuation and the integrated backscattering. The formalism also provides an unique framework for describing Doppler-shifted echoes and allows us to describe flow in the time domain rather than the usual frequency domain.

In principle, any signal can be represented by a set of wavelets that form an orthogonal basis. This is similar to Fourier's idea that a signal can be thought of as comprised of a series of sinusoidal waves of different amplitudes and frequencies. Consequently, a signal can be viewed in either a time or a frequency domain. Wavelet Transformation decomposes a time domain signal into smaller segmented time domain signals known as wavelets. Each wavelet has a different but finite duration while they all have the same shape. By adding or subtracting these wavelets via appropriate weighing factors, or wavelet coefficients, the original time domain signal can be reconstructed. For simplicity, a special basis known as the Harr wavelet is used in this study. Harr wavelets consist of finite pieces of 1's, 0's and -1's with various durations. A time domain signal decomposed by Wavelet Transformation is represented by a set of wavelet coefficients. These weighing factors can then be used to reconstruct the time signal. Like the power spectrum in Fourier Analysis, these wavelet coefficients provide an interesting insight as to the nature of the signal under investigation.

In this paper, both computer simulated and physical data from a number of reticulated foam samples were studied via a method inspired by wavelet analysis. Using envelope-detected signals, we show that there is a definite relationship between the wavelet coefficients and the spatial dependence of a medium's attenuation-slope. A histogram of these wavelet coefficients thus provides a novel and simple means for estimating attenuation and, perhaps, for also characterizing tissue and tissue state.

Using wavelet analysis to estimate attenuation also provides an estimate of the energy lost as a function of depth. Then, after a compensation procedure, in which the detected envelope signal at each depth is divided by the corresponding attenuation factor, the total area under the corrected envelope is then used as a measurement of the exact amount of the backscattered energy present. Thus, wavelet analysis provides a simplified means to measure the integrated backscattering. Conventional integrated backscattering measurements on a number of reticulated foam samples were in excellent agreement with values obtained via wavelet analysis.

Finally, we examine the problem anew of Doppler-shifted ultrasound echoes reflected from moving blood cells, using wavelet analysis rather than conventional Fourier techniques. It will be demonstrated that the initial ultrasound pulse as well as the unprocessed echo signal may be described by a wavelet. Thus, the imaging and measurement of flow is a natural arena for applying wavelet analysis --- not only as a processing tool, but also as a theoretical framework for analyzing the signals.

WAVELET TRANSFORMS

The concept of wavelets appears to have originated with Harr in 1910 and the publication of his classic work, *The Theory of Orthogonal Functions*. However, this theory was not developed, codified, and actually applied until 1984 when Grossman and Morlet applied this formalism to sound waves[1] . They used an analysis window which was short at high frequency and long at low frequency. This corresponds to a frequency response which is logarithmically scaled along the frequency axis.

Formally, the wavelet transform can be defined as the decomposition of a given signal onto a set of basis functions, $\{\phi(t), \psi(t)\}$, where ϕ is the scaling or dilation function and ψ is the primary wavelet function. These basis functions have the following properties:

$$\lim_{t\to\infty} |\psi(t)| = 0 , \qquad \int_{-\infty}^{\infty}\psi(t)dt = 0 , \qquad \int_{-\infty}^{\infty}\phi(t)dt = 1 \tag{1a}$$

with

$$\phi(t) = \phi(2t) + \phi(2t - 1)\ ,\quad \psi(t) = \phi(2t) - \phi(2t - 1) \tag{1b}$$

Given these properties, let X represent the continuous wavelet transform, a the scaling factor and b the time. Then,

$$X(a,b) = \int X(\gamma)\psi_{a,b}{}^{*}(\gamma)\, d\gamma \tag{2a}$$

where

$$\psi_{a,b}(t) = |a|^{(-1/2)}\, \psi[(t - b) / a] \tag{2b}$$

The duration and temporal location of each scaling and wavelet function is specified by two parameters, the dilation coefficient, j, and the shift coefficient, m. For a waveform with N values, j ranges from 0 to j_{max} and m ranges from 0 to m_{max} where

$$j_{max} = \log_2 N - 1\ ,\ m_{max} = 2^j - 1 \tag{3}$$

In general, then, the wavelet transform with resolution J is given by

$$X(n) = \Sigma^{J}{}_{j=0}\Sigma_{m=0} d_j(m)\psi_{j,m}(n - 2^j m) + \Sigma_{m=0}{}^{mMAX}\, a_j(m)\phi_{j,m}(n - 2^j m) \tag{4}$$

where $d_j(m)$ is the wavelet coefficient and $a_j(m)$ is the scaling coefficient. For simplicity, we replace these more general expressions with Harr Wavelets which are finite pieces of 1's, 0's, and -1's with various durations. Figure 1 shows how the Harr Wavelets are selected.

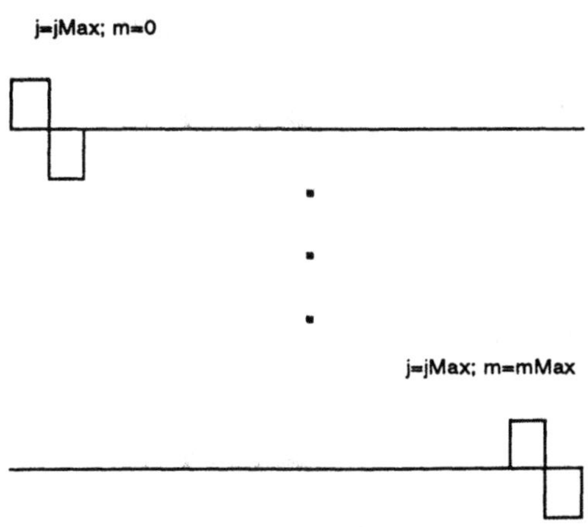

Figure 1. Example of Harr Wavelets

As illustrated in the above figure, the j value, or the dilation coefficient, for a particular wavelet characterizes the resolution associated with that wavelet. While low values of j are responsible for crafting the coarse outline of a signal, high values of j give it its fine structure. From this perspective, low j values are analogous to the low frequency components of a signal's Fourier transformation, while high j values are the counterparts of the transformation's high frequency components. In fact, the resolution represented by the j+1 level wavelets is double that represented by the j level wavelets. The m value is the translation coefficient, and it specifies the location of a wavelet for a particular j. Overall, the major difference between Fourier and Wavelet Transformation is that the latter carries out the entire process in the time domain. Note that wavelets of the same j value are responsible for the same degree of resolution regardless of m.

Using the above formalism much in the same way we would use Fourier analysis, we estimated the attenuation slope from the A-line pulse-echo data for a series of three reticulated foam samples[2]. The results are shown in Table 1.

Table 1. Measured attenuation-slopes for three differentreticulated foams using the wavelet method and the RF method (short term FFT)

Sample	Measure Attenuation-slope (dB/MHz/cm) (wavelet method)	Measure Attenuation-slope (dB/MHz/cm) (RF method)
Foam #1	0.45	0.42
Foam #2	1.07	0.94
Foam #3	0.84	1.01

THE DOPPLER EFFECT

The Doppler effect, in the context of medical ultrasound applications, may be described as the change in the frequency of an ultrasound wave when reflected from a moving interface (or scatterer), as measured by a stationary transducer or hydrophone. Perhaps for historical reasons, the Doppler effect has continued to be described from the vantage point of the Fourier domain. However, it is relatively straightforward to prove that a time domain description is at least as powerful, and arguably more appropriate, for uncovering the structure of the Doppler echo field. Consider an ultrasound pulse, of the form p(t), and with Fourier spectrum $P(\omega)$, with ω denoting the angular frequency, reflected from an interface traveling towards the emitting transducer with velocity V. Each frequency component in the incident pulse is Doppler-shifted to a new value, ω', such that

$$\omega' = (1 + V/C)\omega = \omega/a \tag{5}$$

where C is the wave velocity in the stationary medium filling the space through which the interface moves, and $V/C \ll 1$. The echo spectrum, $P'(\omega')$, is a distorted version of the incident spectrum, $P(\omega)$, and transforms according to the rule which maintains energy conservation:

$$|P'(\omega')|^2 d\omega' = |P(\omega)|^2 d\omega \tag{6}$$

$$\Rightarrow \quad P'(\omega') = \sqrt{a}\, P(\omega) \tag{7}$$

The time-domain structure of the echo signal is expressed as

$$p'(t) = \int P'(\omega') \exp(-i\omega' t)d\omega' \tag{8}$$

$$= \sqrt{a}\int P(\omega) \exp(-i\omega t/a)d\omega = \sqrt{a}\, p(t/a) \tag{9}$$

Thus, the Doppler effect may be seen as a process which modifies both the time-scale and amplitude of the echo, compared to the values for a stationary reflector. This simple, but powerful, result allows the rapid computation of the Doppler-shifted form of any broad-band signal, without recourse to more cumbersome and computationally demanding Fourier methods.

The origin of echoes from flowing blood is thought to be scattering from red blood cells. Each cell gives rise to a time-scaled echo --- depending on the velocity of the cell --- with an amplitude that depends on a number of factors. Assume that we can represent the echo signal from a single stationary scatterer at range R by $Ap(t - 2R/C) = Ap(t - \tau)$, where A denotes the echo amplitude relative to the incident pulse. Given that a large number, N, of scatterers are distributed throughout the resolution volume of the pulse, the signal from the ensemble of moving cells can be written as:

$$X(t) = \Sigma_{i=1}^{N}\ A_i\, p([t - \tau_i] / a_i) \sqrt{a_i} \tag{10}$$

$$\Rightarrow \quad \iint W(a, \tau)\, p_{a,\tau}(t)da d\tau \tag{11}$$

where W is an appropriate weighting factor and the notation $p_{a,\tau}(t)$ is self-explanatory. Equation 11 is the continuum analogue of the discrete Equation 10, and its structure is strongly reminiscent of that of a wavelet transform[3]. Note that the wavelet transform of the signal X(t) can be written as:

$$W_X(a, \tau) = \iint X(t)\bullet h^*_{a,\tau}(t)dt \tag{12}$$

where * denotes the complex conjugate, and the term

$$h_{a,\tau}(t) \equiv h([t - \tau] / a) \sqrt{a} \tag{13}$$

denotes scaled and translated versions of a basic wavelet. Thus, it is essentially the wavelet transform rather than the Fourier transform of the received Doppler echo sequence that encodes information about the distribution of the scatterer velocities in the ultrasound resolution cell.

In order to examine more closely the appropriateness of the Fourier approach towards unraveling the velocity distribution of the scatterers, it is moot to examine the structure of the echo sequence, as described in the Fourier domain. In practice, the Doppler shifted echo has a Fourier transform which is only slightly distorted and shifted relative to that from a stationary reflector. Such small frequency changes are easily swamped by other frequency-modifying effects, such as dispersive attenuation in the tissues overlying the blood vessel being investigated. The measured signal is, in fact, a superposition of the echo signals from a number of scatterers at different locations within the system resolution cell --

with different amplitudes and with different Doppler shifts. The Fourier domain equivalent of Equation 10 may be written as:

$$FT\{X(\omega)\} = \int X(t)\exp(i\omega t)dt = \Sigma_{m=1}^{N} A_m \exp(i\tau_m\omega)\, P_m(\omega) \quad (14)$$

$$\approx [\Sigma_{m=1}^{N} A_m \exp(i\tau_m\omega)]\, P_{ave}(\omega) \quad (15)$$

where $P_{ave}(\omega)$ denotes the average spectrum. Since all the Doppler shifts are so small, $P_{ave} = P$. The magnitude of the first factor in Equation 15 encodes the interference between the constituent echoes that make up the signal, and therefore fluctuates rapidly throughout the bandwidth of P_{ave} . Thus, the power spectrum of the received signal is essentially a corrupted version of the power spectrum of the echo from a reflector moving at a velocity equal to the average for the red blood cell ensemble.

CONCLUSIONS

Wavelet analysis provides an effective means for estimating such acoustical properties as attenuation and the integrated backscattering from pulse-echo data. It also offers a natural and simple means to describe Doppler waveforms and an interesting theoretical framework for analyzing the signals.

REFERENCES

1. A. Grossman and J. Morlet, Decomposition of Hardy functions into square integrable wavelets of a constant shape, *SIAM Journal of Mathematical Analysi, 15*, 723-736 (1984).
2. David M. W. Lee, *Evaluation of Myocardial Fibrosis Using Quantitative Ultrasonic Methods*, Dissertation, University of California Irvine (1998).
3. I Daubechies, The wavelet transform, time-frequency localization and signal analysis, *IEEE Trans. on Information Theory, 36*,961-1005 (1990).

SPATIAL DISTRIBUTION OF EIGENVIBRATION ON THE HEART WALL

Hiroshi Kanai[1] and Yoshiro Koiwa[2]

[1]Department of Electrical Engineering
Graduate School of Engineering
Tohoku University, Sendai 980-8579, Japan
[2]First Department of Internal Medicine
Tohoku University School of Medicine
Seiryou-machi, Aoba-ku, Sendai 980-8574, Japan

INTRODUCTION

We have already developed a novel ultrasonic-based method to accurately track the movement $x_i(t)$ of a point (i) in the heart wall based on both the phase and magnitude of the demodulated signals[1,2]. By this method, the velocity signal, $v(t;x_i)$, of point i at a depth of x_i in the heart wall with small amplitudes, which are less than several micrometers on the large motion resulting from a heartbeat, can be successfully detected with sufficient reproducibility in the frequency range up to at least 100 Hz.

Moreover, by applying time-frequency analysis to the resultant velocity signal, $v(t;x_i)$, of the heart wall, the instantaneous frequency, $f_2(t)$, of the eigenvibration on the left ventricle (LV) is determined. By assuming that the LV wall has an eigenvibration with mode-2 around 30 Hz at the end-diastole as shown in Fig. 1 and using the average thickness, h, and the average radius, r, of the LV, which are determined from the standard B-mode images, the LV-end-diastolic pressure (EDP) is noninvasively estimated[3]. The differences between the estimates of the LV-EDP and the actual values measured by cardiac catheterization were sufficiently small, ranging from -4.4 to +3.6 mmHg for 5 patients. The LV-EDP, the normal value of which lies between 5 and 12 mmHg, cannot be obtained from the blood pressure measured at the brachial artery, and the LV-EDP is usually needed to assess LV function in clinical settings. These *in vivo* experimental results lead to the conclusion that the noninvasively estimated LV-EDP can be an effective parameter for the evaluation of the heart function. Thus, establishment of the above-mentioned noninvasive measurement of the LV-EDP using ultrasound is significant. Though the mode-2 eigenvibration of the LV has been observed in *in vitro* experiments with the isolated canine hearts,[4,5] it has not yet been confirmed in *in vivo* experiments with human hearts.

In this paper, therefore, we propose a new method to control the directions, $\{n_{k,j}\}$, of the ultrasonic beams so that the velocity signals, $\{v(t;x_{kj})\}$, are simultaneously measured at multiple points $\{x_{kj}\}$ on the IVS and the LV posterior wall in order to

Acoustical Imaging, Volume 25, Edited by Halliwell and Wells.
Kluwer Academic/Plenum Publishers, 2000.

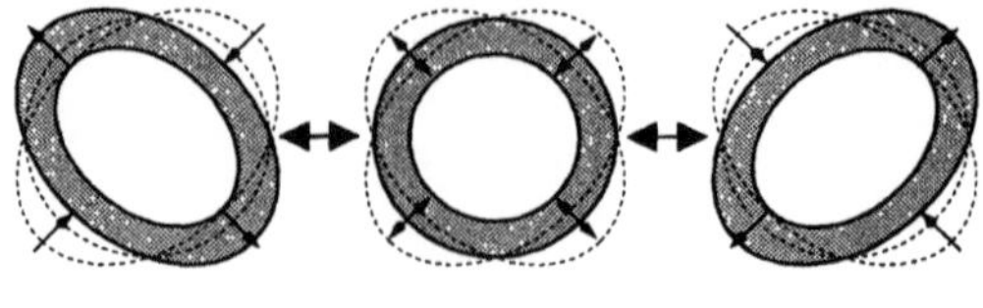

Figure 1. Three phase of a mode-2 vibration of an elastic spherical shell.

image their spatial distribution, where $x_{k,j}$ denotes the the depth of jth points on the ultrasonic beam in the kth direction. By applying this method to the LV wall of a young volunteer, we obtain the spatial distribution of the frequency component around 30 Hz in the resultant vibrations $\{v(t;x_{k,j})\}$ at the end-diastole. From the *in vivo* experimental results superimposed on the B-mode cross-sectional image, we confirm that there is mode-2 vibration on the LV wall at the end-diastole.

MEASUREMENT OF HEART WALL VIBRATIONS

RF pulses with an angular-frequency of $\omega_0 = 2\pi f_0$ are transmitted at a time interval of ΔT from an ultrasonic transducer on the chest surface. The phase difference, $\Delta\theta(x;t)$, between the received signal, $y(x;t)$, and the subsequently received signal, $y(x;t+\Delta T)$, is given by

$$\Delta\theta(x;t) = \theta(x;t+\Delta T) - \theta(x;t) = \frac{2\omega_0}{c_0}\Delta x(t), \tag{1}$$

where $\Delta x(t) = x(t+\Delta T) - x(t)$ is the movement of the object during the period ΔT around a time t and c_0 is acoustic velocity in the human body. By dividing the movement Δx by the period ΔT, the average velocity $v(t+\frac{\Delta T}{2})$ of the object during the period ΔT is given by

$$\widehat{v}\left(t+\frac{\Delta T}{2}\right) = \frac{\Delta x(t)}{\Delta T} = \frac{c_0}{2\omega_0}\frac{\widehat{\Delta\theta}(x;t)}{\Delta T}, \tag{2}$$

where the phase difference $\widehat{\Delta\theta}(x;t)$ is accurately determined by the constraint least-squared approach based on the complex cross-correlation between $y(x;t)$ and $y(x;t+\Delta T)$ for suppressing noise components[1]. By multiplying the resultant velocity $\widehat{v}\left(t+\frac{\Delta T}{2}\right)$ by the period ΔT, the next depth $\widehat{x}(t+\Delta T)$ is estimated by

$$\widehat{x}(t+\Delta T) = \widehat{x}(t) + \widehat{v}\left(t+\frac{\Delta T}{2}\right) \times \Delta T. \tag{3}$$

By moving the depth of the object along the direction of the ultrasonic beam based on the resultant object depth $\widehat{x}(t+\Delta T)$ by $\widehat{v}\left(t+\frac{\Delta T}{2}\right) \times \Delta T$, the displacement of the object is successfully tracked, and then the velocity signal $\widehat{v}\left(t+\frac{\Delta T}{2}\right)$ on the tracked large motion is accurately estimated[1].

CONTROL OF THE DIRECTIONS OF ULTRASONIC BEAMS

In Eq. (2), the phase difference during the pulse repetition interval, ΔT, must be less than π to avoid the aliasing. Therefore, the upper limit in the velocity measurement, $|v_{max}|$, is determined by

$$|v_{max}| < \frac{c_0\pi}{2\omega_0\Delta T} = \frac{c_0}{4f_0\Delta T}. \tag{4}$$

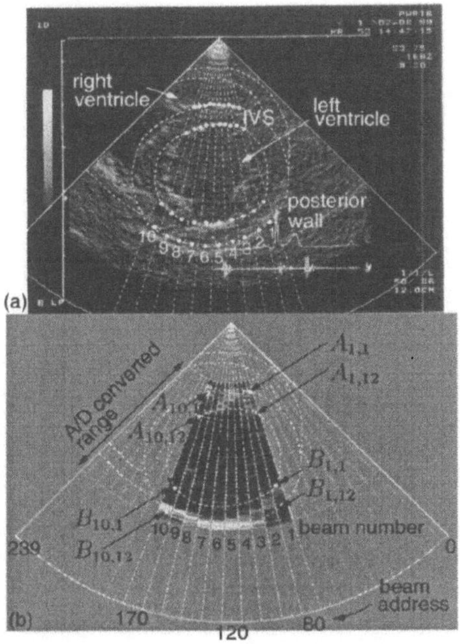

Figure 2. (a) Cross-sectional short-axis image of the LV of a 21-year-old healthy male volunteer at the timing of the R-wave. Forty measurement points on the beams from the first to 10*th* address are also shown by white dots. (b) Reconstructed cross-sectional image of 10 directions of ultrasonic beams.

In a typical case of ΔT =167 μs and f_0=3 MHz, the upper limit $|v_{max}|$ is equal to 0.77 m/s.

To simultaneously measure the velocity signals at multiple points along the surface of the LV, the directions $\{n_k\}$ of the ultrasonic beams in the modified ultrasonic diagnostic equipment are controlled in real time using a micro-computer. In the employed phased-array ultrasonic probe, the direction of the ultrasonic beam in the fan-shape region is identified by inputing the beam address, n, which is numbered from 0 to 239, into the address register, which controls the direction of the ultrasonic beam transmitted from the phased-array ultrasonic probe. If N directions are selected from the above addressable 240 directions, the upper limit, $|v_{max,N}|$, of the measurable velocity is reduced to $|v_{max,N}| = |v_{max}|/N = c_0/(4f_0\Delta TN)$. The maximum value of the velocity at the points in the IVS or the posterior wall of the human heart is about 0.07 m/s[1] in a typical case, which is about 1/10 of $|v_{max}|$ in Eq. (4). Thus, we must select 10 appropriate directions (N=10) from the 240 directions addressable in the phased-array ultrasonic probe.

IN VIVO EXPERIMENTAL RESULTS

The velocity signals at 10 points $\{A_{k,j}\}$ ($k = 1, 2, \ldots, 10$) on the LV side of the IVS and 10 points, $\{B_{k,j}\}$ ($k = 1, 2, \ldots, 10$), on the LV posterior wall of a 21-year-old healthy male volunteer are simultaneously measured. In *in vivo* experiments, the employed pulse repetition interval ΔT and the ultrasonic frequency f_0 are 167 μs and 3 MHz, respectively. That is, the original pulse repetition frequency (PRF=$1/\Delta T$)

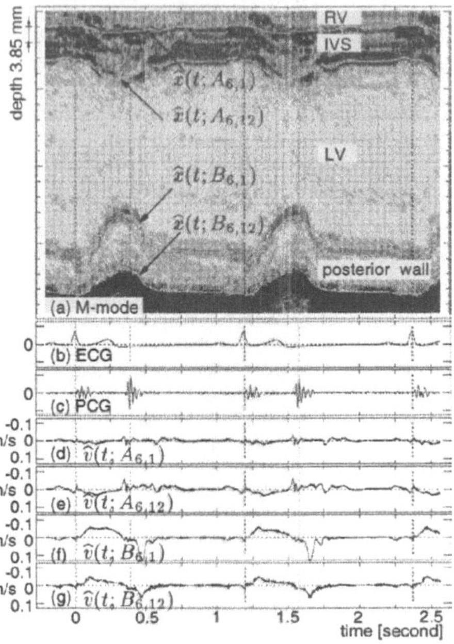

Figure 3. *In vivo* experimental results of the LV vibrations at points $A_{6,1}$, $A_{6,12}$, $B_{6,1}$, and $B_{6,12}$ on the $6th$ ultrasonic beam for the same subject as in Fig. 2. (a) Four Tracking results overlaid on the M-mode image, (b) ECG, (c) phonocardiogram (PCG), (d)(e) the velocity signals $\hat{v}(t; A_{6,1})$ and $\hat{v}(t; A_{6,12})$ of the IVS, (f)(g) the velocity signals $\hat{v}(t; B_{6,1})$ and $\hat{v}(t; B_{6,12})$ of the posterior wall.

is 6 kHz. Figure 2(a) shows the cross-sectional B-mode image along the short axis of the LV at the timing of the R-wave of the electrocardiogram (ECG). The curved broken lines show the surface of the LV wall and that of the right ventricular wall (RV). The ultrasonic beam is transmitted in 10 directions every $10th$ address from n_1=80 to n_{10}=170. These 40 measurement points are shown by white dots on the RV side and LV side of the IVS and endocardial and epicardial sides of the posterior wall in Fig. 2(a). Figure 2(b) shows the cross-sectional image reconstructed from the amplitude of the A/D converted data in 10 directions of ultrasonic beams at the period of 6.7 ms from the R-wave.

Figure 3 shows a typical example of the tracking results and the velocity signals at the two points on the 6th direction of the ultrasonic beam. The 4 points, $A_{6,1}$, $A_{6,12}$, $B_{6,1}$, and $B_{6,12}$, are respectively preset on the RV side and LV side of the IVS and endocardial and epicardial sides of the posterior wall at the timing of the R-wave. The tracking results $\hat{x}(t; A_{6,j})$ and $\hat{x}(t; B_{6,j})$ of these four points are overlaid on the M-mode image as shown in Fig. 3(a). The velocity signals, $\hat{v}(t; A_{6,j})$ and $\hat{v}(t; B_{6,j})$, at these four points are shown in Fig. 3(d), 3(e), 3(f), and 3(g), respectively. The resultant velocity signals are sufficiently reproducible for two heartbeat periods. During systole, the signals $\{\hat{v}(t; A_{k,j})\}$ and $\{\hat{v}(t; B_{k,j})\}$ have velocity components so that the LV lumen contracts. At the beginning of the diastole, on the contrary, they have velocity components because the LV lumen expands. A similar tracking operation is applied to each of the signals received from other directions from n_1 to n_{10}.

The heart chamber contracts during systole and expands during diastole. The

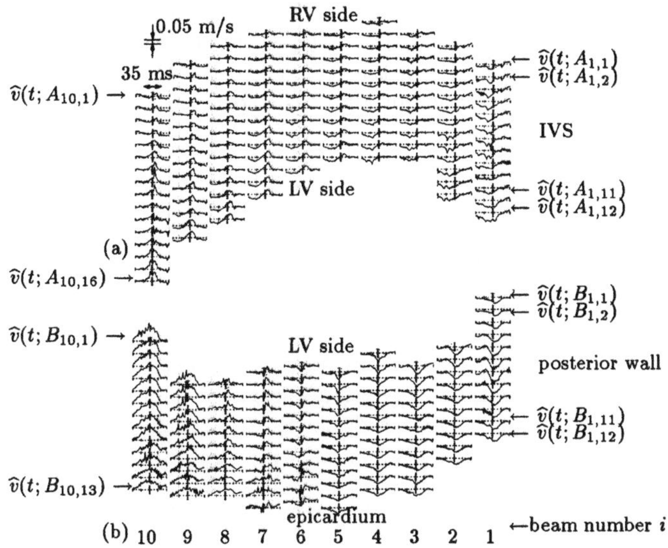

Figure 4. (a) The vibration velocity signals $\{\hat{v}(t; A_{k,j})\}$ $(k = 1, 2, \ldots, 10)$ in the IVS in 10 directions during one cardiac cycle. (b) The vibration velocity signals $\{\hat{v}(t; B_{k,j})\}$ $(j = 1, 2, \ldots, 10)$ in the posterior wall in 10 directions during the same cardiac cycle as that in Fig. (a).

direction of the ultrasonic beam of n_6 is almost perpendicular to the heart wall during one cardiac cycle and the points $A_{6,j}$ and $B_{6,j}$ on the beam trace the movements. For the points other than $A_{6,j}$ and $B_{6,j}$, however, it is difficult to accurately trace the movement of the point on the heart wall because the direction of the ultrasonic beam is always fixed. The point, the depth of which is preset at the R-wave, slips off the ultrasonic beam during systole. Thus, in a strict sense, each of the velocity waveforms $\{\hat{v}(t; A_{k,j})\}$ and $\{\hat{v}(t; B_{k,j})\}$ $(k \neq 6)$ does not continuously reveal the accurate movement of each point for one cardiac cycle. However, at least around the R-wave when the position of each point is preset, these points are in the heart wall. In this study, therefore, the spatial distribution of the instantaneous amplitudes of these velocity signals of these points is analyzed at the end-diastole.

Figures 4(a) and 4(b) show a typical example of the wavelets of the simultaneously measured velocity signals $\{\hat{v}(t; A_{k,j})\}$ $(k = 1, 2, \ldots, 10)$ in the IVS and $\{\hat{v}(t; B_{k,j})\}$ in the posterior wall, respectively, at the end-diastole. The central vertical bar in each wavelet shows the timing 6.7 ms after the R-wave. The waveforms gradually vary according to the directions of the ultrasonic beams. At this timing, the LV chamber vibrates as the vibration mode in the leftmost illustration of Fig. 1.

It has been experimentally shown that the eigenfrequency of the human heart is about 30 Hz at the end-diastole[4,5]. Therefore, in order to detect the spatial distribution of the eigenvibration components, Fourier transform is applied to each wavelet in Figs. 4(a) and 4(b) and the resultant phase values at 30 Hz, $\{\phi(t; A_{k,j})\}$ and $\{\phi(t; B_{k,j})\}$ $(k = 1, 2, \ldots, 10)$, are color-coded and shown in Fig. 5.

Each phase value shows the phase delay of the measured wavelet when the cosine wave of 30 Hz is fitted. Both for the IVS and the posterior wall, the phase values vary from green at the right-hand side, through blue, to red at the left-hand side. That is, the phase of the vibration mode of the LV chamber corresponds to the leftmost of the

Fig. 1 at this timing. Thus, there should be one node of the eigenvibration between the section from $A_{1,j}$ to $B_{1,j}$ and another between that from $A_{10,j}$ to $B_{10,j}$. These nodes exist near the connecting points of the RV wall to the LV wall. From these *in vivo* results, the mode-2 eigenvibration in Fig. 1 is recognized at the end-diastole for each subject.

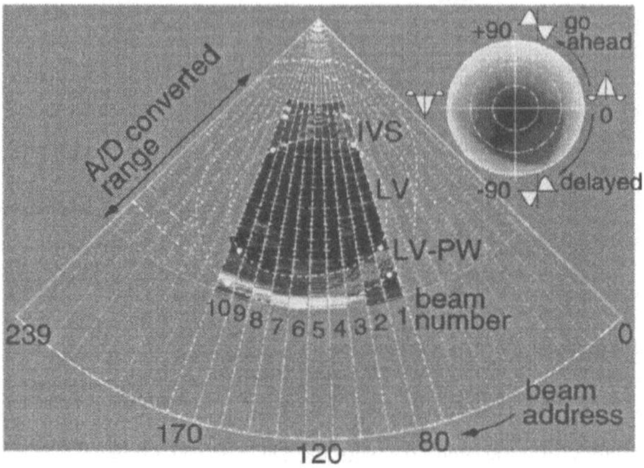

Figure 5. Spatial distribution of color-coded phase value of the spectrum at 30 Hz of the measured wavelets at the end-diastole in Fig. 4.

CONCLUSIONS

In this paper, we have proposed a novel method for controlling the directions of the ultrasonic beams so that the velocity signals are simultaneously measured at multiple points on the LV wall for imaging of their spatial distribution. By applying this method to the LV wall of a young volunteer, we obtained spatial distributions of small vibrations of about 30 Hz at the end-diastole. The results were superimposed on the B-mode image. From the results, we confirmed that the there is mode-2 eigenvibration on the LV wall at the end-diastole. This method offers potential for imaging of the spatial distribution of the small vibrations with high frequency components, which cannot be recognized in the standard B-mode or M-mode images.

REFERENCES

1. H. Kanai, M. Sato, Y. Koiwa, and N. Chubachi, Transcutaneous measurement and spectrum analysis of heart wall vibrations, *IEEE Trans. on UFFC*, 43:791(1996).
2. H. Kanai, H. Hasegawa, N. Chubachi, Y. Koiwa, and M. Tanaka, Noninvasive Evaluation of Local Myocardial Thickening and Its Color-Coded Imaging, *IEEE Trans. on UFFC*, 44:752(1997).
3. H. Honda, S. Nakaya, H. Kamada, H. Hasegawa, J. Demachi, H. Chikama, K. Sugimura, Y. Yamamoto, N. Kumasaka, T. Takita, J. Ikeda, H. Kanai, Y. Koiwa, K. Shirato, Non-invasive Estimation of Human Left Ventricular End-diastolic Pressure, *Medical Engineering & Physics*, 20:485(1998).
4. Y. Koiwa, T. Ohyama, T. Takagi, J. Kikuchi, H. Honda, R. Hashiguchi, Y. Shimizu, J. P. Butler, T. Takishima, The left ventricular vibration mode in the ventricular transfer function method and at the moment of the first heart sound, *Frontiers Med. Biol. Engng*, 1:59(1988).
5. M. Sato, H. Kanai, N. Chubachi, H. Honda, and Y. Koiwa, Method for Noninvasive Estimation of Left Ventricular End Diastolic Pressure Based on Analysis of Heart Wall Vibration, *Electronics Letters*, 32:949(1996).

RADIOFREQUENCY REAL TIME PROCESSING: ULTRASONIC SPECTRAL IMAGES AND VECTOR DOPPLER INVESTIGATION

E. Biagi, L. Masotti, L. Breschi, M. Calzolai,
L. Capineri, S. Granchi, M. Scabia

Ultrasound and Non Destructive Testing Laboratory
Dipartimento di Elettronica e Telecomunicazioni,
Università degli Studi di Firenze, via S. Marta 3, 50139 Firenze, Italia

INTRODUCTION

Novel radio frequency processing techniques for biological tissue characterization with ultrasound are presented in order to improve the diagnostic power of ultrasonic echographic systems.

Spectral images of biological "in-vitro" and "in-vivo" tissue, obtained through the Discrete Wavelet Packet Transform (DWPT) are presented, as well as velocity vector maps of blood flow obtained with 2-D Doppler investigation.

The implementation of the Discrete Wavelet Packet Transform through a digital filter produces, for each acquired frame, real time spectral maps, in different frequency bands. Multi-parametric images are composed by merging these maps using a dedicated "balance image fusion" algorithm. New blood images, obtained without using the Doppler effect, for "in-vivo" and "in-vitro" experiments are presented based on the exploitation of non-linear ultrasound-medium interaction effects [1,2,3,4]. The proposed spectral processing procedure seems to be suitable to perform tissue characterization. Pathological portions inside tissue could be detected thanks to their different echo frequency content which in turns is determined by linear and non-linear ultrasonic-medium interaction. The target of future clinical applications is to investigate the potential of the procedure as a "virtual biopsy".

Vector Doppler multi parametric images are obtained with compound measurements of Doppler shifts along different directions, and superimposing the resulting 2-D velocity vector maps to the conventional morphological B-mode representation [5,6,7]

The results, here presented, were produced by employing a hardware and software platform dedicated to ultrasonic signal and image processing [8,9,10]. The radiofrequency signal for multi parametric calculation and presentation with a multiprocessing digital architecture was used. This platform provides a multi analysis and

Acoustical Imaging, Volume 25, Edited by Halliwell and Wells.
Kluwer Academic/Plenum Publishers, 2000.

interactive image system for those clinical applications where a high image production rate is required. In order to obtain an efficient interactive system, it was necessary also to realize a fast signal processing, as well as to implement a fast visualization tool for managing multiple images. The platform is completely programmable, and for a specific application, can be reconfigured on-line depending on the parameters that need to be evaluated. The platform main features were extensively explained in previous works [9,10]

METHODS

Ultrasonic wave transmission, reflection, diffraction and diffusion associated with frequency selective attenuation are the basic interaction phenomena on which medium characterization is based. For fluid materials, such as biological ones, non-linear interaction effects have also been investigated in this last decade [1, 2]. It should be possible, in the near future, to extract information of biological tissue about their healthy or pathological nature, as physicians now obtain data on vessels hematic flow, as well as cinematic characteristics of the moving parts such as heart structures, vessel walls and foetus.

Ultrasonic waves interact with medium discontinuities along their propagation path according to the Rayleigh law in dependence on discontinuity size-wavelength ratio. Specular reflection and scattering are the two main phenomena involved. For a given scatterer distribution the single echo signal amplitude at each frequency depends on the scatterer size-wavelength ratio. Non linear effects must also be considered as a contribution to the back propagated energy. When the time equivalent mutual distance of scatterers is smaller than the echo signal duration, constructive and destructive interference occurs on the receiving transducer face generating an amplitude modulated echo signal. This amplitude modulation originates from the echo signal phase shift depending on the scatterer mutual distance-wavelength ratio. For the broadband echographic signal this phase shift is responsible for the "scalloping effect" in the frequency domain, i.e. spectral amplitude weighting according to a multicosine function whose period depends on the scatterer mutual distance [11].

These remarks impose radio frequency processing to extract the typical local informative content of the echo signal.

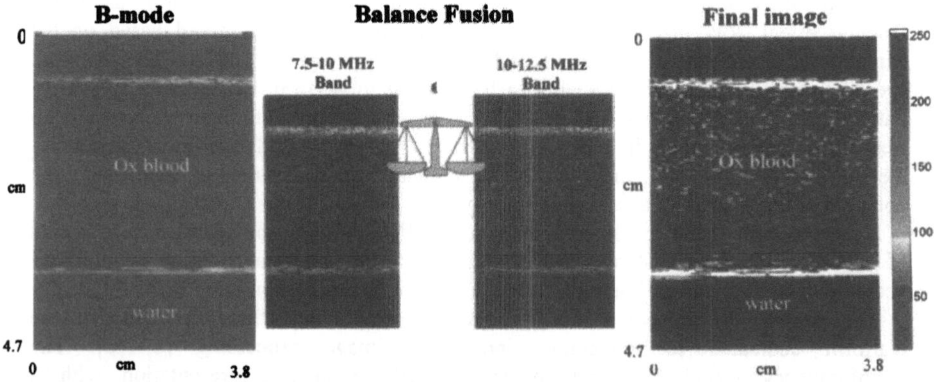

Figure 1: "In-vitro" acquisition of a longitudinal section of a thin wall plastic tube filled with ox blood. Left: B-mode image. Center: spectral maps for Balance fusion algorithm. Right: final image.

Over the last 20 years, a number of investigators [3,12-21] defined processing techniques to extract information about biological tissue microstructure or material properties that are not conveyed in conventional B-mode images. All these techniques use radio-frequency echo signals to preserve information that are otherwise lost.

Finally also the Doppler frequency shift is contained in the radio frequency signal phase.

Radio-frequency information content must be not only extracted by the "most appropriate" signal processing but also represented on-line; the information must exhibit an accurate time/space correlation, a fast rate production and must be easily readable.

The most effective interaction between operator and investigation system is another aspect that must be taken into account.

All these requirements concerning ultrasonic investigation systems, motivated our efforts to develop a novel platform devoted to real-time digital signal processing, of radio-frequency echo [8-10]. The platform architecture was designed to provide an interactive and fast visualization system firstly to adapt the measurement procedure to the medium characteristics to be exploited and also to manage multiparametric images for an immediate data interpretation .

EXPERIMENTAL RESULTS: SPECTRAL IMAGES FOR TISSUE CHARACTERIZATION

This paragraph concerns the radio-frequency spectral analysis for medium microstructure characterization.

The results are obtained by a particular platform configuration [10]. The basic element for this specific application is the *spectral processor module*, based on a Gray Chip GC 2011 digital filter with 64x2 taps, 80 MHz maximum operating frequency [8,9]. The digital filter is used to perform the Discrete Wavelet Transform (DWT) decomposition, in its Packet version DWPT, by successive filtering and decimation on different programmable frequency bands.

ESAOTE AU3 echograph, which uses a 7.5 MHz central frequency linear array probe was employed as front end.

The experimentation was performed for blood investigation through the DWPT spectral analysis of the backpropagated ultrasonic signal. New spectral contributions due

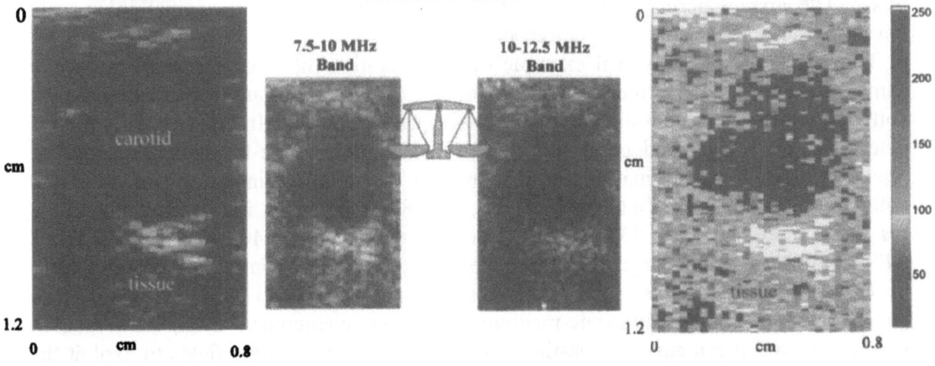

Figure 2: "In-vivo" acquisition of a transversal section of a human carotid. Left: B-mode image. Center: spectral maps for Balance fusion algorithm. Right: final image.

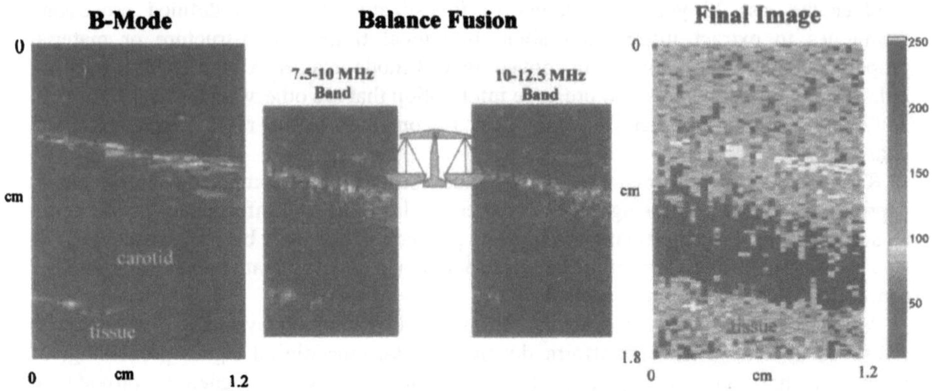

Figure 3: "In-vivo" acquisition of a longitudinal section of a human carotid. Left: B-mode image. Center: spectral maps for Balance fusion algorithm. Right: final image.

to non-linear effects in blood were extracted and presented. For this specific application the digital filter performs an echo signal spectral decomposition into two different bands: [7.5MHz-10MHz] [10MHz-12.5MHz]. The first one corresponds to the higher portion of the probe frequency response while the second one was found to be a typical blood behavior induced by non linear propagation effects. The two produced data frames are then transferred to the PC, where an image fusion, based on a weighted comparison of DWPT coefficient amplitude, is performed by a dedicated software named "balance image fusion" algorithm. In figure 1, the obtained images for an "in vitro" acquisition are reported. The test object was a thin wall plastic tube filled with ox blood immersed in tank with distilled water. The B-mode longitudinal section is reported on the left of figure 1, the two spectral data frames produced in real time by the S*pectral Processor module* appear in the center, while the final image is on the right. This "in vitro" experiment was performed to test the "balance image fusion" algorithm immunity to noise contributions and its capability to recognize the presence of blood.

The results obtained for "in-vivo" acquisition of a transversal section of human carotid vessel are reported in figure 2 with the same image sequence as in figure 1. The images for the carotid longitudinal section are in figure 3. From these images the "balance image fusion" algorithm effectiveness to differentiate the blood with respect to the surrounding tissues appears. Red color, in the original image, was utilized to indicate the presence of blood. The images here reported are the gray level version of the original color scale ones. For these images red color corresponds to level 50.

This application is a typical example of the operative mode of our platform: while some operations are performed by the hardware, other operations are demanded of the PC software in order to obtain the maximum image production rate. In this case only the PC determines the image production rate; in fact it must execute the following procedures,: scan converter, image formation and display. For example, the image visualization and software scan converter of a 640x480 pixels, with 8 bits per pixel, image takes about 25.4 ms, on an Intel Pentium II 233 MHz PC, corresponding to about 40 frames/s. Note that a 640x480 image is much larger than those typically provided by commercial echographs.

The spectral analysis potential is enhanced when a very broad band signal is used to cover a wider class of ultrasonic-medium interaction phenomena. Skin, intravascular, endoscopic and percutaneous investigation appear to be appropriate fields to exploit the broad band- spectral analysis synergism [26,27].

EXPERIMENTAL RESULT: 2-D VECTOR DOPPLER

A second example of application for the platform is 2-D vector Doppler for blood flow investigation. Vector Doppler [5,6,7,28] is performed by using a modified version

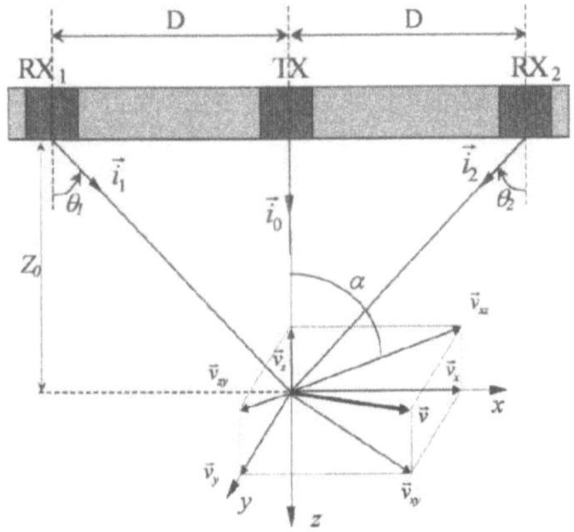

Figure 4: 2D vector Doppler using a linear array probe.

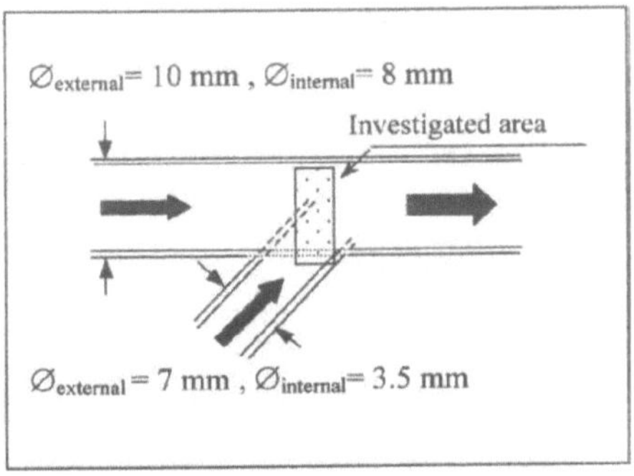

Figure 5: Schematic diagram of the Doppler flow phantom and details of the investigated area.

of a commercial echograph, ESAOTE AU3 echograph, based on a 5 MHz linear array probe (figure 4). A *Tx/Rx & Scan Control* module of the platform controls the selection of the active elements of the ultrasonic array probe and programs the beam steering. The 2-D vector Doppler estimation is thus obtained as a combination of two 1-D Doppler measurements carried out by two different receiving sub-apertures in the array, selected by the module, with different beam steering.

The *Tx/Rx & Scan Control* first selects a central sub-aperture that transmits, with 0° beam steering, the 5 MHz ultrasonic burst; a second subaperture (Rx1), with +12° beam steering, is then selected in order to receive the backscattered burst. This process is repeated for a certain number of bursts (up to 256), producing the Doppler measurement relative to the path Rx0-Rx1 in figure 4. The same measurement is performed with the second receiving sub-aperture Rx2, with 12° beam steering.

This allows estimation of the projection on the probe scan plane of the velocity vector in the investigated voxel. This process is repeated for all the voxels that must be investigated, thus producing a vector velocity map.

The platform configuration devoted to 2-D Doppler is equipped with a digital demodulation module, D*igital Demodulator that* is responsible for the demodulation of the radio frequency signal. Such module is followed by a *Spectral Processor* to execute the Doppler shift centroid extraction.

The velocity vector calculation is performed, over all the elementary volumes of the investigated area, by the *Velocity Estimator* module and then transferred to the PC. A dedicated software module carries out image interpolation and visualization of the 2-D Doppler map. Figure 5a and 5b show 2-D Doppler maps obtained on a Doppler flow phantom, whose schematic diagram is shown in figure 5; the phantom contains a solution of water, glycerol and Sephadex. In particular the investigated area shown in figures 6a and 6b consists of the confluence of two flows into a single tube.

In these figures, gray level representations of the original color scale images are reported. The confluence creates a reverse flow which in the figure appears in black,

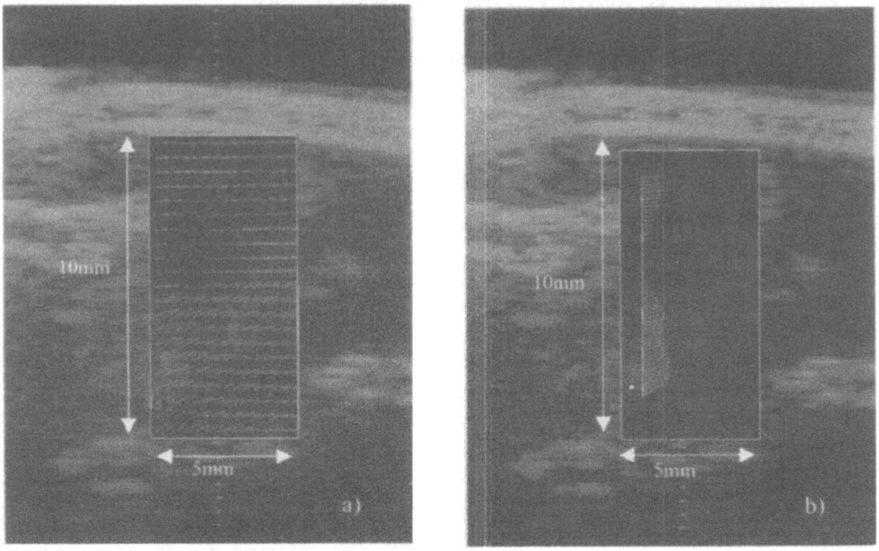

Figure 6: Vector velocity maps of the investigated area of the Doppler flow phantom represented in figure 5. (a) Raw image, showing vectors for every voxel. (b) Interpolated image, with vectors represented along a chosen line.

meaning flow away from the probe, while the rest of the flow is gray, meaning flow towards the probe.

Figure 5b represents the interpolated version of the Doppler map, where velocity vectors indicate the direction of the represented flow along a selected line. Another version of the same velocity map, with no interpolation and with velocity vectors applied to each voxel, is shown in figure 5a.

CONCLUSIONS

In this work we have presented two applications of a hardware-software platform developed for applications where multiparametric and multi image on-line continuous representation of data is required.

Results obtained with two different platform configurations have been presented not only as novel signal acquisition, processing and representation but also to clarify the meaning of multiparametric images.

The implementation of the Discrete Wavelet Packet Transform through a digital filter led to produce, for the same acquired frame, real time spectral maps, in different frequency bands. Merging these maps using a "balance fusion" algorithm multi-parametric images are provided. "In-vivo" and "in-vitro" images of blood were obtained, without Doppler effect, using non-linear effects that generate new spectral contributions. The proposed spectral processing procedure seems to be suitable to perform tissue characterization. Pathological portions inside tissue could be detected thanks to their different echo frequency content which in turn is determined by linear and non-linear ultrasonic-medium interaction. The target of future clinical applications is to investigate the potential of the procedure as a "virtual biopsy". Prostate investigation will be the first clinical platform job to differentiate pathological from inflamed tissues. Other clinical applications will be aimed to differentiate calcific from lipidic plaques inside vessels as well as to investigate the thyroid tissue.

Vector Doppler multi parametric images were also obtained by compound measurements of Doppler shifts along different directions, and superimposing the resulting 2-D velocity vector maps to the conventional morphological B-mode representation.

REFERENCES

1. Siemens Medical System, Inc., Ultrasound Group, "Tissue Harmonic Imaging", United States Patent, n° 5,632,277. September 23, 1998.
2. Averkiou, et al., "Ultrasonic diagnostic imaging of response frequency differing from transmit frequency", ATL Ultrasound, United States Patent, n° 5,879,303, March 9, 1999.
3. L. Masotti et alii, "L'assorbimento selettivo degli ultrasuoni nell'ecografia clinica", Atti del 1° Congresso della Società Italiana per lo Studio degli Ultrasuoni in Medicina, Ed. Minerva Medica, pp. 35-55, 1976.
4. L. Masotti et alii, "Caratterizzazione dei tessuti biologici per ultrasonografia clinica", Atti del 1° Congresso della Società Italiana per lo Studio degli Ultrasuoni in Medicina, supplemento a RAYS-International Jurnal of Radiological Sciences, 1977.
5. G. Bruni, M. Calzolai, L. Capineri, A. Fort, L. Masotti, S. Rocchi, M. Scabia, "Measurement and imaging of a velocity vector field based on a three transducer Doppler system", Acoustical Imaging. vol. 22. New York and London, Plenum Press, pp. 431-438, 1995.
6. M. Calzolai, L. Capineri, A. Fort, L. Masotti , S. Rocchi, M. Scabia, "A 3-D PW Ultrasonic Doppler Flowmeter: Theory and Experimental Characterization", IEEE Transaction on UFFC, Vol.46, n° 1, pp.108-113 January 1999.
7. M. Scabia, M. Calzolai, L. Capineri, L. Masotti, A. Fort. "A 'Real Time' two-dimensional pulsed wave Doppler system". Ultrasound in Medicine & Biology , Vol. 26, n° 1, pp 121-131. 2000.
8. E. Biagi, M. Calzolai, L. Masotti, "Method for enhancing the diagnostic power of ultrasonic scanning system by using real-time spectral maps, and device operating by said method" ESAOTE S.p.A., International Patent pending", FI 97A 141.
9. E. Biagi, M. Calzolai, S. Granchi, L. Masotti, M. Scabia, "Real-time processing of the radiofrequency ultrasonic echo signal for on-line spectral maps", 24th International Symp. On Acoustical Imaging, Santa Barbara, 1998.
10. E. Biagi, M. Calzolai, L. Capineri, S. Granchi, L. Masotti, M. Scabia, "FEMMINA: a Fast Echographic Multiparametric Multi Imaging Novel Apparatus, Invited paper 1999 UFFC, International Ultrasonic Symposium, 17-20 October, Lake Tahoe.
11. E. Biagi, G. Castellini, L. Masotti, S. Rocchi, "Ultrasonic spectroscopy of composite materials", Ultrasonic International, Butterworth Scientific Ltd, pp. 56-64, June 1989.
12. F.L. Lizzi, M. Ostromogilsky, E.J. Feleppa, M.C. Rorke, M.M. Yaremko, "Relationship of ultrasonic spectral parameters to features of tissue microstructure", IEEE Trans. Ultrasonic Ferroelectric Frequency Control, vol.UFFC-34, pp. 319-329, 1987.
13. E. Biagi, A. Corvi, L. Masotti, "Ultrasonic nondestructive evaluation of damage evolution in composite materials", Acoustical Imaging, vol. 19, Plenum Press, New York, pp. 595-600, 1992.
14. L. Masotti et al., "Spectral maps by AR models as a support to conventional ultrasonic imaging", EUSIPCO, Brussels, Elsevier Science Pub., Amsterdam, 1992.
15. E. Biagi, L. Masotti, E. Borchi, L. Masi, " Ultrasonic testing for oxide high temperature superconductors", Acoustical Imaging, vol. 20, Plenum Press, pp. 33-40, New York, 1993.
16. E. Biagi, L. Masotti, A. Fort, L. Ponziani, "Ultrasonic high resolution images for defect detection in ceramic materials", Research in Non Destructive Evaluation, n° 6, pp. 219-226, 1995.
17. F.L. Lizzi, M. Astor, A. Kalisz, T. Liu, D.J. Coleman, R.H. Silverman, R. Ursea, M.J. Rondeau, "Ultrasonic spectrum analysis of different scatterer morphology; theory and very high frequency clinical results", IEEE Ultrasonic Symposium, pp. 1155-1154, 1996.

18. S.L. Bridal, V. Roberjot, P. Laugier, G. Berger, "Attenuation and backscatter coefficient measurements from 2 to 60 MHz using backscattered signals from a tissue-mimicking phantom", IEEE Ultrasonic Symposium, pp. 1151-1154, 1996.
19. H.J. Huisman, J.M. Thijssen, "An *in vivo* Ultrasonic Model of liver Parenchyma", IEEE Ultrasonic Symposium, pp. 1143-1148, 1997.
20. F.L. Lizzi, "Ultrasonic scatter-property images of the eye and prostate", IEEE, Ultrasonic Symposium, pp.1109-1117, 1997.
21. P. Laugier, P. Giat, C. Chappard, Ch. Roux, G. Berger, "Clinical assessment of the backscatter coefficient in osteoporosis", IEEE Ultrasonic Symposium, pp. 1101-1104, 1997.
22. J.H. Letcher, " Methods and apparatus for reflective ultrasonic imaging", United States Patent, n°5,253,530, 19 October 1993.
23. A.A. Abdel-Malek, K. W. Rigby, "Enhanced method for reducing ultrasound speckle noise using wavelet transform", General Electric Company, United States Patent, n° 5,619,998, 15 April 1997.
24. Nikolaj Hess-Nielsen and Mladen Victor Wickerhauser, "Wavelets and Time-Frequency Analysis", Proceedings of the IEEE, 1995.
25. D. M. Healty, Jr.Jian Lu, J.B. Weaver, "Two Application of Wavelet on Real-time Techniques in Medical Imaging", in Annals of Biomedical Engineering, Vol. 23, pp.637-665, 1995.
26. E. Biagi, M. Brenci, S. Fontani, L. Masotti, M. Pieraccini, "Photoacoustic generation: optical fiber ultrasonic sources for non-destructive evaluation and clinical diagnosis",Optical Review, Vol. 4, n°.4, pp.481-483, 1997.
27. E. Biagi, L.Masotti, "Fiber optic ultasonic sources", Patent n. FI96A 216
28. J.A. Jensen, P. Munk, "A new method for estimation of velocity vectors", IEEE Transactions on Ultras Ferr Freq Contr, UFFC 45, pp.837-851, 1998.

ULTRASOUND STORAGE CORRELATOR ARRAYS FOR REAL-TIME BLOOD FLOW VELOCITY VECTOR IMAGING

Jian-yu Lu, Shiping He, and Khalil F. Dajani

Ultrasound Laboratory, Department of Bioengineering
The University of Toledo
Toledo, Ohio 43606, U.S.A.
Email: jilu@eng.utoledo.edu

ABSTRACT

Conventional speckle tracking method can be used for blood flow velocity vector imaging. However, it requires a large amount of computations. In this paper, storage correlator array is used to perform the correlations for blood flow velocity vector imaging at a high frame rate. In the method, sliding correlations are used for speckle tracking and matching. Previous speckle patterns are stored as reference signals in the storage correlator arrays, waiting for tracking current speckle patterns. The reference signals stored can be easily programmed, rewritten, or erased at any time. A blood flow velocity vector image can be obtained from peaks of correlations between the previous and the current speckle patterns.

INTRODUCTION

Storage correlator array[1-4] is a key component in high frame rate[5-6] blood flow velocity vector imaging due to their short processing time. They can be used effectively as a programmable matched filters to improve the signal-to-noise ratio (SNR) of signals. The programmable feature of storage correlator arrays allows speckle patterns stored to be updated for each frame of constructed radio-frequency (RF) image (an RF ultrasound image is one before an envelope detection). Real-time tracking and matching of previous and current speckle patterns make high speed blood flow velocity vector imaging practical. The high-speed processing also increases the maximum measurable velocities.

Conventional speckle tracking techniques[7-9] estimate the motion of objects based on speckle pattern matching and correlation algorithms. Velocity vector images are obtained by comparing the speckle pattern in a small kernel region in the current image to each

Acoustical Imaging, Volume 25, Edited by Halliwell and Wells.
Kluwer Academic/Plenum Publishers, 2000.

possible match in a larger surrounding search region in the previous one. The best match indicates the most likely location to which the speckle pattern has moved. The displacement of the best match and the time between the acquisition of the kernel and search regions are used to compute blood flow velocity vectors, thereby overcoming the limitations of conventional Doppler-based velocity estimators: angle dependence and aliasing. However, the conventional speckle tracking method is limited by several factors. One of them is that it takes a long time to calculate the correlation between the kernel and search regions and to search for the best matching speckle pattern. The long searching time may reduce the correlation between the kernels and the possible matches in search regions producing fault results.

To overcome the problem, in this paper, storage correlator arrays are used to perform the integrations of correlations for real-time speckle tracking. The operations of storage correlator arrays can be categorized into two models depending on how the integration is carried out, i.e., time-domain[10-12] and space-domain processing models[13]. In time-domain processing, integrations of correlated signals accumulate stored charges over time and the time bandwidth product is determined by the product of the transducer bandwidth and the storage time of the storage correlation arrays. Because storage correlation arrays have a long storage time, the processing gain of the signal is increased greatly. In space-domain processing, signal integration is carried out across a one- or two-dimensional space. The time bandwidth product of the signals is determined by the transducer bandwidth and the propagation time of ultrasound waves across a diode array.

Although the time bandwidth product of space-domain processing model is smaller than that of the time-domain processing, the space-domain model is better suited for real-time speckle tracking and matching. In the processing, a RF image constructed from a previous data acquisition is stored in a storage correlator array as a reference. Then, cross-correlations between the previously stored image and the multiple kernels obtained from the second RF image are processed with the storage correlator array in parallel. The outputs of the storage correlator array produce correlation peaks that indicate the location of possible matches. Velocity vector images of motions can thus be obtained. In addition, storage correlator arrays have a very large storage capacity and images stored can be easily overwritten or erased. The size of correlator arrays is small (usually about tens of millimeters in signal length and a few micrometers for each storage element).

STORAGE CORRELATOR ARRAYS

A storage correlator array is a device that can store a snapshot of the waveform of a surface acoustic wave (SAW) while the wave propagates from one finger transducer to another[11]. A simple storage correlator array has three ports (two acoustical and one electrical) (Fig. 1). Storage elements are usually a diode array. The storage principle is based on a nonlinear interaction between the waves and the diodes. Two systems consisting of such devices are shown in Figs. 2 and 3, respectively. Internal structure of a storage correlator array is shown in Fig. 4.

The device shown in Fig. 2 has a center frequency of 76 MHz (typical frequency is between 10MHz to a few GHz) with an interaction distance (where waveforms can be stored) of 20 mm, a diode size of 4 μm $\times$ 4 μm, and an inter-diode distance of 4 μm. The speed of sound of this device is close to 4000 m/s. The diode array is situated above an ultrasound piezoelectric delay line with 3-phase unidirectional transducers.

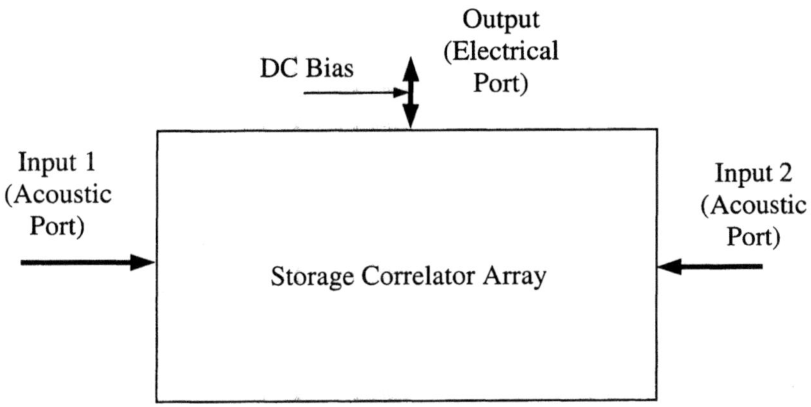

Figure 1. Simplified block diagram of a storage correlator array.

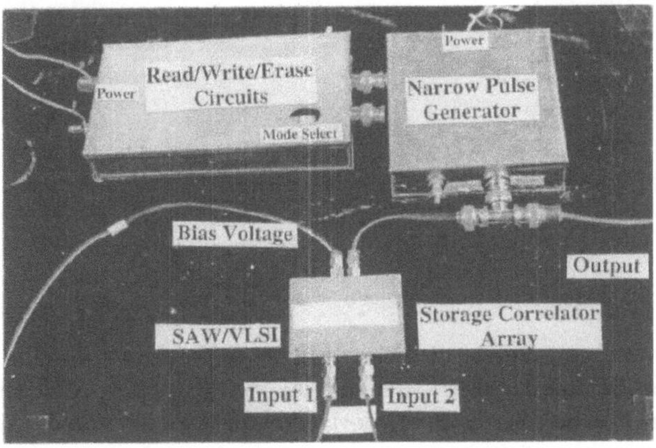

Figure 2. A subsystem consisting of a SAW/VLSI storage correlator array.

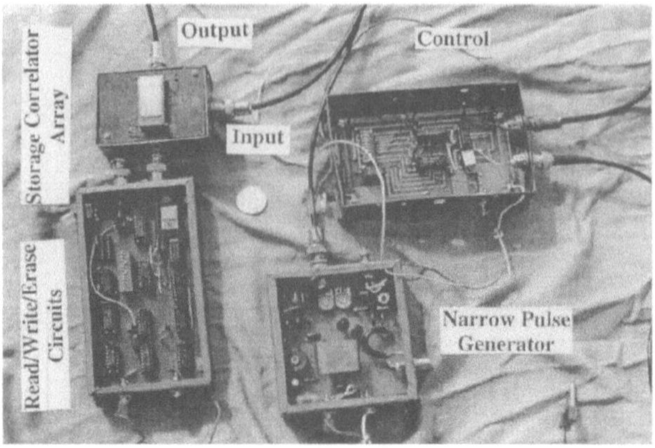

Figure 3. A subsystem with a high efficiency GaAs storage correlator array.

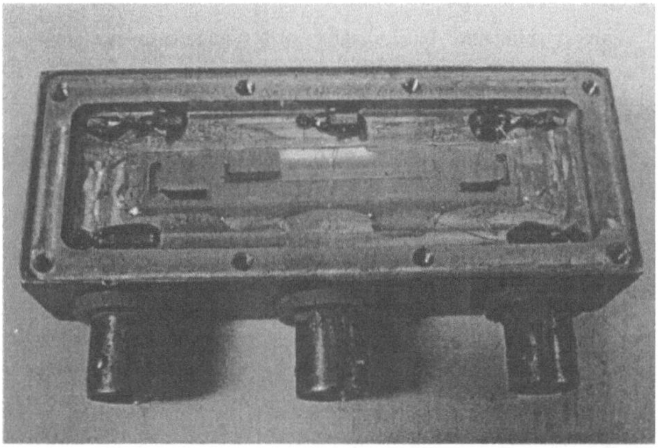

Figure 4. Internal structure of a storage correlator array.

To write a waveform into the diode array (write-in process), a short pulse of a width of a few nanoseconds can be applied through the electrical port. After storage, a signal that is proportional to the correlation or convolution between the stored waveform and the waveform input from one of the two acoustic ports can be read out from the electrical port (readout process). A DC bias of about 10V can be applied to increase the amplitude of output signals. The waveform storage time of the diodes is from hundreds of milliseconds to a few seconds and is a function of the DC bias voltage.

An experiment result of the autocorrelation with the storage correlator array in Fig. 2 is shown in Fig. 5. The waveform in the first row represents a signal that was stored in the diode array. The signal in the second row represents an autocorrelation of the stored signal.

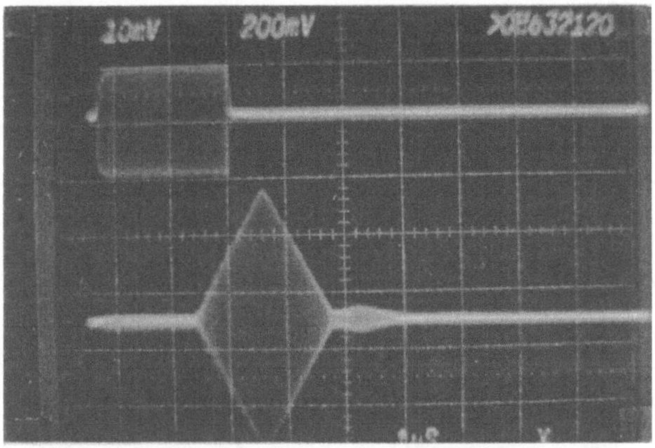

Figure 5. Autocorrelation of a tone burst with the storage correlator array shown in Fig. 2.

A simulation of the autocorrelation of a 31–chip pseudorandom (PN) code is shown in Fig. 6 and its corresponding experiment using the storage correlator array in Fig. 2 is shown in Fig. 7.

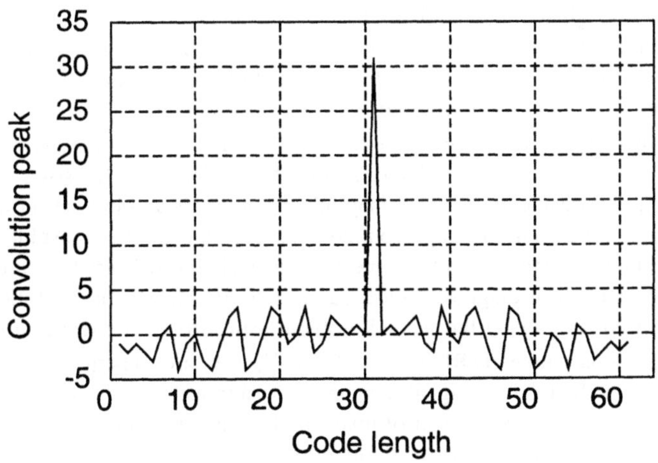

Figure 6. Simulation of autocorrelation of a 31–chip PN code.

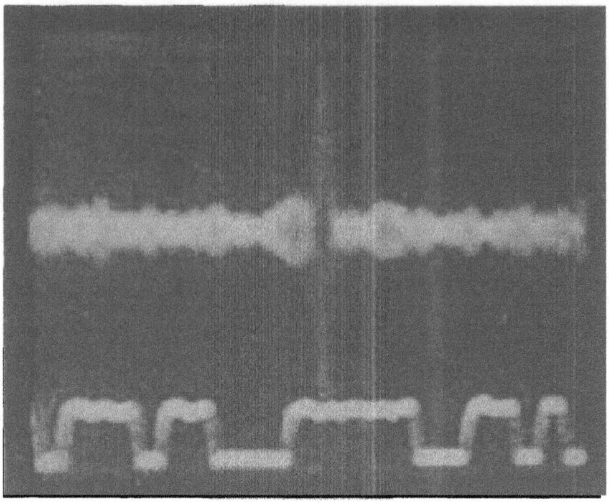

Figure 7. Experiment of autocorrelation of the same PN code as in Fig. 6.

VELOCITY VECTOR ESTIMATIONS

Speckle tracking techniques[7–9,12] can be used to obtain a blood flow velocity vector by comparing the speckle pattern in a small kernel region of the second image to each possible match in a larger surrounding search area of the first. The best match indicates the most likely location to which the speckle pattern, and thus the target, has moved. The location of the best match and the time between the acquisitions of the two images are used to estimate the velocity vector of the kernel.

To implement the speckle tracking techniques in practice, there are two major problems. The first is pattern decorrelation due to low frame rate of ultrasound images, and the second is a large amount of computation required.

Recently, a high frame rate imaging method that could be used to construct images at a frame rate up to 3750 frames or volumes/second for biological soft tissues at a depth of about 200 mm was developed[5,6]. This may address the first problem because the time between two consecutive frames is only about 267 μs. During this time, a fast blood stream at 5 m/s moves about 1.3 mm. Therefore, speckle decorrelation will not be a problem for frequencies used in medical diagnostic ultrasound (1 to 10 MHz). In the following, an approach is proposed for solving the second problem using storage correlator arrays.

Assuming two consecutive 2D RF images of size $N_x \times N_z$ are constructed with the high frame rate imaging method[5,6] from signals acquired with a transducer of center frequency of about 3.5 MHz, where N_x is the number of pixels in the scan direction (around 128 pixels) and N_z is the number of pixels in the depth direction (at 40 MHz sampling rate, there will be about 10680 pixels for a depth of 200 mm in biological soft tissues). The first image can be subdivided into multiple search areas, each of which has the size of $N_{sx} \times N_{sz}$, where N_{sx} and N_{sz} are the number of pixels along the scan and depth directions, respectively. In the second image, kernels of the size of $N_{kx} \times N_{kz}$ centered at the corresponding search areas of the first image can be obtained, where $N_{kx} < N_{sx}$ and $N_{kz} < N_{sz}$. The sizes of the search and the kernel areas are determined by the anticipated maximum flow velocity and resolution of the constructed velocity vector images. At a high image frame rate, the difference between the sizes of the search and kernel areas can be small (because the maximum displacement of an object may be less than 2 mm between two consecutive frames of images acquired). A small kernel size will increase the resolution in constructed velocity vector images. However, a kernel that is too small may lose the feature of the kernel leading to false correlations.

With conventional method where a digital computation is used to perform correlation integrations, it is difficult to complete all 2D correlations within a short period of time, say, 267 μs, because there will be at least $(N_x \times N_z)/(N_{sx} \times N_{sz})$ 2D correlations. This problem may be solved with a storage correlator array modified from that shown in Fig. 1. To perform a parallel processing, more electrical and acoustical ports are added and the diode array is extended from 1D to 2D (Fig. 8).

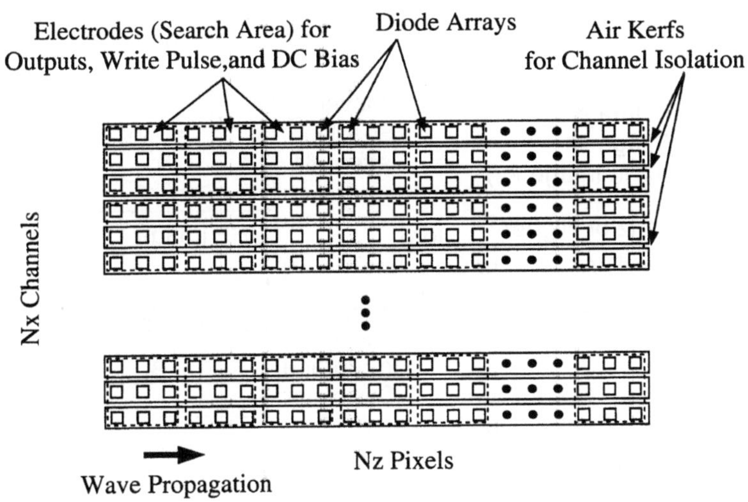

Figure 8. 2D diode array for parallel 2D correlations in a modified storage correlator array.

To perform correlations, the first RF digital image constructed with the high frame rate method is read out from a random access memory (RAM) and converted to high frequency signals that match the frequency, say, 76 MHz, of the storage correlator array via N_x digital-to-analog (D/A) converters simultaneously. The analog signals are sent to the acoustic ports (Fig. 9) and stored (by applying a short strobe pulse) in parallel when they reach the end of the area of the storage diodes (mapping the entire RF image in the diode array). The kernels in the second image are then converted also to analog signals via D/A converters. The analog signals of the kernels are sent in parallel to the acoustic ports of the storage correlator array. Correlations are performed when the signals of the kernels are within their corresponding search areas of the first image stored. Correlation peaks are obtained simultaneously via the multiple electrical ports. From the positions of correlation peaks, motion of each kernel along the depth direction can be obtained. To obtain the motions of kernels along the scan direction, the analog signals of the kernels are shifted step by step in this direction for $N_{step} = N_{sx} - N_{kx}$ steps. The position (in both x and z coordinates) of the highest correlation peak among all steps for each kernel indicates the best possible match within the corresponding search area. From the position of the peak, the velocity vector of the kernel can be calculated.

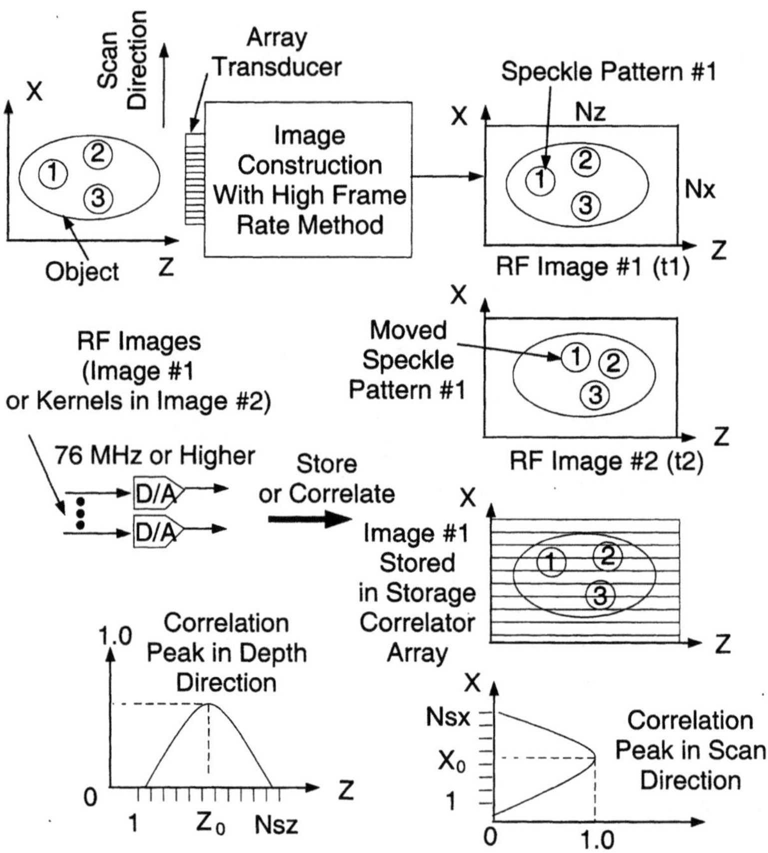

Figure 9. A schematic for parallel speckle tracking with a 2D storage correlator array.

Apparently, with the approach described above, the number of velocity vectors obtainable is equal to that of the search areas or the number of electrical ports. To increase the number of velocity vectors or improve the resolution of the constructed flow images, new kernels that lie between existing kernels and new search areas centered around these kernels can be selected. New velocity vector images can be obtained by repeating the correlation procedure above and interlacing the results with existing images.

With a higher frequency storage correlator array and a higher diode density, it is possible to obtain a high resolution velocity vector image within the time between two successive RF images (267 μs in the example above). Assuming the depth of biological soft tissue is 200 mm, with 3.5 MHz center frequency, there will be about 933 cycles within the round trip distance of the ultrasound. After D/A conversion, the same number of cycles will be obtained but at a higher frequency, say, 76 MHz. If the high frequency signal propagates at about 4000 m/s within the storage correlation arrays, the wavelength will be 52.6 μm and the 933 cycles will occupy a a total length of about 49.12 mm. Therefore, for each correlation process, it takes about 12.28 μs to store the first image, and additional 12.28 μs for each step of correlation. To obtain velocity vector image of a higher resolution, the total time will be $12.28 \times (N_{step} + 1) \times N_{off}$ μs, where N_{off} and N_{step} are number of offsets and number of steps, respectively. N_{off} indicates the number of velocity vector images interlaced. N_{off} can be smaller if an interpolation is used to increase the pixel density of velocity vector images.

To reduce the total time required for the correlation processing, the frequency of the D/A converters can be further increased. The total length of the acoustic pathway of the storage correlator array can be reduced at a higher frequency for the same number of signal cycles. This reduces the propagation time (from 12.28 μs in the example above). The number of diodes within the total length should be kept the same without degrading performance. The minimum number of diodes required should satisfy the Nyquist sampling rate, i.e., at least 2 diodes per cycle. For 933 cycles, the minimum of 1866 diodes are needed. With a 49.12 mm total length, the distance between the diodes must be smaller than 26 μm. A device shown in Fig. 2 has an inter-diode distance of 8 μm.

DISCUSSION

In this paper we have presented an approach to perform correlations with storage correlator arrays using the space-domain processing model. The described system has the ability to construct velocity vector images in real-time.

The extension of 1D to 2D storage correlator arrays require good acoustic separation between channels. This can be accomplished by using an air kerf between channels. A pair of finger transducers is required for each channel.

Because the high frame rate imaging method developed recently can also be used to construct 3D images at an ultrahigh frame rate (a few thousand volumes per seconds), the method developed in this paper can be extended to 3D for volumetric velocity vector imaging by using more than one multi-input/output storage correlator array. Such arrays can be stacked to form a 3D storage matrix that stores an entire volume of RF image. Multiple kernels can be obtained from the second volumetric image for real-time volumetric velocity vector imaging.

CONCLUSION

A real-time speckle tracking method using storage correlator arrays has been proposed in this paper. Previous RF speckle patterns are stored in a storage correlator array as references and current speckle patterns can be tracked in real time. The storage correlator arrays can be easily programmed, rewritten, or erased at any time. Blood flow velocity vector images can be obtained from the correlations between the speckle patterns of previous and current RF images.

ACKNOWLEDGMENTS

This work was supported in part by a grant HL60301 from the National Institutes of Health.

REFERENCES

1. S. He and B. Wu, SAW MZOS storage correlators/convolvers, *Acta Electronica Sinica.* 18(4):114-117 (1990).
2. J. H. Yin and S. He, Modified SAW diode convolver, *Ectronics Letter.* 28(2):172-174 (1992).
3. S. He, H. Hong, and B. Wu, An investigation of image spectrum systems using SAW storage correlators, *Proc. of 1989 International Symposium on Electromagnetic Compatibility.* Nagoya, Japan, 913–916 (1989).
4. B. Wu and S. He, A new development of SAW correlator and it's application in an experimental spread spectrum system, *Proc. of 1991 IEEE International Symposium on Circuits and Systems.* Cape Cod, USA, 2609-2612 (1991).
5. Jian-yu Lu, 2D and 3D high frame rate imaging with limited diffraction beams, *IEEE Trans. Ultrason. Ferroelec. Freq. Contr.* 44(4):839-856 (1997).
6. Jian-yu Lu, Experimental study of high frame rate imaging with limited diffraction beams, *IEEE Trans. Ultrason. Ferroelec. Freq. Contr.* 45(1):84-97 (1998).
7. L. N. Bohs, B. H. Friemel, B. A. McDermott, and G. E. Trahey, Lateral Velocity Profile and volume Flow measurements via 2-D speckle tracking, *Acoustical Imaging.* 21:503-507 (1995).
8. L. N. Bohs, B. J. Geiman, M. E. Anderson, S. M. Breit, and G. E. Trahey, Ensemble tracking for 2D vector velocity measurement: Experimental and initial clinical results, *IEEE Trans. Ultrason. Ferroelec. Freq. Contr.* 45:3-10 (1998).
9. L. N. Bohs, B. H. Friemel, and G. E. Trahey, Experimental velocity profiles and volumetric flow via two-dimensional speckle tracking, *Ultrasound Med. Biol.* 21(7):885-898 (1995).
10. S. He, Y. Shui, and H. Hong, A larger TBP time domain processing spread spectrum systems, *Proc. of IEEE International Symp. on Electromagnetic Compatibility.* Japan (1994).
11. S. He, T. K. Y. Lo, and J. Litva, A spread spectrum system with a time domain processing device, *IEEE Trans. on Circuits and Systems-II: Analog and Digital Signal Processing.* 44(6):447-452 (1997).
12. S. D. Foster, P. M. Embree, and W. D. O'Brian, Jr, Flow velocity profile via time-domain correlation error analysis and computer simulation, *IEEE Trans. Ultrason., Ferroelect., Freq. Contr.* 37(2):164-175 (1990).
13. S. He, H. Hong, and B. Wu, Analysis of interference rejection of spread spectrum systems using SAW storage correlators, *Journal of China Institute of Communication.* 12(3):76-82 (1991).

SIMULATION OF POWER DOPPLER ANGIOGRAPHIC IMAGES OF A SEVERE STENOSIS

Damien Garcia [1], Louis-Gilles Durand [1], Zhao Qin [1], Champlain Landry [1], and Guy Cloutier [1,2]

[1] Laboratory of Biomedical Engineering, Clinical Research Institute of Montreal, and [2] Laboratory of Biorheology and Medical Ultrasonics, University of Montreal Hospital, Quebec, Canada.

ABSTRACT

Objectives — This study sought to clarify the mechanisms leading to a signal loss observed in *in vitro* experiments when reconstructing the 3D geometry of a stenosis with power Doppler angiography (PDA).

Methods and Results — *In vitro* experiments were performed in a vessel model containing a severe stenosis (80 % area reduction). The PDA cross-sectional images of the stenosis, reconstructed in 3D by longitudinally scanning the artery, were compared with theoretical binary images obtained from finite-element modeling (FEM) simulations. Predictions of the PDA image characteristics were performed under steady and pulsatile physiological flows. The FEM simulations showed that the wall filter induces a loss of signal, which can significantly affect the assessment of the stenosis severity. Under steady flow, the FEM simulations reproduced the *in vitro* experimental images with a good accuracy. Dynamic changes in the position and importance of the signal loss within the vessel were observed under pulsatile flow.

Conclusions — The loss of signal on the PDA images is attributed to the wall filter and is not related to the backscattering properties of the fluid. When applied clinically, a significant underestimation of the stenosis severity may occur with this method. However, it is shown that the signal loss under pulsatile flow may be of diagnostic value.

INTRODUCTION

Power Doppler angiography has been developed to overcome some of the limitations of color Doppler imaging to map the blood flow field[1]. This technique is more sensitive than color Doppler, has a better signal-to-noise ratio, and is less affected by aliasing and Doppler angle. *In vitro* experiments performed by our group in a vessel model containing a severe stenosis showed that a flow-dependent loss of signal can occur[2]. We hypothesized

Acoustical Imaging, Volume 25, Edited by Halliwell and Wells.
Kluwer Academic/Plenum Publishers, 2000.

that this loss of signal was mainly due to the wall filter cutoff frequency, which affects the detection of low flow velocities. To test this hypothesis, we compared theoretical PDA images computed from FEM simulations to those obtained from *in vitro* experiments.

METHODS

In vitro model — A wall-less agar phantom with a rectilinear vessel containing an axi-symmetric stenosis of 80% area reduction was used. The diameter of the non-obstructed lumen was 7.96 mm and the cosine shaped stenosis had a length of 20 mm. The inlet length of the flow phantom was long enough to have a fully developed flow before the stenosis, and the length of the tube after the stenosis ensured the complete re-establishment of the flow after the jet. A blood mimicking fluid containing water (1/3), glycerol (2/3) and sephadex particles (50 μm) was circulated in the model under steady and pulsatile flows varying between 100 and 800 mL/min. The viscosity and density of the fluid were 3.5 cP and 1000 $kg.m^{-3}$, respectively. The flow rate was measured with an electromagnetic flowmeter (Cliniflow II, Carolina Medical Electronics, King, NC).

Power Doppler angiography — PDA measurements were performed at 4 MHz with an ATL-HDI Ultramark 9 system. The wall filter cutoff frequency was set to 25 or 100 Hz and the Doppler angle was 70°. The highest persistence was used for steady flow measurements whereas no persistence was selected for pulsatile flow. 3D images were reconstructed by longitudinally scanning the artery over a distance of 9.6 cm. For steady flow experiments, the cross-sectional image acquisition and 3D reconstruction were performed with a workstation commercialized by Life Imaging System Inc. For pulsatile flow, an in-house LabView (National Instruments) acquisition program allowing flow gating was used.

FEM simulations — FEM simulations of the flow in the vicinity of the stenosis were performed with Fidap (ver. 8, Fluent Inc.). The dimension and the shape of the stenosis simulated by FEM had the same characteristics as those of the *in vitro* model. The problem was axi-symmetric and nonlinear, and the flow was considered laminar. The mesh of the flow region was composed of 12,431 nodes and 12,000 four-node quadrilateral elements. It was refined near the wall (Fig. 1). The solver used was the successive substitution owing to its large radius of convergence and the relative simplicity of the flow region geometry. A relaxation factor of 0.5 was used to improve the rate of convergence. The fluid viscosity and density were the same as the mimicking fluid. A null velocity was imposed at the wall and the radial component of the velocity was fixed at zero along the central axis.

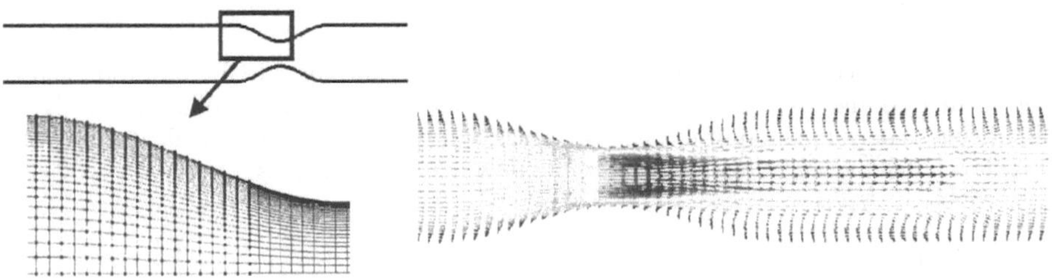

Figure 1. Schematic representation of the mesh used for the FEM simulations (left), and example of the velocity field along the vessel (right). The velocities were reconstructed by symmetry for a flow rate of 100 mL/min.

The boundary conditions at the inlet were imposed as follows :

Steady flow : A parabolic velocity profile defined by the following equation was imposed at the inlet :

$$V_z(r) = 2\frac{Q}{\pi R^2}\left[1 - \left(\frac{r}{R}\right)^2\right] \tag{1}$$

where r is the radial coordinate, z is the longitudinal coordinate, Q is the flow rate, and R is the vessel radius.

Pulsatile flow : A time-dependent longitudinal component of the total normal stress was imposed at the inlet. It is defined as [3] :

$$\sigma_{ZZ}(r,t) = -P(r,t) + 2\mu\frac{\partial V_Z(r,t)}{\partial z} \tag{2}$$

where t is the time, P is the static pressure, μ is the fluid viscosity, and V_z is the velocity.

Because the entrance length is long enough, the term $\mu\, \partial V_z/\partial z$ is negligible at the inlet and P is independent of r. Consequently,

$$\sigma_{ZZ\,\text{inlet}}(r,t) \approx -P_{inlet}(t). \tag{3}$$

Assuming that the pressure drop due to turbulence created by the stenosis is negligible, that the pressure is considered to be zero at the outlet, then the inlet pressure can be written as a function of the pressure gradient as :

$$P_{inlet}(t) \approx L\,\Delta P(t) \tag{4}$$

where L is the length of the vessel and ΔP is the pressure gradient.

The pressure gradient $\Delta P(t)$ can be estimated from the flow rate $Q(t)$ by using the Womersley method [4,5] : Because the flow is periodic, both the pressure gradient and the flow rate can be decomposed as a Fourier series :

$$Q(t) = \sum_{n=-N}^{N} Q_n e^{-in\omega t} \quad \text{and} \quad \Delta P(t) = \sum_{n=-N}^{N} \Delta P_n e^{-in\omega t} \tag{5}$$

where i is the imaginary number $\sqrt{(-1)}$, and $\omega = 2\pi / T$ (T is the period of the flow).

The relationship between ΔP_n and Q_n can be written as :

$$\Delta P_n = \frac{n\omega\rho}{i\pi R^2} Q_n \left[\frac{J_0(i^{3/2}\alpha_n)}{J_0(i^{3/2}\alpha_n) - 2J_1(i^{3/2}\alpha_n)/(i^{3/2}\alpha_n)}\right], \text{ where } \alpha_n = R\sqrt{\frac{n\omega\rho}{\mu}}, \tag{6}$$

ρ is the fluid density, and J_0 and J_1 are the Bessel functions of order zero and one, respectively.

Because $Q_n = \frac{1}{T}\int_0^T Q(t)e^{-in\omega t}dt$, we obtain:

$$\Delta P(t) = \sum_{n=-N}^{N} \frac{n\omega\rho}{i\pi R^2}\left[\frac{1}{T}\int_0^T Q(t)e^{-in\omega t}dt\right]\frac{J_0(i^{3/2}\alpha_n)}{J_0(i^{3/2}\alpha_n) - \frac{2}{i^{3/2}\alpha_n}J_1(i^{3/2}\alpha_n)} e^{-in\omega t}. \tag{7}$$

The time-dependent flow rate used for the FEM simulations was that measured by the electromagnetic flowmeter during the *in vitro* experiments. In the above equation, N is the number of harmonics chosen in such a way that the correlation coefficient between $Q(t)$ and his Fourier series was greater than 0.999 ($N = 12$ in our case).

Theoretical PDA imaging — Based on the velocities obtained with the FEM simulations, the Doppler shifted frequencies were calculated at each point of the mesh by using the following equation:

$$F(r,z) = \frac{2F_0}{c}\left(V_r \times \sin\theta_D \times \cos\varphi + V_z \times \cos\theta_D\right) \quad (8)$$

where F_0 is the ultrasound transmitted frequency (4 MHz), c is the sound velocity (1540 m/s), r is the radial coordinate, z is the longitudinal coordinate, V is the velocity, θ_D is the Doppler angle, and φ is the angle between the image plan used for the FEM visualization and the vertical axis (Fig. 2). In the present study, we choose the vertical plane ($\varphi = 0$). Only the points whose frequency F was greater than the wall filter cutoff frequency (25 or 100 Hz) contributed to the theoretical reconstruction of the stenosis geometry (theoretical PDA image).

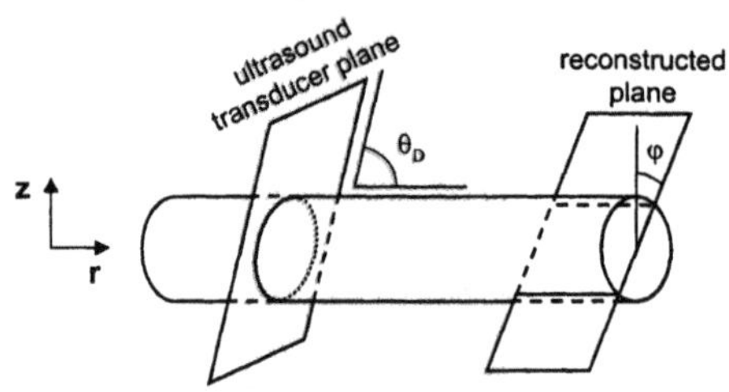

Figure 2. Definition of the angles θ_D and φ used for the theoretical PDA image reconstruction.

RESULTS

Under steady flow, the theoretical PDA images based on the FEM simulations showed that the wall filter induces a loss of signal in the recirculation zone, downstream of the stenosis, or near the wall due to the presence of low velocities. As expected, Figs. 3 and 4 show that the signal loss increased with decreasing flow rate. At low steady flow rates (100 and 250 mL/min), the cross-sectional area of the artery looks smaller than it is in reality. Depending on the situation, this tends to underestimate the severity of the stenosis when considering the percentage of area reduction, and/or overestimate the length of the stenosis. This phenomenon is more emphasized for a wall filter frequency of 100 Hz. Note that the stenosis cannot be detected at 100 mL/min for a wall filter of 100 Hz. Interestingly, similar results were found for the *in vitro* 3D reconstruction of the PDA images. At a steady flow rate of 250 mL/min and a wall filter of 25 Hz, the recirculation zone was visible on both theoretical and experimental PDA images.

Under pulsatile flow, the FEM simulations predicted dynamic changes in the position and severity of the signal loss within the vessel (Fig. 5). The complex fluid dynamics and reverse flow explain these image characteristics (results not shown). During the acceleration of the flow, the wall filter affected the PDA images downstream and upstream of the stenosis. An almost complete disappearance of the image was predicted. At peak systole, the shape of the stenosis was almost entirely recovered because of the presence of high velocities, when projected on the ultrasound transducer axis. During deceleration, reverse flow close to the wall resulted in signal loss by the wall filter. Although the image acquisition program allowed flow gating, the frame rate of the ATL instrument and the absence of flow synchronization in the image formation did not allow us to confirm these simulations experimentally.

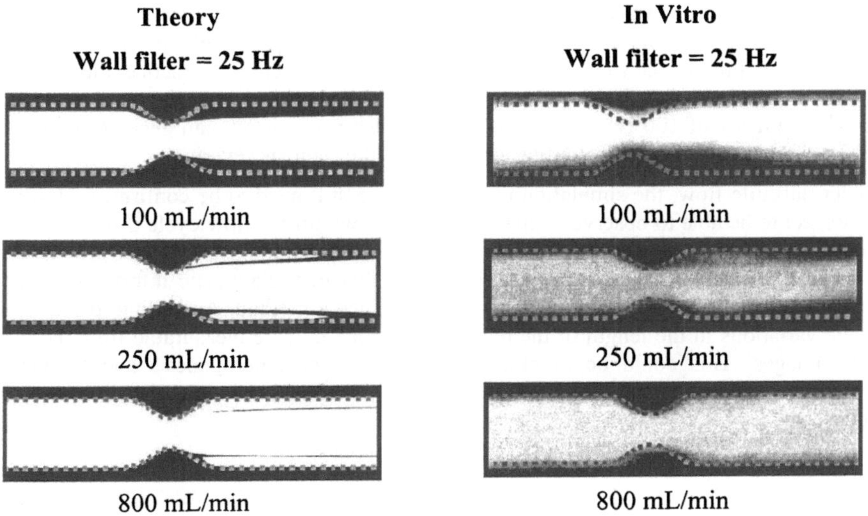

Figure 3. PDA angiographic images simulated by FEM (theory) and experimental results obtained *in vitro* with the ATL instrument. These results correspond to steady flow and the use of a 25 Hz cutoff frequency. The dotted lines represent the real dimension of the stenosis.

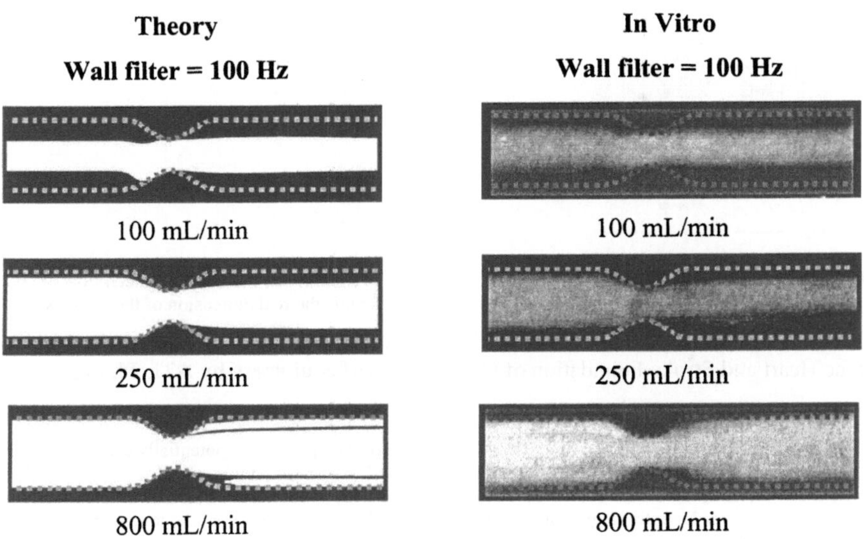

Figure 4. PDA angiographic images simulated by FEM (theory) and experimental results obtained *in vitro* with the ATL instrument. These results correspond to steady flow and the use of a 100 Hz cutoff frequency. The dotted lines represent the real dimension of the stenosis.

DISCUSSION

For steady flow, there was a good agreement between the simulated results and those of the experimental study. This confirms the validity of the hypothesis relating the PDA signal loss in the vicinity of the stenosis to the wall filter effect. In clinical practice, the wall filter is used to attenuate the flash artifacts due to the wall motion within the cardiac cycle. A cutoff frequency of 100 Hz is commonly used. Such a frequency can generate significant PDA signal loss, which may alter the evaluation of the stenosis severity.

Under pulsatile flow, the simulation results of Fig. 5 still need to be confirmed. However, we expect to be able to observe similar PDA images within the flow cycle if one uses ECG gated image formation and no persistence. Modern ultrasound systems can provide such processing of PDA images. It is thus anticipated that new diagnostic information on the stenosis severity may be made available by using such a method. According to Fig. 5, the cyclic variations in the length of the recirculation zone may be measurable from the gated PDA images. This could be of clinical relevance since atherogenesis and thrombosis preferentially occur in regions of blood flow recirculation, oscillating shear stresses, and flow stasis[6].

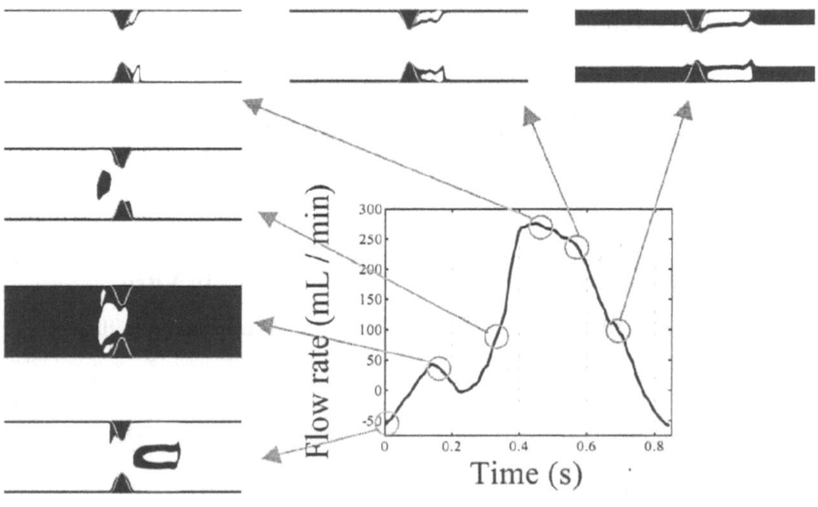

Figure 5. Theoretical PDA images under pulsatile flow. These simulations considered a mean flow of 100 mL/min and a 25 Hz cutoff frequency. The gray line represents the real dimension of the stenosis.

ACKNOWLEDGEMENT: Grants from the Medical Research Council of Canada (MA-14985) and the Heart and Stroke Foundation of Québec; loan of equipment by ATL-Philips.

REFERENCES

1- Rubin JM, Bude RO, Carson PL, Bree RL, Adler RS, Power Doppler US: a potentially useful alternative to mean frequency-based color Doppler US. Radiology, 190 (3): 853-6, 1994.
2- Cloutier G, Qin Z, Garcia D, Soulez G, Oliva V, Durand L.G, Assessment of arterial stenosis in a flow model with power Doppler angiography: Accuracy and observations on blood echogenicity, submitted for publication.
3- Munson BR, Young DF, Okiishi TH., Fundamentals of fluid mechanics, 2nd edition. John Wiley & Sons, 1994.
4- Jaffrin MY, Goubel F., Biomécanique des fluides et des tissus. Masson, 1998.
5- Mc Donald DA, Blood flow in arteries, 2nd edition. Arnold, 1974.
6- Malek AM, Alper SL, Izumo S, Hemodynamic shear stress and its role in atherosclerosis. JAMA, 282: 2035-2042, 1999.

DOPPLER BEAM STEERING FOR BLOOD FLOW VELOCITY VECTOR IMAGING

Shiping He, Khalil F. Dajani, and Jian-yu Lu

Ultrasound Laboratory, Department of Bioengineering
The University of Toledo
Toledo, Ohio 43606, U.S.A.
Email: jilu@eng.utoledo.edu

ABSTRACT

Angle independent blood flow velocity vector measurements could be of a great value for the diagnoses of many vascular and cardiac diseases, such as stenotic carotid and coronary lesions. Although many methods such as speckle tracking and multiple transceivers have been proposed, they have limitations. In this paper, a blood flow velocity vector measurement method called Doppler beam steering is proposed for either 2D or 3D imaging. In the method, a narrow ultrasonic beam is steered at three or more angles in transverse directions. The changes in frequency of the returned echoes provide radial components of blood flow velocity along these directions. Blood flow velocity vector is then calculated from these components and images of the velocity vector distribution can be obtained. Both *in vitro* and *in vivo* experiments were carried out to verify the method. Velocities measured with this method agree well with those measured directly with a flowmeter. Results show that this method is capable of estimating blood flow velocity vectors and thus may provide useful diagnostic information that is difficult to obtain with conventional Doppler imaging method.

INTRODUCTION

Blood flow velocity measurement is very important in diagnosing various cardiac and vascular diseases. Although conventional Doppler blood flow imaging methods are widely used, they have a major limitation: only the velocity components along the lines of sights of ultrasound beams are measured [1,2]. Because of this limitation, conventional Doppler methods are not only unsuitable for quantitative evaluation, but also prone to misinterpretation of complicated blood flow patterns presented in curves, branches, bifurcations, and tortuous vessels where the direction of flow velocity may be helical

Acoustical Imaging, Volume 25, Edited by Halliwell and Wells.
Kluwer Academic/Plenum Publishers, 2000.

in nature [3,4]. In addition, the presence of diseases and its associated turbulence in the arteries cause the blood to move in many different directions, e.g. non-radial jets. These complicated changes in the velocity vectors are not obvious and cannot be detected by conventional Doppler methods including color Doppler imaging.

To overcome these problems, various techniques for measuring two- or three-dimensional (2D or 3D) velocity vectors (true velocity) have been proposed. These techniques fall roughly into three categories: speckle tracking, multiple transceivers, and projection computed velocimetry. In speckle tracking [5-10], the direction and magnitude of the displacement of local blood speckle pattern in consecutive B-mode images are measured. Because the time between image acquisitions is known, velocity vectors can be obtained. This technique overcomes the angle dependence and aliasing limitations of current Doppler velocity measurement methods [11]. Multiple transceiver method uses multiple piezoelectric transducers mounted on the top surface of a lens to measure blood flow velocity vector [12]. Ultrasound beams from the transducers are focused to produce multiple parallel and closely spaced beams. Motion of blood cells is tracked with RF correlation techniques as the cells move along the beams and from beam to beam, and thus 3-D flow velocity vectors can be estimated. Finally, projection computed velocimetry can also be used for blood flow velocity vector measurement [13]. Target velocity components both along and perpendicular to the beam axis are obtained with this method in two steps. The first is to perform the Fourier transform of received signals in time domain (pulse-echo signals). The second is to get projection integration in polar direction. These two steps provide two velocity components along and transverse to the beam axis, respectively.

The methods mentioned above for flow velocity vector measurement suffer from several limitations. The Speckle tracking techniques have problems due to relative motion of ultrasonic scatters within a sample volume in a low frame rate flow measurement system, the pixel-based pattern correlating and matching algorithms employed, and a low signal-to-noise ratio (SNR) of acquired images. Multiple transceiver techniques complicate the scanning system, and may lead to an unjustifiable cost increase. Although the projection computed velocimetry method has a merit of separating the composite velocity vectors with a linear processing, it requires a lot of computation time for either 2D or 3D Fourier transforms for each of the measured points.

In this paper, we propose a blood flow velocity vector measurement method called Doppler beam steering for 2D or 3D imaging. In this method, a narrow ultrasonic beam is produced and steered at either three or four angles. The changes in frequency of the returned echoes provide radial components of the blood flow velocity and are used to calculate velocity vector. Selecting sample volumes over the entire region of interest, 2D or 3D velocity vector images can be constructed.

PRINCIPLE

In conventional pulse and color Doppler methods, velocities at different positions of an image are obtained from the frequency shift with the equation:

$$|\vec{v}| = \frac{f_d c}{2 f_0 \cos\theta}, \tag{1}$$

where $|\vec{v}|$ is the velocity magnitude, c is the speed of sound in tissues, f_d is the Doppler frequency shift, f_0 is the center frequency of the transducer, and θ is the angle between transducer axis and the direction of velocity. Because θ is usually unknown, conventional color Doppler method can not obtain a true velocity (both magnitude and direction) and only the radial component of the velocity can be measured.

To overcome this problem, a Doppler beam steering method is proposed (Fig. 1). In this method, three focused beams (the principle is the same for four or more beams) are steered at steering angles, γ_i, $(i = 1, 2, 3)$, and azimuthal angles, ϕ_i, $(i = 1, 2, 3)$. The center of the aperture of the ultrasonic transducer is located at the origin of the coordinates, (0,0,0). The radial components of the velocity at three points of a volume (a small volume that covers the three points) of blood are measured.

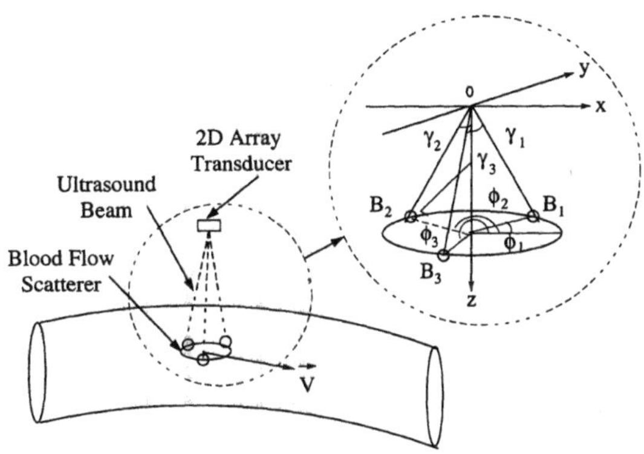

Figure 1. An illustration of Doppler beam steering method.

In Figure 1, it is assumed that the volume of blood have no relative random motions. If $\gamma_i = \gamma$, the three components and the magnitude of the velocity vector of blood volume can be obtained with the matrix [14,15]:

$$\begin{bmatrix} \sin\gamma\cos\phi_1 & \sin\gamma\sin\phi_1 & \cos\gamma \\ \sin\gamma\cos\phi_2 & \sin\gamma\sin\phi_2 & \cos\gamma \\ \sin\gamma\cos\phi_3 & \sin\gamma\sin\phi_3 & \cos\gamma \end{bmatrix} \begin{bmatrix} v_x \\ v_y \\ v_z \end{bmatrix} = \begin{bmatrix} v_{r_1} \\ v_{r_2} \\ v_{r_3} \end{bmatrix}, \tag{2}$$

where v_{r_i} $(i = 1, 2, 3)$ are radial velocities along different beams. Because each velocity component is a simple linear function of the radial components of the velocity, velocity calculations could be performed in real time.

IN VITRO AND IN VIVO EXPERIMENTS AND RESULTS

Both *in vitro* and *in vivo* experiments were conducted at The Jobst Vascular Center at Toledo Hospital, Toledo, OH. *In vitro* experiment was performed with a peripheral vascular Doppler Flow Controller and Pumping System (Model 700, ATS Laboratories, Inc., Bridgeport, CT) as shown in Figure 2. The system consists of a test fluid reservoir, a positive displacement variable-speed pump, a "flow integrator" to control the amount of flow rate, two in-line flowmeters, a pressure gage, and a flow phantom of 13.8 cm wide × 21 cm long × 10 cm deep. The system has the capability to detect flow at varying depths. The phantom is made of rubber-based tissue-mimicking material. Its speed of sound is in the range between 1450 and 1473 m/s, and the frequency-dependent attenuation is about 0.5dB/cm/MHz at room temperature (23° C). Four cylindrical flow channels, drilled through the phantom material, were available for flow rate measurement. The diameters of the channels are 2, 4, 6, and 8 mm, respectively, and their centers are located at 25 mm from the top surface of the phantom. Experiments were conducted for all channels. ATS 707-G Doppler fluid was used in the experiment. The system can provide steady-state flow rates in the range of 20 to 950 ml/minute. In the experiment, the pump speed was set manually to provide flows around 200 ml/minute. Actual flow rate was read from the flowmeters display during data collection. Using these in-line flowmeters, continuous monitoring of the flow rate during the testing was performed to guarantee a stable flow measurements. Standard ultrasound gel was placed over the phantom for acoustic coupling.

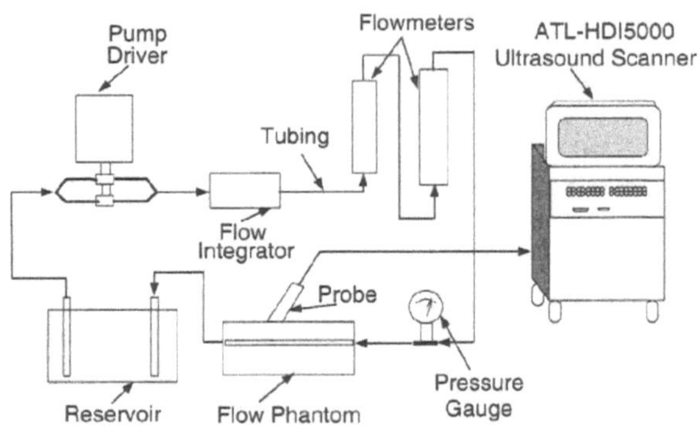

Figure 2. *In vitro* experiment system for flow measurement with both flow meters and the Doppler beam steering methods.

A wideband (4-7 MHz) ultrasound probe (L7-4) was used. It was connected to an ATL (Advanced Technology Laboratories, Bathell, WA) HDI 5000 ultrasound scanner. The probe was manually held to insonify the flow channel. To guarantee that flow sample was taken at the center of the channel without lateral distortion, B-mode images were used to measure the channel diameter to match the actual diameter. The diameter of the beam was about 1.0 mm, and the steering angle, γ, was set to 5° off the vertical axis in four directions.

The result of the *in vitro* experiment with three Doppler beams was obtained from the radial velocities measured within the channels and with Eq. (2). The mean value of the magnitude of the velocity vector of a small volume of the Doppler test fluid was calculated. The result (29.76 mm/s) is compared well with that measured with the flow meter (26.24 mm/s).

In vivo experiment was conducted on the carotid artery of a patient using the same ultrasound scanner as that used for the phantom. B-mode Doppler flow carotid ultrasonography was performed with the patient supine. The head was extended and rotated away from the side to be studied. The anatomic portion of the study was performed first. A cursory survey of the internal and external carotid arteries was done prior to store digital images of the study. This was done in a relative short time period to familiarize us with the patient anatomy and to assess the scanning approach (anterior, lateral or posterial) that the patient anatomy and to assess the scanning approach (anterior, lateral or posterial) that provided optimal access to the carotid artery. Using a low viscosity gel as acoustic coupling agent, the artery was scanned in cross-sectional plane, proximal to the distal internal carotid artery, using the same probe used in the *in vitro* experiment. Using conventional duplex scan, a pulsed Doppler was superimposed on a two dimensional grey-scale image and a single adjustable sample volume was used to obtain quantitative and qualitative information regarding the blood flow in that particular segment. Such spectral information allows us to calculate maximum velocities and use this information to ascertain the measurements. Scanning sequence was recorded from different angles and stored digitally. Results were obtained for both maximum (65.62 mm/s) and minimum (37.63 mm/s) velocities during a cardiac cycle.

DISCUSSION

The results obtained from both *in vitro* and *in vivo* experiments have shown that the Doppler beam steering technique is capable of estimating 3-D blood flow velocity vector with either three or four beams. This technique can obtain the true velocity that current Doppler blood flow system can not.

In the *in vitro* experiment, there is a small difference between velocities obtained with the experiment and flowmeters. This difference is due to errors in the estimation of steering angles. Using an accurate control of steering angle such as electronic steering, the accuracy can be improved. If the errors in Doppler beam steering measurements are random and independent, the standard deviation can be estimated[16]. If the steering angles are reduced to minimize the sampling volume and increase the number of velocity vectors measured within a region of interest, the errors of the three components of blood flow velocity vectors will be higher.

To compare the Doppler beam steering with the multiple transceiver method, the radial velocity components of one point in center of the channel of the phantom was measured with four beams incident from different angles. The results are also compared well with those obtained with the Doppler beam steering method. The advantage of the Doppler beam steering method is that it is easier to implement and thus may simplify imaging systems.

Because the speed of sound is finite in tissues, it takes time to get echoes for each transmit beam. Therefore, image frame rate will be reduced if more beams are used in the Doppler beam steering method to increase velocity accuracy.

CONCLUSION

Conventional Doppler ultrasound methods do not provide complete information about blood flow velocity vectors. In addition, these methods are prone to errors when the beam-to-flow angles is near 90°. To overcome these limitations, a new method for measuring 2D or 3D blood flow velocity vector distribution has been proposed in this paper. The principle of this method for measuring velocity vectors of moving blood flow cubes in three dimensions has been described. Both *in vitro* and *in vivo* experiments have been carried out with the method. Three and more steered beams have been used to measure the velocity vector of a small cube of blood volume. Results show that the new method can be used to obtain flow velocity vector information that are unobtainable with conventional methods. The usefulness of the proposed technique and its potential for practical use in 3D blood flow velocity vector estimation are demonstrated. The method is especially useful for large blood vessels where the flow velocity in a small volume of blood is approximately uniform. Combining standard vascular and cardiac imaging techniques with the proposed method for the analysis of flow velocities in the arteries, the diagnosis and management of diseased lesions related to blood flow can be improved.

ACKNOWLEDGMENTS

This work was supported in part by grant HL60301 from the National Institutes of Health.

REFERENCES

1. P. Atkinson, J. P. WoodcockP. *Doppler Ultrasound and Its Use in Clinical Measurements.* Academic Press, London (1982).
2. W. D. Barber, J. W. Eberhard, and S. G. Karr, A new time domain technique for velocity measures using Doppler ultrasound, *IEEE Trans. Biomed. Eng.* 32:213-229 (1985).
3. D. N. Ku, D. P. Giddens, D. J. Phillips, Hemodynamics of the normal human bifurcation: *In vitro* and *in vivo* studies, *Ultrasound Med. Biol.* 11:13–16 (1985).
4. D. J. Phillips, F. M. Greene, Y. Langlois, Flow velocity patterns in the normal carotid bifurcation of young presumed normal subjects, *Ultrasound Med. Biol.* 9:39–49 (1983).
5. L. N. Bohs, B. H. Friemel, B. A. McDermott, and G. E. Trahey, A real time system for quantifying and displaying two-dimensional velocities using ultrasound, *Ultrasound Med. Biol.*, 19:751–761 (1993).
6. L. N. Bohs, B. J. Geiman, K. R. Nightingale, C. D. Choi, B. H. Friemel, and G. E. Trahey, Ensemble tracking: a new method for 2D vector velocity measurement, *Ultrasonics Symposium Proceedings*, 2:1485-1488 (1995).
7. L. N. Bohs, B. H. Friemel, B. A. McDermott, and G. E. Trahey, Lateral Velocity Profile and volume Flow measurements via 2-D speckle tracking, *Acoustical Imaging.* 21:503-507 (1995).
8. L. N. Bohs, B. J. Geiman, M. E. Anderson, S. M. Breit, and G. E. Trahey, Ensemble tracking for 2D vector velocity measurement: Experimental and initial clinical results, *IEEE Trans. Ultrason. Ferroelec. Freq. Contr.* 45:3-10 (1998).
9. B. H. Friemel, L. N. Bohs, G. E. Trahey, Relative performance of two-dimensional speckle-tracking techniques: normalized correlation, non-normalized correlation and sum-absolute-difference, *Ultrasonics Symposium Proceedings.* 2:1481-1484 (1995).
10. F. Yeung, S. F. Levinsonn, K. J. Parker, Multilevel and motion model-based ultrasonic speckle tracking algorithms, *Ultrasound Med. Biol.* 24:427-441 (1998).
11. L. N. Bohs, B. H. Friemel, and G. E. Trahey, Experiment velocity profiles and volumetric flow via two-dimensional speckle tracking, *Ultrasound Med. Biol.* 21:885-898 (1995).
12. I. A. Hein, Triple-Beam lens Transducers for three-dimensionmal ultrasonic fluid flow estimation, *IEEE Trans. Ultrason., Ferroelect., Freq. Contr.* 42(5):854-869 (1995).
13. K. Katakura and M. Okujima, Ultrasonic vector velocity measurement by projection computed velocimettry, *IEEE Trans. Ultrason., Ferroelect., Freq. Contr.* 42(5):889-895 (1995).
14. T. Shiina, *Proc. 61th Meet. JSUM. Kobe, 1992, JPN. J. Med. Ultrason.* 61:43 (1992).
15. B. H. Brigga and R. A. Vincent, Spaced-antenna analysis in the frequency domain, Radio Science, *Radio Science.* 27(2)117-129 (1992).
16. S. G. Foster, P. M. Embree and W. D. O'brien, Flow velocity profile via time-domain correlation error analysis and computer simulation, *IEEE Trans. Ultrason., Ferroelect., Freq. Contr.* 37(2):164-175 (1990).

DYNAMIC EFFECTS OF ACOUSTIC RADIATION FORCE ON MICROBUBBLES

Piero Tortoli[1], Vittorio Michelassi[2], Francesco Guidi[1], Massimo Corsi[1]
Masatsugu Hori[3], Ken Ishihara[4], Fuminori Moriyasu[5], Yasuhito Takeuchi[6]

[1] Electronics and Telecommunications Department, University of Florence
[2] Dept. of Mechanical and Industrial Engineering, University of Roma Tre, Italy
[3] Osaka University School of Medicine [4] Ehime University School of Medicine,
[5] Kyoto University School of Medicine [6] Kagoshima University, Department of Information and Computer Science, Japan

INTRODUCTION

The movement of compressible, encapsulated microbubbles is known to be influenced by the application of an ultrasound field. Dayton et al.[1] have demonstrated through optical methods that Albunex® (Molecular Biosystems, San Diego, USA) bubbles are appreciably displaced toward the far wall of a tube by ultrasound radiation force. Doppler spectra shown by Tortoli et al.,[2] originated by Levovist® (Schering AG, Berlin, Germany) bubbles suspended in distilled water, exhibited considerable changes in bandwidth and shape when pressure amplitudes of a few hundreds kPa were used. Also this phenomenon was attributed to radiation force pushing the microbubbles away from the transducer.

The bubble motion was recently described as the result of the combined action of radiation pressure and drag force of the fluid where the bubble is suspended[3]. Because of their small inertia the bubbles react very rapidly to the ultrasound field and follow a flow path that may considerably deviate from the axial flow. The resulting relative velocity between bubbles and fluid produces a drag force. When the drag force equals the ultrasound force the bubble trajectory is stabilized. The validity of the proposed model was demonstrated by following an indirect approach, where experimental PW Doppler spectra are compared with the simulated spectra.

In this paper, we discuss at which extent the phenomenon is affected by a) the transmitted ultrasound beam, b) the physical properties of the bubbles, c) the characteristics of the fluid. The results obtained with three different contrast agents are presented and compared to each other. We also show that, for a given bubble, different displacements are obtained in CW or PW conditions, even though the transmitted intensity is the same in the two modes.

Acoustical Imaging, Volume 25, Edited by Halliwell and Wells.
Kluwer Academic/Plenum Publishers, 2000.

FORCES ACTING ON MOVING BUBBLES

The following sections describe the two forces that simultaneously drive a microbubble suspended in a moving fluid insonified by an ultrasound beam.

Radiation Force

The instantaneous primary force produced by an ultrasound field acting on a compressible spherical bubble can be evaluated following Dayton et al.[1]. The expression of the ultrasound force, F_{us}, results:

$$F_{us}(\omega)=\frac{\pi P_a^2 D}{\rho_o c\omega}\cdot\frac{\delta_{tot}\omega_o/\omega}{\left[\left(\frac{\omega_o}{\omega}\right)^2-1\right]^2+\left(\frac{\delta_{tot}\omega_o}{\omega}\right)^2} \tag{1}$$

where:

- P_a (local ultrasound pressure) and ω (transmitted frequency) depend on the transmitted ultrasound beam;
- D (diameter), ω_o (pulsing eigenfrequency) and δ_{tot} (damping coefficient) depend on the physical properties of the microbubble;
- c (ultrasound propagation speed), ρ_o (fluid density) and the same δ_{tot} depend on the fluid where the bubble is suspended.

Drag Force

The drag force is generated when a relative velocity exists between the bubble and the viscous fluid. In absence of radiation force, the bubbles move at exactly the same velocity of the fluid due to their vanishingly small mass. In presence of radiation force, the bubble is accelerated and diverted from the fluid flow path and a drag force develops. The drag force is proportional to the relative bubble-fluid velocity and increases until it equals the ultrasound force: at this stage the acceleration of the bubble tends to zero. Under the assumption of Newtonian fluid and spherical shape of the bubble, the size of which oscillates around a mean diameter D, the drag force, F_D, can be easily computed as follows[4]:

$$F_D = C_D \,\mathrm{Re}\cdot\frac{\pi D \upsilon \rho_o}{8}\cdot\left|V_f - V_b\right| \tag{2}$$

where V_f and V_b are the fluid and bubble velocity, respectively, and υ is the fluid kinematic viscosity. Re and C_D (Reynolds number and the drag coefficient respectively) are to be computed as:

$$\mathrm{Re}=\frac{\left|V_f - V_b\right|}{\upsilon}\cdot D \qquad\qquad C_D=\frac{24}{\mathrm{Re}}+\frac{6}{1+\sqrt{\mathrm{Re}}}+0.4 \tag{3}$$

The definition of C_D is strictly valid only for spherical bubbles and is accurate in the Reynolds number range [0-10^5] under the hypothesis of steady flow conditions.

Balance of forces

The trajectory of the bubble can be traced by solving the vector motion equation:

$$\vec{F}_{us}(t) - \vec{F}_D(t) = m \cdot \frac{d\ \vec{V}_b(t)}{dt} \tag{4}$$

where m is the bubble mass, which, according to Leighton[5], for low Reynolds numbers can be estimated equivalent to one half the mass of the displaced surrounding fluid. Equation (4) is decomposed into its radial and axial components, thereby producing two scalar equations that are marched in time by a simple Euler one-step method[6]. Due to their very small mass, the bubbles react instantly to the ultrasound field and can consequently deviate from the fluid core flow path[3]. An accurate knowledge of the parameters affecting both the ultrasound and drag forces of eq. (4) is crucial for the correct description of the bubble trajectory.

FACTORS AFFECTING THE BUBBLE DYNAMICS

Influence of the ultrasound beam characteristics

According to eq. (1) the radiation force is proportional to the squared value of the pressure, P_a, experienced by the bubble. Since the sound field spatial distribution is in general not uniform, moving bubbles are exposed to different pressure levels, depending on their instantaneous position. For example, a bubble crossing the axis of a beam will subsequently experience increasing pressures while approaching the beam axis, and decreasing pressures while receding from the beam axis. The extent to which the bubble is affected by radiation force will also depend on the beamwidth: large beams will deviate the bubble from its original path more consistently than narrow beams.

Since eq.(1) reports the radiation force that is instantaneously applied to a bubble, the transmission modality (Continuos Wave - CW, or Pulsed Wave - PW) plays an important role. In CW mode, the two forces are continuously applied to the bubble. In PW mode the bubble is subject to radiation force only when the ultrasound pulse travels over it. In the remaining part of the repetition interval, it is only subject to the drag force. Hence, even though the ultrasound intensity is the same in the two cases, different bubble trajectories have to be expected due to the non-linearity of the acting forces.

Influence of the bubble physical properties

As stated above, the radiation force acting on a bubble depends on its diameter, resonance frequency, and damping coefficient.

For a given contrast agent, the size distribution of the absolute volume fraction of gas contained in stabilized microbubbles is usually non-flat around a value that is typically presented as the bubble median diameter, D_0. For each transmitted frequency, only those bubbles having a specific diameter actually resonate.

The relation between a bubble diameter and the corresponding resonant frequency, f_r, depends on the bubble mechanical properties. For free air bubbles, the resonant frequency can be computed by the following equation[7]:

$$f_r^2 = \frac{1}{(\pi D)^2}\left[\frac{3}{\rho_o}\left(p_o + \frac{8\sigma}{3D}\right)\right] \tag{5}$$

where the fluid pressure, p_O, is assumed 100 kPa, and the surface tension, σ, is 72×10^{-3} N/m for bubbles in water. For encapsulated bubbles one should consider[8]:

$$f_r^2 = \frac{1}{(\pi D)^2}\left(\frac{3p_o}{\rho_o}\right) + \frac{2S_p}{m_e\pi} \tag{6}$$

where p_O is the fluid pressure, m_e is the effective mass of the bubble and S_p is the so-called shell parameter (including the shell elasticity and thickness). Equation (6) indicates that the stiffer is the bubble (i.e., the higher is its shell parameter), the higher is its resonance frequency. Conversely, for a given transmission frequency, stiffer contrast agents have larger resonant diameters. The presence of bubbles with different diameters implies a not uniquely defined resonant frequency for each contrast agent: for each possible frequency transmitted over a relatively wide range, there may be a corresponding group of bubbles whose diameter leads them to resonance. According to the second part of eq.1, these resonating bubbles experience the maximum radiation force, and are therefore those of major interest.

The damping coefficient, δ_{tot}, including every radiation energy loss, depends on the bubble diameter, too. However, this parameter is less critical than the others, and an average value is usually acceptable. The damping coefficient becomes important especially when two contrast agents have approximately the same resonant diameter. In this case, the larger δ_{tot}, the lower the ultrasound force.

Table1 shows the typical values of D_O, S_p and δ_{tot} related to Levovist®, Albunex® and to a contrast agent mimic, F-4E® (Matsumoto Yushi-Seiyaku Co. Ltd., Osaka Japan). The latter consists of thermoplastic shell hollow microspheres containing low-molecule-number hydrocarbon gas (C_3H_8), and will be referred here as "toy" bubbles[9]. In Table 1, S_p was evaluated through a method based on attenuation measurements[8], while the coefficient δ_{tot} is related to bubbles suspended in distilled water, resonating at 4 MHz. For Levovist, the shell parameter is not reported because we found that a model based on eq.(5) (free bubble) is most suitable in our case.

Table 1. Main parameters influencing the radiation force for 3 contrast agents

	Levovist	Albunex	Toy Bubbles
Median diameter (D_o - μm)	3	4	4
Shell Parameter (S_p - kg/s)	-	8	12
Damping coeff (δ_{tot})	0.2	0.65	0.45

Fig.1 shows the radiation force evaluated for the three contrast agents as a function of the bubble diameter, when the transmission frequency is set at 4 MHz. The behavior is clearly resonant, with different resonant diameters in the three cases. In no case these are coincident with the median diameter. It may be observed that Albunex® and toy bubbles have similar resonant diameters, but radiation force is larger for toy bubbles because damping is lower.

Influence of the fluid characteristics

As suggested by eq. (4), equilibrium is achieved when the drag force, F_D, becomes equal to the primary ultrasound force, F_{us}. As shown in eq. (2), F_D depends on the relative velocity, $|V_f - V_b|$, between the fluid and the bubble, as well as on the fluid viscosity. For a given F_{us}, the F_D module that equals F_{us} requires smaller relative velocities for higher values of fluid viscosity. As an example, it has to be expected that the relative velocities reached by the bubbles due to radiation force are lower in blood than in water, since blood kinematic viscosity is 4 times higher.

Moreover, the fluid viscosity also influences the damping coefficient. For example, for free bubbles δ_{tot} can be considered almost proportional to ν.

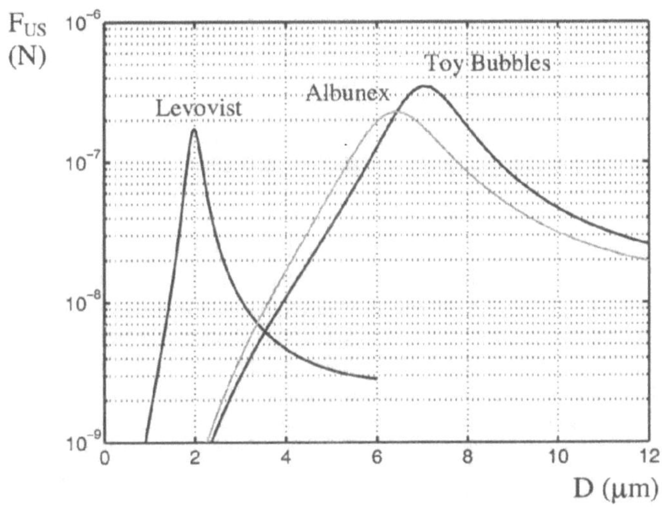

Figure 1. Radiation forces acting on bubbles of different diameters for 3 contrast agents suspended in distilled water (f = 4MHz, P_a = 500 kPa)

RESULTS

The combined effects of the two forces described above were simulated by considering both PW and CW transmission modalities, in the hypothesis that the bubbles initially move at constant velocity along the axis of a laminar flow. In PW mode, the radiation force is applied to the bubble only during the passage of the travelling ultrasound burst. Fig.2 shows the path followed by the resonating bubbles of 3 contrast agents when insonified by a single 5.2 μs-long pulse directed at 45° to the fluid stream. It may be observed that, despite being subject to a lower radiation force, the Levovist bubble is radially displaced more than the others. This is consistent with the fact that the Levovist bubble resonating diameter is the lowest one, as shown in Fig.1. Hence, the other bubbles, which have larger diameters, are subject to higher drag forces that finally limit their radial displacement. In particular, the toy bubble is displaced more than the Albunex® bubble because of its minor damping.

When bubbles cross the acoustic field produced by a transducer, the entire pressure profile must be considered. Figure 3 (right) shows the total displacements imposed to a resonating Levovist bubble by 4 MHz PW and CW ultrasound fields having the pressure profile

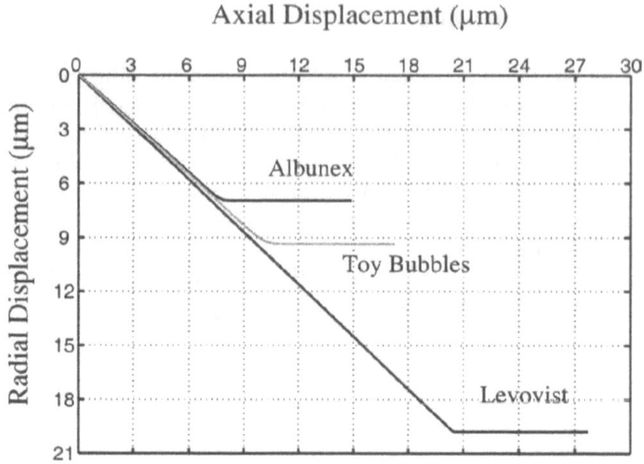

Figure 2: Simulated displacements of resonating Levovist, Albunex and Toy bubbles in response to a single burst of ultrasound (f = 4 MHz, P_a = 500 kPa)

shown on the left. It may be observed that in both cases the original trajectory is appreciably displaced away from the transducer in the region closer to the beam axis, where the highest pressure levels are experienced. Although the intensity is the same (≈ 280 mW/cm^2), the displacement is more evident in CW mode than in PW mode.

In PW mode, the estimated bubble trajectory can be used to evaluate the inter-pulse phase shifts that have to be expected between subsequent echoes backscattered from the moving bubble[10]. Differently from the case of scatterers moving at constant velocity (yielding constant phase shifts), here the echo-Doppler signal is typically modulated in both amplitude and phase. Its spectrum is therefore correspondingly changed, turning out to be typically asymmetrically enlarged[3]. Since a same contrast agent contains bubbles of different sizes, we have considered the contributions produced by a finite number of bubbles with diameters distributed over a range including the median diameter. For each bubble, the displacement due to radiation and drag force was evaluated. The individual echoes were then computed and scaled according to a factor that takes into account the size-dependent scattering cross-section[8]. Finally, all the echoes are summed together, to produce the simulated spectra.

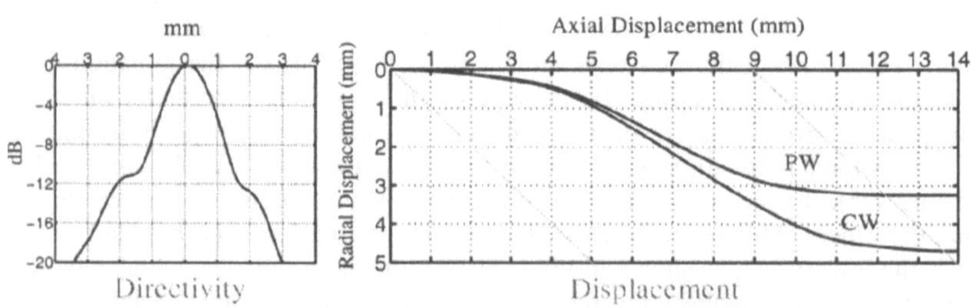

Figure 3: Left: pressure profile experienced by a resonant Levovist bubble suspended in a fluid moving at constant velocity (V_f = 12 cm/s); Right: Corresponding simulated trajectories obtained when CW and PW transmission modes are used. The beam-to-flow angle is 45° (i.e, it is contained within the grey oblique lines) and the ultrasound intensity is the same in the two cases

An indirect demonstration of the validity of this model was achieved by comparing the Doppler spectra estimated from simulations, to those obtained from tests made with the experimental set-up described by Tortoli et al[2]. Fig.4 shows the simulated and experimental spectra obtained for Toy Bubbles suspended in water (left) and in a mixture of 60% water and 40% glycerol (right) at different pressure peak levels. It has to be remarked that in both cases there is an excellent agreement between simulations and experiments.

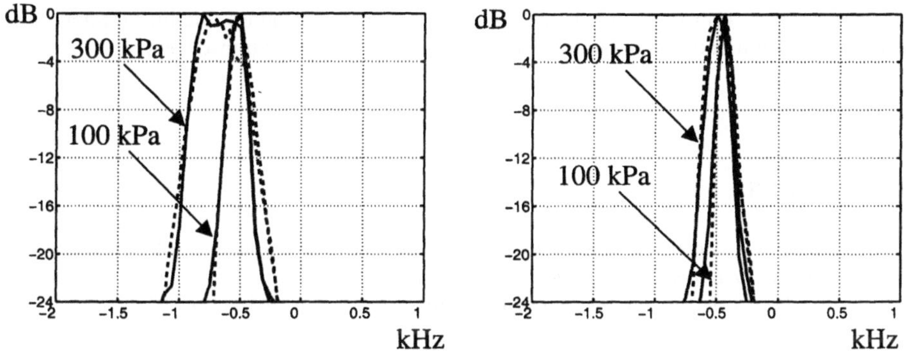

Figure 4: Simulated (continuous line) and experimental (dotted line) spectra obtained for Toy Bubbles suspended in water (left) and in a mixture of 60% water and 40% glycerol (right) (f = 4 MHz). The central streamlines of a laminar flow in a 10-mm-diameter tube were interrogated at 45°.

DISCUSSION AND CONCLUSION

This paper has discussed the leading factors that influence the trajectory of microbubbles crossing an ultrasound field. These may be summarized in transmitted intensity, transmission modality, bubble elasticity and fluid viscosity.

Dependence on transmission intensity, I, is relatively trivial: the higher is I, the larger are the displacements. However, the comparison between displacements obtained in CW and PW modes demonstrated that the phenomenon is not linearly dependent on I, because in PW mode two distinct mechanisms are yielded, during the passage of the ultrasound pulse and during the remaining part of each pulse repetition interval, respectively.

For a given contrast agent, a different resonant diameter, D_{res}, corresponds to each possible transmission frequency. In general, the larger D_{res}, the larger the corresponding radiation and drag forces. Hence, similar displacements have to be expected[3].

When different contrast agents are insonified at the same frequency, the resonant diameter depends on the bubble elasticity. The stiffer the bubble, the larger its D_{res}. Hence, the drag force is larger, too, and, unless damping reverts the situation (as for Albunex® compared to toy bubbles - see Fig.2), the bubble radial displacement is generally lower.

Finally, the fluid viscosity may drastically limit the bubble displacement. This is evident in Fig.4, which shows that the enlargement of the Doppler bandwidth is lower in a fluid having the same viscosity as blood. This fact, together with the intensity limitations imposed by FDA, may explain why the phenomenon has not been routinely reported by clinical tests.

In conclusion, there are different possible reasons yielding spectral modifications of the echoes produced by contrast agents. They are (1) resonant (harmonic) and non-resonant (chaotic) nonlinear microbubble echogenicity due to size changes, (2) echogenicity change (scintillation) appearence, growth, shrinkage and disappearence of gas bubble on and after the shell breakage, and (3) displacement of gas bubble due to transmitted ultrasound field, as discussed in this paper. The former ones typically produce repetition or symmetrical broadening of the fundamental Doppler spectrum, while the latter seems to be the only one giving rise to shifts like those visible in Fig.4.

Acknowledgements

The authors wish to acknowledge valuable help by Alessandro Gubbini (Esaote spa, Firenze) and Emiliano Maione in the calibration of the ultrasound transducers.

REFERENCES

1. P.A. Dayton, K.E. Morgan, A.L. Klibanov, G. Brandenburger, K.R. Nightingale and K.W. Ferrara, A preliminary evaluation of the effects of primary and secondary radiation forces on acoustic contrast agents, *IEEE Trans. on Ultrason., Ferroelect., Freq. Contr.*, 44:6, p.1264-1277 (1997).

2. P.Tortoli, F.Guidi, E.Maione, F.Curradi, V.Michelassi, Radiation force Doppler effects on contrast agents, in: *Acoustical Imaging* vol.24, H.Lee Ed., Plenum Publishing Company, New York (1999).

3. P.Tortoli, M.Pratesi and V.Michelassi, Doppler spectra from contrast agents crossing an ultrasound field, *IEEE Trans. on Ultrason., Ferroelect., Freq. Contr.*, 47:3 (2000).

4. C. Benocci, J.-M.Buchlin, V.Michelassi, P.Weinacht, Numerical modelling of gas-droplet flows for industrial applications, *Third International Conference on Computational Methods and Experimental Measurements*, Porto Carras, Greece (1986).

5. T.G. Leighton, *The Acoustic Bubble*, Academic Press, New York (1994).

6. C. Hirsch, *Numerical Computation of Internal and External Flows*, John Wiley & Sons, New York (1988).

7. B. Ran, J. Katz, The response of microscopic bubbles to sudden changes in ambient pressure, *J. Fluid Mech 224*, pp.91-115 (1991).

8. N.de Jong, *Acoustic properties of ultrasound contrast agents*, Ph.D. dissertation, de Erasmus Universiteit Rotterdam (1993).

9 Y.Takeuchi, Industrial use thermoplastic microballon to mimic the contrast agent and its in-vitro behavor including released gas dynamics, *IEEE Ultrasonics Symposium Proceed.*, pp.1579-1582 (1997).

10 J.A. Jensen, An analysis of pulsed wave ultrasound systems for blood velocity estimation, in: *Acoustical Imaging* vol.22, P.Tortoli & L.Masotti Eds., Plenum Publishing Co., New York, pp.377-384 (1996).

IMAGING OF TISSUE ELASTIC MODULUS DISTRIBUTION BASED ON ESTIMATED 3-D DISPLACEMENT VECTOR

Tsuyoshi Shiina, Naotaka Nitta

Institute of Information Sciences and Electronics,
University of Tsukuba
1-1-1 Tennodai Tsukuba, 305-8573, Japan
TEL:+81-298-53-5504 / FAX:+81-298-53-5206
E-mail: shiina@is.tsukuba.ac.jp

INTRODUCTION

Imaging of tissue elastic properties is expected to be a new modality by which to assess tissue diseases such as cancer. The first approach to the evaluation of tissue elastic properties is strain imaging, which estimates elasticity directly by ultrasonic echo. However, to obtain more quantitative information, the elastic modulus image must be reconstructed by using the estimated strain data. Some reconstruction techniques based on 1-D strain data require the imposition of many boundary conditions. In order to reconstruct under less restricted conditions, it is necessary to fully evaluate the mechanical equilibrium equations using all components of the displacement vector induced by tissue deformation.

In this work, we employ a method, which we previously proposed for blood flow measurement,[1,2] as a 3-D displacement vector estimator. The method can attain real-time processing by the use of a 2-D array transducer and the simple linear operation, though the upper boundary of measurable displacement is restricted due to aliasing. Therefore, to make the proposed method applicable to the measurement of strain for large displacement (i.e., beyond wavelength), a phase unwrapping procedure is incorporated into the above method as the phase shift detector.[3]

In addition, we developed a method of reconstructing the elastic modulus distribution based on the equilibrium equations. The feasibility of the proposed methods were investigated by numerical simulation using the ultrasonic echo signals based on the 3-D finite element tissue model. Herein, we describe the theoretical foundation of our methods and present the results of the numerical simulation.

METHODS

Measurement of 3-D displacement vector

It is assumed that the successive pulses under pre- and post-deformation are transmitted from the circular 2-D array aperture as illustrated in Fig. 1. The phase shift

Acoustical Imaging, Volume 25, Edited by Halliwell and Wells.
Kluwer Academic/Plenum Publishers, 2000.

$\phi(x,z)$ between successive scattering waves received by each element is converted to a function $p(x,z)$ as eq.(1).

$$p(x,z) = (R+L)u_y - \frac{\lambda}{2\pi}R\phi(x,z)$$
$$= u_x \cdot x + u_z \cdot z \qquad (1)$$

where, u_y is the axial displacement along the beam direction and can be obtained by the phase shift near the center of the 2-D array aperture as follows,

$$u_y = \frac{\lambda}{4\pi}\phi(0,0) \qquad (2)$$

Equation (1) indicates that $p(x,z)$ constitutes a plane with respect to two variables, that is, the lateral and elevational coordinates. The gradient of the function on the receiving aperture corresponds to the lateral (u_x) and elevational (u_z) displacement components.[1,2]

The above method of processing implies that three components of the displacement vector are obtained by simple linear operation; that is, this technique is capable of real-time processing.

In order to obtain the high S/N strain image and to reconstruct the elastic modulus image accurately, it is necessary that a large deformation should be added to the tissue surface. Under the condition of a large deformation, the phase shift at each element becomes to be affected by aliasing, and therefore cannot be measured by the conventional Doppler technique. To redress this difficulty, we incorporated a phase unwrapping procedure into the above displacement estimator.[3]

Methods of elastic modulus reconstruction

Assume that tissues comprise an isotropic, compressible, linear elastic body under static compression, and that Poisson's ratio is constant throughout the body. It follows that the elastic properties of the body are described by Young's modulus. Based on the measured 3-D displacement vector, all components ε_{ij} of strain tensor can be obtained as follows,

$$\varepsilon_{ij} = \frac{1}{2}(u_{i,j} + u_{j,i}) \quad (i,j=1,2,3) \qquad (3)$$

where u_1, u_2 and u_3 are the lateral, axial and elevational displacements. By substituting the relationship between the stress tensor and strain tensor into the equilibrium equations, simultaneous equations are obtained as Eq.(4).

$$(l\varepsilon_{kk}\delta_{ij} + \varepsilon_{ij})\frac{E_{,j}}{E} = -(l\varepsilon_{kk,i} + \varepsilon_{ij,j}) \qquad (4)$$

Equation (4) include the unknown parameters about derivatives of Young's modulus ($E_{,1}/E$, $E_{,2}/E$, $E_{,3}/E$).

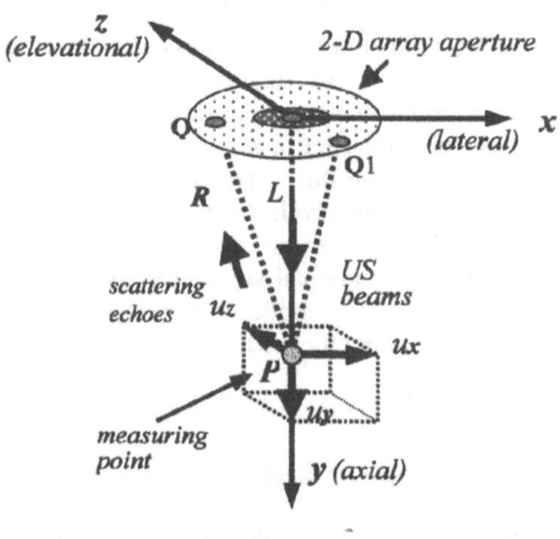

Figure 1. Measurement coordinate system on the 2-D array aperture

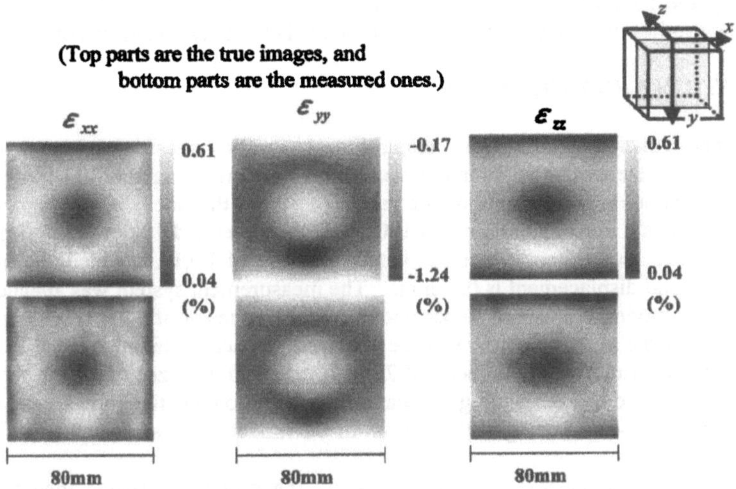

Figure 2. Estimated normal component images of strain tensor

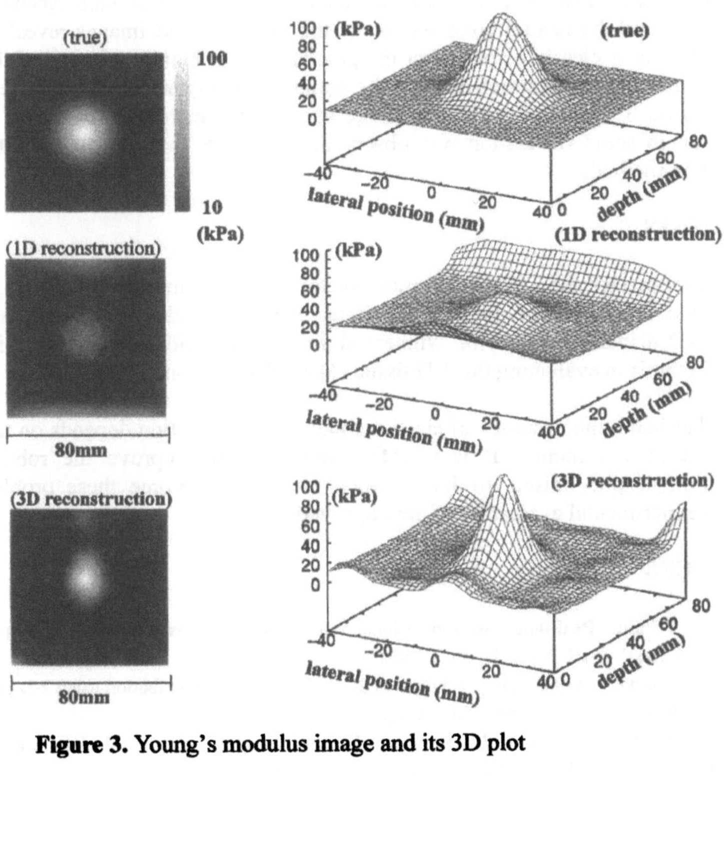

Figure 3. Young's modulus image and its 3D plot

After solving the equations, the elastic modulus distribution can be reconstructed by integrating the solutions from the tissue surface. The above method of reconstruction processing implies that the surface Young's modulus is the only boundary condition.

SIMULATION AND DISCUSSION

The performance of the 3-D displacement vector measurement and elasticity reconstruction were evaluated by numerical simulation involving the 3-D finite element model. This model is a cube with a side length of 80mm that includes the hard spherical inclusion and has a Young's modulus of 100kPa at its center and 10kPa in the surrounding area. Under these conditions, when the surface pressure of 100Pa is added, the maximum value of the axial displacement is 0.65mm. The measurement system was attached on the bottom of this model. Ultrasonic pulses with a center frequency of 3.75MHz were radiated vertically, then the scattering waves were received at each element that had a pitch of 6.5mm on the aperture with a diameter of 20 to 70mm, and digitized at the rate of 10MHz. Rf signals, produced by convoluting the scatterer distribution and the point spread function, were acquired by 2-D scanning at the bottom of the model, then white noise was added to these signals.

The displacement vectors were measured by the proposed method then the elastic modulus (Young's modulus) distribution was reconstructed using strain tensor obtained by the measured displacement vectors. The error on the horizontal displacement measurement was evaluated with respect to the receiving aperture diameter and SNR of rf signals. These results show that all errors are below 5% at noise levels over 10dB. Figure 2 shows the normal components of strain tensor estimated using the measured 3-D displacement vector distribution around the central cross-section of the model. These images reveal that each strain distribution is clearly displayed at the practical level. Finally a image of the elastic modulus distribution reconstructed using the estimated components of strain tensor was shown in figure 3 compared with true image and 1D reconstruction based on the axial strain. Although some fluctuation was observed, the results indicate the validity of this reconstruction method.

CONCLUSION

Herein, we present the 3-D displacement vector measurement method utilizing the 2-D array and the reconstruction method of tissue elastic modulus distribution using the measured 3-D displacement vector. Numerical simulation validated the feasibility of the proposed methods in evaluating the 3-D tissue elastic distribution.

On the other hand, the accuracy of elastic modulus reconstruction depends on that of the displacement measurement. It is therefore important to improve the robustness of reconstruction despite noise. In future work, we will overcome these problems, and conduct an experimental evaluation of the improved method.

REFERENCES

1. N.Nitta and T.Shiina, Real-time three-dimensional velocity vector measurement using weighted phase gradient method, Jpn. J. Appl. Phys., 37, 3058 (1998).

2. T.Shiina and N.Nitta, A new imaging technique of 3-D velocity vector distribution using 2-D phased array probe, *proc. of IEEE Ultrason.* sympo., 1509(1998).

3. T.Shiina, M.M.Doyley and J.C.Bamber, Strain imaging combined RF and envelope autocorrelation, proc. Of IEEE Ultrason. Sympo., 1331(1996).

ELASTICITY IMAGING OF INCOMPRESSIBLE BIOLOGICAL SOFT TISSUE FROM INCOMPLETE ULTRASONIC DISPLACEMENT DATA

Yasuo Yamashita[1], Hitoshi Kaneko[1], and Mitsuhiro Kubota[2]

[1]Department of Industrial Engineering and Management,
Nihon University College of Industrial Technology,
Narashino 275-8575, Japan
[2]Department of Surgery, Tokai University School of Medicine,
Isehara 259-0011, Japan

ABSTRACT

In this paper, we present a formulation and numerical simulation for the determination of the elasticity parameter of an isotropic, inhomogeneous, incompressible elastic medium subject to external mechanical forces. Given a knowledge of the displacements within the medium, a variational formulation is used to determine the shear modulus throughout the tissue. A finite-element based model for static deformation is proposed for solving the distribution of the relative shear modulus of the tissue. The feasibility of the proposed method is demonstrated using the simulated deformation data of the simple three-dimensional inclusion problem. The results show that the relative shear modulus may be reconstructed from the displacement data measured locally in the region of interest within an isotropic, incompressible medium, and that the relative shear modulus can be recovered to some degree of accuracy from only one-dimensional displacement data.

INTRODUCTION

Palpation has been the clinical diagnostic modality to detect the changes in soft tissue elasticity, because the elasticity is usually related to some abnormal, pathological process. Internal tissue deformation induced by externally applied mechanical sources or by the primary cardiac pulsation has been evaluated to characterize tissue elasticity. Under the assumption that the tissue is elastic, isotropic, and subject to a constant uniform stress field, the strain field could be interpreted as a relative measure of elasticity distribution. Although the tissue elasticity is ultimately correlated with strain field of internal deformation, deformational geometry as well as the pattern of external force/displacement sources can greatly affect the strain distribution as well. Consequently, for a quatitative elasticity imaging, material elasticity parameter must be reconstructed from estimates of internal displacement and strain (Emelianov et al., 1995).

In this paper, we present a formulation and numerical demonstration of the determination of the shear modulus of an isotropic, inhomogeneous, incompressible elastic soft tissue, subject to external static force. Given a knowledge of the displacement within the medium, a new implementaion of a variational formulation is used to determine the shear modulus throughout the medium.

In the theory section, the general theoretical approach to determine the deformation inside tissue based on the common model of an elastic, isotropic, incompressible medium is presented. A variational formulation for the forward problem is presented in the next section. Practical method to estimate the elastic modulus using displacement data is formulated by inversely solving the forward problem. The feasibility of the proposed method is validated using the simulated deformation data of a simple 3-D inclusion problem. The result indicates that the method may be applied locally within an inhomogeneous, isotropic, incompressible medium. Finally, the paper concludes with a discussion of the results.

THEORY

Governing Equations for Static Deformation

Consider a three-dimensional (3-D) elastic medium occupying a volume Ω bounded by a surface Γ (cf. Fig. 1(a)). When the soft tissue is deformed very slowly under the external mechanical source outside the region of interest, the governing equation is simply the static equilibrium equation which may be expressed in indicial notation as:

$$\sigma_{ij,j} = 0, \qquad i = 1,2,3, \tag{1}$$

where σ_{ij} is the cartesian component of the stress tensor, $\sigma_{ij,j}$ denotes the partial derivative of σ_{ij} with respect to the cartesian coordinate x_j, and summation is assumed over repeated indices.

Assuming linear elasticity, the constitutive equation in an isotropic medium is

$$\sigma_{ij} = \lambda\epsilon_{kk}\delta_{ij} + 2\mu\epsilon_{ij}, \qquad i,j = 1,2,3, \tag{2}$$

where δ_{ij} is the Kronecker delta and ε_{ij} is the cartesian components of the strain tensor. Denoting the elastic displacement field within the medium by $\boldsymbol{u}^t$=(u_1, u_2, u_3), this strain tensor is described as:

$$\epsilon_{ij} = \frac{1}{2}(u_{i,j} + u_{j,i}), \qquad i,j = 1,2,3. \tag{3}$$

In Eq. (2) the parameters $\lambda = E\nu/(1+\nu)(1-2\nu)$ and $\mu = E/2(1+\nu)$ are the Lamè coefficients in which E is the Young's modulus and ν is the Poisson's ratio. Thus, the static deformation for the isotropic continuous medium is characterized by two elastic parameters.

Under the assumption of infinitesimal deformation for an isotropic elastic material, Eq. (2) can be stated as follows using two elastic constants:

$$\frac{1}{3}\sigma_{kk} = \lambda\epsilon_{kk}, \tag{4}$$

$$\sigma_{ij} - \frac{1}{3}\sigma_{kk}\delta_{ij} = 2\mu(\epsilon_{ij} - \frac{1}{3}\epsilon_{kk}\delta_{ij}), \qquad i = 1,2,3. \tag{5}$$

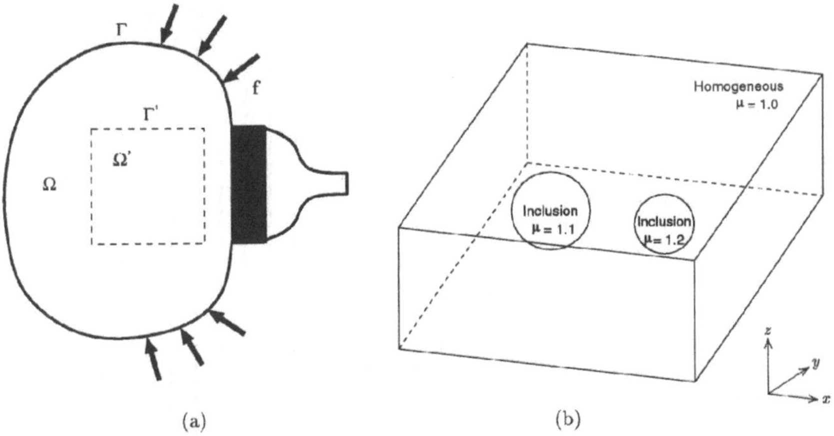

Figure 1. Biological tissue model being subject to external forces (a) and a image of the shear modulus distribution (b) used for numerical experiments. The tissue elasticity distribution contains two spherical hard inclusions. The inclusions are 1.2 and 1.1 harder than its surrounding medium, respectively.

The biological soft tissues is incompressible (Sarvazyan, 1975), namely the Poisson's ratio approaches 0.5, so that the volume change due to deformation must be zero:

$$\epsilon_{kk} = 0 \tag{6}$$

Thus, for incompressible medium, the longitudinal Lamè constant λ approaches infinity. Under these conditions, the stress-strain relation for static deformation reduces to

$$\sigma_{ij} = p\delta_{ij} + 2\mu\epsilon_{ij} \qquad i,j = 1,2,3, \tag{7}$$

where p is the static internal pressure, defined as

$$\lim_{\lambda/\mu\to\infty,\ \epsilon_{kk}\to 0} [\lambda\epsilon_{kk}] = p. \tag{8}$$

which, from Eqs. (6) and (7), is equivalent to the mean normal stress:

$$p = \frac{1}{3}\sigma_{kk}. \tag{9}$$

Therefore, the static deformation of an isotropic, incompressible medium can be completely characterized by a single elastic parameter, μ or E (which equals 3μ for an incompressible medium). Since the stress tensor is not determined from the strain tensor, we cannot obtain the constitutive equation for an incompressible medium in the usual sense. Note that a pressure term must be included in Eq. (7) to fully describe deformations in incompressible media.

Variational Formulation of the Forward Problem

An equivalent variational form of the problem is obtained in the following way.

In the forward problem, the equilibrium equation (1) along with the constitutive equation (7) is solved for the unknown displacement and static internal pressure, when

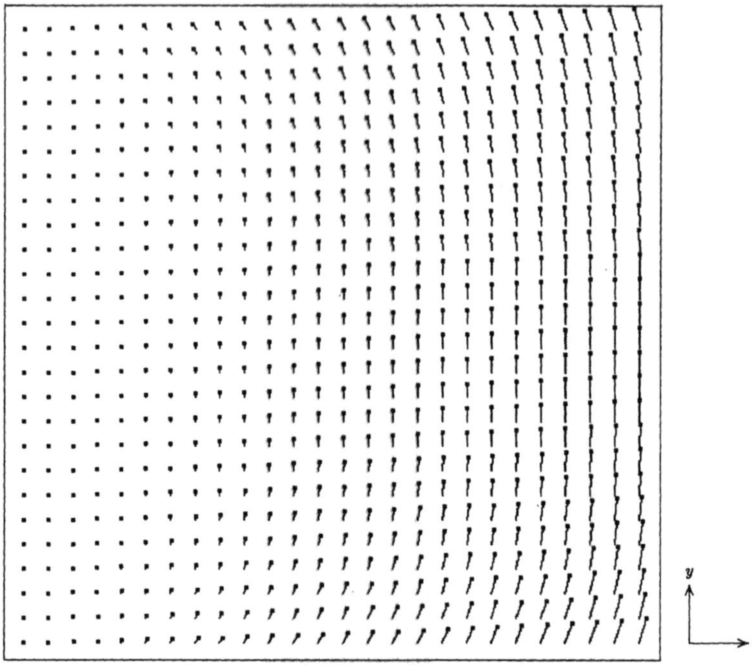

Figure 2. Displacements $(u_x and u_y)$ in the plane at z=5 intersecting the centers of two spherical inclusions when a shear force in addition to small uniform compression is applied at the right edge of the medium. The small dot of the displacement vector indicates the direction of displacement and the displacements are magnified by a factor of 5.

the material elasticity and the boundary conditions are known (see Fig. 1(a)). The medium is presumed to be acted upon by a traction $\boldsymbol{f}^t=(f_1, f_2, f_3)$ applied on Ω. The requirement of continuity of normal forces is expressed in the boundary condition:

$$\sigma_{ij}n_j = f_i \qquad i = 1, 2, 3, \tag{10}$$

where $\boldsymbol{n}^t=(n_1, n_2, n_3)$ is the unit normal vector at the surface Γ directed toward the exterior of the medium Ω.

The variational principle states that the energy functional (Zienkiewicz, 1989)

$$\Pi \equiv \frac{1}{2}\int_\Omega \sum_{ij}\left[(\sigma_{ij} - \frac{1}{3}\sigma_{kk}\delta_{ij}) + p\delta_{ij}\right]\epsilon_{ij}d\Omega - \int_\Gamma \sum_i u_i f_i d\Gamma \tag{11}$$

gives, on minimization, the satisfaction of the equilibrium equation along with the boundary condition $\sigma_{ij}n_j = f_i$. By considering p in addition to ϵ_{ij} as independent variables, the minimization of Eq. (11) results in

$$\int_\Omega \delta\boldsymbol{\epsilon}^t \left[2\mu(I - \frac{1}{3}\boldsymbol{m}\boldsymbol{m}^t)\boldsymbol{\epsilon} + \boldsymbol{m}p\right] d\Omega - \int_\Gamma \delta\boldsymbol{u}^t\boldsymbol{f}d\Gamma = 0 \tag{12}$$

where δ stands for a variation symbol, I is the identity matrix and

$$\boldsymbol{m}^t = (\ 1,\ 1,\ 1,\ 0,\ 0,\ 0\ ),$$
$$\boldsymbol{\epsilon}^t = (\ \epsilon_{11},\ \epsilon_{22},\ \epsilon_{33},\ \epsilon_{12},\ \epsilon_{23},\ \epsilon_{31}\ ).$$

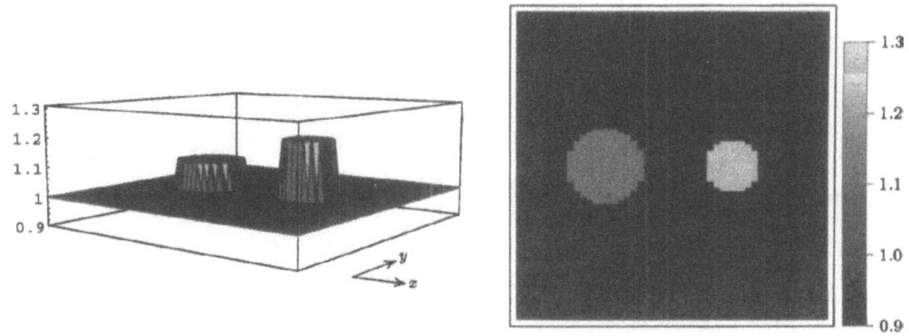

Figure 3. Shear modulus distribution reconstructed from the 3-D displacement fields (u_x, u_y, u_z).

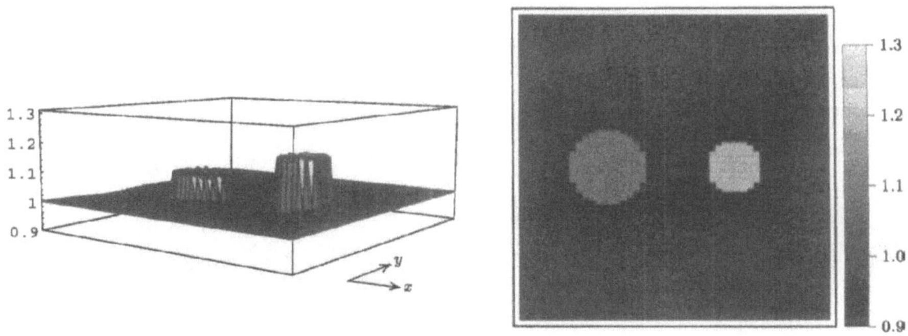

Figure 4. Shear modulus distribution reconstructed from the 2-D displacement fields (u_x, u_y) given in Fig. 2.

Furthermore, since the static internal pressure p cannot be definitely related to the strain, we require for sufficiently large λ as a weak form of Eq. (4):

$$\int_\Omega \delta p \left[\boldsymbol{m}^t \boldsymbol{\epsilon} - \frac{p}{\lambda} \right] d\Omega = 0. \tag{13}$$

In applying the finite element method, the region of interest, Ω', is divided into a number of subdivision called elements. The true functions of displacement and static internal pressure within each element is approximated using their nodal values as well as the interpolation function. The interpolation function will be chosen to be linear in the nodal values but may be nonlinear with respect to space coordinates. Let the strain tensor $\boldsymbol{\epsilon}$ to be approximated using the nodal displacement $\vec{u}$ such as $\boldsymbol{\epsilon} \cong B\vec{u}$. The minimization of Eqs. (12)(13) results in a set of linear equations for the unknown values of displacement and static internal pressure:

$$\begin{bmatrix} W[\vec{\mu}] & Q \\ Q^t & -V \end{bmatrix} \begin{bmatrix} \vec{u} \\ \vec{p} \end{bmatrix} = \begin{bmatrix} \vec{f} \\ 0 \end{bmatrix} \tag{14}$$

where

$$W[\vec{\mu}] = \int_\Omega 2\mu B^t \left(I - \frac{1}{3} \boldsymbol{m}\boldsymbol{m}^t \right) B d\Omega,$$

$$Q = \int_\Omega B^t \boldsymbol{m} d\Omega,$$

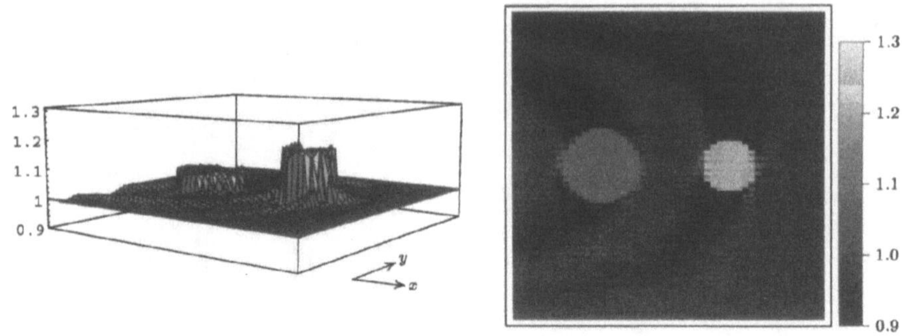

Figure 5. Shear modulus distribution reconstructed from the 1-D displacement field u_x given in Fig. 2.

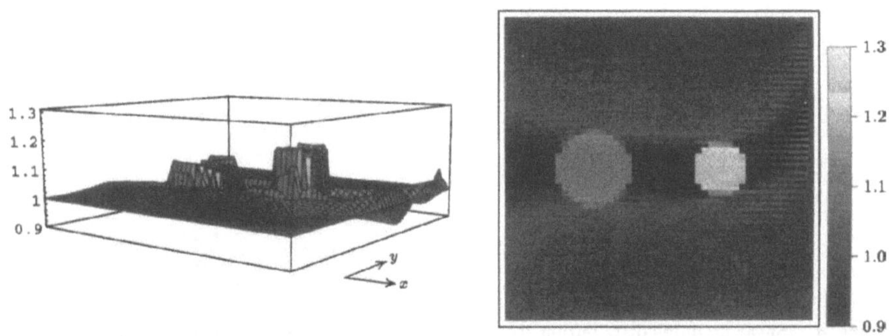

Figure 6. Shear modulus distribution reconstructed from the 1-D displacement field u_y given in Fig. 2.

$$V = \int_\Omega \left(\frac{d\Omega}{\lambda} \right) .$$

A sparse symmetric matrix $W[\vec{\mu}]$ is known as the stiffness matrix which is a function of the shear modulus, $\vec{\mu}$, of each element. In Eq. (14), $\vec{u}$ denotes the unknown displacements at the nodal points, $\vec{p}$ denotes the unknown static internal pressure in the element, and $\vec{f}$ defines the applied external forces. Thus, the displacements of all nodes as well as the static internal pressure in all elements are determined from Eq. (14).

Reconstruction of Shear Modulus from Measured Displacements

The variational formulation for the forward problem leads to a system of equations of the form $W[\vec{\mu}]\,\vec{u}+Q\vec{p}=\vec{f}$, where $\vec{u}$ is the unknown displacements and $\vec{p}$ is the unknown static internal pressure. Since $W[\vec{\mu}]$ of Eq. (14) is also linear for the shear modulus $\vec{\mu}$ of each element, $W[\vec{\mu}]\,\vec{u}$ can be rearranged with the shear modulus as $C[\vec{u}]\,\vec{\mu}$ and, moreover, the internal pressure, $\vec{p}$, can be eliminated by using the 2nd matrix equation of (14). This yields a different linear system of equations

$$C[\vec{u}]\,\vec{\mu} = \vec{f} - QV^{-1}Q^t\vec{u} \tag{15}$$

where $\vec{\mu}$ is the shear mudulus of all elements. Note that the element of the matrix $C[\vec{u}]$ is now a function of the measured displacements $\vec{u}$.

Since the matrix $C[\vec{u}]$ tends to be ill-conditioned, the solution for Eq. (15) is generally obscured by wildly oscillating noise and is usually very unstable. The regularized least squares technique to stabilize the solution yields

$$\vec{\mu} = (C^tC + \gamma I)^{-1}C^t(\vec{f} - QV^{-1}Q^t\vec{u}) \tag{16}$$

where the scalar quantity γ is the regularizing parameter whose optimal value depends on the noise level included in the measured displacements.

NUMERICAL SIMULATION

The biological soft tissue model consists of a 3-dimensional finite incompressible elastic medium, which contains two bonded spherical inclusions whose shear modulus differ from that of the surrounding medium. In this study, the region of interest is a size of 500×500×200 which has two spherical hard inclusions of radius 40 and 60 whose shear modulus is 1.2 and 1.1, respectively, whereas the shear modulus of the surrounding media is 1.0 (see Fig. 1(b)). Here, we assume that the model is subject to a linear shear force in addition to small uniform compression at the right edge of the medium.

The displacement fields were computed by solving the forward problem using the finite element method. We assume that we can measure the 3-dimensional displacement vector at arbitrary points in the region of interest ultrasonically by using a speckle tracking method. Fig. 2 shows the computed displacements u_x and u_y in the plane at z=100 intersecting the centers of two inclusions.

Based on the displacement fields given in Fig. 2, we can solve Eq. (15) for the unknown shear modulus. Generally speaking, since we may not exactly know the boundary conditions imposed at the surface of the medium or it is difficult to measure the exact external mechanical force applied at the surface, we exclude the equilibrium equation at the surface node at which the unknown external force is acting. The resulting equation become linearly dependent:

$$D[\vec{u}]\,\vec{\mu} = -QV^{-1}Q^t\vec{u} \tag{17}$$

so that we can only reconstruct the relative value of the shear modulus. In the reconstruction process we assigned 1.0 to the shear modulus of some elements in the right edge of the region of interest as the reference element.

Fig. 3 shows the relative shear modulus reconstructed from the 3-D displacement data. The result shows a perfect reconstruction of the relative shear modulus distribution.

Fig. 4 shows the relative shear modulus reconstructed from the 2-D displacement data, (u_x, u_y), given in Fig. 2. Two hard inclusions are well reconstructed but the relative shear modulus has extremely small perturbation at surrounding medium around the spherical hard inclusions.

The reconstruction was carried out for the case in which only the 1-D displacement data was measured. Since 1-D displacement measurement can be accomplished more easily by ultrasonic means than 3-D or 2-D displacement measurement, this reconstruction problem seems to have practical interest (Sumi et al., 1995). We reconstructed the relative shear modulus only from the displacement data in the x-direction, in which, since only the equilibrium equation in x-direction is used to solve for the unknown relative shear modulus, the number of the matrix equation of (16) reduces to one third. Fig. 5 shows the relative shear modulus reconstructed using only the x-component of the displacement data, u_x, in which the relative shear modulus gives fairly good estimates compared to that shown in Fig. 3, although erroneous perturbation around inclusions can be observed. The relative shear moduli at the central part of the inclusions have the mean value of 1.17 and 1.08, respectively. Fig. 6 shows the relative shear modulus reconstructed using only the y-component of the displacement, u_y, in which the relative shear modulus gives good estimates, although some artifact again exists around inclusions.

DISCUSSION

A variational formulation was presented to solve the inverse problem of elasticity imaging of an isotropic, incompressible biological soft tissue. A system of equations, $D[\vec{u}]\,\vec{\mu} = -QV^{-1}Q^t\vec{u}$, for reconstructing the relative shear modulus, $\vec{\mu}$, from the measured displacements, $\vec{u}$, was derived by rearranging the forward equation obtained from the variational formulation for the static deformation. Numerical simulation was carried out to validate the proposed method using the simulated deformation data of the simple 3-D inclusion problem. Simulation results demonstrated that the inversion algorithm can be applied to reconstruct the relative shear modulus distribution of isotropic, incompressible biological soft tissue.

In the present study, only 1-D component of the displacement vector was also used to reconstruct the relative shear modulus. The result showed good estimates comparable to those obtained from 3-D or 2-D displacement data. It should be pointed out that the relative shear modulus can be recovered to some degree of accuracy from only the displacement data in one direction.

ACKNOWLEGEMENTS

This work was supported in part by the Ministry of Education under Grant 11680852. The authors wish to thank T. Sai for developing the computer programs during this work, and H. Abe for preparing the manuscript.

REFERENCES

Emelianov, S.Y., Skovoroda, A.R., Lubinski, M.A., and O'Donnell, M., 1995, Reconstructive elasticity imaging, *Acoustical Imag.*, 21:241.

Kallel, F., and Bertrand, M., 1996, Tissue elasticity reconstruction using linear perturbation method, *IEEE Trans. Med. Imag.*, 15:299.

O'Donnell, M., Skovoroda, A.R., Shapo, B.M., and Emelianov, S.Y., 1994, Internal displacement and strain imaging using ultrasonic speckle tracking, *IEEE Trans. Ultrason. Ferroelect. Freg. Contr.*, 41:314.

Raghavam, K.R., and Yagle, A., 1994, Forward and inverse problems in imaging the elasticity of soft tissue," *IEEE Trans. Nucl. Sci.*, 41:1639.

Sarvazyan, A.P., 1975, Low frequency acoustic characteristics of biological tissues, *Mechnics of Polymers*, 4:691.

Skovoroda, A.R., Emelianov, S.Y., and O'Donnell, M., 1995, Tissue elasticity reconstruction based on ultrasonic displacement and strain images, *IEEE Trans. Ultrason. Ferroelect. Freq. Contr.*, 42:747.

Sumi, C., Suzuki, A., and Nakayama, K., 1995, Estimation of shear modulus distribution in soft tissue from strain distribution, *IEEE Trans. Biomed. Eng.*, 42:193.

Yamashita, Y., and Kubota, M., 1994, Ultrasonic characterization of tissue hardness in the *in vivo* human liver, *Proc. IEEE Ultrason. Symp.*, 1449.

Zienkiewicz, O.C., and Taylor, R.L., 1989, *The Finite Element Method*, MaGraw-Hill, London.

Romano, A.J., Shirron, J.J., and Bucaro, J.A., 1998, On the noninvasive determination of material parameters from a knowledge of elastic displacemens: Theory and Numerical simulation, *IEEE Trans. Ultrason. Ferroelect. Freq. Contr.*, 45:751.

RECONSTRUCTING YOUNG'S MODULUS DISTRIBUTIONS WITHIN SOFT TISSUES FROM FREEHAND ELASTOGRAMS

M. M. Doyley[1, 3], J. C. Bamber[1], P. M. Meaney[2],
F. G. Fuechsel[1], N.L. Bush[1], and N. R. Miller[1]

[1]Institute of Cancer Research and Royal Marsden NHS Trust, Downs Road, Sutton Surrey, UK
[2]Thayer School of Engineering, Dartmouth College, Hanova, NH, USA
[3]Presently at the Throaxcentre, Erasmus University, Rotterdam The Netherlands

INTRODUCTION

Elastography is an ultrasonic method that depicts tissue elasticity by imaging internal tissue strain that is induced using an externally applied strain. Various specialised methods and equipment have been used to apply the stress to organs such as the breast, with the view of obtaining the best possible estimate of internal tissue strain. Unfortunately, these limit the clinical usefulness of the technique by restricting the anatomical region that can be imaged or by interfering with the normal procedure of an ultrasound examination. A further potential limitation of elastography is that strain images are generally not suitable for clinical applications that require quantitative measure of tissue elasticity. This is because they do not take account of how the stresses are distributed in the medium. Potential clinical examples where a quantitative assessment of tissue elasticity would be required may include monitoring the response of a tumour to therapy and the characterisation of different tissue types. Furthermore, if no account is taken of the stress distribution then stress inhomogeneities will appear as artifacts in the resulting elastograms.

We have recently proposed an iterative inverse reconstruction method for computing Young's modulus within soft tissues from ultrasonically measured internal tissue displacements[1]. It was demonstrated by way of computer simulations that reconstruction based on knowledge of the displacement boundary conditions produces modulus images that are

Acoustical Imaging, Volume 25, Edited by Halliwell and Wells.
Kluwer Academic/Plenum Publishers, 2000.

superior, in terms of spatial and contrast resolution, compared to those produced when only the stress boundary conditions are known. The latter however, provides a quantitative method capable in principle of measuring Young's modulus *in vivo*, whereas the former does not.

In addition to developing methods for imaging Young's modulus we have also developed methods for imaging internal tissue strain induced using a hand-held transducer[2]. It would appear from the results of experiments conducted in elastically homogeneous phantoms that only a small loss in performance was incurred when hand-induced transducer motion is used as the source of tissue displacements. This paper reports the result of a preliminary study, which was conducted to explore the feasibility of producing clinically useful relative Young's modulus images from freehand elastograms.

INVERSE RECONSTRUCTION METHOD

The iterative scheme for recovering the Young's modulus distribution was derived directly from the forward elasticity problem. For this implementation, the finite element method was chosen over other numerical techniques such as the finite difference method because of its flexibility in allowing the analysis of structures with complex geometries, in-homogeneity, and boundary conditions. For the purpose of this paper, soft tissue was modelled as a two-dimensional linear, isotropic, incompressible medium. The Galerkin finite element representation of the governing elasticity equations was derived using the three-step approach described by Reddy[3], to give

$$[K]\{U\} = \{F\} \tag{1}$$

where [K] is the global stiffness matrix, {F} is the global force vector, and {U} is the nodal displacement vector, which contains both x and y components of displacement, i.e.

$$\{U\} \equiv \{X_1 \quad Y_1 \quad .. \quad X_N \quad Y_N\}^T \tag{2}$$

where the superscript T stands for transpose. Since [K] is generally sparse (i.e. a large proportion of the matrix away from the main diagonal is zero) a banded storage technique was employed to reduce the computational overhead. The rank of the matrix is 2N, due to the two degrees of freedom at each node, where N is the total number of nodes in the mesh. All internal forces such as gravity were neglected for all problems studied in this paper when required in the design of our study. Once the displacement boundary conditions were imposed, equation (1) was solved directly using the LU decomposition method[4].

Modified Newton Raphson iterative scheme

The general basis of our reconstruction technique is to minimise the squared error between the measured and the computed axial displacements. In this work the goodness-of-fit was calculated using the following expression

$$\Phi(\{E\}) = \|U\{E\} - \{U\}\|^2 \tag{3}$$

where {U} and {U{E}} represents the measured and calculated displacements, respectively. Differentiating equation (3) with respect to the Young's modulus distribution {E} gives

$$\frac{\partial\Phi(\{E\})}{\partial\{E\}} = 2(U\{E\} - U) \qquad (4)$$

Since there is a non-linear relationship between Young's modulus and displacement, the forward model is linearised by expanding equation (1) about an arbitrary Young's modulus distribution, $\{E_0\}$ using a truncated Taylor series to give

$$\{U\{E\}\} = \{U_0\} + \frac{\partial\{U\{E\}\}}{\partial\{E\}}\{\Delta E\} \qquad (5)$$

where $\{\Delta E\}$ is the Young's modulus residual (i.e. the difference between the true and trial Young's modulus distribution), and $\{U_0\}$ is the true displacement. Substituting equation (5) into equation (4) and setting the resulting equation to zero gives

$$[J]\{\Delta E\} = -\{\Delta U\} \qquad (6)$$

where the vector $\{\Delta U\}$, represents the residual between the measured and estimated displacements at each nodal point, and the Jacobian matrix, [J], represents the derivative of equation (1) with respect to the Young's modulus of elements in the mesh, i.e.$[J] = \frac{\partial U\{E\}}{\partial\{E_0\}}$. The normal equation, which describes the least squared problem[4] was formulated by multiplying both sides of equation (6) by the transpose of the Jacobian to give

$$[J^T J]\{\Delta E\} = -[J^T]\{\Delta U\} \qquad (7)$$

A new trial solution is generated at each iterate by first solving for $\{\Delta E\}$ in equation (7) then updating the Young's modulus values as follows:

$$\{E^{i+1}\} = \{E^i\} + \{\Delta E\} \qquad (8)$$

where the superscript refers to the iteration number. Because the Hessian (i.e. the matrix on the left-hand side of equation (7)) was ill *posed* and *ill conditioned*, equation (7) was solved using the Marquardt method [5].

Reconstruction begins using the assumption (i.e. the trial solution) that the tissue under investigation is elastically homogeneous. Once the trial solution have been constructed the reconstruction procedure can be summarized in the following steps:

1) Compute the displacements at each node in the finite element mesh based on the current Young's modulus distribution,$\{E\}$, by solving equation (1)
2) Compute the displacement residual vector, $\{\Delta U\}$, by subtracting the calculated displacements from the measured displacements at each nodal point.
3) Compute the displacement error, $\Phi(\{E\})$, between the computed and ultrasonically measured tissue displacements.
4) Construct the Jacobian matrix, [J], using the method described by Kallel and Bertand [6].
5) Compute the Young's modulus perturbation vector, $\{\Delta E\}$, by solving equation (7) using the Marquardt method.
6) Update the values of $\{E\}$, at the $(i+1)^{\text{Th}}$ iteration based on $\{\Delta E\}$ using equation (8).
7) Repeat steps 1 to 6 multiple times until a suitable stopping criterion is reached.

Note that solving for $\{\Delta E\}$ in step 5 using the Marquardt method involved scaling both the Hessian matrix and the right hand side of equation (7), then adding a positive definite scalar, λ, to the diagonal of the scaled Hessian. This has the net effect of improving the condition of the Hessian prior to matrix inversion. Because of this feature, i.e. weighting the diagonal of the Hessian, the Marquardt method is sometimes perceived as a regularisation method [7].

Spatial filter

The performance of iterative methods is known to deteriorate rapidly when measurement noise is significant[2]. One approach of solving this problem is of course to improve the displacement measurement method. There will however be a limit to how much improvement is possible in this respect. In addition, it may be possible to alter the reconstruction method to render it less sensitive to noisy input data. Consequently, we have recently developed a noise reduction technique to minimise the effect displacement measurement noise on the quality of reconstructed Young's modulus images. The general idea is to filter the Young's modulus updates at each iterate by computing the weighted average of each element in the finite mesh and the neighbouring elements.

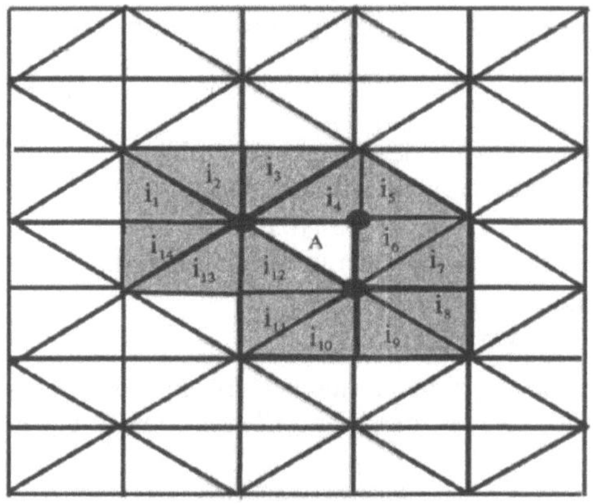

Figure 1 Illustrates the spatial filter, which was developed to suppress the effect of measurement noise. Showing an isolated element and its 14 surrounding elements.

To illustrate the technique, consider an isolated element A, and its neighbouring elements (Fig. 1). At the $K^{th.}$ iteration, the Young's modulus at element A is filtered as follows

$$E_k^{new}\{A\} = (1-\theta)E_K^{old}\{A\} + \frac{\theta}{N}\sum_{i=1}^{N} E_K^{old}\{A_i\} \qquad (9)$$

where the superscripts old and new represents the un-filtered and the filtered Young's modulus value, respectively, and θ is a value between 0 and 1 which controls the filtering weight, and N is the number of neighbouring elements (i.e. the spatial range of the filter, in this case 14). It is important to realise that the spatial range of the filter is significantly reduced at the edges because of the manner in which filter is constructed. As a result of this there is likely to be some edge effects because of in adequate filtering. Since equation (9) is a linear operation, it can be re-written in matrix form as follows

$$\{E_K^{new}\} = [F]\{E_k^{old}\} \qquad (10)$$

where the filtering matrix [F] is sparse. Because the filtering matrix depends solely on θ, it need be constructed once prior to the iterative process, with the filtering at each iterates being accomplished by a simple matrix-vector multiplication.

MODULUS AND STRAIN IMAGING

Experiments were conducted on an elastically inhomogeneous phantom (68.5 mm (lateral) by 68.5 mm (axial) and 90 mm (elevational)) to demonstrate the feasibility of reconstructing relative Young's modulus images from freehand elastograms. The phantom was manufactured entirely from porcine skin gelatine (Type A, approximately 175 bloom, Sigma chemical co. P.O. Box 14508, St. Louis, MO USA). The background gel was made from 6 % by weight gelatine suspension. A circular hole was made in the centre of the phantom. This was filled with 20 % by weight porcine skin gelatine to simulate a 29-mm diameter by 70-mm long hard inclusion. Both regions of the phantom were loaded with polyethylene granules (~119μm diameter, 0.36 % by weight,) to provide acoustic scattering centres for tracking.

Elasticity measurements were first conducted using the mechanical elastography imaging system. This was conducted by placing the phantom between two rigid Perspex plate (150 mm^2) as illustrated in Fig. 2. The lower plate was secured to an elevating state (Time and Precision, Hampshire, UK). This was used to deform the phantom under investigation by 5 % in steps of 0.2 % against a rigid base plate (i.e. the upper compressor) that was secured to a fixed support. A thin slit was cut in the base plate, to a perfect fit for an Acuson L7 linear transducer array to provide a window of echo imaging using an Acuson 128-XP ultrasound scanner.

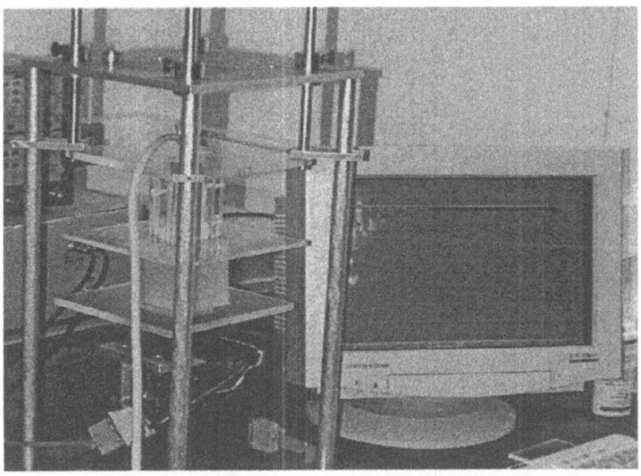

Figure 2 Photograph of mechanical compression system used to create mechanical elastograms, showing the inhomogeneous phantom mounted in pre-deformed state in between the two compressing plates

Elasticity measurements were also conducted using hand-held transducer. The phantom was also deformed by 5 %, however in this case it was deformed in one continuous motion at a strain rate of 2.96 % s^{-1}. Sequences of intermediate frequency (IF) echo frames were captured continuously from the scanner at full frame rate (15 Hz) using a modified MV-1000 frame grabber (Mu-tech Corp., MA).

IF echo-signals were collected in both sets of measurements (i.e. freehand and mechanical elastography) from a 70 mm (deep) by 38 mm wide region of interest using a 7 MHz Acuson (L7) linear transducer array. During all elasticity measurements, the scanner was set to operate with a dynamic receive focus and a single transmit focal zone, which was centred at 35 mm. All echo signals were digitised to 10 bits at a sampling rate of 20 MHz and stored on a 166

MHz Pentium based personal computer for off-line processing using MATLAB™ programming environment. Figure 3 shows examples of sonograms obtained when the phantom was in a pre-deformed state during mechanical and freehand elastographic imaging. There appear to be an acoustic shadowing in both images, which suggest that there is an increase in attenuation in the inclusion due to the difference in gelatin concentration. Both sonograms were computed by complex demodulation of the analytical radio frequency (RF) echo signals, which was reconstructed from the digitized IF echoes using a signal recovery technique that were based on heterodyning[2].

Freehand and mechanical elastograms were created using a thee-step approach. First, a two-dimensional normalized cross-correlation tracking technique was applied to consecutive pair of RF echo frames to create a sequence of high-resolution axial displacement images. Second, the resulting axial displacement images were median filtered (4-point by 4-point) and averaged to produced a composite axial displacement image. Echo tracking was performed in both cases using a 1.5 mm by 2.7 mm kernel, which was allow to overlap by 50 % and 99 % in the axial and lateral dimension respectively. Third, the composite displacement image was differentiated using 10-point LSQSE strain estimator[8], to produce an elastogram.

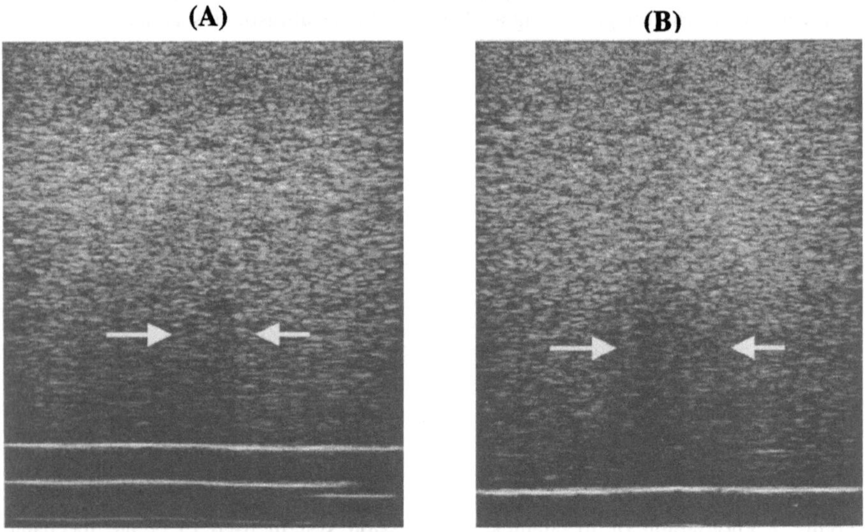

Figure 3 Sonogram of phantom in pre-deformed state during (a) Freehand, and (b) standard elastographic imaging. Arrows donates acoustic shadowing.

Figure 3 shows example of freehand and mechanical elastograms that were created during this experiment. Note that although the inclusion was not discernible in sonograms they are highly visible in the elastograms, which demonstrates that both freehand and standard elastography is capable of providing new diagnostic information that are generally not available in standard ultrasound images. The freehand elastogram is however appear noisier compare to the mechanical elastograms. The difference in image quality is likely to be due to effects of out-off plane transducer motion, which the authors were unable to compensate for during this study. Nevertheless the contrast resolution of the freehand elastograms were sufficiently high for visualising the inclusion and the characteristic bi-directional shadowing due to stress concentration at the tissue-inclusion interface.

(A) (B)

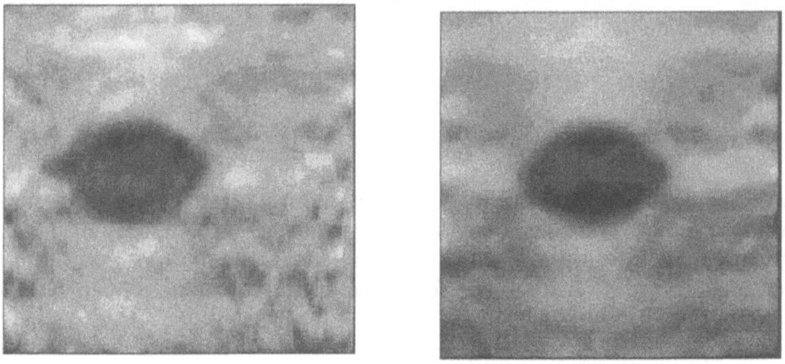

Figure 4 Elastograms produced using (a) hand-induced transducer motion, and (b) mechanical induced transducer motion.

Relative Young's modulus images were computed from the composite axial displacement images. For the purpose of this study the isonified region was represented by 70-mm long by 38-mm wide finite element model that contained 961 nodes and 1800 elements. Young's modulus was computed using a homogenous trial solution that had a Young's modulus of 20 kPa. A filtering weight of 0.2 was used in the spatial filter and λ was set to 10^{-3} at the start of all reconstruction. In both cases the reconstruction process was terminated when either a) the maximum number of iteration exceeded 20 iterations, b) the RMS displacement error was less than 10^{-8}, or c) the RMS Young's modulus error between any two iteration was less than ten.

(A) (B)

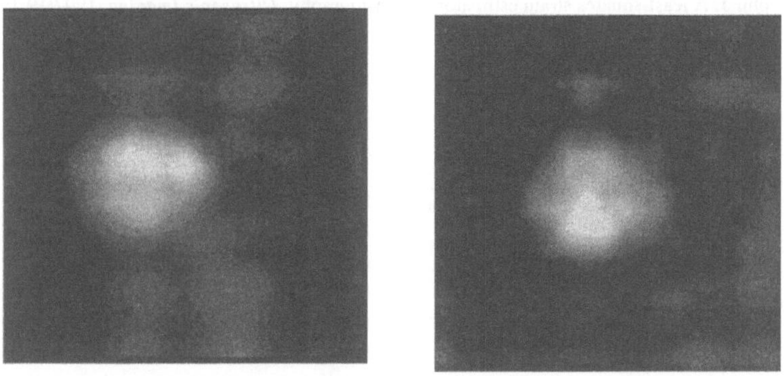

Figure 5 Relative Young's modulus images produced using (a) hand-induced transducer motion, and (b) mechanical induced transducer motion. Note that in these images white and black represents hard and soft material, respectively.

Figure 5 shows examples of freehand and mechanical modulus images that were produced. It would appear from these images that substantial improvement in image quality is achieved through the reduction of stress-concentration artefacts. Notice that the characteristic bi-directional artefacts due to stress concentration at the tissue-inclusion interface is apparent in

Fig. 4 but is not present in the reconstruction images. It is also interesting to note that despite the difference in quality of the elastograms the quality of the freehand and mechanical modulus elastograms were comparable.

CONCLUSIONS

We have demonstrated that despite the loss in performance incurred when hand-induced transducer motion are used as the source of tissue displacement, the quality of the elastograms are sufficiently high for creating clinically useful relative modulus images. Further work is currently been conducted to evaluate the performance of our quantitative freehand elastography system *in vivo.*

REFERENCES

1. Doyley MM, P.M.; M, Bamber JC. Evaluation of an iterative reconstruction method for quantitative elastography. *Physics in Medicine and Biology.* 2000;in press.
2. Doyley. Methods for improving the clinical usefulness of elastography. Ph.D. dissertation London: University of London; 1999.
3. Reddy JN. *An Introduction to the finite element method.* Hightstown, NJ 08520: McGraw-Hill,Inc; 1993.
4. Press WH, Teukolsky SA, Vetterling WT, Fannery BP. *Numerical Recipes in C : The Art of Scientific Computing.* Cambridge: Cambridge University press; 1988.
5. Marquardt DW. An Algorithm for Least-squares Estimation of Nonlinear Parameters. *J.Soc.Indust.Appl.Math.* 1963:431-441.
6. Kallel F, Bertrand M. Tissue elasticity reconstruction using linear perturbation method. *Ieee Transactions On Medical Imaging.* 1996;15:299-313.
7. Meaney PM, Paulsen D, Ryan TP. Two-dimensional Hybrid element image reconstruction for TM illumination. *IEEE transactions on antennas and propagation.* 1995;43:239-247.
8. Kallel F, Ophir J. A least-squares strain estimator for elastography. *Ultrasonic Imaging.* 1997;19:195-208.

ELASTIC CONTRAST DETECTION: A COMPARISON OF PERFORMANCE FOR ELASTOGRAPHY AND FOR THE DIRECT OBSERVATION OF B-MODE MOVIES DURING PALPATION

Naomi R. Miller and Jeffrey C. Bamber

Joint Department of Physics
Institute of Cancer Research / Royal Marsden Hospital
Downs Road, Sutton, Surrey SM2 5PT
United Kingdom

INTRODUCTION

Manual palpation has traditionally been used to detect abnormal regions of tissue elasticity such as may indicate the presence of a tumour. Since palpation cannot be used to detect small lesions or those located deep in the body, there has been considerable interest in methods of elasticity imaging (EI), which depict tissue stiffness by using conventional imaging modalities to evaluate tissue motion in response to an applied force (Hein and O'Brien, 1993; Céspedes et al., 1993; Ophir et al., 1991; Gao et al., 1996). Breast imaging by "elastography", for example, uses cross-correlation to track segments of echo data in radio frequency (RF) ultrasound images (Céspedes et al., 1993). The axial displacement of each tissue region is estimated by finding the best match between similarly sized windows in RF images of pre- and post-distorted tissue. The map of tissue displacements is then processed to generate an axial strain distribution, which is displayed as a grey-scale image known as an "elastogram". As implemented hitherto, elastography has required substantial modification to standard ultrasound breast imaging procedures and equipment. The purpose of this study, therefore, was to compare elastography with an alternative method for assessing tissue stiffness that is already in clinical use. This method, which involves the direct observation of tissue motion seen in B-mode image movies while palpating with the ultrasound probe, is referred to as relative motion assessment (RMA). The aim of this study was to perform a series of perception experiments to measure Young's modulus contrast thresholds for the detection of positive contrast elastic lesions by means of both RMA and elastography.

Acoustical Imaging, Volume 25, Edited by Halliwell and Wells.
Kluwer Academic/Plenum Publishers, 2000.

THEORETICAL BACKGROUND

Contrast in RMA and elastography

Hall et al. (1993) used the term *object contrast* to describe the inherent contrast in an object independent of the imaging system. In the present study, the fundamental object contrast was the Young's modulus contrast, defined as

$$C_Y = \frac{Y_L - Y_B}{Y_B} \qquad \text{(Eq. 1)}$$

where Y_L and Y_B were the values of Young's modulus in the lesion and in the background respectively. To describe this quantity we have used the term ***modulus contrast***.

The *image contrast* is the contrast that is displayed in the image (Hall et al., 1993). For both RMA and elastography, the displayed contrast is a strain contrast, defined as

$$C_S = \frac{S_L - S_B}{S_B} \qquad \text{(Eq. 2)}$$

where S_L , S_B are average displayed tissue strain values in the lesion and in the background. Note that for RMA, the strain in the moving B-mode image is not displayed as an image brightness, but as a change in the local dimensions of echo structure. Strain contrast may be perceived as a difference in this property from one image region to another.

In order to examine the relationship between object contrast and image contrast, we note that when the stress distribution within the tissue is uniform such that Hooke's law ($S \propto 1/Y$) is valid, the strain contrast can be related to the modulus contrast according to

$$C_S = \left(\frac{1}{C_Y + 1}\right) - 1 \qquad \text{(Eq. 3)}$$

This relationship between strain contrast and Young's modulus contrast is plotted in Figu 1. Note that while the modulus contrast might theoretically be varied from zero to +∞, the corresponding variation in the magnitude of strain contrast is from zero to 1. This non-linear relationship had a number of implications for the present study, as explained below.

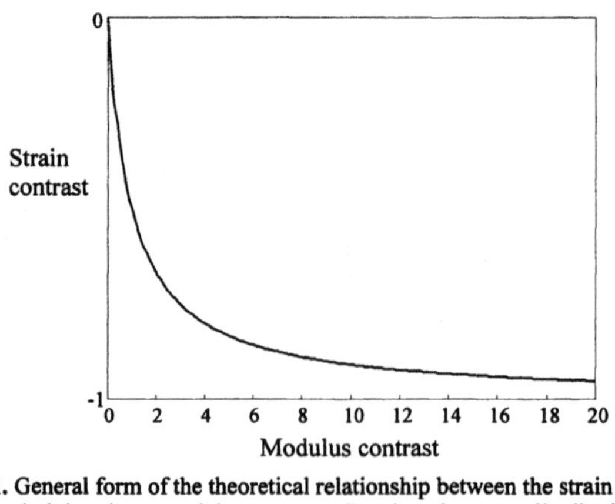

Figure 1. General form of the theoretical relationship between the strain contrast, C_S, and the underlying tissue modulus contrast, C_Y, when the stress distribution is uniform.

Contrast thresholds and psychometric curves

Contrast thresholds describe the ability of the imaging system and observer in combination to detect objects that lie at the threshold of detectability by virtue of their contrast and size, for a given noise level. The contrast threshold may be defined as the contrast at which the presence of an object will be detected with a given level of probability. In the present study, thresholds were obtained by conducting a two-alternative forced choice (2-AFC) experiment in which observers were required to identify which of two images contained a target, knowing that the other contained only noise. A curve known as a psychometric curve was derived by fitting a cumulative normal distribution to a plot of the proportion of correct responses, P_C, as a function of image contrast. The contrast threshold was then extracted from this curve as the image contrast for which $P_c = 0.75$.

In order to fit the psychometric curves, the contrast quantity plotted on the abscissa was the magnitude of the strain contrast. However, as described below, the simulation required that the lesion contrast be specified in terms of a Young's modulus contrast. Thus in order to fit the curves, the modulus contrast values were converted to strain contrasts using Eq. 3. Then, having determined the contrast thresholds in terms of strain, Eq. 3 was used to convert them to modulus contrast thresholds.

METHOD

Simulation

Simulated images were used since this allowed complete control over the Young's modulus contrast in the object, its relationship to the strain contrast in the image, the speckle noise due to the ultrasound system, and the lesion size. The two-dimensional (2-D) simulation began by setting up a finite element model of the tissue and specifying the Young's modulus contrast. In order to choose the boundary conditions, it was assumed that the ultrasound coupling gel at the face of the transducer generates slip conditions, while the chest wall constitutes a non-slip boundary. The applied axial strain was optimised separately for each technique (7% for RMA, 2.5% for elastography). The forward elasticity problem was solved to yield the distribution of internal tissue displacements.

The lateral and axial displacement distributions were used to distort a model of the tissue's ultrasonic backscattering impulse response that consisted of a fine distribution of scatterers designed to generate speckle. The undistorted and distorted tissue models were imaged by convolution with the point-spread function of a simulated ultrasound scanner, resulting in RF images of the pre- and post-distorted tissue. The tissue model and method of image formation were based on the simulation of Bamber and Dickinson (1980). Identical simulation parameters were employed for RMA and elastography.

The B-mode images for the RMA study were obtained by envelope detection of the RF images. These were assembled sequentially and played as a movie. There was no ultrasound echo contrast between the lesion and the background, so the lesion could only be detected by observing the speckle motion. Speckle dimensions were measured from the full width half maximum of normalised autocovariance functions in the lateral and axial directions (Belaid et al., 1994). The average lateral dimension was 13.7 pixels and the average axial dimension was 9.8 pixels, where the size of the B-mode image was 420 x 420 pixels. Fig. 2 shows pre- and post-distorted B-mode images of tissue containing a high contrast elastic lesion.

Figure 2. Simulated B-scans before (left) and after (right) tissue distortion. Note that in the experiment, these are not displayed side-by-side as above, but sequentially as a movie. The lesion diameter is 0.24 of the axial image size. The modulus contrast is 0.99, which makes the lesion easily visible when the B-scans are played as a movie.

The elastograms were generated by processing the RF images of the pre- and post-distorted tissue. The aim was to produce images of the highest possible quality within a reasonable amount of processing time. This was achieved by implementing some of the noise reduction techniques proposed in the literature. 2-D global companding (Chaturvedi et al., 1998) reduces signal decorrelation by warping the pre-distorted RF image prior to displacement estimation. In this study, the displacement distributions used to warp the image were those obtained from finite element analysis for homogeneous tissue. In order to reduce decorrelation due to lateral motion even further, the residual displacements were measured using a cross-correlation method that searched in both the axial and lateral directions (Hein and O'Brien, 1993). Axial strain was computed from the axial displacement image using a least squares strain estimator (Kallel and Ophir, 1997). Finally, as proposed by Ophir et al. (1991), a correction was applied to compensate for non-uniform stress due to the boundary conditions. The amplitude histograms in the corrected images were found to be a close approximation to a Gaussian distribution, as also observed by Belaid et al. (1994). The mean-to-standard deviation ratio was approximately 26. Fig. 3 shows a strain image containing a lesion at the highest contrast level for the particular size.

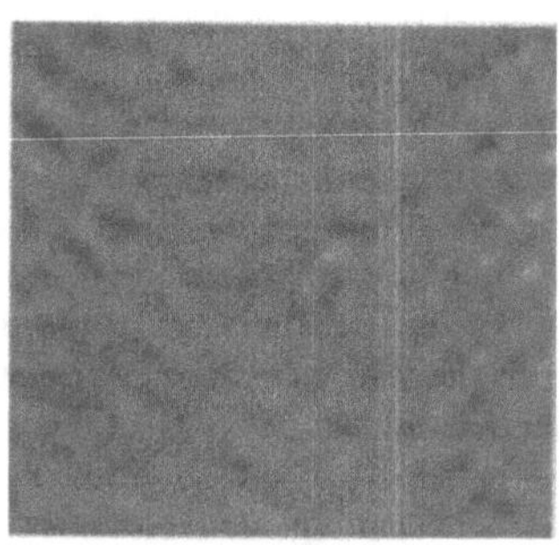

Figure 3. Simulated strain image containing a lesion whose diameter is 0.18 of the axial image size. The modulus contrast is 0.11, which is the highest contrast level at this size.

Image Preparation

The lesion sizes for RMA were chosen as described previously (Miller et al., 2000). The size of the smallest lesion was limited by the non-linear relationship between object contrast and image contrast. Since the image contrast required to perceive a lesion is inversely related to its size, for small lesions it was not possible to produce sufficient contrast in the image for the lesion to be detected, no matter how high the modulus contrast. For the simulation parameters employed, this phenomenon imposed a lower limit for the lesion size of about 42 speckle cells. In pixels, the lesion diameters were 75, 102, and 137, corresponding to physical dimensions on the display screen of 10.3 mm, 14.0 mm and 18.8 mm. The average number of speckle cells within the lesions was 42, 77, and 139.

Since contrast thresholds were considerably lower for elastography than for RMA, it was decided that the elastography study would include an additional experiment at a fourth, smaller lesion size. The diameter of this lesion was 43 pixels, corresponding to a physical diameter of 6.0 mm on the display screen. The lesion contained an average of 14 speckle cells. As the study progressed, it became apparent that there would be little benefit in acquiring data for elastography at the largest lesion size, since in this region of variable-space, the contrast threshold varied relatively slowly with lesion size.

For both RMA and elastography, 12 contrast levels were employed each lesion size. These were chosen such that there would be equal intervals in the strain contrast. Twenty statistically independent images were generated at each lesion contrast for a given size.

Detection Task and Image Display

The two alternatives were displayed side-by-side on the monitor. For each lesion size, the experiment was divided into 20 viewing sessions, each containing 12 observations. These corresponded to the 12 different contrast levels and were presented in a random order. For elastography, a viewing session took approximately 5 minutes to complete, compared to 15-20 minutes for RMA. At each lesion size, threshold estimates were obtained for between 4 and 6 observers.

The image display and viewing conditions were identical for RMA and elastography, with the exception of the mean grey level in the test images, which was higher for elastography. This was due to the fact that the amplitude histogram for elastography is a Gaussian distribution, whilst for echography, it is a Rayleigh distribution. The background grey level of the display screen was set to be equal to the mean grey level in the test images. The viewing distance was 75 cm and the size of 1 pixel was 0.14 mm on the display screen. The contrast and brightness controls of the monitor were fixed throughout all the experiments.

RESULTS

Contrast thresholds

Table 1 shows the strain contrast thresholds and modulus contrast thresholds, averaged over all the observers, for both RMA and elastography. The modulus contrast thresholds are also plotted in Fig. 4 as a function of lesion size, where size is expressed as the square root of the average number of speckle cells within the lesion, $M^{1/2}$. The standard errors on the contrast threshold values were obtained as described previously (Miller et al., 2000).

It is clear that elastography substantially reduced contrast thresholds for lesion detection compared with RMA. For example, for the lesion containing an average of 77 speckle cells, the average modulus contrast threshold for elastography was 35 times lower than that measured for RMA. The differences between the modulus contrast thresholds measured for RMA and elastography were found to be highly significant ($p < 0.001$).

Table 1. Average contrast threshold values for RMA and elastography

Lesion size, $M^{1/2}$	Mean strain contrast threshold for RMA	Mean modulus contrast threshold for RMA	Mean strain contrast threshold for elastography	Mean modulus contrast threshold for elastography
3.74	-	-	0.046 ± 0.015	0.048 ± 0.016
6.48	0.551 ± 0.077	1.318 ± 0.460	0.026 ± 0.006	0.027 ± 0.006
8.78	0.437 ± 0.070	0.806 ± 0.232	0.023 ± 0.004	0.023 ± 0.005
11.8	0.315 ± 0.056	0.459 ± 0.127	-	-

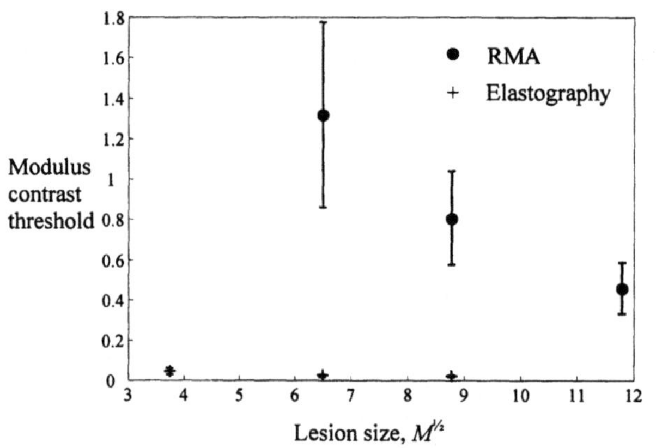

Figure 4. Contrast thresholds measured for all lesion sizes for RMA and elastography. Lesion size is expressed as the square root of the average number of speckle correlation cells (M) within the lesion.

DISCUSSION AND CONCLUSIONS

Contrast and size thresholds

The modulus contrast thresholds for lesion detection by both RMA (0.46-1.32) and elastography (0.02–0.05) appear to be within the range of contrast values that have been measured in clinically relevant tissues. For example, Sarvazyan's (1993) *in vitro* measurements showed that the tissue of malignant breast tumours is between 3 and 12 times stiffer than normal breast tissue (i.e., $C_Y = 2 - 11$). However, the substantial reduction in contrast thresholds provided by elastography is likely to be extremely valuable for visualising low-contrast lesions and for assessing small differences in elasticity, such as those that may arise from normal anatomical variation.

Lesions that were below the size threshold for detection by RMA could be easily visualised using elastography, suggesting that the latter could be of great benefit for the early detection of tumours. It is of interest to know what the lesion sizes investigated in the present study would correspond to in a practical imaging situation. A typical breast scanner operating with a 7.5 MHz linear array transducer might have a lateral resolution of 1 mm and axial resolution of 0.5 mm. The speckle cell area would then be approximately

0.39 mm^2 and the lesion diameters studied would correspond to values of about 2.7, 4.6, 6.2 and 8.4 mm. Such tumour sizes would be highly relevant to the early detection of breast cancer although, as mentioned below, there are many reasons as to why the present results cannot be used to make accurate predictions concerning lesion detectability in a clinical situation.

Implications of the simulation employed

There were a number of differences between the present simulation and the conditions that would exist in clinical practice. For example, a 2-D simulation was employed limiting all displacements to be within the scan plane. Whilst both the human observer and the computational algorithm were able to track lateral in-plane speckle motion, they would not have been able to compensate for speckle motion orthogonal to the scan plane. Thus it is likely that the 2-D simulation enhanced lesion detectability compared with a simulation in which structures are able to move out of the scan plane. A second way in which the simulation can be considered ideal is that the Young's modulus in the background tissue was uniform, so noise due to the elastic heterogeneity of normal breast anatomy was neglected.

In other senses, the simulated conditions were less favourable than those that would exist in practice. In breast imaging, for example, the B-mode image is not limited to speckle noise, which decorrelates rapidly as a function of strain; it also contains structural echoes that can, unlike speckle, be tracked over large displacements.

Although it would be essential to consider the above limitations if one wished to apply the results of the present study to making anything other than very rough predictions concerning lesion detectability in a clinical situation, this was not our aim. Rather, we wished to compare contrast thresholds for elastography and RMA under conditions that were both identical and reproducible. For this purpose the present simulation was adequate.

Comparison of clinical RMA and elastography

RMA presents a number of advantages over elastography in its present form. It is a real-time technique that is already a part of the breast ultrasound examination procedure. The radiologist can rapidly explore a suspicious area using different forcing functions and scan planes. The backscatter and the elasticity information can be viewed simultaneously, and a variety of dynamic features can be assessed including axial strain, lateral strain, mobility, and rotational motion.

As implemented hitherto, elastography has required the use of non-standard techniques and equipment, such as a mechanical compressor and the facility to store RF data. The elastograms have only depicted axial strain and have been computed off-line. However, these deficiencies could be overcome in the future. For example, Doyley (1999) has demonstrated the feasibility of freehand elasticity imaging. His approach was to develop ways of compensating for undesired transducer motions. Chen et al (1995) produced images of lesion mobility by applying a shear force to the breast and then using speckle tracking to measure the resulting lateral displacements.

Acknowledgments

This work was supported by the Medical Research Council and the Cancer Research Campaign.

REFERENCES

Bamber J. C. and Dickinson R. J., 1980, Ultrasonic B-scanning: a computer simulation, *Phys. Med. Biol.* 25:463-479.

Belaid N., Céspedes I., Thijssen J. M., and Ophir J., 1994, Lesion detection in simulated elastographic and echographic images: a psychophysical study, *Ultrasound Med. Biol.* 20:877-891.

Céspedes I., Ophir J., Ponnekanti H., and Maklad N., 1993, Elastography: elasticity imaging using ultrasound with application to muscle and breast in vivo, *Ultrason. Imag.* 15:73-88.

Chaturvedi P., Insana M. F., and Hall T. J., 1998, 2-D companding for noise reduction in strain imaging, *IEEE Trans. Ultrason. Ferroelec. Freq. Contr.* 45:179-191.

Chen E. J., Adler R. S., Carson P. L., Jenkins W. K., and O'Brien W. D., 1995, Ultrasound tissue displacement imaging with application to breast cancer, *Ultrasound Med. Biol.* 21:1153-1162.

Doyley M. M., 1999, An investigation into methods for improving the clinical usefulness of elasticity imaging, *PhD thesis*, University of London.

Gao L., Parker K. J., Lerner R. M., and Levinson S. F., 1996, Imaging of the elastic properties of tissue - a review *Ultrasound Med. Biol.* 22:959-977.

Hall T. J., Insana M. F., Soller N. M., and Harrison L. A., 1993, Ultrasound contrast-detail analysis: a preliminary study in human observer performance, *Med. Phys.* 20:117-127.

Hein I. A. and O'Brien W. D., 1993, Current time-domain methods for assessing tissue motion by analysis from reflected ultrasound echoes - a review, *IEEE Trans. Ultrason. Ferroelec. Freq. Contr.* 40:84-102.

Kallel F. and Ophir J., 1997, A least squares strain estimator for elastography, *Ultrason. Imag.* 19:195-208.

Miller N. R. and Bamber J. C., 2000, Thresholds for visual detection of Young's modulus contrast in simulated ultrasound image movies, *Phys. Med. Biol.* (submitted).

Ophir J., Céspedes I., Ponnekanti H., Yazdi Y., and Li X., 1991, Elastography: A quantitative method for imaging the elasticity of biological tissues, *Ultrason. Imag.* 13:111-134.

Sarvazyan A. P., 1993, Shear acoustic properties of soft biological tissues in medical diagnostics, *J. Acous. Soc. Am. Proceedings of the 125th Meeting* 93 (part 2) 2329.

2D TRANSIENT ELASTOGRAPHY

Laurent Sandrin, Stefan Catheline, Mickaël Tanter, and Mathias Fink

Laboratoire Ondes et Acoustique
E.S.P.C.I.
Université Paris VII
U.M.R. 7587, C.N.R.S. 1503
10 rue Vauquelin
75231 Paris cedex 05
France

INTRODUCTION

Relative stiffness imaging is of great interest since the pathological state of soft tissues is often correlated with changes in their stiffnesses. Palpation, which is a standard medical practice, relies on qualitative estimation of the low frequency (LF) stiffness of tissue. But despite the difference in stiffnesses a pathological lesion may not be detectable by palpation if it is located deep in the body or if it is too small. Moreover the echogenicity and the stiffness of tissues are generally uncorrelated and the lesion may not be ultrasonically detectable.

Different elastographic techniques have been developed in order to image tissue stiffness. These techniques differ by the kind of perturbation which is applied to the medium and by the imaging system used to estimate displacements within the medium. Transient elastography was introduced by Catheline et al. (1999). They measured the longitudinal component of the displacements induced by a piston that transiently vibrates perpendicularly to the surface of a half space viscoelastic medium. The longitudinal component of the displacements was measured versus time on the axis of the piston using a standard cross-correlation technique. The different waves could be separated. The first displacements to be observed after the LF excitation was applied were due to the fast compressional wave. The second observed wave was the slow (~ 4 $m.s^{-1}$) shear wave whose longitudinal component is non-zero in the near field of the vibrator. They showed that diffraction biases could be avoided with a transient excitation and that both shear elasticity and shear viscosity could be measured.

Sandrin et al. expanded the work of Catheline et al. (1999) on transient elastography to the study of 2D shear wave propagation. We propose here results obtained in gel-based homogeneous and heterogeneous phantoms with 2D transient elastography. The technique involves an ultra-fast ultrasonic imaging electronic system that can produce up to 10000

Acoustical Imaging, Volume 25, Edited by Halliwell and Wells.
Kluwer Academic/Plenum Publishers, 2000.

frames per second. With such an imaging system one single LF pulse excitation is necessary to acquire the full data set. Acquisition times are considerably reduced compared with MRI or Doppler techniques.

METHODS

The experimental setup is shown in Figure 1. Shear waves are generated using a circular piston built on a vibrator. The piston vibrates perpendicularly to the surface of the medium. The shape and the frequency of the LF vibrations are controlled with a function generator. The entire system is operated with a personal computer (Pentium 300 MHz). An electronic system is used to produce ultrasonic images of the medium. The ultrasonic transducer array may be situated in position 1 or position 2 of Figure 1.

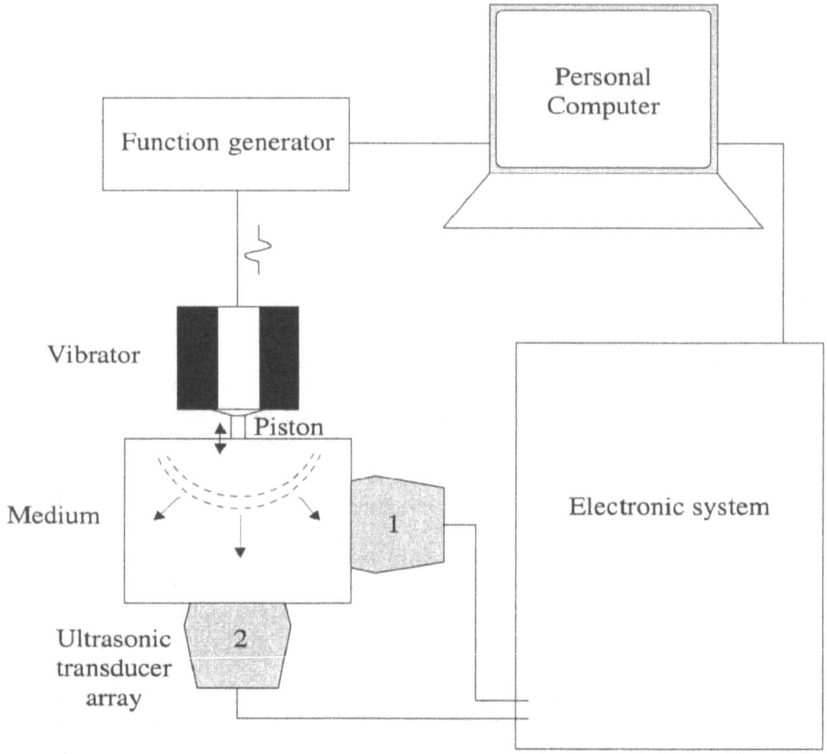

Figure 1. Experimental setup.

Ultra-fast ultrasonic imaging

The electronic system is presented in Figure 2. Standard ultrasonic scanners are limited to about 100 frames per second which is not sufficient to study the propagation of the low speed shear wave (~ 4 $m.s^{-1}$). Higher frame rates can be obtained when no focusing is performed in transmit and during reception. Thus the only limitation is the travel time of the ultrasounds within the medium. Parallel beam forming (PBF) is carried out afterwards numerically. As shown in Figure 3, a plane wave is used to illuminate the medium and each

(a)

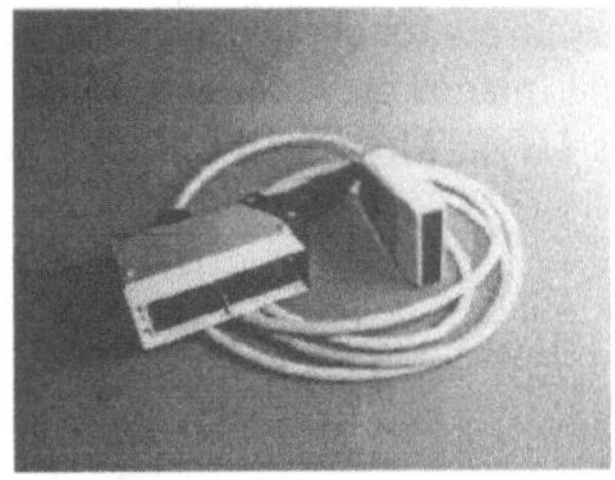

(b)

(c)

Figure 2. (a) The ultra-fast imaging electronic system. (b) 128 elements 3.5 MHz transducer array. (c) Gel-based phantom in a plexiglas cylinder.

element of the array has its own memory that is used to store the back-scattered (BS) signals or RF signals. As soon as reception is completed, the transducer can be used to emit a new signal.

We use a linear array of 64 transducers working at 3.5 MHz. Each transducer is connected to a fully programmable channel having 128 Kbytes memory. Sampling frequency is 30 MHz. Since the duration of the response that must be stored on each channel in order to construct a speckle image is typically about 100 μs (3000 points), the maximum number of frames that can be stored is 43.

Parallel beam forming (PBF) is applied to the BS (RF) signals received and stored by each of the 64 transducer elements. PBF consists in delaying and adding responses from the different elements with respect to the positions in the image in order to construct a focused speckle image. The Figure 4(a) shows a 5 μs long region of the 64 elements RF responses obtained during an experiment on a gel-based phantom and Figure 4(b) a 27×4 mm^2 region of the computed speckle image. The method is very similar to the one used with standard ultrasonic scanner. The difference is that beam forming is performed afterwards and that the illuminating wave is plane. The consequence of a plane illumination is a degradation of the speckle image dynamic and signal to noise ratio. However high dynamic range is not needed in our experiments since we intend to perform cross-correlation between consecutive speckle images.

Axial displacement estimation

Many speckle tracking algorithms (O'Donnel, 1994) have been developed in order to estimate the axial or even lateral displacement or strain in soft tissue. In our study speckle lines are segmented versus depth z into about 2 mm slices. The axial displacement is estimated in each of the segments using a cross-correlation technique between successive speckle images. The precision on axial displacement is about 1 μm. Figure 5 shows the full process used to obtain displacements. After PBF of N unfocalized speckle data set, N speckle images are obtained. Plane waves are emitted with a delay of T_{prf}. Thus the frame rate is $1/T_{prf}$. The ith speckle image is obtained by beam forming the ith rough RF signals stored.

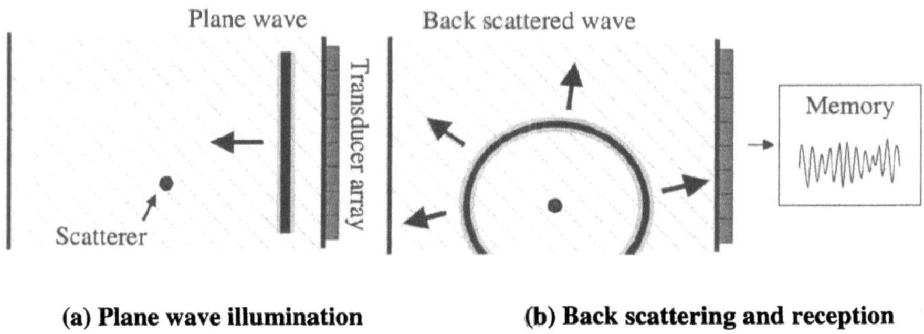

Figure 3. The ultra-fast imaging sequence. (a) The medium is illuminated using a plane ultrasonic wave. (b) The BS signals are received by the transducer array and stored in the channel memories.

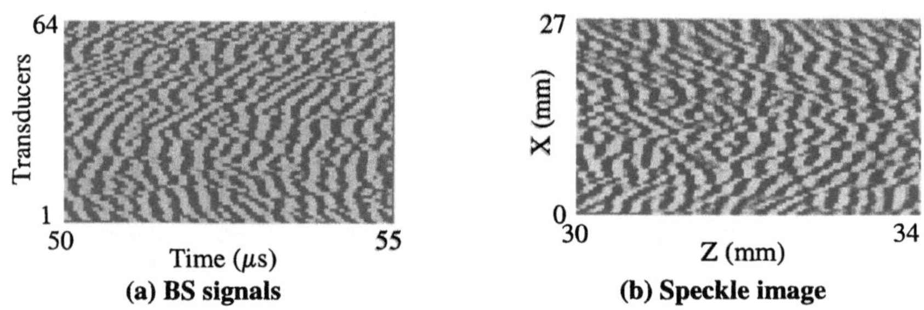

Figure 4. Parallel beam forming process. (a) 5 μs long BS (RF) signals sampled and stored by each of the 64 transducer elements. (b) Part of the speckle image computed by delaying and adding responses from the different elements with respect to the positions in the image.

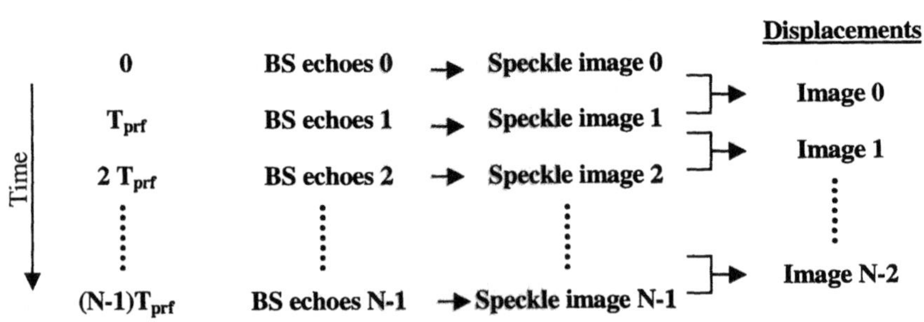

Figure 5. From *N* speckle data set, *N* speckle images are obtained which give *N-1* displacement images.

EXPERIMENTAL RESULTS AND DISCUSSION

Experiments were conducted in tissue mimicking gel-based homogeneous and heterogeneous phantoms. The first phantom is homogeneous of size 100×100×200 mm. The second phantom is cylindrical of size 70×∅100 mm and is composed of two layers. The top layer is stiffer than the bottom layer. Stiffness contrast is obtained using different gelatine concentration (typically between 3 and 10%). Phantoms contain agar powder as echogenic particles.

In all the experiments the excitation central frequency is 200 Hz, its duration is 1 period and in the case of the heterogeneous phantom and 3 periods in the case of the homogeneous phantom. The amplitude of the axial displacement of the piston is about 1 mm. The piston diameter is 10 mm. It vibrates perpendicularly to the surface. The frame rate is 1000 frames per second. Acquisition time is very short : about 50 ms. The post-acquisition procedure (to get displacement animation) takes about 5 minutes.

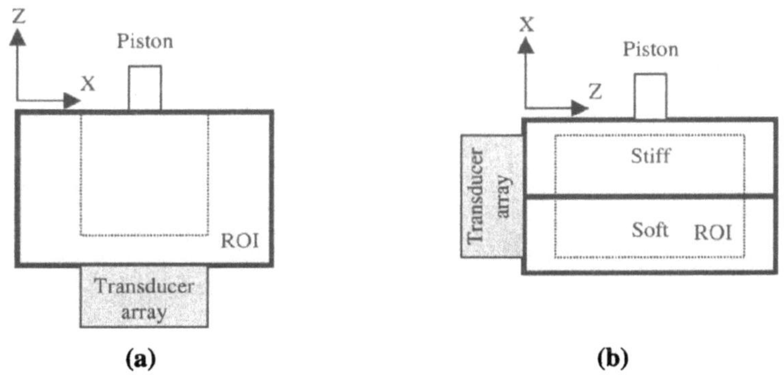

Figure 6. (a) In the case of the homogeneous phantom the transducer array is placed in front of the vibrator. (b) In the case of the bilayered phantom the transducer array is placed on the side of the medium while the piston vibrates on the top.

As shown in Figure 6(b), in the case of the bilayered phantom the transducer array is placed on the side of the phantom while the piston vibrates on the stiff top layer. The region of interest (ROI) covers the two layers equally. Since displacements are measured on the horizontal axis, due to symmetry versus the vertical axis of the piston we expect the displacement distribution to be antisymmetric versus this axis. In the case of the homogeneous phantom (Figure 6(a)), the piston and the array are placed face to face. The displacements are measured on the vertical axis and thus due to symmetry we expect the displacement distribution to be symmetric versus this axis.

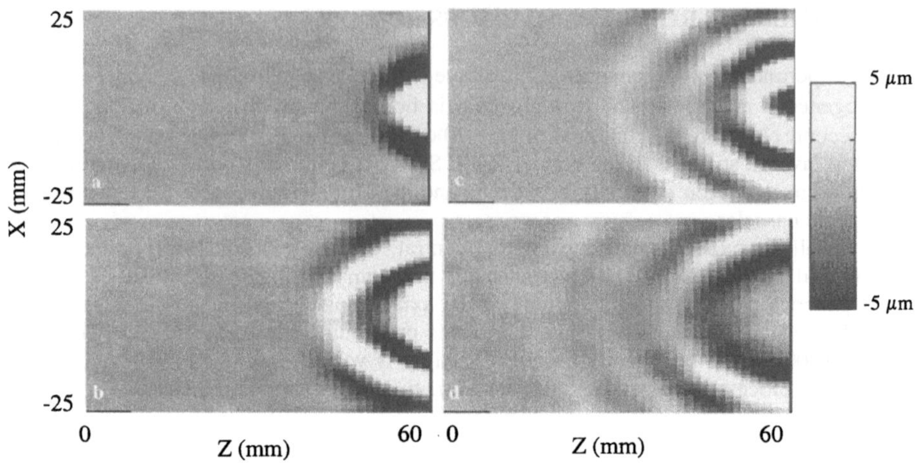

Figure 7. Displacement images obtained at time 9 (a), 15 (b), 21 (c), and 27 ms (d) in an homogeneous phantom. LF excitation is composed of 3 periods of a sinusoid at 200 Hz.

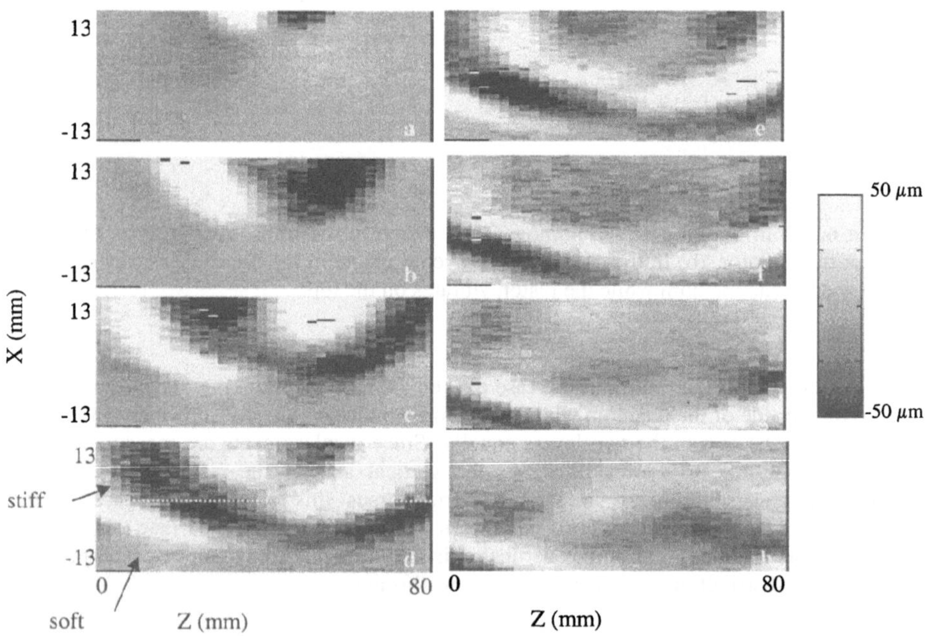

Figure 8. Displacement images obtained at time 6 (a), 8 (b), 10 (c), 12 (d), 14 (e), 16 (f), 18 (g) and 20 ms (h) on a bilayered phantom. The interface between the two layers is shown by a dotted white line on image (d). The top layer is stiffer than the bottom layer.

Homogeneous phantom case

In Figure 7 the axial displacements induced in the homogeneous phantoms after a LF excitation of central frequency 200 Hz and duration 3 periods has been sent to the vibrator are shown at different times : 9, 15, 21 and 27 ms. These displacements images are represented every 6 ms. The frame rate is 1000 frames per second (T_{prf} = 1 ms). The speed of the 200 Hz transient shear wave is estimated of about 2 m/s. The displacement amplitude attenuates as the wave goes deeper and deeper. The wavelength $\lambda = c/f_c$ and the shear modulus are respectively about 1 cm and 4000 Pa. As previously explained the longitudinal displacement due to the shear wave on the axis of the vibrator is non-zero. This is particularly true close to the vibrator. By contrast the longitudinal component disappears as the wave reaches the far field where the shear wave is known to be purely transverse.

Bilayered phantom case

To investigate the bilayered phantom a frame rate of 1000 frames per second (T_{prf} = 1 ms) was used and 43 frames could be stored. Results are shown in Figure 8. Displacements are measured over the whole area with 1 μm precision and low noise level. 42 displacement images were obtained. The maximum displacement between two frames is about 125 μm. Images are antisymmetric with respect to the axis of the vibrator (vertical) as expected from symmetry considerations. The wavelength, λ, decreases from ~2 cm in the stiff layer to ~1 cm in the soft layer. The speed, c, is calculated using the central frequency of the excitation (f_c = 200 Hz), $c = \lambda f_c$. It decreases from 4 m/s in the stiff layer to 2 m/s in the soft layer. The shear modulus, $\mu = \rho c^2$ where ρ is the density, is deduced assuming attenuation can be neglected and is respectively of 16 kPa in the stiff layer, and only of 4 kPa in the soft layer. Refraction can also be observed at the interface between the two layers: the wave front rotates as the wave penetrates into the soft layer (Figure 8 (d)).

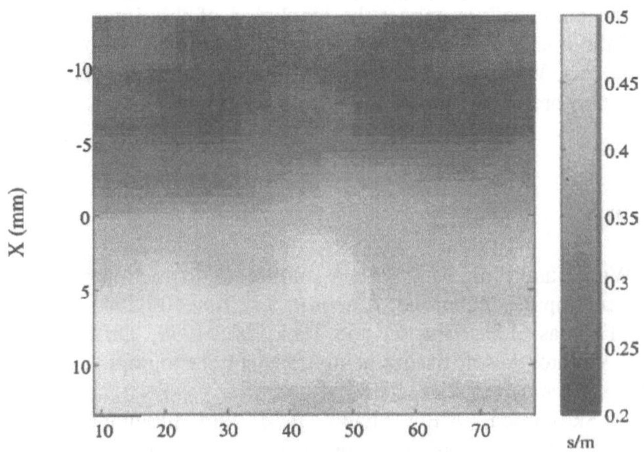

Figure 9. Experimental shear wave slowness ($s.m^{-1}$) cartography of the gel-based bilayered medium.

Data inversion

A very simple data inversion has been carried out by assuming the medium to be incompressible. In that case, the displacement vector u verifies $div\ \boldsymbol{u} = 0$ and each component u_x and u_z verifies the wave equation :

$$\Delta u_i - \frac{1}{c^2(x,z)} \frac{\partial u_i}{\partial t^2} = 0 \qquad i = x, z$$

where $c(x,z)$ is the shear wave speed. By simply calculating the second order derivatives in time and space of the axial displacement u_x in the nonzero displacements areas, we obtain the local shear wave speed $c(x,z)$ at each frame sampling time. This shear wave speed is then averaged over all frames. As seen on Figure 9, the data inversion clearly allows to reconstruct the spatial elasticity distribution of the bilayered medium. The boundary between the two layers is well defined and the shear wave speeds deduced from inversion are in good agreement with the previous wavelength measurement.

CONCLUSION

The important result presented in this paper is the possibility of imaging the propagation of a transient low frequency shear wave in biological tissues by means of an ultra-fast ultrasonic imaging system which is described in this paper. This imaging system can produce up to 10,000 frames per second in the medium. The shear wave propagation images were obtained with 1000 and 2000 frames per second from homogeneous and heterogeneous phantoms. From these images the wavelength and the speed of the shear wave are qualitatively estimated and thus the shear modulus is obtained. Since the displacements induced by the propagating centimeter wavelength shear wave are measured locally inside the medium, we expect the resolution of this imaging technique to be the same as that of the image of the displacements. In the future, we plan to measure also the lateral displacements. With two components of the displacement vector we should be able to solve the inverse problem without the strong assumption of a perfectly incompressible medium.

BIBLIOGRAPHY

Catheline, S., Wu, F. and Fink, M., 1999, A solution to diffraction biases in sonoelasticity : The acoustic impulse technique, *J. Acoust. Soc. Am.* 105:2941.

Catheline, S., Thomas, J.L., Wu, F. and Fink, M., 1999, Diffraction field of a low-frequency vibrator in soft tissues using transient elastography, *IEEE Trans. Ultrason. Ferroelectr. Freq. Control* 46:1013.

O'Donnel, M., Skovoroda, A.R., Shapo, B.M. and Emelianov, Y., 1994, Internal displacement and strain imaging using ultrasonic speckle tracking, *IEEE Trans. Ultrason. Ferroelectr. Freq. Control* 41:314.

Sandrin, L., Catheline, S., Tanter, M., Hennequin, X. and Fink, M., submitted, Time-resolved pulsed elastography with ultra-fast ultrasonic imaging, *Ultrasonic Imaging*.

ADVANCING INTRAVASCULAR PALPOGRAPHY TOWARDS CLINICAL APPLICATION

M. M. Doyley[1, 2], C. L. de Korte[1], F. Mastik[1],
S.G. Carlier[1], and A. W. F. Van der Steen[1, 2]

[1]Thoraxcentre, Erasmus University, PO Box 1738, 3000DR Rotterdam, The Netherlands
[2] Inter-university Cardiology Institute of the Netherlands

INTRODUCTION

Therapeutic techniques for treating atherosclerotic lesions, such as angioplasty, stenting and atherectomy are predominantly mechanical in nature. Therefore the morphology and mechanical properties of the atherosclerotic lesion could influence the outcome of these procedures. Intravascular ultrasound (IVUS) provides real-time, high-resolution cross-sectional images of the arterial wall that accurately depict plaque morphology. However, it is generally difficult to characterise the composition of plaques using IVUS. Nevertheless, using IVUS it is possible to obtain information regarding the compressibility of vascular tissues, which is related to their elastic properties. Consequently, there have been considerable recent interests in developing IVUS based elasticity imaging methods for visualising the elasticity properties of vascular tissues[1,2,3,4]. The main limitation with the vast majority of these techniques is that very complex signal processing techniques are employed, with the view of obtaining the best possible estimate of internal tissue strain. Unfortunately, this prevents real-time application, and thus reduces the likelihood of rapid acceptance and widespread utilisation of elasticity imaging in the catherization laboratory. Although, it is necessary to pursue complex signal processing schemes when solving the inverse elasticity problem[5], it would be of immediate practical value to develop a simple real-time imaging system, which is capable of visualising tissue elasticity. Consequently, we have developed a prototype elasticity imaging technique based on IVUS palpography[6]. Although this technique only provides one-dimensional elasticity information, it is simple enough to be implemented in real-time using relatively in-expensive hardware, and should be robust compared to previously reported techniques. This paper reports the results of experiments conducted on phantoms to compare the quality of single-step and multi-step palpograms produced using the prototype system. All elasticity measurements were performed under dynamic conditions in order to simulate the motion incurred when the intra-coronary pressure is used as the source of mechanical stimulation.

THEORY

Various measures have been used for characterising the elastic properties of vascular tissues. These include compliance, distensibility (i.e. change in lumen area) and pressure-strain elastic modulus[7, 8,9]. Unfortunately, these measures are usually inaccurate because they do not take into consideration the thickness of the arterial wall[6]. A more representative measure of vascular elasticity can be computed from radial strain (ε) using the following expression[6]

$$\varepsilon = \frac{\Delta R_2 - \Delta R_1}{R_2 - R_1} \tag{1}$$

where ΔR_1 represents the displacement of a region of tissue encompassing the lumen vessel interface, ΔR_2 represents the displacement of a deeper layer of tissue within the arterial wall.

EQUIPMENT

Figure 1 shows a schematic diagram of the prototype intravascular palpography system. It consists of an InVision intravascular ultrasound scanner (EndoSonics Corp., CA.) that was equipped with a 64-element phase array catheter (EndoSonics Corp., CA), which was operating at 20 MHz. A fast data acquisition system (DAS) was developed based on a modified PCI bus Coreco Viper digital frame grabber (Coreco Inc. Quebec, Canada) and incorporated on the scanner. This was used to capture long sequences RF frames from the ultrasound scanner at full-frame rate (i.e. 30 Hz) in order to cope with motion artifacts and decorrelation noise sources, which is likely to be incurred when the inter-coronary pressure is used as the source of tissue displacement. The low frequency physiological signals (i.e. ECG and pressure) were digitised to 12 bits at a sampling frequency of 5 kHz using a multi-channel input-output (I/O) PCI card (National instruments). The entire data acquisition was controlled using in-house software, which was running under WINDOWS NT™ operating system.

MATERIALS AND METHODS

Phantom fabrication: An elastically homogeneous vessel phantom (15 cm long), which had inner and outer diameters of 3 mm and 20 mm, respectively was manufactured from 15 % by weight porcine skin gelatin (Type A, approximately 175 bloom, Sigma chemical St. Louis, MO), 1 % by weight Agar (Boom BC, Meppel, The Netherlands), and 1 % by weight carborundum particles (3-10 μm mean diameter). The acoustic and elasticity properties of the phantom was chosen to be representative of vascular tissues.

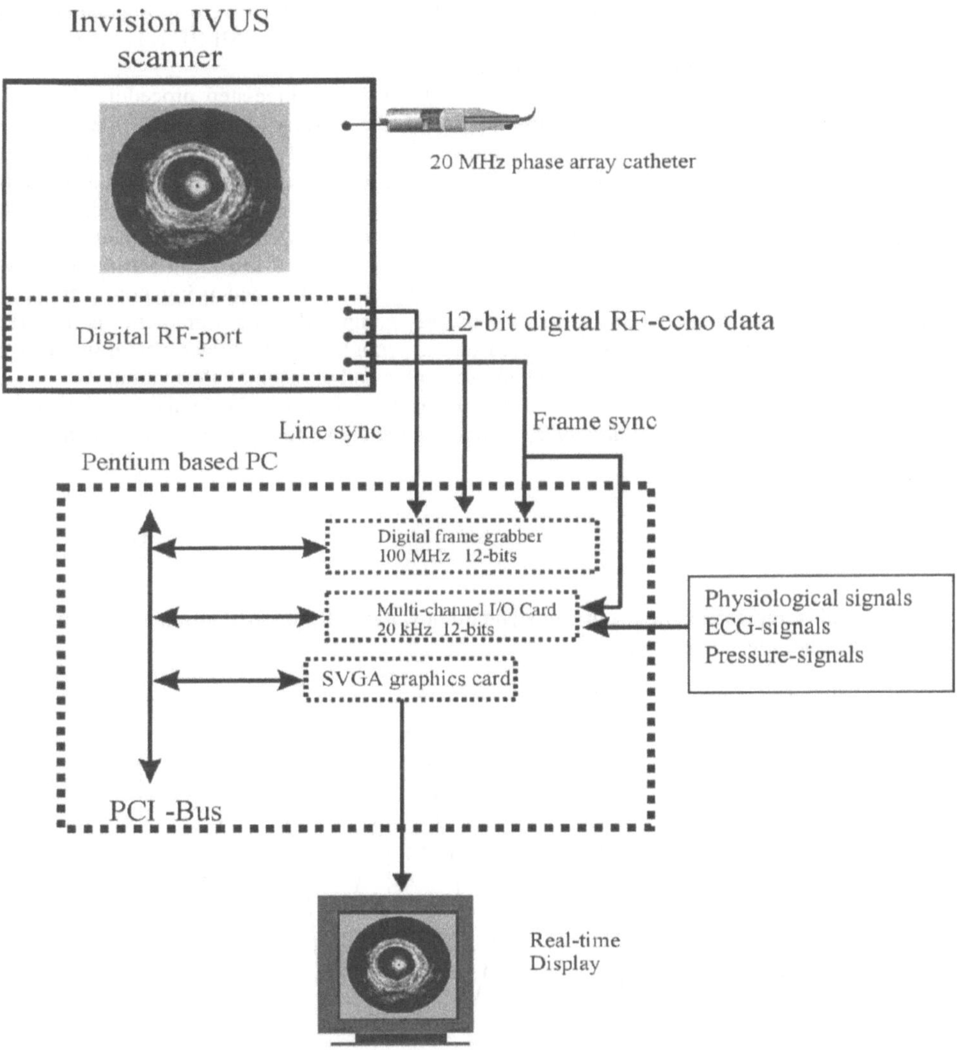

Figure 1 Schematic diagram of prototype intravascular elasticity imaging system.

Elasticity measurements: All measurements were performed in a water tank (18 cm (long) by 8 cm (high) by 10 cm (wide)), which was equipped with two sheaths (8 F), one was located at each side of tank. The phase array catheter was inserted in the distal sheath and positioned approximately in the centre of the lumen. The phantom was subjected to time-varying intra-lumen pressure (1Hz) that was generated using a programmable peristaltic pump (Watson and Marlow 5000), which was connected to the proximal sheath. The pressure within the lumen vessel phantom was measured using a pressure sensor (DTX plux, Ohmeda, The Netherlands) which was connected to the distal sheath. A total of four independent acquisitions were made.

Each acquisition contained a total of 121 RF echo frames (4 s), which were subsequently stored on a Pentium III based PC for off-line analysis using MATLAB™. All RF echo frames were digitised to 12 bits at a sampling rate of 100 MHz.

Signal processing: Palpograms were created using a three-step procedure. First, RF echo frames corresponding to the same phase of the cardiac cycle were identified by differentiating the pressure curve. Figure 2 shows an example of a pressure curve, which was obtained during the elasticity measurements. Second, the lumen vessel interface was detected by applying a simple edge detection method to the digitised RF echo signals. Third, the induced displacements of tissue layer near the lumen vessel interface were estimated using either a single-step or multi-compression strain estimation strategy. A one-dimensional normalised cross-correlation tracking method. Furthermore, all tracking was performed using non-overlapping kernels. The induced tissue strains were subsequently computed using a gradient operator (i.e. Eq. 1). These were colour-coded and encoded on the sonogram at the lumen vessel interface to provide a compound image containing both elastographic and ultrasonic information.

Performance evaluation: In addition to visually assessing the images, the quality of palpograms was assessed quantitatively by computing the signal to noise ratio of the palpogram, which was defined as

$$SNR = \frac{\mu_s}{\sigma_s} \qquad (2)$$

where μ_s and σ_s represents the mean and standard deviation of the radial strain estimated within an elastically homogeneous vessel.

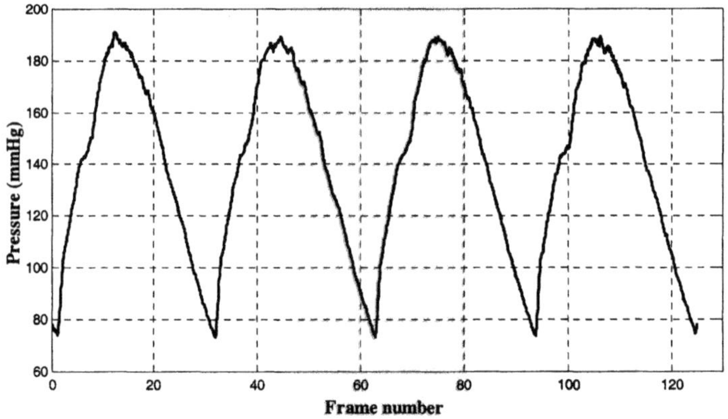

Figure 2 Example of the pressure curve obtained during the elasticity measurement.

RESULTS AND DISCUSSION

Figure 3 shows examples of elasticity images (i.e. elastograms and palpograms) that were produced using the single step approach. Note that the elastograms are included here solely for the purpose of visual comparison with the palpograms. Note that the stress decay artefacts, which is present in the elastograms gives the impression that phantom consists of two materials, i.e. an elastically soft inclusion which surrounded by an elastically hard material. One approach to removing this artefact would be to reconstruct the Young's modulus using an inverse reconstruction approach. This can however, be very time consuming and is not necessary in situations the elasticity properties of vessels are assessed using intravascular palpography because elasticity measurements are performed very close to the lumen-vessel interface where stress decay is negligible. It is important to note that because palpography does not suffer stress decay artifacts, palpograms can be interpreted as Young's modulus images using the assumption of uniformity.

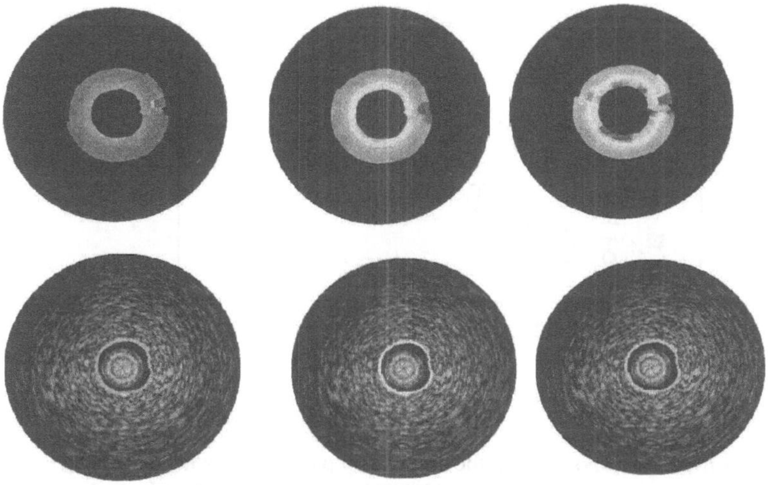

Figure 3 Examples of (top panel) elastograms and (bottom panel) palpograms created using the single-step strain estimation strategy. In these figures green, yellow and red represents high, medium and low strains, respectively.

Figure 4 shows the mean SNR of the palpograms plotted as a function of mean internal tissue strain when different window size were employed in the single-step strain estimation approach. Single-step palpograms were computed, in a given phase of the cardiac cycle, by comparing the frame obtained at the lowest pressure with all the remaining frames. It is important to realise that these plots represent experimentally defined strain filters. To our knowledge this represents the first such use of the strain filter concept[10] to characterise the performance of an IVUS based elasticity imaging system. It should also be noted that the strain filters defined in Fig. 4 are not fully defined, since relatively small internal strains were generated in this study. Nevertheless, a number of important observations are apparent from this figure. Firstly, the SNR of the palpograms generally increases with increasing applied intraluminal strain up to 0.8 % however, because of decorrelation effects further increase in the intraluminal strain results in a degradation in image quality (i.e. decreases the SNR). Secondly, the quality and the repeatability (i.e. variance) of the palpograms improve significantly with increasing window size for small strains (i.e. < 0.8 %). This confirms the observation of previously reported work[11, 12] that estimating internal tissue displacements in situation of small internal strain using small observation windows generally increases the displacement estimation error, since this increases the likelihood of peak hopping. It is also interesting to note that the effect of decorrelation becomes more pronounced when internal tissue strain is estimated using large observation windows. It would also appear from Fig. 4 that the prototype system has a strain sensitivity* of approximately 0.2 %.

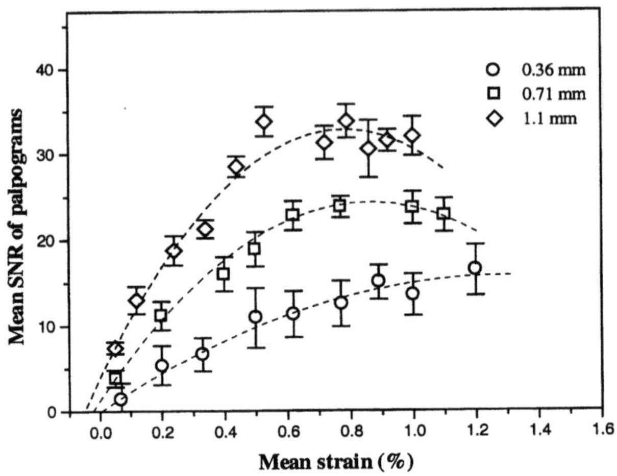

Figure 4 Strain filter corresponding to the single step palpogram, with increasing window size. The error bars represent ± 1 STDEV estimated our 16 independent measurements (4 cycles per measurements).

Figure 5 shows examples of elasticity images (i.e. elastograms and palpograms) that were produced using the multi-step approach. Note that the visual appearance of the elastograms was improved considerable when the multi-step averaging approach was employed. Figure 6 show examples of strain filters, which were computed from palpograms, produced using two different inter-frame strains in the multi-step approach. All computations were performed in this example using the optimum window size (i.e. 1.1-mm) as derived using the single step approach. Notice the remarkable improvement in image quality (i.e. peak SNR) achieved when the multi-step method was employed. However, it is apparent from this figure that performance of the technique is highly on the inter-frame strain, and thus experiments must be conducted to optimise the technique. An obvious possibility would be to select the optimum inter-frame strain based on strain filters derived using a single-step approach.

---

* Strain sensitivity was measured in this work using a half-maximum criterion (i.e. 50 % down on the peak strain SNR).

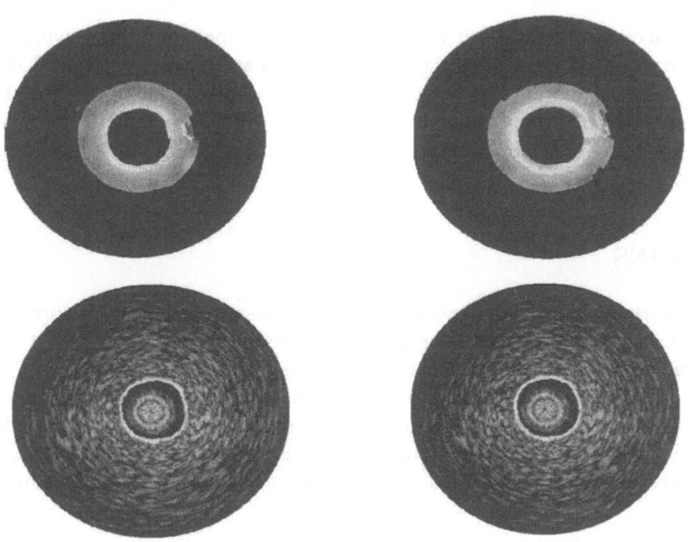

Figure 5 Examples of (top panel) elastograms and (bottom panel) palpograms created using a multi-step strain estimation strategy. In these figures green, yellow and red represents high, medium and low strains, respectively.

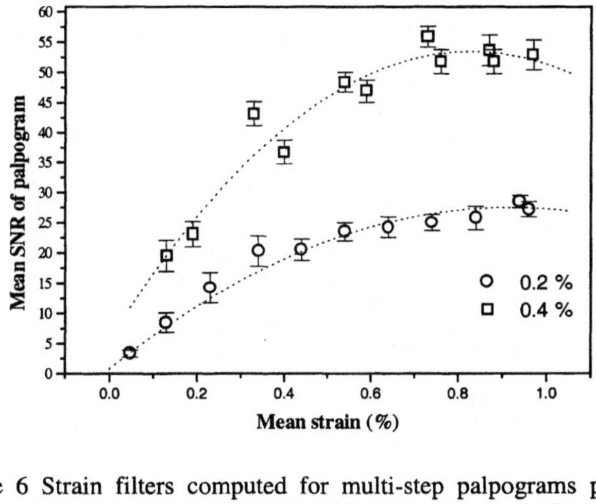

Figure 6 Strain filters computed for multi-step palpograms produced using an inter-frame strain of 0.2 % and 0.4 % respectively. The error bar represents ± 1 STDEV estimated over 16 independent estimates.

Despite these encouraging results it is important to note that multi-step averaging may prove ineffective in practice because of in plane and out-off plane catheter motion. Consequently, we are developed a global motion compensation method to correct for in-plane catheter motion[13], and are currently conducting further experiments to study the benefit of multicompression averaging under more realistic clinical conditions; the results of these investigations will be presented later.

CONCLUSIONS

The quality of palpograms produced using multi-step averaging is superior compared to those created using a single-step approach. The length of the kernel used in both signal-processing strategies should be chosen carefully, since this will have a direct bearing on the quality of the elasticity images that are produced. Despite the apparent advantage gained when multi-step averaging is employed, this strain estimation approach has the distinct disadvantage of been computationally very demanding, therefore we recommend that in situations were real-time assessment of vessel elasticity is desired that palpograms should be created using the single step approach.

REFERENCES

1. De Korte, C.L.; Pasterkamp G.; van der Steen, A.F.W.; Woutman H.A.; Bom N. Characterisation of plaque components using intravascular ultrasound elastography in human femoral and coronary arteries in vivo. Circulation: in press; 2000
2. Shapo, B.M; Crowe, J.R.; Skovoroda, A.; Eberle, M.J.; Cohen, N.A.; and O'Donnell, M. A new technique for imaging tissue imaging of coronary arteries with intraluminal ultrasound. IEEE Trans. On UFFCS (43), pp. 234-246, 1996
3. Cespedes E.I.; De Korte C.L.; van der Steen, A.F.W.; von Birgelen C.; Lancee, C.T. intravascular elastography: prinicples and potentials. Semin Interv Cardiol, (2)1, pp 55-62, 1997.
4. Talhami, H.E.; Wilson, L.S. and Neale, M.L. Spectral tissue strain: A new technique for imaging tissue strain using intravascular ultrasound. Ultrasound in medicine and Biology 20(8), 759-772.
5. Doyley, MM; Meaney, P.M.; Bamber, J.C. Evaluation of an iterative reconstruction method for quantitative elastography. Physics in medicine and biology: in press; 2000
6. Cespedes, E.I.; de Korte, C.L. and van der Steen, A.F.W. Intravascular ultrasonic palpation: assessment of local wall compliance. IEEE ultrasonics Symposium Proceedings, 1079-1082.
7. Hoeks, A.P.G; Brands, P.J., and Reneman, R.S Assessment of the arterial distension waveform using Doppler signal processing. Journal of Hypertension 10(6), pp 19-22,1992.
8. Stadler, R.W., Karl, W.C., and Lees, R.S. The application of echo-tracking methods to endothelium-dependent vasoreactivity and arterial compliance measurements. Ultrasound in Med. & Biol. 22(1), pp. 35-42,1996.
9. Loree, H.M.; Gordzinsky, A.J.; Park S.Y.; Gibson, L.J.; and Lee, R.T. Static circumferential tangential modulus of human atherosclerotic tissue. Journal of Biomechanics (27)2, pp195-204,1994.
10. Varghese, T. and Ophir, J. A theoretical frame work for performance characterisation of elastography: the strain filter. IEEE trans. On UFFC, (44), pp 164-172,1997.
11. Lubinski, M.A., Emelianov, S.Y., O'Donnell, M.O. Speckle tracking methods for ultrasonic elasticity imaging using short-time correlation. IEEE Trans. On UFFCS 46(1), pp82-96,1999
12. Céspedes, E.I.; de Korte, C.L.; van der Steen, A.F.W. Echo decorrelation from displacement gradients in elasticity and velocity estimation. IEEE Trans. On UFFCS 46(4), pp791-801,1999
13. Doyley, M.M.; de Korte, C.L., van der Steen, A.F.W., Céspedes, E.I.A noise and artifact reduction strategy for intravascular palpography. IEEE symposium of UFFCS, 1999

MINERALIZED TOOTH TISSUE STUDY USING SCANNING ACOUSTIC MICROSCOPY

R.Gr. Maev,[1] L.A. Denisova,[2] E.Yu. Maeva[3], A.F. Denisov[4]

[1,3]Department of Physics, University of Windsor
Windsor, N9B 3P4 Canada
[2,4]Institute of Biochemical Physics
Russian Academy of Sciences
Moscow, 117334, Russia

INTRODUCTION

The first paper dedicated to the application of ultrasound in dental research was published in 1963 by Baum et al.[1], since that time, barely thirty papers on that subject have appeared (detailed review in Maev et.al.,[2]). The objective of most investigators in that field was to use ultrasound in dental research for pathology diagnostics on the whole tooth in vivo. However, several factors such as small dimensions, complicated multilayered structures and the curved surface of teeth made it difficult to overcome problems imaging the microstructure of the whole tooth. With all this, there is no explanation why an excellent opportunity to study tooth microstructure with an acoustic microscope in flat-parallel slices without fixation, decalcination, dehydration and staining has been ignored for such a long period. We have not been able to find any detailed information concerning the hystological structure in acoustic images of teeth except for a brief description of enamel caries by Peck and Briggs[3,4].

The aim of the present study was to investigate various samples of normal and pathological tooth tissues using flat-parallel slices and longitudinally cut halves of teeth, and to compare obtained results with literature and data on tooth morphology. Fine hystological structure of human tooth had been well studied with conventional light and electron microscopy and described in numerous papers, monographs and manuals. On the basis of that knowledge we attempted to identify some well-known details in acoustic images of teeth.

Additionally, flat samples of some dental restorative materials were examined with special consideration of their microstructure and resemblance with tooth tissues in physico-mechanical properties.

SAMPLES

Human teeth were extracted for orthodontic reasons and stored in a saline solution with tymol. Flat-parallel slices were prepared by grinding extra parts. Acoustic images were obtained with ELSAM acoustic microscope[5] at a pulse frequency of 200, 400, 600 MHz and with a Wide Field Pulse Scanning Acoustic Microscope, fabricated in the Center for Acoustic Microscopy of the Russian Academy of Sciences[6], at a pulse frequency of 50 and 100 MHz.

Acoustic images of healthy and carious teeth display a distinct interface between dentine and enamel and showed boundaries of pulp chamber (Figure1, 2).

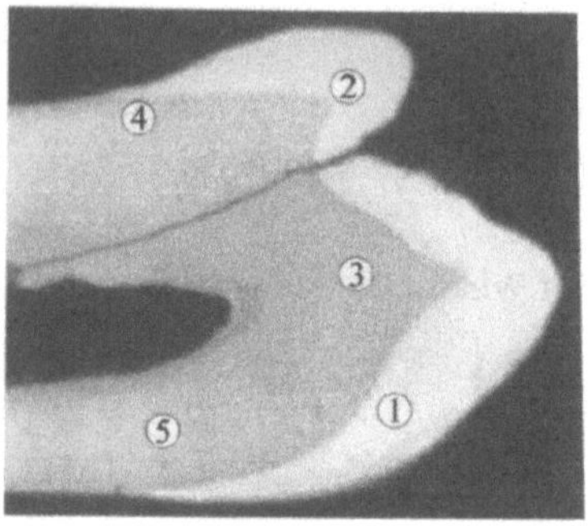

Figure 1. Acoustic image of human premolar. Scan size 11×17 mm, frequency 50 MHz, Numbers show spots used for sound velocity measurements

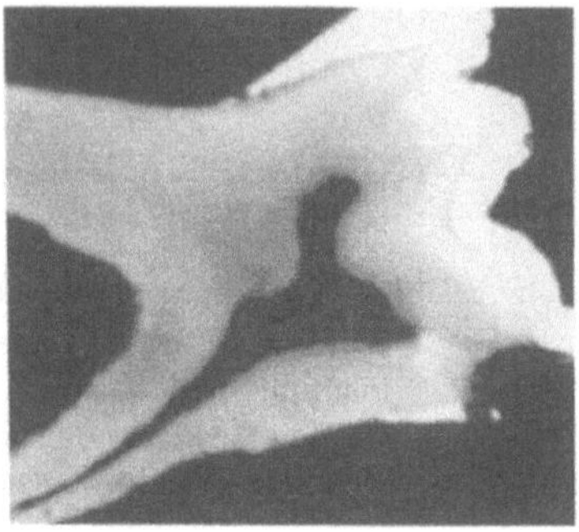

Figure 2. Human molar with a deep carious cavity. Scan size 10×12 mm 50 MHz

Detailed investigation reveals that dentin is not a homogeneous tissue and at least 3 zones with different acoustic properties can be observed: more light dentine in bulk, dark area around the pulp chamber and dark layer underneath enamel, so called mantle dentin. Elements of interglobular dentin can be seen as black spots beneath mantle dentin.

The sound velocity in different dental tissues was measured using several flat-parallel sections 1.2 to 2 mm thick. Figure 1 shows, as example, set of points where the measurements were taken. Delays of *L*- and *LT*-impulses were measured in enamel (points 1 and 2) and dentine (points 3, 4, and 5). Values for longitudinal and horizontal velocities are presented in Table 1.

Table 1. Sound velocities and impedances in dental tissues.

Parameter	Point Number					Water	Silver
	Enamel		Dentin				
	1	2	3	4	5		
c_L, km/s	5.9 ± 0.3	6.2 ± 0.3	3.9 ± 0.2	3.9 ± 0.2	3.6 ± 0.2	1.49	3.7
c_T, km/s	3.1 ± 0.2	3.4 ± 0.2	1.6 ± 0.1	1.6 ± 0.1	1.6 ± 0.1	–	–
Z, 10^6 g/cm²·s	1.8 ± 0.2	1.8 ± 0.2	0.8 ± 0.2	0.8 ± 0.2	0.8 ± 0.2	0.15	3.9

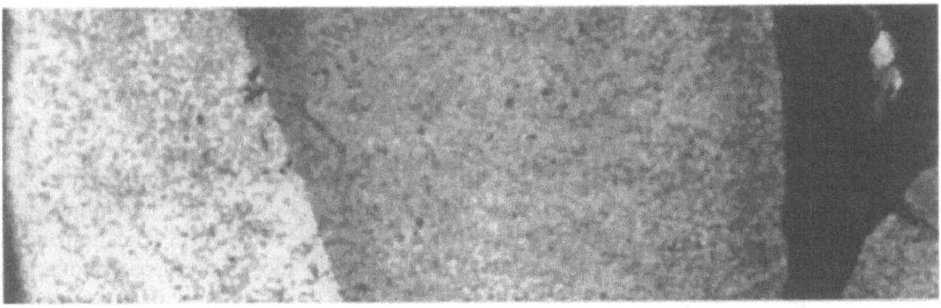

Figure 3. Enamel (left), dentin (center) and pulp chamber (right) in ELSAM acoustic microscope. scan size 1.3×4.0 mm, 200 MHz

For quantitative measurements in damaged regions, samples with caries in a spot stage were chosen. This is an early stage, which is characterized by change in enamel color (a white stain is visible on a dry surface), while process of its destruction has not yet begun. In dentine, the affected regions appear darker than healthy dentine tissue.

The dark zone around the pulp is 200–300 μm in width. It is less homogeneous (Figure 3). Areas with low acoustic impedance between spherical structures with higher impedance can be distinctly distinguished. It is known that in the circumpulpar area, production of new dentin occurs during the entire period of tooth life. Initially, an organic substrate is formed, and later on mineralization begins. Hydroxyapatite crystals form spherical structures called calcospherites, which grow and merge with each other, creating a mass of substance that is irregularly formed and has areas of lower density between merged globules. Mineral content of this immature dentin is lower than in bulk.

A narrow dark zone of dentin adjoining the enamel (width about 100 μm) is more homogeneous than dentin in the bulk (Figure 3). Formation of that dentin starts before teething. It features a special orientation of collagen fibers and hydroxyapatite crystals: both of them are parallel to dentin tubules and perpendicular to dentin-enamel interface.

In the bulk of dentin, the majority of collagen fibers are oriented tangentially to dentin tubules, and hydroxyapatite crystals are partially parallel to collagen fibers and partially form calcospherites.

Dentin tubules can be revealed as thin dark lines in longitudinal sections and as small orifices, they can be several microns in diameter in transversal sections (Figure 4). In enamel, rather weakly expressed dark and light slanting strips (width of 100–150 μm) in Figure 3 resemble the size and layout of Hunter-Schreger bands. Due to the S-shaped bend

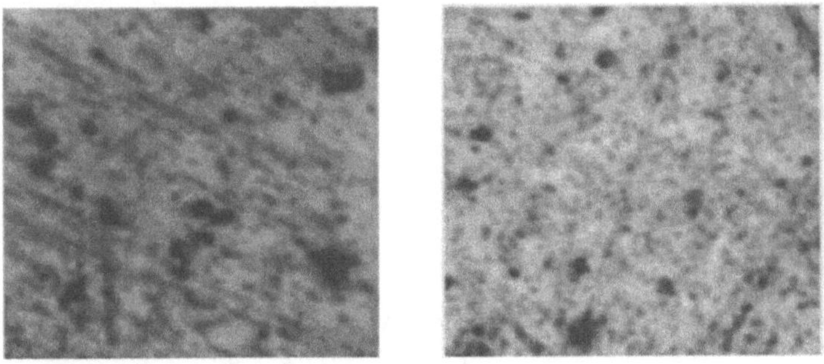

Figure 4. Dentin tubules at the acoustic images of transversal (left) and longitudinal (right) tooth slices. Scan size 300×300 μm, 600 MHz

of enamel prisms, Hunter-Schreger bands appear in longitudinal sections because some of them have been cut transversely and some of them along the axis. They look like more and less dense zones – parazones and diazones. Hunter-Schreger bands include 10–15 layers of prisms. There is no difference between parazones and diazones in mineral content, and variation of acoustic properties can be caused just by structural organization.

In a molar with deep carious cavities, (Figure 2) darker zones of dentine under the enamel and around the pulp chamber can be distinguished. Next to that, a complicated organization of dentine structure near a carious cavity can be revealed. The light area, spreading from the bottom of a carious cavity to the pulp chamber, is transparent dentine, which has higher density due to extra mineralization and obliteration of dentine tubules. Strengthening the dentine prevents distribution of the carious process into the pulp. Between the area of transparent dentine and healthy tissue, there is a small zone of dark dentine, which can be caused by rarefying of destroyed tissue. An interesting phenomenon is the layer with increased impedance, which is distinctly distinguished under the occlusive surface, which is beneath the dark mantle dentine. Width of this layer is 650–700 μm. We have not found any published data about differences of structure or density in the bulk of dentine, which resembles this revealed layer, and this fact requires additional investigation. However, it is possible to suppose that such hardening could develop to provide more strength under the occlusal surface and protect the harder but more fragile enamel from cracks.

Enamel attrition develops due to mechanical wear of enamel. In the acoustic image of a human incisor with deep attrition, (Figure 5) beside the dark zones around the pulp chamber and in the mantle dentin we can see light layers of dentin, 600–700 μm width, under vestibular and oral surfaces of the tooth. Under the area of complete enamel attrition, there is a central streak about 600 μm in width that spreads from the pulp chamber to the external surface. In the pulp chamber under attrition, there is a mass of reparative dentin, which has lower acoustic impedance. In the root we can see obliteration of the pulp channel and acoustic impedance in that part is higher than in the bulk.

The presented results demonstrate that, examining human teeth sections with ultrasound scanning allows us to obtain new information on dentin microstructure. It has

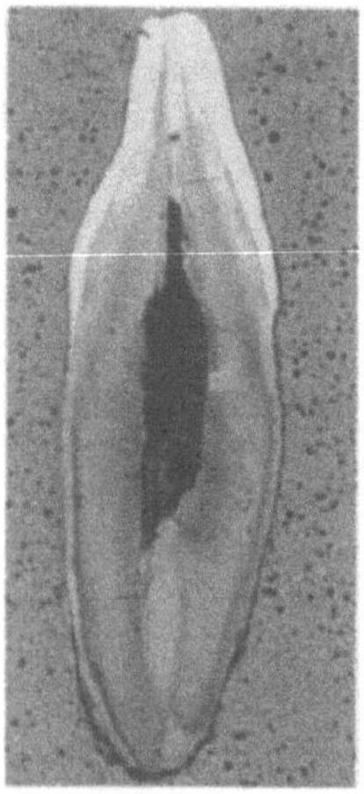

Figure 5. Human premolar acoustic image. Scan size 11×17 mm, 50 MHz

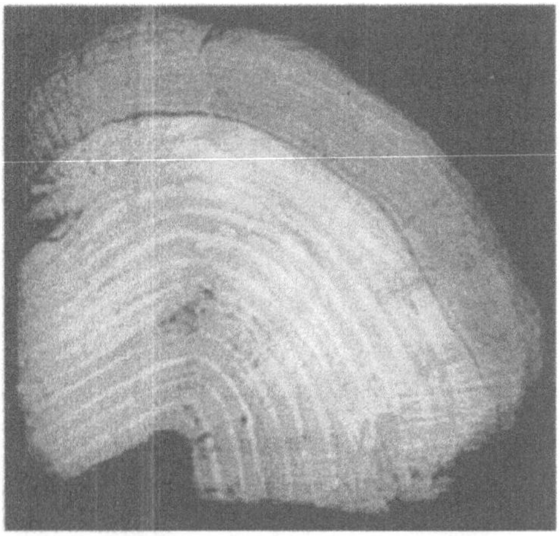

Figure 6. Acoustic image of a tooth section of northern elephant seal Mirounga angustirostris. Scan size 11×11 mm, 50 MHz. Fragment of the tooth has been kindly provided by Dr. Klevezal, Institute for Animals Morphology, Moscow

been shown that human dentin has at least 3 layers, which differ with their acoustic properties. In the case of decreasing tooth strength, due to carious lesion or attrition, additional local hardening of dentin occurs under the destroyed surface.

In mammals, layered structure of dentin is more distinct and is caused by seasonal changes of dentinogenesis. Alteration of lighter and darker zones in the tooth's acoustic images reveals a distinct picture of incremental lines (Figure 6). The number of incremental lines allows us to precisely determine that the animal was 12 years old. The width of circles reflects the animal's state in each year of its' existence.

Thus, we can see that structural visualization of dental flat-parallel sections with an acoustic microscope gains new and useful information about spatial distribution of tooth tissue elements, practically the same way as we can do with an optical or electron microscope. However, the most important feature of acoustic visualization is that we can study the unfixed, undemineralised, and unstained specimen. Moreover, acoustic images reflect physico-mechanical properties of tissue that is of special value in cases of mineralized tissue investigation.

At present, one of the most important branches of dental research is design of new dental materials for tooth restoration and denture modeling. Dental material must meet numerous standard requirements including its resemblance and compatibility to dental tissue in hardness, durability, elasticity, and compression resistance et.al. Acoustic microscopy provides us with a new nondestructive method for the evaluation of dental materials microstructure and physico-mechanical properties.

Acoustic image of a longitudinal slice of the tooth with restoration (Figure 7A) reveals inhomogeneity of the cement and composite, there are some micropores in the bulk. Bonding of the composite with the dentin is not complete. Under the restoration, dentine tubules look wider and more distinct; this may be due to acid etching. Ultrasonic imaging enables us to determine location, size and shape of pores and areas with bonding defects at the dentin-composite interface, as well as evaluate structural and physico-mechanical features of the material. It is important that we can examine composite consistency after polymerization and to detect any structural changing caused by shrinkage, aging degradation, oxidation or moisturising.

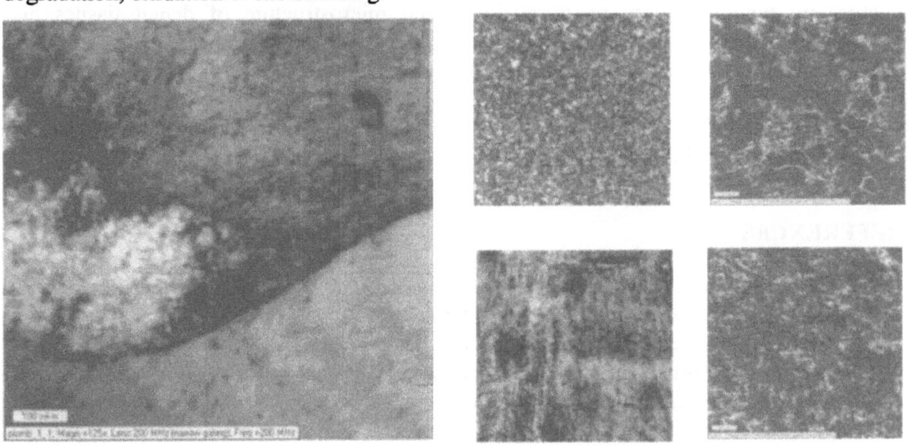

(a) (b)

Figure 7. (a) Prismafil restoration, 200 MHz, the sample has been kindly provided by Dr. I. Poyurovskaya, Central Institute for Dental Research, Moscow. (b) Distribution of hydroxyapatite (HAP) in polymer matrix of dental composites: *top right* – PMMA + 30% HAP, 50 MHz, 1×1 mm; *bottom right* – PE + 25% HAP, 50 MHz, 500×500 μm; *top left* – PE + 30% HAP, 400 MHz, 800×800 μm; *bottom left* – modified PE + 30% HAP, 400 MHz, 800×800 μm. Composite samples have been kindly provided by Dr. A.Volozhin, Moscow Medical Institute for Dentistry

(a) (b) (c)

Figure 8. Features of crack formation in materials, used for dentures: (a) shape of a crack on the surface of a white basic polymer, 200 MHz, 800×800 μm; (b) shape of a crack on the surface of etacrylic specimen, reinforced with 2 layers of threads installed at 45° to each other; (c) the same specimen, armouring threads at the depth of 1 mm beneath the surface, 50 MHz, 10×10 mm

Ultrasound scanning of dental composite samples (Figure 7b) produces images, which reflect the microstructure of the material, in detail, and allows us to evaluate the spatial distribution of hydroxyapatite grains in organic matrix, shape and size of inclusions, their ability to form aggregations and other morphometric indices. Acoustic images give us opportunity to find defects in material and simultaneously to visualise microstructure of the substance. Therefore we can observe how structural elements such as grains, granules, and micropores can affect crack or other defect formation (Figure 8). If multilayered material is studied, it is possible to control coordination of the internal reinforcing layer and its interaction with the developing crack (Figure 8b, 8c).

CONCLUSIONS

The results presented in our study, demonstrate that application of scanning acoustic microscopy methods allow us to evaluate fine microstructure of dental tissues and restorative materials. Acoustic images reflect, in detail, morphological features and provide the investigator with valuable data on local physico-mechanical properties. Acoustic microscopy is a powerful instrument, which can provide us with new knowledge about mineralized biological tissues and bring us closer to understanding the mechanical properties of teeth and the methods for restorative treatment.

REFERENCES

1. G.Baum., J.Greenwood., S.Slawski., R.Smirnov, Observation of internal structures of teeth by ultrasonography. *Science*, N.Y., v.139, pp.495-496 (1963).
2. R.Maev, Yu. Maximovsky, L. Denisova, E. Maeva, A. Denisov, T. Chirkova, D. Domyshev, Acoustic microscopy – new method for teeth investigation "Stomatologia", 2000, No. 5-6. (submitted).
3. S.D.Peck, G.A.D.Briggs, A scanning acoustic microscope study of the small caries lesion in human enamel, *Caries Res.* v.20, pp.356-360 (1986).
4. S.D.Peck, J.M.Rowe, G.A.D.Briggs, Studies on sound and carious enamel with the quantitative acoustic microscope, *J.Dent.Res.*, v.68, No.2, pp.107-112 (1989).
5. A.Thaer, M.Hoppe, W.J.Patzelt, The ELSAM Acoustic microscope, *Leitz Mitteilungen fur Wissenschaft und Technik*, v.2, No.3/4, pp.61-67 (1982).
6. K.I.Maslov, Acoustic Scanning microscope for investigation of subsurface defects. *"Acoustical Imaging"*, Plenum Press, N.Y.-London, v.19, pp.645-649 (1992).

BONE IMAGING BY LOW FREQUENCY ULTRASONIC REFLECTION TOMOGRAPHY

P. Lasaygues, JP. Lefebvre

CNRS Laboratoire de Mécanique et d'Acoustique
31 Chemin Joseph Aiguier
13402 Marseille cedex 20
France
Tel: 33 491 164 277 - email : lasaygues@lma.cnrs-mrs.fr

INTRODUCTION

This investigation deals with the reconstruction of object by ultrasonic reflection tomography (URT). URT borrows from echography its physical concepts (exploitation of scattered echoes) and from X-ray tomography most of its image formation process (numerical reconstruction from projections). The method is based on a linear approximation of the inverse scattering problem, the Inverse Born Approximation (IBA), which allows to reconstruct small perturbations from a known reference problem,[1,2,3].

For media with weak inhomogeneities, such as biological tissues (weak scatterers), the method works straightforward. This leads to a "Constant Background" IBA method, whose practical solution results in regular angular scanning with broad-band pulses, allowing one to cover slice-by-slice the spatial frequency spectrum of the imaged object,[4,5,6]. This leads to "Reconstruction-From-Projections" algorithms like those used for x-ray Computed Tomography.

For media with strong heterogeneities as bone (strong contrasts, large objects with respect to wavelength), the problem is quite non-linear and there is in general no single solution. However, for example, in the non-destructive evaluation domain, one is generally concerned only by flaws, which appear to be strong (but small and localised so that the result is a small disturbance) inhomogeneities in well known media, the part of component to be inspected. In this case, one can use a "Variable Background" IBA method - the reference background being the water-specimen set - to reconstruct the perturbation,[7,8].

Since characterization of human bone is a new domain of application for ultrasonics imaging (for osteoporosis diagnosis or orthopaedic research), one is interested in extending the method.

In this case, one would guess that Low Frequency Tomography (100 kHz < frequency < 1 MHz, leading to a better accuracy of the Born approximation) will have a larger domain of validity than the classical one. Moreover, when one turns down the frequency of the

Acoustical Imaging, Volume 25, Edited by Halliwell and Wells.
Kluwer Academic/Plenum Publishers, 2000.

transducer, one increases the penetration length of the wave. But this is done at the expense of resolution and the usual algorithm leads to poor resolution images, inappropriate for bone imagery.

To improve resolution, we propose to apply some signal processing prior to apply the reconstruction procedure. The chosen signal processing procedure is a deconvolution procedure of Papoulis type (extension of the generalised inversion in the complementary bandwidth of the electroacoustic set-up),[9]. The resolution enhancement procedure was first tested on various academic targets,[10]. We shown that the resolution of the image can be improve.

In this paper, we tried to improve the resolution of images of various kind of animal and human bones and we shown the possibility to obtain high-resolution images with low-frequency transducers.

BASIC PRINCIPLE OF ULTRASONIC REFLECTION TOMOGRAPHY (URT)

Basic principles of URT are now well established for imaging smoothly media as biological structures. Because of the almost homogeneous (from mechanical point of view) nature of biological tissues, one can linearise the scattering problem with using a Born approximation leads to a simple relation between the back-scattering frequency response for plane waves and the spatial Fourier transform of an "object", the impedance fluctuations. That relation states that, in frequency domain, broadband pulses give slices of the spectrum of the object. When working with the amplitude of the echoes like in echography, the object we are dealing with is called the "reflectivity", and it is that object we are interested to image in a first time.

The philosophy of the imaging method is simple inverse of the forward scattering problem. The result is called the Inverse Born Approximation (IBA), more precisely, for biological structures, the IBA with "constant background", that background being the propagation medium, that is the uniform medium equivalent (for propagation) to the tissue, that is nothing but water.

The technique used for image formation is a "reconstruction from projections" technique as those used in X-ray computed tomography. It is the natural way, since we acquire slices in the frequency domain. The algorithm we use is an improved version of standard "summation of filtered back-projections" algorithm that takes in account the relative sparse nature of the data.

To bone imaging, the problem is quite non-linear: cancellous or cortical bone is quite inhomogeneous and porous. High frequency waves used in classical echography are very perturbed by this kind of strong contrast media. To limited attenuation and scattering effects of the wave by heterogeneities, the low frequency tomography (< 1 MHz) is gradually becoming an interesting tool with a larger domain of validity.

With low frequency transducers, the penetration lengths of the wave increase. But the bandwidth of the transducer is then limited and the usual algorithm leads to poor resolution images. We then tried now to operate an improvement of the spectral bandwidth by Papoulis deconvolution,[10] based essentially on a non-linear adaptive extrapolation of the Fourier domain.

The general architecture of the mechanical system is that of a first generation X-ray tomograph : a main symmetric arm holds two transverse arms allowing the parallel translation of two transducers. Angular scanning is allowed by the rotation of either the main arm or the object holder. The transducers used for data acquisition were Imasonic broadband videoscan transducers, with center frequencies of 500 kHz and 1 MHz.

PAPOULIS DECONVOLUTION

The method proposed by Papoulis,[10] to determine h(t) as a function of its Fourier transform within a band, is a non-linear adaptive modification of an extrapolation method. It takes advantage of the finite width of impulse responses in both time and frequency.

We consider a function h(t) and its Fourier transform $H(\upsilon)$. If we define the frequency band $B = [\upsilon_1 , \upsilon_2]$ of X , we assume that $H(\upsilon)$ is known only in the segment B, i.e. that only the function $W_1(\upsilon)$ is known.

$$W_1(\upsilon) = H(\upsilon)P(\upsilon) = G(\upsilon)Y(\upsilon) \text{ where } P(\upsilon) = \begin{bmatrix} 1 & \text{if } \upsilon \in B \\ 0 & \text{if } \upsilon \notin B \end{bmatrix} \qquad (1)$$

The problem is to compute $H(\upsilon)$ for all frequencies, i.e. to compute $H(\upsilon)$ from the knowledge of $W_1(\upsilon)$. We build the different functions $W(\upsilon)$ and the inverse Fourier transforms w(t) by an iterative processing,[11]. The main problem is the estimation of B and T ; i.e. the level under which the signal is considered as noise.

The "true" impulse response of the object is given after two deconvolutions : deconvolution by the mechanical response of the apparatus determined for each angles and deconvolution by the impulse response of the transducer given by air-water interface echoes.

POROUS AND LONG BONE IMAGING

Since many years, scientists and physicians exploit the potentiality of the low frequency ultrasonic waves (500 kHz) to bone characterization. For example, Laugier and al, have developed an apparatus to control in vivo the osteoporosis of woman. The apparatus is based on celerity and attenuation imaging of specified bone (Broadband Ultrasound Attenuation Method),[12]. On clinical aspects, the system detects osteoporosis only on calcaneum that is an excellent indicator of skeletal status. But, analysis of femur, vertebra or wrist appears better sites to follow the evolution of the illness. With the geometry of these sites, the tomography is better adapted to a good diagnostic,[13].

In biomechanic, scientists are interested by a systematic characterisation of bone structures,[14] or by the knowledge of elastic properties of bone,[15,16]. The quantitative imaging is an objective. In orthopedic surgery, we can speak about biocompatibility or prosthesis and in this case, the low frequency ultrasonic tomography becomes an interesting tool.

This approach to generation of high-resolution low frequency tomography was the validated experimentally.

Cancellous bone : vertebra

This example concern analysis of L2 lumbar vertebra, with visibly external spinal body diameter of approximately 30 mm. The specimen containing a 4 mm circular metallic rod.

For this kind of porous and heterogeneous body, the classical imagery with high frequencies ultrasonics used in medical echography (1 to 10 MHz) don't allow visualise internal structure. In fact, multiple reflections and interference phenomena disturb considerably waves. Attenuation of the wave is very important.

For our application, the nominal frequency of the transducer is 500 kHz (λ = 3 mm in water) and the sinogram is constitute by 180 projections with 1024 samples. The size of the image is 255 x 255 pixels. The first image reconstructed was pulse echo reflection tomography without treatment (Figure 1).

Figure 2 present low frequency tomogram after deconvolution with frequential threshold at -30 dB, with frequency range [0.16 – 1,17] MHz, the contour of bone is well mark and the dimension of the rod is well reconstructed.

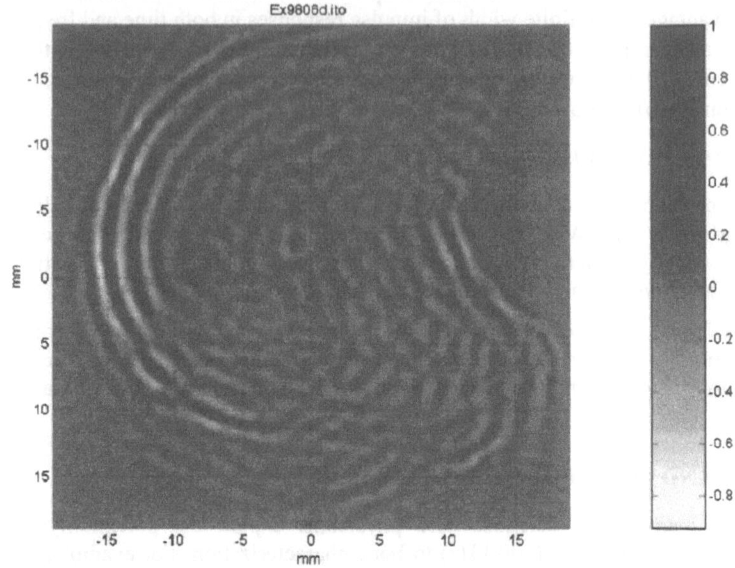

Figure 1. Low Frequency (500 kHz) URT of cancellous human vertebra
($\phi \approx$ 30 mm -180 projections - 1024 samples - image 255 x 255 pixels)

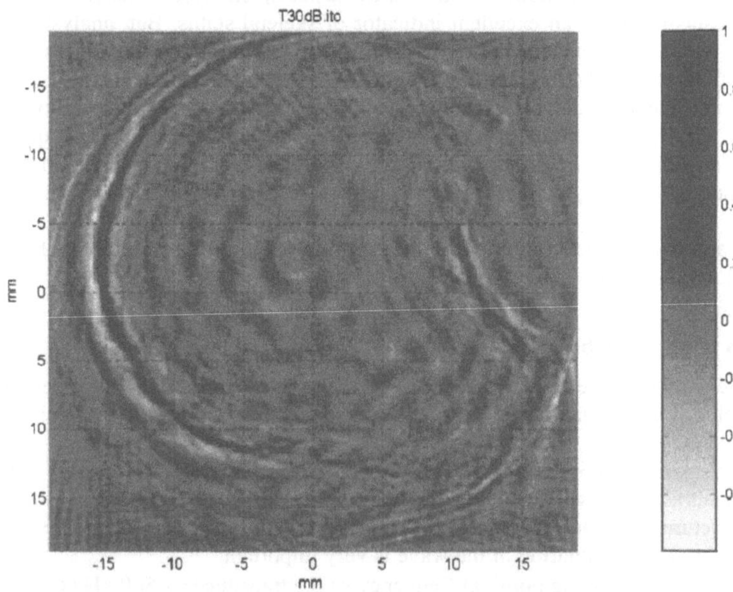

Figure 2. Low Frequency (500 kHz) URT of cancellous human vertebra
Papoulis deconvolution of each projections (frequency range [0.16 – 1,17] MHz , -30 dB)

Cortical human femur

The second example concern the non-invasive measurement of bone mass. X-ray computed tomography is the main method to visualise structural characteristic of bones (Figure 3). Nevertheless in contrast with X-rays which are propagating according to straight lines, acoustical ray are diffracted by the medium in which they propagate.

The tomogram of the figure 4 is obtained after Papoulis deconvolution out the frequency bandwidth [330 – 760] kHz (–15 dB), the boundaries are correctly improvement.

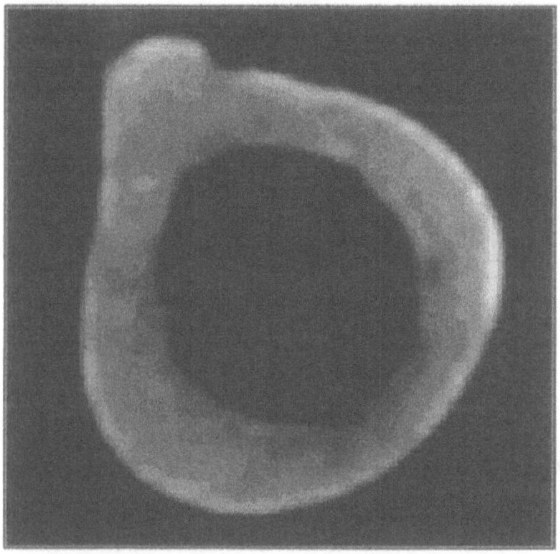

Figure 3. X-ray tomogram of cortical human femur

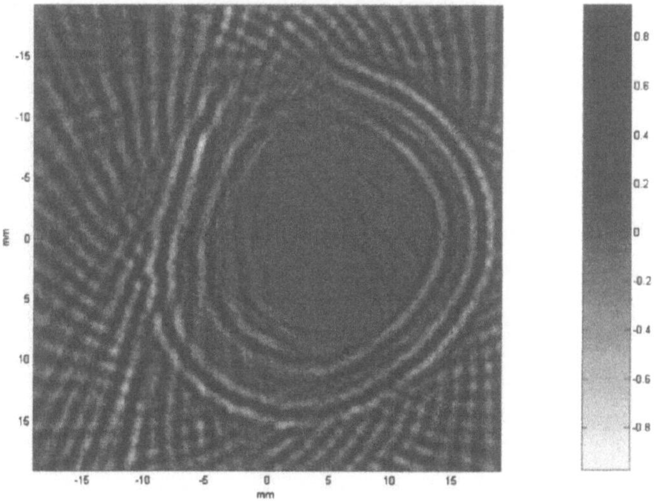

Figure 4. Deconvolued ultrasonic tomogram of cortical human femur
(60 projections, 1024 samples; 255 x 255 pixels, frequency range [330 – 760] kHz, –15 dB)

Cortical bovine femur

Now we show the results of employing low frequency ultrasonic tomography with cortical bovine femur (Figure 5). The femur, whose mean apparent thickness is 8 mm, is placed at the center of the bench. The target is insonified at 90 angles with 2048 samples for each projection. The nominal frequency of the transducer is 1 MHz. After deconvolution, we improve the resolution of the image (Figure 6), at contour and thickness.

Figure 5. Picture of bovine femur

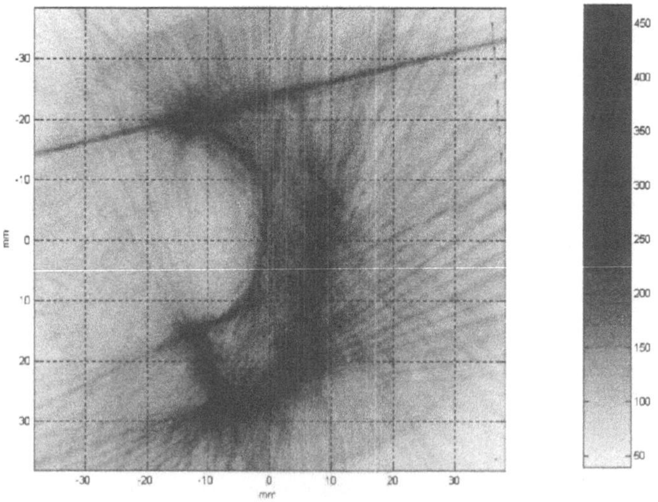

Figure 6. Deconvolued Low Frequency (1 MHz) URT of cortical bovine femur (90 projections, 2048 samples; 255 x 255 pixels, frequency range [330 – 760] kHz, –15 dB)

CONCLUSION

An implementation scheme for bone ultrasonic imaging by means of reflected tomography has been presented.

Ultrasonics is used in medical engineering with high frequencies (> 1MHz) and the results is very successful in imaging applications as echography.

Echography allows analysing only small contrast objects as biological tissues. In this case, the linear theory is available as Born approximation.

If the medium present strong heterogeneities or strong contrast as bones, the problem is more complicated and high frequency techniques, as classical echography are then very limited and not outstanding.

For this kind of medium, low frequency ultrasonics (< 1 MHz) are more interesting because the penetration length of the wave increase.

Scientists and physicians used low frequency ultrasonics to bone imaging and characterisation. For example, to diagnostic the osteoporosis of bone, one can applied angular tomography to cylindrical sites as vertebra, femur or wrist. But the useful of low frequencies in conjuncture with tomography technique decrease the image resolution.

In this paper, we have exposed a solution to imaging porous and long bones by used of low frequency ultrasonic reflection tomography and we have presented an improvement of the resolution by Papoulis deconvolution.

This approach demonstrates potential for increasing the image obtainable about bone.

REFERENCES

1. A.C. Kak, B.A. Roberts, Reconstruction from Projections: Applications in Computerized Tomography, *Handbook of Pattern Recognition and Image Processing,* pp. 649-693, (1986).

2. S.J. Norton, M. Linzer, Ultrasonic reflectivity tomography: Reconstruction with circular transducer arrays, *Ultrasonic Imaging,* 1 154 – 184, (1979).

3. J.P. Lefebvre, Progress in linear inverse scattering imaging: NDE application of Ultrasonic Reflection Tomography, in *Inverse Problem in Engineering Mechanics*, pp 371-375, A.A.Balkema/ Rotterdam/Brookfield, (1994).

4. J.F. Greenleaf. Computerized transmission Tomography, *in methods of Experimental Physics,* vol 19, Ultrasonics, pp 563-589, (1981).

5. A.J. Devaney, A filtered backpropagation algorithm for diffraction tomography, *Ultrasonic Imaging,* 4, 336-350, (1982).

6. S. Mensah, Tomographie ultrasonore en réflexion, application à l'imagerie médicale, *PhD, University of Aix-Marseille II,* (1990).

7. O.S. Haddadin, Ultrasound inverse scattering for tomographic imaging and self-focussing arrays, *PhD, University of Michigan,* (1998).

8. P. Lasaygues, Etude de l'écho d'une fissure, traitement de signaux par analyse en ondelettes en contrôle non destructif, *PhD, University of Aix-Marseille II,* (1992).

9. P. Lasaygues, J.P. Lefebvre, S. Mensah, High Resolution Low Frequency Ultrasonic Tomography, *Ultrasonic Imaging*, vol. 19, p 278, (1997).

10. A. Papoulis, A new algorithm in spectral analysis and band-limited extrapolation, in *IEEE Trans. Circuits Syst.*, vol. Cas-22, 9, pp. 735-742, (1975).

11. P. Lasaygues, J.P. Lefebvre, High-Resolution Process in Ultrasonic Reflection Tomography, *24th International Symposium on Acoustical Imaging*, Santa Barbara, California, USA September 23-26, (1998).

12. P. Laugier, P. Giat, G. Berger, New ultrasonic methods of quantitative assessment of bone status, *European Journal of Ultrasound* 1, pp 23-38, (1994).

13. E. Ouedraogou, P. Lasaygues, J.P. Lefebvre, G. Berger, M. Gindre, P. Laugier, Transmission Ultrasonic Tomography of Human Femur: Limits and Application, *25th International Symposium on Acoustical Imaging*, Bristol, United Kingdom, March, (2000).

14. R.B. Martin, D.B. Burr, N.A. Sharkey, Skeletal Tissue Mechanics, *Springer Verlag*, (1988).

15. J.L. Katz, H.S. Yoon, S. Lipson, R. Maharidge, A. Meunier, P. Christel, The Effects of Remodeling on Elastic Properties of Bone, *Calcified Tissue International*, 36, pp 31-36, (1984).

16. M. Pithioux, P. Lasaygues, P. Chabrand, M. Jean, Characterisation of mechanical properties of compact bone using an ultrasonic method, *Journal of Physiology and Biochemistry*, (1999).

ULTRASONIC TRANSMISSION TOMOGRAPHY OF THE HUMAN FEMUR : APPLICATION AND LIMITS

E. Ouedraogo,[1] P. Lasaygues,[2] J.-P. Lefebvre,[2] G. Berger,[1] M. Gindre,[1] and P. Laugier[1]

[1]Laboratoire d'Imagerie Paramétrique CNRS UMR 7623
15 rue de l'école de Médecine 75270 Paris Cedex 06 France
[2]Laboratoire de Mécanique et d'Acoustique CNRS
31 Chemin Joseph Aiguier 13402 Marseille Cedex 20 France

INTRODUCTION

In recent years, quantitative ultrasound (QUS) measurements has played a growing role in the assessment and management of osteoporosis and is gradually becoming an integrated part of the diagnosis of osteoporosis[1]. This development is attributable to the now wide availability of ultrasonic equipment which provides equivalent fracture risk assessment compared to conventional X-ray absorptiometric techniques[2,5]. Currently available technologies are based on measurements in transmission of the slope of the frequency-dependent attenuation and the velocity[6]. However, QUS is still in its infancy. While currently available devices are limited to the measurement in transmission of peripheral skeletal sites (calcaneus, finger phalanges[6]), several investigations are being conducted at our laboratory to develop innovative QUS techniques to determine and utilise the full potential of this technology for the benefit of detecting pathological conditions that affect bone strength : QUS transmission imaging has the potential for enhancing the reproducibility in repetitive measurements, for monitoring disease progression or response to treatment[7]; reflection techniques have the potential for assessing bone properties at non peripheral skeletal sites[8], scattering techniques may be useful for the quantitative analysis of bone micro-architecture[9]. Transmission ultrasound tomography is another approach that should provide useful information based on quantitative images of bone at skeletal sites that are currently assessed by X-ray quantitative computed tomography, e.g., the wrist. In transmission tomography, the local velocity in the studied material can be evaluated. Sound velocity is a good index of bone mineralisation and can be used to predict the risk of fracture for osteoporotic women[6]. To the best of our knowledge, ultrasound tomography has not been applied successfully to the assessment of bone. One of the main difficulty in

Acoustical Imaging, Volume 25, Edited by Halliwell and Wells.
Kluwer Academic/Plenum Publishers, 2000.

applying a such technique to the investigation of bone relates to the high mismatch in acoustical properties between bone and soft tissue. Transmission ray tomography is a class of highly non-linear, ill-posed inverse problems due to well-known sources of ray bending and incomplete and noisy data. We use the Born approximation[10-11] to linearize the wave equation of the direct problem.

We have developed a transmission tomographic technique providing sound velocity images of cortical bone sections in which we compensate refraction by *a priori* knowledge of the wave path for a particular ray orientation in the material. This compensation is based on *a priori* knowledge of the geometrical and acoustical properties of the studied objects. In order to obtain parallel propagation of the acoustical rays in the test material we calculate the incident and refracted angles which determine the adequate position and orientation of the transducers.

THEORY

Direct problem

The direct problem is solved with the Born approximation : we assume weak scattering characteristics in the two media, surrounding fluid and object, and therefore we consider only first order term scattering. The solution to the single frequency inverse scattering problem is given by the following equation[12] :

$$\frac{1}{c_0}\int_0^{|R-T|} n(s)ds \equiv \tau \qquad (1)$$

where τ is the time-of-flight (TOF) along a straight line of the acoustic wave between points T and R along the path s. This time-of-flight is due to the disturbance n which is the weak variation of the refractive index in a background medium of velocity c_0. In our work τ represents the reconstruction parameter.

In an inhomogeneous medium, the time-of-flight may be written as a function of the local velocity $c(s)$:

$$\tau_{\mathrm{TR}} = \int_{\mathrm{TR}} \frac{1}{c(s)} ds \qquad (2)$$

where TR is the ray path from the transmitter (T) to the receiver (R). In transmission the TOF is modified by the acoustical properties of the object. Therefore, the delay term $\Delta\tau$ in equation (3) is the TOF difference due to the object.

$$\Delta\tau = \int_{RT} \left(\frac{1}{c(s)} - \frac{1}{c_o}\right) ds \qquad (3)$$

An image of velocity $c(s)$ can be obtained from the reconstruction of the TOF difference. This time is obtained from the cross-correlation between the signal resulting from propagation through the object and the surrounding fluid and the reference signal resulting from propagation in the surrounding fluid only.

Inverse Problem

Tomography is a multiview imaging technique. Each view corresponds to an incidence angle θ of the transmitter with respect to the object. For each view, the object is illuminated with an incident wave and the transmitted wave is measured. After recording several data sets, the views are assembled together to form the total image. Our image reconstruction technique uses a filtered backpropagation (FBP) algorithm based on Radon's theorem[13]. This algorithm assumes that the wave propagates along straight ray paths (Fig. 1). If $p(d,\theta)$ (d: axis of translation) represents all the projection data around the objet, the inverse Radon transform R provides the object reconstruction parameter $\tau(x,y)$ according to:

$$R^{-1}(p(d,\theta)) = \tau(x,y) \tag{4}$$

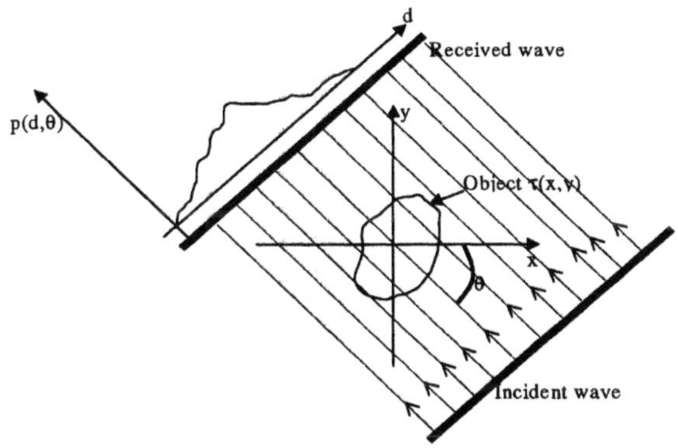

Figure 1. The classical configuration in transmission tomography. The different views are recorded by changing the angle of incidence θ around the object.

Using the FBP algorithm, the backpropagation can be formalized by :

$$\tau(x,y) = \frac{1}{2\pi}\int_0^{2\pi} \hat{p}(t,\theta)d\theta = \frac{1}{2\pi}\int_0^{2\pi} p(\omega,\theta)|\omega|e^{i\omega t}d\omega \tag{5}$$

where $\hat{p}$ represents the filtered data and ω the frequency domain filter. The idea is to backpropagate the measured data for each view and form a partial image[12,13]. These partial images are subsequently added to give the total tomographic image.

COMPENSATION

In transmission mode and for objects of high acoustical contrast compared to the surrounding medium, only a small proportion of the incident energy penetrates into the object. In addition the signal is highly attenuated, especially at high frequency, and the transmitted wave is deviated from its initial direction because of refraction, which results in an additional loss of data. To take into account the particular geometry of the object, we propose a more advanced compensation and correction procedure. This work is based on the usual assumption required for transmission tomography: propagation along straight parallel rays in the object to be reconstructed. Using Snell's law and *a priori* knowledge on the geometry of the object (i.e. the shape of the fluid-object interface), we determine the positions and orientations of the transducers which are required to obtain straight ray propagation (Fig. 2) through the object. Our first approximation consisted in assuming a simple cylindrical shape. Therefore, calculation of the position d and orientation α_1^* of the transducers was performed with solid cylinders whose position is known *a priori*.

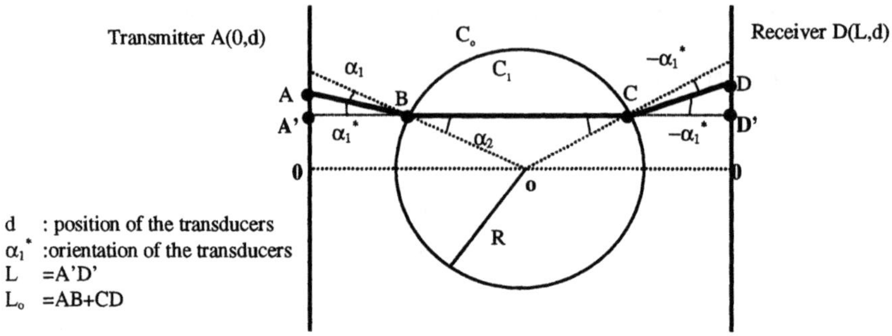

Figure 2. Geometrical configuration used in the compensation algorithm in the case of solid cylinder.

The diameter of the object was obtained from ultrasonic measurements in the echo mode. The position d and the orientation α_1^* are given by:

$$d = R.\sin\alpha_2 + (\frac{L}{2} - R.\cos\alpha_2).tg\alpha_1 * \qquad (6)$$
$$\alpha_1 * = -(\alpha_2 - \alpha_1)$$

The ultrasound waves propagated along the path A-B-C-D and the time-of-flight τ of the signal recorded at point D in Figure 2 was corrected to provide the time-of-flight τ' (equation 7) along the straight line A'-B-C-D' according to the FBP reconstruction algorithm.

$$\tau' = \tau - L_o(d).\frac{1-\cos\alpha_1^*(d)}{c_0} \qquad (7)$$

However the solid cylinder approximation is too far from the actual structure of long bones. The diaphysis of long bones consists of an approximately cylindrical cortical shell surrounding an approximately cylindrical central cavity filled with a fluid-like medium (bone marrow) and its projection may be considered as projection of hollow cylinder. In this case we used the solid cylinder correction everywhere except in the projection of hollow part of the object which was considered as a multi-layer medium with parallel and plane interfaces. For this approach the incident angle α_1^* may be closer to the normal incidence (0°) in the hollow part.

EXPERIMENTAL SET-UP

The experimental set-up consisted of two focused transducers, one transmitter and one receiver, mounted on the same mechanical axis. The central frequency of the transducers was 1 MHz, which represented a good compromise between the depth of penetration of ultrasonic waves in bone (the attenuation in bone increases with frequency) and resolution. At 1 MHz the lateral resolution of the transducer in the reference medium (water) was 5 mm. The object was placed in the center of the system and 90 tomographic views were recorded, each consisting of measurements taken at 128 equally spaced points (0.22 mm). Measurements of the reference signal (water without object) were also performed for the same 90 views.

Validation of the technique was first performed on a solid cylinder made of Plexiglas (ϕ=30 mm). We then used one hollow cylinder made of Plexiglas (external diameter $\phi_{ext.}$ = 20 mm and internal diameters $\phi_{int.}$= 8) filled with water in order to test the ability of the method to reconstruct the internal cavity of the object. The external diameter of the bone specimen is not constant; therefore we have tested the "robustness" of the method by introducing in Equation (6) extreme estimates of the solid cylinder diameter, 10 mm and 50 mm, which are beyond the true value of 30 mm. The method was then applied to a human femur specimen *in vitro*. Each object was tested with and without the compensation procedure. According to tabulated and reported values[14,15], velocities of 2700 $m.s^{-1}$ for the Plexiglas cylinders, 3400 $m.s^{-1}$ for the bone specimen and 1478 $m.s^{-1}$ for water were used in the model for the calculation of the positions of the transducers.

RESULTS AND DISCUSSION

The tomographic velocity images are shown in grey levels (from black to white as the value increases). The field of view is 32 mm × 32 mm.

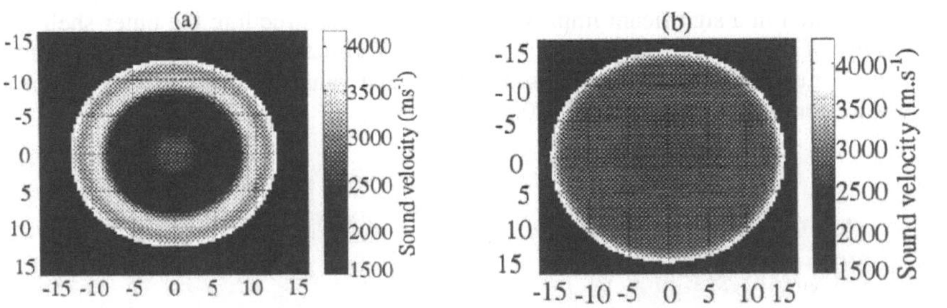

Figure 3. Tomographic images of the solid cylinder (ϕ=30 mm),(a) image obtained without compensation procedure, (b) image obtained with compensation procedure

Figures 3a and 3b show respectively the reconstructed image of the solid cylinder (ϕ=30 mm) without correction and with the correct orientation of the transducers. In Figure 3a the size was smaller than it should be (the estimated diameter was 83 % of the actual size). The corrected image was also more homogenous except on the edge where Gibbs phenomena occurred. The mean value of the velocity (± SD) in Figure 3b was 2710 ± 40 $m.s^{-1}$ and the size (29 mm) was close to the actual size whereas inaccurate values of sound velocity were observed in Figure 3a (the velocity varied from 1675 $m.s^{-1}$ to 3490 $m.s^{-1}$). The compensation procedure leads to a correct reconstruction in this case. Figures 4a and 4b show the results for the same cylinder obtained by introducing in the correction procedure *a priori* inaccurate values of the diameter, 10 mm and 50 mm respectively. Both images have a smaller size than the actual size (82 % of the actual diameter), however the velocity value fluctuates less than in Figure 3a. The mean velocity is 2200 ± 130 $m.s^{-1}$ in Figure 4a and 2000 ± 15 $m.s^{-1}$ in Figure 4b. These values illustrate that an inaccurate *a priori* knowledge of the diameter of the cylinder results in inaccurate reconstruction of both the shape and estimated velocity values. Further studies are clearly warranted to define the range of variation of the diameter that could lead to reasonable reconstruction.

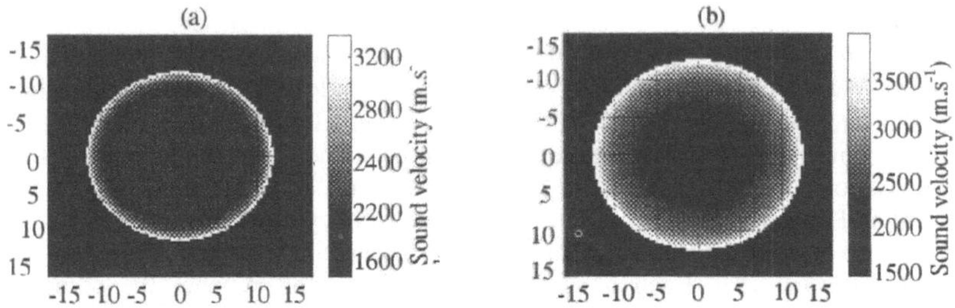

Figure 4. Images of the solid cylinder (ϕ=30 mm), (a) image obtained with compensation procedure with inaccurate value of diameter : 10 mm , (b) image obtained with compensation procedure with inaccurate value of diameter : 50 mm

Figures 5a and 5b show the images of a hollow cylinder (external diameter 20 mm, internal diameter 8 mm), respectively without and with the correction procedure. The data acquired without refraction compensation (Fig.5a) completely failed in reconstruction of the structure, size and sound velocity values: the inner hollow part of the cylinder is not reconstructed, the external diameter was 15 mm instead of 20 mm and the velocity values vary from 1730 $m.s^{-1}$ to 3350 $m.s^{-1}$). In contrast, using the local refraction correction (Fig.5b) resulted in a significant improvement in the reconstruction: the outer shell and the inner cavity of the cylinder were clearly visible, the estimated values of the inner and outer diameters (7 mm and 19 mm) were closed to the actual values and the velocity values in the cavity were correctly reconstructed (1500 $m.s^{-1}$).

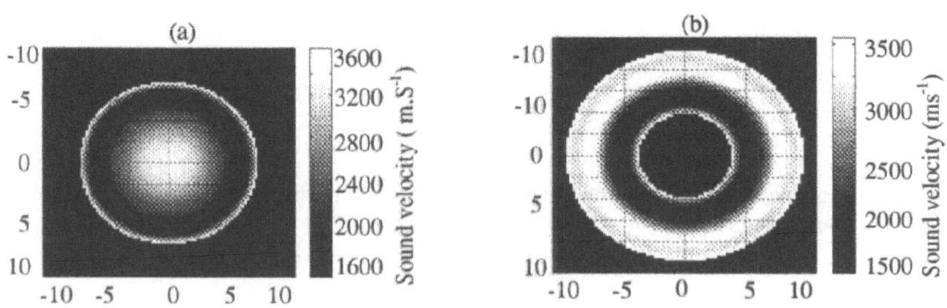

Figure 5. Images of the hollow cylinder (ϕ_{int}=20 mm, ϕ_{ext}=8 mm), (a) image obtained without compensation procedure, (b) image obtained with compensation procedure

However, some problems were encountered in the estimation of the velocity values in the shell showing two different zones: outer part 4200 ± 400 $m.s^{-1}$, inner part 2500 ± 250$m.s^{-1}$.

Reasons for this discrepancy are not fully elucidated yet, but several potential sources of inaccuracy can be anticipated including: limited precision of the mechanical acquisition device, "two-zones" compensation procedure too simple and difficulties encountered in the signal detection.

The reconstructed image without compensation of the bone specimen (Figure 6a) shows a reconstructed central cavity which velocity values were 3000 ± 100 $m.s^{-1}$ instead of expected mean value of 1500 $m.s^{-1}$. The reconstructed image with compensation shows that one part of the fluid in the internal shape was reconstructed with the correct values of velocity (1500 $m.s^{-1}$). Mean velocity values in cortical shell were respectively 4000 ± 600 $m.s^{-1}$ and 4100 ± 200 $m.s^{-1}$ in Figure 6a and Figure 6b. The reported values of the longitudinal velocity in cortical bone[14,15] vary from 3480 $m.s^{-1}$ to 4200 $m.s^{-1}$. The tomographic image of bone shows important distortion in Figure 6a than in Figure 6b. The inversion algorithm is based on the Born approximation according to which the object should be a weak scatter. But bone is an inhomogeneous scattering medium and in addition the measured signal is highly attenuated and complex for the time-of-flight determination. This may explain the distortion and the incomplete reconstruction of the internal shape.

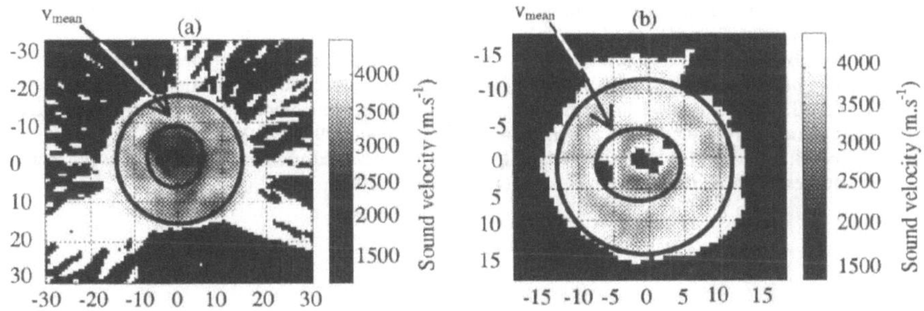

Figure 6. Images of the bone specimen (Femur), (a) image obtained without compensation procedure, (b) image obtained with compensation procedure

CONCLUSION

This new method which corrects locally the refraction in the object during data acquisition give a good result for cylindrical objects. The reconstructed images show that the shape and the velocity values were in good agreement with the actual object. "Two-zones" compensation resulted in a significant improvement in the reconstruction of a hollow cylinder. This type of compensation might enhance the efficiency of FBP reconstruction algorithm in case of bone. However the preliminary experimental results in bone requires algorithms for better time-of-flight determination.

REFERENCES

1 C. C. Glüer, Quantitative ultrasound techniques for the assessment of osteoporosis: expert agreement on current status, *J. Bone Miner. Res.* **12**, 1280-1288, (1997).

2 D. Hans, P. Dargent-Moline, A. M. Schott, J. L. Sebert, C. Cormier, P. O. Kotski, P. D. Delmas, J. M. Pouilles, G. Breart, P. J. Meunier, Ultrasonographic heel measurements to predict hip fracture in elderly women: the Epidos prospective study, *Lancet* **348**, 511-514, (1996).

3 D. C. Bauer, C. C. Glüer, J. A. Cauley, T. M. Vogt, K. E. Ensrud, H. K. Genant, D. M. Black, Broadband ultrasound attenuation predicts fractures strongly and independently of densitometry in older women, *Arch. Intern. Med.* **157**, 629-634 (1997).
4 P. W. Thomson, J. Taylor, R. Oliver, A. Fisher, Quantitative ultrasound (QUS) of the heel predicts wrist and osteoporosis-related fractures in women age 45-75 years, *J. Clin. Densitometry* **1**, 219-225 (1998).
5 E. W. Gregg, A. M. Kriska, L. M. Salamone, M. M. Roberts, S. J. Anderson, R. E. Ferrell, The epidemiology of quantitative ultrasound: a review of the relationships with bone mass, osteoporosis and fracture risk, *Osteoporosis Int.* **7**: 89-99 (1997).
6 Njeh, D. Hans, T. Fuerst, C. C. Glüer, H. K. Genant, Assessment of osteoporosis and bone status in : *Quantitative Ultrasound*, Edited by C. F. Martin Dunitz, london, (1999)
7 P. Laugier, B. Fournier, G. Berger, Ultrasound parametric imaging of the calcaneus:*in vivo* results with a new device, *Calcified Tissue Int.***58**:326-331 (1996).
8 E. Camus, M. Talmant, G. Berger, P. Laugier, Experimental study of lateral wave in human cortical bone, *IEEE 1999 Ultrasonics Symposium proceedings*, Eds SC Schneider, M Levy, BR McAvoy, IEEE, Piscataway, in press, (2000).
9 S. Chaffaï, F. Peyrin, G. Berger G, P. Laugier, Variation of human cancellous bone ultrasonic properties with density and micro-architecture, *IEEE 1999 Ultrasonics Symposium*, proceedings, Eds SC Schneider, M Levy, BR McAvoy, IEEE, Piscataway, in press (2000).
10 N. Sponheim, I. Johansen, Experimental results in ultrasonic tomography using a filtered backpropagation algorithm, *Ultrasonic Imaging* **13** : 56-70 (1991).
11 N. Sponheim, L.-J. Gelius,I. Johansen,J.J. Stamnes. Ultrasonic tomography of biological tissue. *Ultrasonic Imaging* **16** : 19-32, (1994).
12 J.F. Greenleaf, Computirized transmission tomography, in *: Method of Experimental Physics* vol.19, Ultrasonics Academic Press edition, NY, (1981)
13 A.C. Kak, B.A. Roberts, Reconstruction from projection : Application in computerized tomography, in : *Handbook of Pattern Recognition and Image Processing*, Academic Press edition, NY, (1986)
14 G Lowet, G. Van der Perre, Ultrasound velocity measurement in long bones :measurement method and simulation of ultrasound wave propagation . *J. Biomechanics* **29** :1255-1262, (1996)
15 S.C. Lee, B.S. Coan, M.L Bouxsein, Tibia ultrasound velocity measurement in situ predicts the material properties of tibial cortical bone, *Bone* **21(1)** : 119-125, (1997).

SHAPE AND SIZE RETRIEVAL OF OBSTACLES BY NONLINEAR INVERSION OF MEASURED SCATTERED FIELD DATA IN RESPONSE TO AUDIBLE PROBE RADIATION

Erick OGAM and Armand WIRGIN

Laboratoire de Mécanique et d 'Acoustique, UPR 7051 du CNRS,
31 chemin Joseph Aiguier,
13402 Marseille cedex, France

Abstract. Shape and size retrieval of an obstacle by inversion of the scattered audible acoustic field, and a method for acquiring this field in an anechoic chamber, are described. An optimization algorithm, artificial neural networks, and ambiguity diagrams, all appealing to the ICBA description of the scattered field, are employed for solving this ill posed inverse problem.

INTRODUCTION

This work is concerned with the inverse scattering problem of shape and size retrieval as arises in applications such as non-destructive evaluation, underwater imaging, medical ultrasonic imagery, robotics (artificial vision), etc. In many of previous investigations on this topic[1,2] the scattered field data was simulated numerically and no limitations were placed on the duration of the computations involved in the inversion. The emphasis herein is on a fast (for the near real time inversions covered by the above-mentioned applications) and fairly-accurate method (thus linearization schemes are avoided) that is capable of inverting measured scattered field data. The obstacles under consideration (located in an anechoic chamber) are cylinders (of length much greater than the wavelength) whose cross-section shape and size (comparable to the wavelength) are to be determined. The scattered field, to which audible probe radiation gives rise, is measured on a circle Γ_b circumscribing the cylinder in a plane perpendicular to the (z-) axis of the latter, so that, in the first approximation, the problem has 2D (cylindrical) symmetry. Only objects with star-shaped boundaries Γ_τ that can be described by the parametric equation $r = \tau(\theta)$, wherein $\tau(\theta)$ is a single-valued, continuous function of θ, are considered.

Acoustical Imaging, Volume 25, Edited by Halliwell and Wells.
Kluwer Academic/Plenum Publishers, 2000.

THE FORWARD PROBLEM

The object to be studied, occupying the bounded domain D in the x-y cross-section plane E, and whose boundary is $\Gamma_\tau = \partial D$, is insonified by a monochromatic acoustic wave (implicit time factor $\exp(-i\omega t)$, with ω the angular frequency and $i = \sqrt{-1}$). The host medium is air in which the sound speed is c, the total pressure field p, the incident wavefield p^i and the scattered wavefield p^s. From the experimental viewpoint, the scattering problem is to determine p^s from the measured fields p^i and p.

The objects treated herein are essentially sound-hard, so that the mathematical boundary-value problem, related to the physical forward scattering problem, is expressed by :

$$(\Delta + k^2)p(\mathbf{x}) = 0; \quad \mathbf{x} \in \mathbf{R}^2 \setminus \overline{D} \quad (\overline{D} \text{ the closure of } D,\ \mathbf{x} \equiv (x, y),\ k = \omega / c), \tag{1}$$

$$\partial p(\mathbf{x}) / \partial \nu = 0; \quad \mathbf{x} \in \partial D \qquad (\partial / \partial \nu \text{ the normal derivative}), \tag{2}$$

$$p(\mathbf{x}) = p^i(\mathbf{x}) + p^s(\mathbf{x}) \qquad (p^i(\mathbf{x}) = p_0 \exp(-ikr\cos\theta),\ (r,\theta) \text{ the polar coordinates in } E), \tag{3}$$

$$(\partial / \partial r - ik)p^s(\mathbf{x}) \sim o(r^{-1/2}); \quad r \to \infty, \text{ uniformly in } \theta . \tag{4}$$

This problem has been solved in closed form for circular cylinders [3]. The exact solution for scattering from such a canonical body is obtained by separation of variables and is expressed by :

$$p(\mathbf{x}) = p^i(\mathbf{x}) + \sum_{m=0}^{\infty} B_m H_m(kr)\cos m\theta; \quad \forall r \geq a , \tag{5}$$

wherein :

$$B_m = -p_0 \varepsilon_m i^m J_m'(ka) / H_m'(ka), \tag{6}$$

a is the cylinder radius, J_m, H_m the m-th order Bessel functions and Hankel functions of the first kind respectively, J_m', H_m' the derivatives of these functions with respect to their arguments, and $\varepsilon_0 = 1$, $\varepsilon_{m>0} = 2$.

For cylinders of more general (albeit star) shapes, a fairly-accurate solution is obtained by the Intersecting Canonical Body Approximation (ICBA) [4] and expressed by :

$$p(r,\theta) \approx p^i(r,\theta) + \sum_{m=0}^{M} F_m(\theta) H_m(kr)\cos m\theta; \quad \forall r \geq d \quad (d \equiv \underset{\theta \in [0,2\pi[}{Max}\, \tau(\theta)), \tag{7}$$

$$F_m(\theta) = -p_0 \varepsilon_m i^m J_m'(k\tau(\theta)) / H_m'(k\tau(\theta)), \quad M = O(kd) . \tag{8}$$

This method assumes that the response of the actual object to the incident wave, at points (r,θ), beyond the smallest circumscribing circle *r=d*, is approximately the same as the response (to the same incident wave) of the circular (canonical) body (center on the z-axis) whose radius is $\tau(\theta)$.

THE GEOMETRY OF THE SCATTERING PROBLEM AND THE EXPERIMENTAL SETUP

We use a commercial loudspeaker as an acoustic source. The latter is placed at one end of an anechoic chamber, with the object that is to be studied suspended at its center. Raw data is obtained by measuring: 1) the reference pressure field captured by a microphone placed near the mouth of the loudspeaker, 2) the total pressure field detected by a microphone moving along a circle Γ_b in a plane perpendicular to the axis of the cylindrical object, the latter being located in the far-field zone of the loudspeaker. This data is preprocessed, via Hilbert transforms, to obtain the amplitude and phase of the total field as a function of angular position (of the microphone). The same process is then repeated in the absence of the object, thereby yielding the incident field and subsequently (by subtraction) the scattered field. The output signals of the two microphones are acquired with an HP3566A (Paragon) signal analyzer.

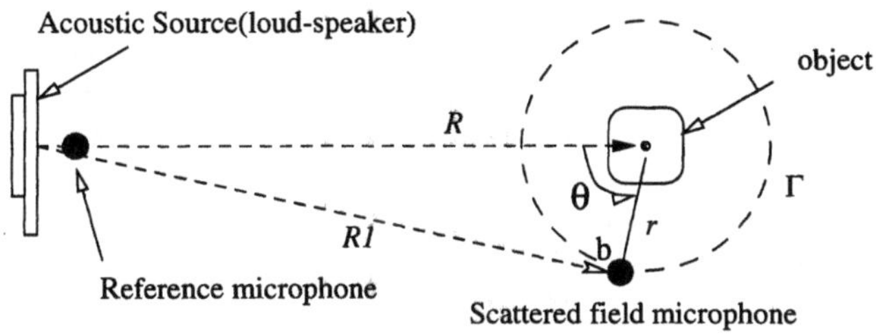

Figure 1. Geometry of the scattering problem

For the geometric configuration in figure 1, the measured acoustic pressure $P(b,\theta;t)$ at point (b,θ), in the absence of the object, is a spherical wave, which, due to the large distance between the loudspeaker and the center of rotation of the microphone, behaves like a plane wave, this being expressed by :

$$P^i(b,\theta;t) \sim [A/R_1(\theta)]\exp[-i(\omega t - kR_1(\theta)+\phi)] \approx [A/R]\exp[-i(\omega t - kR_1(\theta)+\phi)]; \quad kR >> 1, \tag{9}$$

wherein $R_1(\theta) = \sqrt{R^2 - 2bR\cos\theta + b^2}$, and A, ϕ are the amplitude and phase respectively at the center of the loudspeaker.

THE INVERSE PROBLEM

The inverse problem consists in determining the boundary $\Gamma_\tau = \partial D$ from knowledge of the measured scattered acoustic data. Several inversion schemes are employed, all appealing to the ICBA model[4] of the sound-obstacle interaction. Since the theoretical response of the canonical body (intersecting circular cylinder at each point of the obstacle) is known in closed-form (see (7) and (8)), the time-consuming resolution of the direct problem is bypassed during the inversion. Basically, the local radius $\tau(\theta)$ at measurement angle θ is determined by minimizing the discrepancy between the measured ($\hat{p}^s$) and ICBA model of the scattered field at (b,θ), i.e., by seeking $\tau(\theta)$ which minimizes the cost function (see (7) and (8)) :

$$J_l(\tau(\theta)) = \left| \hat{p}^s(b,\theta) - \sum_{m=0}^{M} F_m(\theta) H_m(kb) \cos m\theta \right|^l \qquad (10)$$

for some integer $l \geq 1$. If this is done at a rather dense set of angles in $[0,2\pi[$, then, in principle, the shape and size of the complete boundary of the object can be so determined. The use of the ICBA amounts to finding the minima of a single non-linear function (due to the non-linear nature of $F_m(\theta)$ as a function of $\tau(\theta)$) in one unknown (i.e., $\tau(\theta)$) for each measurement angle. The fact that $\tau(\theta_1)$ is obtained in fully uncoupled manner from $\tau(\theta_2)$, and that only single-parameter cost functions have to minimized, is a unique feature of the ICBA model.

Usually, many values of $\tau(\theta)$ correspond to a minimum of the cost function at angle θ. This is an oft-occurring phenomenon in shape identification problems[1] . When necessary, to overcome this difficulty, we employ data at two frequencies and single out the sought-for shape by a summation and filtering process. It turns out that this procedure is most successful for the insonified portion of the obstacle, so that it is advisable to rotate the object for fixed loudspeaker and microphones, rather than rotate the microphone, for fixed object and loudspeaker. The diversity of the previously-mentioned inversion schemes derives from the manner of seeking the minima of the cost functions. We employ two classes of methods: neural networks, and iterative minimization of a non-linear least squares cost function.

The first inversion scheme is implemented with the help of the Stuttgart neural network simulator and the MATLAB neural networks toolbox [5], and employs 6 input neurons, 1 output neuron and 3-7 hidden neurons. The target output of the network for each input is the local radius of the object. At each training phase we verify that the mean square error (MSE)

$$MSE = (1/N)\sum_{j=0}^{N-1} \left(T_j - O_j\right)^2 , \qquad (11)$$

(in which T_j is the target value and O_j is the corresponding solution obtained by the network for the *j*-th learning pattern, with *N* the total number of patterns presented to the network) is reasonably small. This error is minimized via the back-propagation algorithm inherent in the network.

The second inversion scheme appeals to the Nelder-Mead simplex method[6] for minimizing a function of several variables. Actually, we employ this method for one variable at each observation angle. Rather than displaying the minima of the cost function, one can give a more global representation of the object by simply tracing a polar plot of the cost function, the minima of which indicate possible candidates for points on the object boundary.

RESULTS

Alignment and proper shielding from stray objects was assured by comparison between scattered fields obtained synthetically (via (5) and (6)) and the ones measured experimentally for a circular cylinder test obstacle insonified by a plane wave (see figure 2). This aspect of the experimental procedure is crucial in order to reduce mismatch between the

experimental data and the theoretical prediction of the latter employed in the inversion.
Figure 3 depicts the degree of success of the neural network and Nelder-Mead inversion schemes. The latter fares better as concerns shape, and both are quite accurate as concerns size (in fact, the errors are those that can be expected from the evaluation of the measurement error (see figure 2).

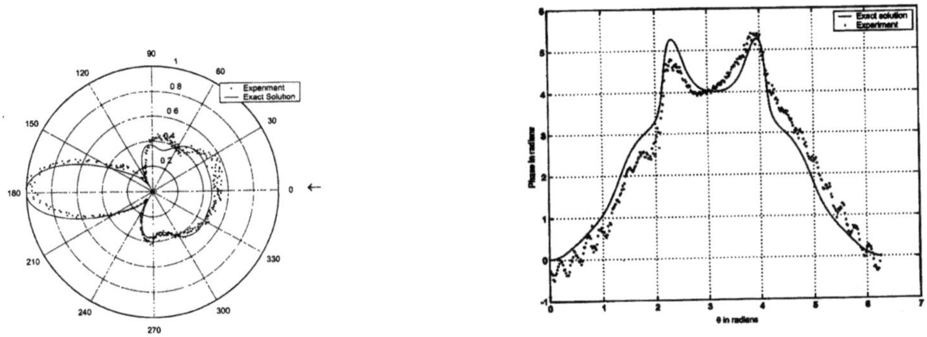

Figure 2. Amplitude and phase of the scattered field at 3500Hz for a solid PVC cylinder *a*=0.06m, *b*=. 19m

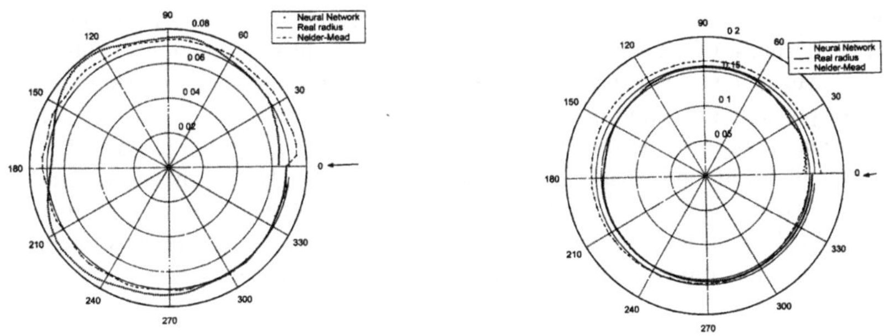

Figure 3. Reconstruction for a steel cylindrical shell (radius a=0.07m) and hollow PVC cylinder (radius a=0.1575m) using neural network . . . and Nelder-Mead simplex minimization . _ . _

As mentioned previously, it can be useful to employ a second method for acquiring the scattered field which (for fixed loudspeaker) consists in rotating the object and measuring the field with a fixed microphone placed at an azimuthal angle of $\pi/2$. This was employed for boundary identification of two objects : four long laminated wood panels taped together at their edges so as to form a hollow square-shaped cylinder(size 0.20m x 0.20m), and a hexagonal-shaped hollow metallic cylinder. The polar diagram of the cost function $J_1(\tau(\theta)$ (which is often termed « ambiguity diagram ») is given in figure 4, wherein it can be seen that this representation gives directly the object's shape, size and position (with respect to the axis of rotation). Of course, secondary minima introduce ambiguities in the identification, which can be resolved by the two-frequency operations described previously. Note should be taken of the fact that the corners of the square cylinder are not well-resolved, probably due to the non-rigid scotch-tape bond at these locations (which, of course, is not accounted-for in the ICBA model).

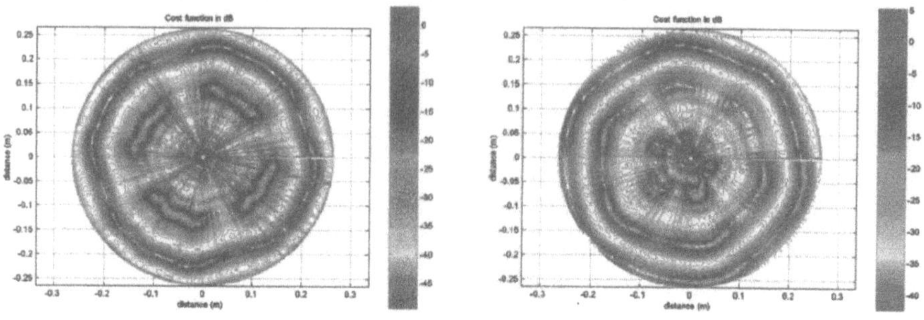

Figure 4. Cost function for rotating square and hexagonal-shaped cylinders at 3.5KHz for a fixed loudspeaker and fixed microphone at $\pi / 2$

CONCLUSIONS

A simple experimental method for acquiring and preprocessing the scattered acoustic field was shown to give rise to data that accurately matches theoretical predictions (both in amplitude and phase) for circular cylinder objects. A slight modification of this experiment, which consists in revolving the object instead of the microphone, was shown to be more suitable for providing the kind of information required by the wave-object interaction model. The latter was chosen to be the ICBA, which provides a good compromise between high computational speed and shape reconstruction accuracy. Neural networks and standard minimization schemes were shown to give rise to comparable images of the object, whereas the ambiguity diagram technique provides a useful alternative which reveals the existence of image artifacts while obviating the explicit search for minima of the cost function.

REFERENCES

1. COLTON D. and R. KRESS: 1992, Inverse Acoustic and Elecromagnetic Scattering theory. Berlin Heidelberg New York: Springer.
2. KLEINMAN R.E. and P.M.VAN DEN BERG: 1994, 'Two-dimensional location and shape reconstruction'. In: Radio Science, **29**, 1157-1169.
3. MORSE P. and H. FESHBACH: 1953, Methods of Theoretical Physics. International series in Pure and Applied Physics. New York: McGraw-Hill Book Company Inc.
4. SCOTTI T. and A. WIRGIN: 1995, 'Shape reconstruction using diffracted waves and canonical solutions'. Inverse Probs., **11**, 1097-1111.
5. Neural Network Toolbox User's Guide. Mathworks Inc.
6. BERTSEKAS D.P : 1995, Nonlinear Programming. New York : Athena Scientific.

PERFORMANCE OF ULTRASOUND ECHO DECOMPOSITION USING SINGULAR SPECTRUM ANALYSIS: MONTE CARLO APPROACH

Carlos Dias Maciel[1] and Wagner Coelho de Albuquerque Pereira[1]
[1] Biomedical Engineering Program - COPPE/UFRJ – PO Box. 68510
Zip 21945-970 Rio de Janeiro - RJ, Brazil - maciel@peb.ufrj.br
wagner@peb.ufrj.br

INTRODUCTION

The problem of quantitative ultrasound (US) characterization of media is still a main topic for research. In particular, in the field of biological tissue identification, statistical signal processing tools have been explored using basically the spectral analysis approach to quantify parameters like mean particle diameters and mean scatterer spacing (MSS), etc [2,4].

Among those techniques Singular Spectrum Analysis (SSA) seems to be a powerful tool for time-series analysis different of the common approaches and not yet well explored for biological tissue characterization. It has the capability to identify periodic structures from the time signal even in a noisy enviroment.

The objective of this work is to investigate the potential of SSA when applied to simulated US backscattered echoes, submitted to a Monte Carlo process. An experimental example of bovine liver is also shown and discussed.

THEORY

The SSA is a tool commonly used on nonlinear dynamics studies [1,6,7] and several signal processing areas [5]. It is based on the classical Principal Component Analysis (PCA) and gives the principal axes of a data vector ($\boldsymbol{x}_i$, $1\leq i \leq N$) which can be expanded with respect to an orthonormal basis (E^k, $1\leq k \leq M$):

$$\boldsymbol{x}_{i+j} = \sum_{k=1}^{M} a^k{}_i E^k{}_j \,,\ 1\leq j \leq M, \text{ where } M<N. \qquad (01)$$

The projection coefficient a^k_i are called the Principal Components (PCs), and the basis vector E^k, the empirical orthonormal function (EOFs). The vectors E^k are the eigenvectors of the cross-correlation matrix (C_x) of the sequence ($\boldsymbol{x}_i$). This procedure is a direct application of the general Karhunen-Loève bi-orthogonal expansion.

The presence of eigenelements whose values are close enough to be identified as a "pair" can be associated with a signal periodic component. The frequency of this periodic component can be found by taking the average of the peak of the amplitude spectrum of the 2 associated eigenvectors. Thus, SSA provides a way to characterize separately the periodic and non-periodic structures of the medium. The next challenge is to define a rule to establish what is a pair of eigenvalues and how many of them should be accepted to compose the periodic subspace of the data.

For the present analysis we will use a visual criterion to identify a pair of eigenvalues. Then we will consider one or two pairs if they represent at least 65% of all variance and the associated frequency of their eigenvectors are no more than 2.5 % apart of each other.

RESULTS

It will be shown two results of this analysis. The first one is an example of a real phantom signal using a set of parallel nylon lines (spaced of 3.0 mm) immersed into water. The inspection pulse is 5.9 MHz (duration 0.5 μs). The received signal was digitized in oscilloscope (TDS 420-Tektronics), sampling frequency of 25 MHz. Figure 1A shows the RF echo backscattered from the phantom. The two arrows show the signal window used in the analysis. Figure 1B shows a zoom of this window.

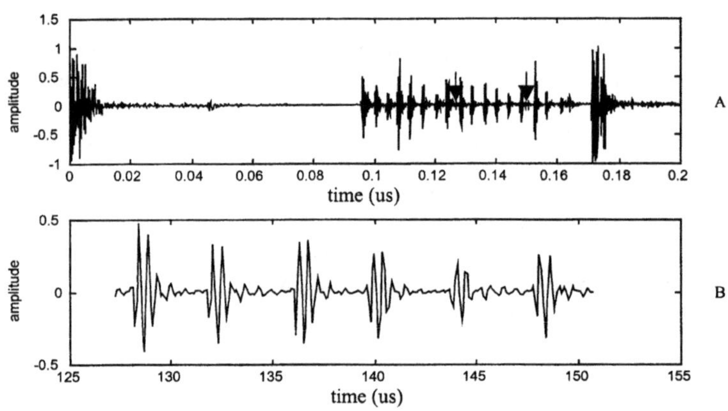

Figure 1 – Ultrasound RF line (A) arrows show the signal window to be analyzed (B) zoom of the window marked between two arrows.

Figure 2A shows the Singular Spectrum with the eigenvalues in a descending order. It is observed that it has two eigenvalue pairs (arrows) that may reproduce a quasi-periodic behavior. The continuous line represents the accumulated variance. In Figure 2B it is presented the power spectrum (as a gray scale) of the eigenvectors in the Y-axis and in the X-axis is the eigenvector component number. Each pair of eigenvectors has the peak response in the same frequency (arrows). This indicate the presence of periodicity.

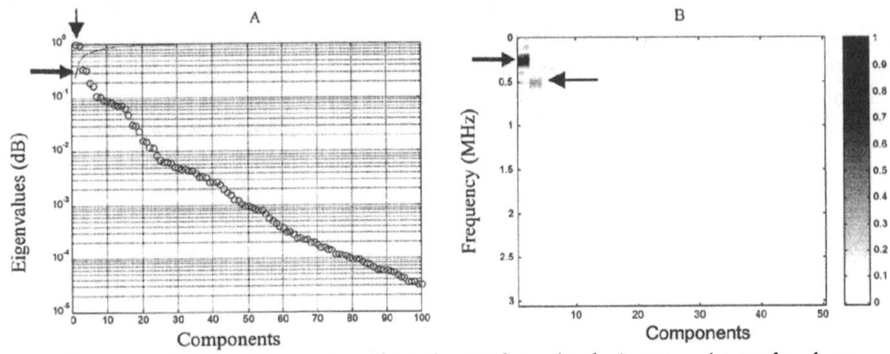

Figure 2 - (A) Singular spectrum from the envelope signal. Arrows point to the chosen eigenvalue pairs. Continuos line represents accumulated variance. (B) Frequency response of eigenvector pairs that indicate the presence of periodic structures. Amplitude is in gray scale.

In Figure 3A, it can be observed the reconstructed signal using these two pairs of eigenvectors and, in Figure 3B, its power spectrum. The non-periodic signal (figure 3C) is obtained by subtracting the periodic signal from the original one. In figure 3D its power spectrum. It is important to notice that the peak in figure 3B is in 0.254 MHz which corresponds to a periodicity of 3.03 mm (the real spacing is 3 mm).

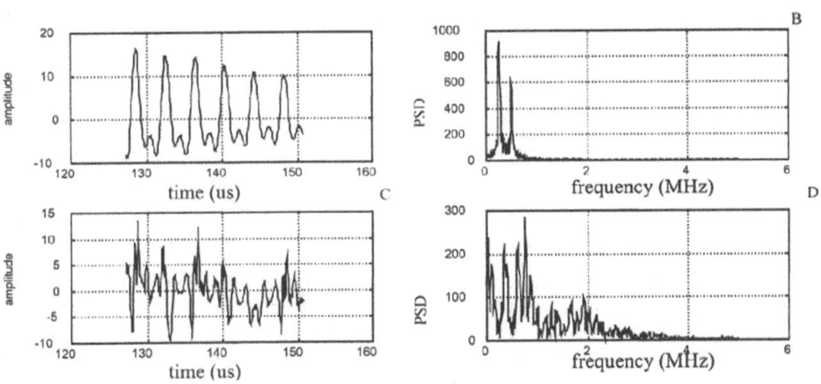

Figure 3 – (A) Reconstructed periodic signal, (B) power spectrum of the reconstructed periodic signal, (C) non-periodic signal, (D) power spectrum of the non-periodic signal.

The second result is a Monte Carlo analysis using simulated backscattered echoes to see how this kind of process behaves in different scenarios. The simulation model adopted is a one dimension RF echo from tissue, using a convolution of tissue scattering amplitude and a RF pulse inspection as described in [2,4]. The simulation parameters are Signal-to-Noise Ratio (SNR) in five different levels (12, 5, 0, -3 and –5 dB) and jitter of the regular structure (1%, 5%, 10%, 20% and 30%). SNR is estimated from the ratio: regular structure backscattered echoes to diffuse backscattered echoes. It is important to note that both signals represent information from biological tissue. Five hundred simulated signals were generated for each case. As an example, in Figure 4 there are the histograms (in two different forms) where the X-axis represents MSS expressed in MHz (Frequency=c/2*MSS, with c as the velocity of ultrasound propagation) and Y-axis is the number of events. In Figures 4(C) and 4(D), both with lower SNR, it is possible to observe that one cluster is being built around the first harmonics of the expected response. It can also be seeing that, for the case of higher jitter the estimates spread around the original clusters, occupying new histogram classes.

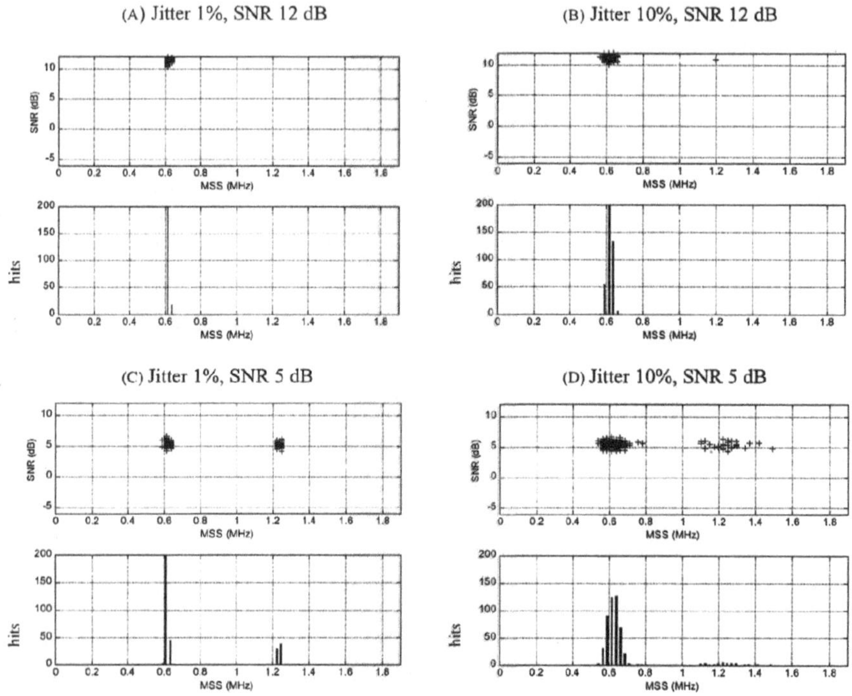

Figure 4 – Examples of the histograms of the Monte-Carlo Analysis from 500 simulated signals with 256 samples (5 periods). In each item the histograms are presented in two different ways. (A) Jitter = 1% & SNR = 12 dB. (B) Jitter = 10% & SNR = 12 dB. (C) Jitter = 1% & SNR = 5 dB. (D) Jitter = 10% & SNR = 5 dB. Observe the occurrence of a second cluster in items (c) and (d).

Figure 5 shows the consolidated results of the performance for the SSA method applied to the simulated ultrasound signals, including all the five levels of SNR and jitter. Here a correct estimate (called "hit") is defined as the one whose value falls within a ±5% window around the real value.

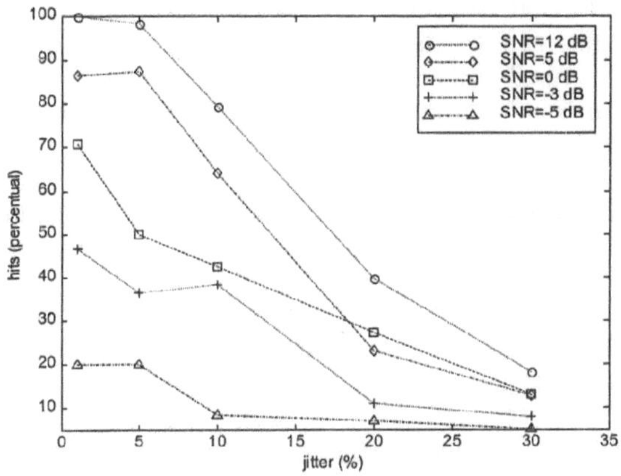

Figure 5 – Monte-Carlo Result from simulated US signals for five different levels of jitter and SNR. A correct estimate is considered when falling in the range of ± 5% of the real value. 500 simulated signals were used for each combination of (jitter,SNR).

REAL LIVER SIGNAL ANALYSIS

This analysis was performed with an experimental set-up with a 3.3 MHz transducer with low Q Factor. The bovine liver was placed in water in a precision rotatory support. It was collected 500 samples [3]. These results (figure 6) show a MSS estimate around 3.7 mm. It is notable the spread of the estimates around the peak of the histogram (most os the values fall in the range 3.1 mm to 4.8 mm). The value of ultrasound velocity propagation experimentally estimated was 1,630 m/s.

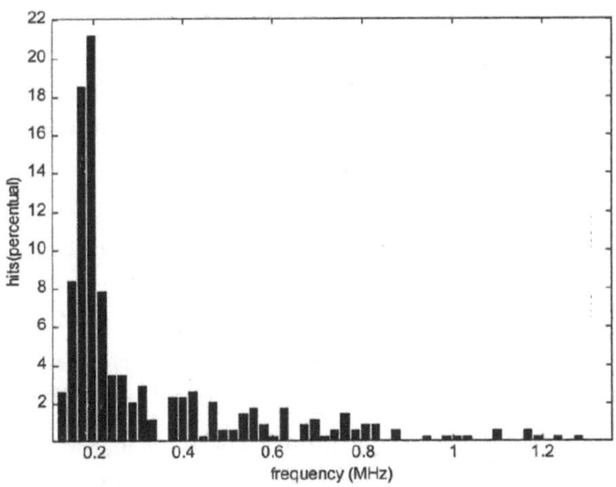

Figure 6 – Real bovine liver MSS estimation using SSA technique. The highest peak points to a MSS estimate of 3.7 mm. Notice the spreading of hits around the peak value.

DISCUSSION AND CONCLUSION

This application of SSA shows a simple way to study the decomposition of RF ultrasound signals for possible characterization of biomedical ultrasound data. In this study we have made an analysis to quantify the repeatability of the process. From this study 4 main conclusions can be taken:

a) As the jitter level grows (keeping the same level of noise), the effect seems to be the broadening of the estimate clusters (fig.4), fulfilling the adjacent histogram classes;

a) on the other hand, as the noise level grows (keeping the same level of jitter), the effect seems to be the appearance of new estimate clusters centered around the harmonics of the main MSS frequency (fig.4);

c) for the over all performance (fig.5) in the presence of noise, the process tends to generate consistent estimates (at least 45% of hits) for jitter levels up to 10%, in SNR scenarios as low as 0 dB. Below this SNR level the performance degrades fast;

d) considering the general behavior as jitter levels grow, the performance goes down. After 20% jitter the process can be seen as not repeatable, but one can raise the question of until what level of jitter a periodic structure can be understood as such.

The spread around the highest peak for the liver estimates (fig.6) can be understood as a natural spread of values for a real experiment. If we consider that both, the US beam and the tissue, have a 3-dimensional nature, then different incident angles from the beam reach the periodic structure, "looking at" slightly different MSS values all over the insonified region.

We have also made an initial US phantom signal analysis and the preliminary results show good agreement between theory and experiments. The next step is to devise a methodology to optimize the separation between periodic and non-periodic subspaces so one can extract a unique periodic signature of the medium. Once that is reached SSA can be used to automatically localize intermittent oscillations.

REFERENCES

[1] D. S. BROOMHEAD and G. KING (1986), "Extracting Qualitative Dynamics from Experimental Data", Physica 20D p. 217-236 North-Holland, Amsterdam.
[2] C.D. MACIEL and W. C. A. PEREIRA (1998), "Rf Ultrasound Echo Decomposition Using Singular-Spectrum Analysis" 24th Acoustical Imaging vol 24, in press.
[3] F. R. PEREIRA (1999), "Ultrasonic wave speed measurement using the time-delay profile of RF-backscattered signals", D.Sc. Thesis [*in portuguese*], Biomedical Eng. Program, Federal Univ. of Rio de Janeiro, Brazil.
[4] C. SIMON, J. SHEN, R. SEIP and E. S. EBBINI (1997), "A Robust and Computationally Efficient Algorithm for Mean Scatterer Spacing Estimation", IEEE Trans. on UFFC, vol. 44, n. 4.
[5] C. THERRIEN (1992), "Discrete Random Signals and Statistical Signal Processing" Prentice Hall Processing Series, Englewood Cliffs, New Jersey.
[6] R. VAUTARD and M. GHIL (1989), "Singular Spectrum Analysis in Nonlinear Dynamics With Applications to Paleoclimatic Time Series", Physica D, 35 pp. 395-424 North-Holland, Amsterdam.
[7] R. VAUTARD, P. YOIU and M. GHIL (1992), "Singular-Spectrum Analysis: a Toolkit for Short, Noisy Chaotic Signals", Physica D 58 pp. 95-126 North-Holland, Amsterdam.

IMAGE FORMATION WITH THE AMPLITUDE-STEERED ARRAY USING TIME-FREQUENCY DISTRIBUTIONS

Catherine H. Frazier and William D. O'Brien, Jr.

Bioacoustics Research Laboratory
Department of Electrical and Computer Engineering
University of Illinois
405 N. Mathews Avenue
Urbana, IL 61801

INTRODUCTION

The amplitude-steered array, introduced by Hughes and Thompson in 1976 [1], was originally designed to steer a monochromatic signal to a particular direction. The array is now being considered for a fast imaging system [2]. The amplitude-steered array spatially separates frequencies in one direction, which allows a sector to be scanned with a single pulse. Images can be formed by calculating the spectrogram, which is the magnitude-squared of the short-time Fourier transform (STFT). The temporal position of the processing window gives information about the range of the target, and the frequencies contained within the window give information about the lateral position of the target. Although the spectrogram is conceptually simple and easy to program, there are several drawbacks that make it possibly undesirable as the means for forming the image. In particular, there is an inherent tradeoff between time and frequency resolution, so that an improvement in one implies a degradation in the other.

Many time-frequency distributions have been studied for various purposes, each having its own benefits and drawbacks. It is generally accepted that the best distribution for an application must be chosen based on the properties of the signal and the criteria for the result. We compare five distributions for a test signal, that is a simulated reflection from seven point targets. The distributions we have chosen are the spectrogram, the constant-Q spectrogram [3], the Wigner distribution [4], the smoothed pseudo-Wigner distribution [5], and the Choi-Williams distribution [6]. In sonar applications, time-frequency representations and time-frequency filtering have been used to identify targets [7], [8]. In our application, we do not identify targets based on characteristics of the distribution, but rather, we use the time-frequency distribution to form an image of the target.

In the next section, we describe how the amplitude-steered array is used to collect data. Then we discuss properties of each of the distributions. Next, we describe our test signal,

Acoustical Imaging, Volume 25, Edited by Halliwell and Wells.
Kluwer Academic/Plenum Publishers, 2000.

and the basis for comparison of each of the distributions. Finally, we give results and conclusions. In the following discussion, the terms time-resolution and frequency-resolution are used; however the terms range-resolution and lateral-resolution, respectively, could easily be substituted.

ARRAY FIELD PATTERN

The array pattern for an amplitude-steered array with an even number of elements can be written as [1],[2]

$$H(\theta) = \frac{2}{N}[\cos\phi\cos u + \cos 3\phi\cos 3u + \ldots + \cos((N-1)\phi)\cos((N-1)u)] \quad (1)$$
$$+\frac{2}{N}[\sin\phi\sin u + \sin 3\phi\sin 3u + \ldots + \sin((N-1)\phi)\sin((N-1)u)]$$

where θ is measured from the array broadside direction, N is the number of elements in the array, ϕ is a constant equal to $(k_0 d\sin\theta_0)/2$, and u is equal to $(kd\sin\theta)/2$. θ_0 is the design steering angle, k_0 is the wavenumber at which the beam will be steered to θ_0, and d is the interelement spacing. In Equation (1), $\cos((2n-1)\phi)$ and $\sin((2n-1)\phi)$ terms are constants which we interpret as amplitude weights on the elements. The beam is steered by weighting the elements. The $\cos((2n-1)u)$ and $\sin((2n-1)u)$ terms represent combinations of pairs of elements on opposite sides of the center of the array in-phase or 180° out of phase, respectively. Equation (1) can be interpreted as the sum of the outputs of two arrays.

If the array described above is excited by a frequency different from the design frequency used to calculate ϕ, the maximum response will occur at an angle different from the designed steering direction, θ_0.

$$\phi = \frac{k_0 d\sin\theta_0}{2} = \frac{k_f d\sin\theta_f}{2} \quad (2)$$

$$\theta_f = \sin^{-1}\left(\frac{f_0}{f}\sin\theta_0\right) \quad (3)$$

The subscript, f, was added to k and θ to emphasize that the new steering direction is calculated for a specific frequency, f.

An example of several beams from a 9.76-cm-length array is shown in Figure 1. The array is designed to steer to 5° at 5.6 MHz. Beams are shown for 5.6 MHz (5°), 4.5 MHz (6.23°), 3.4 MHz (8.25°), 2.3 MHz (12.25°), and 1.2 MHz (24°).

TIME-FREQUENCY PROCESSING FOR IMAGE FORMATION

In this section, we briefly describe the properties of the time-frequency distributions used to form images. Expressions are given for calculating only the discrete distributions. More detailed discussions and expressions for calculating the continuous distributions can be found in the references for the distributions provided in the introduction.

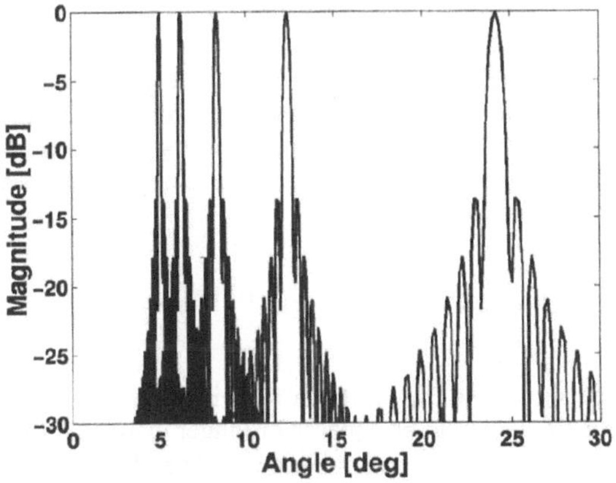

Figure 1. Beams from a 9.76-cm aperture steered to 5° at 5.6 MHz. Beams are shown for 5.6 MHz (5°), 4.5 MHz (6.23°), 3.4 MHz (8.25°), 2.3 MHz (12.25°), and 1.2 MHz (24°).

Spectrogram

The spectrogram, which is the magnitude-squared of the STFT, is the classic method for studying signals whose spectra vary with time. The spectrogram has the advantage that oscillating cross-terms, which are present for some distributions, are reduced to zero as long as the signal components do not overlap in time. And the spectrogram is a nonnegative definite distribution which leads to easy interpretation as an image.

The time and frequency resolution of the spectrogram are determined by the length of the window used in the calculation. Different tradeoffs can be made by changing the window shape. However, the window cannot be narrow in both the time and frequency domain, so there is an inherent tradeoff in resolution for the spectrogram.

Constant-Q Spectrogram

The constant-Q distribution is a special case of the spectrogram, where the window length is chosen based on the frequency sample being calculated. The number of cycles in the window is kept constant; therefore, a constant ratio between center frequency and bandwidth, can be maintained.

The expression for the constant-Q spectrogram is given by

$$X[k] = \frac{1}{N[k]} \sum_{n=0}^{N[k]-1} W[k,n]x[n]e^{-j2\pi Qn/N[k]} \tag{4}$$

where x is the digitized signal, $N[k]$ is the number of samples used to evaluate the expression for frequency sample k, $W[k,n]$ is a window function set to a Hanning window, for example, and Q is defined as $f/\Delta f$, where f is the frequency.

In the constant-Q spectrogram, cross-terms are reduced to zero unless the signal components overlap in time, as for the conventional spectrogram. However, there remains a tradeoff between time and frequency resolution that is based on the window length, which is quantified by a single value, the length of the window in terms of number of cycles.

Wigner Distribution

The Wigner distribution produces a time-frequency representation with the greatest resolution, but the representation also has very high cross-terms. The distribution is real, but it is not necessarily zero when the signal is zero, due to the cross-terms. It is zero before the signal starts and after the signal finishes, and it can have negative values.

The discrete Wigner distribution is given by

$$D(n,k) = \frac{1}{\pi}\sum_{-L+1}^{L-1} s^*(n-\tau)e^{-j4\pi k\tau/N}s(n+\tau) \tag{5}$$

where *s(n)* is the digitized signal and the limits on τ are included for practical evaluation of the sum. The total length, *2L-1*, should be equal to or longer than the duration of the signal.

The Wigner distribution is periodic with a period π. Therefore, to produce a representation that is free from aliasing, the signal must either be oversampled by a factor of two, or the analytic signal must be used. Using the analytic signal is preferable because it will reduce the number of cross-terms. Cross-terms appear when two frequency components interfere with each other, including interference between positive and negative frequencies. With the analytic signal, negative frequency components are no longer present.

Smoothed Pseudo-Wigner Distribution

The smoothed pseudo-Wigner distribution (SPWD) is an attempt to reduce the oscillating cross-terms in the Wigner distribution. Smoothing is applied independently in the time and frequency directions; therefore, there is no undesirable connection between the two resolutions as with the spectrogram. Smoothing degrades the resolution in the direction in which it is applied.

The SPWD is a filtered version of the Wigner distribution,

$$D(t,\omega) = \int L(t-t',\omega-\omega')D^{Wigner}(t',\omega')dt'd\omega' \tag{6}$$

where in our case, *L* is a Gaussian filter in each direction, $L(t,\omega)=(1/\alpha\beta)exp(-t\text{^}2/\alpha-\omega\text{^}2/\beta)$.

Choi-Williams Distribution

The Choi-Williams distribution was created to reduce cross-terms while preserving other desirable properties. The discrete version of the Choi-Williams distribution is given by Equation (20) of [6].

$$\begin{aligned} D(n,k) = 2\sum_{\tau=-\infty}^{\infty} W_N(\tau)e^{-j2\pi k\tau/N} \times \\ \sum_{\mu=-\infty}^{\infty} W_M(\mu)\frac{1}{\sqrt{4\tau^2/\sigma}}\exp\left(-\frac{\mu^2}{4\tau^2/\sigma}\right)s(n+\mu+\tau)s^*(n+\mu-\tau) \end{aligned} \tag{7}$$

where $W_N(t)$ is a symmetric window that is nonzero in the range, $t \in [-N/2, N/2]$, and $W_M(t)$ is a rectangular window that is nonzero in the range, $t \in [-M/2, M/2]$, effectively reducing the limits on the sums. σ is the parameter used to control the properties of the distribution. Large σ implies more smoothing and reduction of the cross-terms. However, this also leads to greater loss of resolution. The loss of resolution is not independently controlled for the time and frequency directions because only one parameter is used. The Choi-Williams distribution also has the drawback that it is difficult to reduce cross-terms if two signal components occur at the same time or the same frequency [9].

TEST SIGNAL AND BASIS FOR COMPARISON

The operation of the linear amplitude-steered array has been simulated using the Field II program, developed by J. A. Jensen [10], [11]. The test signal is a simulated reflection from a set of seven point targets received by the amplitude-steered array operated in pulse-echo mode. The point targets are located at (6°, 4.0 m), (9°, 4.01 m), (12°, 4.02 m), (15°, 4.03 m), (18°, 4.04 m), (21°, 4.05 m), and (24°, 4.06 m). The amplitude weighting is determined so that the main beam is steered to 5° at 5.6 MHz. The transmitted signal is a linear FM chirp with frequency swept from 1.2 MHz to 5.6 MHz. The speed of sound is assumed to be 1500 m/s and, attenuation is not included. The transducer we simulate has a low-Q impulse response. The main effect of the transducer's transfer function is to reduce the amplitude of targets away from the resonance frequency of the transducer, not to effect resolution. Therefore, we replace the transducer's impulse response with an impulse.

The resolution of different distributions is measured as the -6-dB axial and lateral widths of each point target. The highest level of a cross-term relative to the maximum target amplitude in the image is also observed. Each distribution is optimized within the limits of the parameters available so that the axial and lateral resolutions of the point target at 6° are approximately equal and close to 1 cm and so that the highest level of a cross-term is at least 20 dB below the peak target. The criterion for highest level of a cross-term was chosen based on the appearance of the images. Lowering cross-terms causes a blurring of auto-terms. In most cases we do not have the degrees of freedom to control all of these criteria, so subjective evaluation is used to produce the best image.

RESULTS

The respresentations of the signal on the time-frequency plane are taken as images of a target. In the cases where the distribution has negative values, the absolute value is taken before logarithmic compression and display of the images.

Measured lateral and axial widths are presented in Table 1 and Table 2. Because of the high cross-terms for the Wigner distribution, measurements for targets at 12° to 21° are made using images with only a single target present at a time. The target at 24° was not measured because it appeared at the edge of the image. The target at 24° was not measured for the Choi-Williams distribution because it was not visible above the noise in the image.

The images produced with the spectrogram and constant-Q distributions are shown in Figure 2 and Figure 3. As expected, the spectrogram and constant-Q distributions did not have any cross-terms. The constant-Q distribution achieved better lateral resolution than the spectrogram, although the spectrogram generally produced better axial resolution. The constant-Q distribution maintained equal lateral and axial resolution for each individual target, while the spectrogram produced much better axial resolution than lateral resolution for the targets at lower frequencies.

An image formed using the Wigner distribution is shown in Figure 4. The Wigner distribution has no parameters for smoothing, and therefore the level of the cross-terms could not be reduced. In the image, only the targets at 6 and 9 degrees could be distinguished from the cross-terms. The Wigner distribution is still of interest as a measure of the achievable resolution.

Table 1. Measured lateral widths of the point spread function in mm.

Angle (deg)	6	9	12	15	18	21	24
Spectrogram	11.8	19.5	30.6	45.7	66.1	81.8	102
Constant-Q	11.1	17.2	23.2	31.0	37.6	42.6	48.7
Wigner[a]	11.1	16.6	21.5	28.4	34.7	38.5	--
SPWD	11.6	17.1	22.9	30.9	38.9	45.8	53.9
Choi-Williams	16.8	24.9	34.8	51.6	88.9	87.1	--

[a]Values for Wigner distribution were measured using images with one target present at a time.

Table 2. Measured axial widths of the point spread function in mm.

Angle (deg)	6	9	12	15	18	21	24
Spectrogram	11.2	11.2	14.9	16.8	18.7	18.7	24.3
Constant-Q	11.0	18.3	23.8	29.3	34.8	40.3	45.8
Wigner[a]	4.76	8.02	10.2	12.7	14.4	17.7	--
SPWD	7.07	9.41	12.2	14.1	16.0	19.0	21.4
Choi-Williams	7.10	9.89	13.0	14.8	15.6	20.6	--

[a]Values for Wigner distribution were measured using images with one target present at a time.

The smoothed pseudo-Wigner distribution, shown in Figure 5, performed the best of the distributions compared here in terms of combined lateral and axial resolution. Both lateral and axial resolutions were comparable to those for the Wigner distribution. And the cross-terms were reduced to 30 dB below the peak target signal.

The Choi-Williams distribution, shown in Figure 6, was not able to match the performance of the smoothed pseudo-Wigner distribution. As mentioned previously, cross-terms are difficult to suppress if signal components overlap in time or frequency. In our case, the two targets at the lower frequencies overlap enough in time and frequency that the cross-term between them could only be lowered below 20 dB at the cost of great loss in resolution. Therefore, some of the cross-terms were allowed to remain in the image at higher levels. Still, with only one parameter to control time and frequency smoothing and to control the cross-terms, we were unable to effectively trade more axial resolution for better lateral resolution.

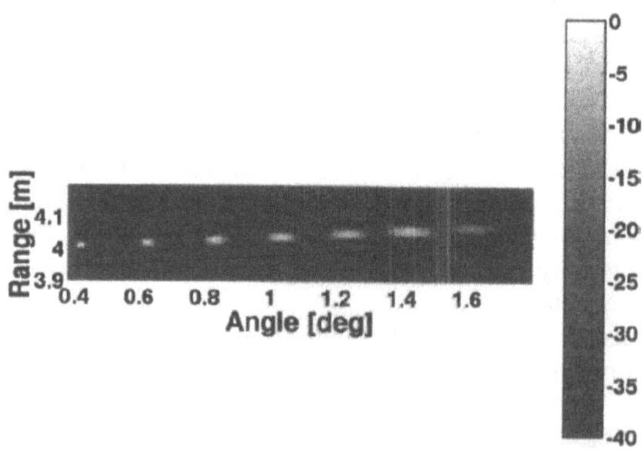

Figure 2. Image produced using the spectrogram. A 590-point (28.8 μs) Hanning window was used for the calculation.

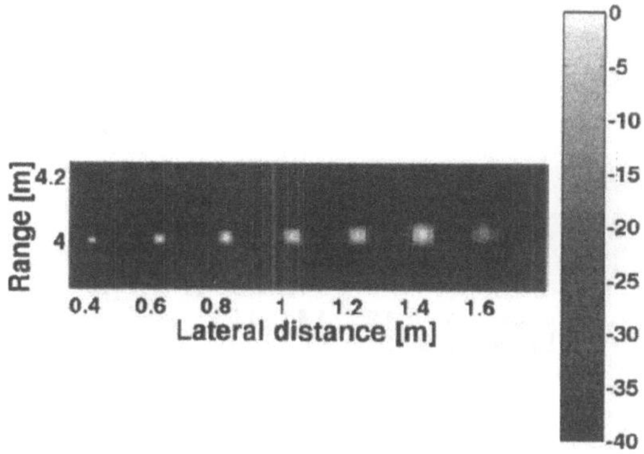

Figure 3. Image produced using the constant-Q spectrogram. The processing window contained 141 cycles (Q=141).

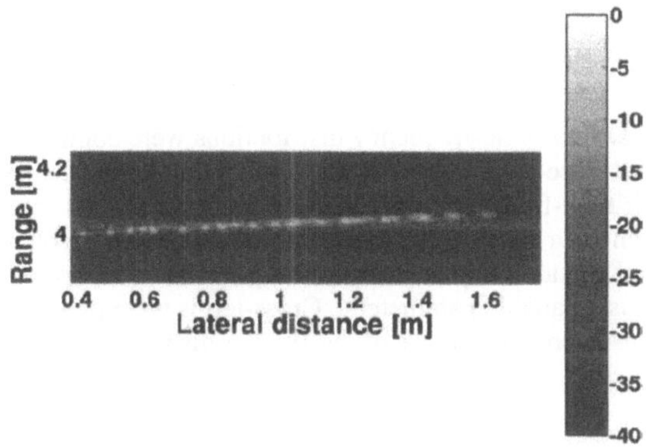

Figure 4. Image produced using the Wigner distribution. There were no parameters available for optimization.

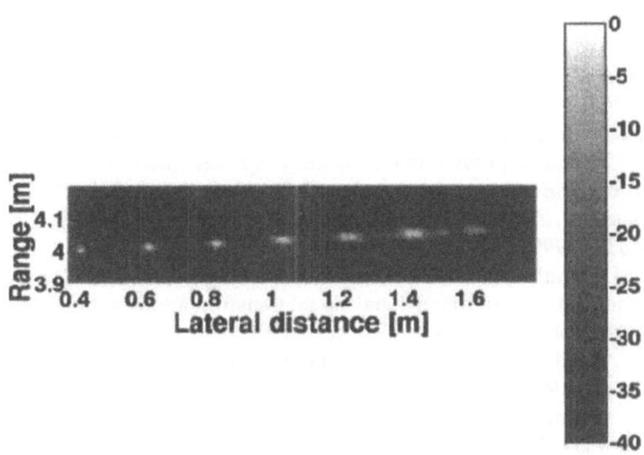

Figure 5. Image produced using the smoothed pseudo-Wigner distribution. The filter was Gaussian in both time and frequency directions. α=11.4E-12 s^2 and β=15.8E9 (rad/s)^2.

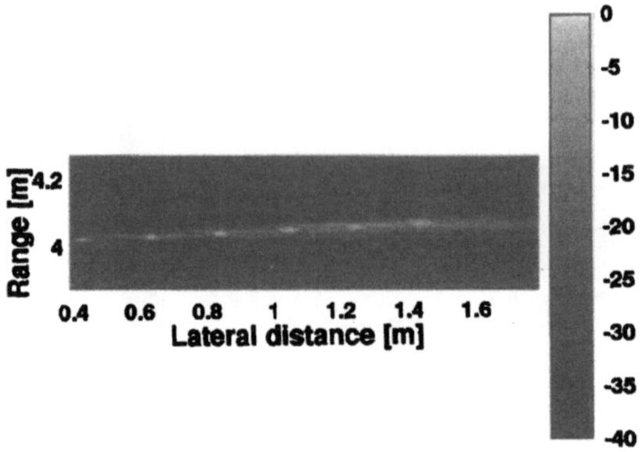

Figure 6. Image produced using the Choi-Williams distribution. W_M was a rectangular window with 256 points. W_N was a rectangular window with 513 points. σ was set to 0.75.

CONCLUSIONS

The performance of several time-frequency distributions were compared using a test signal, which represented a pulse-echo signal from seven point targets. By moving from the spectrogram to other time-frequency distributions, we have shifted from a tradeoff in axial and lateral resolution to a tradeoff in axial resolution, lateral resolution, and cross-term level. The smoothed pseudo-Wigner distribution gave the best overall results most likely because it has the most available parameters. Cross-terms were present at low levels. Axial and lateral resolutions approached those of the Wigner distribution.

REFERENCES

1. W. J. Hughes and W. Thompson, Jr., Tilted directional response patterns formed by amplitude weighting and a single 90° phase shift, *J. Acoust. Soc. Am.* **59**(5), 1040-1045 (1976).
2. C. H. Frazier, W. J. Hughes, and W. D. O'Brien, Jr., Analysis of resolution for an amplitude-steered array, *J. Acoust. Soc. Am.*, accepted (2000).
3. J. C. Brown, Calculation of a constant-Q spectral transform, *J. Acoust. Soc. Am.* **89**(1), 425-434 (1991).
4. T. A. C. M. Claasen and W. F. G. Mecklenbrauker, The Wigner distribution – A tool for time-frequency signal analysis, *Philips J. Res.* **35**(4/5) 276-300 (1980).
5. L. Cohen, Time-frequency distributions – A review, *Proc. of the IEEE* **77**(7), 941-981, (1989).
6. H.-I. Choi and W. J. Williams, Improved time-frequency representation of multicomponent signals using exponential kernels, *IEEE Trans. Acoust. Speech Sig. Proc.* **37**(6), 862-971 (1989).
7. P. Chevret, N. Gache, and V. Zimpfer, Time-frequency filters for target classification, *J. Acoust. Soc. Am.* **106**(4), 1829-1837 (1999).
8. L. R. Dragonette, D. M. Drumheller, C. F. Gaumond, D. H. Hughes, B. T. O'Connor, N.-C. Yen, and T. J. Yoder, The application of two-dimensional signal transformations to the analysis and synthesis of structural excitations observed in acoustical scattering, *Proc. of the IEEE* **84**(9), 1249-1263 (1996).
9. F. Hlawatsch and G. F. Boudreaux-Bartels, Linear and quadratic time-frequency representations, *IEEE Sig. Proc. Mag.* **9**, 21-67, 1992.
10. J. A. Jensen, FIELD: A program for simulating ultrasound systems, *Med. Biol. Engr. Comp.* **34** (1, Suppl.1), 351-353 (1996).
11. FIELD II program is available at the web site http://www.it.dtu.dk/~jaj/field/field.html.

VIBRO-ACOUSTOGRAPHY: SPECKLE FREE ULTRASONIC C-SCAN

James F. Greenleaf, Mostafa Fatemi and Randall R. Kinnick

Department of Physiology and Biophysics
Mayo Clinic and Foundation
Rochester, MN 55905 USA

INTRODUCTION

It is well known that changes in elasticity of soft tissues are often related to pathology. Traditionally, physicians use palpation as a simple method for estimating mechanical properties of tissue. The dynamic response of soft tissue to an oscillating or impulsive force is also valuable in medical diagnosis. For instance, rebound of tissue upon sudden release of pressure exerted by the physician's finger on the skin provides useful diagnostic information about the tissue. In vibro-acoustography ultrasound is used to produce an oscillating radiation force which then produces an acoustic response from tissues. Specifically acoustic energy is emitted from tissues in response to an oscillatory radiation force produced by interfering focused beams of ultrasound. Frequency spectra of ultrasound stimulated acoustic emission exhibit object resonances. Raster scanning the radiation force over the object and recording the amplitude and phase of the emitted sound results in data from which speckle free images related to elastic compositions of the acoustically emitting objects can be computed. We show here that intact and broken struts on prosthetic disk heart valves (convexo-concave Bjork-Shiley) can be distinguished by their resonant response to vibro-acoustography.

BACKGROUND

Quantitative measurement of the mechanical properties of tissues and their display in a raster format is the aim of a class of techniques generally called elasticity imaging or elastography (Gao et al., 1996). The general approach is to measure tissue motion caused by an external (or, in some methods, internal) force or displacement and use it to reconstruct the elastic parameters of the tissue. The excitation stress can be either static, or dynamic (vibration). Dynamic excitation is of particular interest because it provides more comprehensive information about tissue properties in a spectrum of frequencies. Because the motion is dynamic, derivatives of displacement in both space and time can be measured allowing the computation of elastic properties with low dependence on boundary condi-

Acoustical Imaging, Volume 25, Edited by Halliwell and Wells.
Kluwer Academic/Plenum Publishers, 2000.

tions. Magnetic resonance elastography is a recently developed method that employs a mechanical actuator to vibrate the body surface and then measures the strain waves with phase sensitive Magnetic Resonance Imaging resulting in a method termed Magnetic Resonance Elastography (Muthupillai et al., 1995). The majority of elasticity imaging methods are based on an external source of force, resulting in a spatially wide stress field distribution. This requires the stress field to pass through the superficial portion of the object before reaching the region of interest within the interior. This requirement can complicate estimation of stiffness because the stress field patterns vary with depth. An alternative strategy is to apply a stress directly to the localized region of interest. One way to accomplish this is to use the radiation pressure of ultrasound. An example is the use of ultrasound radiation pressure to vibrate the tissue and the use of MRI to measure the displacement (Sarvazyan et al., 1998 and Tau et al., 2000). The approach taken in this paper is to estimate the mode of vibration of an object from the acoustic emission resulting from the cyclic displacement.

Acoustic radiation force is the time average force exerted by an acoustic field on an object. This force is produced by a change in the energy density of an incident acoustic field (Beyer, 1978), for example, due to absorption or reflection. Several advantages may result from using ultrasound radiation force for evaluating tissue stiffness, including: a) acoustic (ultrasound) energy is a noninvasive means of applying force, b) current ultrasound technology and devices can be readily modified for this purpose, thus eliminating the need for developing a new technology, c) radiation force can be applied remotely inside tissue without disturbing its superficial layers, d) the radiation stress field can be localized, thus providing for precise positioning of the excitation point, and e) radiation force can be applied in a wide range of frequencies or temporal shapes. These features make radiation force methods highly attractive for creating remote stress fields within objects.

VIBRO-ACOUSTOGRAPHY THEORY

Study of radiation force and radiation pressure dates back nearly one century, to the time of Rayleigh (1902 and 1905). A historical review of radiation force and radiation pressure is presented in Beyer (1978), and a critical review of the subject can be found in Chu and Apfel (1982). Some recent developments of radiation force/pressure in attenuating medium, which may be applicable to biological tissues, are presented in Rudenko et al. (1996) and Jiang and Greenleaf (1996). Acoustic radiation force is a universal phenomenon in any wave motion that introduces some type of unidirectional force on absorbing or reflecting targets in the wave path. Radiation force in fluids is often studied in the context of radiation pressure. Definitions and theoretical developments of radiation force depend on the boundary conditions. Simple explanations of these definitions can be presented by considering a sound traveling inside, and along the axis of, a cylindrical container toward a wall (Beyer, 1978). Rayleigh radiation pressure is the excess pressure produced on the wall when the container's side wall is confining the fluid inside. Langevin radiation pressure is the excess pressure on the wall when the confining side wall is absent, so that the fluid is free to move. Here we will focus on Langevin radiation pressure because the conditions for which this pressure is defined apply to our experimental situation.

It can be shown that the Langevin radiation pressure of a plane wave impinging normally on a perfectly absorbing wall is equal to the total energy density $\langle E \rangle$, where $\langle \rangle$ represents the time average. If the wall is partially reflecting, this pressure would be equal to $(1+R)\langle E \rangle$, where R is the power reflection coefficient (Beyer, 1978). Thus, in general, we can write the radiation force of a normally impinging sound beam on a wall as $F = d_r \langle E \rangle$, where F is the force in the beam direction, and d_r is the "radiation force function" or the drag coefficient. This dimensionless coefficient is defined per unit incident energy density and unit projected area. For a planar object, F is numerically equal to the force on the object. Physically, the drag coefficient represents the scattering and absorbing properties of the object (Westervelt, 1951). For a perfectly absorbing object, $d_r = 1$ and for a perfectly reflecting object $d_r = 2$. In the case of oblique incidence, the radiation force will have a normal as well as a transverse component. A more detailed description is presented in Westervelt (1951). To produce a time varying radiation force, the intensity of the incident beam can be modulated in various ways. For example, a short ultrasound pulse can produce a transient pulsed radiation force, and a sinusoidally modulated beam can result in a sinusoidally varying force.

METHODS

We interrogate the object by arranging the intersection of two focused continuous wave (CW) ultrasound beams of different frequencies to occur at a selected point on the object. Interference in the intersection region of the two beams produces sinusoidal modulation of the ultrasound energy density. Modulation of the energy density creates an oscillatory force, effectively vibrating the object at the selected region. The resulting vibration of the object produces an acoustic field (Fatemi and Greenleaf, 1998) that can be measured some distance away.

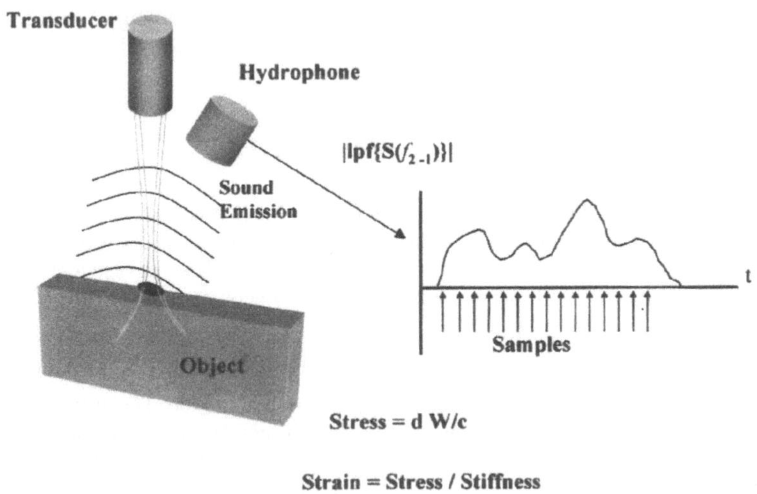

Figure 1. Ultrasound is made to interfere on the region of interest and the resulting acoustic emission is detected with a hydrophone or microphone to estimate the response of the object. The signal is digitized for analysis.

Ultrasound beams can be constructed in a variety of ways for this purpose (Fatemi and Greenleaf, 1999). We used two coaxial, confocal transducer elements of a spherically focused annular array (consisting of a central disc and an outer annulus) driven by two CW signals at slightly different frequencies f_1 and f_2 (Figure 1). The energy density at a point in this ultrasound field, say at the focus, is proportional to the square of the sum of the ultrasound fields from the two elements. Squaring the sum of two sines gives rise to sum and difference frequency terms. Thus, high frequency and low frequency variations in energy density result at the intersection of the two beams produced by the two elements. Ultrasound-stimulated acoustic emission results from the energy term that produces a low-frequency vibration. The low-frequency force on a target at the focal point can be computed with the following integration: $F_l = d_r \int_S \langle E \rangle da$, S is the area over which $\langle E \rangle$, the total energy density in the focal plane, has significant value, and $\langle \ \rangle$ represents a short-term time average. For focused beams, the intersection region can be small enough that F_l, the low frequency force, can be considered to be an oscillating point force applied to the object at the focal intersection of the beams. To produce an ultrasound stimulated vibro-acoustic emission spectrogram, we vibrate a small region of the object with an oscillating radiation force of varying frequency. The complex amplitude of the resulting acoustic emission field, $\Phi(\Delta f)$, can be written: $\Phi(\Delta f) = d_r H(\Delta f) Q(\Delta f)$, where $Q(\Delta f)$ is a complex function representing the mechanical frequency response, or admittance, of the object at the selected point and $H(\Delta f)$ represents the combined frequency response, or transfer function, of the propagation medium and receiver and is assumed to be fixed and known (Fatemi and Greenleaf, 1999). Recording $\Phi(\Delta f)$ allows us to obtain $Q(\Delta f)$ for each point, within a constant multiplier. We raster scan the radiation force over the object to produce data that can be mapped into a pictorial format. The spatial resolution of the resulting image is determined by the region in which significant interference between the ultrasound beams occurs and is of the order of a few wavelengths at the ultrasound frequency. In a slight alteration of vibro-acoustography called vibrography we attach an accelerometer to the object to measure its vibration. The scan of the vibration inducing ultrasound beam remains the same. We show both types of images here.

RESULTS

To test the feasibility of using the technique to image mechanical properties of tissues, we measured the amplitude of acoustic emission from intact and broken struts on Bjork-Shiley convexo-concave (BSCC) prosthetic heart valves. The valves were scanned in a plane perpendicular to the ultrasound beam axis. Vibro-acoustography of the valve at 40 kHz resulted in an image of $Q(40k)$ and is illustrated in Figure 2. Also illustrated in Figure 2 are vibrography images representing $Q(7k)$ of a valve in which one leg of the small strut is broken. One can see that the strut vibrates differently when broken. The amplitude images are highly detailed and exhibit variations in acoustic emission from mechanical objects having different vibration characteristics. These differences are due to variations in the product of the reflection properties, d_r, and the effective mechanical vibration admittance properties of the object. Thus, both vibro-acoustography and vibrography are similar to conventional pulse echo ultrasound imaging, which is sensitive to the ultrasonic parameters of the object, but these two new imaging methods have the advantage of also being sensitive to the mechanical admittance at low frequencies.

DISCUSSION

Several implications can be deduced from the results of this study. The results shown here imply that fractured struts could be found in similar types of valves implanted in patients assuming an appropriate clinical instrument can be built. Vibration modes of complex objects could be characterized without having to touch the object. Motion induced by ultrasound and measured with ultrasound pulse echo has been used previously to study "hardness" (Dutt et al., 2000). However, the sensitivity of ultrasound pulse echo to motion at common ultrasound frequencies is limited to several micrometers. The advantage of ultrasound stimulated vibro-acoustic emission is its high displacement sensitivity. Cyclic displacement of 100 nm at 10 kHz produces an acoustic intensity of about 3.0 x 10^3 W/cm^2. Hydrophones similar to the one used in these experiments are sensitive to as little as 10^{-15}W/cm^2, and therefore, can detect very small cyclic displacements. The method will be more sensitive for higher frequency sound because acoustic power is proportional to the square of frequency for constant displacement amplitude.

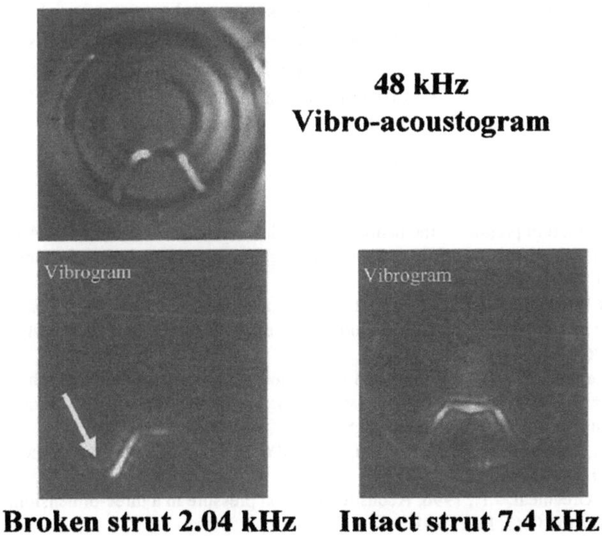

Figure 2. Upper; vibro-acoustography image of artificial heart valve (BSCC) showing high detail and no speckle. Highlights on strut are due to reflection of ultrasound into the hydrophone which was at the right. Lower; images obtained with vibrography at the resonance of the broken (left) and intact (right) struts. The strut is broken at the arrow and the free end vibrates strongly when driven at frequency of 2.04 kHz, much lower that the 7.4 kHz resonance of the intact strut on the right.

Ultrasound stimulated vibro-acoustic spectrography may have application in at least two general areas. The first is nondestructive evaluation of materials, where material characteristics and structural flaws can be identified by measuring changes in the mechanical response to vibration at a point. The object under test could be remotely probed by beams propagating and interfering in either water or air. Beams propagating within the object could also be used to produce acoustic emission from flaws. The second general area of application is medical imaging and detection. Vibro-acoustography appears particularly suitable for noninvasive detection of hard tissue inclusions, such as imaging arteries with calcification, detecting breast micro-calcifications, visualizing hard tumors, and detecting foreign objects. Because changes of stiffness alter the vibration frequency response or damping of tissue, vibro-acoustography can potentially provide a noninvasive, remote, high-resolution, speckle free "palpation" technique that can reach small abnormalities that are otherwise untouchable by conventional methods.

SUMMARY

In an *in vitro* experiment the acoustic emission characteristics of broken and intact struts on BSCC valves exhibited different frequency responses and different vibro-acoustography images. The images indicated that the vibration modes of the object could be deduced in addition to the structural integrity of the object. Structural integrity of such prostheses could be deduced using vibro-acoustography *in vivo*.

ACKNOWLEDGMENTS

The authors gratefully acknowledge the secretarial assistance of Elaine C. Quarve and the provision of heart valves by Edmond Rambod at the California Institute of Technology. This work was supported in part by a grant from the National Institutes of Health HL 61451 and a grant from the US Department of Defense DAMD 17-98-1-1821.

REFERENCES

Beyer, R.T., 1978, Radiation pressure - the history of a mislabeled tensor, J Acoust Soc Am 63(4):1025.

Chu B-T., and Apfel, R.E., 1982, Acoustic radiation pressure produced by a beam of sound. J Acoust Soc Am 72(6):1673.

Dutt, V., Kinnick, R.R., Muthupillai, R., Oliphant, T.E., Ehman, R.L., and Greenleaf, J.F., 2000, Acoustic shear-wave imaging using echo ultrasound compared to magnetic resonance elastography, Ultrasound Med Biol 26(2) (to appear in Feb issue).

Fatemi, M., and Greenleaf, J.F., 1998, Ultrasound-stimulated vibro-acoustic spectrography, Science 280:82.

Fatemi M., and Greenleaf, J.F., 1999, Vibro-acoustography: an imaging modality based on ultrasound-stimulated acoustic emission, Proc Natl Acad Sci USA 96:6603.

Gao, L., Parker, K.J., Lerner, R.M., and Levinson, S.F., 1996, Imaging of the elastic properties of tissue - a review, Ultrasound Med Biol 22(8):959.

Jiang, Z-Y., and J.F. Greenleaf, J.F., 1996, Acoustic radiation pressure in a three-dimensional lossy medium. J Acoust Soc Am 100(2, Pt. 1):741.

Lord Rayleigh, 1902, Philos Mag 3:338; also see Lord Rayleigh, 1905, Philos Mag 10:364.

Muthupillai, R., Lomas, D.J., Rossman, P.J., Greenleaf, J.F., Manduca, A., and Ehman, R.L., 1995, Magnetic resonance elastography by direct visualization of propagating acoustic strain waves, Science 269:1854.

Rudenko, OV., Sarvazyan, A.P., and Emelianov, S.Y., 1996, Acoustic radiation force and streaming induced by focused nonlinear ultrasound in a dissipative medium, J Acoust Soc Am 99(5):1.

Sarvazyan, A.P., Rudenko, O.V., Swanson, S.D., Fowlkes, J.B., and Emelianov, S.Y., 1998, Shear wave elasticity imaging: a new ultrasonic technology of medical diagnostics, Ultrasound Med Biol 24(9):1419.

Wu, T., Felmlee, J.P., Greenleaf, J.F., Riederer, S.J., and Ehman, R.L., 2000, MR imaging of shear waves generated by focused ultrasound. Magn Reson Med 43:111.

Westervelt, P.J., 1951, The theory of steady force caused by sound waves, J Acoust Soc Am 23(4):312.

LASER-ULTRASOUND FOR INTERFACE MONITORING WITHIN THE EYE

Ali Sadr[1], Peter A. Payne[1], Emanuel S Rosen[2] and Richard J. Dewhurst[1]

[1] Department of Instrumentation and Analytical Science
University of Manchester Institute of Science and Technology (UMIST)
Manchester M60 1QD, UK
[2] The Rosen Eye Surgery Centre, Alexandra Hospital-Victoria Park,
Manchester M14 5QH, UK

INTRODUCTION

Generation of ultrasound energy by nanosecond laser pulses was introduced about twenty years ago for high resolution characterisation of materials[1,2]. This field became known as laser-ultrasound.[3] Following significant advances in both the science and engineering of the subject, it has become established as a non-contact measurement technique for solid materials. It offers remote and non-invasive measurement, and has important applications within the industrial field of non-destructive evaluation (NDE).

In the late 1980s, Payne and Dewhurst conducted experiments to show that biological tissue could also generate ultrasonic signals when irradiated with laser pulses[4]. Since then, a number of research groups around the world have investigated the phenomenon and now a broad range of biolaser-ultrasound applications is being studied[5-9]. One early application was concerned with the potential imaging of arteries in the human body using probe designs suitable for intra-arterial applications[10-13]. An important consideration in laser-generated ultrasound was that the optical fibre used to introduce the laser energy into the artery or other region could also be used to guide high-energy laser irradiation for the purposes of tissue destruction or ablation. Hence, any probe system used for diagnostic purposes and imaging using laser-generated ultrasound could also be used for some form of guided surgery by switching from low average-power laser levels to those with a higher power level. Descriptions of potential intra-arterial laser imaging have already been published[14,15].

More recently, we have begun to study the use of laser-generated ultrasound to detect interfaces within the eye, which is the subject of this paper. Presently, in ophthalmology, conventional pulse echo ultrasound is used to conduct dimensional measurements of the position of various interfaces within the eye. The conditions that make these diagnostic measurements possible are: (i) ocular tissues do not have too much effect on the

Acoustical Imaging, Volume 25, Edited by Halliwell and Wells.
Kluwer Academic/Plenum Publishers, 2000.

propagation of ultrasound; (ii) the organ is superficially located; and (iii) healthy eye's anatomy is quite simple. High-resolution measurements are needed to provide on-axis dimensions of the eye, and pulse-echo signals depend on small changes in acoustic impedance at each internal eye interface. Similarly, laser-generated ultrasound can probe the eye for interfaces, but this time the signals depend on differences in optical absorption at each eye interface. Laser-generated ultrasound is potentially capable of delivering high-resolution measurements in one shot for all the dimensional measurements required in the eye. Since it is sensitive to optical changes, it is potentially useful for monitoring any optical degradation that may be associated with ageing or from eye disease. It may offer vital supplementary aid in the solution of pre- and post-operative examination, as well as in specialist diagnostic problems.

In this paper, we indicate how measurements may be undertaken using laser-ultrasound. We describe the experimental procedures employed and present results for bovine eyes suspended in normal saline solutions. For validation of the origin of ultrasound features, evidence is also presented for a simple plastic lens mounted in saline.

BIOLASER-ULTRASOUND MEASUREMENTS IN AN EYE

The experimental arrangement used for examining bovine eye samples is shown in Figure 1. The laser source consisted of a Quanta-Ray type GCR-150 laser system. Based on Nd:YAG, it was Q-switched to produce 8 ns laser pulses at a wavelength of 532 nm, with a repetition rate of 2.5 Hz. It was retained at this low repetition level to permit thermal relaxation within the tissue volume, allowing the tissue to recover to its original temperature before absorption of subsequent laser pulses. Energy outputs from this type of laser were normally in excess of 100 mJ, so that neutral density glass filters were used for laser beam attenuation. Laser energies of the order of 2 mJ were delivered from the tip of the fibre. This energy level ensured that laser-ultrasound signals were generated in the thermoelastic-heating regime. Laser pulses were launched into a 600 μm multi-mode optical fibre of the Ocular Laser-Ultrasound Probe (OLUP) via an SMA connector. Transducers were based on a 28 μm PVDF polymer film integrated onto the head of the probe. The head of the OLUP was located just in front of the cornea, with saline separation from the sample of between 2.5 and 13 mm.

Partial absorption of laser light takes place as the laser beam propagates through the eye to the retina. At an interface, any small change in absorption creates a pressure pulse that propagates in two opposite directions. The piezoelectric transducer detects signals propagating back towards the probe. Acoustic impedance changes are comparatively small, so that ultrasound reverberations within the eye are not as significant for laser-ultrasound signals.

A simplified diagram illustrating the nature of the laser-ultrasound probe employed is shown in Figure 2. As can be seen, a piezoelectric polymer ultrasound transducer is built with a concentric optical fibre. Details of the performance of this probe are given by Dewhurst *et al*[13]. For experiments with bovine eyes, samples were obtained fresh from the School of Biological Sciences, University of Manchester, and the eyeballs were stored at 5°C until required. Experimental work was normally conducted within 36 hours of removal of the eyes. The samples were mounted in an aluminium ring inside a small water tank filled with normal saline solution at room temperature.

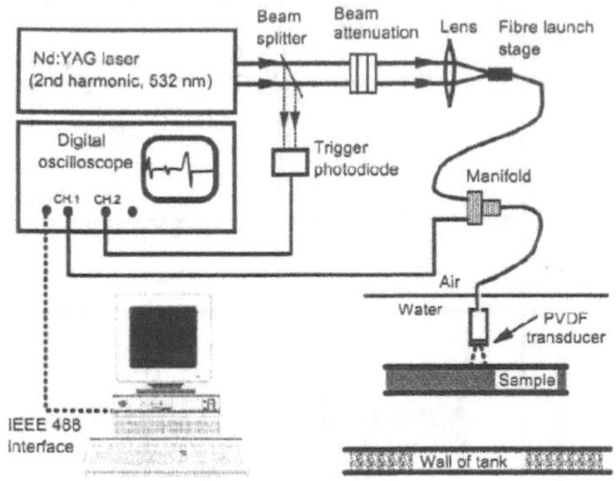

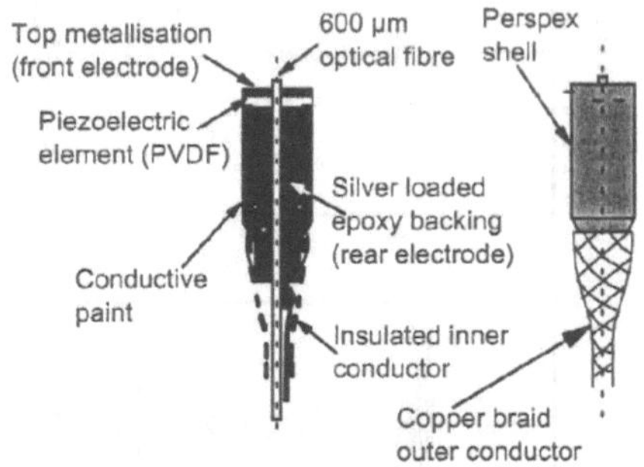

Figure 1. Experimental arrangement for examination of bovine eyes.

Figure 2. Constructional details of the sensor head.

Figure 3 shows a diagram of the eye and associated with it are the corresponding photoacoustic signals that are expected from the various interfaces. This assumes that the laser light is introduced down the main axis of the eye. Laser-ultrasound signals may arise from the anterior cornea (AC), the posterior cornea (PC), the anterior lens (AL), the posterior lens (PL), and features from the region of the retina and choroid (AR) and (PR).

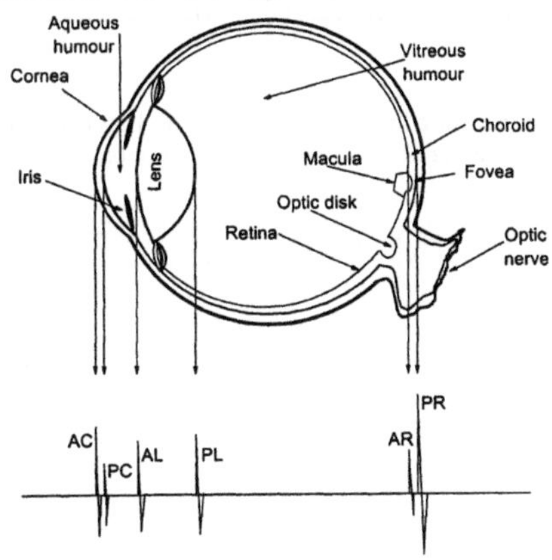

Figure 3. Outline diagram of a bovine eye, together with the acoustic signals generated at optical interfaces.

Figures 4 and 5 show typical results for two bovine eye specimens, arising from waveform averaging over 1,024 laser shots. Figure 4 shows waveforms on shorter time-scales, with T designating the laser trigger time when the laser pulse was initially transmitted from the probe tip. Electronic noise following this event was associated with probe design. With a probe stand-off distance of a few millimetres, following fast transient in each waveform are associated with signals from the eye that reached the probe. Relatively small amplitude signals arise from the front of the cornea, AC, and the posterior cornea, PC. There was a large amplitude signal from the lens capsule and lens surface, AL. In Figure 5, waveforms on longer timescales revealed additional signals that arose from the region of the retina. Signal structure was complicated due to the physiology of that region of the eye. We note that the interface, PL, shown in Figure 3, has not been conclusively identified. Identification of signal features was assisted by estimates of eye dimensions obtained by freezing the eye, followed by a section along the axis, Figure 6.

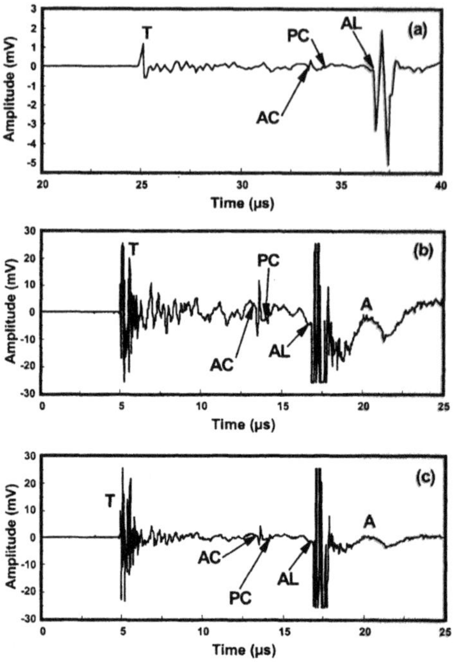

Figure 4. Laser-ultrasound signals measured from bovine eyes on short timescales (A – artefact).

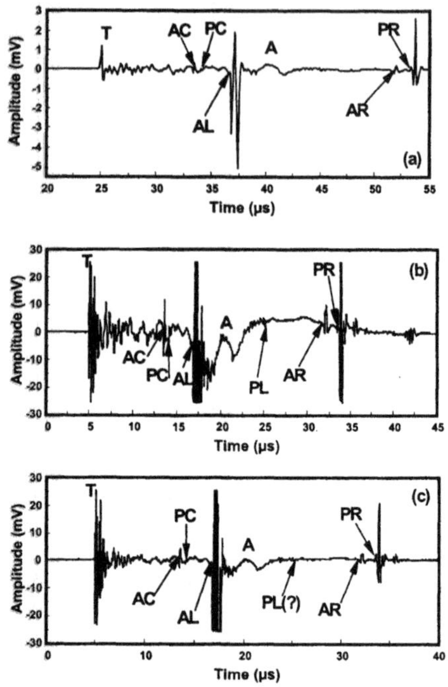

Figure 5. Laser-ultrasound signals measured from bovine eyes on long timescales (A – artefact).

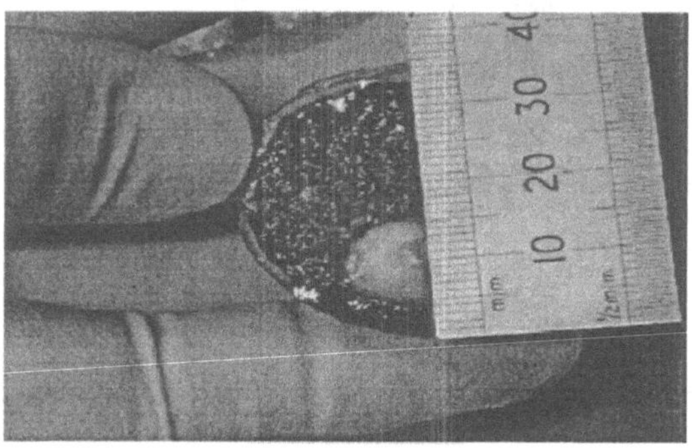

Figure 6. Cross-section of a frozen bovine eye.

In order to understand why the laser-ultrasound signal from the rear surface of the lens might be absent, we have assessed known optical data for the optical and acoustic properties of the various tissues within the eye. Table 1 shows some typical optical transmission percentages for a laser wavelength of 532 nm, together with typical ultrasound velocities[16].

On examination of the optical absorption coefficient data, it is clear that between the lens and the vitreous humour there is little or no optical absorption change. Thus at 532 nm, we might expect little or no laser-ultrasound generation. This situation might be quite different should there be any eye disease.

Table 1. Typical optical and acoustic properties of the eye

(See note)	Transmission[(a)]	Absorption Coefficient[(b)] (cm^{-1})	Ultrasound Velocity[(c)] (m.s^{-1})
Cornea	63%	3.08	1,588
Aqueous humour	60%	0.49	1,503
Lens	47%	0.18	1,673
Vitreous humour	37%	0.17	1,532
Retina	1%	36.10	1,532

Note: [(a)]Percentage transmission through the human eye reaching the anterior surface of a particular structure (estimated).
[(b)]These values were calculated using Beer's law.
[(c)]Typical values quoted for human eyes.

Cataract Evaluation

We are at a very early stage in the work of using laser-generated ultrasound measurements for the assessment of cataract formation. Microwave techniques to induce simulated degradation of the bovine lens have not yet been successful. However, one of the experiments conducted so far has been to use a plastic lens held in a normal saline solution similar to the set-up shown in Figure 1. Initially, the lens was examined in its normal transparent state. The surface of the rear of the lens was then roughened using fine emery paper until it became translucent to simulate a posterior subcapsular opacity in the human lens. Measurements were then repeated on the modified lens. The two sets of signals are shown in Figure 7. There are clear differences in the amplitudes of the two signals recorded for the translucent lens compared with the transparent one, suggesting that relative opacification of the crystalline lens may be measured by means of this method. Interestingly, the reduction in thickness caused by the abrasion of the rear surface also shows up in small timing differences for the two ultrasonic signals. It may also be an aid to the surgeon to determine whether phakoemulsification is a viable option.

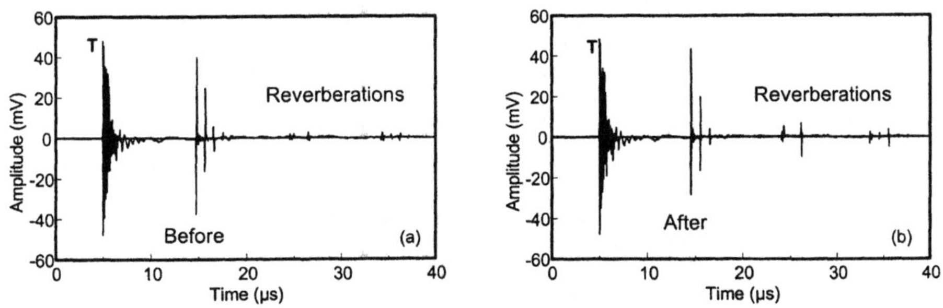

Figure 7. Laser-ultrasound signals from a plastic lens in saline, (a) before and (b) after abrasion.

CONCLUSIONS

We have described an ophthalmic laser-ultrasound probe and have shown that, in its early prototype form, it is capable of generating dimensional information for both the anterior and posterior chambers of the eye. Laser pulse energies of about 2 mJ in 8 ns were required to produce laser-ultrasound signals, and these were delivered from a fibre optic probe to bovine samples mounted in saline solution. As expected, there were large amplitude signals arising from the anterior lens surface and the retina, with smaller signals from the other physiological boundaries. From time-of-flight measurements, we were able to identify many of the other interfaces in the eye. These techniques therefore offer a basis for a new form of interrogation system.

ACKNOWLEDGEMENTS

We wish to acknowledge studentship support from the Iran University of Science and Technology and The Burdens Charitable Foundation. The original work on the intra-arterial probe was funded and is currently being funded by a UK charity, the National Heart Research Fund. We are also grateful to the School of Biological Sciences, University of Manchester, for the supply of bovine samples, and Miss A.M. Richardson for her assistance with the production of this paper.

REFERENCES

1. C.B. Scruby, R.J. Dewhurst, D.A. Hutchins and S.B. Palmer, Quantitative studies of thermally-generated elastic waves in laser-irradiated metals, *J. Appl. Phys.* 51:6210 (1980).
2. R.J. Dewhurst, D.A. Hutchins, S.B. Palmer, and C.B. Scruby, Quantitative measurements of laser-generated acoustic waveforms, *J. Appl. Phys.* 53:4064 (1982).
3. C.B. Scruby and L.E. Drain. *Laser Ultrasonics: Techniques and Applications*, Adam Hilger, Bristol (1990).
4. P.A. Payne and R.J. Dewhurst, *Ultrasonic Probe*, UK Patent Application No. 8727875, 27th November 1987.
5. Q. Shan, A. Kuhn, P.A. Payne and R.J. Dewhurst, Characterisation of laser-ultrasound signals from an optical absorption laser within a transparent fluid, *Ultrasonics* 34:629 (1996).
6. A.A. Oraevsky, S.L. Jacques and F.K. Tittel, Measurement of tissue optical properties by time-resolved detection of laser-induced transient stress, *Appl. Opt.* 36:402 (1997).
7. A.A. Oraevsky, R.O. Esenaliev and A.A. Karabutov, Laser optoacoustic tomography of layered tissue: signal processing, *Proc. SPIE* 2979:59 (1997).
8. R.O. Esenliev, A.A. Karabutov, F.K. Tittel, B.D. Fornage, S.L. Thomsen, C. Stelling and A.A. Oraevsky, Laser optoacoustic imaging for breast cancer diagnostics: limit of detection and comparison with X-ray and ultrasound imaging, *Proc. SPIE* 2979:71 (1997).
9. P.C. Beard and T.N. Mills, Characterisation of post mortem arterial tissue using time-resolved photoacoustic spectroscopy, *Phys. Med. Biol.* 42:177 (1997).
10. Q.X. Chen, G.A. Davies, R.J. Dewhurst and P.A. Payne, Photo-acoustic probe for intra-arterial imaging and therapy, *Electron. Lett.* 29:1632 (1993).
11. Q.X. Chen, R.J. Dewhurst, P.A. Payne and B. Wood, A new laser-ultrasound transducer for medical applications, *Ultrasonics* 32:309 (1994).
12. P.C. Beard and T.N. Mills, An optical fibre sensors for the detection of laser generated ultrasound in arterial tissues, *Proc. SPIE* 2331:112 (1994).
13. R.J. Dewhurst, Q.X. Chen, G.A. Davies, A. Kuhn, K.F. Pang, P.A. Payne and Q. Shan, Developments in a photoacoustic probe for potential use in intra-arterial imaging and therapy, in: *Acoustical Imaging, Volume 21*, J. P. Jones, ed., Plenum Press, New York (1995).
14. K.A Roome, R.F. Caller, P.A. Payne and R.J. Dewhurst, Laser-ultrasound tissue characterisation methods for potential use in laser angioplasty procedures, in: *Materials Science Forum: Nondestructive Characterization of Materials VIII*, R.E. Green Jr, ed., Plenum Press, New York (1998).
15. K.A Roome, P.A. Payne, and R.J. Dewhurst, Towards a sideways looking intravascular laser-ultrasound probe, Sensors *and Actuators A*, 76:197 (1999).
16. F.A. Duck. *Physical Properties of Tissue: A Comprehensive Reference Book*, Academic Press, London, (1990).

AUTHOR INDEX

SUBJECT INDEX

The items in the index include those taken from a list of key terms provided by each author. The referenced page number indicates the page on which the item occurs.

MIX
Papier aus verantwortungsvollen Quellen
Paper from responsible sources
FSC® C105338

If you have any concerns about our products, you can contact us on
ProductSafety@springernature.com

In case Publisher is established outside the EU, the EU authorized representative is:
Springer Nature Customer Service Center GmbH
Europaplatz 3, 69115 Heidelberg, Germany

Printed by Libri Plureos GmbH
in Hamburg, Germany